Die Grundlehren der mathematischen Wissenschaften

in Einzeldarstellungen
mit besonderer Berücksichtigung
der Anwendungsgebiete

Band 190

Carl Faith

Algebra: Rings, Modules and Categories I

Springer-Verlag Berlin Heidelberg New York 1973

Carl Faith

Rutgers University
New Brunswick, N. J., U. S. A.

AMS Subject Classifications (1970):
12-01, 13-01, 15-01, 16-01, 18-01, 20-01

ISBN 3-540-05551-7 Springer-Verlag Berlin Heidelberg New York
ISBN 0-387-05551-7 Springer-Verlag New York Heidelberg Berlin

Library of Congress Catalog Card Number 72-96724.

For my three muses:

Mickey, Heidi, and Cindy

Preface

This book is designed to introduce students to the basic ideas and operations of rings, modules, and categories as patiently and as thoroughly as time and space permit, and then bring them to the frontiers of research as rapidly and as comprehendingly as their abilities permit. Therefore, the reader is forewarned not to be deceived by the elementary beginning, including a foreword on set theory and a first chapter on rudimentary group theory and category theory.

This book is a survey of aspects of ring theory since Jacobson's Colloquium volume published in 1956, and revised in 1964, and all of Volume II is devoted to ring theory. Many of the chapters are based on papers in areas where there is intense research activity — the authors of these papers are mentioned briefly at the end of this preface, and the papers are discussed in the introduction — so students will encounter plenty of motivation in the current literature should they wish to continue the study!

Although a rapid intellectual growth is desireable, it is intellectual growth (period) which excites the teacher, and so ideas ought not be given short shrift in any serious study of them. For this reason, many important theorems are presented in the text in more than one aspect: I have included three independent proofs of the Goldie and Lesieur-Croisot theorems on rings having semisimple right quotient rings, two independent proofs of the Johnson-Utumi maximal right quotient rings, three proofs of one of the most important Morita theorems, viewed tensor products both as adjoint functors and as traditionally constructed (ditto for the Gabriel localization theory and localization in commutative rings); and in many other instances for paedagogical reasons I have offered the reader a *déja vu*, and, thus often a *preview*! This accounts both for the dual nature of the text—interwoven as much of it is with parts purely ring theoretical and parts purely categorical, and for the size—two volumes in six parts. [All the category theory appears in Volume I, however, and, as I have said, Volume II is almost wholly (not holy!) ring theory.]

The first systematic study of the category mod-A of right modules over a ring A was done, I believe, by Morita in 1958. In his University of Oregon lectures in 1962, Bass gave simplified proofs of a number of "Morita Theorems", incorporating ideas of Chase and Schanuel. One of the Morita theorems characterizes when there is an equivalence of categories mod-$A \approx$ mod-B for two rings A and B. Morita's solution organizes ideas so efficiently that the classical Wedderburn-Artin theorem is a simple consequence, and moreover, a similarity class $[A]$ in the Brauer group $Br(k)$ of Azumaya algebras over a commutative ring k consists of all algebras B such that the corresponding categories mod-A and mod-B consisting of k-linear morphisms are equivalent by a k-linear functor. (For fields, $Br(k)$ consists of similarity classes of simple central algebras, and for arbitrary commutative k, this is subsumed under the Azumaya [51][1] and Auslander-Goldman [60] Brauer group.) Numerous other instances of a wedding of ring theory and category (albeit a shotgun wedding!) are contained in the text. Furthermore, in my attempt to further simplify proofs, notably to eliminate the need for tensor products in Bass's exposition, I uncovered a vein of ideas and new theorems lying wholely within ring theory. This constitutes much of Chapter 4 —the Morita theorem is Theorem 4.29—and the basis for it is a correspondence theorem for projective modules (Theorem 4.7) suggested by the Morita context. As a by-product, this provides foundation for a rather complete theory of simple Noetherian rings—but more about this in the introduction.

Category theory has long passed the confines of a "theory" and truly is one of the fundamental mathematical tools, unifying many disciplines much the way group theory does. The universality of the concepts of morphism, functor, and natural transformation has changed the intellectual basis of mathematical knowledge, irrevocably I believe, from the 19th century's dependence on point-sets and functions to the 20th century. (It is rather surprising to find that categories and functors were not introduced until the 20th was middle aged[2].)

Nevertheless, I have eschewed a functorial foundation for set theory, mainly, I suppose, because I do not think about rings and modules functorially. Rather than solve problems in ring theory, category theory raises interesting questions which deepen our understanding of the subject[3]. Thus, diverse theorems of Wedderburn, Artin, Jacobson, Johnson, Goldie, Lesieur, Croisot, Utumi, Morita, and Azumaya could

---

[1] Azumaya [51] refers to a paper in the Bibliography published by Azumaya in 1951. If more than one paper or book appears in one year, then small letters, such as Jacobson [45a], [45b], and [45c], will distinguish them.

[2] In 1945, by Eilenberg and MacLane; cf. MacLane [63, p. 34].

[3] "The notions of category and functor provide not profound theorems but a convenient language." Ibid., p. 33. (A theorem in a recent study by Mitchell [72] may date this quotation!)

be subsumed under the Grothendieck-Gabriel theory of quotient abelian categories and localizing functors in Gabriel's thesis. (More about this in the introduction.)

It is a deceptive practice to list authors whose works are being presented in a text, not only because some authors inevitably will be slighted (for perhaps no other reason than that the importance of their work has not yet been perceived by the author), but also because the anthologized may be represented by less than his best. Nevertheless, we have anthologized the following immortals: E. Artin, G. Burnside, R. Dedekind, H. Fitting, G. Frobenius, C. F. Gauss, D. Hilbert, L. Kronecker, J. Levitzki, E. Maschke, T. Nakayama, E. Noether, O. Ore, I. Schur, J. von Neumann, and J. H. M. Wedderburn.

The major contributors to ring theory and module theory represented in the text are: S. A. Amitsur, K. Asano, M. Auslander, G. Azumaya, R. Baer, H. Bass, J. E. Björk, N. Bourbaki, R. Brauer, H. Cartan, S. U. Chase, C. Chevalley, J. H. Cozzens, R. Croisot, B. Eckmann, A. W. Goldie, O. Goldman, M. Ikeda, N. Jacobson, R. E. Johnson, I. Kaplansky, G. Köthe, W. Krull, J. Lambek, L. Lesieur, E. Matlis, K. Morita, H. Prüfer, A. Schopf, Y. Utumi, C. L. Walker, and E. A. Walker.

The major contributors to the category theory are: D. Buchsbaum, S. Eilenberg, P. Freyd, P. Gabriel, A. Grothendieck, D. Kan, S. Lubkin, S. MacLane, B. Mitchell, N. Popesco, C. Watts, and N. Yoneda.

We hasten to remark that these divisions are hardly exact! We have commented on the categorical aspects of Morita's work, but one could also trace the influence on category theory of ring theory, for example the structure of rings without chain conditions of Jacobson, or the invention of the injective hull of Baer and Eckmann-Schopf. Similarly, the contributions to ring theory of Eilenberg and MacLane are inestimable.

The second and third groupings are contemporary and/or recent authors, but here again there are wide variations. Perhaps it is senseless to make distinctions which omit two of the most influential mathematicians in the subject, namely P. M. Cohn and I. N. Herstein. But I can take this opportunity to note the excellent exposition of the work of Herstein, and of his students, in the Carus Monograph [68]; and, moreover, the work of P. M. Cohn on free ideal rings, and coproducts of rings, is included in his book Cohn [71].

Princeton, February 1973 Carl Faith

Acknowledgements

No book ever is written by the efforts of one man alone. I wish to thank my wife for this observation, and the amount of help she gave me which prompted it. I have to thank my teen-age daughters Heidi and Cindy for bearing my great despair, forgiving my inattentions, and denying me pity. All of which served me to get on with it.

It is a pleasure to acknowledge other dependencies as well. Research which eventually became a part of the text was supported by a number of fellowships. In 1959—1960, I held a NATO Postdoctoral Fellowship at Heidelberg, and an NSF Postdoctoral Fellowship at the Institute for Advanced Study in 1960—1961, followed by a grant-in-aid in 1961—1962. The main writing of the text was aided by two Rutgers Faculty Fellowships, one taken to the University of California, Berkeley, in 1965—1966, and one taken in Princeton, 1969—1970. I have to thank the Rutgers Research Council for these.

I am deeply aware of the privileges such generous support for my work bestows, and the concomittent obligations it exacts. The writing of this book, the love and labor I have lavished on it, is my wholehearted attempt to discharge these obligations, and to show my appreciation. I owe the faculties and staffs of the Institute for Advanced Study, and the Mathematical Institute of Heidelberg, too many favors to acknowledge individually, but Deane Montgomery and Friedrich Kasch have my eternal gratitude for their help.

In addition to my former students Ann K. Boyle, John H. Cozzens, and Barbara L. Osofsky, I am obliged to many people who read parts of the manuscript in its various editions. For many, many suggestions, which I often incorporated in the text without specific acknowledgement, I wish to thank Paul M. Cohn, Kent R. Fuller, Nathan Jacobson, Arun Jategaonker, Serge Lang, William R. Scott, and Wolmer Vasconcelos. I want to thank Hyman Bass for his generous permission to use portions of his Tata Lecture Notes [67]. The material in Chapter 12 on the Morita theorems and the Picard group, and the study of separable algebras and the Brauer group, Chapter 32, are with minor changes, his work. I am indebted to Richard Brauer for sending me a requested copy of a chapter of his on Frobenius algebras, and for an inspiring communication from Balestrand. Saunders MacLane raised a point on my definition of categories. I hope that I have resolved it correctly. For other help on categories, I have leaned on my colleagues Myles Tierney and Barry Mitchell, especially the excellent text of Mitchell [64]. My friend Joe Rosenstein pointed out and corrected as many logical errors in the Foreword as I would allow! I also want to thank Bill Scott for some heart-felt advice, reaped from his own experiences in book-writing (which I took).

I am grateful for the full-hearted cooperation of my students, colleagues, and mathematical friends, at Rutgers and elsewhere, but, of course, they are not to be blamed for errors! John H. Cozzens and John Koehl can not be given too much credit for diligence in picking out countless typos. They have my great thanks.

Mathematics typing requires a special kind of skill, not to mention intelligence, or bravery. For courage under fire, and beyond the call of duty, I wish to praise Ms. "Taffy" Griffith, Ms. "Euthie" Anthony, and Ms. Caroline Underwood. Ms. Underwood has helped legions of visitors at the Institute for Advanced Study, but what has helped me most in my work there is her great sunniness and kindness.

The problem of assembling the bibliography accurately from hundreds of index cards was promptly solved by my friend and neighbor, Frank Crawford, President of Princeton Microfilm Company. He has my finest *Cordon Rouge* '66.

Mathematicians have a theory about everything—I have a theory about mathematicians: so intent are we on our scribbles and our equations, we are likely to be adjudged too detached, and therefore too cold and unemotional. My theory is that it is *because* mathematicians are so emotional that they *can* become mathematicians. They, at least, do have the capacity to be moved by the austere beauty mathematics possesses—the "white goddess" the poet Robert Graves spoke of as the muse of poetry—the muse of us all.

Introduction

Grothendieck's paper [57] placed homological algebra in the setting of abelian categories with exact direct limits (Chapter 14). Although most of the text requires only a modicum of preparation from category theory, and almost none from abelian categories, the far-reaching Popesco-Gabriel theorem [62] depends not only of the foundations of abelian categories (Chapters 5 and 6), but also very heavily on Gabriel's thesis [62]. We give an account of this in Chapters 15 and 16, preceded by preliminaries of Freyd's adjoint functor theorem (Chapter 5), abelian categories (Chapter 6), and Grothendieck categories, including the Freyd and Mitchell embedding theorems (Chapter 14). The adjointness of the functors $\mathrm{Hom}_R(M,\ )$ and $\otimes_R M$, for any R-module (Kan [58]) is taken up in Chapter 5, and again in Chapter 11, where the basic properties of tensor products are developed in the traditional mode, and flat modules introduced.

These 5 chapters on categories, nevertheless, represent only about a sixth of the text, and despite the fact that the Morita theorems, and the theorems on quotient rings, are special cases of Gabriel's localizing functors, we prove these theorems ring-theoretically. For example, Morita's theorem characterizing the similarity of two rings A and B, that is, when mod-$A \approx$ mod-B, is Theorem 29 of Chapter 4. This happens iff there exists a finitely generated projective module P which generates the category mod-B, and for which there is a ring isomorphism $A \approx \mathrm{End}\, P_B = \mathrm{Hom}_B(P,P)$. This condition is right-left symmetric, and we write $A \sim B$. (The commonest example is $B \sim B_n$, the $n \times n$ matrix ring over B.) This is preceded in Chapter 4 by a correspondence theorem for projective modules generalizing the Morita correspondence to the situation where U is a finitely generated projective faithful left module over a ring B, and $A = \mathrm{End}_B U$. Then, there is a lattice isomorphism between the set of (right) A-submodules of U, and the right ideals of B of the form $I = IT$, where T is the trace ideal of U in B. The inverse mapping sends $I = IT$ onto IU, and (B,A)-submodules of U onto ideals of the form $I = IT$. Moreover, the correspondence $S \mapsto S'$ of the set of ideals of A and the set of ideals of B of the form $S' = TS'T$ is a lattice and semigroup isomorphism such that $S'U = US$. The main applications characterize when A is semiprime, prime, simple, right Artinian or Noetherian, has uniform or lateral right ideals.

Simple right Noetherian rings are thereby characterizable as rings of endomorphisms $A = \mathrm{End}_B U$ as above, where B is a right Ore domain[1] with at most one nontrivial ideal T, and satisfying the a.c.c. on the right ideals of the form $I = IT$.

Chapter 3 contains some sixty-odd propositions which I consider preliminary to any deep understanding of the structure of rings and modules. How preliminary this is can be determined by the fact that the Chevalley-Jacobson density theorem is taken up not here but in Chapter 19 as a consequence of a more general theorem of Jacobson and Johnson-Wong on the double annihilator condition (d.a.c.) for quasinjective modules. For semisimple modules, this latter theorem is a theorem of Tate (cf. Artin [50]). The last half of Chapter 3 is devoted to injective modules, Baer's criterion for injectivity, and the Eckmann-Schopf notion of the injective hull of a module. These concepts are the most basic tools in the study of modules, and it is important, I believe, that they come first, not last, in a course on the subject. A more detailed list of topics is offered in the introduction to the chapter. We postpone any reference to Chapters 1 and 2 to the Introduction to Volume I.

Part II is largely devoted to elementary presentations of the Wedderburn-Artin theorems on the structure of semiprime rings (rings with no nontrivial nilpotent ideals) satisfying the descending chain condition on right ideals (right Artinian rings). These are semisimple rings, that is, rings over which every module is a direct sum of simple modules; and when simple, they are full $n \times n$ matrix rings over fields (Chapters 7 and 8). There follows the structure of rings with semisimple right quotient rings, including the Goldie and Lesieur-Croisot theorems, and Levitzki's theorem (Chapter 9), and the Faith-Utumi theorem for orders in semilocal matrix rings (Chapter 10).

Chapter 7 requires just the minimum prerequisites from Part I listed in the introduction to Part II, and contains a number of theorems on simple rings, including those of Hart and Cozzens. Poynomial and differential polynomials rings are defined, and the Hilbert basis theorem proved. Amitsur's theorem on the simplicity of the ring of differential polynomials $A[y,D]$ over a simple ring A of characteristic 0, and outer derivation D, is proved, and Cozzens's examples of a simple domain with a unique simple module, which is injective, is discussed. (One of Cozzens's domains requires the existence of a universal differential field with respect to a derivation D, as established by Kolchin [53], but not con-

[1] A right Ore domain B is an integral domain having a right quotient field D (Ore [31]). In the case at hand, Goldie's theorem and the Lesieur-Croisot theorem (discussed below) imply that A has a (classical) right quotient ring isomorphic to the ring of all $n \times n$ matrices over D, for a unique integer n.

structed here.) All these examples of (simple) domains originated with Ore [31, 33], Littlewood [33] and Jacobson [42].

We begin Part II (Chapter 7) with an elementary proof of an important theorem of Morita on balanced modules, employing just linear and matricial algebra. (A more sophisticated proof is offered in Chapter 4 involving ideas behind the correspondence theorem.)

A right R-module M is balanced if the canonical homomorphism of R into the biendomorphism (bicentralizer, bicommutator) ring of M is surjective. The classical Wedderburn-Artin Theorem states that every minimal right ideal M of a simple right Artinian ring R is balanced, in which case R has a representation as a full ring of linear transformations on a finite dimensional vector space M over the (generally noncommutative) field $D = \text{End}\, M_R$. The theorem of Morita (4.1.3, 7.1, and 7.3) applies to any ring R: *any generator M of the category* mod-R *is balanced, and moreover, M is finitely generated projective as a canonical left module over* $B = \text{End}\, M_R$. Now, if R is any simple ring, then it is a triviality to verify that any nonzero right ideal M is a generator of mod-R, so $R \to \text{End}_B M$ is surjective. Moreover, simplicity of R implies the kernel is zero, $R \approx \text{End}_B M$, and so Morita's theorem implies the General Wedderburn Theorem of Rieffel [65]. Moreover, finite projectivity of M over B yields an idempotent e in the full $n \times n$ matrix ring B_n, for an appropriate integer n, and a ring isomorphism $R \approx eB_n e$. Moreover, $B_n e B_n = B_n$ so Morita's theorem implies that of Hart [67]. In the case M is a minimal right ideal of R, then B is a field by Schur's lemma, so then M is a finite dimensional (by the finite generation of M over B) vector space over B. In this case, M is free over B, say $M \approx B^n$, and then R is isomorphic to the $n \times n$ matrix ring B_n so Morita's theorem implies the Wedderburn [08]—Artin [27] Theorem. Moreover, if M is a uniform right ideal (in the sense that every nonzero submodule is directly indecomposable), then B is a right Ore domain with a right quotient field D, and R has a full right quotient ring $Q(R) \approx D_m$, for some integer m (Theorem of Goldie [58] and Lesieur-Croisot [59]).

The endomorphism theorem for a right Noetherian simple ring R states R is isomorphic to the endomorphism ring of a torsionfree module U of finite rank over a right Ore domain K, where U can be any uniform right ideal (Faith [64]). [In fact, R can be any simple ring with a uniform right ideal U (l.c.).] The theorems of Goldie, Hart, Morita just cited imply that U is finitely generated projective over a right Ore domain $B = \text{End}\, U_R$. The converse question of when a finitely generated projective faithful module U over a right Ore domain B has simple (Noetherian) endomorphism ring A is fully discussed in Chapter 4, using the correspondence theorem. In this case, the center C of B is a field isomorphic to the center of A, the subring $B_0 = C + T$, has at most one nontrivial ideal, namely, $T = \text{trace}_B U$, and U is also finitely generated projective over B_0. Furthermore, $A \approx \text{End}_{B_0} U$ canonically. *This shows that much of the structure theory for simple Noetherian rings can be reduced to the structure of right Ore domains with at most one nontrivial ideal.* It is unknown whether or not B_0 (and B) can be chosen to be a simple domain (see conjectures on simple rings, pp. 381—2), however, every such B which arises this way is simple only if R is right hereditary (Theorem 4.40).

Chapter 8 is on semisimple rings and modules, the Wedderburn-Artin theorems, and global dimension of rings. Auslander's theorems on global dimension are proved, namely, the right global dimension is the supremum of the projective dimensions of the cyclic right modules; when the ring is not semisimple, this is just the sup of the projective dimensions of the right ideals. Thus, a ring has gl. dim ≤ 1 if every right ideal is projective. (Such a ring is called a right hereditary ring.) We prove the Hilbert Szyzgy Theorem in the form given it by Eilenberg-Rosenberg-Zelinsky, namely the polynomial ring over R in n variables has global dimension n plus that of R. Chapter 8 concludes with a theorem stating that any simple Noetherian ring of global dimension $n \leq 2$ is similar to a right Ore domain. (The classical Wedderburn-Artin theorem for a right Artinian semisimple ring is the $n = 0$ case.)

In Chapter 9, we prove the theorems of Goldie [58, 60] and Lesieur-Croisot [59] characterizing a ring R which has a (classical) full right quotient ring $Q(R)$ which is semisimple Artinian. Briefly, a right quotient ring Q is a ring $\supseteq R$ such that: (1) Each regular element (non zero divisor) of R is a unit of Q; (2) Each element $q \in Q$ can be written $q = ab^{-1}$, for suitable $a, b \in R$. Then R is said to be a right order in Q. The existence of Q implies the Ore condition which states that given $a, b \in R$, and b regular, then there exists $a_1, b_1 \in R$, with b_1 regular such that $ab_1 = ba_1$. The Ore condition is necessary and sufficient for the existence of a right quotient ring, and then the quotient ring may be constructed as pairs (a, b), with $a, b \in R$, and b regular, much the way the rational number field is constructed out of the integers. The necessary and sufficient conditions of Goldie and Lesieur-Croisot for the existence of a semisimple quotient ring $Q(R)$ is that R be a ring without nilpotent ideals $\neq 0$ (a semiprime ring) satisfying two ascending chain conditions: $(\text{acc})^{\perp}$, namely the ascending chain condition on annihilator right ideals; and (acc) $\oplus$, the a.c.c. on direct sums of right ideals contained in R. (The latter condition is equivalent to the requirement that the injective hull $\hat{R}$ of R be finite direct sum of indecomposable modules.) A ring satisfying the latter two conditions is said to be a right Goldie ring.

Specializing to the case when $Q(R)$ is a simple Artinian ring, this yields historically the first finite representation theory of arbitrary prime right Noetherian rings: namely, R has a right quotient ring $Q(R) \approx D_n$, the full $n \times n$ matrix ring over a field D, iff R is a prime ring satisfying (acc)$^\perp$ and (acc)$\oplus$. Thus, the elements of R are then representable as $n \times n$ matrices over D. This was proved by Goldie [58] for the case $Q(R)$ is the left and right quotient ring, by Lesieur-Croisot for $Q(R)$ the right quotient ring, but R prime, and by Goldie [60] for the theorem as stated above for R semiprime. We give three independent proofs of this important theorem: the one in Chapter 9 of Goldie [69]; the one in Chapter 16 of Gabriel [62]; and the one in Chapter 19 of Faith [72]. Goldie's proof uses the idea of a singular ideal of a ring of R. E. Johnson, which is related to the existence of the maximal right quotient rings of Johnson [51] and Utumi [56] (see below). Gabriel's proof uses the Grothendieck-Gabriel concept of a quotient abelian category, and localizing functors. Gabriel's method also yields the Johnson maximal right quotient ring of a ring with vanishing singular ideal. The latter is taken up again from a purely ring-theoretical point of view in Chapter 19. The third proof explicitly uses the Johnson maximal right quotient ring, in the mode of the author's original proof of Goldie's theorem in 1961–62 (published in [67]), and moreover, prime right Goldie rings are characterized by the existence of an injective faithful module E, containing no fully invariant submodules other than the trivial ones, and finitely generated over endomorphism ring. When this is so, then $D = \operatorname{End} E_R$ is a field, and $Q = \operatorname{End}_D E$ is the full right quotient ring of R. A semiprime right Goldie ring has a faithful injective module F which is a finite direct sum of modules of the type E just described, $B = \operatorname{End} F_R$ is a finite product of fields, and $\operatorname{End}_B F$ is the full right quotient ring.

As defined by Findlay-Lambek [58], a module M is a rational extension of a submodule N provided it is true that for any intermediate module P we have $\operatorname{Hom}_R(P/N, M) = 0$. This is equivalent to the requirement that to any $y \in M$ and nonzero $y \in M$ there corresponds $r \in R$ such that $xr \in N$, and $xr \neq 0$. Any rational extension is thereby an essential extension. Every module N has a maximal rational extension $\overline{N}$, unique up to isomorphism over N, and in fact, if E denotes the injective hull of N, and $S = \operatorname{End} E_R$, then $\operatorname{ann}_E \operatorname{ann}_S N$ is a maximal rational extension. The maximal rational extension of R is a ring $\overline{R}$ containing R as a subring, and is called the (Johnson [51], Utumi [56]) maximal right quotient ring of R.

The zero right singular ideal of R is the set of all elements the right annihilator of which is an essential right ideal, and R is said to be right neat when this ideal vanishes. In this case, $\overline{R}$ is a right selfinjective (von Neumann) regular ring (Johnson [51a] and Johnson-Wong [59]). Moreover, $\overline{R}$ is a semisimple ring iff R satisfies (acc)$\oplus$, and then R

satisfies $(\mathrm{acc})^{\perp}$. In this case, $\bar{R}$ is the (classical) full right quotient ring of R iff R is semiprime. Thus, the maximal right quotient ring $\bar{R}$ coincides with $Q(R)$ for a semiprime right Goldie ring. (Cf. Chapter 19.)

Chapter 10 contains the Faith-Utumi theorem [65a] in the form generalized by Faith [71b]. The theorem characterizes when R has a right quotient D_n for the respective cases D is a semisimple ring, and semilocal. (The latter is the case D is semisimple modulo radical.) The theorem is the same in all three cases, namely, changing matrix units in D_n if necessary, R contains the $n \times n$ matrix ring F_n over a right order F of D. (If D_n is also the left quotient ring of R, the change of matrix units is not necessary.) As noted in the original paper [65a], the Goldie-Lesieur-Croisot theorem is not required for the proof.) A corollary is a theorem of Goldie [62] which states that any prime principal right ideal ring is a full $n \times n$ matrix ring F_n over a right Ore domain F.

I have just given very brief introductions to Parts I, II, and IV; I now will do the same for the other parts. Part III starts with the classical development of properties of tensor products, although in one or two places references are made to more general properties of left adjoints in Chapter 5. The most important part of Chapter 11 gives the basic properties of flat modules. These include: the characterization by Lazard [64] of flat modules as direct limits of finitely generated projective modules; the characterization of (von Neumann) regular rings by Auslander [57] and Harada [56] as the rings of zero weak global dimension, that is, when *all* modules are flat; and Chase's characterizations [60] of what are now called left coherent rings (defined by the property that finitely generated left ideals are finitely presented).

The Morita theorems, and applications to the Picard group, appear in Chapter 12, employing tensor products. Chapter 13 gives the applications of the Wedderburn-Artin theorem to finite dimensional algebras over a field, including algebraically closed fields. Wedderburn's factor theorem for separable algebras is proved, and so is Maschke's theorem for the semisimplicity of group algebras of finite groups over fields of characteristic 0. The theorems of Skolem-Noether and Brauer are sketched in exercises in Chapter 13, and developed over commutative rings in Chapter 32.

Part V, the biggest part of the text, is almost too big to give a short sketch of its contents. It embraces a wide variety of theorems on the structure of modules over nonsemisimple rings. Chapter 17 is on modules of finite length, and theorems of Kurosch-Ore (for modular lattices), Jordan-Hölder, Schreier, and Fitting.

Chapter 18 introduces the Jacobson radical of a ring and module, namely the intersection of the maximal submodules. The radical is characterized à la Perlis [42]—Jacobson [45] as the maximal quasi-regular ideal, where an ideal J is quasi-regular provided that $1+x$ is a unit $\forall x \in J$. A ring R is semilocal, provided that $R/\text{rad}\, R$ is semisimple, and local if $R/\text{rad}\, R$ is a field. An object M of an abelian category C has a decomposition into a finite direct sum of objects with local endomorphism rings if and only if $\text{End}_C M$ has the same property. This happens if and only if $\text{End}_C M$ is semilocal and idempotents lift modulo the radical. Other theorems include the theorem of Hopkins and Levitzki stating that right Artinian rings are right Noetherian, Asano's theorem on primary decomposable rings, and the Chinese remainder theorem for pairwise comaximal ideals.

Chapter 19 is a study of quasinjective (QI) modules, and their endomorphism rings. The main theorems are those of Wong-Johnson characterizing QI modules as fully invariant in the injective hull, and Utumi's theorem stating that the radical of their endomorphism ring is the set of those endomorphisms with essential kernels. (Moreover, the endomorphism ring is a von Neumann regular ring *modulo* radical.) This characterizes the radical of right selfinjective rings. Many theorems on modules finite over endomorphism ring are cited. Thus, every right QI module finite over endomorphism ring is injective modulo annihilator. Moreover, every right QI module is finite over endomorphism ring if and only if R is right Artinian.

Also given in Chapter 19 is the Johnson-Utumi right quotient ring, and Burnside's theorem. Chapter 20 is on the subject of direct sum decompositions of modules, including theorems of Faith on Σ-injective modules, Matlis-Papp on decomposability of injective modules into indecomposable modules, and Faith-Walker on the decomposability of modules into a direct sum of modules belonging to a set. We defer Chase's theorem characterizing when a direct product of projectives is projective to Chapter 22, but the preliminary lemmas appear here. In particular, Baer's observation that a direct product of copies of $\mathbb{Z}$ is not free is deduced from Chase's lemmas.

In Chapter 22, the classical theory of Artinian nonsemisimple rings is subsumed under the study of rings perfected by Bass. The gist of the theory is to achieve a duality with injective hulls by providing right R-modules with projective covers. Bass [60] determined that such rings satisfied the d.c.c. on principal left ideals (and conversely). The terminology perfect rings reflects the many properties enjoyed by these rings. For example, every projective module (on the same side of the covers) is isomorphic to a direct sum of indecomposable modules, and each indecomposable projective module has the simplest possible structure in that it is isomorphic to a principal indecomposable right ideal. The basic module of a semiperfect ring is a summand of every generator. We prove Björk's theorem stating that the d.c.c. on principal submodules of a module implies that on the finitely generated modules, and conclude the chapter with Brauer's theory of blocks generalized to perfect rings.

In Chapters 23 and 24, we study Morita's duality theory for a category mod-A of right A-modules. Dual to the idea of equivalence is that of a duality $C \rightsquigarrow D$ between categories: namely, $TS \approx 1_D$ and $ST \approx 1_C$ for contravariant functors $T: C \rightsquigarrow D$ and $S: D \rightsquigarrow C$. A theorem of Morita determines that a duality between a Serre class $\mathscr{S}_A$ of mod-A, which contains A, and a Serre class ${}_B\mathscr{S}$ of B-mod, which contains B, is actually a U-duality, that is, is induced by $T = \mathrm{Hom}_A(\ ,U)$ and $S = \mathrm{Hom}_B(\ ,U)$ for an appropriately chosen (B,A)-bimodule U. Necessary and sufficient conditions for a (B,A)-module U to define a U-duality are that U be an injective cogenerator both in mod-A, and in B-mod, such that $A \approx \mathrm{End}_B U$ and $B \approx \mathrm{End}\ U_A$ canonically. In this case, A must be semiperfect, and when A is right Artinian, then U_A must be finitely generated. Conversely, when A is right Artinian, and U_A is a finitely generated injective cogenerator, then U defines a U-duality $\mathscr{S}_A \rightsquigarrow {}_B\mathscr{S}$, where $B = \mathrm{End}\ U_A$. Quasifrobenius rings, studied in Chapter 24, have an A-duality

$$\text{fin. gen. mod-}A \rightsquigarrow \text{fin. gen. } A\text{-mod}$$

and numerous other interesting characterizations.

Chapter 25 is on uniserial rings, and more generally rings over which every module is a direct sum of cyclic modules. These are called Σ-cyclic rings.

Chapter 26 is on semiprimitive rings, or rings with zero Jacobson radical. A ring A is (right) primitive if it has a simple faithful right module. In this case A is isomorphic to a dense subring of the complete ring L of linear transformations in the vector space consisting of V over the (Schur) field of endomorphisms of V. (Dense in the finite topology on L [Chevalley-Jacobson Density Theorem 19.22].) A ring R is semiprimitive iff R is isomorphic to a subdirect product of primitive rings (Jacobson [45]). We prove Amitsur's theorem on the semiprimitivity of group algebras over uncountable fields. This necessitates Amitsur's theorems on the radical of polynomial rings. The (Brown—McCoy) prime radical is characterized à la Levitzki.

Before I conclude this introduction with references to Part VI, I ought to say that Part VI is a victim of the limitations of a book's dimensions: Not every chapter can be Chapter 2, or Chapter 3, etc. Therefore, some part has to be last. This notwithstanding, Part VI contains much of the classical theory of commutative rings needed for a proper background and motivation of the other parts, including unique factorization domains, principal ideal rings, Dedekind rings, Prüfer rings, the Noetherian ideal theory for Noetherian rings, hereditary rings, separable algebras and the Brauer group. The Hilbert Nullstellensatz, the Artin-Rees Lemma, the Krull Intersection Theorem, and the Krull-Akizuki Theorem are proved. Azumaya algebras are studied, and the Skolem-Noether Theorem proved.

Chapter 31 contains Kaplansky's theorems on the structure of projective modules: they are direct sums of countably generated modules. And over a local ring, every projective module is free. Over a right hereditary ring, every projective module is isomorphic to a direct sum of right ideals. We also prove Albrecht's theorem: over a right semihereditary ring, a projective module is isomorphic to a direct sum of finitely generated right ideals. Moreover, a theorem of Bass then states that every projective left module is isomorphic to a direct sum of duals of finitely generated right ideals (31.15).

Chapter 32 on separable algebras and the Brauer group over commutative rings, as I have stated, is extracted from Bass's Tata Lecture Notes [67]. But more about this in Volume II.

Contents

Part IV

Structure of Abelian Categories

Contents of Volume II:

Logical Dependencies of the Chapters (Rough)

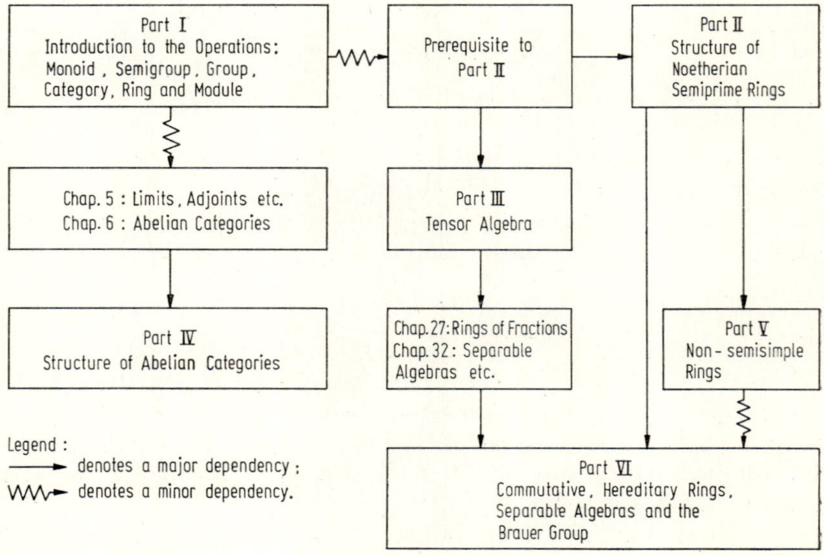

Suggestions for Reading this Book

The first thing to be said is that Part II does not depend upon all of Part I. Part II consists of parts of classical ring theory, from a largely traditional point of view, although there are a number of innovations in Chapter 7. Much of the material of Part I has been included for general knowledge, or for use in parts of the text following Part II. Therefore, anyone who wishes only the specific knowledge of rings contained in Part II should not bother reading Part I. The introduction to Part II lists prerequisites from Part I for a reading of Part II; these include Chapter I.

A slavish following of the text should be similarly avoided in other instances as well. A good rule for reading any book is to skip portions that do not interest you, and space out the more difficult sections with easier material. For example, Chapter 17 (also Chapter 26) is more elementary than Chapter 4.

The following are other examples:

A. After a brief introduction to projective modules, one can prove Kaplansky's theorem in Chapter 31 on hereditary rings. (Indeed, this appears in Cartan-Eilenberg, Chapter 1.) With this at hand, a look at the basis theorem for abelian groups might be profitable. Although this is given in Chapter 31 for modules over principal ideal rings, by simplifying the proof for modules over $\mathbb{Z}$, the result could be introduced very early in a course (cf. Chapter 1 of Lang [65]).

B. When reading Chapter 3 on injective hulls in module type category, the reader might pursue the subject in Grothendieck categories in Chapter 14.

C. Goldie's theorem in Chapter 9, localizing functors in Chapters 15 and 16, and localization in commutative rings in Chapter 27, are connected.

D. Chapter 10 is independent of most of Part II.

E. Chapter 11 on tensor products can be read profitably in conjunction with Chapter 5 on adjoints.

F. Chapter 32 on separable algebras can be read immediately following Chapter 13. The localization theory required can be administered ad hoc. (Indeed, one can read it like a telephone directory, as Bass has suggested.)

G. Much of Chapter 30 on Prüfer and Dedekind rings is elementary.

Numerous other examples can be given, and have been given in cross references sprinkled throughout the text. Furthermore, difficult exercises at the end of each chapter often will lead the solver to proofs found in following chapters. The exercises also serve as a way to draw attention to related results in the literature. In these cases I have tried to give full bibliographic credit, but unlike some authors I am not attempting to relegate to mere exercises important theorems, such as the ones by P. M. Cohn and G. Hochschild on the freedom of right ideals in a free algebra over a field (cited in Chapter 8). I am, instead, encouraging the reader to think about related theorems which fall beyond the range of the text. [It is not impossible, I believe, for some student to solve them, but in the case of the subgroup theorem (Chapter 2), it is highly unlikely!]

I might offer my personal experience with the textual material. As a basis for a one-year undergraduate algebra course, I have used the Foreword on Set Theory, and much of Chapters 1, 3, and 7, supplemented by additional material (such as the first chapter of Lang [65]) on the Sylow theorem, and the basis theorem for abelian groups. In addition, I have included in the first-year graduate course in algebra, the chapters on abelian categories, tensor products, the Morita theorems, and Grothendieck categories, namely Parts I and III, and Chapter 14. In seminars and reading courses parallel to, or following, the graduate algebra course, we have studied the theorem of Gabriel-Popesco (Chapter 15), Gabriel's localizing functors (Chapter 14) applied to quotient rings and the theorems of Goldie, Johnson, and Utumi (Chapter 16), Bass's structure of perfect rings (Chapter 22), Morita duality (Chapter 23) and Quasi-frobenius rings (Chapter 24), Prüfer and Dedekind rings (Chapters 30, 31), separable algebras and the Brauer group over commutative rings (Chapter 32). □[1]

In the Fall semester 1970—1971, in a course in the theory of rings, I covered Chapters 1—4, differential polynomial rings in Chapter 7, global dimension in Chapter 8, Dedekind rings in Chapter 30, and the Kaplansky theorems on projective modules, and hereditary rings, including the basis theorem for abelian groups.

---

[1] The symbol □ is used to conclude mathematical discussions. It is most frequently found at the end of proofs of theorems.

Special Symbols

	Name	Example
&	ampersand	—
$\forall$	universal quantifier	$\forall\, a \in A$
$\exists$	existential quantifier	$\exists\, a \in A$
$\in$	membership	$a \in A$
$\notin$	nonmembership	$a \notin A$
$\subset$	proper containment	$A \subset B$
$\subseteq$	containment	$A \subseteq B$
$\Rightarrow$	implication	$A \Rightarrow B$
$=$	equals	$A = B$
$\varnothing$	empty set	—
$\mathbb{N}$	natural numbers	$1, 2, \ldots$
$\mathbb{Z}$	integers	$0, \pm 1, \pm 2, \ldots$
$\mathbb{Q}$	rational numbers	$a/b,\ a,\ b \in \mathbb{Z},\ b \neq 0$
$\mathbb{R}$	real numbers	$\sqrt{2}$, pi
$\mathbb{C}$	complex numbers	$a + bi,\ i^2 = -1,\ \ a, b \in \mathbb{R}$
$\langle a, b\rangle$	ordered pair	—
$\cup$	union	$A \cup B$
$\cap$	intersection	$A \cap B$
$+$	plus	$a + b$
$-$	minus	$a - b$
$\circ$	composition	$f \circ g$
$\times$	Cartesian product	$\alpha \times \beta$
$\to$	mapping	$A \to B$
$f: A \to B$	mapping	—
$\mapsto$	mapping	$a \mapsto b$
X	cross	$\mathsf{X}_i A_i$
Π	product	$\Pi_i A_i$
$\amalg$	coproduct	$\amalg_i A_i$
$\approx$	equivalence	$A \approx B$
$\wedge$	wedge	$A \wedge B$
$\vee$	cup	$A \vee B$
$>$	greater than	$a > b$
$<$	less than	$a < b$
$\rightsquigarrow$	functor	$A \rightsquigarrow B$
$\triangledown$	wedge	$A \triangledown B$
$\triangleright$	wedge	$A \triangleright B$
$\blacktriangledown$	solid wedge	$A \blacktriangledown B$
$\perp$	perpendicular ("perp")	$A^{\perp}$ and ${}^{\perp}A$
$\bigcirc$	circle	$A \bigcirc B$
Σ	summation	$\Sigma_{i \in I} A_i$
$\oplus$	direct sum	$A \oplus B$ $\Sigma_{i \in I} \oplus A_i$
$\otimes$	tensor product	$A \otimes_R B$

Introduction to Volume I

A foreword supplies the necessary background in the theory of sets and cardinal (ordinal) numbers, and the concepts of group (and the Noether homomorphism theorems), semigroup, category, functor are introduced in Chapter 1. Products, coproducts, and the free object functors in various categories constitute Chapter 2. Rings, modules, ideals, and the Baer-Eckmann-Schopf notions of injectivity, essential extension, and injective hull are introduced in Chapter 3. Chapter 4 contains an elementary presentation (without tensor products) of the Morita theorems characterizing category equivalence of categories of modules over two rings. The basis for this is the correspondence theorem for projective modules, from which also is derived most of the known structure theory of noncommutative Noetherian simple rings. Chapters 5 and 6 survey limits, limit preserving functors, additive and abelian categories, free algebras, (semi) group and polynomial algebras, and the adjoint functor theorems of Eilenberg, Freyd, Gabriel, Kan, Mitchell, and Watts. These six chapters comprise Part I.

Part II, largely independent of Part I, consists of four chapters on the structure theory of semiprime Goldie rings of Wedderburn-Artin, Hopkins, Levitzki, Goldie, Lesieur-Croisot, Faith-Utumi, Hart, and Robson in which Ore domains play a central role[1].

Part III (Chapters 11—13) is the tensor algebra of Kronecker, Whitney, Morita, and applications to the Morita theorems, the Picard group of a category of modules, the Wedderburn theorems on finite dimensional algebras over fields. The Brauer group and Skolem-Noether theorems are presented only in exercises (to Chapter 13), but the generalizations of Azumaya and Auslander-Goldman to separable algebras over commutative rings are presented in Volume II. The basic results on flat modules are in Chapter 11.

Part IV, which completes Volume I, is the theory of Grothendieck categories, the Gabriel theorems on localizing, idempotent, and topologizing functors and quotient categories, the Mitchell-Gabriel embedding theorems, and the Gabriel-Popesco theorem. The Goldie quotient ring (developed in Part II), and the Johnson-Utumi maximal quotient ring (developed ring theoretically in Volume II) are obtained as examples of Gabriel's localization.

---

[1] Amitsur's, Cozzens's, and Littlewood's examples of simple Ore domains are constructed in Chapter 7.

Foreword on Set Theory

The building blocks of mathematics are sets, yet no one should say what a set is. An axiomatic treatment of set theory postulates the existence of certain undefined or primitive objects, called sets, together with symbols and axioms governing their use. A graphic analogy can be made with geometry where there are given undefined objects called points, lines, and planes, together with a collection of axioms relating these objects to each other. An axiom of projective geometry is $\mathscr{P}$: *If L_1 and L_2 are distinct lines, there is one and only one point P on both L_1 and L_2.* Here the relation "P is on L_1" is undefined. Another axiom is $\mathscr{P}^*$: *If P_1 and P_2 are distinct points, there is one and only one line L on both P_1 and P_2.* Any collection of objects which, when properly named as "points" and "lines", satisfy the axioms of geometry, serve as points and lines, equally as well as any other such collection. *It must be possible to replace in all geometric statements the words point, line, plane by table, chair, mug* (David Hilbert, quoted by Weyl [44, p. 635]). Nevertheless, from the axioms we soon discover that some objects in a geometry are not "points" and some objects are not "lines". Analogously in the theory of sets, after we have named our candidates for sets, and listed the undefined symbols and the axioms governing them, we find that not everything in sight is a set. For example, as Cantor observed, there is no such set as "the set of all sets".

It is not our purpose to present here a complete account of set theory. There are many excellent texts, some listed in the bibliography, which do this, and it requires a goodly amount of time to master the parts which every mathematician should know. On the other hand, it is not our purpose to discuss set theory merely from a utilitarian aspect. This would not do justice to a beautiful and rewarding part of mathematics. Futhermore, topics such as the axiom of choice, the well ordering theorem, and the continuum hypothesis always provoke controversy whenever they are discussed and, therefore, are stimulating introductions to mathematics. As a result of this controversy, the student learns that mathematics is not the isle of tranquillity that calculus and the theory of complex functions led him to believe. I believe that it is important that he should learn this.

The minimal aim of this Foreword is to supply the reader with convenient references to terminology, notation, and theorems of set theory used in this book. It is expected that many readers will skip this; others, however, may find the brief review helpful.

Basic Notions of Set Theory

The following symbols are used in mathematical discussions

$\sim$	&	$\vee$	$\Rightarrow$	$\Leftrightarrow$
not	*and*	*or*	*implies*	*if and only if*[1]

$\forall$	$\exists$	$=$	$(,)$	$\{ , \}$
for all	*there exists*	*equals*	*parentheses*	*braces*

The symbols in the first row are called propositional connectives. The symbol $\forall$ is called a *universal* quantifier, and $\exists$ is called an *existential* quantifier. The parentheses and braces group expressions to limit the scope of the quantifiers and to form grammatically correct statements in mathematical language. The rules for forming valid statements are known as the Propositional Calculus. These rules, like those of English grammar, are in constant use without conscious knowledge of this fact. Formal language, as set forth in P. J. Cohen's excellent text, "Set Theory and the Continuum Hypothesis" (pp. 8—11), makes usage explicit for the Propositional Calculus as a subsystem of the Predicate Calculus. However, for the most part, we use the traditional informal, or naive, mathematical language and set theory.

The natural (or counting) numbers

$$1, 2, 3, \ldots, n, \ldots$$

and the integers

$$0, \pm 1, \pm 2, \pm 3, \ldots, \pm n, \ldots$$

are familiar examples of sets. The common fractions a/b, where a and b are integers, are called rational numbers. Common designations of these and other sets are

$\mathbb{N}$	$\mathbb{Z}$	$\mathbb{Q}$	$\mathbb{R}$	$\mathbb{C}$
natural	*integral*	*rational* *numbers.*	*real*	*complex*

The symbol

$$\in$$

element sign

[1] Also abbreviated as "iff".

(also called membership sign) is used to relate sets x and y, as in

$$x \in y$$

x is an element of y

(*x belongs to y, x is a member of y, x lies in y,* all mean the same thing).

Axioms for Zermelo-Fraenkel Set Theory

Axioms for set theory, given by Zermelo [08] and modified by Fraenkel [22], usually are referred to as ZF-set theory. (We follow Cohen [66, pp. 51 ff].) These are listed as axioms ZF_1 to ZF_9. The undefined primitive notions in ZF are set, $\in$, and $=$.

ZF_1. Axiom of Extensionality

$$\forall A, B\,(\forall x (x \in A \Leftrightarrow x \in B) \Rightarrow A = B).$$

In other words, two sets A and B are **equal** in case they have precisely the same elements, or equivalently, a set is determined by its elements. Intuitively, if A and B have the same elements, then one can be substituted for the other in any statement without affecting the truth or falsity of the statement. The symbol $A \neq B$ designates the negation of $A = B$, that is, $\sim(A = B)$; the symbol

$$a \notin B$$

denotes the negation of $a \in B$, that is, $\sim(a \in B)$.

The symbol

$$A \subseteq B$$

is defined by

$$A \subseteq B \Leftrightarrow \forall x (x \in A \Rightarrow x \in B).$$

In this case we say that A **is a subset of** B (A is *contained in* B, A *is included in* B, or B *includes* A, all of which mean the same). If $A \subseteq B$, but $A \neq B$, then A is a **proper** subset of B. If a set A is not a subset of B, we write $A \nsubseteq B$, or $B \nsupseteq A$.

ZF_2. Axiom of the Empty Set

$$\exists x \forall y (y \notin x).$$

This axiom asserts the existence of at least one set. The set x defined is called the **empty** (or **null**) set. Then ZF_1 implies that there is a unique empty set, and we let $\varnothing$ denote this.

ZF_3. Axiom of Unordered Pairs

$$\forall A, B \exists C \forall x (x \in C \Leftrightarrow x = A \vee x = B).$$

The set C so defined is called the **unordered pair** $\{A, B\}$. It is a set consisting of elements A and B, where A and B are sets. The symbol $\{A\}$ denotes $\{A, A\}$ for any set A, called **singleton** A.

The **ordered pair** $\langle A, B\rangle$ is defined by

$$\langle A, B\rangle = \{\{A\}, \{A, B\}\}.$$

One easily deduces:

$$\langle A, B\rangle = \langle C, D\rangle \Rightarrow A = C \text{ and } B = D.$$

(This implication frequently is used to define ordered pair.)

We now use the concept of ordered pair to define a function. Let A and B be sets. A **function** f is a set consisting of ordered pairs $\langle x, y\rangle$, where $x \in A$, $y \in B$, one for each $x \in A$, such that

$$\langle x, y\rangle \in f \;\&\; \langle x, z\rangle \in f \Rightarrow y = z.$$

Mapping is synonymous with function. The set A is called the **domain** of f, the set B is called the **range**, or **target**, of f, and the set[1] of all $y \in B$ such that $\exists x \in A$ with $\langle x, y\rangle \in f$ is called the **image** of f, denoted $\operatorname{im} f$. We let $f: A \to B$ denote that f is a function with domain A and target B. The symbol

$$f \begin{cases} A \to B \\ x \mapsto y \end{cases}$$

means the same, and, in addition, that $\langle x, y\rangle \in f$. Equivalently, we say that $f: A \to B$ is the function defined by $f: x \mapsto y$. The symbol $y = f(x)$ is used synonymously with $\langle x, y\rangle \in f$, and then we say *$y$ is the image or value of f at x.*

A function $f: A \to B$ is also called a **family of elements** (or **sets**) **indexed by** A. Intuitively, we think of this as the set elements of B of the form $f(a)$ for some $a \in A$. The symbol $\{x_a\}_{a \in A}$, or $\{x_a \mid a \in A\}$, will denote a family of elements of B, where $x_a = f(a) \in B$ for all $a \in A$. A **sequence of elements of** B is a function $f: \mathbb{N} \to B$. Thus, setting $x_i = f(i)$, we have

$$x_1, x_2, \ldots, x_n, \ldots.$$

---

[1] An axiom is required for the existence of a set having this property. See axiom ZF_{6n}. The terms domain and target will be employed in the more general context of categories defined in Chapter 1, p. 58.

If $b \in B$, and if A is any set, then there is a mapping

$$b_{\text{const}} \begin{cases} A \to B \\ a \mapsto b \end{cases}$$

called the **constant mapping** defined by b.

$\mathbf{ZF_4}$. Axioms of Unions

$$\forall A \exists B \forall x \big(x \in B \Leftrightarrow \exists X (x \in X \;\&\; X \in A)\big).$$

If A is a set, then the set B is the set whose elements are elements of some set belonging to A. B is called the **union** of the sets in A. This set is denoted by $\bigcup_{X \in A} X$, or if A is a family of sets indexed by a set I, then

$$\textstyle\bigcup_{i \in I} X_i, \quad \text{or} \quad \displaystyle\bigcup_{i \in I} X_i,$$

means the same thing. Let

$$Z = X \cup Y,$$

the union of X and Y, denote the special case $A = \{X, Y\}$. Thus, $t \in Z \Leftrightarrow t \in X \vee t \in Y$.

Intersection is defined by:

$$\forall A \exists B \forall x \big(x \in B \Leftrightarrow \forall X \in A\, (x \in X)\big).$$

The set B consists of those elements that are members of every member of A. The existence of the intersection of given sets follows from axiom ZF_6. This set is denoted by

$$\textstyle\bigcap_{X \in A} X$$

and is called **the intersection of the family of sets** A. If $\{X_i\}_{i \in I}$ is a family of sets indexed by a set I, then

$$\textstyle\bigcap_{i \in I} X_i, \quad \text{or} \quad \displaystyle\bigcap_{i \in I} X_i,$$

denotes the intersection. The special case $A = \{X, Y\}$ is denoted by $Z = X \cap Y$. Thus, $t \in Z \Leftrightarrow t \in X \wedge t \in Y$. Note that

$$\forall X \forall Y (X = X \cap Y \Leftrightarrow X \subseteq Y).$$

A family of sets $\{X_i\}_{i \in I}$ is said to be (**pairwise**) **disjoint** in case $X_i \cap X_j$ is empty $\forall i \neq j \in I$.

Intersections are dual to unions in a sense formulated by the duality principle (Theorem 10).

Let P be a property that may or may not be possessed by certain elements of a set A, and such that it can be determined whether a given element $x \in A$ has property P. (Property is defined by the axiom ZF_{6_n}, and is discussed at the close of the axioms for ZF.) Then, if there is a set consisting precisely of those elements of A having property P, we let the symbol

$$\{x \in A \mid x \text{ has property } P\}$$

denote this set. In other words, the symbol | is read "such that". For example, if B is a subset of A, then

$$\{x \in A \mid x \notin B\}$$

is a subset of A called the **complement of B in A** (or **relative complement of B**). This subset is denoted by $A - B$. Actually, $A - B$ makes sense even if B is not contained in A. Thus,

$$A - B = \{x \in A \mid x \notin B\}.$$

The **de Morgan formulas** for a family $\{X_i\}_{i \in I}$ of subsets of a set U state that

$$U - \bigcap_{i \in I} X_i = \bigcup_{i \in I} (U - X_i)$$

and

$$U - \bigcup_{i \in I} X_i = \bigcap_{i \in I} (U - X_i).$$

These are easily verified illustrations of the duality between intersection and union.

ZF_5. Axiom of Infinity

$$\exists x (\varnothing \in x \;\&\; \forall y (y \in x \Rightarrow y \cup \{y\} \in x)).$$

This implies the existence of a set having infinitely many elements. (An infinite set may be defined to be a set equivalent to a proper subset. Equivalence is used in the sense of Theorem 3. A countably infinite set is one equivalent to $\mathbb{Z}$.) Then ZF_2 and ZF_3 imply the existence of infinitely many sets.

In order to describe the next axiom, the reader will be referred to Cohen [65, p. 7] for the definition of "free variables" in order to enumerate countably many formulas $A_n(x, y; t_1, \ldots, t_k)$ with at least two free variables where k depends on n. The exclamation point ! is reserved to mean "unique".

$\mathbf{ZF_{6_n}}$. Axiom of Replacement

$$\forall t_1, \ldots, t_k \big(\forall x \exists ! y \big(A_n(x, y; t_1, \ldots, t_k) \Rightarrow \forall u \exists v B(u, v)\big)\big), \qquad \text{where}$$

$$B(u, v) \equiv \forall r \big(r \in v \Leftrightarrow \exists s \big(s \in u \,\&\, A_n(s, r; t_1, \ldots, t_k)\big)\big).$$

This axiom can be phrased in the following way. If for fixed $t_1, \ldots, t_k$, it is true that $A_n(x, y; t_1, \ldots, t_k)$ defines y uniquely as a function of x, say $y = f(x)$, then for each u, the image of f on u is a set. (The role of the $t_1, \ldots, t_k$ is merely that of "parameters" in the definition of the function f.) ZF_{6_n} is a powerful axiom in that it allows forming a new set by putting together sets chosen from a universe of sets (defined directly).

A weaker axiom would be the following axiom.

$\mathbf{ZF_{6_n}}$. Axiom of Separation

$$\forall t_1, \ldots, t_k \forall x \exists y \forall z \big(z \in y \Leftrightarrow z \in x \,\&\, A_n(z; t_1, \ldots, t_k)\big).$$

This axiom can be paraphrased by the statement that for every x and property defined by formulas $A_n(z; t_1, \ldots, t_k)$ ranging over all formulas with at least one free variable, there is a set y consisting of all z in x satisfying the given property. (As we stated under ZF_4, this axiom permits the formation of the intersection of any family of subsets of a set.)

$\mathbf{ZF_7}$. Axiom of Powers

$$\forall A \exists X \forall B (B \in X \Leftrightarrow B \subseteq A).$$

The set X so defined is the set of all subsets of A and is denoted $\operatorname{Pow} A$, called the **power set of** A. The axiom of powers implies the existence of uncountable sets because Cantor's theorem (Theorem 5) states that $\operatorname{Pow} A$ is "bigger" than A. Thus, if A is any countable infinite set (e.g., $\mathbb{N}$), then $\operatorname{Pow} A$ is not countable. (Cf. Theorem 43 which compares $\operatorname{Pow} \mathbb{N}$ and $\mathbb{R}$.)

$\mathbf{ZF_8}$. Axiom of Choice

Let I be a nonempty set, and let $\{X_i\}_{i \in I}$ be a nonempty family of nonempty sets. Then the axiom of choice is the statement that there is a mapping

$$f : I \to \bigcup_{i \in I} X_i$$

such that $\forall i \in I$ it is true that

$$f(i) \in X_i .$$

In other words, given a mapping of the nonempty set I, namely $\{\langle i, X_i\rangle\}_{i\in I}$, such that X_i is a nonempty set $\forall i \in I$, then there is another mapping $\{\langle i, f(i)\rangle\}_{i\in I}$ such that $f(i) \in X_i \forall i \in I$. This axiom implies that the choice of a single element $x_i \in X_i$ can be made simultaneously for each $i \in I$.

The **cartesian product** $\mathsf{X}_{i\in I} X_i$ of the family $\{X_i\}_{i\in I}$ of sets is defined to be the set of all mappings

$$f : I \to \bigcup_{i\in I} X_i$$

such that

$$\forall i \in I\big(f(i) \in X_i\big).$$

In other words, *the axiom of choice asserts that the cartesian product of a nonempty family of nonempty sets is nonempty.*

If $X_i = X_j = X$ for all $i \in I$, then we let X^I denote $\mathsf{X}_{i\in I} X_i$. In this case, $X^I \neq \varnothing$ without assuming the choice axiom because X^I contains x_{const}, where $x \in X$.

ZF_9. Axiom of Regularity

This axiom prohibits the existence of infinite decreasing sequences $\{x_n\}_{n\in\mathbb{N}}$ such that $x_{n+1} \in x_n \forall n \in \mathbb{N}$, and it implies that $\forall x (x \notin x)$.

This completes the axioms for ZF-set theory. The axiom of infinity is necessary for classical mathematics because there exists a model M for set theory that satisfies the other axioms but that contains only finite sets. Such an example is the set M of all finite sets generated by $\varnothing$, its successor $\varnothing \bigcup \{\varnothing\}$, and so on.

At first glance, the axioms of ZF are finite in number. Actually, the axiom of replacement requires an infinite number of statements. Essentially, the axiom of replacement states that if $y = f(x)$ is a function defined by some property A, then the range of f on any set u is a set. An infinite number of statements, however, are required to axiomatize the notion of "property". Neverthess, we can finitely axiomatize classes by GB_1—GB_{18} following.

Axioms of Gödel-Bernays

This will be called the GB system (or class theory). The primitive objects are called classes.

GB. Axioms for Sets

$$Y \in X \Rightarrow Y \ \textit{is a set}.$$

This states that every element of a class is a set. If the only primitive notions are those of class, then this axiom defines "set".

In addition, there are axioms GB_1 to GB_8 for classes just as ZF_1 to ZF_7 are described in ZF set theory with the word "class" replacing "set".

GB_8. Axiom for Replacement

$$\forall X (\forall u \exists !\, v \langle u, v\rangle \in X \Rightarrow \forall u \exists v)$$

(v is the image of the function defined by X on u).

This axiom states that if $y = \varphi(x)$ is a function defined by some property, that is, is a class, then for any set u, the image of φ on u exists.

GB. Axioms for Class Formations

GB_9. $\exists X \forall a \big(a \in X \Leftrightarrow \exists b \exists c (a = \langle b, c\rangle \,\&\, b \in c)\big)$.

GB_{10}. $\forall X \forall Y \exists Z \forall u (u \in Z \Leftrightarrow u \in X \,\&\, u \in Y)$.

GB_{11}. $\forall X \exists Y \forall u (u \in Y \Leftrightarrow u \notin X)$.

GB_{12}. $\forall X \exists Y \forall u \big(u \in Y \Leftrightarrow \forall v (\langle v, u\rangle \in X)\big)$.

GB_{13}. $\forall X \exists Y \forall u \big(u \in Y \Leftrightarrow \exists r \exists s (u = \langle r, s\rangle \,\&\, s \in X)\big)$.

GB_{14}. $\forall X \exists Y \forall a \big(a \in Y \Leftrightarrow \exists b, c (\langle b, c\rangle = a \,\&\, \langle c, b\rangle \in X)\big)$.

GB_{15}. $\forall X \exists Y \forall u \big(u \in Y \Leftrightarrow \exists a, b, c (\langle a, b, c\rangle \in X$
$\& \langle b, c, a\rangle \in Y \,\&\, \langle b, c, a\rangle = u)\big)$.

GB_{16}. $\forall X \exists Y \forall u \big(u \in Y \Leftrightarrow \exists a, b, c (\langle a, b, c\rangle \in X$
$\& \langle a, c, b\rangle \in Y \,\&\, \langle a, c, b\rangle = u)\big)$.

These axioms are the $\in$-relation (GB_9), domain (GB_{12}), and cartesian product of two classes (GB_{13}). The remaining axioms are axioms of "inversion".

GB_{17}. Axiom of Choice

$$\exists X\, \forall a\big(a \neq \varnothing \Rightarrow \exists!\, u(u \in a \,\&\, \langle a, u\rangle \in X)\big).$$

Thus, there is a class that consists of one member of each set.

GB_{18}. Axiom of Regularity

$$\forall X\big(X \neq \varnothing \Rightarrow \exists u\big(u \in X \,\&\, \forall y(y \in X \Rightarrow y \notin u)\big)\big).$$

Remarks.

1. Every theorem of ZF is a theorem of GB, and every theorem of GB that speaks only about sets is a theorem of ZF (cf. Cohen [66, p. 77]).
2. The axioms of ZF are infinite in number because of the axiom of replacement; however, the GB axioms are finite in number (cf. Cohen [66, pp. 13 and 83]).
3. Gödel [40] proved that the axiom of choice (AC) (and also the generalized continuum hypothesis (GCH), which is stated following Theorem 39 in the Foreword) is consistent with the GB axioms for classes.
4. Cohen [63, 64] proved that the axiom of choice (also the GCH) is independent of the GB axioms.
5. Sierpinski [47] proved that GCH implies AC.

Set Versus Class

Mathematics developed traditionally without distinguishing between set and class; however, Russell's paradox (p. 18) changed this. The group-valued embedding theorem for abelian categories 14.21, Freyd's adjoint functor theorem 5.50 and a theorem of Walker and the author, 20.8, are theorems distinguishing set and class[1].

Universe

A **universe** is a nonempty set U that satisfies the following conditions:

1. If $X \in U$, then $X \subseteq U$.
2. If $X \in U$, then $\operatorname{Pow} X \subseteq U$.

[1] The notation 20.8 refers to Theorem 8 in Chapter 20.

3. If $\{X_i | i \in I\}$ is a family of sets, where $I \in U$ and $X_i \in U \; \forall \, i \in I$, then $\bigcup_{i \in I} X_i \in U$.

4. If $X, Y \in U$, then the ordered pair $\langle X, Y \rangle \in U$.

An axiom asserting that every set is a member of some universe, leads to the existence of infinite sets (cf. Cohn [65, p. 8]).

Exercises. Let U be a universe.

1. If $X \in U$, and if Y is a set contained in X, then $Y \in U$.

2. If $X \in U$ and $Y \in U$, then $X \cup Y \in U$, $\{X, Y\} \in U$, and $X \times Y \in U$.

3. If $\{X_i | i \in I\}$ is a family of sets, such that $I \in U$ and $X_i \in U \; \forall \, i \in I$, then $\bigtimes_{i \in I} X_i \in U$.

Diagrams

The diagram

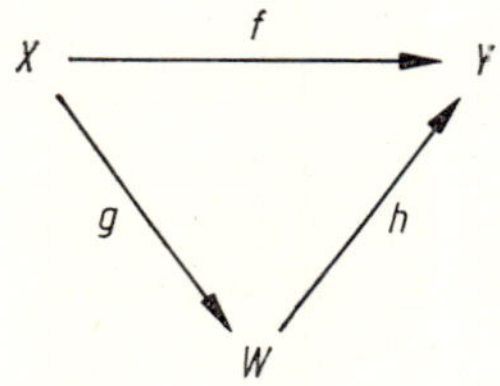

signifies that X, Y, and W are sets and that $f : X \to Y$, $g : X \to W$, and $h : W \to Y$ are mappings. The diagram **commutes**, or is said to be **commutative**, in case $hg = f$. An arbitrary (polygonal) diagram commutes if each triangular sector commutes.

Completed Diagrams

Statements I and II are defined to have the same meaning:

I. The diagram

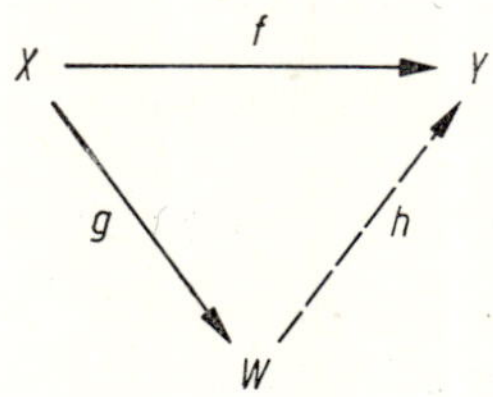

can be completed (by $h : W \to Y$).

II. There exists $h: W \to Y$ and a commutative diagram

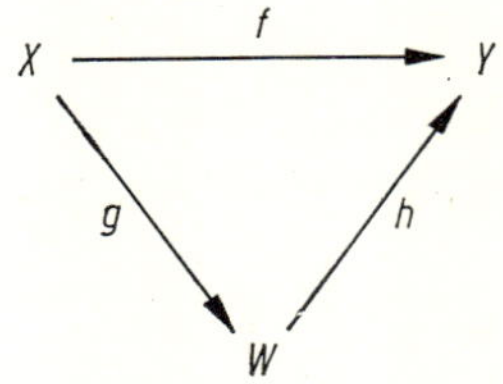

Similarly, to say that a diagram

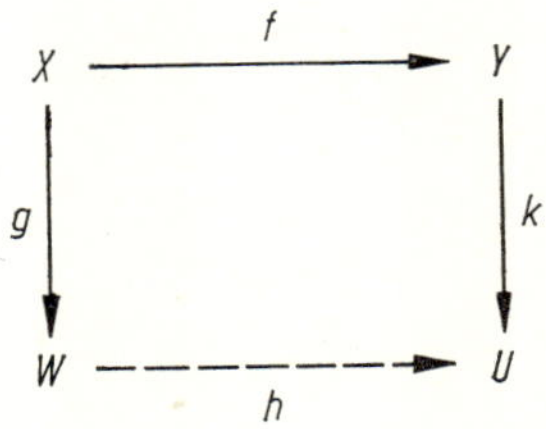

can be completed is defined to mean that there exists a mapping $h: W \to U$ making this diagram commutative.

The terminology for diagrams will be adopted without comment for diagrams consisting of objects and morphisms in abstract categories encountered in the text.

Composition and Restriction

Let X, Y, U, and V be sets, and let

$$f: X \to Y, \quad g: Y \to U, \quad h: U \to V$$

be mappings. Then the mapping

$$gf \begin{cases} X \to U \\ x \mapsto g(f(x)) \end{cases}$$

is called the **composite** or **compose** of f and g, denoted gf, or $g \circ f$. (Proof of existence of composite?) Composition of mappings is **associative** in the sense that

$$(hg)f = h(gf).$$

(The set of mappings $X \to X$ provides an example of a semigroup defined in Chapter 1.)

The notation gf for the compose of the two maps just given may be confusing because the first mapping f follows g in gf. Writing the compose in the reverse order, however, introduces confusions of its own. For example, $fg(x)$ would have to equal $g(f(x))$. For reasons of symmetry we sometimes write $(x)f$ for the value of f at $x \in X$ instead of the conventional symbol $f(x)$. When this happens, the symbol for the compose is the more satisfactory appearing fg. Thus, $(x)fg = ((x)f)g$ in this case.

Let $f: X \to Y$ be a mapping, and let U be a subset of X. Then there is a mapping

$$\begin{cases} U \to Y \\ u \mapsto f(u) \end{cases}$$

called the **restriction** of f to U, denoted $f|U$. (Proof of existence of restriction?) If $h: U \to Y$ and $f: X \to Y$ are mappings such that $h = f|U$, then f is said to be an **extension** of h to X, and we say that f **extends**, or **induces**, h. The mapping

$$1_X \begin{cases} X \to X \\ x \mapsto x \end{cases}$$

is the **identity** mapping. Thus, $1_X(x) = x \; \forall \, x \in X$. The mapping $i_U^X : U \to X$ induced by the identity mapping $1_X : X \to X$ is called the **inclusion** mapping of U in X. We also denote the inclusion by 1_U^X. Note that any mapping $h: U \to Y$ can be extended to a mapping of $f: X \to Y$; for example, pick any element $y \in Y$, and choose f so that $f|(X - U) = y_{\text{const}}$. In general, h may be extended in many ways.

If $f: A \to B$ is a mapping, then $\operatorname{im} f$ is a subset of B consisting of all $b \in B$ for which $b = f(a)$ for some $a \in A$. If $\operatorname{im} f \neq B$, then there is another function $h: A \to \operatorname{im} f$ defined by $h(a) = f(a) \; \forall \, a \in A$. We shall distinguish between the mappings f and h. In fact, if $C = \operatorname{im} f$, then

$$f = i_C^B h,$$

where $i_C^B : C \to B$ is the inclusion mapping, the diagram commutes

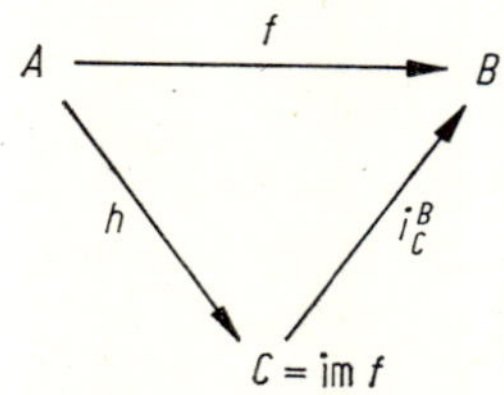

and we say that $h: A \to \operatorname{im} f$ is **induced** by f.

A mapping $f: X \to Y$ is said to be:

surjective in case $\operatorname{im} f = Y$,

injective in case $f(x) = f(x') \Rightarrow x = x' \; \forall \, x, \; x' \in X$,

bijective in case f is both injective and surjective.

The f is said to be, respectively, a **surjection**, an **injection**, and a **bijection**. A bijection $X \to X$ is called a **permutation.** The symbol $f: X \leftrightarrow Y$ will denote that f is a bijection. Note that if $f: X \to Y$ is injective, there is a unique mapping $h: \operatorname{im} f \to X$ such that $hf = 1_X$ (and in fact, if $g: Y \to X$ is any mapping which induces h, then $gf = 1_X$). Thus, an injection $f: X \to Y$ induces a bijection $X \to \operatorname{im} f$.

1. Proposition. *Let $f: A \to B$ be a mapping of nonempty sets. Then*

1.1 *f is injective if and only if there exists a mapping $g: B \to A$ such that $gf = 1_A$;*

1.2 *f is surjective if and only if there exists a mapping $g: B \to A$ such that $fg = 1_B$;*

1.3 *f is bijective if and only if there exists a mapping $g: B \to A$ such that $gf = 1_A$ and $fg = 1_B$.*

Proof of 1.1. Suppose $gf = 1_A$ for some $g: B \to A$. If $a, a' \in A$, and if $f(a) = f(a')$, then

$$a = 1_A(a) = gf(a) = gf(a') = 1_A(a') = a'.$$

Thus, f is injective. Conversely, if f is injective, and if $C = \operatorname{im} f$, then there is a mapping $h: C \to A$ defined by $f(a) \mapsto a \; \forall \, a \in A$. If $g: B \to A$ is any mapping that extends h, then $gf = 1_A$.

Proof of 1.2. If $g: B \to A$ is as given, and if $b \in B$, then

$$b = 1_B(b) = f(g(b)) \in \operatorname{im} f,$$

thus f is surjective. (This shows that the map g in 1.1 is surjective, and 1.1 shows that g in 1.2 is injective.) For each $b \in B$, let $A_b = \{a \in A \mid f(a) = b\}$. Each set A_b is nonempty, and in each A_b choose one element b'. Thus, there is a mapping $g: B \to A$, where $g(b) = b' \; \forall \, b \in B$. Clearly, $fg = 1_B$.

Proof of 1.3. By 1. and 2., f is bijective if and only if there exist mappings $g: B \to A$ and $h: A \to B$ such that $gf = 1_A$ and $fh = 1_B$.

In this case, $\forall\ b \in B$, we have

$$g(b) = g(fh(b)) = gf(h(b)) = h(b),$$

consequently, $g = h$. Together with 1. and 2. this proves 3. ∎

The axiom of choice is defined elsewhere in the foreword. Does the proof require this axiom?

Inverses

Let $f: A \to B$ be a mapping. A **left inverse** of f is a mapping $g: B \to A$ such that $gf = 1_A$, and a **right inverse** of f is a mapping $h: B \to A$ such that $fh = 1_B$. An **inverse** of f is a mapping $B \to A$ that is both a left and right inverse. The proof of 1.3 shows that a mapping f has at most one inverse. If f has an inverse, we denote it by f^{-1}. Whether or not a mapping $f: A \to B$ has an inverse, if C is any subset of B, then we define

$$f^{-1}(C) = \{a \in A \mid f(a) \in C\}.$$

In this terminology the proposition can be rephrased.

2. **Proposition.** *Let $f: A \to B$ be a mapping of nonempty sets. Then*

2.1 *f is injective if and only if f has a left inverse;*

2.2 *f is surjective if and only if f has a right inverse;*

2.3 *f is bijective if and only if f has an inverse.*

The proposition establishes that the set of bijections $X \to X$ of a nonempty set X is an example of a group (defined in Chapter 1) with respect to the operation of composition of mappings. If X is a set containing not less than three elements, then this group is noncommutative in the sense that there exist f and g (bijections) such that $fg \neq gf$. (Prove this, and show that X must have at least three elements for this.)

Equivalent Sets

Two sets A and B are **equivalent** if there exist mappings $f: A \to B$ and $g: B \to A$ such that $gf = 1_A$ and $fg = 1_B$. In this case, we write $A \approx B$; otherwise, A and B are said to be **inequivalent.** It follows immediately from the previous proposition that $A \approx B$ if and only if there exists a bijection $h: A \to B$.

The following theorem is attributed to Bernstein by Hausdorff [49, 57], and to Schröder-Bernstein by Halmos [60]; it is sometimes called the Cantor-Schröder-Bernstein theorem.

3. **Theorem.** *If A and B are sets, and if there exist injections $g : A \to B$ and $h : B \to A$, then $A \approx B$.*

Proof. First suppose that $B \subseteq A$; that is, $h = 1_B^A$. If $X_1, X_2, \ldots, X_n, \ldots$ are subsets of A, let $X_1 + X_2 + \cdots + X_n + \cdots$ denote their set union if and only if they are mutually disjoint, that is, if and only if $X_i \cap X_j = \varnothing\ \forall\, i \neq j$. For example, $A = B + C$, where C is the complement of B in A. Let Q denote the union of the sets $g(C), g(g(C)), \ldots$. Since g is injective, the sets $g(C), g(g(C)), \ldots$ are mutually disjoint; therefore, $Q = g(C) + g(g(C)) + \cdots$. Let $P = g(g(C)) + g(g(g(C))) + + \cdots$. Then $Q = g(C) + P$, and $P = g(Q)$. Since g is injective, $g|Q$ is a bijection $Q \to P$. Similarly, $f = g|C$ is a bijection $f : C \to g(C)$. Thus, $C \approx g(C)$ and $Q \approx P$. Since $Q = g(C) + P$ (disjoint union), it follows that $C + Q \approx Q$. (In fact, a bijection $k : C + Q \to Q$ is simply $k = g|(C + Q)$.) Now $B = Q + K$, where K is the complement of Q in B, and then $A = Q + K + C$. Since $Q + K \approx (C + Q) + K$, it follows that $B \approx A$. (An explicit bijection $p : A \to B$ is given by the formula: if $a \in A$, and if $a \in Q + C$, then $p(a) = k(a) = g(a)$; if $a \in K$, then $p(a) = a$.)

The proof for the general case: Let $B' = h(B)$. Then, there is an injection $q : A \to B'$, where $q = h'g$, where $h' = h|g(A)$. By the preceding proof, $A \approx B'$. But $h' : B \to B'$ is a bijection, thus $B \approx B'$, and therefore $A \approx B$, completing the proof. ▯

Characteristic Functions

Let the symbol 2 denote the set $\{0, 1\}$ consisting of two elements. For any set A, then 2^A denotes the totality of mappings of A into the set $\{0, 1\}$. If X is any nonempty subset of A, the **characteristic function** $\chi_A(X)$ of X (with respect to A) is the mapping $\chi : A \to \{0, 1\}$ defined as follows:

$$\begin{cases} \chi(a) = 1, & \text{if } a \in X, \\ \chi(a) = 0, & \text{if } a \notin X. \end{cases}$$

Thus, each nonempty subset X of A determines an element $\chi_A(X) \in 2^A$. Conversely, if $f \in 2^A$, and if $Y_f = \{a \in A \mid f(a) = 1\}$, then f is the characteristic function of Y. Thus since different subsets of A have different characteristic functions, the next theorem is proved.

4. **Theorem.** *If A is a set, then* Pow A *is equivalent to* 2^A. ▯

If A is a finite set consisting of n elements, then Pow A consists of 2^n elements. (Proof?) Since $2^n > n$, we conclude that Pow A is inequivalent to A. This is a special case of the following theorem.

5. **Cantor's Theorem.** *If A is any nonempty set, then A is not equivalent to* Pow A.

Proof. Assume a bijection $f: A \to \operatorname{Pow} A$. The proof consists of showing that this assumption leads to a contradiction. To do this, let $X = \{x \in A \mid x \notin f(x)\}$. Now $X \in \operatorname{Pow} A$, and, since f is surjective, there exists $a \in A$ such that $f(a) = X$. If $a \in X$, then by the definition of X, $a \notin f(a) = X$. This contradiction forces a to lie outside of X. But, again by the definition of X, $a \in f(a) = X$. In summary, $a \in X$ and $a \notin X$, a contradiction that proves Cantor's theorem.

Russell's Paradox

There are sets that are not members of themselves. Let A denote the set of all those sets that are not members (or elements) of themselves; that is, $A = \{B \mid B \text{ is a set and } B \notin B\}$. Is $A \in A$? Suppose so. From the definition of A we deduce that $A \notin A$, a contradiction. Then certainly $A \notin A$ is the case. But if this is true, then the definition of A implies that $A \in A$, another contradiction. This example of a "set", and the contradiction it causes, is known as Russell's paradox (after its discoverer, Bertrand Russell).

Cantor's Paradox

Cantor's theorem leads to the following statement.

6. **Cantor's Paradox.** *There does not exist a set A having the property that for any set B there exists an injective mapping $f: B \to A$.*

Proof. Deny the statement. Then there exists an injection $f: \operatorname{Pow} A \to A$. On the other hand, there is an injection

$$g \begin{cases} A \to \operatorname{Pow} A \\ x \mapsto \{x\}. \end{cases}$$

From the (Cantor-)Schröder-Bernstein theorem, deduce that $\operatorname{Pow} A \approx A$, which contradicts Cantor's theorem. ▯

The statement shows that there is no such thing as a "largest" set; that is, there is no set into which every other set can be injected. The contradiction that arises from assuming the existence of such a largest set is known as Cantor's paradox.

The paradox also can be interpreted as follows. Given "collections" of sets, we may form their union. Therefore, consider the set A, which is the union of the "collection" of all sets. Then A is a "largest" set in the sense just stated, since every set is a subset of A, and yet we have shown that A cannot be a set.

The moral of this paradox, and others such as Russell's paradox, is that some "things" are not sets, or rather that certain statements that purport to define sets, in fact, do not. This is only an apparent paradox since there is no axiom that permits the formation of, for example, the "set of all sets".

These paradoxes led to a reorganization of the axioms of set theory in which "classes", "sets", and "membership" are primitive notions, and one can speak of "the class of all sets" (see axioms of Gödel-Bernays).

It is interesting to note that Russell's paradox implies that of Cantor: Suppose that A is a set such that if X is a set, then $X \in A$. Then, by the axioms of powers and separation there exists a set B consisting of those sets that are not members of themselves. But Russell's paradox shows that B cannot be a set; hence, A is not a set.

Sometimes it is convenient to dispense with subscripts and write $A \times B$, $A \times B \times C$, and so on, for nonempty sets $A, B, C, \ldots$. We define $A \times B = \mathsf{X}_{i \in I} X_i$, where $I = \{1, 2\}$, $X_1 = A$, and $X_2 = B$. (Note in general that $A \times B = B \times A$ if and only if $A = B$.) If $I = \{1, \ldots, n\}$, for brevity write $\mathsf{X}_{i=1}^n X_i$ instead of $\mathsf{X}_{i \in I} X_i$. Thus $A \times B = \mathsf{X}_{i=1}^2 X_i$. Define $A \times B \times C = \mathsf{X}_{i=1}^3 X_i$, where $X_1 = A$, $X_2 = B$, and $X_3 = C$. Note that $(A \times B) \times C \neq A \times B \times C$, since $f \in (A \times B) \times C$ is a mapping whose domain has two elements, whereas g is a mapping whose domain has three elements.

7. *Exercises.*

7.1 Prove that $(A \times B) \times C \approx A \times B \times C$ and $A \times (B \times C) \approx A \times B \times C$ for sets A, B, and C. Generalize. Compare $A \times B$ with the set of all ordered pairs $\langle a, b\rangle$, $a \in A$, $b \in B$.

7.2 A mapping $f : X \to Y$ is said to be **epic** in case it is true that for every pair of mappings $g_i : Y \to Z$, $i = 1, 2$, the following implication holds: $g_1 f = g_2 f \Rightarrow g_1 = g_2$. Show that a mapping of sets is epic if and only if it is surjective. The mapping $f : X \to Y$ is said to be **monic** if it is true that for every pair of mappings $h_i : U \to X$, $i = 1, 2$, the following implication holds: $f h_1 = f h_2 \Rightarrow h_1 = h_2$. Show that a mapping is monic if and only if it is injective. (Cf. Epics and monics in Chapter 2.)

Binary Relations

If A is a set, a **binary relation** in A is a subset $\mathscr{R}$ of the cartesian product $A \times A$. If $(a, b) \in A \times A$, then we write $a \mathscr{R} b$ if and only if $(a, b) \in \mathscr{R}$. If $\mathscr{R}'$ denotes the complement $A \times A - \mathscr{R}$ of $\mathscr{R}$ in $A \times A$, then $\mathscr{R}'$ is also a binary relation. Also, there is an inverse binary relation $\mathscr{R}^{-1}$ defined by the formula $a \mathscr{R}^{-1} b$ if and only if $(b, a) \in \mathscr{R}$. (What subset of $A \times A$ does $\mathscr{R}^{-1}$ correspond to? Why is this binary relation designated $\mathscr{R}^{-1}$?)

If B is any subset of A, and if $\mathscr{R}$ is a binary relation in A, then $\mathscr{R}$ induces a binary relation in B. This is the relation in B corresponding to the subset $\mathscr{R} \cap (B \times B)$ of $B \times B$.

One of the most general concepts in mathematics is that of an **equivalence relation.** An equivalence relation in a set A is a binary relation $\mathscr{R}$ in A satisfying the three conditions for any $a, b, c \in A$:

Reflexivity: $a \mathscr{R} a$.

Transitivity: If $a \mathscr{R} b$ and $b \mathscr{R} c$, then $a \mathscr{R} c$.

Symmetry: If $a \mathscr{R} b$, then $b \mathscr{R} a$.

If $\mathscr{R}$ is an equivalence relation in A, then for $a \in A$, $\mathscr{R}_a = \{b \in A \mid b \mathscr{R} a\}$ is called the **equivalence class determined by** (or **containing**) a. The class of all equivalence classes of A under $\mathscr{R}$ is a set denoted by $\mathscr{R}_A$, and there is a surjective mapping

$$\begin{cases} f : A \to \mathscr{R}_A \\ f : a \mapsto \mathscr{R}_a \end{cases}$$

called **canonical.** Conversely, any mapping of A determines an equivalence relation as in the following statement.

8. Statement. *If $R : A \to X$ is a mapping, then there is an equivalence relation $\mathscr{R}$ in A defined by*

$$b \mathscr{R} a \Leftrightarrow R(a) = R(b).$$

The mapping

$$\begin{cases} \mathscr{R}_A \to X \\ \mathscr{R}_a \mapsto R(a) \end{cases}$$

is injective and is said to be ***induced*** *by R, and there is a commutative diagram:*

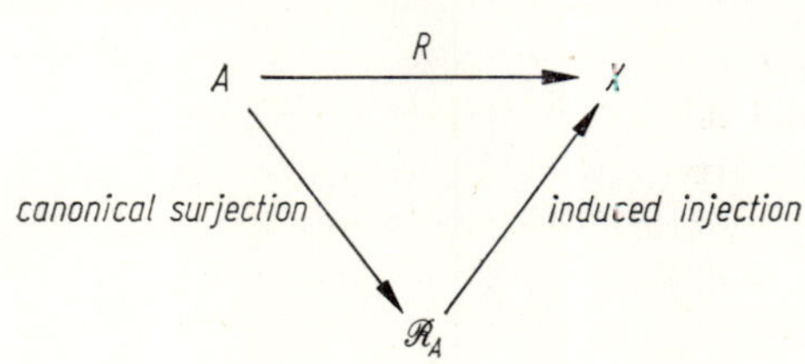

If Y is a set such that

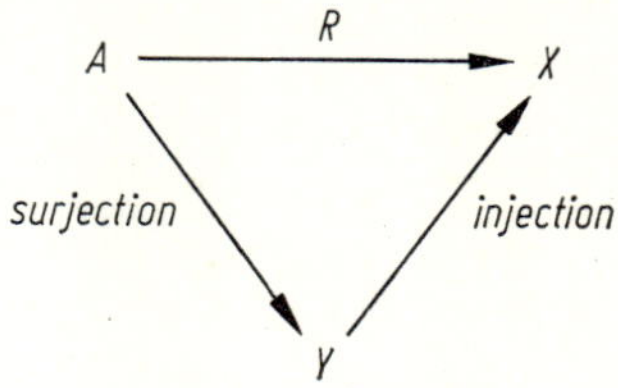

commutes, then there exists a unique equivalence $Y \to \mathscr{R}_A$ *such that*

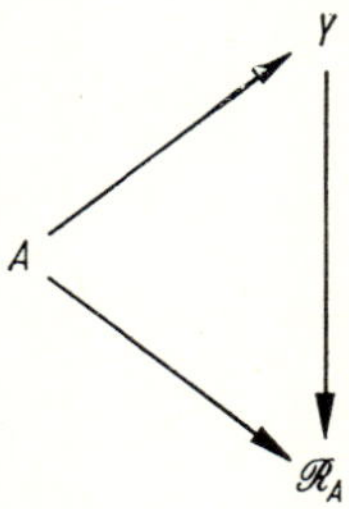

commutes. $\square$

Ordered Sets

A relation defined on a set A is said to be **antisymmetrical** in case it satisfies the law:

A. **Antisymmetry**: $a \mathscr{R} b$ and $b \mathscr{R} a \Rightarrow a = b$.

An **order** in a set A is a reflexive, transitive, antisymmetrical relation. We then call A an **ordered set.** If $\mathscr{R}$ is an order in a set A, then ordinarily the relation $\mathscr{R}$ is denoted by the symbol $\geq$. Thus, A is ordered by $\geq$ provided that it satisfies the conditions:

(1) $a \geq a$,

(2) $a \geq b$ and $b \geq c \Rightarrow a \geq c$,

(3) $a \geq b$ and $b \geq a \Rightarrow a = b$.

The symbol $a \geq b$ is read "a greater than or equal to "b". Define $b \leq a$ to mean $a \geq b$, and read "b less than or equal to a".

The notation $(A, \geq)$ indicates that A is a set ordered by a relation denoted by $\geq$. This notation is convenient when several orderings of the same set are to be distinguished.

The chief (and in some sense the only—see the next proposition) example of order is that defined in Pow X, for any set X, by the rule:

$$A \geq B \Leftrightarrow A \supseteq B.$$

This is called **order by inclusion.** This order is valid for the class of all sets, that is, we need not restrict order by inclusion to the set of subsets of some particular set.

If A and B are ordered sets, then a mapping $f: A \to B$ is an **order homomorphism** provided that

$$a \geq b \Leftrightarrow f(a) \geq f(b).$$

$\forall\, a, b \in A$. Then f is an **order injection** (resp. **surjection, bijection**) in case f is an order homomorphism and an injective (resp. surjective, bijective) mapping.

An **order isomorphism** $f: A \to B$ is a bijection such that both f and f^{-1} are order homomorphisms.

Note that every order injection $f: A \to B$ induces an order isomorphism $f: A \to \operatorname{im} f$, hence, any bijective order homomorphism is an order isomorphism.

9. Proposition. *If A is an ordered set, then there exists an order injection $A \to \operatorname{Pow} A$, where $\operatorname{Pow} A$ is ordered by inclusion.*

Proof. The function

$$f \begin{cases} A \to \operatorname{Pow} A \\ a \mapsto \{b \in A \mid b \leq a\} \end{cases}$$

is an order homomorphism. If $f(a) = f(b)$, then $a \leq b$, and $b \leq a$, so that $a = b$. Thus, f is an injection. □

Intuitively, this proposition says that every order in a set is "induced" by order by inclusion in some set. Every mapping $f: X \to Y$ of sets induces an order homomorphism $\operatorname{Pow} X \to \operatorname{Pow} Y$ sending $A \mapsto f(A)\ \forall\, A \in \operatorname{Pow} X$.

Duality

A mapping $f: A \to B$ of ordered sets A and B **reverses** order, or is an order **antihomomorphism**, provided that

$$a \geq b \quad \text{in} \quad A \Leftrightarrow f(b) \geq f(a) \quad \text{in} \quad B.$$

A **duality**, or order **anti-isomorphism**, is an antihomomorphism which is bijective. For every ordered set A there is an ordered set A^* and a duality $A \to A^*$. In fact, if $(A, \geq)$ is an ordered set, there is a new order $\geq^*$ defined by the implication

$$a \geq b \Leftrightarrow b \geq^* a.$$

This order is called **reverse**, or **dual**, order. The ordered set $(A, \geq^*)$ is called the ordered set dual to A and is denoted by A^*. Thus, the identity mapping $1_A : A \to A$ is a duality $A \to A^*$. The very simple fact of the existence of A^* for each ordered set A has useful consequences which seem to belie the triviality of this fact. The most important of these is the duality principle which asserts that for every theorem about ordered sets there is a dual theorem obtained simply by reversing order.

Let U be a universe of sets. Then U is ordered by inclusion. We call the reverse order **exclusion.** Then, there is a duality

$$d \begin{cases} U \to U \\ X \mapsto U - X \end{cases}$$

which maps intersections onto unions and unions onto intersections. Thus, for any family $\{X_i\}_{i \in I}$ of sets of U,

$$U - \bigcup_{i \in I} X_i = \bigcap_{i \in I} (U - X_i)$$

and

$$U - \bigcap_{i \in I} X_i = \bigcup_{i \in I} (U - X_i).$$

For any statement S about ordered sets, there is a **dual statement** S^* obtained from S by reversing order, that is, by interchanging $\geq$ and $\leq$ throughout. The statement S is **self-dual** provided that $S = S^*$. A set $\{S_i\}_{i \in I}$ of statements about ordered sets is **self-dual** provided that there exists a permutation p of I such that $S_i^* = S_{p(i)}\ \forall\, i \in I$. A set of statements $\{S_i\}_{i \in I}$ are **axioms** for a class C of ordered sets provided that $A \in C$ if and only if A satisfies $S_i\ \forall\, i \in I$. (Here it is assumed that it is possible to determine whether or not a given ordered set A satisfies a given axiom, or condition, S_i.)

The same terminology applies to classes of sets. If C is a class of ordered sets, *closed under order isomorphism*, then C is said to be **closed.** Let

$$C^* = \{A^* \mid A \in C\}.$$

Then C^* is called the class **dual** to C. Since C is closed under order isomorphism, and since A is order isomorphic to $(A^*)^*$ for every $A \in C$, it follows that $C = C^{**}$, where X^{**} denotes $(X^*)^*$. We say that C is **self-dual** provided that $C = C^*$. Thus, C is self-dual whenever, $\forall\, A \in C$, it is true that $A \in C$ implies $A^* \in C$.

10. Duality Principle.

10.1 *If S is a theorem about ordered sets, then S^* is a theorem about ordered sets, called the theorem dual to S.*

10.2 *Let C be a closed class of ordered sets axiomatized by a set S of statements. If S is self-dual, then C is self-dual.*

10.3 *Let C be a closed class of ordered sets. Then the following two conditions are equivalent:*

(a) *C is self-dual;*

(b) *A statement S is a theorem about the class C if and only if the dual statement is a theorem about the class C.*

Proof.

1. The first statement is a special case of the second, because the three axioms (reflexivity, transitivity, and antisymmetry) for order are each self-dual.

2. Clearly, an ordered set A belongs to C if and only if A^* belongs to C^*, that is, $A \in C$ if and only if A^* satisfies the axioms S^*, but since S is self-dual, this is equivalent to the requirement that A^* satisfy the axioms S, that is, that $A^* \in C$.

3. Assume that C is self-dual, and let S be a theorem. If $A \in C$, then $A^* \in C$, hence A^* satisfies S, or equivalently, A satisfies S^*. Thus, S^* is a theorem about C. Conversely, let S denote the statement:

S: *If* $A \in C$, *and if there exists a duality* $A \to B$, *then* $B \in C^*$. Since we are assuming that C, and hence C^*, is closed under order isomorphism, then S is a true statement about C, that is, a theorem. The dual statement S^* is the assertion:

S^*: *If* $A \in C$, *and if there exists an order isomorphism* $A \to B$, *then* $B \in C^*$. By the assumption (b), S^* must be true. Thus, $A \in C$ implies $A \in C^*$. Writing $A = B^*$, with $B \in C$, we see that $A^* = (B^*)^*$ is order isomorphic to $B \in C$, hence $A^* \in C$. Thus, C is self-dual. □

Chains and Totally Ordered Sets

A **chain** is an ordered set $(A, \geq)$ such that $\forall\, a, b \in A$ either $a \leq b$ or $b \leq a$. (A chain is sometimes called a **totally**, **linearly**, or **simply ordered** set.) The set $\mathbb{R}$ of all real numbers is a chain under the natural order (by magnitude). If $(A, \geq)$ is any ordered set, then a **chain in** A is a subset C of A such that if $a, b \in C$, then either $a \leq b$ or $b \leq a$.

An element a of an ordered set $(A, \geq)$ is a **minimal** element of A in case $b \leq a$ implies $b = a\ \forall\, b \in A$. (Maximal elements are dually defined.) A is said to have a **least** (resp. **greatest**) element if there is an element $a \in A$ satisfying $a \leq b$ (resp. $b \leq a$) $\forall\, b \in A$. It follows from the definition of order that there can be at most one least element in an ordered set A, but in general there are many minimal elements. If B is a subset of an ordered A, then a least element of B is an element b satisfying $b \leq c \,\forall\, c \in B$. Such an element may or may not exist. For example, there is no least positive real number.

Well Ordered Sets

A **well ordered set** is a chain $(A, \geq)$ such that every nonempty subset contains a least element. Then we say A is **well ordered by** $\geq$. For example, N is well ordered by its natural ordering (by magnitude).

We say that *a set A can be well ordered in case there is an order $\geq$ defined on A such that the resulting ordered set $(A, \geq)$ is well ordered.*

Since an ordered set $(A, \geq)$ induces an ordering of any subset of A, any subset of a well ordered set can be well ordered. Furthermore, if A is a well ordered set, and if $f: B \to A$ is an injection, then B can be well ordered, where $b \geq c$ if and only if $f(b) \geq f(c)$. It follows that any denumerable set (a set equivalent to a subset of $\mathbb{N}$) can be well ordered.

In their natural orderings (by magnitude), the following sets are not well ordered: $\mathbb{Z}$, $\mathbb{Q}$, $\mathbb{R}$, the rational and real numbers on the interval $[0, 1]$ (cf. the well ordering theorem discussed later).

11. *Exercises.*

11.1 If $(A, \geq)$ is an ordered set, and if every non-empty subset has a least element, then A is a well ordered set.

11.2 If A and B are well ordered sets, show that $A \cup B$ and $A \times B$ can be well ordered in a "natural" way.

Minimum and Maximum Conditions

An ordered set A satisfies the **maximum** (resp. **minimum**) condition in case every nonempty subset X of A contains an element that is maximal (resp. minimal) in X.

Any well ordered set A satisfies the minimum condition, and A in its reverse order satisfies the maximum condition. An example of this is $\mathbb{N}$, which in its natural order satisfies the minimum but not the maximum condition. In its reverse order the opposite is true.

An ordered set satisfies the **ascending chain condition** (a.c.c.) (resp. **descending chain condition**, d.c.c.) provided the following holds: If $a_1 \leq a_2 \leq \cdots \leq a_n \leq \cdots$ (resp. $a_1 \geq a_2 \geq \cdots \geq a_n \geq \cdots$) holds for any countable sequence $\{a_n \mid n = 1, 2, \ldots\}$ of elements of A, then there exists an integer k such that $a_n = a_k \; \forall \, n \geq k$.

12. **Proposition.** *In an ordered set A, the maximum (resp. minimum) condition is equivalent to the ascending (resp. descending) chain condition.*

Proof. Assume the maximum condition, and let $a_1 \leq a_2 \leq \cdots \leq a_n \leq \cdots$ be a countable chain of elements of A. Let a be a maximal element in the subset $\{a_n \mid n = 1, 2, \ldots\}$ of A. Now $a = a_k$ for some k, and hence $a_n = a = a_k \; \forall \, n \geq k$.

Conversely, assume the a.c.c., and let X be a nonempty subset of A. Then X has an element a_1, and if a_1 is not maximal in X, there is an element $a_2 \in X$ such that $a_1 < a_2$. Hence, assume there exist $a_1, \ldots, a_n \in X$ such that $a_1 < a_2 < \cdots < a_n$. Then either a_n is maximal in X, or else there exists $a_{n+1} \in X$ such that $a_n < a_{n+1}$. We conclude that either X has a maximal element or else there exists an infinite chain $a_1 < a_2 < \cdots < a_n < \cdots$ of elements of X. Only the first alternative is allowed by hypothesis. The principle of duality 10 now suffices for the parenthetical part of the theorem. □

Lattices

Let $(A, \geq)$ be an ordered set, and let X be a nonempty subset of A. If a is an element of A such that $a \geq x$ (resp. $a \leq x$) $\forall \, x \in X$, then a is called **an upper bound** (resp. **lower bound**) **of** X **in** A; a is a **least upper bound** (l.u.b.) in case a is the least element in the set of upper bounds of X in A. Thus, if X has a l.u.b. in A, it is unique. There is the dual notion of **greatest lower bound** (g.l.b.). A **lattice** is an ordered set A in which each subset $\{a, b\}$ consisting of two elements $a, b \in A$ has a l.u.b., $a \vee b$, and a g.l.b., $a \wedge b$ in A. The class of all lattices is closed and self-dual (see 10).

13. *Examples and Exercises.*

13.1 Let $A = \text{Pow}\, Y$, where $Y \neq \varnothing$. Then A is ordered by inclusion, and for $a, b \in A$, $a \vee b = a \cup b$ (set union) and $a \wedge b = a \cap b$ (set intersection). Hence, Pow Y is a lattice. Furthermore, for any set A, there is a duality

$$d \begin{cases} \text{Pow}\, A \to \text{Pow}\, A \\ X \mapsto A - X. \end{cases}$$

This duality maps intersections of subsets of A onto unions

$$d(\bigcap_{i\in I} X_i) = \bigcup_{i\in I} dX_i$$

and maps unions onto intersections

$$d(\bigcup_{i\in I} X_i) = \bigcap_{i\in I} dX_i.$$

13.2 Let $\mathbb{Z}$, as usual, denote the set of integers. If $a, b \in \mathbb{Z}$, write $a \leq b$ if and only if $a \mid b$ (a divides b). Let (a, b) = greatest common divisor of a and b, and let $[a, b]$ = least common multiple of a and b. Then $\mathbb{Z}$ is a lattice, since $a \vee b = [a, b]$ and $a \wedge b = (a, b)$.

13.3 Any chain A is a lattice, for, given $a, b \in A$, either $a \geq b$ or $b \geq a$. In the former case $a = a \vee b$ and $b = a \wedge b$.

13.4 Any "geometric" lattice is a lattice. By geometric lattice we mean a configuration, oriented as in the diagram, consisting of two classes of parallel lines that intersect each other. The points of intersection are called lattice points. The set A of lattice points is an ordered set when we write $a_1 \geq a$ in case a can be joined to a_1 by a polygonal path whose segments starting from a are always rising. Thus, $b_1 \geq b$ in the figure; but a, b, and a_2 are not comparable. Neither are a_2 and b_1, but $b_1 \geq a$. The ordered set A is a lattice: for $a, b \in A$ and $a \neq b$, the lines on a and the lines on b meet in precisely two points c, d, which are comparable. If $c \geq d$, then $c = a \vee b$ and $d = a \wedge b$.

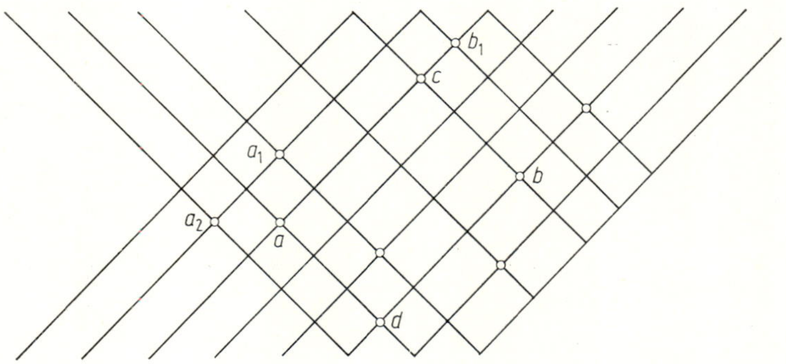

13.5 If G is a group, then the set of subgroups (resp. set of normal subgroups) is a lattice. Here $G_1 \wedge G_2 = G_1 \cap G_2$, and $G_1 \vee G_2 = G_1 \cup G_2$ (group union), for any two (normal) subgroups G_1, G_2.

13.6 The set A of all real-valued functions defined on the unit interval $[0, 1]$ is an ordered set, where $f \geq g$ means $f(x) \geq g(x) \forall x \in \mathbb{R}$.

A is also a lattice, where $f \vee g$ and $f \wedge g$ are the functions: $f \vee g(x) = \max\{f(x), g(x)\}$ and $f \wedge g(x) = \min\{f(x), g(x)\} \forall x \in \mathbb{R}$.

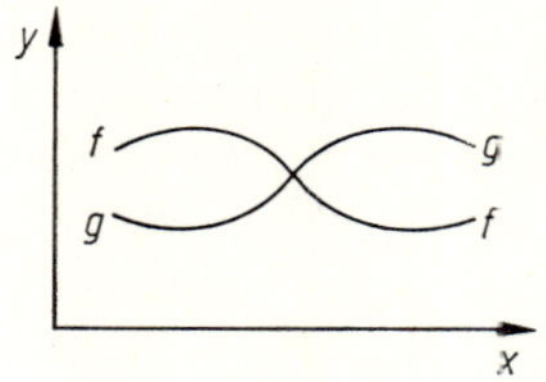

13.7 An ordered set is **complete** if every subset has a least upper bound and is **conditionally complete** if every nonempty subset bounded above has a least upper bound. Prove that a set A is conditionally complete if and only if every nonempty subset bounded below has a greatest lower bound. Examples 1 and 5 are complete lattices.

13.8 If A is a conditionally complete ordered set, if $f: A \to A$ is a nondecreasing function [in the sense that $x \leq y \Rightarrow f(x) \leq f(y) \forall x, y \in A$], and if there exist a and b in A such that

$$a \leq f(a) \leq f(b) \leq b,$$

then there exists $c \in A$ such that $a \leq c \leq b$ and $f(c) = c$. (c is called a **fixpoint** of f.)

13.9 Any nondecreasing function $f: A \to A$ of a complete ordered set A has at least one fixpoint.

13.10 If A is an ordered set with a.c.c. on subsets, then any surjective mapping $f: A \to A$ is bijective. Dually, for injective mappings.

13.11 (a) There are just two nonisomorphic ordered sets of two elements, each of which is self-dual.

(b) There are five nonisomorphic ordered sets of three elements, three of which are self-dual.

(c) The number of nonisomorphic ordered sets of four elements is 16, but there are 219 different orders possible (cf. Birkhoff [48, p. 4]).

13.12 (a) There are just five nonisomorphic lattices of five elements, three of which are self-dual.

(b) There are just four nonisomorphic lattices of less than five elements.

(c) There are just 15 lattices of six elements, of which exactly seven are self-dual (cf. Birkhoff [48, p. 17]).

13.13 (a) (Kaplansky) Every lattice of more than six elements contains a sublattice of exactly six elements.

(b) The analogous assertion for seven elements is false.

Inductive Sets

There is one consequence of the axiom of choice that is particularly suited for use in algebra, namely Zorn's lemma. Let M be a set. Recall that a chain in M is a collection C of subsets of M such that for each pair $X, Y \in C$ either $X \supseteq Y$ or $Y \supseteq X$. A collection P of subsets of M is said to be **inductive** in case it has the property: *If C is any chain in M such that each $X \in C$ belongs to P, then the union of the sets of C belong to P.* We are now able to formulate Zorn's lemma.

14. **Zorn's Lemma.** *If P is a collection of nonempty subsets of a set M, and if P is inductive, then P contains a maximal element S; that is, P contains an element S that is contained in no element of P other than S.*

For highly readable accounts of Zorn's lemma and its connection with the axiom of choice, see Kelley [56, pp. 30—36], Kurosch [56 pp. 26—29], or Halmos [60, p. 59ff.]. (In Kurosch, Zorn's lemma in an equivalent form is called the Kuratowski-Zorn theorem.)

Well Ordering Theorem

The well ordering theorem states that any set A can be well ordered. This theorem is a controversial one in mathematics. The controversy is typified by the following question: If $\mathbb{R}$ *can* be well ordered, what *is* the ordering? The point is that a well ordering of $\mathbb{R}$ is difficult, if not impossible, to visualize. Expressed otherwise, there is no known effective procedure that will determine in a well ordering of $\mathbb{R}$ when $a \geq b$ for any pair $a, b \in \mathbb{R}$. For this reason, a minority of mathematicians prefer not to use this theorem. Since the well ordering theorem is equivalent to the axiom of choice, this is tantamount to an avoidance of the axiom, which results in the loss of a number of theorems and eliminates many branches of mathematics that most mathematicians find interesting. In any case, mathematical economy, if nothing else, would require that the axiom be used only when necessary. (Axiom of necessity?) This raises the valid question (often ignored): when is the use of this axiom necessary?

The work of Gödel and Cohen has established that the axiom of choice is independent of the other axioms of set theory. Most mathematicians accept the axiom of choice without quibble and frequently use it, or its equivalents, without comment. For the most part, hereafter, so shall we.

A school of mathematics that rejects the axiom of choice is the so-called intuitionalistic mathematics. Ironically, a leader of this school is

L. E. J. Brouwer the author of a fix-point theorem in topology, which is proved using the axiom of choice!

15. Well Ordering Theorem. *Any set A can be well ordered.*

Proof. The proof uses Zorn's lemma. Let W denote the collection of all well ordered subsets of A. We (partially) order W by the rule: If $(X, \geq_1)$ and $(Y, \geq_2)$ are well ordered subsets of A, then $(X, \geq_1) \geq (Y, \geq_2)$ if and only if: (1) $X \supseteq Y$; (2) $\geq_1$ induces $\geq_2$; and (3) if $x \in X$ and $y \in Y$, and if $y \geq_1 x$, then $x \in Y$. W is nonempty since W contains the empty set. If $C = \{(X_i, \geq_i) \mid i \in I\}$ is any chain in W, then $(X, \geq) \in W$, where $X = \bigcup_{i \in I} X_i$ and $\geq$ is an ordering of X that induces $\geq_1$ $\forall i \in I$. Specifically, if $x, y \in X$, then $x, y \in X_j$ for some $j \in I$, and we define $x \geq y$ if and only if $x \geq_j y$. If also $x, y \in X_k$, then $x \geq_j y$ if and only if $x \geq_k y$. Since each X_i is a chain, $(X, \geq)$ is a chain. Furthermore, if Y is any nonempty subset of X, then $Y \cap X_i$ is nonempty for some i. Let a be a least element of $Y \cap X_i$ with respect to the ordering $\geq_i$. If $b \in Y$, then $b \in Y \cap X_j$ for some j. Suppose $b \leq a$. Then $b \leq_j a$, so $b \in X_i$ by (3). Then $b \in Y \cap X_i$, and $b \leq_i a$, so $b = a$ by the choice of a. This shows that a is the least element of Y, so $(X, \geq)$ is well ordered; that is, $(X, \geq) \in W$. Since W is therefore inductive, by Zorn's lemma, W has a maximal element which we denote by $(X, \geq)$. Assume momentarily that $A \neq X$. If $a \in A$ and $a \notin X$, let $X' = X \cup \{a\}$. Define an ordering $\geq'$ of X' as follows: If $x, y \in X'$, and if $x = a$, then $x \geq' y$; the remaining possibility is when both $x, y \in X$, in which case let $x \geq' y$ if and only if $x \geq y$. Then $(X', \geq')$ is well ordered, and $(X', \geq') > (X, \geq)$ in the ordering of W. This contradiction establishes that $X = A$, whence A is well ordered. ▯

16. Theorem. *The well ordering theorem implies the axiom of choice.*

Proof. Let $\{X_i\}_{i \in I}$ be a nonempty family of nonempty sets. Assuming that each set X_i is well ordered, then there exists a least element $x_i \in X_i$ $\forall i \in I$. Thus, $\times_{i \in I} X_i \neq \varnothing$. ▯

Product of Sets

Let $\{X_i \mid i \in I\}$ denote a family of sets indexed by a set I. A set P is called a **product** of the family $\{X_i \mid i \in I\}$ in case there exists a family $\{\pi_i : P \to X_i\}$ of mappings with the following property: If X is any set, and if $\{g_i : X \to X_i\}$ is any family of mapping, then there exists a unique mapping $\prod_{i \in I} g_i : X \to P$ such that

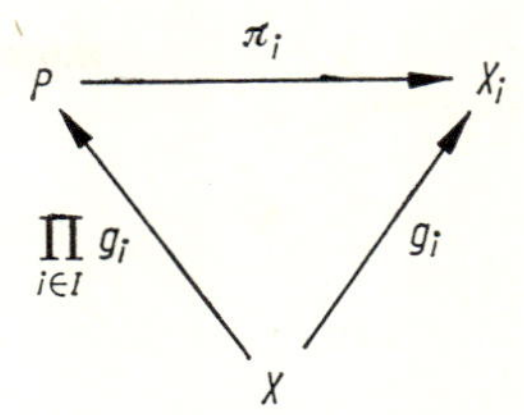

commutes $\forall i \in I$. When this is so, we denote P by $\prod_{i \in I} X_i$.

The mapping $\prod_{i \in I} g_i : X \to P$ is called the **product of the family of mappings** $\{g_i\}_{i \in I}$. The mapping $\pi_i : P \to X_i$ is called the ***i*th projection** from the product.

17. **Proposition.** *Let $\{X_i \mid i \in I\}$ be any family of sets.*

17.1 *The cartesian product $\times_{i \in I} X_i$ is a product, with projections*

$$\begin{cases} p_i : \times_{i \in I} X_i \to X_i \\ p_i : f \mapsto f(i) \end{cases}$$

$\forall i \in I,\ f \in \times_{i \in I} X_i$.

17.2 *If $\{\pi_i : P \to X_i\}$ is any product of the family $\{X_i\}_{i \in I}$, then*

$$\prod_{i \in I} p_i : \times_{i \in I} X_i \to P$$

is the unique equivalence such that

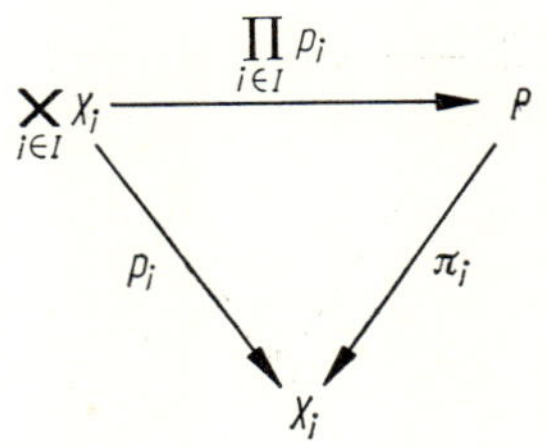

commutes $\forall i \in I$.

Proof. $\times_{i \in I} X_i$ is a product because given $\{g_i : A \to X_i\}$, then

$$g = \prod_i g_i \begin{cases} A \to \times_{i \in I} X_i \\ a \mapsto g(a), \end{cases}$$

where $g(a) : i \mapsto g_i(a) \forall i \in I$, is the unique mapping such that

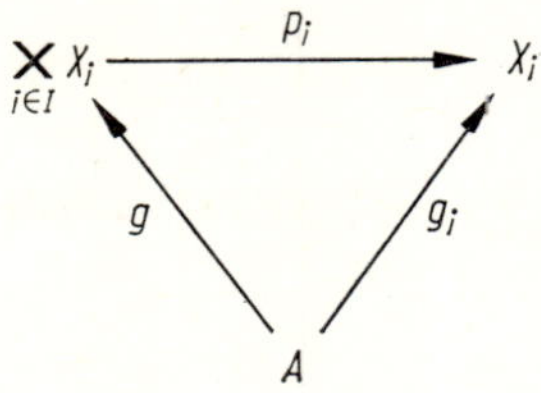

commutes. Let $F = \coprod_{i\in I} p_i$ denote the unique mapping such that the diagram in the statement of the proposition commutes, and let $G : P \to \times_{i\in I} X_i$ denote the corresponding mapping such that $p_i G = \pi_i \forall_i \in I$, given by the fact P is a product. Then

$$p_i GF = \pi_i F = p_i$$

$\forall i \in I$. But, by the definition of product, the identity mapping 1_X is the unique mapping $H : X \to X$, where $X = \times_{i\in I} X_i$, such that $p_i H = p_i \forall i \in I$. Thus, $GF = 1_X$. By symmetry, $FG = 1_P$. Therefore, $F = \coprod_{i\in I} p_i$ is an equivalence and is unique as asserted. □

Coproduct and Disjoint Union

Let $\{f_i : A_i \to A\}_{i\in I}$ be a family of mappings indexed by a set I. Then A is said to be a **coproduct** of this family of mappings, provided that the following condition holds:

If $\{g_i : A_i \to X\}_{i\in I}$ is any family of mappings, then there exists a unique mapping $g : A \to X$ making the diagram

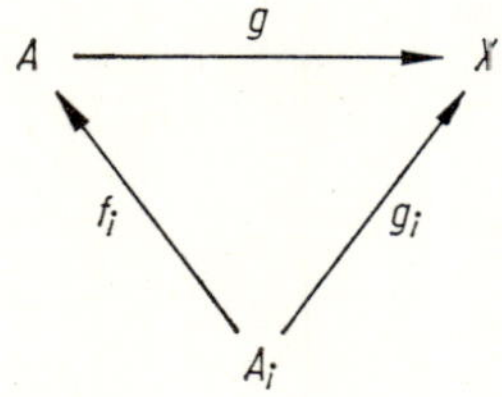

commutative.

The proposition on the existence of coproducts (Proposition 19) states that any two coproducts for the family $\{f_i : A_i \to A\}$ are equivalent. Denote the coproduct by $\coprod_{i\in I} A_i$, and denote the mapping g by $\coprod_{i\in I} g_i$, the **coproduct of the family of mappings** $\{g_i\}_{i\in I}$.

In addition to the duality principle for ordered sets, there is a duality principle for mappings. A formal statement of this for categories appears in Chapter 1 (1.6) and will not be given here. Intuitively, this principle can be applied to any statement about mappings of sets to derive another statement by interchanging domain and range and reversing arrows. In this sense coproduct is dual to product, and surjection is dual to injection.

Define a **disjoint union** of the family $\{A_i \mid i \in I\}$ to be a set B together with a family of injections $\{f_i : A_i \to B\}$ such that:

(1) $$\operatorname{im} f_i \bigcap \operatorname{im} f_j = \varnothing \text{ when } i \neq j,\ \forall i, j \in I;$$

(2) $$B = \bigcup_{i\in I} \operatorname{im} f_i.$$

18. Lemma. *If $\{A_i \mid i \in I\}$ is a family of sets, the disjoint union exists.*

Proof. Let B be the union of the family $\{C_i \mid i \in I\}$ of sets, where $C_i = \big\{\{a\} \times \{i\} \mid a \in A_i\big\}\ \forall i \in I$. Thus, the sets $\{C_i \mid i \in I\}$ are mutually disjoint, and the injections

$$f_i \begin{cases} A_i \to B \\ a \mapsto \{a\} \times \{i\} \end{cases}$$

show that B is the disjoint union. ▯

19. Proposition.

19.1 *The disjoint union $\{f_i : A_i \to B\}$ is a coproduct of the family $\{A_i\}_{i\in I}$ of sets.*

19.2 *If $\{X_i\}_{i\in I}$ is any family of sets, and if $\{h_i : X_i \to X\}_{i\in I}$ and $\{g_i : X_i \to Y\}_{i\in I}$ are coproducts, then there is a unique equivalence $f : X \to Y$ making the diagram*

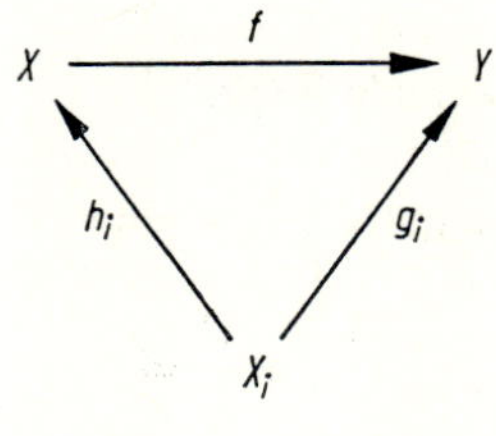

$\forall i \in I$ commutative. In particular, the mappings $\{h_i : X_i \to X\}_{i\in I}$ are injections. Furthermore, if $\{k_i : X_i \to Y_i\}_{i\in I}$ is any family of equivalences of sets, then the coproduct $\coprod_{i\in I} k_i$ of the mappings is an equivalence $\coprod_{i\in I} X_i \approx \coprod_{i\in I} Y_i$.

Proof. Exercise. ▯

20. *Exercises.*

20.1 If $\{X_i \mid i \in I\}$ and $\{Y_j \mid j \in J\}$ are two families of sets, if there exists an equivalence $p: I \to J$, and if there exist equivalences $X_i \approx Y_{p(i)}$ for every $i \in I$, then $\prod_{i \in I} X_i$ is equivalent to $\prod_{j \in J} Y_j$. Also, $\coprod_{i \in I} X_i \approx \coprod_{j \in J} Y_j$.

20.2 For sets X, I, and J, an equivalence $I \approx J$ implies an equivalence $X^I \approx X^J$. Conversely.

20.3 The projection mapping $\prod_{i \in I} X_i \to X_j$ is surjective for each $j \in I$. Demonstrate an injection $X_j \to \prod_{i \in I} X_i$. Dually for coproduct.

20.4 If $\{X_i\}_{i \in I}$ is a family of ordered sets, and if I is well ordered, then the **lexicographical ordering** of $\times_{i \in I} X_i$ is defined by: If f and g lie in the product, and if a is the least element of I such that $f(a) \neq g(a)$, then $f \leq g$ if and only if $f(a) \leq g(a)$.

(a) Show that if each X_i is well ordered, the lexicographical ordering of the product is a well ordering.

(b) Demonstrate a well ordering of the coproduct of a family $\{X_i\}_{i \in I}$ of well ordered sets indexed by a well ordered set I.

20.5 Prove that the well ordering theorem implies Zorn's lemma.

20.6 Prove that the axiom of choice implies Zorn's lemma.

20.7 Fully justify the assertion that the notions of injection and surjection are dual.

20.8 Prove Proposition 19 by "dualizing" the proof of Proposition 17.

Ordinals

The rest of the Foreword is on the subject of ordinal and cardinal numbers, following the development of Cohen [66]. We omit the proofs which may be found in Cohen's book.

Let A be a well ordered set. A **segment** of A is a subset B such that if $a \in A$ and if there exists $b_1, b_2 \in B$ such that $b_1 < a < b_2$, then $a \in B$. A segment of A that contains the least element of A is an **initial** segment. Also, for any $c \in A$, $\{a \in A \mid a \leq c\}$ is an initial segment of A, denoted by A_c.

If A is well ordered, and if $C \subseteq A$, then $x = \sup C$ denotes the least element $a \in A$ such that $c \leq a \; \forall \, c \in C$. (Sup abbreviates *supremum* another term for l.u.b.)

An injection $f: A \to B$ of well ordered sets is called an **initial order injection** if im f is an initial segment of B and f is an order homomorphism.

21. **Lemma.** *Let A and B be well ordered sets. There can be at most one initial order injection $f: A \to B$.*

22. **Theorem.** *If A and B are well ordered sets, then there exists a unique initial order injection of one into the other.*

23. **Corollary (Comparability of Sets).** *If A and B are sets, then either there exists an injection $A \to B$ or there exists an injection $B \to A$.*

If A and B are well ordered sets, then define

23.1 $\tilde{A} \leq \tilde{B}$ in case there exists an initial order injection $A \to B$;

23.2 $\tilde{A} = \tilde{B}$ in case there is an order isomorphism $A \to B$;

23.3 $\tilde{A} < \tilde{B}$ in case $\tilde{A} \leq \tilde{B}$, but $\tilde{A} \neq \tilde{B}$.

24. **Proposition.** *If A and B are well ordered sets, then precisely one of the following holds:* $\tilde{A} < \tilde{B}$, $\tilde{A} = \tilde{B}$, or $\tilde{B} < \tilde{A}$.

A set A is **transitive** in case

$$\forall x, y, z \in A\,(y \in x \,\&\, z \in y \Rightarrow z \in x).$$

The $\in$**-relation** is a relation $\mathscr{R}$ defined in A by:

$$x \mathscr{R} y \Leftrightarrow (y \in x) \vee (y = x).$$

Thus, if A is transitive, then

$$y < x \Leftrightarrow y \in x$$

and

$$y \leq x \Leftrightarrow (y < x) \vee (y = x).$$

In other words, if A is transitive, then the $\in$-relation orders A. An **ordinal** is defined to be a transitive set in which the $\in$-relation is a well ordering.

25. **Proposition.** *The following properties of ordinals hold:*

25.1 *Any initial segment of an ordinal is an ordinal.*

25.2 *If α is an ordinal, and if $x \in \alpha$, then $x = \{y \in \alpha \mid y < x\}$, and x is an ordinal.*

25.3 *If α is an ordinal, then $\alpha \cup \{\alpha\}$ is an ordinal, the successor of α.*

26. *Examples.* The following examples of ordinals are deduced from 25.3

$$\varnothing,$$

$$\{\varnothing\} = \varnothing \cup \{\varnothing\},$$

$$\{\varnothing, \{\varnothing\}\} = \{\varnothing\} \cup \{\{\varnothing\}\},$$

and so forth.

27. *Exercise.* $\{\varnothing\}$ is the unique ordinal with precisely one element.

28. Theorem. *If A is any well ordered set, there exist a unique ordinal α and order isomorphism $f: A \to \alpha$.*

29. Corollary. *If α and β are ordinals, then*

$$\tilde{\alpha} = \tilde{\beta} \Rightarrow \alpha = \beta$$

and

$$\tilde{\alpha} < \tilde{\beta} \Rightarrow \alpha \in \beta .$$

30. Corollary. *If α is an ordinal, then*

$$\alpha = \{\beta \mid \tilde{\beta} < \tilde{\alpha}\} . \quad \square$$

Corollary 30 expresses the fact that *an ordinal is the set of all ordinals that precede it.*

31. Corollary. *If α is an ordinal, then $\alpha \cup \{\alpha\}$ is the least ordinal greater than α and is denoted by $\alpha + 1$.* $\square$

32. Corollary. *If S is any nonempty set of ordinals, there exists a least ordinal in S.*

33. Theorem. *If S is a set of ordinals, there exists a least ordinal α such that $\beta \leq \alpha \forall \beta \in S$, and we write $\alpha = \sup S$.*

An ordinal α is a **successor** if there exists an ordinal β such that $\alpha = \beta + 1$. An ordinal α is a **limit ordinal** in case $\alpha \neq 0$ (the least ordinal) and α is not a successor.

An ordinal α is an **integer** (or natural number) if $\beta \leq \alpha$ implies that β is a successor or 0.

34. Theorem. *There exists a limit ordinal.*

35. Theorem. *Let x be any limit ordinal. The least limit ordinal $\omega \leq x$ is the set of all integers (natural numbers).*

36. Theorem (Mathematical Induction). *Let $x \subseteq \omega$, let $\varnothing \in x$, and assume:*

$$\forall n (n \in x \Rightarrow n + 1 \in x) .$$

Then, $x = \omega$. $\square$

37. Transfinite Induction Theorem. *Let $A_n(x, y; t_1, \ldots, t_k)$ denote all formulas in ZF with at least two free variables x and y, let $t_1, \ldots, t_k$ be given, and assume that*

$$\forall x \exists ! y \big(A_n(x, y; t_1, \ldots, t_k)\big) .$$

Let $y = \varphi(x)$ *denote the function so defined. Then for every ordinal* α *and set* z *there is a unique function* f *defined on the set* $\alpha + 1 = \{\beta \mid \beta \leq \alpha\}$ *such that* $\forall \beta \leq \alpha$, $f(0) = z$ *and* $f(\beta) = \varphi(h)$, *where* h *is* f *restricted to* β.

The simplest consequence of transfinite induction is this: *Let* A *be a class of ordinals containing* 0, *containing* $a + 1$ *whenever* $a \in A$, *and containing* sup X *whenever* X *is a subset of* A. *Then* A *contains all ordinals.* As illustrations of transfinite induction, one may define ordinal sum, product, and exponentiation.

38. **Ordinal Sum.** For fixed α, define $\alpha + 0 = \alpha$ and $\alpha + \beta = \sup\{\alpha + \gamma \mid \gamma < \beta\}$, if α is a limit ordinal. Define $\alpha + (\beta + 1) = (\alpha + \beta) + 1$.

Ordinal Product. For fixed α, define $\alpha \cdot 0 = 0$ and $\alpha \cdot (\beta + 1) = \alpha \cdot \beta + \alpha$, and if β is a limit ordinal, define $\alpha \cdot \beta = \sup\{\alpha \cdot \gamma \mid \gamma < \beta\}$.

Exponentiation. For fixed α, define $\alpha^0 = 1$, $\alpha^{\beta+1} = \alpha^\beta \cdot \alpha$, and if β is a limit ordinal,

$$\alpha^\beta = \sup\{\alpha^\gamma \mid \gamma < \beta\}.$$

Cardinals

Let A and B be sets. Write $|A| \leq |B|$, if there exists an injection $A \to B$, and write $|A| = |B|$ in case there is a bijection $A \to B$. An earlier theorem implies that if A and B are sets, then either $|A| \leq |B|$ or else $|B| \leq |A|$. The Cantor-Schröder-Bernstein theorem asserts the implication:

$$|A| \leq |B| \;\&\; |B| \leq |A| \Rightarrow |A| = |B|.$$

We say that A and B have the same **cardinality** (numerosity) when $|A| = |B|$.

A **cardinal** is defined to be an ordinal α such that the following implication holds for ordinals β:

$$\beta < \alpha \Rightarrow |\beta| < |\alpha|.$$

This defines a cardinal to be the least ordinal of a given cardinality.

Since every set can be well ordered by the Well Ordering Theorem 15, and since every well ordered set is order isomorphic to an ordinal by Theorem 28, it follows that every set A is equivalent to a unique cardinal $|A|$. In this case, we say that A **has cardinal** $|A|$, and it follows that sets A and B are equivalent if and only if they have the same cardinal, that is, if and only if $|A| = |B|$. (The notation $|A|$ for cardinal therefore is consistent with its use for cardinality.) Cantor's theorem (Theorem 5) implies that $|\mathrm{Pow}\, A| > |A|$, for any set A, and shows that there exist arbitrarily large cardinals (and ordinals).

39. Theorem. *There is a bijection*

$$\begin{cases} \text{ordinals} \to \text{cardinals} \\ \alpha \mapsto \aleph_\alpha, \end{cases}$$

where $\aleph_\alpha$ *is the least infinite cardinal greater than* $\aleph_\beta$ *for all* $\beta < \alpha$.

The cardinality of ω is $\aleph_0$. An ordinal is **countable** if its cardinal is $\aleph_0$. Thus, ω is the least countable ordinal, that is, $\omega = \aleph_0$. An ordinal α is **finite** if $\alpha \in \omega$. Thus, the finite ordinals are cardinals (integers).

Since $|\mathrm{Pow}\, \omega| > \omega = \aleph_0$, the question arises as to which $\aleph_\alpha$ is $|\mathrm{Pow}\, \omega|$. The **continuum hypothesis** states that

$$c = |\mathrm{Pow}\, \omega| = \aleph_1,$$

that is, that the cardinality of Pow ω is the first uncountable cardinal. The **generalized continuum hypothesis** states that, for every ordinal α,

$$|\mathrm{Pow}\, \aleph_\alpha| = \aleph_{\alpha+1}.$$

The set of real numbers on the closed interval $[0, 1]$ is defined to be the **continuum**. The continuum hypothesis (CH) receives its name from the fact that the continuum has cardinality $|\mathrm{Pow}\, \omega|$ (Theorem 43). In ZF-set theory, the generalized continuum hypothesis implies the axiom of choice (Sierpinski [47]), but the axiom of choice does not imply the generalized CH (Cohen [63, 64]). Furthermore, CH does not imply AC, nor do AC and CH imply the generalized CH (Solovay [63]).

If α and β are countable ordinals, then $\alpha + \beta$, $\alpha\beta$, and α^β are all countable, e.g., $|2^\omega| = \omega$. This illustrates the obvious fact that ordinal exponentiation is not related to cartesian product, except notationally. Formerly A^B represented the cartesian product, namely the set of all mappings $B \to A$. For example, the power set Pow ω is equivalent to the cartesian product 2^ω (Theorem 4), and Pow ω is not equivalent to ω by Cantor's theorem. Thus, $|2^\omega| \neq \omega$. Ordinal exponentiation is so infrequently used, that no confusion should result.

Cardinal Arithmetic

Ordinal sum $\alpha + \beta$ and product $\alpha\beta$ of two ordinals have been defined. Cardinals are ordinals, yet neither ordinal sum nor ordinal product of two cardinals need be a cardinal. For example $\omega + 1 > \omega$, yet $|\omega + 1| = |\omega|$. Thus, since a cardinal is the least ordinal of given cardinality, it follows that $\omega + 1$ is not a cardinal.

40. Theorem and Definition.

40.1 *For any two ordinals α and β,*

$$\big||\alpha| + |\beta|\big| = |\alpha + \beta| = |\alpha \amalg \beta| = \max\{|\alpha|, |\beta|\}$$

and

$$|\alpha|\,|\beta| = |\alpha\beta| = |\alpha \times \beta| = \max\{|\alpha|, |\beta|\},$$

where $\alpha \times \beta$ denotes the cartesian product and $\alpha \amalg \beta$ denotes the disjoint union (or coproduct) of the sets α and β.

40.2 *Cardinal sum and product are defined for ordinals $\{\alpha_i\}_{i\in I}$ by*

$$\sum_{i\in I} |\alpha_i| = \Big|\coprod_{i\in I} \alpha_i\Big|$$

and

$$\prod_{i\in I} |\alpha_i| = \Big|\bigtimes_{i\in I} \alpha_i\Big|.$$

Proof. Exercise. □

41. *Exercises.*

41.1 For any limit ordinal α, prove that $1 + \alpha = \alpha \neq \alpha + 1$ and that $n\alpha \neq \alpha n$ for any integer $n \neq 0$.

41.2 Assuming the generalized continuum hypothesis, prove that, for any set A, the set of all mappings $A \to A$ is equivalent to Pow A, that is, $A^A \approx 2^A$. (Hint: $A \approx 2^B$, for some set B.)

41.3 Prove that the continuum is equivalent to the set $\mathbb{R}$ of real numbers, that is, $c = |\mathbb{R}|$.

41.4 For any sets A and B, show that the set of bijections of A is equivalent to the set of bijections of B if and only if A is equivalent to B. If A has infinite cardinality α, show that the set of bijections of A has cardinality α^α. Assuming the generalized continuum hypothesis, prove that the set of bijections of A has cardinality 2^α, where α is the cardinality of A.

41.5 Prove that $\alpha\beta$, $\alpha + \beta$, and α^β are countable when α and β are countable.

41.6 Show that cardinal sum and product are commutative, associative, and distributive (see Chapter 1 for definitions).

41.7 *Let $\{a_i \mid i \in I\}$ be a family of cardinal numbers, where $a_i = a$ $\forall i \in I$. Then $\sum_{i\in I} a_i = a\,|I| = \max\{a, |I|\}$.*

41.8 *For any infinite cardinal number a and natural number n,*

$$a^n = a \quad \text{and} \quad na = a.$$

41.9 *Let X and Y be infinite sets, and let $f : X \to Y$ be a surjection. For each $y \in Y$, let $a_y = |f^{-1}(y)|$. Then*

$$|X| = (\textstyle\sum_{y \in Y} a_y) \cdot |Y|.$$

Furthermore, if $|f^{-1}(y)| \leq |Y| \; \forall y \in Y$, then

$$|X| = |Y|.$$

(In particular, $|X| = |Y|$ in case $|f^{-1}(y)| < \aleph_0 \forall y \in Y$.)

41.10 Start with a countable sequence $\{F_n \mid n = 1, 2, \ldots\}$ of finite sets, and for each n, assume that there is a map Φ_n of F_n into Pow F_{n+1}. In order to simplify notation, denote Φ_n by Φ, $\forall n$, and the union of the given family of finite sets by $\mathscr{F}$. A **path** in (the ordered pair) $(\mathscr{F}, \Phi)$ is a finite or infinite sequence of elements $b_1, \ldots, b_n, \ldots$ of $\mathscr{F}$ such that $b_i \in F_i$ and $b_{i+1} \in \Phi(b_i)$, $i = 1, 2, \ldots$. The length of a finite path $b_1, \ldots, b_m$ is m; the length of the infinite path $b_1, b_2, \ldots$ is infinite. Prove: **König's Graph Lemma.** *If $\{\mathscr{F}, \Phi\}$ has paths of ever greater length, then it has a path of infinite length.*

The continuum hypothesis implies that c is uncountable. This can be proved without recourse to the continuum hypothesis.

42. Theorem. *c is uncountable.*

Proof. Any real number in $[0, 1]$ has an essentially unique decimal representation $0 \cdot a_1 a_2 \cdots a_n \cdots$, where a_i is a natural number between 0 and 9; essentially unique, since for example,

$$0.1575999\ldots$$

also has the representation

$$0.1576000\ldots$$

This ambiguity can be avoided, for example, by always choosing the first representation, that is, by not allowing any decimal to terminate (in zeros).

If c is countable, then we can arrange the decimals $0.a_1 a_2 \cdots$ in a countable sequence:

$$\begin{array}{l} 0.a_{11} a_{12} \cdots a_{1n} \cdots \\ 0.a_{21} a_{22} \cdots a_{2n} \cdots \\ \vdots \quad \vdots \quad \vdots \qquad \vdots \\ 0.a_{m1} a_{m2} \cdots a_{mn} \cdots \\ \vdots \quad \vdots \quad \vdots \qquad \vdots \end{array}$$

Let $b = 0.b_1 b_2 \cdots b_n \cdots$ be a decimal whose n-th digit b_n is a natural number (between 0 and 9) not equal to a_{nn} or 0. Then b is a number not in the list, since its n-th digit differs from that of any decimal in the list. But we had assumed that we could list all the decimals in such a countable sequence. This contradiction proves that c is uncountable. $\square$

The method of proof is **Cantor's diagonal argument** and is pervasive in infinite mathematics. (The proof of Cantor's theorem, Theorem 5, employs Cantor's diagonal theorem. Cf. Abian [65, p. 255]. Also, see Cohen [66, pp. 39] for Godel's diagonalization methods).

43. **Theorem.** $2^{\aleph_0} = c = |\mathbb{R}|$.

44. **Corollary.** For any natural number n,

$$c = 2^{\aleph_0} = 3^{\aleph_0} = \cdots = n^{\aleph_0} = \cdots.$$

45. *Exercises.*

45.1 If a and b are cardinal numbers, and if n is an integer, then (a) $a + n = b + n \Rightarrow a = b$ and (b) $na = nb \Rightarrow a = b$.

45.2 Use $c = 2^{\aleph_0}$ to derive $c = c^2 = c^3 = \cdots = c^{\aleph_0}$. Conclude that all real vector spaces of finite or countable dimensions have the same cardinal.

45.3 Prove that $c^c = 2^c$.

45.4 Prove that $\aleph_0^c = 2^c$.

45.5 Prove König's theorem: The sum of an infinite series of cardinal numbers different from zero is less than their product.

45.6 If a, b, c, and d are cardinals, and if $a \leq c$ and $b \leq d$, then $a^b \leq c^d$.

45.7 If 2 and 3 are factors of a cardinal, then 6 is a factor.

General References

Abian, A.: The Theory of Sets and Transfinite Arithmetic. Philadelphia: Saunders 1965.

Birkhoff, G.: Lattice Theory, Colloquium Publication. Vol. 25 (revised). Amer. Math. Soc., Providence 1948, 1967.

Bourbaki, N.: Theorie des Ensembles. Actualités Scientifiques et Industrielles, Nos. 1212, 1243. Paris: Hermann 1960, 1963, Chapters 1–3.

Cantor, G.: Contributions to the Founding of the Theory of Transfinite Numbers. New York: Dover Publ. 1942.

Cohen, P. J.: Set Theory and the Continuum Hypothesis. New York: Benjamin 1966.

— The independence of the continuum hypothesis, I, II. Proc. Nat. Acad. Sci. U.S.A. **50**, 1143–1148 (1963); **51**, 105–110 (1964).

Cohn, P. M.: Universal Algebra. New York: Harper & Row 1965.

Fraenkel, A.: Über den Begriff „definit" und die Unabhängigkeit des Auswahl axioms. Sitzungsber. Preuss. Akad. wiss. Phil.-math. **1922**, 253—257.
Bar-Hillel, Y.: Set Theory. Amsterdam: North-Holland Publ. 1958.
Gödel, K.: The Consistency of the Continuum Hypothesis. Princeton: Princeton University Press 1940; 1958.
Halmos, P.: Naive Set Theory. New York: Van Nostrand 1960.
Hausdorff, F.: Set Theory. New York: Chelsea 1957.
— Mengenlehre. Leipzig 1914.
Kamke, E.: Theory of Sets. New York: Dover Publ. 1950.
Kelley, J.: General Topology. New York: Van Nostrand 1955.
Kurosch, A.: Theory of Groups, Vols. 1 and 2. New York: Chelsea 1955, 1956.
— Lectures on General Algebra. New York: Chelsea 1963.
Lawvere, F.: Elementary theory of the category of sets. Proc. Nat. Acad. Sci. U.S.A. **52**, 1506—1511 (1964).
— The category of categories as a foundation for mathematics. Proc. Conf. Categorical Algebra. Amsterdam: North-Holland Publ. 1965, pp. 1—20.
Sierpinski, W.: L'Hypothèse généralisée du continu et l'axiome du choix. Fund. Math. **24**, 1—5 (1947).
— Cardinal and Ordinal Numbers. Monographs of the Polish Academy, Warsaw, 1958.
Solovay, R.: Independence results in the theory of cardinals (Preliminary Report), Notices of the Amer. Math. Soc. **10**, 595, (1963).
Zermëlo, E.: Beweis, daß jede Menge wohlgeordnet werden kann. Math. Ann. **59**, 514—516 (1904).

I. Introduction to the Operations: Monoid, Semigroup, Group, Category, Ring, and Module

Chapter 1

1. Operations: Monoid, Semigroup, Group, and Category

In addition to the topics in the title, this chapter presents the operations of direct product and sum, quotient group, the Noether homomorphism theorems, the Zassenhaus butterfly lemma, and exercises on permutation and alternating groups, solvable and nilpotent groups, simple groups, the Sylow theorems, and the basic theorems for abelian groups.

Operation

The concept of an operation is the basis for the definitions of monoid, semigroup, group, category, ring, and module presented in the first two chapters, and, in turn, these are the subject of the book. If G and S are sets, then G **operates** on S in case there is a mapping $f: G \times S \to S$. The mapping f is called a (binary) **operation**, and the elements of G are **operators**. When f is fixed, then gx denotes $f(g, x)$, or when there are two operations, then variously $g \cdot x$, $g \circ x$, or $g + x$, denote the image of (g, x) under the second operation. If $x \in S$, then the set $\{gx \mid g \in G\}$ is the **orbit** of x under G. A subset T of S is **stable** under G provided that T contains the orbit of each of its elements.

Monoids and Semigroups

A special case of an operation occurs when a set S operates on itself. An operation $S \times S \to S$ is **associative** provided that the formula

$$(xy)z = x(yz)$$

is valid $\forall\ x, y$, and $z \in S$. A **monoid** is defined to be an ordered pair consisting of a set S and an associative operation $S \times S \to S$. An **identity** element of a monoid S is an element $1 \in S$ satisfying the formula

$$1x = x1 = x$$

$\forall x \in S$. A monoid can have at most one identity element, since if e is another, then $1 = e1 = e$. A **semigroup** is a monoid with an identity element. A **multiplicative (additive)** monoid or semigroup is one such that xy (resp. $x + y$) denotes the image of (x, y) under the operation, and then the operation is called **multiplication (addition)**. In an additive semigroup, the symbol 0 denotes the identity element, called **zero**, and in this case $0 + x = x + 0 = x$ for all x.

Suboperation

If $S \times S \to S$ is an operation, then a **suboperation** is an induced operation $T \times T \to T$, where T is a subset of S. If S is a monoid, then T is also a monoid, called a **submonoid** of S. If S is a semigroup, however, a submonoid T of S may not be a semigroup since T may not have an identity element. If S is a semigroup with identity 1, then a **subsemigroup** T is a submonoid with identity element e. Thus, a subsemigroup is a semigroup with identity e possibly $\neq 1$. The intersection of any number of submonoids is a submonoid,, and the intersection of any number of subsemigroups having the same identity element is also a subsemigroup.

A **unit** in a semigroup S is an element x for which there exists a $y \in S$ such that $xy = yx = e$, where e is the identity. y is unique denoted x^{-1}, and called x **inverse**. The element x is then said to be **invertible**. If S is additive, then $-x$ denotes x inverse and is called **minus** x. Thus, $0 = x + (-x) = -x + x$ in this case. For any $y \in S$, set $y - x = y + (-x)$.

Group

A **group** is defined to be a semigroup in which every element is a unit. For any semigroup S, the set consisting of all units of S is denoted units S, and is a subsemigroup of S. This subsemigroup is a group, called the **units group** of S. A **subgroup** T is a suboperation which is a group. If $T = \{e\}$, then T is the **identity subgroup**.

If $x_1, \ldots, x_n$ are elements of a monoid S, the **product** $x_1 \cdots x_n$ is defined inductively:

$$x_1 \cdots x_n = \prod_{i=1}^{n} x_i = (x_1 \cdots x_{n-1})\, x_n .$$

If S is additive, then the **sum** is

$$x_1 + \cdots + x_n = \sum_{i=1}^{n} x_i = (x_1 + \cdots + x_{n-1}) + x_n .$$

(These formulas are unambiguous because S is associative.) When $x = x_1 = \cdots = x_n$, then x^n denotes the product $x_1 \cdots x_n$ (and nx denotes the sum $x_1 + \cdots + x_n$). If x is a unit, then x^{-n} is defined to be $(x^{-1})^n$. In a semigroup S with identity 1, define $x^0 = 1$. Then x satisfies the **exponents law** $x^n x^m = x^{n+m}$, which holds for all integers when x is a unit. The **generalized associative law** is proved inductively:

$$\prod_{i=1}^{m} x_i \prod_{j=1}^{n} x_{m+j} = \prod_{k=1}^{m+n} x_k .$$

Abelian Monoid

A monoid S is **commutative**, or **abelian**[1], in case the following identity holds: $xy = yx \; \forall \, x, y \in S$. If S is a semigroup, then, for any nonempty subset X of S,

$$\text{centralizer } X = \{a \in S \mid ax = xa \; \forall \, x \in X\}$$

is a subsemigroup of S. The **center** of S is defined to be centralizer S. Thus, center S is an abelian monoid (semigroup) when S is a monoid (semigroup).

If the semigroup is commutative, then usually the additive notation $x + y$ for the semigroup operation is employed. Of course, this convention is variable. For example, under multiplication the set $\mathbb{N}$, consisting of the natural (or counting) numbers 1, 2, 3, ..., is a commutative semigroup. Under addition, $\mathbb{N}$ is again a commutative semigroup. Similarly, $\mathbb{R}$, $\mathbb{Q}$, and $\mathbb{Z}$ are additive and multiplicative semigroups.

Maps X

Maps X denotes the set of all mappings $X \to X$, where X is a nonempty set. (Of course, this is the cartesian product X^X, but this no-

---

[1] Abelian groups are named after Niels Henrik Abel, a Norwegian mathematician. He discovered that not every equation $a_0 x^n + a_1 x^{n-1} + \cdots + a_n = 0$ can be solved by a formula obtained from the "coefficients" $a_0, \ldots, a_n$ by rational operations and the operation of extracting mth roots. However, equations with abelian Galois groups are solvable. An account of this may be found in Artin [59], Jacobson [64], or Kaplansky [69a].

tation is reserved for direct product defined presently.) Maps X is a semigroup with respect to the operation $(f, g) \mapsto fg$, where fg is defined by **composition of mappings**:

$$fg(x) = f(g(x))$$

$\forall\, x \in X$. The identity mapping 1_X is the identity of Maps X. For any two sets X and Y, for each $y \in Y$, the mapping

$$y_{\text{const}} \begin{cases} X \to Y \\ x \mapsto y \end{cases}$$

is called **constantly** y. This mapping satisfies the identities

$$fy_{\text{const}} = f(y)_{\text{const}}$$

and

$$y_{\text{const}}f = y_{\text{const}}$$

for all $y \in X$, $f \in$ Maps X. It follows that Maps X is abelian if and only if X consists of a single element. An **idempotent** in a monoid is an element e such that $e^2 = e$. In a group, the identity element is the only idempotent. In Maps X, the element y_{const} is idempotent for any $y \in X$, hence Maps X is a group if and only if X consists of a single element. If e is an idempotent in a monoid S, then the subset

$$eSe = \{eae \mid a \in S\}$$

is a submonoid which is a semigroup with identity e. Conversely, if T is a subsemigroup of a semigroup S, then the identity e of T is an idempotent, and $T \subseteq eSe$.

Symmetric Group

If X is a set, then **biject** X will denote the set of bijections $X \to X$. Thus, by Foreword Proposition 2, biject $X =$ units Maps X is a group with respect to composition of mappings. If $f: X \to Y$ is an equivalence of sets, then the mapping

$$\begin{cases} Y^Y \to X^X \\ h \mapsto f^{-1}hf \end{cases}$$

is an equivalence which induces an equivalence

$$\text{biject } Y \to \text{biject } X.$$

For each integer $n > 0$, the set of bijections of n, denoted S_n, is called the **symmetric group on** n, and also the symmetric group on n letters, objects, or words.

1.1 *Exercises.*

1.1.1 The cardinality of S_n is $n! = n(n-1) \cdots 2 \cdot 1$.

1.1.2 S_n is noncommutative if $n > 2$.

1.1.3 Determine the cardinality of biject X, for a set X.

1.1.4 Prove that biject $X \approx$ biject Y implies $X \approx Y$.

1.1.5 If S is a monoid and A and B are submonoids of S, define $AB = \{ab \mid a \in A, b \in B\}$. If S is abelian, then the set of submonoids of S is a monoid, with respect to the product AB defined.

1.1.6 Let S be a monoid, A and B submonoids, and let $A \circ B$ be the smallest submonoid containing both A and B. Show that the set of submonoids of S is a monoid with respect to the operation $A \circ B$ and every element is idempotent. Dualize with $A \circ B = A \cap B$ instead.

Direct Product and Sum

If $\{X_i\}_{i \in I}$ is a nonempty family of monoids, then the (cartesian) product $\prod_{i \in I} X_i$ is a monoid[1] with respect to the operation $(f, g) \mapsto fg$, defined $\forall\, i \in I$ by

$$fg(i) = f(i)\, g(i)$$

called the (**direct**) **product** of the family $\{X_i\}$ of monoids. The operation is called **pointwise multiplication** of functions. If the additive notation for the monoids is employed, then the operation is defined by

$$(f + g)(i) = f(i) + g(i)$$

and is called **pointwise addition** of functions. (Synonyms for pointwise are **coordinatewise** and **componentwise**.) Then the set X_i is called a **component**, in fact, the **i-th component**, of the product. Associativity of the product $(fg)h = f(gh)$ follows immediately from that in every component. In the multiplicative notation, the components are called the (direct) **factors** of the product and are called **summands** in the additive notation. The product $\prod_{i \in I} X_i$ is a semigroup if and only if each component is a semigroup. Then, if 1_i is the identity element of X_i, the

[1] See product of sets, Foreword Proposition 17.

function

$$\begin{cases} I \to \prod_{i \in I} X_i \\ i \mapsto 1_i \end{cases}$$

is the identity element of the product. The **support** of an element f in the product is defined to be the set of all $i \in I$ such that $f(i) \neq 1_i$. The support of f is **finite** in case it is a finite subset of I and nonempty. The set of all functions in the product having finite support is again a semigroup with respect to pointwise multiplication (or addition) of functions. This semigroup is denoted by $\sum_{i \in I} \oplus X_i$, or $\bigoplus_{i \in I} X_i$, and is called the **direct sum** of the family $\{X_i\}_{i \in I}$ of semigroups. The direct product equals the direct sum when I is finite. The direct product or direct sum is a group if and only if every component is a group and is abelian if and only if every component is abelian. Many properties of the direct product or sum are inherited from, or reduced to, corresponding properties of the components X_i, which is a way of saying that frequently it is true that the direct product or sum belongs to a category of semigroups if and only if each of the components do.

A special case of the direct product is X^I, the set of all mappings $I \to X$, where X is a semigroup and I is a nonempty set. In this case, the direct sum[1] is denoted by $X^{(I)}$. The identity element of X^I (also of $X^{(I)}$) is 1_{const}, where 1 denotes the identity element of X. (In case X is additive, then 0_{const} is the identity.) Note that when X is a semigroup, there are two ways to make X^X into a semigroup, namely, the direct product and Maps X.

1.2 *Exercise.* For any cardinal number A show that there exists a group G such the set G has cardinal A.

Homomorphisms

A mapping $f: A \to B$ of monoids is a **homomorphism**, or **map**, in case

$$f(ab) = f(a) f(b) \tag{1}$$

for all $a, b \in A$. Intuitively, a homomorphism of any algebraic system is a mapping that preserves the operation. If the operation in A is written additively $a + b$, and that in B multiplicatively $x \circ y$, then a homo-

[1] Some authors employ the symbol ${}^I X$ to denote the direct sum. I have rejected it for purely optical reasons.

morphism $f: A \to B$ satisfies the identity

$$f(a+b) = f(a) \circ f(b) \tag{2}$$

for all $a, b \in A$. The set $\mathbb{R}^+$ of all positive real numbers is a group under the operation of multiplication of real numbers; and for a fixed real number $r > 0$ and $r \neq 1$, there is a homomorphism

$$\exp \begin{cases} \mathbb{R} \to \mathbb{R}^+ \\ a \mapsto r^a \end{cases}$$

of the additive group $\mathbb{R}$ of real numbers into $\mathbb{R}^+$. The mapping exp satisfies (2) since this is just the exponents law $r^{a+b} = r^a r^b$. A property of real numbers is that $\exp: \mathbb{R} \to \mathbb{R}^+$ has an inverse $\log: \mathbb{R}^+ \to \mathbb{R}$. (For $r = 10$, this is just the common logarithm, and when $r = e$, this is called the natural log.) Thus:

$$\log xy = \log x + \log y$$

and

$$\log(\exp a) = a$$

$$\exp(\log x) = x$$

for all $a \in \mathbb{R}$ and $x \in \mathbb{R}^+$.

Another example, provided by any abelian semigroup A and a fixed positive integer n, is the power mapping $x \mapsto x^n$ (or $x \mapsto nx$ in the additive notation). If A is an abelian group, this also works for negative integers because a^{-n} can be defined to be $(a^{-1})^n$ [or $(-nx) = n(-x)$ in an additive group]. In particular, for an abelian group, the mapping sending every a into its inverse a^{-1} is a bijective homomorphism. Moreover, every subgroup N of an abelian group A determines a group A/N and a homomorphism $A \to A/N$; this will be presently discussed. In brief: homomorphisms of abelian groups are abundant!

Hom (X, Y)

If X and Y are monoids and Y is abelian, then the set Hom (X, Y), consisting of all homomorphisms $X \to Y$, is a submonoid of Y^X, that is, for all $f, g \in \text{Hom}(X, Y)$ and $a, b \in X$

$$\begin{aligned} fg(ab) = f(ab)g(ab) &= f(a)f(b)g(a)g(b) \\ &= f(a)g(a)f(b)g(b) \\ &= fg(a)fg(b). \end{aligned}$$

If, in addition, Y is an abelian (semi) group, then Y^X, hence Hom (X, Y), is an abelian (semi) group.

For any monoid X, if $f, g \in \text{Hom}(X, X)$ and $a, b \in X$, then

$$fg(ab) = f(g(a)g(b)) = fg(a)fg(b),$$

and therefore Hom (X, X) is a semigroup, a subsemigroup of Maps X, which we denote by End X and call the **Endomorphism semigroup** of X. The elements of End X are called **endomorphisms.**

Image and Kernel

For a homomorphism $f : A \to A'$ of groups A and A', the set

$$\{b \in A' \mid \exists\, a \in A \;\&\; b = f(a)\}$$

is a subgroup of A', called the **image** of f and denoted im f. If e denotes the identity element of A, and if e' is the identity element of A', then since a homomorphism preserves idempotents, and since e' is the only idempotent in a group A', it follows that $f(e) = e'$. Thus e' lies in the image of every homomorphism $f : A \to A'$. This includes the special case when A is a subgroup of A' and $f : A \to A'$ is the inclusion, that is, the identity element e' of group lies in every subgroup. Note also that $f(a)^{-1} = f(a^{-1})$ for every $a \in A$. It follows that im f is a subgroup of A'. The **kernel of** f is defined to be

$$\ker f = \{a \in A \mid f(a) = e'\}.$$

This is a subgroup of A with the property that, for any $x \in A$, $x^{-1}ax \in \ker f \;\forall\, a \in \ker f$. Any subgroup N of A such that $x^{-1}ax \in N$ for all $x \in A$ is called a **normal subgroup** of A. If A is abelian, then every subgroup is normal[1].

Quotient Group

If N is any subgroup of A, and if $x \in A$, then we define

$$xN = \{xa \mid a \in N\}$$

to be the **left coset** of N **containing** x. Thus,

$$y \in xN \Leftrightarrow x^{-1}y \in N.$$

---

[1] There is a class of nonabelian groups, called Hamiltonian groups, with this property. See Chapter 1 Exercises 19.

The right coset Nx is defined symmetrically. Then N is a normal subgroup if and only if $xN = Nx$ for every $x \in A$. For any subgroup N of A, the relation

$$x \mathscr{R} y \Leftrightarrow xN = yN$$

is an equivalence relation. If $xN = yN$, we write $x \equiv y \pmod N$ and say x **is congruent to** y **modulo** N.

Let A/N denote the set $\{xN \mid x \in A\}$. When N is a normal subgroup (and only then), A/N is a group under the operation

$$(xN)(yN) = xyN.$$

The group A/N is said to be the **factor**, or **quotient**, **group of** A **modulo** N, and

$$\begin{cases} A \to A/N \\ x \mapsto xN \end{cases}$$

is a homomorphism, called **canonical.** Note that N is the kernel of the canonical homomorphism $A \to A/N$.

If A is an abelian group, then in the additive notation, $x + N$ denotes the coset determined by x. Thus, in A/N, the operation reads

$$(x + N) + (y + N) = (x + y) + N.$$

For example, if n is any integer and $n\mathbb{Z}$ is the set of integer multiples of n, then $\mathbb{Z}/n\mathbb{Z}$ is a group which is denoted $\mathbb{Z}_n$. The additive quotient group $\mathbb{Q}/\mathbb{Z}$ is called the **rationals mod 1**, and $\mathbb{R}/\mathbb{Z}$ is called the **reals mod 1**.

Epimorphisms and Monomorphisms

A surjective homomorphism $f: A \to A'$ of monoids is called an **epimorphism**, and then A' is called an **epimorph** of A. Thus, for any normal subgroup N of a group A, the canonical mapping $A \to A/N$ is an epimorphism (cf. First Noether Isomorphism Theorem 1.10.) An injective homomorphism $f: A \to A'$ of monoids is called a **monomorphism**, or an **embedding** of A in A'. When A is a group, and A' is a semigroup, then f is a monomorphism if and only if the kernel of f consists of the identity element e of A, notationally, $\ker f = e$ (or $\ker f = 0$ when A is additive). The inclusion $1_B^A : B \to A$ of any submonoid B of a monoid A is an embedding which also is denoted in: $B \to A$, $B \xrightarrow{\text{in}} A$, or $B \subseteq A$.

Projections and Injections

If $\{X_i\}_{i\in I}$ is a family of monoids, then for each $j \in I$, the jth **projection** mapping is defined to be the mapping

$$p_j \begin{cases} \prod_{i\in I} X_i \to X_j \\ \qquad f \mapsto f(j) \end{cases}$$

and is an epimorphism of monoids from the product to the jth component which induces an epimorphism

$$\sum_{i\in I} \oplus X_i \to X_j$$

called the jth *projection from the direct sum.*

When $\{X_i\}_{i\in I}$ is a family of semigroups, then, for each $j \in I$, there is a monomorphism

$$u_j : X_j \to \prod_{i\in I} X_i$$

defined by

$$\begin{cases} u_j(x)\,(j) = x & \forall x \in X_j \\ u_j(x)\,(i) = \text{identity when } i \neq j. \end{cases}$$

The monomorphism is called the **canonical injection** (or **embedding**) of X_j in the product. Since u_j has finite support, then the image of u_j is contained in the direct sum, and the induced mapping $X_j \to \sum_{i\in I} \oplus X_i$ is called the *canonical injection of* X_j *into the direct sum.* In either case, for all $i \neq j \in I$ and all $x \in X_i$ and $y \in X_j$,

$$u_i(x)\,u_j(y) = u_j(y)\,u_i(x).$$

Exact Sequences

If M, N, and P are groups, then a sequence

$$M \xrightarrow{f} N \xrightarrow{g} P$$

denotes that $f : M \to N$ and $g : N \to P$ are homomorphisms. The sequence is said to be **exact** in case $\operatorname{im} f = \ker g$. In this terminology, $f : M \to N$ is a monomorphism if and only if

$$0 \to M \xrightarrow{f} N$$

is exact, and f is an epimorphism if and only if

$$M \xrightarrow{f} N \to 0$$

is exact. Here 0 represents the identity subgroup.

If $\{M_i\}$ is a family of groups, then a sequence

$$\cdots \xrightarrow{f_{i+2}} M_{i+1} \xrightarrow{f_{i+1}} M_i \xrightarrow{f_i} M_{i-1} \xrightarrow{f_{i-1}} M_{i-2} \to \cdots$$

is **exact** in case $\operatorname{im} f_i = \ker f_{i-1}$ for every i.

Generators

If N is a submonoid of a monoid M, then a subset S of M is said to **generate** N, notationally $N = (S)$, provided that the following two equivalent conditions hold:

(1) N is the intersection of all submonoids of M which contain S.

(2) S is a subset of N, and every element $x \in N$ is a product of elements of S.

Here, any element $s \in S$ is by definition "a product of elements of S". Thus, for any nonempty subset T the set of all finite products of elements of T is a submonoid $(T) \supseteq T$. In particular, N is generated by its elements. If M is a semigroup, and N is a subsemigroup, then a subset S of M generates N provided that N is the submonoid of M that S generates. If M is a group and N is a subgroup, then a subset S generates N, notationally $N = (S)$, provided that the two equivalent conditions hold:

$(S-1)$ N is the intersection of all subgroups of M which contain S.

$(S-2)$ S is a subset of N, and N is the submonoid of M generated by the set $S \bigcup S^{-1}$, where $S^{-1} = \{s^{-1} \mid s \in S\}$.

A subgroup, subsemigroup, or submonoid N is **finitely** (resp. **countably**) **generated** provided that N is generated by a finite (resp. countable) subset. A monoid, semigroup, or group G is **cyclic** if it is generated by a single element a, that is, $G = (a)$.

Index and Order

The **order** of a monoid M is its cardinality $|M|$. If $a \in M$, then the order of a is defined to be the order of the submonoid (a) generated by a. If a has finite order >1, then there exist integers $0 \leq k < n$ such that

$a^n = a^k$ (cf. bicycle semigroup in the Exercises 1.8). If n is chosen to be the least integer with this property, then the order of a is $n - k$. If a has infinite order, then

$$\begin{cases} \mathbb{N} \to (a) \\ n \mapsto a^n \end{cases}$$

is a bijective homomorphism. On the other hand, if G is a group, and if $a \in G$ has infinite order, then

$$\begin{cases} \mathbb{Z} \to (a) \\ i \mapsto a^i \end{cases}$$

is a bijective homomorphism. A semigroup S is a **cancellation semigroup** if

$$ax = ay \Rightarrow x = y$$

$$x'b = y'b \Rightarrow x' = y'$$

$\forall a, x, y, b, x', y' \in S$. In a cancellation semigroup, $a^n = a^k$ implies $a^t = 1$, where $t = n - k$, and thus, a is a unit with $a^{-1} = a^{t-1}$.

1.3 **Proposition.** *If H is a subgroup of a group G, then G is the disjoint union $G = \coprod_{a \in G} aH$ of the inclusions $\{aH \to G\}$, where aH ranges over all left cosets of H. The set of left cosets of H in G is denoted $G : H$. The cardinality $|G : H|$ equals the cardinality of the set of right cosets of H in G, and is called the* ***index*** *of H in G. Thus,*

$$|G| = |G : 1| = |G : H| \, |H : 1|;$$

that is, the order of a subgroup divides the order of the group (Theorem of Lagrange).

Proof. Every element $g \in G$ belongs to a left coset, namely, gH. Furthermore, two left cosets aH and bH are equal if and only if $b^{-1}a \in H$, (or equivalently $a^{-1}b \in H$). Hence, if $g = ah_1 = bh_2 \in aH \cap bH$, where h_1 and $h_2 \in H$, then $b^{-1}a = h_2h_1^{-1} \in H$, proving that $aH = bH$. Thus, different cosets are disjoint, and this proves that G is the disjoint union as asserted. The mapping

$$\begin{cases} G : H \to \text{set of right cosets} \\ aH \mapsto Ha^{-1} \end{cases}$$

is a bijection which proves that $|G : H|$ is the cardinal of the set of right cosets. ∎

In this notation, if H is a normal subgroup, then $|G/H| = |G : H|$.

Divisible Groups

For each additive abelian group G and integer $n > 0$, the set $nG = \{na \mid a \in G\}$ is a subgroup. An abelian group G is said to be **divisible** if $nG = G \forall n > 0$. Examples of divisible groups include $\mathbb{R}$, $\mathbb{Q}$, and the multiplicative group of all nth roots of unity in $\mathbb{R}$. Any epimorph of a divisible group is divisible, for example, the rationals mod 1 or the reals mod 1. A divisible group is necessarily infinite or zero.

If $\{X_i\}_{i\in I}$ is a chain of monoids, then the set union $\bigcup_{i\in I} X_i$ is a monoid and is a group when each X_i is. For each prime p, and integer $n > 1$, there is an embedding $f_n : \mathbb{Z}_{p^n} \to \mathbb{Q}/\mathbb{Z}$ (exercise), and hence $\mathbb{Q}/\mathbb{Z}$ contains a chain

$$\mathbb{Z}_p \subset \mathbb{Z}_{p^2} \subset \cdots \subset \mathbb{Z}_{p^n} \subset \cdots.$$

The unions of any two such chains of subgroups (in any two groups) are isomorphic (exercise), and the union of the chain above is denoted $\mathbb{Z}_{p^\infty}$ (cf. Exercises 1.8.13).

Torsion and Torsionfree Groups

A group G is **torsion** if every element has finite order. If there exists a prime p such that every element x has order equal to some power of p, then G is a **p-group**. Thus, $\mathbb{Z}_{p^n}$ is a finite p-group, and $\mathbb{Z}_{p^\infty}$ is an infinite p-group. The multiplicative subgroup of $\mathbb{R}$ consisting of all roots of unity, the rationals mod 1, and the direct sum of the family $\{\mathbb{Z}_n\}_{n\in N}$ are examples of torsion groups which are not finite. (Note that the direct product of the family $\{\mathbb{Z}_n\}_{n\in N}$ is not torsion.)

A group is **torsion of bounded order** in case there exists an integer $m > 0$ such $mG = 0$. The **exponent** of G is then the l.u.b. of all such m. Such a group may be infinite, as the infinite direct sum of countably many groups each equal to $\mathbb{Z}_2$ shows. (In this case, the direct product is also torsion and of bounded order.) Any subgroup and any quotient group of a torsion group (of bounded order) have the same property.

The **torsion subgroup** of an abelian group G is the subgroup consisting of all elements of G of finite order. The torsion subgroup of G is a torsion group. A group is **torsionfree** provided that the only torsion element is the identity element. For any abelian group G, if torsion G denotes the torsion subgroup, the quotient group G modulo torsion G is torsionfree. A **free abelian group** is one isomorphic to $\mathbb{Z}^{(A)}$ for some set A. Any subgroup of a free abelian group is torsionfree (in fact, free. See end-of-Chapter Exercises 27).

1.4 *Exercises.*

1.4.1 (a) Let A be a semigroup. For each $a \in A$, let a_s denote the mapping

$$a_s \begin{cases} A \to A \\ x \mapsto ax, \end{cases}$$

and let a_d denote the mapping

$$a_d \begin{cases} A \to A \\ x \mapsto xa. \end{cases}$$

Then $a_d b_s = b_s a_d$ for all $a, b \in A$.

(b) Consider the mappings

$$s_A \begin{cases} A \to \text{Maps } A \\ a \mapsto a_s \end{cases} \quad \text{and} \quad d_A \begin{cases} A \to \text{Maps } A \\ a \mapsto a_d. \end{cases}$$

Then $s_A : A \to \text{Maps } A$ is an embedding ($d_A : A \to \text{Maps } A$ is an anti-embedding, i.e., an injection f such that $f(xy) = f(y)f(x) \; \forall x, y$).

(c) If A is a group, then $\text{im } s_A \subseteq \text{biject } A = \text{units Maps } A$, and also $\text{im } d_A \subseteq \text{biject } A$.

(d) If A is a group of finite order n, then there is an embedding $A \to S_n$ of A into the symmetric group on n.

1.4.2 Show that $\mathbb{Z}_{p^\infty}$ has no subgroups except a cyclic subgroup of order p^n for each integer n.

1.4.3 Show that any abelian cancellation monoid S can be embedded in a group G such that $G = \{st^{-1} \mid s, t \in S\}$.

1.4.4 Every subgroup of a cyclic group is cyclic, and a cyclic group contains exactly one subgroup of order d for each integer d dividing the order of the group.

1.4.5 Prove the converse of 1.4.4, namely, if G is a finite group of order n such that the mapping

$$\begin{cases} \text{set of subgroups} \to \text{divisors of } n \\ S \to |S| \end{cases}$$

is a bijection, then G is cyclic.

1.4.6 Two cyclic groups A and B have the same order if and only if there is a bijective homomorphism $A \to B$.

1.4.7 If H is a subgroup of a group G, then there exists a subset of elements $\{a_i\}_{i\in I}$ of G such that $G:H = \{a_i H\}_{i\in I}$ and such that the mapping

$$\begin{cases} G:H \to \text{set of right cosets of } H \\ a_i H \mapsto H a_i \end{cases}$$

is bijective.

1.4.8 If H is a subgroup of G (of finite index), then

$$\bigcap_{x\in G} xHx^{-1}$$

is a normal subgroup (of finite index).

1.4.9 The **normalizer** $\text{norm}_G A$ of a subgroup A of G is defined to be the intersection of all subgroups B of G such that A is normal in B. Thus, $\text{norm}_G A = \{x \in G \mid x^{-1}Ax = A\}$. A subgroup of G of the form $x^{-1}Ax$ is said to be a **conjugate** of a subgroup A. Show that the number of conjugates of A in G is the index of the normalizer of A in G.

1.4.10 Let f be an element of the symmetric group S_n of order 2. If r is the number of elements x such that $f(x) = x$, then $n - r$ is even. Show that every finite group G of even order contains an element of order 2.

1.4.11 If every element of a group G has order ≤ 2, then G is abelian.

1.4.12 If N is a subgroup of G of index 2, then N is normal.

1.4.13 Show that S_3 contains exactly six subgroups of which three are normal and three are not.

1.4.14 Let H and K be proper subgroups of a group G containing at least two elements each. Show that G is not the set union of H and K.

*1.4.15 (a) Any finitely generated torsionfree abelian group is embeddable in a free abelian group (and hence is a free abelian group. Cf. Chapter 1 Exercises 27).

(b) Any abelian group N is an epimorph of a free abelian group. [Hint: Let A be a set of generators of N. For each $a \in A$ there is a unique homomorphism $\mathbb{Z}^{(A)} \to N$, sending 1 in the a-th component onto a. More generally, for each finite sum $n_1 a_1 + \cdots + n_t a_t \in N$, with $n_i \in \mathbb{Z}$, $a_i \in A$, there is a unique homomorphism $h: \mathbb{Z}^{(A)} \to N$ sending 1 in the a_i-th component onto a_i, $i = 1, \ldots, t$.]

Category

A **category** C consists of a class obj C, the elements of which are called **objects**, and a class Mor C, consisting of ordered triples $f: A \to B$, where A and B are objects. An element $f: A \to B$ of Mor C is called a

morphism; the **source** of f is A, and the **target** of f is B. For each ordered pair $\langle A, B\rangle$ it is required that the class $\mathrm{Mor}_C(A, B)$ of all morphisms $A \to B$ be a set[1]. Furthermore, for every ordered triple $\langle A, B, C\rangle$ of objects, there is an operation

$$\begin{cases} \mathrm{Mor}_C(B,C) \times \mathrm{Mor}_C(A, B) \to \mathrm{Mor}_C(A,C) \\ \qquad\qquad\qquad\qquad (g, f) \mapsto gf \end{cases}$$

which is associative in the sense that $h(gf) = (hg)f$ whenever hg and gf are defined. (Note that hg is defined only if target $g =$ source h.) Finally, for each object A, $\mathrm{Mor}_C(A, A)$ is a semigroup with **identity** 1_A satisfying $f1_A = f$ and $1_A g = g$ for all morphisms $f: A \to B$ and $g: C \to A$.

For each object A of a category C, the semigroup $\mathrm{Mor}_C(A, A)$ is called the **endomorphism semigroup** of A, denoted $\mathrm{End}_C A$. An element of $\mathrm{End}_C A$ is called an **endomorphism** of A.

If C and D are categories, if $\mathrm{obj}\, D \subseteq \mathrm{obj}\, C$, if $\mathrm{Mor}\, D \subseteq \mathrm{Mor}\, C$, and if the operation $\mathrm{Mor}\, D \times \mathrm{Mor}\, D \to \mathrm{Mor}\, D$ is induced by the operation defined by Mor C, then D is said to be a **subcategory** of C. Any subclass S of objects of C defines a subcategory S', where morphisms $A \to B$ between objects are the same as in C, and morphisms in S' are composed as in C. Then S' is a category, is a subcategory of C, and is called the **full subcategory of C on the objects S.**

Functors

If C and D are categories, a **covariant functor** $T: C \rightsquigarrow D$ is a pair of mappings

$$T\begin{cases} \mathrm{obj}\, C \to \mathrm{obj}\, D \\ \quad X \mapsto TX \end{cases} \quad \text{and} \quad T\begin{cases} \mathrm{Mor}\, C \to \mathrm{Mor}\, D \\ \qquad f \mapsto Tf, \end{cases}$$

which preserves composition of morphisms and identities:

$$Tfg = TfTg, \qquad \forall f, g \in \mathrm{Mor}\, C,$$

$$T1_A = 1_{TA}, \qquad \forall A \in \mathrm{obj}\, C.$$

[1] Other designations for $\mathrm{Mor}_C(A, B)$ are $\mathrm{Hom}_C(A, B)$ and $\mathrm{Ar}(A, B)$. (A morphism is called an arrow in Gabriel-Zissman [65].) Not everyone makes the requirement that $\mathrm{Mor}_C(A, B)$ be a set. For example, a semigroup may be defined to be a class, rather than a set, with an associative operation. All the interesting examples, however, are sets.

Thus, if $f : X \to Y$ and $g : Y \to Z$ are morphisms of C, then

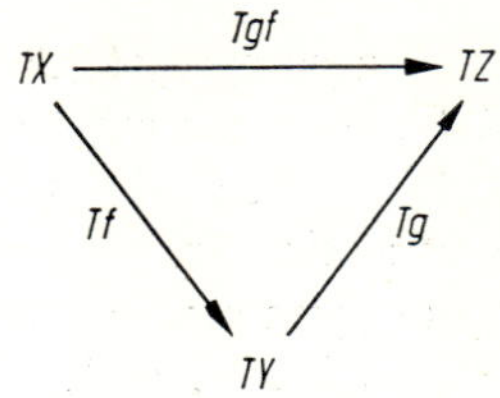

commutes.

A **contravariant functor** $S : C \rightsquigarrow D$ is a pair of mappings

$$S \begin{cases} \text{obj } C \to \text{obj } D \\ A \mapsto SA \end{cases} \qquad S \begin{cases} \text{Mor } C \to \text{Mor } D \\ f \mapsto Sf \end{cases}$$

preserving identities but reversing composition:

$$Sgf = SfSg$$
$$S1_A = 1_{SA}.$$

Thus, if $f : X \to Y$ and $g : Y \to Z$ in C, then

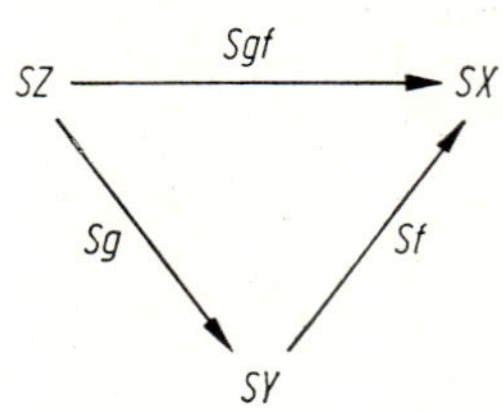

commutes when S is contravariant.

The term **functor** will designate a covariant functor. If $T : C \rightsquigarrow D$ is covariant, and if X and Y are objects of C, then the mapping

$$T_{X,Y} \begin{cases} \text{Mor}_C(X, Y) \to \text{Mor}_D(TX, TY) \\ f \mapsto Tf \end{cases}$$

is said to be **induced** by T. In particular, for each object X of C, the functor T induces a homomorphism $\text{End}_C X \to \text{End}_D TX$ of semigroups. The functor $T : C \rightsquigarrow D$ is said to be **faithful** provided that the induced mappings $T_{X,Y}$ are injective $\forall X, Y \in \text{obj } C$. In this case, for each X, the induced mapping $\text{End}_C X \to \text{End}_D TX$ is an embedding. If, on the other hand, the induced mappings $T_{X,Y}$ are surjective $\forall X, Y \in \text{obj } C$, then T is said to be **full.** In particular, then the induced mappings $\text{End}_C X \to \text{End}_D TX$ are semigroup epimorphisms $\forall X \in \text{obj } C$.

Duality

If C is a category, there is a category C^*, called the **dual**, or **opposite, category**, or **cocategory**, defined as having the same objects and morphisms as C but reversing the source and target and reversing composition of morphisms. To distinguish an object A and morphism f of C when considered in C^* denote them by A^* and f^*. Thus, for objects A^* and B^*, $\mathrm{Mor}_{C^*}(A^*, B^*) = \mathrm{Mor}_C(B, A)$, and, $f^* g^* = (gf)^*$. Thus, $f^* g^*$ is defined if and only if gf is defined, and the functor

$$\begin{cases} C \rightsquigarrow C^* \\ A \mapsto A^* \quad \text{(on morphisms and objects)} \end{cases}$$

is a contravariant functor, called the **duality functor.**

For every property P about categories, described by diagrams and morphisms and the completions of diagrams, there is a dual property, or coproperty, P^*, described by the dual diagrams obtainable via the duality functor. In this sense, the dual of one of the axioms for category is one of the axioms of category, which is another way of stating that the dual category C^* exists. The duality principle for ordered set (10 in the Foreword) applies in the same way to categories, namely, the dual S^* of any theorem S about categories is a theorem about categories. Furthermore, if $\mathscr{C}$ is a class of categories axiomatized by a self-dual set of axioms, and if $\mathscr{C}$ is closed under equivalence of categories, then the dual S^* of any theorem S about $\mathscr{C}$ is a theorem about $\mathscr{C}$. (Equivalence of categories will be defined presently.)

1.5 *Examples.*

1.5.1 **Composite functors.** If $T: C \rightsquigarrow D$ and $S: D \rightsquigarrow E$ are functors, then the **composite functor** is the functor

$$ST \begin{cases} C \rightsquigarrow E \\ A \mapsto S(TA) \quad \text{(on objects and morphisms)}. \end{cases}$$

If S and T are both covariant, or both contravariant, then ST is covariant. Sometimes $S \circ T$ denotes ST.

1.5.2 **Identity functor.** The **identity functor** $1_C : C \rightsquigarrow C$ of a category C sends each object and morphism onto itself.

1.5.3 **Inclusion functor.** If D is a subcategory of C, then the **inclusion functor** $1_D^C : D \rightsquigarrow C$ is the functor induced by the identity functor 1_C. Thus, 1_D^C is a faithful functor which is full if and only if D is a full subcategory of C.

1.5.4 **Sets.** The category C such that obj C is the class of all sets and Mor C is the class of all mappings of sets is denoted SETS.

1.5.5 **Forgetful functor.** If C is a subcategory of SETS, then C is said to be **concrete**, and the inclusion functor $C \rightsquigarrow$ SETS is called the **forgetful functor.**

1.5.6 **Semigroups.** There is a subcategory C of SETS such that obj C is the class of all semigroups, and for each pair X, Y of semigroups, $\mathrm{Mor}_C(X, Y)$ is the set of all homomorphisms $X \to Y$. This category is denoted SEMIGROUPS, and the image of an object under the forgetful functor is called the **underlying set.**

Any semigroup S with identity is a category C with one object S, and Mor C is the semigroup S. If C and D are categories with a single object X and Y, respectively then any covariant functor $T : C \rightsquigarrow D$ induces a homomorphism

$$\begin{cases} \mathrm{Mor}\, C \to \mathrm{Mor}\, D \\ \quad f \mapsto Tf, \end{cases}$$

and conversely any homomorphism $\mathrm{Mor}\, C \to \mathrm{Mor}\, D$ defines a covariant functor. *Thus the theory of categories with a single object coincides with the theory of semigroups*[1]. The same symbol S often represents the semigroup, the corresponding category, the unique object, and also the set of morphisms. Thus, $\mathrm{Mor}\, S = \mathrm{Mor}_S(S, S) = \mathrm{End}_S S$.

1.5.7 **Groups.** A full subcategory of SEMIGROUPS is GROUPS the objects of which are the class of all groups.

1.5.8 **Abelian groups.** A full subcategory of GROUPS is mod-$\mathbb{Z}$ the objects of which are the class of all abelian groups. (Anticipating terminology introduced in Chapter 3, an abelian group is a $\mathbb{Z}$-module.)

1.5.9 **Divisible groups.** A full subcategory of mod-$\mathbb{Z}$ is the DIV mod-$\mathbb{Z}$ the objects of which are the divisible abelian groups.

1.5.10 **The category of propositions.** The objects are sentences, and a morphism $f : A \to B$ is an implication "A implies B".

1.5.11 **The category of ordered sets.** The objects are ordered sets, and the morphisms are order homomorphisms.

1.5.12 **The morphism, or representation, functors.** If A is an object of a category C, then the **first representation, or covariant morphism functor** is the functor defined by

$$h^A \begin{cases} C \rightsquigarrow \text{SETS} \\ X \mapsto \mathrm{Mor}_C(A, X) \qquad \text{(on objects),} \end{cases}$$

[1] This is the traditional (vague) way of saying something quite precise. See Equivalent Categories, p. 65.

such that for each morphism $f: X \to Y$ of C,

$$h^A f \begin{cases} \mathrm{Mor}_C(A, X) \to \mathrm{Mor}_C(A, Y) \\ \qquad g \mapsto fg. \end{cases}$$

Thus, $(h^A f)g = fg$, whenever fg is defined. The functor h^A is also denoted $\mathrm{Mor}_C(A, \ )$, and sometimes $(A, \ )$. If the objects $h^A X = \mathrm{Mor}_C(A, X)$ are objects of (SEMI) GROUPS for every X, then the induced functor $C \rightsquigarrow$ (SEMI) GROUPS is also denoted h^A and is called the **(semi) group valued functor.** Similarly, if D is a subcategory of SETS, that is, if D is a concrete category, such that $h^A X \in \mathrm{obj}\, D \ \forall X \in \mathrm{obj}\, C$, then the induced functor is called the D **valued functor**[1]. (In chapters following Chapter 2, module valued functors are encountered.)

The **second representation**, or **contravariant morphism functor** $h_A : C \rightsquigarrow$ SETS is such that for each morphism $f: X \to Y$ of C,

$$h_A f \begin{cases} \mathrm{Mor}_C(Y, A) \to \mathrm{Mor}_C(X, A) \\ \qquad g \mapsto gf. \end{cases}$$

$\mathrm{Mor}_C(\ , A)$ and $(\ , A)$ are variant forms for this functor.

1.5.13 **Category of sequences over C indexed by a set I.** If C is a category, then there is another category, the objects of which are sequences of objects $\{X_i\}_{i \in I}$ of C indexed by I. This category is denoted C^I. A morphism of C^I is a sequence $\{g_i\}_{i \in I}$ of morphisms of C. The compose of this with another morphism $\{h_i\}_{i \in I}$ is defined if and only if $h_i g_i$ is defined $\forall i \in I$ in which case $\{h_i\}\{g_i\} = \{h_i g_i\}$.

1.5.14 **Category of commutative sequences over C indexed by I.** The objects are sequences $\{f_i : X_i \to Y_i\}_{i \in I}$ of morphisms of C, that is, the objects are the morphisms of C^I. A morphism between this and an object $\{g_i : U_i \to V_i\}_{i \in I}$ is an ordered pair $\{X_i \to U_i\}$ and $\{Y_i \to V_i\}$ of morphisms of C^I such that

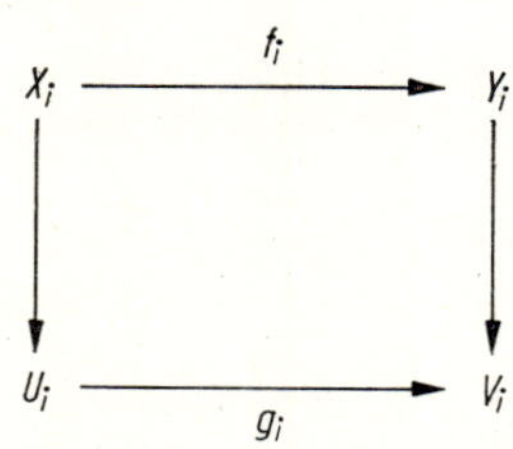

commutes for every $i \in I$. Then the morphisms are composed as in C^I.

1.5.15 **Source X commutative sequences over C indexed by I.** This is the subcategory of the category of commutative sequences over C indexed by I, the objects of which are the sequences $\{f_i : X \to Y_i\}_{i \in I}$

[1] We are assuming that h^A *does* induce a functor $C \rightsquigarrow D$.

with source X for a given object X of C. If $\{g_i : X \to Z_i\}$ is another object, then a morphism is defined to be a sequence $\{h_i : Y_i \to Z_i\}_{i \in I}$ such that

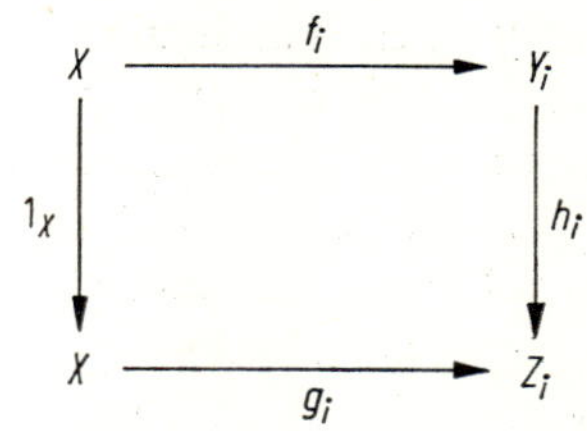

commutes. Thus, the ordered pair $\{1_X : X \to X\}$ and $\{h_i : Y_i \to Z_i\}$, both indexed by I, is a morphism in the category of commutative sequences, that is, for simplicity the sequence of identity morphisms $X \to X$ is suppressed.

1.5.16 **Source** X. If X is an object in a category C, then source X is the category of commutative sequences over C with source X and indexed by 2, that is, an object of source X is a morphism $X \to Y$ of C. A morphism from this object to the object $X \to Z$ is a morphism $Y \to Z$ such that

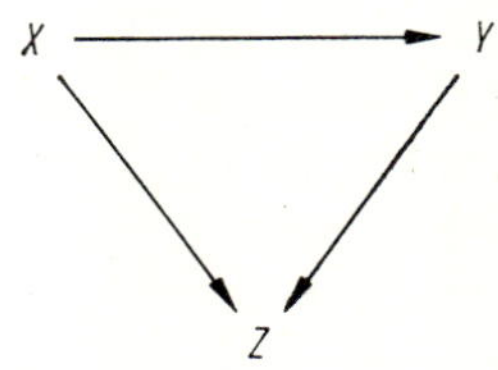

commutes.

1.5.17 **Target** X. This is the category dual to that in Example 1.5.16 namely, the objects are morphisms $Y \to X$. The category defined in Example 1.5.15 has a similar dual.

Equivalences and Automorphisms

A morphism $f : A \to B$ in a category C is said to be an **equivalence** provided that there exists a morphism $g : B \to A$ such that

$$gf = 1_A \qquad \text{and} \qquad fg = 1_B.$$

The morphism g is the unique morphism with this property and is denoted f^{-1}. Equivalence is a reflexive, symmetric, and transitive relation. The symbol $A \approx B$ denotes an equivalence, and B is called a **copy** of A. An equivalence $A \to A$ is an **automorphism** of A. The set of all equivalences $A \to A$ coincides with the group units $\text{End}_C A$ and is denoted $\text{Aut}\, A$. An equivalence in SEMIGROUPS is called an **iso-**

morphism. By Foreword Proposition 1, an isomorphism $A \approx B$ is necessarily a bijection. Then B is an **isomorph** of A.

Two objects are **inequivalent** if they are not equivalent. If D is a class of objects of a category C, then an **equivalence class** of objects of D determined by an object A in D is the subclass of all objects in D equivalent to A; then there is a subclass Equiv D with the property that every object in D is equivalent to an object in Equiv D, and two different objects of Equiv D are inequivalent. If Equiv D is a set, then the cardinal of Equiv D is called the number of inequivalent objects in class D. If C is SEMIGROUPS, then an equivalence class is called an **isomorphy class**. (Cf. p. 20.)

Natural Transformation of Functors

Let $S : C \rightsquigarrow D$ and $T : C \rightsquigarrow D$ be two covariant functors. A **natural transformation** $S \to T$ is a function h that associates to each object $A \in C$ a morphism $h(A) : S(A) \to T(A)$ such that for each morphism $f : A \to A'$ of C, the diagram

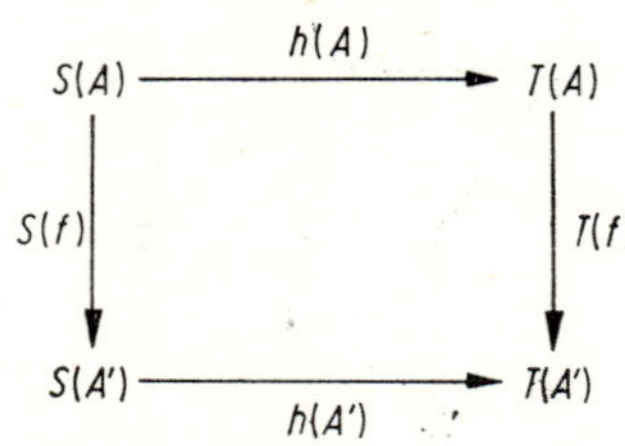

is commutative; that is,

$$T(f)\, h(A) = h(A')\, S(f).$$

A natural transformation $h : S \to T$ between the functors S and T is called a **natural equivalence** of the functors S and T whenever $h(A)$ is an equivalence $\forall A \in \text{obj } C$. In this case we write $S \approx T$. This relation is reflexive, symmetric, and transitive.

If $g : R \to S$ and $h : S \to T$ are natural transformations of functors, then the compose $hg : R \to T$ is a natural transformation defined by composition of morphisms: $hg(A) = h(A)\, g(A)$.

The phrase "ignoring sizes, C is a category ..." means that the requirement that $\text{Mor}_C(X, Y)$ be a set is waived. Ignoring sizes, there is the category CAT of categories, the objects of which are categories, the morphisms are functors $C \rightsquigarrow D$, and the operation is composition of functors. The identity functor $C \rightsquigarrow C$ is denoted 1_C.

Again ignoring sizes, for two fixed categories C and D, there is a category Funct (C, D) whose objects are the class of all functors $C \rightsquigarrow D$, and a morphism between two functors S and T is a natural transformation of functors. The category operation is composition of natural transformations. The term **functor morphism** is used synonymously with natural transformation, and **functor equivalence** means a natural equivalence of functors. Funct (C, D) is also denoted by D^C.

Equivalent Categories

An equivalence $C \rightsquigarrow D$ between two categories is an ordered pair (T, S) of covariant functors $T : C \rightsquigarrow D$ and $S : D \rightsquigarrow C$, and a pair of natural equivalences

$$ST \approx 1_C \qquad \text{and} \qquad TS \approx 1_D$$

of functors. In this case, we say that C and D are **equivalent categories**, we write $C \approx D$, and we say that $T : C \rightsquigarrow D$ is an equivalence. The relation $C \approx D$ is reflexive, symmetric, and transitive. An equivalence $C \rightsquigarrow C$ is called an **autoequivalence.** We now show that an equivalence $C \rightsquigarrow D$ is determined up to an autoequivalence of C (or D). If (T_1, S_1) and (T_2, S_2) are two equivalence pairs $C \rightsquigarrow D$, then

$$T_1 S_2 T_2 S_1 \approx T_1 S_1 \approx 1_D,$$

also $T_2 S_1 T_1 S_2 \approx 1_D$, and so $T_1 S_2 : D \rightsquigarrow D$ is an autoequivalence H of D. Therefore,

$$T_1 \approx T_1(S_2 T_2) = (T_1 S_2) T_2 = H T_2 \approx T_2,$$

and $S_1 \approx S_2$ by symmetry. For each ordered pair (C, D) of equivalent categories C and D, a single equivalence pair (T, T^{-1}) is selected, and T^{-1} is called the **inverse** equivalence. The class of autoequivalences $C \rightsquigarrow C$ is denoted Aut C, and when it is a set, it is a group. In case $ST = 1_C$ and $TS = 1_D$, then T is called a category **isomorphism** $C \rightsquigarrow D$.

We now can make more precise what was vaguely stated in 1.5.6: *the full subcategory of CAT on the categories with single objects is equivalent to **SEMIGROUPS**.*

Examples.

1. **The commutator functor.** If G is a group, then the **commutator**, or **derived subgroup** of G is the subgroup $[G, G]$ generated by $\{aba^{-1}b^{-1} \mid a, b \in G\}$ and is the intersection of all normal subgroups N of G such that G/N is abelian. The **commutator functor**

$$\begin{cases} \text{GROUPS} \rightsquigarrow \text{mod-}\mathbb{Z} \\ \qquad G \mapsto G/[G, G] \qquad \text{(on objects)} \end{cases}$$

maps each homomorphism $h : G \to H$ of groups onto the homomorphism

$$\begin{cases} G/[G,G] \to H/[H,H] \\ g[G,G] \mapsto h(g)\,[H,H] \end{cases}$$

of abelian groups. The canonical homomorphism

$$\text{can}\,(G) : G \to G/[G,G]$$

defines a natural transformation of the identity functor of GROUPS to the commutator functor.

2. **The morphism functor**

$$h^{\mathbb{Z}} \begin{cases} \text{mod-}\mathbb{Z} \rightsquigarrow \text{mod-}\mathbb{Z} \\ \qquad A \mapsto \text{Hom}\,(\mathbb{Z}, A) \end{cases}$$

is naturally equivalent to the identity functor I of mod-$\mathbb{Z}$ under the natural transformation $g^{-1} : I \to h^{\mathbb{Z}}$, where for any $A \in$ mod-$\mathbb{Z}$,

$$g_A \begin{cases} \text{Hom}\,(\mathbb{Z}, A) \to A \\ \qquad\qquad f \mapsto f(1). \end{cases}$$

To see this, note first of all that g_A is surjective: if $a \in A$, then there is a map

$$a_s \begin{cases} \mathbb{Z} \to A \\ n \mapsto na, \end{cases}$$

and $g_A(a_s) = a_s(1) = 1a = a$. This proves surjectivity. If $f(1) = 0$, then $f(a) = af(1) = 0$ and therefore $f = 0$. Thus, $\ker g_A = 0$, proving the isomorphism. Naturality follows forthwith.

The same proof works for the category mod-R of all right R-modules over a ring R defined in Chapter 3, that is, the morphism functor h^R : mod-$R \rightsquigarrow$ SETS induces an autoequivalence of mod-R.

A functor $T : C \rightsquigarrow D$ is called a **duality** or **antiequivalence**, provided that it is contravariant and there exist a functor $S : D \rightsquigarrow C$ and natural equivalences of functors

$$TS \approx 1_D \qquad \text{and} \qquad ST \approx 1_C.$$

The ordered pair (T, S) is a **duality** pair. A duality $C \rightsquigarrow C$ is called an **involution.**

1.6 Proposition and Definition.

1.6.1 *If (T_1, S_1) and (T_2, S_2) are duality pairs $C \rightsquigarrow D$, then $T_1 \approx T_2$ and $S_1 \approx S_2$. For each duality $C \rightsquigarrow D$, a single duality pair (T, T^{-1}) is* *selected, called canonical, and T^{-1} is called the inverse duality. The canonical functor $C \rightsquigarrow C^*$ is called the duality functor.*

1.6.2 *For every functor $T : C \rightsquigarrow D$, the inverse duality functor $C^* \rightsquigarrow C$ composed with T is denoted T^* and is called* ***the first functor dual to*** *T. T composed with the duality functor $D \rightsquigarrow D^*$ is denoted T_*, called the* ***second functor dual to*** *T. Thus, T is covariant (contravariant) if and only if the dual functor, first or second, is contravariant (covariant), and T is a duality (equivalence) if and only if the dual functor is an equivalence (duality).*

1.6.3 *For any category C and object A, the first functor dual to the set-valued functor h^A (resp. h_A) is h_{A*} (resp. h^{A*}).*

Proof. 1.6.1 is proved exactly in the same way as equivalence pairs. For the rest, there is nothing to prove. ▯

If S is a semigroup, then the dual category is a semigroup called the **semigroup opposite** to S, and denoted S^{op}. If S and S' are semigroups, then a mapping $f : S \to S'$ is an **antihomomorphism** provided that it is a contravariant functor. Also f is an anti-isomorphism if it is a duality. An involution $S \to S$ is, then, an anti-isomorphism. Every group G has an anti-isomorphism, namely

$$\text{inverse} \begin{cases} G \to G \\ x \mapsto x^{-1}. \end{cases}$$

It follows that there is an isomorphism $G \approx G^{\text{op}}$ (cf. 1.6.2).

Automorphisms of groups are abundant. If G is abelian, then the mapping $x \mapsto -x$ is an automorphism which is the identity mapping if and only if $x + x = 0$ for every x. Even in this case, G has automorphisms not the identity[1] unless $G \approx \mathbb{Z}/2\mathbb{Z}$.

Assume that G is not abelian and adopt the multiplicative notation. The **center** of G is a normal subgroup consisting of all $x \in G$ such that $xy = yx$ for all $y \in G$. For any $y \in G$, the mapping I_y such that $I_y(x) = yxy^{-1}$, $\forall x \in G$, is an automorphism of G, called the **inner automorphism** of G defined by y. The set InAut G consisting of all inner automorphisms of G is a normal subgroup of Aut G, as indicated by the formula

$$fI_yf^{-1} = I_{f(y)}$$

$\forall y \in G$, $f \in \text{Aut } G$. Furthermore, the mapping

$$\begin{cases} \text{InAut } G \to G/(\text{center } G) \\ \qquad I_y \mapsto y\,(\text{center } G) \end{cases}$$

is an isomorphism of groups.

[1] This is an observation of Kaplansky.

1.7 Proposition (Cayley).

1.7.1 *For any group A, and element $a \in A$, there is a bijection*

$$a_s \begin{cases} A \to A \\ x \mapsto ax. \end{cases}$$

The mapping

$$s_A \begin{cases} A \to \text{biject } A \\ a \mapsto a_s \end{cases}$$

is a monomorphism. Thus, if A has finite order n, there is an embedding $A \to S_n$ of A into the symmetric group S_n on n letters.

1.7.2 Aut A *is a subgroup of* biject A, *and*

$$f a_s f^{-1} = (fa)_s$$

$\forall f \in \text{Aut } A, a \in A$.

1.7.3 *Every automorphism f' of* im s_A *is induced by an inner automorphism of* biject A; *namely, $f'(a_s) = f(a)_s = f a_s f^{-1}$ for some $f \in \text{Aut } A$, and then I_f induces f'.*

1.7.4 *The mapping*

$$d_A \begin{cases} A \to \text{biject } A \\ a \mapsto a_d, \end{cases}$$

where $a_d(x) = xa \; \forall x \in A$, is an injective antihomomorphism.

1.7.5 *The injections $s_A : A \to \text{biject } A$ and $d_A : A \to \text{biject } A$ commute, that is, $a_s b_d = b_d a_s \; \forall a, b \in A$.* □

1.8 *Exercises.*

1.8.1 Determine the number of nonisomorphic semigroups with precisely four elements.

1.8.2 Let n be a positive integer an k be an integer, $0 \leq k < n$. Show that there exists a cyclic additive semigroup $A_{n,k} = (a)$ such that $na = ka$. If A' is any cyclic semigroup, then either A' is infinite, in which case A' is isomorphic to the additive semigroup $\mathbb{Z}^+$ of integers ≥ 0, or else there exist a unique ordered pair (n, k) and an isomorphism $A' \approx A_{n,k}$. The semigroup $A_{n,k}$ is called the **bicycle semigroup**.

1.8.3 The number of nonisomorphic finite semigroups is countable, but there exist a countable semigroup M and uncountably many nonisomorphic epimorphs of M.

1.8.4 If a semigroup M is finitely generated, then the number of nonisomorphic epimorphs of M is countable.

1.8.5 Show that the multiplicative semigroup $\mathbb{N}$ of natural numbers is isomorphic to the direct sum of countably many copies of the additive semigroup $\mathbb{Z}^+$ of non-negative integers.

1.8.6 If G is cyclic, then is Aut G?

1.8.7 If p is a prime number, show that Aut $\mathbb{Z}/p\mathbb{Z} \approx \mathbb{Z}/(p-1)\mathbb{Z}$. Describe Aut $\mathbb{Z}/n\mathbb{Z}$, for any $n \geq 0$ in $\mathbb{Z}$.

1.8.8 Determine necessary and sufficient conditions on positive integers $n_1, \ldots, n_k$ in order that a direct product of the cyclic groups $\mathbb{Z}_{n_i}$ $i = 1, \ldots, k$, be cyclic.

1.8.9 If $\{X_i\}_{i \in I}$ is any family of monoids, and if $J \subseteq I$, then there exists an exact sequence

$$0 \to \prod_{i \in I-J} X_i \to \prod_{i \in I} X_i \to \prod_{j \in J} X_j \to 0.$$

Establish an isomorphism

$$\prod_{i \in I} X_i \approx \Big(\prod_{i \in I-J} X_i\Big) \times \prod_{j \in J} X_j.$$

1.8.10 The direct sum of a family of groups is a normal subgroup of the direct product.

1.8.11 (a) Let $\{X_i \mid i \in I\}$ be a family of subsemigroups of a semigroup G. Show that there is a homomorphism of the direct sum $h: \sum_{i \in I} \oplus X_i \to G$ such that $h: (\ldots, x_i, \ldots) \mapsto \prod_{i \in I} x_i$ (finite product) if and only if the following commuting condition holds: each element x_i of X_i commutes with each element x_j of X_j when $i \neq j$.

(b) Let $\{Y_i \mid i \in I\}$ be a family of semigroups. Show that $G \approx \sum_{i \in I} \oplus Y_i$ if and only if there exists a family $\{X_i \mid i \in I\}$ of subsemigroups of G satisfying the four conditions: (1) there exists a family $\{f_i\}_{i \in I}$ of isomorphisms $f_i : Y_i \approx X_i$; (2) the commuting condition (a) holds; (3) G is the subsemigroup generated by the set $\bigcup_{i \in I} X_i$; (4) for each $i \in I$, the intersection of X_i with the subsemigroup generated by $\bigcup_{j \in I-\{i\}} X_j$ is the identity subgroup. [This implies that each X_i contains the identity element of G.]

When this is so, then each element $x \in G$ has a unique representation as a finite product $x = \prod_{i \in I} x_i$, with $x_i \in X_i$. Moreover, when G is a group, then this implies that X_i is a normal subgroup $\forall i \in I$.

1.8.12 Let $\{X_i \mid i \in I\}$ be a family of groups, and let $X = \prod_{i \in I} X_i$. Show that Maps X has idempotents other than the identity mapping whenever I has cardinality $\neq 1$. If $J \subseteq I$, there is an idempotent mapping

$$\prod_{i \in I} X_i \to \prod_{j \in J} X_j \to \prod_{i \in I} X_i,$$

called a **projection idempotent**. For a general abelian group X, discuss to what extent each idempotent e in End X is a projection idempotent. [Hint: Show that X is isomorphic to the product (im e) $\times$ (ker e).]

1.8.13 Show that $\mathbb{Q}/\mathbb{Z}$ is isomorphic to a direct sum of the p-groups $\mathbb{Z}_{p^\infty}$, one for each prime. Show that every finite subgroup of $\mathbb{Q}/\mathbb{Z}$ is cyclic.

Three Isomorphism Theorems

There are three fundamental isomorphism theorems. The first is an analog of the formula $(A/K)/(B/K) = A/B$ for ordinary fractions of integers. The second is a cancellation law for groups: $(AB)/B \approx (A \cap AB)/A \cap B = A/A \cap B$, where A and B are subgroups of a given group and AB is the group product $\{ab \mid a \in A, b \in B\}$. If the group is noncommutative, then B is assumed to be a normal subgroup; if the group is commutative, then the additive notation is adopted, and the law appears $(A+B)/B \approx A/A \cap B$. The third isomorphism theorem is the Zassenhaus (or Butterfly) lemma 1.12.

1.9 Correspondence Theorem. *Let $f: A \to A'$ be an epimorphism of groups. For a subgroup B of A, let $f(B) = \{f(b) \mid b \in B\}$. Then $B \mapsto f(B)$ is a bijection from the set of subgroups of A containing the kernel of f to the set of subgroups of A'. Let B and K be normal subgroups of A, where $B \supseteq K$. The mapping*

$$\begin{cases} A/K \to A/B \\ aK \mapsto aB \end{cases}$$

is a group homomorphism with kernel $\{aK \mid a \in B\}$, and there is a row-exact commutative diagram:

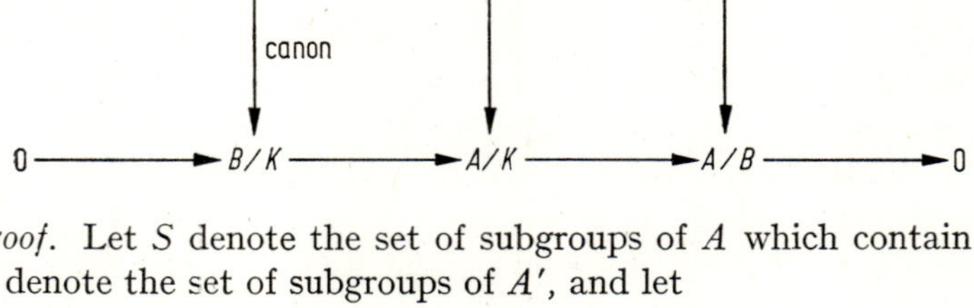

Proof. Let S denote the set of subgroups of A which contain $\ker f$, let S' denote the set of subgroups of A', and let

$$f' \begin{cases} S \to S' \\ B \mapsto B' = f(B). \end{cases}$$

Recall for a subset Y of A' that $f^{-1}Y = \{x \in A \mid f(x) \in Y\}$. Clearly, every subgroup Y of A' satisfies $ff^{-1}Y = Y$, so that f' is surjective. Furthermore, $f^{-1}fB \subseteq B$ for every $B \in S$. But if $b \in B$, then

$f(b) = f(f^{-1}fb)$. If $K = \ker f$, then this proves that $f^{-1}fbK = bK \subseteq B$. Thus, $f^{-1}fb \in B \,\forall b \in B$, whence $f^{-1}fB = B$, which proves that f' is injective. The commutativity of the diagram is obvious. ▯

1.10 First Noether Isomorphism Theorem. *If $f: A \to A'$ is an epimorphism of groups, and if N is a normal subgroup of A containing the kernel of f, then a unique isomorphism $f': A/N \to A'/f(N)$, called* **the isomorphism induced** *by f, completes the diagram:*

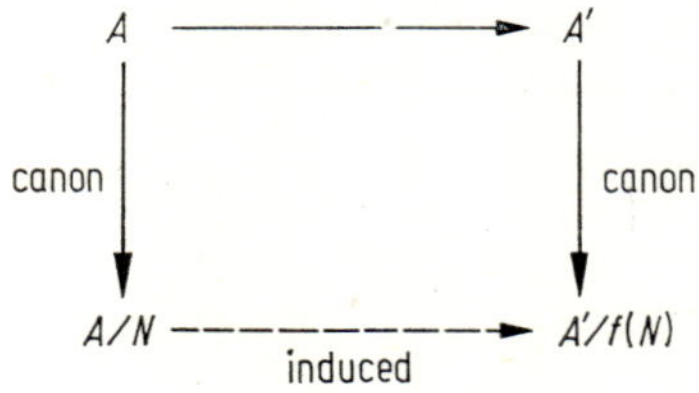

In particular, f induces an isomorphism $A/\ker f \approx A'$. Thus, the set of quotient groups of A contains an isomorphy class for the class of epimorphs of A.

Proof. First, there is a commutative diagram:

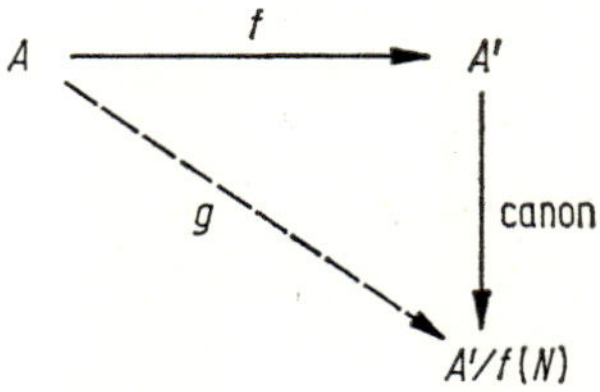

Second, since $N = f^{-1}f(N)$ by the correspondence theorem, the following statements are equivalent: (1) $a \in \ker g$; (2) $f(a) \in f(N)$; (3) $a \in N$. Thus, $\ker g = N$. By the correspondence theorem, there is a commutative diagram:

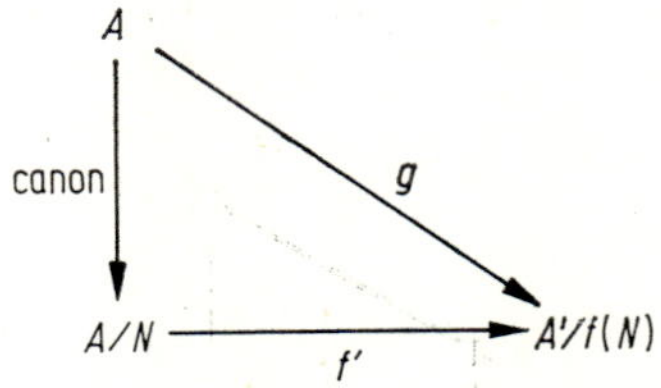

where f' is an isomorphism. The diagrams together yield the stated diagram. ▯

The first Noether isomorphism theorem for exact categories appears in Chapter 5, p. 263.

1.11 Corollary. *Let $A \supseteq B \supseteq N$ be groups with B and N both normal in A. Then there is an isomorphism $A/B \to (A/N)/(B/N)$ that completes the diagram*

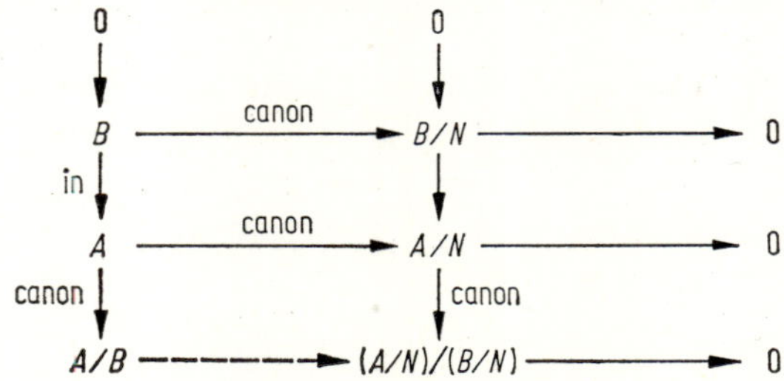

Second Noether Isomorphism Theorem. *If A and B are subgroups of a group G and B is normal in G, then $AB = \{ab \mid a \in A, b \in B\}$ is a subgroup of G, and $A \cap B$ is a normal subgroup of A. The mapping $A \to AB/B$ induced by the canonical mapping $G \to G/B$ is called canonical and has kernel $A \cap B$. Thus,*

$$\begin{cases} A/(A \cap B) \to AB/B \\ a(A \cap B) \mapsto aB \end{cases}$$

is an isomorphism, and there is a row and column exact commutative diagram:

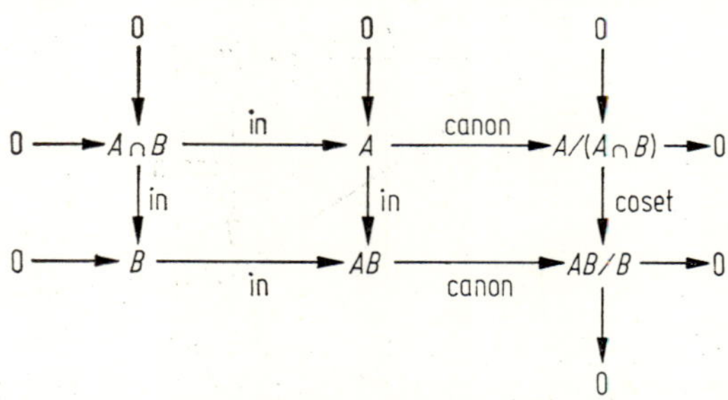

Proof. Every element of AB/B has the form aB for some $a \in A$, so the coset map $A \to AB/B$ is an epimorphism, and the kernel is $A \cap B$. Then the coset mapping $A/(A \cap B) \to AB/B$ completes the diagram. ▯

A mnemonic is the (Hasse) diagram:

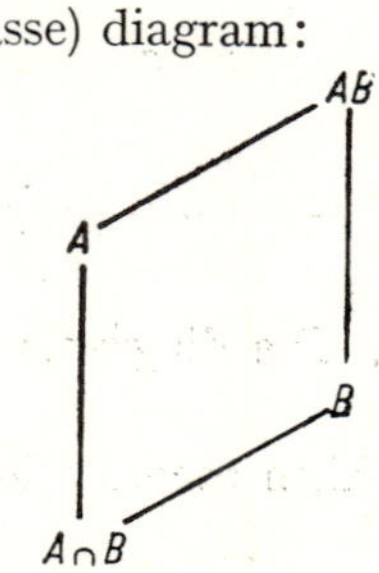

Also $AB/B \approx (A \cap AB)/(A \cap B) = A/(A \cap B)$, that is, intersect both "numerator" and "denominator" of AB/B with A. In the additive notation: $(A + B)/B \approx \big(A \cap (A + B)\big)/(A \cap B)$.

1.12 Butterfly Lemma for Groups. (Zassenhaus) *Let a, A, b, and B be subgroups of a group G such that a is normal in A and b is a normal subgroup of B. Then $a(A \cap b)$ is a normal subgroup of $a(A \cap B)$, and*

$$\frac{a(A \cap B)}{a(A \cap b)} \approx \frac{b(A \cap B)}{b(a \cap B)},$$

or

$$\frac{a + (A \cap B)}{a + (A \cap b)} \approx \frac{b + (A \cap B)}{b + (a \cap B)}$$

in the additive notation.

Proof. The second isomorphism theorem for additive groups states for subgroups C and D, with D normal, that $C/(C \cap D) \approx (C + D)/D$. Let $C = A \cap B$ and $D = a$, then $C \cap D = a \cap B$, so that

(i) $$\frac{A \cap B}{a \cap B} \approx \frac{a + (A \cap B)}{a}.$$

By symmetry,

(ii) $$\frac{A \cap B}{b \cap A} \approx \frac{b + (A \cap B)}{b}.$$

If h denotes the homomorphism $h: A \cap B \to \big(a + (A \cap B)\big)/a$ defined by (i), then

$$h\big((a \cap B) + (b \cap A)\big) = \frac{a + (a \cap B) + (b \cap A)}{a} = \frac{a + (b \cap A)}{a}.$$

By the first isomorphism theorem, e.g., $A/N \approx f(A)/f(N)$,

$$\frac{A \cap B}{(a \cap B) + (b \cap A)} \approx \left(\frac{a + (A \cap B)}{a}\right) \Big/ \left(\frac{a + (b \cap A)}{a}\right).$$

By the corollary to this theorem $\big((M/K)/(L/K) \approx M/L\big)$, the right side can be replaced by $\big(a + (A \cap B)\big)/\big(a + (b \cap A)\big)$. Therefore,

(iii) $$\frac{A \cap B}{(a \cap B) + (b \cap A)} \approx \frac{a + (A \cap B)}{a + (b \cap A)}.$$

By symmetry,

$$\text{(iv)} \qquad \frac{A \cap B}{(b \cap A) + (a \cap B)} \approx \frac{b + (A \cap B)}{b + (a \cap B)}.$$

Since the left side of (iii) is precisely the same as the left side of (iv), the result now follows. ▯

The butterfly lemma gets its name from the Hasse diagram:

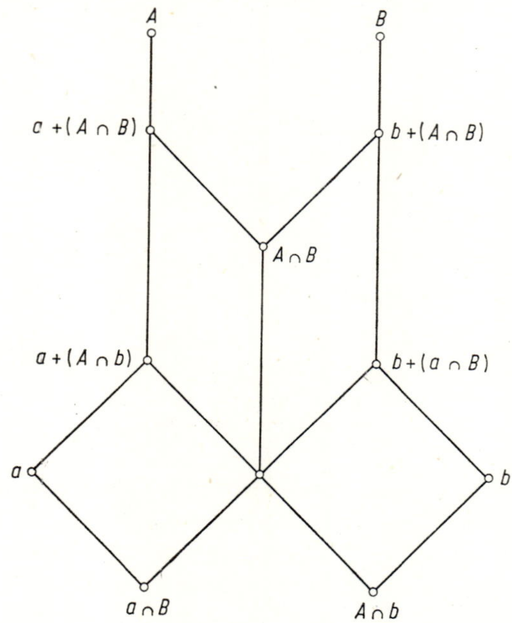

Exercises for Chapter 1

1. Every group of order not exceeding 5 is abelian. How many non-abelian groups are there of order 6? Show that there are just two non-abelian groups of order 8; namely, the *quaternion group*

$$H = \{\pm 1, \pm i, \pm j, \pm k\},$$

where $+x = x$, $+1x = x$, $-1x = -x$, $(-x)(-x) = x^2 \; \forall x \in H$ and

$$\begin{aligned} i^2 = j^2 = k^2 &= -1 \\ ij = -ji &= k \\ ki = -ik &= j \\ jk = -kj &= i; \end{aligned}$$

and the *dihedral group*

$$D = \{a^i b^j \mid i = 0, 1, 2, 3, \text{ and } j = 0,1\},$$

where $a^0 = b^0 = a^0 b^0 = 1$, the identity element, and

$$a^4 = b^2 = 1$$
$$bab^{-1} = a^3.$$

(Since $b^{-1} = b$, the last condition is $bab = a^3$. Cf. Exercise 17.)

2. Every group of order 15 is cyclic.

3. If a subgroup H of a finite group G has index which is a prime smaller than any other prime dividing $|G|$, then H is normal. Find a group G with a subgroup H of prime index which is not normal.

4. Show that S_n contains elements of order n.

5. A group of order p^2 is abelian if p is prime, and the group is either cyclic or isomorphic to a direct sum of two cyclic groups.

6. An abelian group of order pq for primes p and q is cyclic if $p \neq q$.

***7**[1]. Prove the Jordan-Hölder theorem for groups. (See 17.7 and 17.9.4).

***8.** For a group G inductively define the derived series

$$G = G^{(0)} \supseteq G^{(1)} \supseteq \cdots \supseteq G^{(i)} \supseteq G^{(i+1)} \supseteq \cdots,$$

where $G^{(n)}$ is the commutator group of $G^{(n-1)}$ for every natural number n, called the nth **derived** group of G. A group G is said to be **solvable** provided that for some n the nth derived group is the identity subgroup. Show that any subgroup or quotient group of a solvable group is solvable. If a finite group G is not solvable, then G contains a normal subgroup N of order >1 equal to its own derived group. A solvable group of order >1 contains an abelian normal subgroup of order >1. Show that G is solvable if and only if there exists a finite sequence

$$G = N_0 \supseteq \cdots \supseteq N_i \supseteq N_{i+1} \supseteq \cdots \supseteq N_t = e = \text{identity}$$

such that N_{i+1} is a normal subgroup of N_i and N_i/N_{i+1} is abelian, $i = 0, 1, \ldots, t-1$. Prove that the symmetric group S_3 or S_4 is solvable. (It is known that S_n is not solvable for $n > 4$.)

---

[1] The star * denotes exercises falling into one or more of the following categories: (1) more than routine interest; (2) more than routine difficulty; (3) a related result standing in the literature. The latter category is more for convenient references than for the purpose of an exercise, for example, the Sylow theorems (and proof) outlined in Exercise 15, and the Baer-Specker theorems, Exercise 21. The absence of a star, however, does not denote the negation of any of this!

***9.** Let G be a group, and define the **ascending central series**

$$Z_0 = e \subseteq Z_1 \subseteq \cdots \subseteq Z_i \subseteq Z_{i+1} \subseteq \cdots$$

such that Z_i is a normal subgroup of Z_{i+1} and Z_{i+1}/Z_i is the center of G/Z_i, $i = 0, 1, 2, \ldots$. Define the **descending central series**

$$N_0 = G \supseteq N_1 \supseteq \cdots \supseteq N_i \supseteq N_{i+1} \supseteq \cdots$$

such that N_i is a normal subgroup of G and $N_{i+1} = [G, N_i]$ for every i. A group G is defined to be **nilpotent** provided that the two equivalent conditions hold: (a) $Z_n = G$ for some integer $n > 0$; (b) $N_t = e$ for some integer $t > 0$. (In case G is nilpotent, $n = t$.) Every subgroup and quotient group of a nilpotent group is nilpotent. Show that S_3 is not nilpotent. Conclude that a group G which has a nilpotent normal subgroup N such that G/N is nilpotent is not necessarily nilpotent. Every nilpotent group is solvable. Every p-group is nilpotent. (Cf. Exercise 15.)

10. A permutation f of the 2-dimensional plane $P = \mathbb{R} \times \mathbb{R}$ is said to be an **isometry** provided that f preserves the distance P_1P_2 between any two points P_1 and P_2. If $P_i = (x_i, y_i)$, $i = 1, 2$, then

$$P_1P_2 = \sqrt{(x_1 - x_2)^2 + (y_1 - y_2)^2}.$$

A **translation** of P is defined by two real numbers h and k to be a mapping $(x, y) \mapsto (x + h, y + k)$. Show that the set T of translations of P is a normal subgroup of the group G of all isometries of P. Show that G/T contains a normal subgroup of index 2 which is isomorphic to the group of rotations of P around some fixed point. Conclude that G is solvable.

11. A permutation $f \in S_n$ is an r-**cycle**, or a cycle of length r, provided that there exists $i_1, \ldots, i_r$ such that $f(x_j) = x_j$ when j is not one of the i_k, and

$$f(x_{i_1}) = x_{i_2}, \quad f(x_{i_2}) = x_{i_3}, \quad \ldots, \quad f(x_{i_{r-1}}) = x_{i_r}, \quad f(x_{i_r}) = x_{i_1}.$$

In an element of the group S_n, the order of an r-cycle is r. A **transposition** is defined to be a 2-cycle. Show that every cycle is a product of transpositions. Two permutations f and g are **disjoint** in case every element moved by one is fixed by the other, that is, $f(x) \neq x \Rightarrow g(x) = x$ and $g(y) \neq y \Rightarrow f(y) = y$. If f and g are disjoint, then $fg = gf$. Prove for $n \geq 2$ that every permutation in S_n is a product of disjoint cycles of lengths ≥ 2 and that the factorization is unique except for the order in which the cycles are written. Conclude that every element of S_n has order not exceeding n and that the exponent of S_n is n. Furthermore, since every cycle is a product of transpositions, conclude that every permutation is a product of transpositions.

12. A semigroup S is **simple** provided that every homomorphism $f: S \to S'$ which is not the identity (or zero) mapping is an isomorphism. For a group, this is equivalent to the requirement that every normal subgroup not the identity is S itself. Prove that an abelian group G is simple if and only G is isomorphic to a cyclic group $\mathbb{Z}_p$ consisting of p elements, where p is a prime[1].

13. Consider the polynomial in n variables $x_1, \ldots, x_n$:

$$\Delta(x_1, \ldots, x_n) = \prod_{i<j}^{n} (x_i - x_j).$$

A permutation f in the symmetric group S_n acting on the set $\{x_1, \ldots, x_n\}$ is **even** or **odd** according as $(fx_1, \ldots, fx_n)$ is plus or minus $\Delta(x_1, \ldots, x_n)$. Prove that the set A_n of even permutations is a normal subgroup of S_n of index 2, called the **alternating group.** (It is known that A_n is a simple group unless $n = 4$. Cf. Hall [59, p. 61], or Rotman [65, p. 45.] This remains true for the subgroup A_ω of the symmetric group S_ω on ω consisting of all permutations which can be written as a finite product of an even number of transpositions.)

14. Show that there are $n+1$ different embeddings of S_n in S_{n+1}. Show that S_n can be embedded in a normal subgroup of S_{n+2}, in fact S_n embeds in A_{n+2}.

15. Let G operate on a finite set X, and let O_x denote the orbit of x. The stability group Stab x is the group of all elements g of G which fix x. Then

$$|O_x| = |G : \text{Stab}\, x|$$

so that $|O_x|$ divides the order of G. Write X as the disjoint union of orbits

$$X = \coprod_{i=1}^{t} O_{x_i}.$$

Thus, x_i belongs to one and only one orbit, and

$$|X| = \sum_{i=1}^{t} |G : \text{Stab}\, x_i|.$$

This is called the **orbit decomposition formula.** Let $\text{Fix}\, g$ be the set of all x fixed by g. Then

$$t = |G|^{-1} \sum_{g \in G} |\text{Fix}\, g|.$$

---

[1] Feit and Thompson [63] proved that every finite group of odd order is solvable. Via the Jordan-Hölder theorem (see Exercise 7), this implies, and is implied by, the assertion that every noncommutative simple group has even order.

Now let G be any group, and let G operate on itself by the formula $g \cdot h = ghg^{-1}$. The operation is called **conjugacy**, and the orbit of an element of G is called a **conjugacy class**. In this case, Stab x is the normalizer of x (the centralizer of x), and the orbit decomposition formula becomes

$$(\textit{class formula}) \qquad |G| = \sum_{i \in 1}^{t} |G : \text{normalizer}_G x_i|.$$

An element x lies in center G if and only if its normalizer is G. Thus, if the $\{x_i\}_{i \le t}$ are numbered so that center $G = \{x_i \mid i = 1, \ldots, c\}$, where $c = |\text{center } G|$, then the class formula reads:

$$|G| = |\text{center } G| + \sum_{x_i \notin \text{center } G} |G : \text{centralizer } x_i|.$$

Use this formula to prove that every finite p-group $\neq 1$ has center $\neq 1$. Conclude that every finite p-group is nilpotent. Prove that every finite abelian group contains an element of each prime order dividing the order of the group. Use this and the class formula to prove that every finite group contains a subgroup of order p^n for every prime power p^n dividing $|G|$. Call a maximal p-subgroup of G a **p-Sylow subgroup.** Prove the **Sylow theorems**:

(S—1) Every p-subgroup of a finite group is contained in a p-Sylow subgroup.

(S—2) Any two p-Sylow subgroups for the same prime p are conjugate.

(S—3) The number of p-Sylow subgroups for p is $\equiv 1 \pmod{p}$.

16. Prove that a finite group G is isomorphic to a direct product of its p-Sylow subgroups, p ranging over all prime divisors of $|G|$, if and only if every p-Sylow subgroup of G is normal.

17. Let P be a regular polygon with vertices $x_1, \ldots, x_n$. The **dihedral group** D_{2n} is defined to be the group isomorphic to the group of symmetries of P, that is, the group of permutations of P which are rigid motions in the real plane sending vertices into vertices. Let S denote clockwise rotation by one vertex, and let T be reflection about a line of symmetry through x_1 and the center of P. Show that

$$S^n = 1, \qquad T^2 = 1, \qquad \text{and} \qquad TST = S^{-1}.$$

Apply the Sylow theorems to prove that every group of order $2p$, where p is an odd prime, is either cyclic or dihedral.

18. If G is a group of order pq, the product of primes $p > q$, then either G is cyclic or G is generated by elements s and t such that

$$s^p = 1, \qquad t^q = 1, \qquad \text{and} \qquad t^{-1}st = s^r,$$

where $r \neq 1 \pmod p$ and $r^q = 1 \pmod p$. Conclude that G is solvable, and the p-Sylow subgroup is normal[1].

19. A finite nonabelian group G is **Hamiltonian** (in the sense that every subgroup is normal) if and only if

$$G \approx H \times A \times \mathbb{Z}_2,$$

where H is the quaternion group and A is an abelian group every element of which has odd order.

20. Euler's φ-function associates to an integer n the number $\varphi(n)$ of integers less than n and relatively prime to n. Moreover, if $a < n$ is prime to n, then there is a least integer e, the **exponent** of $a \pmod n$, such that $a^e \equiv 1 \pmod n$. If this exponent is $\varphi(n)$, then a is called a **primitive root of** n. Show that Aut $\mathbb{Z}_n$ is cyclic iff n has a primitive root. Show that this happens iff n is 2, 4, p^n, or $2p^n$, where p is an odd prime.

***21.** Show that the direct product $\mathbb{Z}^\omega$ is not free abelian (Baer). Prove that any countable subgroup of $\mathbb{Z}^\omega$ is a free abelian group (Specker). Prove that Hom $(\mathbb{Z}^\omega, \mathbb{Z})$ is free (Specker [50]).

22. A subgroup A is a (direct) **factor** of a group G, provided that there is a subgroup B of G such that the direct sum of the inclusions $A \to G$ and $B \to G$ is an isomorphism. Show that a necessary and sufficient condition for this is that A and B are normal subgroups of G such that $G = AB = \{ab \mid a \in A, b \in B\}$ and $A \cap B = (e)$. In this case, $ab = ba \; \forall \; a \in A$ and $b \in B$. If G is additive, then A is called a (direct) **summand** of G.

23. A homomorphism $f: A \to B$ is a **retraction** provided that there exists a homomorphism $g: B \to A$ such that $fg = 1_B$. The dual statement defines coretraction. (Thus, the mapping g in the definition of retraction is a coretraction.) A retraction is an epimorphism, and a coretraction is a monomorphism. An epimorphism $f: A \to B$ is said to **split** in case it is a retraction. Dually for monomorphism. Prove in SEMIGROUPS that the canonical injections $X_i \to \bigoplus_{i \in I} X_i$ are coretractions, and that the canonical projections $\prod_{i \in I} X_i \to X_i$ are retractions. Show for any retraction (coretraction) $f: A \to B$ in mod-$\mathbb{Z}$ that ker f (resp. im f) is a summand of A (resp. B), and conversely.

---

[1] Burnsides pq theorem asserts that any group of order $p^a q^b$ for primes p and q is solvable. (See M. H. Hall, Jr. [59, p. 291, Theorem 16.8.7].)

24. Show for every abelian group B, there exists a divisible abelian group D and an embedding $0 \to B \to D$. An abelian group A is **injective** provided that A has the following properties which are equivalent:

(D — 1) The image of every monomorphism $A \to B$ in mod-$\mathbb{Z}$ is a summand.

(D — 2) If $0 \to K \to L$ is exact in mod-$\mathbb{Z}$, then $h_A(L) \to h_A(K) \to 0$ is exact in mod-$\mathbb{Z}$ (or what is the same, $h_A(L) \to h_A(K)$ is surjective).

(D — 3) Every monomorphism $f: A \to B$ in mod-$\mathbb{Z}$ is a coretraction.

Prove that (D—1), (D—2), and (D—3) are equivalent. Show that an abelian group is injective if and only if it is divisible. (Hint: Use the embedding of A into a divisible group.) Order mod-$\mathbb{Z}$ by $A \leq B$ if and only if there is an embedding $A \to B$. Prove that in DIV mod-$\mathbb{Z}$ that this order induces an order which is linear in the sense that $A \leq B$ and $B \leq A$ implies $A \approx B$. Show that every A in mod-$\mathbb{Z}$ can be embedded in an injective abelian group D which is minimal in the class of all D' in DIV mod-$\mathbb{Z}$ for which there is an embedding $A \to D'$. Prove that any two minimal injectives D and D' in which A can be embedded are isomorphic by a map $D \to D'$ such that

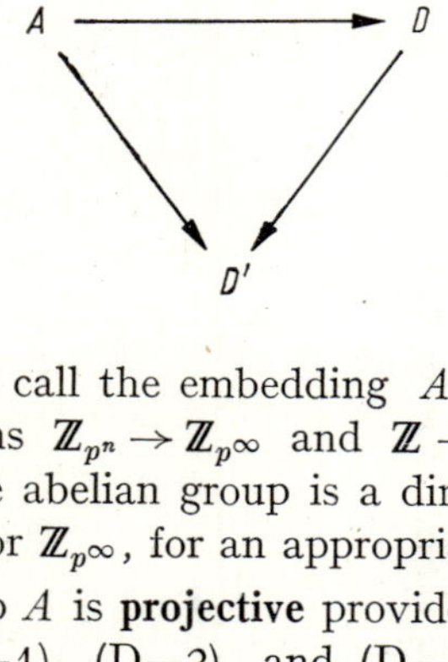

commutes. In this case, call the embedding $A \to D$ an **injective hull.** Show that the inclusions $\mathbb{Z}_{p^n} \to \mathbb{Z}_{p^\infty}$ and $\mathbb{Z} \to \mathbb{Q}$ are injective hulls. Show that any injective abelian group is a direct sum of groups each isomorphic to either $\mathbb{Q}$ or $\mathbb{Z}_{p^\infty}$, for an appropriate prime number p.

25. An abelian group A is **projective** provided that A has the three properties dual to (D—1), (D—2), and (D—3) in Exercise 24, e.g., every epimorphism $B \to A$ in mod-$\mathbb{Z}$ splits. Show that A is projective if and only if A is free, that is, isomorphic to $\mathbb{Z}^{(S)}$ for some set S. Show that any subgroup B of a free abelian group $A \approx \mathbb{Z}^{(S)}$ is a free abelian group $\approx \mathbb{Z}^{(S')}$ for some set S' such that $|S'| \leq |S|$. The concept dual to injective hull in mod-$\mathbb{Z}$ is called **projective cover** (the subject of Chapter 22). Show that A has a projective cover in mod-$\mathbb{Z}$ if and only if A is a free abelian group.

26. Generalize the notion of injective group to the full subcategory mod-$\mathbb{Z}_n$ of mod-$\mathbb{Z}$ consisting of all torsion groups of exponent dividing n, e.g., replace mod-$\mathbb{Z}$ with mod-$\mathbb{Z}_n$ in (D—1). Show that $\mathbb{Z}_n$ is **selfinjective** in the sense that $\mathbb{Z}_n$ is injective in mod-$\mathbb{Z}_n$. If $n = mm'$, and if m and m' have no common factors, show that $\mathbb{Z}_n \approx \mathbb{Z}_m \oplus \mathbb{Z}_{m'}$. Thus reduce the problem to the case $n = p^e$ is a power of a prime. Use selfinjectivity of $\mathbb{Z}_n$ to prove: Any abelian torsion group of bounded order is isomorphic to a direct sum of cyclic groups. If $n = p_1^{e_1} \cdots p_t^{e_t}$ is the factorization of n into a product of powers $p_k^{e_k}$ of distinct primes $p_1, \ldots, p_t$, show that any group A in mod-$\mathbb{Z}_n$ is isomorphic to a direct sum of cyclic p_i-groups, $i = 1, \ldots, t$. (Reduce to the case $t = 1$ by showing that any torsion group of bounded order is a direct sum of p-groups. Any group A of exponent p^e contains a subgroup B isomorphic to $\mathbb{Z}_{p^e}$, and B is a summand of A since $\mathbb{Z}_{p^e}$, whence B, is injective in mod-$\mathbb{Z}_{p^e}$. Write $A = \mathbb{Z}_{p^e} \oplus A'$, and conclude the proof by induction on the number of generators of A if A is finite. Otherwise show that a direct sum of injectives in mod-$\mathbb{Z}_n$ is injective. And, if B is a subgroup of A which is maximal with respect to the property that B is a direct sum of groups isomorphic to cyclic groups $\mathbb{Z}_{p^i}$ for suitable $i \leq e$, then B must be a summand of A, say $A = B \oplus C$. But, then $C \neq 0$ would have a summand $\approx$ some $\mathbb{Z}_{p^j}$, contradicting maximality of B. Thus, $A = B$ is the desired direct sum.)

27. Basis Theorem for Abelian Groups. Any finitely generated abelian group G is a direct sum of cyclic groups, and in fact, torsion G is a summand, whence

$$G \approx G/(\text{torsion } G) \oplus \text{torsion } G.$$

(Cf. Theorem 31.23.) [Hint: Any finitely generated torsionfree abelian group is a free abelian group. First show that such a group can be embedded in $\mathbb{Z}^n$, and then consider the effect on the group of the projections $\mathbb{Z}^n \to \mathbb{Z}$.]

General References

Artin, E.: Galois Theory. South Bend: Notre Dame University 1955.

— Geometric Algebra. New York: Interscience Publishers 1957.

— Nesbitt, E., Thrall, R.: Rings with Minimum Condition. Ann Arbor: University of Michigan Press 1944.

Bass, H.: Algebraic K-Theory. New York: Benjamin 1968.

Birkhoff, G.: Lattice Theory, Colloquium Publication, Vol. 25 (revised). Amer. Math. Soc., Providence 1948, 1967.

Bucur, I., Deleanu, A.: Introduction to the Theory of Categories and Functors. London/New York/Sidney: Interscience Publishers, Wiley 1968.

Cartan, H., Eilenberg, S.: Homological Algebra. Princeton: Princeton University Press 1956.

Chevalley, C.: Fundamental Concepts of Algebra. New York: Academic Press 1956.
Clifford, A. H., Preston, G. B.: Algebraic Theory of Semigroups. Vol. 1. Surveys Amer. Math. Soc., No. 7, Providence 1961.
Cohn, P. M., Morita Equivalence and Duality. University of London, Bookstore, Queen Mary College, Mile End Road, London, 1966.
Feit, W.: Characters of Finite Groups. New York: Benjamin 1967.
Freyd, P.: Abelian Categories. New York: Harper & Row 1964.
Fuchs, L.: Abelian Groups, 2nd Ed. New York: Pergamon Press 1970.
Gorenstein, D.: Finite Groups. New York: Harper & Row 1968.
Hall, M. H., Jr.: The Theory of Groups. New York: Macmillan 1959.
Herstein, I. N.: Topics in Algebra. New York: Harper & Row 1964.
Jacobson, N.: Lectures in Abstract Algebra, Vols. I—III. New York/Princeton: Van Nostrand 1951, 1953, 1964.
Jans, J.: Rings and Homology. New York: Holt, Rinehart & Winston 1964.
Kaplansky, I.: Infinite Abelian Groups, 2nd Ed. Ann Arbor: University of Michigan Press 1969.
Kurosch, A.: Theory of Groups, Vols. 1 and 2. New York: Chelsea 1955, 1956.
— General Algebra. New York: Chelsea 1963.
Lang, S.: Algebra. Cambridge: Addison-Wesley 1965.
MacLane, S.: Homology. Berlin/Göttingen/Heidelberg: Springer 1963.
— Categories for the Working Mathematician, Berlin/Heidelberg/New York: Springer 1971.
—, Birkhoff, G.: Algebra. New York/London: MacMillan 1967.
Mitchell, B.: Theory of Categories. New York: Academic Press 1965.
Northcott, D. G.: Homological Algebra. Cambridge: Cambridge University Press 1960.
Pontryagin, L.: Topological Groups. Princeton: Princeton University Press 1939.
Rotman, J.: Theory of Groups. Boston: Allyn & Bacon 1965.
Waerden, B. L. van der: Modern Algebra, Vols. 1 and 2. New York: Ungar 1948.
Weyl, H.: The Classical Groups, their Invariants and Representations (revised). Princeton: Princeton University Press 1946.
Wielandt, H.: Finite Permutation Groups. New York: Academic Press 1964.
Zassenhaus, H.: The Theory of Groups, 2nd Ed. New York: Chelsea 1958.

Chapter 2

2. Product and Coproduct

The concepts introduced and explored in this chapter are so fundamental that scarcely any of them can be dispensed with hereafter.

A functor $T: C \rightsquigarrow D$ is faithful (resp. full) if T induces an injection (surjection) $\mathrm{Mor}_C(X, Y) \to \mathrm{Mor}_D(TX, TY)$, for every pair X, Y of objects of C. An object $A \in C$ is a generator provided that the first morphism functor $h^A: C \rightsquigarrow \text{SETS}$ is faithful. (An object A of mod-$\mathbb{Z}$ is a generator iff every object is the homomorphic image of a direct sum of copies of A or equivalently, iff A has a summand isomorphic to $\mathbb{Z}$. An object A is said to be projective if the set valued functor h^A preserves epics; an object A of mod-$\mathbb{Z}$ is projective iff it is a free abelian group. Cf. Exercises 2.6.) The dual concept is that of an injective object. An abelian group A is injective in mod-$\mathbb{Z}$ iff A is a divisible group 2.6.5.

For each concrete category C the free object functor $F: \text{SETS} \rightsquigarrow C$ is defined, and when it exists, is faithful if C contains objects of ever greater cardinal. (In the terminology of Chapter 5, the free object functor is left adjoint to the forgetful functor $C \rightsquigarrow \text{SETS}$.) We construct F for the three categories of mod-$\mathbb{Z}$, SEMIGROUPS, and GROUPS.

In this chapter product and coproduct in categories are defined and realized in several concrete categories. In all of them, product coincides with direct product, coproduct in SETS is disjoint union, and in mod-$\mathbb{Z}$, coproduct is direct sum. In mod-$\mathbb{Z}$, the coproduct of any family of homomorphisms $\{\mathbb{Z} \to A\}$, indexed by a set S, is the direct sum mapping $\mathbb{Z}^{(S)} \to A$ of the family, that is, the family of injections $\{\mathbb{Z} \to \mathbb{Z}^{(S)}\}$ is a coproduct in mod-$\mathbb{Z}$, called the free abelian group on the set S. Similarly, in GROUPS there is a family of monomorphisms $\{\mathbb{Z} \to F(S)\}$ into the free group $F(S)$ on a set S which is a coproduct.

The Yoneda functions are defined and they demonstrate the fidelity of the representation functors.

Zero

An object T in a category C is **initial** in case for every object X there is a unique morphism $T \to X$. An object F is **final** in case it has the dual property, namely, for every object X there is a unique morphism $X \to F$. A category may have more than one initial (or final) object, but if T_1 and T_2 are both initial, then the compose of the unique morphisms $T_1 \to T_2$ and $T_2 \to T_1$ must coincide with the unique morphisms $T_1 \to T_1$ which is necessarily the identity morphism 1_{T_1}. By symmetry, composing $T_2 \to T_1$ and $T_1 \to T_2$ yields 1_{T_2}. This proves the following proposition.

2.1 Proposition and Definition. *If T_1 and T_2 are initial objects in a category C, then there is a unique equivalence $T_1 \to T_2$. An object of C is said to be a* **zero object** *if it is both initial and final. A zero object is unique up to a unique equivalence and is denoted* 0.

The formula $X = 0$ symbolizes that X is a zero object, but $X = 0$ and $Y = 0$ does not imply that $X = Y$. (A better representation would be $X \approx 0$ and $Y \approx 0$, in which case $X \approx Y$.) A morphism $f : X \to Y$ is said to **factor through** an object Q in case there is a commutative diagram

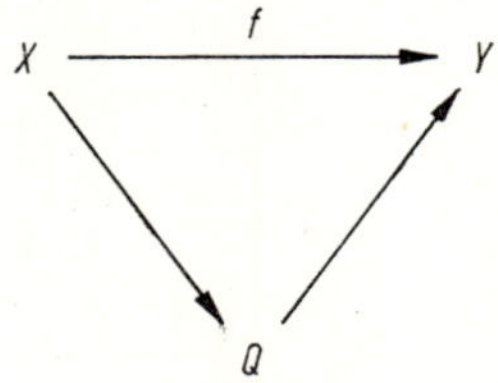

In this case, one also says that the morphism f factors through the morphism $X \to Q$ and the morphism $Q \to Y$ and that these morphisms **allow** f. Thus, in GROUPS, a morphism $f : X \to Y$ factors through the canonical epimorphism $X \to X/K$ if and only if kernel f contains K.

A morphism is said to be **zero** in case it factors through a zero object. Although $\mathrm{Mor}_C(X, Y)$ and $\mathrm{Mor}_C(X', Y')$ are disjoint if $\langle X, Y\rangle \neq \langle X', Y'\rangle$, we denote the zero morphism in each by the same 0 symbol. [Thus, $f = 0$ in $\mathrm{Mor}_C(X, Y)$ and $f' = 0$ in $\mathrm{Mor}_C(X', Y')$ does not mean necessarily that $\mathrm{Mor}_C(X, Y) \cap \mathrm{Mor}_C(X', Y') \neq \varnothing$.] If $f = 0$, then $fg = 0$ and $hf = 0$, for any morphisms h and g such that fg and hf are defined. In GROUPS, for every group G the unique subgroup consisting of a single element e, the identity subgroup, is a zero object and a homomorphism $f : H \to G$ is a zero morphism if and only if $\mathrm{im}\, f = e$. Thus the zero morphism maps everything onto the identity element. This does not hold in SEMIGROUPS, for if f is an idempotent not equal to the identity element e of a semigroup S, and if (e) and (f) denote the semigroups consisting of e and f respectively, then $(e) \approx (f)$, hence there exist at least two different morphisms $(e) \to S$, so (e) is not an initial object.

Preservation Properties

Let P be a property of categories defined for morphisms, objects, or diagrams of objects and morphisms. Then a functor $T : C \rightsquigarrow D$ **preserves** P in case for every morphism, object, or diagram Q it is true that $T(Q)$ has property P whenever Q does. Trivially, every functor, covariant or contravariant, preserves equivalences. (If this were not so, one of the two concepts-equivalence or functor-would have to be discarded!) The study of any algebraic system inevitably is reduced to the determination of what properties are preserved by a prescribed class of morphisms or functors and the dual problem of what class of morphisms or functors preserves a given class of properties. (This principle was first stated by Klein for geometry and was referred to as Klein's *Erlanger program*. In this program, the morphisms were groups, including the group of rigid motions in a plane.) A property of categories is **categorical** provided that it is preserved by category equivalences (see 2.4).

The functor $T : C \rightsquigarrow D$ **reverses** a property P provided that the dual functor $T_* : C^* \rightsquigarrow D$ (or, what is the same, $T^* : C \rightsquigarrow D^*$) preserves P. We then say T **converts** P into P^*. The functor $T : C \rightsquigarrow D$ **reflects** a property P in case for every object, morphism, or diagram Q it is true that Q has property P whenever $T(Q)$ does. A **zero preserving** functor is one which preserves zero objects. Since a zero object A is characterized by the statement that 1_A is a zero morphism, any zero preserving functor also preserves zero morphisms.

Retractions

A morphism $f : A \to B$ is a **retraction** if there exists a morphism $g : B \to A$ such that $fg = 1_B$. The dual property is **coretraction.** Thus, the morphism g such that $fg = 1_B$ is a coretraction. By Foreword Proposition 1, a surjection (injection) in SETS is a retraction (coretraction). In any category C, a morphism f is an equivalence if and only if f is both a retraction and a coretraction. An exact sequence

$$0 \to A \to B \to A' \to 0$$

splits in case $B \to A'$ is a retraction. The sequence is then said to be **split exact.** Retractions and coretractions are preserved by covariant functors and reversed by contravariant functors. Thus, retractions are categorical properties. If $A \to B$ is a coretraction, then A is said to be a **retract** of B.

The same terminology applies the functors. A functor $T : A \rightsquigarrow B$ is a **retraction** provided that there is a functor $S : B \rightsquigarrow A$ and a functor equivalence $TS \approx 1_B$. A functor $T : A \rightsquigarrow B$ is **representative** provided that for each object Y of B, there exist an object X of A and an equivalence $Y \approx TX$ in B. Thus, any retraction $T : A \rightsquigarrow B$ is representative, since $TS \approx 1_B$ implies $Y \approx TSY$.

Full and Faithful Functors

A functor $T : C \rightsquigarrow D$ is **faithful** (resp. **full**) in case $\forall\ X, Y$ in C, the induced mapping

$$\begin{cases} \mathrm{Mor}_C(X, Y) \to \mathrm{Mor}_D(TX, TY) \\ \qquad\qquad f \mapsto Tf \end{cases}$$

is injective (resp. surjective). Any coretraction $T : A \rightsquigarrow B$ is faithful, because if $S : B \rightsquigarrow A$ is such that $ST \approx 1_A$, and if $Tf_1 = Tf_2$ for morphisms $f_i : X \to Y$, $i = 1, 2$, then the commutative diagram

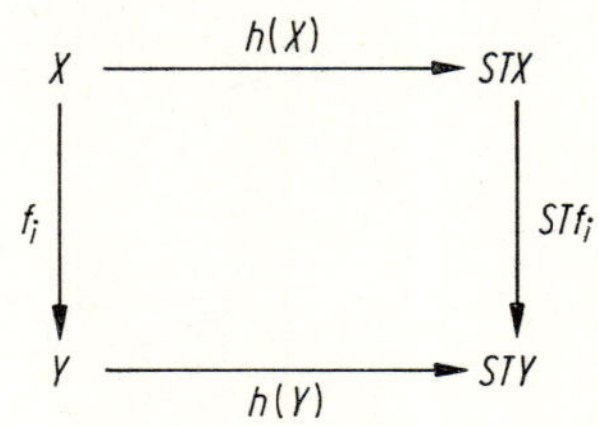

supplied by the natural equivalence of functors $h : 1_C \to ST$ implies that $h(Y) f_1 = h(Y) f_2$, and then $f_1 = f_2$ since $h(Y)$ is an equivalence. (It easily follows that a coretraction T reflects every property definable as a limit or colimit in the terminology of Chapter 5; in particular T reflects exact sequences, kernels, and cokernels, products, and coproducts.) The same argument shows that a retraction $S : D \to C$ induces surjections

$$\mathrm{Mor}_D(TX, TY) \to \mathrm{Mor}_C(STX, STY)$$

and, hence, induces surjections

$$\mathrm{Mor}_D(TX, TY) \to \mathrm{Mor}_C(X, Y).$$

Note that any faithful functor reflects nonzero morphisms and, hence, nonzero objects. (It will be seen that for abelian categories, a faithful functor reflects inexact sequences (see 3.49 and also Chapter 5).

An **embedding** $C \rightsquigarrow D$ is another term for faithful functor. If C is a subcategory of D, then the inclusion functor is an embedding, denoted 1_C^D. (The identity functor 1_D^D is denoted simply 1_D.) A subcategory C of D is a full subcategory if and only if the inclusion functor is full. The terms full and faithful can be inflected, thus: fully faithful, faithfully exact, and so on.

2.2 Equivalence Criterion. *A covariant functor $T : A \rightsquigarrow B$ is an equivalence if and only if T is fully faithful and representative.*

Proof. Any retraction is representative, any coretraction is faithful, and an equivalence $T : A \rightsquigarrow B$ is full, since, as just noted, $T^{-1} : B \rightsquigarrow A$ induces surjections $\mathrm{Mor}_B(TX, TY) \to \mathrm{Mor}_A(T^{-1}TX, T^{-1}TY)$. (Actually, this shows that T^{-1} is full.)

Conversely, assume $T : A \rightsquigarrow B$ is a fully faithful and representative covariant functor. Thus, for each $Y \in \mathrm{obj}\, B$, there corresponds an object $SY \in \mathrm{obj}\, A$ and an equivalence $h(Y) : Y \to T(SY)$. This defines a covariant functor

$$\begin{cases} U : B \rightsquigarrow B \\ U : Y \mapsto TSY & \text{(on objects)} \\ U : f \mapsto h(Y')fh(Y)^{-1} & \text{(on morphisms } f : Y \to Y'\text{).} \end{cases}$$

The following diagram shows that $h : 1_B \to U$ is a natural equivalence of functors; that is, $U \approx 1_B$.

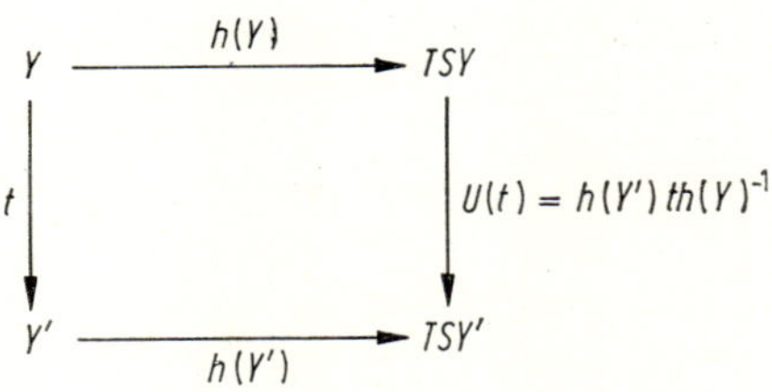

Now T induces a bijection $\mathrm{Mor}_A(X, Y) \to \mathrm{Mor}_B(TX, TY)$, so T^{-1} is defined on Mor (TX, TY) for any pair $X, Y \in \mathrm{obj}\, A$. We now define a covariant functor

$$\begin{cases} S : B \rightsquigarrow A \\ S : Y \mapsto SY & \text{(objects)} \\ S : f \mapsto T^{-1}U(f) & \text{(morphisms)} \end{cases}$$

which satisfies

$$TS = U \approx 1_B.$$

Thus, in order to prove that T is an equivalence we need only prove that $ST \approx 1_A$. Let $X, X' \in \mathrm{obj}\, A$, and suppose that in B there exists an equivalence

$$t : TX \approx TX'.$$

This means that there exists a morphism $t' : TX' \to TX$ such that $t't = 1_{TX}$ and $tt' = 1_{TX'}$. Since T is bijective on morphisms, we can apply T^{-1}:

$$T^{-1}(t')T^{-1}(t) = T^{-1}(t't) = T^{-1}(1_{TX}) = 1_X.$$

Similarly, $T^{-1}(t)T^{-1}(t') = 1_{X'}$, so the equivalence (2) implies an equivalence:

$$T^{-1}(t) : X \approx X'.$$

Now for each $Y \in \operatorname{obj} B$, we have the equivalence $h(Y) : Y \approx TSY$. For $Y = T(X)$, let

$$g(X) = T^{-1}h(TX)$$

denote the corresponding equivalence

$$g(X) : X \approx STX$$

which exists because $h(TX) : TX \approx TS(TX)$. For simplicity, let $h = h(TX)$, $h' = h(TX')$, $g = g(X)$, and $g' = g(X')$. Then

$$g = T^{-1}h \quad \text{and} \quad g' = T^{-1}h'.$$

We now compute the functor ST at a morphism $f : X \to X'$. By the diagram, we can compute TS at $T(f) : TX \to TX'$; namely,

$$TS(Tf) = U(Tf) = h' \cdot Tf \cdot h^{-1}$$

or

$$(TSTf)h = h' \cdot Tf$$

or

$$T(STf \cdot T^{-1}h) = T(T^{-1}h' \cdot f)$$

or

$$STf \cdot T^{-1}h = T^{-1}h' \cdot f$$

or

$$STf \cdot g = g' \cdot f.$$

Thus, we have

$$STf = g' \cdot f \cdot g^{-1}$$

and a commutative diagram

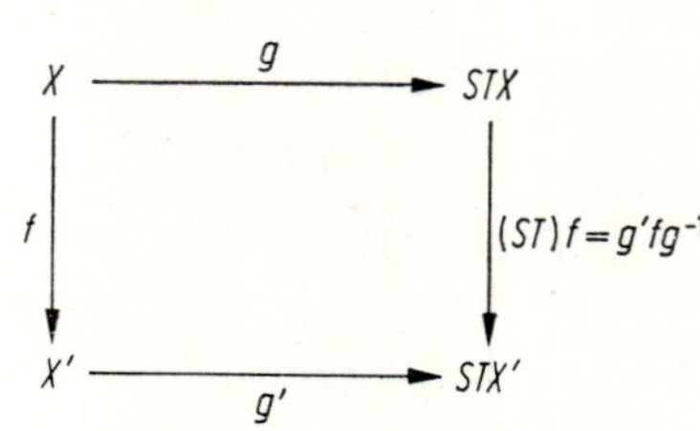

which shows that we have a natural equivalence $g : 1_A \rightsquigarrow ST$, whence $ST \approx 1_A$. □

2.3 Proposition (Yoneda [54]).

2.3.1 *Let C be a category, let A be an object of C, and let $T : C \rightsquigarrow$ SETS be a covariant functor. Then, the* ***Yoneda function***

(1)
$$y_{A,T} \begin{cases} [h^A, T] \to TA \\ \qquad \eta \mapsto \eta(A)\,(1_A) \end{cases}$$

on the set $[h^A, T]$ of all natural transformations $h^A \dot{\to} T$ is bijective and natural in both A and T.

2.3.2 *The functor*

$$\begin{cases} C^{\mathrm{op}} \rightsquigarrow \mathrm{Funct}\,(C, \mathrm{SETS}) \\ A^* \mapsto h^A \end{cases}$$

is a fully faithful functor, called the ***representation functor***.

2.3.3 *If A and B are objects of C, then the Yoneda function*

$$[h^A, h^B] \to \mathrm{Mor}_C(B, A)$$

is bijective. The set-valued functors h^A and h^B are naturally equivalent if and only if the objects A and B are equivalent in C.

Proof. Define a function

$$y' : TA \to [h^A, T]$$

by the rule

(2)
$$y'(x)\,(B)\,(f) = T(f)\,(x),$$

where $B \in \mathrm{obj}\, C$, $x \in TA$, and $f \in h^A(B) = \mathrm{Mor}_C(A, B)$. Thus, $y'(x) : h^A \to T$ is the natural transformation that assigns to each B the function $y'(x)\,(B) : h^A(B) \to TB$, which maps $f \in h^A(B)$ onto $T(f)\,(x)$. It follows from the functorial property $T(gf) = T(g)\,T(f)$, that $y'(x)$ is indeed natural. Naturality of y implies commutativity of

I.
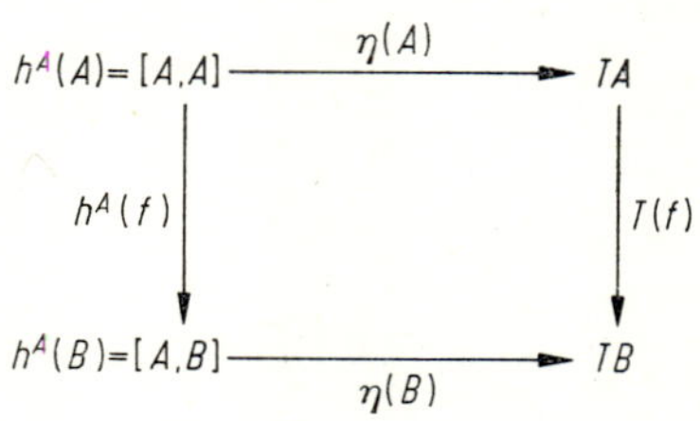

Evaluating clockwise at $1_A \in [A, A]$ yields

(3)
$$T(f)\,\eta(A)\,(1_A) = T(f)\big((y)\,(\eta)\big) = y'\big(y(\eta)\,(B)\big)\,(f),$$

whereas the evaluation counterclockwise provides

(4) $$\eta(B)\,h^A(f)\,(1_A) = \eta(B)\,(f).$$

Hence, commutativity of the diagram yields $y'y(\eta) = \eta$ for all $\eta \in [h^A, T]$; that is, $y'y =$ identity. If $x \in TA$, then

$$y(y'(x)) = y'(x)\,(A)\,(1_A) = T(1_A)\,(x) = x,$$

so $yy' =$ identity, also. This proves that y is bijective as claimed. [If C is an additive category, and T is an additive functor (sup. 3.32), then the definition (2) of y' shows that $y'(x)\,(B)$ is a group homomorphism. Since y is obviously a group homomorphism by (1), then the preceding result implies y is an isomorphism. (This will be used later on.)]

To show that y is natural in A, we must show for any morphism $f: A \to B$ that

II.

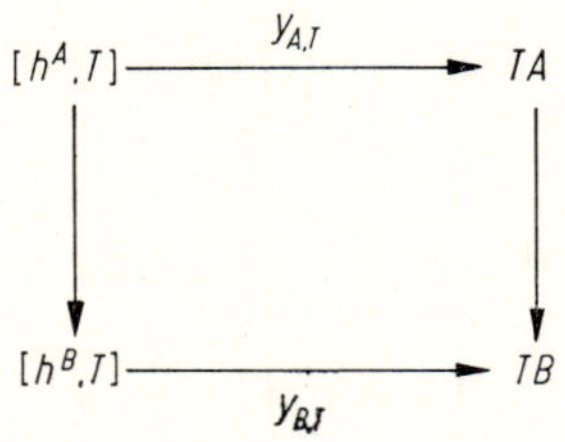

commutes. Starting with $\eta \in [h^A, T]$ and going clockwise, we obtain $T(f)\,\eta(A)\,(1_A)$, and going counterclockwise, we obtain $\eta(B)\,(f)$. Thus, commutativity of II follows from that of I, which established the equality (3) and (4).

To show that y is natural in T, we must show for any natural transformation of functors $\varrho: T \to S$ that

III.

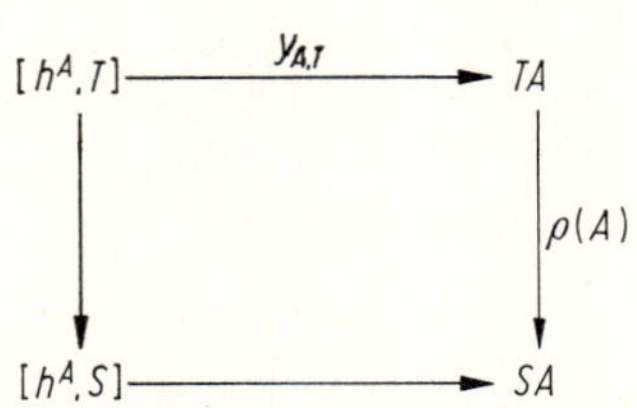

commutes. Starting with $\eta \in [h^A, T]$, clockwise we obtain $\varrho(A)\,(\eta(A)\,(1_A))$, and counterclockwise we obtain $(\varrho\eta)\,(A)\,(1_A)$. By composition of natural transformations, these are equal, so III commutes. This proves (1).

For $T = h^X$, for any X of C, let

$$y'_{X;A} = (y_{A,T})^{-1}$$

be defined by the rule (2) in the proof of (1). Then the representation functor is defined by:

$$H \begin{cases} C^{\text{op}} \rightsquigarrow \text{Funct}(C, \text{SETS}) & \\ A^* \to h^A & \text{(on objects)} \\ t^* \mapsto y'_{X,A}(t) & \text{(on morphisms } t : X \to A). \end{cases}$$

Thus, for any object B

$$H(t)^* \begin{cases} h^A(B) \to h^X(B) \\ \qquad f \mapsto h^X(f)t = ft. \end{cases}$$

If, also $s : Y \to X$, then the representation functor maps

$$H(s^*t^*) = H((ts)^*) = y'_{Y,A}(ts) = y'_{Y,X}(s)\, y'_{X,A}(t) = H(s^*)\, H(t^*),$$

that is, the representation functor is covariant and fully faithful by (1). This proves (2), and (3) is a corollary. □

Generators and Cogenerators

An object A of C is a **generator** provided that the set-valued covariant morphism $h^A = \text{Mor}_C(A, \ )$ is faithful. An object A is a **cogenerator** of C in case A^* is a generator in C^*, or equivalently, provided that the dual functor $(h^A)^* = h_{A^*} : C^* \rightsquigarrow \text{SETS}$ is faithful. Thus, A is a generator (cogenerator) if and only if it is true for each pair of distinct morphisms $f_i : X \to Y$, $i = 1, 2$, of C, that there exists a morphism $g : A \to X$ (resp. $g : Y \to A$) such that $f_1 g \neq f_2 g$ (resp. $g f_1 \neq g f_2$). The truth of this proposition is readily apparent when one observes that

$$h^A(f)g = fg \quad \text{and} \quad h_A(f)g = gf$$

for morphisms f and g of C, when fg or gf are defined. In summary, an object A is a generator provided that for any object X there are enough morphisms $A \to X$ to distinguish between distinct morphisms $X \to Y$ for any object Y. Dually for cogenerator. In SETS the concept is trivial, since any nonempty set is a generator, and any set containing at least two elements is a cogenerator. If A is a generator (cogenerator), and if $B \to A$ is epic (resp. $A \to B$ is monic), then B is a generator (cogenerator).

2.4 Proposition and Definition. *Let C be a category. The following properties in any section, I or II, are either equivalent or defined to be equivalent and are categorical.*

I. Categorical Properties of a Morphism f.

(1) (a) *f is **monic**.*
(b) *For all objects A of C, that is, for all set-valued functors $h^A = \mathrm{Mor}_C(A, \ )$, $h^A(f)$ is an injective mapping.*
(c) $f g_1 = f g_2 \Rightarrow g_1 = g_2$ *for all morphisms g_1 and g_2 of C.*
(d) *f is left cancelable.*

(2) (a) *f is **epic**.*
(b) *For all $A \in \mathrm{obj}\, C$, $h_A(f)$ is a surjection, where $h_A = \mathrm{Mor}_C(\ , A)$.*
(c) $g_1 f = g_2 f \Rightarrow g_1 = g_2$.
(d) *f is right cancelable.*

(3) (a) *f is a **proper monic**.*
(b) *f is a monic, but not an equivalence.*

(4) (a) *f and g are monics such that $f = gh$ for some morphism h.*
(b) $f \leq g$.

(5) *Dualize* (3) or (4) *for epics.*

II. Categorical Properties of an Object A.

(1) (a) *A is **projective**.*
(b) *The covariant set-valued morphism functor h^A preserves epics.*
(c) *If $f: X \to Y$ is epic, then $h^A(f): h^A(X) \to h^A(Y)$ is surjective.*

(2) (a) *A is **injective**.*
(b) *The contravariant set-valued morphism functor h_A converts monics into epics.*
(c) *If $f: X \to Y$ is monic, then $h_A(f): h_A(Y) \to h_A(X)$ is surjective.*

(3) (a) *A is a **generator of** C.*
(b) *The set-valued functor h^A is faithful.*
(c) *If $g_i: X \to Y$, $i = 1, 2$, are distinct morphisms of C, then there exists a morphism $f: A \to X$ such that $g_1 f \neq g_2 f$.*

(4) (a) *A is a **cogenerator of** C.*
(b) *The set-valued functor h_A is faithful.*
(c) *Dualize* 3(c). □

2.5 *Examples.*

2.5.1 In SETS, surjection, retraction, and epic are equivalent conditions for a mapping $f: X \to Y$ (see Foreword Proposition 1). Dually, injection, coretraction, and monic are equivalent conditions in SETS. This explains (b) $\Leftrightarrow$ (c) of (1) and (2) of II.

2.5.2 In any category, a retraction is epic, and a coretraction is monic.

2.5.3 In SEMIGROUPS, the embedding $\mathbb{Z} \to \mathbb{Q}$ of the multiplicative semigroup $\mathbb{Z}$ in the group $\mathbb{Q}$ is an epic which is not surjective. This is true because any map $f : \mathbb{Q} \to A$ is determined by its value on $\mathbb{Z}$, since $f(ab^{-1}) = f(a)\,f(b)^{-1}\ \forall\, 0 \neq b,\, a \in \mathbb{Z}$.

More generally, let $f : B \to A$ be a morphism in SEMIGROUPS such that A is a group, and let A be generated as a group by im f. Then f is epic in SEMIGROUPS.

2.5.4 The embedding of any dense subspace of a topological space is an epic in the category C, the objects of which are topological spaces and the morphisms of which are continuous functions, e.g., the embedding of $\mathbb{Q}$ in $\mathbb{R}$ as topological spaces.

2.5.5 In the category mod-$\mathbb{Z}$, a homomorphism $f : A \to B$ is an epic (resp. monic) if and only if f is an epimorphism (resp. a monomorphism).

That epimorphisms (resp. monomorphisms) are epic (resp. monic) follows from Example 1. Conversely, let $f : A \to B$ be not epimorphic, that is, assume $C = \operatorname{im} f$ is not equal to B. If $g : B \to B/C$ is canonical, then $g \neq 0$, whereas $gf : B \to B/C$ is the zero map; that is,

$$gf = 0 = 0f,$$

so f is not right cancelable, hence f is not epic. The proof for monics is dual.

2.5.6 If f and g are epics (resp. monics) in a category C, and if fg is defined, then fg is epic (resp. monic). fg is obviously right (resp. left) cancelable.

2.5.7 A faithful functor $F : A \to B$ is monic in the category of categories, and a full functor is epic.

2.5.8 An **ordered category** C is a category such that $\operatorname{Mor}_C(A, B)$ is either empty, or else there is a unique morphism $A \to B$. In the latter case write $A \leq B$. The relation $A \leq B$ is an order. Conversely, any ordered set is an ordered category C in which case $A \to B$ is the inclusion. In an ordered category, every morphism $A \to B$ is both epic and monic, but of course, $A \to B$ is an equivalence if and only if $A = B$. Thus, a morphism which is both epic and monic need not be an equivalence. Furthermore, if C denotes the subcategory of SETS, with the same objects, and a morphism $A \to B$ is inclusion, then $A \to B$ is an epic which is surjective if and only if $A = B$. Thus epics in concrete categories need not be surjections. □

2.6 *Exercises.*

2.6.1 Let D be the category of divisible abelian groups, let U be the (multiplicative) group of all roots of unity, and let $f : U \to U$ be the map $f(u) = u^2\ \forall u \in U$. Show that f is monic but not injective.

2.6.2 In SEMIGROUPS, show that the embedding $\mathbb{Z}^+ \to \mathbb{Z}$ of additive semigroups is epic but not surjective.

2.6.3 If A is an injective object in C, then any monic $A \to B$ is a coretraction. Dually for projectives.

*2.6.4 In mod-$\mathbb{Z}$, for any set I, the direct sum $\mathbb{Z}^{(I)}$ is projective and so is any subgroup of $\mathbb{Z}^{(I)}$.

*2.6.5 (a) An abelian group G is a generator in mod-$\mathbb{Z}$ iff $G \approx \mathbb{Z} \oplus A$, for some $A \in$ mod-$\mathbb{Z}$. (This is the assertion made in the introduction to this chapter about generators in mod-$\mathbb{Z}$.)

(b) An abelian group A is injective in mod-$\mathbb{Z}$ if and only if A is divisible. Show that $\mathbb{Q}/\mathbb{Z}$ is an injective cogenerator in mod-$\mathbb{Z}$.

(c) Show that an object E of mod-$\mathbb{Z}$ is a cogenerator iff $\mathbb{Q}/\mathbb{Z}$ embeds in E.

2.6.6 For any abelian group A, the group $A' = \operatorname{Hom}(A, \mathbb{Q}/\mathbb{Z})$ is the **character group** of A. The **bicharacter functor** is defined to be the functor

$$h^2_{\mathbb{Q}/\mathbb{Z}} \begin{cases} \text{mod-}\mathbb{Z} \rightsquigarrow \text{mod-}\mathbb{Z} \\ A \mapsto A'' \end{cases}$$

composed of $h_{\mathbb{Q}/\mathbb{Z}}$ with itself. Thus, $A'' = (A')'$. The transformation

$$\pi \begin{cases} 1_{\text{mod-}\mathbb{Z}} \to h^2_{\mathbb{Q}/\mathbb{Z}} \\ A \mapsto A'' \end{cases}$$

such that $\pi_A(a)f = f(a)$, $\forall a \in A$, $f \in A'$, is natural. Furthermore, $\pi_A : A \to A''$ is monic $\forall A \in$ mod-$\mathbb{Z}$, and

$$h_{\mathbb{Q}/\mathbb{Z}}(\pi_A) : (A'')' \to A'$$

is a splitting epic. Conclude that every finite abelian group A is isomorphic to its character group A'. [Hint: Every finite abelian group A is a direct sum of p-groups (Chapter 1, Exercises 25). Assume A is a p-group of order p^n. Show A' is a p-group of order $\leq p^n$. Then, $|A''| \leq |A'| \leq |A|$. But A embeds in A'' so $|A| = |A'|$. Etc.]

*2.6.7 In GROUPS, monics are injections because a monic must have zero kernel. Show that epics are surjective.

2.6.8 Investigate all the concepts introduced in this chapter for an ordered category C. Also for $C =$ SETS.

The Free Object Functor

Let C be a concrete category. For a set S, consider the category (S, C) the objects of which are $f : S \to A$, where A is an object of C and f is a mapping. A morphism of (S, C) is defined to be a morphism $A \to B$ in C

such that

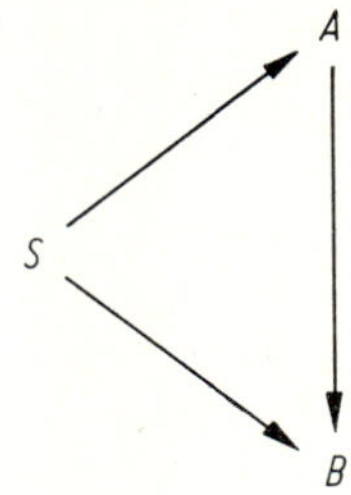

commutes for given objects $S \to A$ and $S \to B$ of (S, C). Composition of morphisms is defined in the obvious way. An initial object $f_S : S \to F(S)$ in (S, C) is said to be the **free object in** C **on the set** S. By the uniqueness of initial objects in categories, if $F_1(S)$ and $F_2(S)$ are two objects of C free on S, then there is a unique equivalence $F_1(S) \to F_2(S)$ of C such that

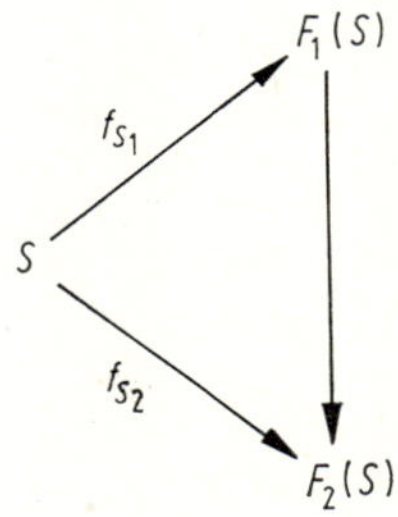

commutes.

The **free object functor**

$$F \begin{cases} \text{SETS} \rightsquigarrow C \\ f_S : S \mapsto F(S) \end{cases}$$

is defined to be the covariant functor which assigns to each set the object $F(S)$ of C free on S and to each mapping $h : S \to T$ of sets the unique morphism $F(h) : F(S) \to F(T)$, given by the free object $f_S : S \to F(S)$, such that $F(h) f_S = f_T h$. This definition depends on the existence of free objects for every set S.

2.7 Free Object Functor Proposition. *Let C be a category such that the free object functor $F : \text{SETS} \rightsquigarrow C$ exists.*

2.7.1 *Then, for each ordinal α, F induces a faithful functor on the subcategory of* SETS *of cardinality $\leq \aleph_\alpha$ if and only if C has an object of cardinality at least $\aleph_\alpha$. Specifically, $f_S : S \to F(S)$ is an injection if and only if $|S| \leq |X|$ for some object X of C. Thus, $F : \text{SETS} \rightsquigarrow C$ is faithful, provided that C contains objects of ever greater cardinal.*

2.7.2 *If*

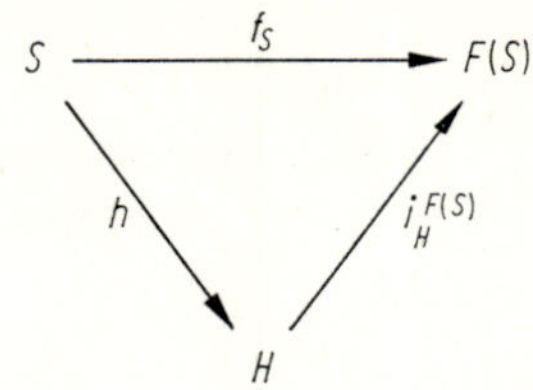

commutes, where H *is an object of* C *and* $i_H^{F(S)}$ *is the inclusion mapping, then* $H = F(S)$, *that is,* $f_S : S \to F(S)$ *factors through no inclusion mapping* $H \to F(S)$ *of an object* $H \neq F(S)$.

Proof of 2.7.1. Assume that C contains an object H of cardinality $\geq \aleph_\alpha$. Then, if S is a set of cardinality $\aleph_\alpha$ and $h : S \to H$ is an injection, then there exists a morphism $t : F(S) \to H$ such that $h = t f_S$. It follows that f_S is injective. Next assume that C contains no objects of cardinal $> \aleph_\alpha$, and let $r : R \to S$ be a surjective mapping. Then $f_S r : R \to F(S)$ is free on the set R, because any mapping $R \to T$, where T is an object of C, factors through $r : R \to S$, and hence, factors uniquely through $f_S r : R \to F(S)$. This shows that the mapping $f_R : R \to F(R)$ of a free object need not be injective; it fails to be injective, however, only if C fails to have objects of cardinal $\geq |R|$.

Now let $h_i : S \to T$, $i = 1, 2$, be mappings. Then $F(h_i) f_S = f_T h_i$, $i = 1, 2$. If $|T| \geq \aleph_\alpha$, then f_T is injective by the preceding arguments, hence $h_1 = h_2$. This proves that F is faithful. Conversely, if $|T| < \aleph_\alpha$, that is, if f_T is not injective, and if $h_i : S \to T$ are distinct mappings such that $f_T h_1 = f_T h_2$, then by the definition of free object, there is a unique morphism $g : F(S) \to F(T)$ such that $g f_S = f_T h_1$. Thus, $F(h_1) = F(h_2)$, and F is therefore not faithful. This proves 1.

Proof of 2.7.2. There is a morphism $g : F_S \to H$ such that $g f_S = h$, and then

$$i_H^{F(S)} g f_S = i_H^{F(S)} h = f_S.$$

This implies that $i_H^{F(S)} g = 1_{F(S)}$, because $1_{F(S)}$ is the unique morphism $p : F(S) \to F(S)$ such that $p f_S = f_S$. It follows that $i_H^{F(S)}$ is surjective, whence $H = F(S)$. □

2.8 **Theorem.** *If* S *is a nonempty set, then there is a free semigroup*

$$f_S \begin{cases} S \to F(S) \\ s \mapsto (s), \end{cases}$$

where $F(S)$ *is the set of all finite sequences* $(a_1, \ldots, a_n)$, *or* n*-tuples, of elements of* S, *the semigroup operation of* $F(S)$ *is composition*

$$(a_1, \ldots, a_n)(b_1, \ldots, b_m) = (a_1, \ldots, a_n, b_1, \ldots, b_m)$$

of finite sequences, and the empty sequence E is the identity element of $F(S)$. The mapping $f_S : S \to F(S)$ is injective, and im f_S *generates $F(S)$.*

Proof. If $h : S \to H$ is a mapping of S into a semigroup H, then

$$g \begin{cases} F(S) \to H \\ (a_1, \ldots, a_n) \mapsto h(a_1) \ldots h(a_n) \end{cases}$$

is the unique homomorphism such that

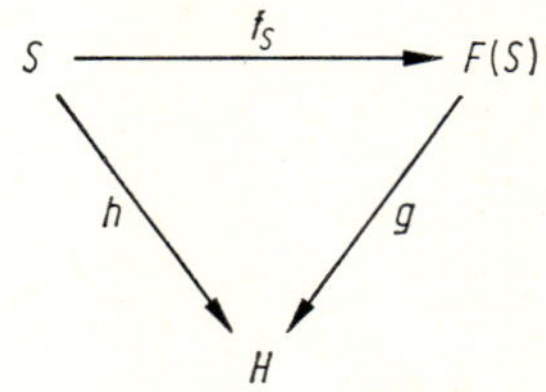

commutes. The last sentence is a consequence of 2.7.2 of the Free Object Functor Proposition. ▯

2.9 Free Group Theorem. *For each set S there exists a free group $f_S : S \to F(S)$. The mapping f_S is injective, and* im f_S *generates $F(S)$. The* ***rank*** *of $F(S)$ is defined*[1] *to be $|S|$.*

Proof (Chevalley [56, p. 40]). Let S' be a set equivalent to S under a mapping $a \mapsto a'$, and let $f_0 : S \coprod S' \to M$ be the free semigroup on the disjoint union of S and S'. Let X denote the set of all epimorphisms $p_Q : M \to Q$, where Q is a semigroup having the property for all $a \in S$ that $p_Q(f_0(a))$ is a unit of Q, and

$$(1) \qquad p_Q(f_0(a))^{-1} = p_Q(f_0(a')).$$

Since M is the subsemigroup generated by im f_0 (Theorem 2.8), that is, since every element of M is a product of elements in the image of f_0, then every element of Q is a product of elements in the image of $p_Q f_0$, so that the condition (1) implies that Q is a group and, hence, so is the direct product $P = \prod_{Q \in X} Q$. Let $g : M \to P$ denote the product of the $\{p_Q\}$, let w_Q denote the canonical projection $P \to Q$, where Q ranges over X, let $F = \operatorname{im} g$, and let $f : S \to F$ be induced by the restriction of $g f_0$ to S. Since

$$w_Q(g(f_0(a))) = p_Q(f_0(a))$$

is the inverse of the element defined by replacing a by a' in this expression, it follows that

$$g(f_0(a')) = g(f_0(a))^{-1}.$$

---

[1] Rank is unique. See end of chapter Exercise 8.

Since the image of $f = gf_0$ generates F, this shows that F is also a group. Let H be any group and $s : S \to H$ a mapping. Then s can be extended to a mapping $s_0 : S \coprod S' \to H$ in such a way that

$$s_0(a') = s_0(a)^{-1} = s(a)^{-1},$$

and there exists a unique homomorphism $h : M \to H$ of the free semigroup M on $S \coprod S'$ such that $hf_0 = s_0$. Let Q denote the image of h, and let $p_Q : M \to Q$ be the induced epimorphism. Since for any $a \in S$, $s_0(a') = s_0(a)^{-1}$, it follows that

$$p_Q(f_0(a')) = p_Q(f_0(a))^{-1},$$

and consequently $Q \in X$. Furthermore, if t denotes the restriction of $1_Q w_Q$ to F, then $t : F \to H$ is the unique homomorphism such that $tf = s$. □

2.10 Free Abelian Group Theorem. *The free object functor*

$$\begin{cases} \text{SETS} \rightsquigarrow \text{mod-}\mathbb{Z} \\ \quad S \to \mathbb{Z}^{(S)} \end{cases}$$

exists, where for each $t \in S$ the function $f_S(t)$ sends t onto 1 and sends $s \neq t$ onto 0.

Proof. If $f : S \to A$ is a mapping of S into an abelian group, then there is a unique homomorphism $g : \mathbb{Z}^{(S)} \to A$ such that $gf_S = f$. □

2.11 Proposition and Definition.

2.11.1 *If S is a set of generators of a (semi) group M, then there exists a unique epimorphism $g : F(S) \to M$ of the free (semi) group such that*

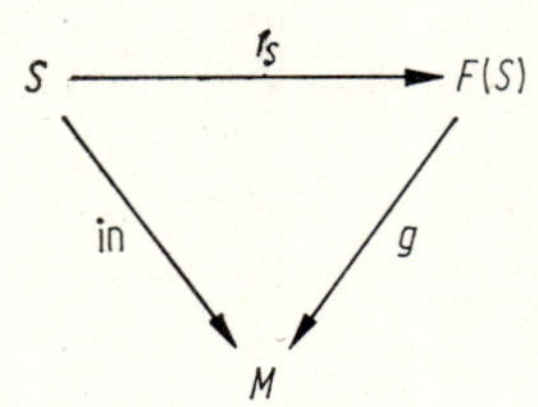

commutes.

2.11.2 *In case M is a group, and the kernel K of g is the normal subgroup generated by $\{r_i\}_{i \in I}$, then M is the group with* ***generators*** *S and* ***relations*** *$\{r_i = 1\}_{i \in I}$. Then, if Δ denotes the set of relations, $(S \mid \Delta)$ is a* ***presentation*** *of M. If $K = 1$, then M is the group generated by S and* ***no relations***. □

2.12 *Exercises.*

2.12.1 The cyclic group of order 6 has generators x and y and relations $x^2 = 1$, $y^3 = 1$, and $x^{-1}y^{-1}xy = 1$. The group with presentation $\{x, y \mid x^2 = 1, y^3 = 1\}$ is infinite.

2.12.2 Determine the set of generators of the free abelian group on a set S.

2.12.3 Determine generators and relations for the dihedral group D_n, the quaternion group, and any of the groups cited in the text thus far (cf. Exercises for Chapter 1).

2.12.4 The infinite cyclic group is a generator of GROUPS. Generalize.

2.12.5 A group M is isomorphic to the free group on a set S iff M is the group generated by S and no relations.

2.12.6 **(The Word Group.)** Let S be a set, which for convenience we suppose is finite, $S = \{a_1, \ldots, a_n\}$, and let $S' = \{a_1', \ldots, a_n'\}$ be a set disjoint from S, so the union $X = S \cup S'$ is a disjoint union. We form the set F of all **words** (with letters in X), namely, all finite sequences $b_1 b_2 \cdots b_t$, with $b_i \in X$, $i = 1, \ldots, t$. The empty word is denoted by 1. A word is **reduced** if no pair of consecutive letters $b_i b_{i+1}$ has $b_{i+1} = b_i'$. Two words w_1 and w_2 are **adjacent** if they have the form $w_1 = u a_j a_j' v$ and $w_2 = uv$, for some j, and words $u, v \in F$. Finally, two words u and v are **equivalent** if there exists a sequence $w_1, \ldots, w_m$ of words, such that $w_i w_{i+1}$ is adjacent for every $i \leq m - 1$, and such that $w_1 = u$ and $w_m = v$. Prove that the class $[w]$ of words equivalent to a given word w contains one and only one reduced word. Prove that the set W of classes of equivalent words is a group with respect to the operation $[w_1][w_2] = [w_1 w_2]$. W is called the **word group** on the set [of letters] S. Prove that the mapping $S \to W$ sending $s \mapsto [s]\ \forall s \in S$ is the free group on the set S.

Other exercises on the free group are placed at the end of this chapter.

Product and Coproduct

Let C be any category, I a set in the universe, and $\{A_i\}_{i \in I}$ a family of objects of C. Consider the category whose objects are families $\{p_i : A \to A_i\}_{i \in I}$ of morphisms of C. A morphism between $\{p_i : A \to A_i\}_{i \in I}$ and $\{f_i : B \to A_i\}_{i \in I}$ is a morphism $h : A \to B$ in C such that $f_i h = p_i$ for all $i \in I$. A final object $\{p_i : A \to A_i\}_{i \in I}$ in this category is called a **product** of the family $\{p_i : A \to A_i\}_{i \in I}$ of morphisms and is denoted $\prod_{i \in I} A_i$. When $\prod_{i \in I} A_i$ is written, the morphisms $p_i : \prod_{i \in I} A_i \to A_i$ are understood. The morphism p_i is then called the ith**-projection from**

the product. Sometimes, we call $\prod_{i\in I} A_i$ the product of the family $\{A_i \mid i \in I\}$ of objects, with the projections understood. Also, for any family $\{f_i : B \to A_i\}$ of morphisms, the unique morphism $h : B \to \to \prod_{i\in I} A_i$ such that $f_i = p_i h$ is denoted $\prod_{i\in I} f_i$ and is called the **product of the morphisms** $(f_i)_{i\in I}$, or the **product morphism**. If A_i is the same object A for all i, we let A^I denote the product $\prod_{i\in I} A_i$. Then the morphism $f^I : B \to A^I$ denotes the product morphism $\prod_{i\in I} f_i$, where $f_i : B \to A$ is the same morphism f for all i. In this case, the **diagonal morphism**

$$\triangle : A \to A^I$$

is defined by $p_i \triangle = 1_A$ for all $i \in I$. The diagonal morphism is monic.

A coproduct $\coprod_{i\in I} A_i$ is defined dually. If the coproduct exists, and if $\{g_i : A_i \to B\}$ is any family of morphisms, then the unique morphism $k : \coprod_i A_i \to B$ such that $k u_i = g_i$, for all i, is called the **coproduct of the family** $\{g_i\}$ **of morphisms**. If $A_i = A$ for all $i \in I$, let $A^{(I)}$ denote the coproduct, and then $g^{(I)} : A^{(I)} \to B^{(I)}$ denotes the coproduct morphism $\coprod_{i\in I} g_i$, where $g_i = g \;\forall i \in I$. Dual to the diagonal morphism is the **codiagonal morphism** $\nabla : A^{(I)} \to A$ defined to be the coproduct $1_A^{(I)}$, that is, $\nabla u_i = 1_A$ for all $i \in I$. The codiagonal morphism is epic.

The product of a finite family $A_1, \ldots, A_n$ is denoted by

$$\prod_{i=1}^{n} A_i, \quad \text{or} \quad A_1 \times A_2 \times \cdots \times A_n,$$

whereas

$$\coprod_{i=1}^{n} A_i, \quad \text{or} \quad A_1 \oplus \cdots \oplus A_n,$$

denotes the coproduct. In this case, for morphisms $f_i : B \to A_i$, $i = 1, \ldots, n$,

$$(f_1, \ldots, f_n) : B \to A_1 \times \cdots \times A_n$$

denotes the product map, and for morphisms $g_i : A_i \to B$, $i = 1, \ldots, n$,

$$(g_1 \oplus \cdots \oplus g_n) : A_1 \oplus \cdots \oplus A_n \to B$$

denotes the coproduct (or direct sum) map. If, in the first case, $A_i = A$, and $f_i = f$, $i = 1, \ldots, n$, then A^n or $A \times \cdots \times A$ denotes the product; we also let A^n, or $A \oplus \cdots \oplus A$, denote the coproduct when coproduct is understood. If $f_i = f$, $i = 1, \ldots, n$, then $f^n : B \to A^n$ denotes the product morphism. Similarly, $g^n : A^n \to B$ denotes the coproduct of morphisms when $g_i = g$, $i = 1, \ldots, n$. The term **finite product** (or coproduct) denotes the product (or coproduct) of a finite family of morphisms.

2.13 Proposition and Definition. *Let C be a category with zero.*

2.13.1 *If $\{p_i : A \to A_i\}$ is a product in C, then $p_i : A \to A_i$ is a retraction, and there exist coretractions $u_i : A_i \to A$ which uniquely satisfy the relations*

$$p_j u_i = \delta_{ij} 1_{A_i}$$

$\forall i, j \in I$.

2.13.2 *Dually, if $\{u_i : A_i \to A\}$ is a coproduct, then u_i is a coretraction, and there exist unique retractions $p_i : A \to A_i$ such that*

$$p_i u_j = \delta_{ij} 1_{A_i}$$

$\forall i, j \in I$.

In either case, we say that A is a ***product (coproduct) with the canonical injections*** *u_i* ***and projections*** *p_i, $i \in I$, and that $A_i \overset{u_i}{\to} A \overset{p_j}{\to} A_j$ is a* ***(co)product representation of*** *A.*

Proof. For each $i \in I$, there exists a unique morphism $u_i : A_i \to A$ such that

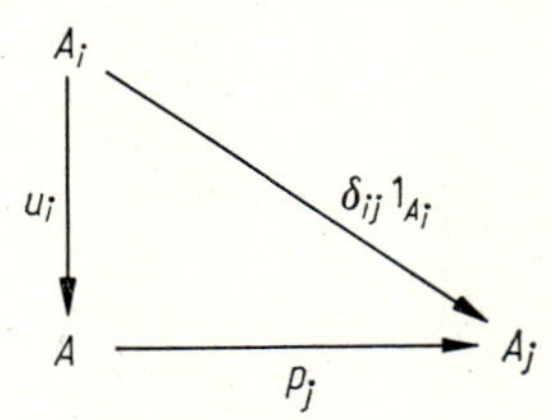

commutes $\forall j \in I$. Since $p_i u_i = 1_{A_i}$, necessarily p_i (u_i) is a (co)retraction. 2.13.2 is dual. ▯

An $I \times J$ matrix over a set S is a mapping $f : I \times J \to S$. Letting, $f(i, j) = f_{ij}$, then (f_{ij}) also denotes the matrix f. Thus, a matrix is simply an element of $S^{I \times J}$. If S has two distinguishable elements 0 and 1, then the **Kronecker-δ** is the matrix $\delta : I \times I \to S$ such that $\delta_{ii} = 1$, and $\delta_{ij} = 0$ when $i \neq j$.

2.14 Matrix Proposition. *Let $\{A_j\}_{j \in J}$ and $\{B_i\}_{i \in I}$ be families of objects in a category C. If f is a morphism $\coprod_{j \in J} A_j \to \prod_{i \in I} B_i$, then the morphism*

$$p_i f u_j : A_j \to B_i,$$

defined by the injections u_j of A_j into the coproduct and the projections p_i onto B_i from the product, is called a ***coordinate morphism*** *and is an element of $\mathrm{Mor}_C(A_j, B_i)$ which is denoted f_{ij}, $\forall i \in I, j \in J$. The matrix (f_{ij}) is an element of the cartesian product $\times_{(i,j) \in I \times J} \mathrm{Mor}_C(A_j, B_i)$, a mapping $I \times J \to \bigcup_{(i,j) \in I \times J} \mathrm{Mor}_C(A_j, B_i)$. The mapping*

$$\begin{cases} \mathrm{Mor}_C\left(\coprod_{j \in J} A_j, \prod_{i \in I} B_i\right) \to \underset{(i,j) \in I \times J}{\times} \mathrm{Mor}_C(A_j, B_i) \\ f \mapsto (f_{ij}) \end{cases}$$

is bijective. If C has a zero, then there is a morphism

$$\delta : \coprod_{j\in I} A_j \to \prod_{i\in I} A_i$$

such that δ is the Kronecker-δ, that is, $\delta_{ii} = 1_{A_i}$ and $\delta_{ij} = 0$ when $i \neq j$. The coproduct is called a **biproduct** *when δ is an equivalence.*

Proof. Two morphisms f and g are equal if and only if they have the same coordinate morphisms $g_{ij} = f_{ij}$ $\forall i, j \in I$. The rest is left as an exercise. □

A matrix $f : I \times J \to S$ into an additive semigroup is **row-finite** in case the subset of I consisting of all i for which there exists j such that $f_{ij} \neq 0$ is finite, that is, there exist only finitely many $i \in I$ such that (i, j) lies in the support of f. Dually for **column-finite**. (Cf. 2.17.4.)

If S is a semigroup, each element a of S determines a matrix $\operatorname{diag}(a) : I \times I \to S$, where

$$\operatorname{diag}(a) = (a\,\delta_{ij})$$

(Kronecker-δ), called a **diagonal matrix**. The mapping $S \to S^{I\times J}$ sending $a \mapsto \operatorname{diag}(a)$ is an embedding.

2.15 Proposition and Definition. *Let C be a category with zero. Let $\{A_i\}_{i\in I}$ be a family of objects of C, and write $I = \bigcup_{k\in K} I_k$ as a disjoint union of sets.*

2.15.1 *For each $k \in K$, let $\{p_{kj} : B_k \to A_j\}_{j\in I_k}$ be a product, and let $\{t_k : B \to B_k\}_{k\in K}$ be a product. Then $\{p_{kj} t_k : B \to A_j\}_{j\in I_k, k\in K}$ is a product for the family $\{A_i\}_{i\in I}$. Hence, if $\prod_{i\in I} A_i$ is any product for this family, there is an equivalence*

$$\prod_{i\in I} A_i \approx \prod_{k\in K} \Big(\prod_{j\in I_k} A_j \Big).$$

2.15.2 *Dually,*

$$\coprod_{i\in I} A_i \approx \coprod_{k\in K} \Big(\coprod_{j\in I_k} A_j \Big).$$

2.15.3 *If J is a subset of I and if the following products exist, then there exists an epic p_{JI}, called the* **canonical projection**, *such that*

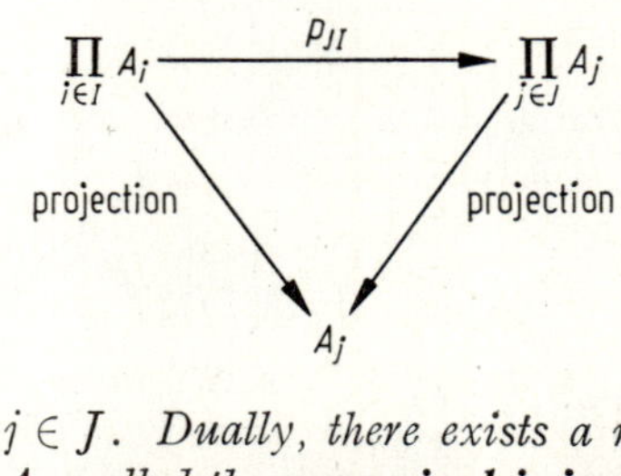

commutes for every $j \in J$. Dually, there exists a monic $u_{IJ} : \coprod_{j\in J} A_j \to \coprod_{i\in I} A_i$ called the **canonical injection**.

Proof of 2.15.1. If $j \in I_k$, let $p'_j = p_{kj} t_k$, all $j \in I$. If $\{f_i : X \to A_i\}_{i \in I}$ is a family of morphisms, then for each $k \in K$, the product morphism $F_k = \prod_{j \in I_k} f_j$ exists. Here $F_k : X \to B_k$, and the product morphism $F : X \to B$ exists, $F = \prod_{k \in K} F_k$. Clearly, $\{p'_i : B \to A_i\}_{i \in I}$ is a product $\prod_{i \in I} A_i$, and $F = \prod_{i \in I} f_i$. Since any two products for the same family are equivalent, we have

$$\prod_{i \in I} A_i \approx \prod_{k \in K} \Big(\prod_{j \in I_k} A_j\Big)$$

as stated.

Proof of 2.15.3. Define p_{JI} to be the product $\prod_{j \in J} p_j$, where, for all $i \in I$, $p_i : \prod_{i \in I} A_i \to A_i$ is the projection. Then, p_{JI} is an epic

$$\prod_{i \in I} A_i \to \prod_{j \in J} A_j$$

as defined in (1). For example, if $K = \{1, 2\}$, and if $I_1 = J$, $I_2 = I - J$ in (1), then p_{JI} corresponds to $t_1 : B \to B_1$. It is epic because projections are. □

2.16 *Examples of Product and Coproduct.*

2.16.1 In SETS, the family of projections $\{\prod_{i \in I} X_i \to X_i\}_{i \in I}$ from the cartesian product of the family of sets $\{X_i\}_{i \in I}$ is a product in SETS, and the family of injections $\{X_i \to \coprod_{i \in I} X_i\}_{i \in I}$ into the disjoint union is a coproduct. (See Foreword Propositions 17 and 19.)

2.16.2 In GROUPS, the family of projections $\{\prod_{i \in I} X_i \to X_i\}_{i \in I}$ from the direct product of a family of groups is a product. The coproduct $\{X_i \to \coprod_{i \in I} X_i\}$ exists in GROUPS. (This is also called the free product in the literature.) If $f_S : S \to F(S)$ is the free group on a set S, then $\{(f_S(t)) \to F(S)\}_{t \in S}$ is a coproduct. Here $(f_S(t))$ is the cyclic group generated by $f_S(t)$ in $F(S)$. Thus, in other notation, if F is a group generated by S and no relations, then $\{(t) \overset{in}{\to} F\}_{t \in S}$ is a coproduct in GROUPS. [This is often expressed by the statement that a free group is a coproduct of infinite cyclic groups. The existence of arbitrary coproducts is analogous to the proof of the existence of arbitrary free groups; the proof of this is left to the interested reader (see 2.17.1).]

2.16.3 In mod-$\mathbb{Z}$, the family of injections into the direct sum is a coproduct.

2.17 *Exercises.*

2.17.1 If $\{S_i\}_{i \in I}$ is a family of groups, if S_i is generated by a set X_i with relations Δ_i, and if F is the group generated by $X = \coprod_{i \in I} X_i$ with relations $\coprod_{i \in I} \Delta_i$, where $\coprod$ denotes disjoint unions of sets, then $S_i = (X_i)$ is the subgroup of F generated by $X_i \forall i \in I$, and the family of inclusions $\{S_i \to F\}_{i \in I}$ is a coproduct in GROUPS. Generalize: when is the coproduct of "cyclic" objects (s), each generated by an element $s \in S$, the free object $F(S)$ on S, where $f_S : S \to F(S)$ is the disjoint union of the maps $\{s\} \to F(S)$?

2.17.2 Show that $\mathbb{Z}$ is a generator in GROUPS.

2.17.3 Show that arbitrary coproducts exist in SEMIGROUPS.

2.17.4 If $(f_{ij}) : I \times J \to S$ is a matrix over an additive monoid, then $S^{I \times J}$ is an additive monoid under matrix addition $S^{I\times J} \times S^{I\times J} \to S^{I\times J}$ defined by

$$(f_{ij}) + (g_{ij}) = (f_{ij} + g_{ij}).$$

(This is just the operation in the direct product $S^{I\times J}$. Thus $S^{I\times J}$ is abelian, a group, or a semigroup when S is.) If S is also a multiplicative monoid and K is another set, define matrix multiplication

$$S^{I\times J} \times S^{J\times K} \to S^{I\times K}$$

by

$$(f_{ij})\,(g_{ij}) = (h_{ij})$$

in case the sum

$$h_{ik} = \sum_{j\in J} f_{ij} g_{jk}$$

is defined or finite [e.g., when (f_{ij}) is row-finite or (g_{ij}) is column-finite]. Under what conditions on S is $S^{I\times I}$ or the set of row-finite or column-finite matrices a monoid (semigroup) under matrix multiplication?

2.17.5 (a) For any categories C and L, the constant functor

$$c_{C,L} \begin{cases} C \rightsquigarrow \operatorname{Funct}(L, C) \\ A \mapsto c(A) \end{cases}$$

is fully faithful. Here $c(A)(F) = A$ and $c(A)(f) = 1_A$, for any functor $F : L \rightsquigarrow C$ and any natural transformation $f : F \to G$ of functors.

(b) If F is an object of $\operatorname{Funct}(L, C)$, and if $\overleftarrow{F}$ is an object of C such that for every A there is a bijection $\operatorname{Mor}_C(A, \overleftarrow{F}) \to [c(A), F]$ natural in the variables A and F, then $\overleftarrow{F}$ is called the **limit** of the functor F. Any object Y of C is called a limit, if $Y \approx \overleftarrow{F}$ for some functor $F : L \rightsquigarrow C$. Dually, for colimits. Show that a product $\prod_{i\in L} X_i$ in a category is a limit of a functor $F : L \rightsquigarrow C$. Show that a final object is a limit of the empty functor. In mod-R, for any ring R, show that kernels and intersections are limits. Discuss "limit preserving" functors, and prove that $h^A : \text{mod-}R \rightsquigarrow \text{mod-}\mathbb{Z}$ is limit preserving for every object A of mod-R.

(c) In an ordered category C, show that $\overleftarrow{F} = \underset{X\in \operatorname{obj} L}{\text{g.l.b}} \{F(X)\}$, for any functor $F : L \rightsquigarrow C$. Conclude that limits, intersections, and products coincide in C, e.g., the family of morphisms $\{\text{g.l.b.}\{A_i\}_{i\in I} \to A_i\}_{i\in I}$ is a product in C.

For the definitions of left and right adjoint functors in the next exercises, consult Chapter 5.

2.17.6 A right (left) adjoint to the constant functor $C \rightsquigarrow \text{Funct}\,(L,C)$ is called the **limit functor** $\lim : \text{Funct}\,(L, C) \rightsquigarrow C$ (resp. colimit functor). Show that $\lim : \text{Funct}\,(L, C) \rightsquigarrow C$ exists if and only if $\bar{F}$ exists for every functor $F : L \rightsquigarrow C$, and then there are equivalences $\lim F \approx \bar{F}$ natural in F.

*2.17.7 Prove that the limit functor $\lim : \text{Funct}\,(L,C) \rightsquigarrow C$ is limit preserving.

*2.17.8 The free object functor SETS $\rightsquigarrow C$ is left adjoint to the forgetful (or inclusion) functor $C \rightsquigarrow$ SETS.

*2.17.9 The free object functor SETS $\rightsquigarrow C$ is colimit preserving, and in particular, preserves coproducts [cf. 2.17.5 (a) and 2.17.7].

Commuting Homomorphisms and Structure of the Direct Sum

Two homomorphisms $h : X \to S$ and $g : Y \to S$ into a semigroup S are said to **commute** provided that

$$h(x)\,g(y) = g(y)\,h(x)$$

$\forall x \in X, y \in Y$. A family of homomorphisms $\{h_i : X_i \to S\}_{i \in I}$ into a semigroup S are said to **commute** provided that for each $i \neq j$ it is true that h_i and h_j are commuting homomorphisms. Thus, the canonical injections $u_i : X_i \to \sum_{i \in I} \oplus X_i$ are commuting. If each X_i is a subsemigroup of S and if the inclusion mappings $\{X_i \to S\}_{i \in I}$ commute, then $\{X_i\}_{i \in I}$ is defined to be a set of **commuting subsemigroups**. (This says nothing about the semigroup operation in each of the X_i, which may be noncommutative.) The **direct sum** of a family $\{h_i : X_i \to S\}$ of commuting homomorphisms is defined to be the unique mapping

$$\sum_{i \in I} \oplus X_i \xrightarrow{\oplus h_i} S$$

such that

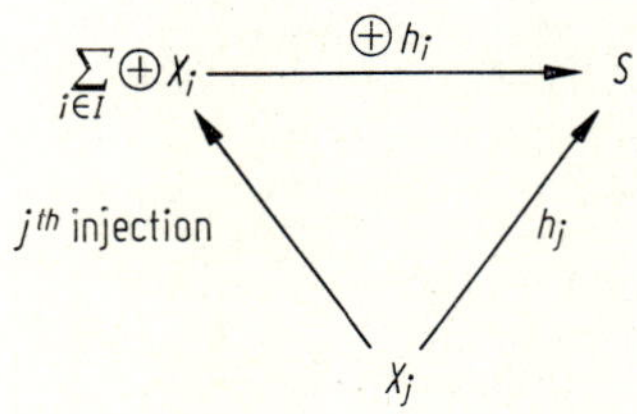

commutes $\forall j \in I$.

2.18 Proposition. *If $\{X_i\}$ is a family of commuting subsemigroups of an additive semigroup S, the set $\sum_{i\in I} X_i$ consisting of all elements of S which can be written as a (finite) sum*

$$s = \sum_{j=1}^{t} s_{i_j} = s_{i_1} + \cdots + s_{i_t},$$

where t depends on s, $i_j \in I$, and $s_{i_j} \in X_{i_j}$, $j = 1, \ldots, t$, is a subsemigroup of S, called the **sum** *of the $\{X_i\}_{i\in I}$. If S is a group, and if $\{X_i\}_{i\in I}$ is a family of commuting subgroups, then $\sum_{i\in I} X_i$ is the image of the direct sum of the inclusion mappings $\{X_i \to S\}_{i\in I}$ and is the subsemigroup of S generated by $\bigcup_{i\in I} X_i$.*

In case S is a group and $\{X_i\}_{i\in I}$ is a family of commuting subgroups, then $\sum_{i\in I} X_i$ is a group.

Proof. It is clear that $\sum_{i\in I} X_i$ is a semigroup generated by the set union of the X_i. In particular, $\sum_{i\in I} X_i$ is contained in the image T of the direct sum of the inclusions $\{X_i \to S\}_{i\in I}$. On the other hand, every element in T is a finite sum of elements in the various X_i, and therefore $T = \sum_{i\in I} X_i$. In case S and X_i, $\forall\, i \in I$, are groups, then the direct sum $\sum_{i\in I} \oplus X_i$ is a group, and hence so is the image $T = \sum_{i\in I} X_i$ of the direct sum of the inclusions. $\square$

2.19 Structure of the Direct Sum Proposition.

2.19.1 *The following conditions* (A) *and* (B) *on a set $\{X_i\}_{i\in I}$ of subsemigroups of an additive semigroup S are equivalent:*

(A) *$\{X_i\}_{i\in I}$ is a commuting family, and the direct sum of the inclusion homomorphisms $\{X_i \to S\}_{i\in I}$ is an isomorphism.*

(B) *$\{X_i\}_{i\in I}$ is a commuting family of subsemigroups each containing 0, and every element $s \in S$ can be written in one and only one way as a finite sum of elements*

$$s = s_{i_1} + s_{i_2} + \cdots + s_{i_t},$$

where t depends on s, $i_j \in I$, and $s_{i_j} \in X_{i_j}$, for $j = 1, \ldots, t$.

2.19.2 *When conditions* (A) *and* (B) *hold, then*

(C) *The $\{X_i\}_{i\in I}$ are commuting subsemigroups such that $S = \sum_{i\in I} X_i$, and such that $\forall\, i \in I$,*

$$X_i \cap \sum_{j\in I-\{i\}} X_j = 0.$$

2.19.3 *If S is a group, and if each X_i is a group, then, conversely,* (C) *implies conditions* (A) *and* (B).

Proof of 2.19.1. (A) $\Rightarrow$ (B). Because the direct sum mapping is assumed to be an isomorphism, it suffices to prove the statement (B) for the direct sum and the family $\{X_i'\}_{i\in I}$, where X_i' is the image of the canonical injection u_i of X_i in the direct sum. Clearly, u_i maps the identity element of X_i onto 0, so 0 is the identity of X_i', for every $i \in I$. Next, if f is an element of the direct sum, then

$$\begin{aligned} f &= \sum_{i\in I} u_i(f(i)) \\ &= \sum_{i\in I} u_i p_i f, \end{aligned}$$

where for any $i \in I$, p_i is the projection from the direct sum. (Compare with 2.14.) The sum is finite because f has finite support. The two functions represented on the right- and left-hand sides of the equality are equal because when evaluated at an arbitrary element $k \in I$ they each yield the value $f(k)$. This also implies the uniqueness of the representation of f as a finite sum

$$f = \sum_{i\in I} f_i'$$

with $f_i' \in X_i' \,\forall\, i \in I$, since necessarily $f_i' = u_i(f(i)) \,\forall\, i \in I$.

(B) $\Rightarrow$ (A). Let $h_i : X_i \to S$ be the inclusion mapping $\forall\, i \in I$. Let f and g be elements of the direct sum such that

$$\bigoplus_{i\in I} h_i(f) = \sum_{i\in I} f(i) = \sum_{i\in I} g(i) = \bigoplus_{i\in I} h_i(g).$$

Since $f(i)$ and $g(i)$ are elements of X_i, then by the uniqueness of the representation of $\bigoplus_{i\in I} h_i(f) = \bigoplus_{i\in I} h_i(g)$ as a finite sum of elements one from each X_i, necessarily $f(i) = g(i) \,\forall\, i \in I$, that is, $f = g$. This proves that the direct sum $\bigoplus_{i\in I} h_i$ mapping is an isomorphism.

Proof of 2.19.2. (B) $\Rightarrow$ (C). The condition $S = \sum_{i\in I} X_i$ follows from the fact that every element $s \in S$ has the representation as a finite sum as indicated in (B). If $s \in S$ belongs to the intersection of some X_{i_j} with the sum of X_{i_k}, where $k \neq j$, then s is a sum of elements belonging to each X_{i_k}, $k \neq j$. For brevity assume that $j = 1$, and write $s = s_{i_2} + \cdots + s_{i_t}$. Then

$$s = 0 + s_{i_2} + \cdots + s_{i_t} = s + 0 + \cdots + 0$$

violates the uniqueness of the representation of s as a sum of elements one from each X_{i_q}, $q = 1, 2, \ldots, t$, unless $s = 0$. This proves that (B) $\Rightarrow$ (C).

Proof of 2.19.3. (C) $\Rightarrow$ (A). Now assume that S and each X_i are groups. Then the sum $\sum_{i\in I} X_i$ is a group. Assume the notation of the proof of (B) $\Rightarrow$ (A), and let f be an element in the kernel of $\bigoplus_{i\in I} h_i$. Then, for each j,

$$f(j) = -\sum_{k\neq j} f(k)$$

is an element of X_j intersected with the sum of the X_k, $k \neq j$. It follows from the assumption C, that $f(j) = 0\ \forall\, j \in I$, that is, that $f = 0$, whence the kernel of the direct sum mapping is zero. This proves that (C) $\Rightarrow$ (A). □

2.20 Structure of the Direct Sum Corollary. *If $\{Y_i\}_{i\in I}$ is a set of semigroups, let $\{u_i : Y_i \to \sum_{i\in I} \oplus Y_i\}$ denote the canonical injections into the direct sum, and let $X_i = \operatorname{im} u_i\ \forall\, i \in I$. Then the direct sum of the inclusions $\{X_i \to \sum_{i\in I} \oplus Y_i\}$ is an isomorphism.* □

A family $\{X_i\}_{i\in I}$ of subgroups of an abelian group is **independent** provided that the canonical map $\sum_{i\in I} \oplus X_i \to \sum_{i\in I} X_i$ is an isomorphism.

Exercises for Chapter 2

1. In any category C, if $\{X \to X_i\}_{i\in I}$ is a product, and if X_i is injective, $\forall\, i \in I$, then X is injective. If C, is a category with zero, then the converse holds. (This is expressed by the statement: In any category with zero, a product of objects is injective if and only if each object is injective. Dually for projectives.)

2. Discuss injectives, projectives, cogenerators, and other categorical concepts for GROUPS and mod-$\mathbb{Z}$ (also mod-$\mathbb{Z}_n$), and compare. (E.g., does GROUPS contain any projective or injective objects.)

3. Discuss the structure of $\operatorname{End}_C M$ for any object M so far introduced in the text for the various categories $C =$ mod-$\mathbb{Z}$, GROUPS, SEMIGROUPS.

4.1 If K is a normal subgroup of a group G, and if A is a group $\approx G/K$, then G is said to be an **extension** of K by A. If Q is a subgroup of G such that $G = KQ$, and $K \cap Q = 1$, then G is called the **semidirect product** of K by Q. (Then G is an extension of K by Q.) The following conditions are equivalent:

(a) G is the semidirect product of K by a subgroup.

(b) The canonical map $p : G \to G/K$ splits.

(c) There exists an idempotent endomorphism $f : G \to G$ such that K is the normal subgroup generated by $\{gf(g^{-1}) \mid g \in G\}$.

When this is so, and if $q: G/K \to G$ is such that $pq = 1_{G/K}$, then:

(a′) G is the semidirect product of K by im q.

(c′) The endomorphism $qp: G \to G$ has the property ascribed to f in (c).

4.2 A group G of order pq, where $p > q$ are primes, is an extension of its p-Sylow subgroup by a group of order q. (Chapter 1, Exercises 18.)

5. A group F is a free group on some set S if and only if every group epimorphism $G \to F$ splits, equivalently, every extension of a group K by F is a semidirect product of K by a subgroup.

6. Every subgroup of a free abelian group $\mathbb{Z}^{(S)}$ is (isomorphic to) a free abelian group $\mathbb{Z}^{(S')}$, and $|S'| \leq |S|$. Conclude that $\mathbb{Z}^{(S)} \approx \mathbb{Z}^{(T)}$ for two sets S and T iff $|S| = |T|$, that is, a free abelian group has unique rank $|S|$, and every subgroup is free of rank $\leq |S|$. Cf. Exercise 13.

***7.** (Nielsen-Schreier Subgroup Theorem). Every subgroup of a free group on a set S is isomorphic to a free group on a set S'. If the free group has finite rank n and the subgroup has index j, then it is free of rank $jn - j + 1$.

8. A free group $F(S)$ has unique rank. (Cf. Exercises 6 and 11.)

9. The free group on one element is isomorphic to $\mathbb{Z}$.

10. The free group on a set S of cardinal > 1 has center $= 1$ and just one element of finite order.

11. If $f_S: S \to F(S)$ is a free group, then the inclusion mapping of the set $F(S):[F(S), F(S)]$ of left cosets of the commutator subgroup of $F(S)$ into $F(S)/[F(S), F(S)]$ is the free abelian group on the set of left cosets isomorphic to $\mathbb{Z}^{(S)}$. Furthermore, for any subset T of S, the group generated by S with relations $\{t = 1\}_{t \in T}$ is the free group on the set of left cosets in $F(S)$ of the normal subgroup generated by T.

***12.** If F is free of rank > 1, then $[F, F]$ is free of infinite rank.

***13.** If S is a subgroup of a free group F of finite rank, such that F/S is finite but not cyclic, then rank $S >$ rank F.

14. The **Subgroup Theorem** states that any subgroup $\neq 1$ of the free group is free (Exercise 7). Show that no such theorem exists for subsemigroups of a free semigroup. (Cf. P. M. Cohn [62].)

General References

See General References for Chapter 1, especially Bass [68], Bucur-Deleanu [68], Chevalley [56], Freyd [64], Fuchs [70], Kurosch [55, 56], Lang [65], Mitchell [65], and Rotman [65].

Chapter 3

3. Ring and Module

A list of topics in this chapter includes ring, module, endomorphism ring, biendomorphism ring (= bicentralizer) of a module, left ideals, nilpotency, simplicity in categories (e.g. simple groups, simple modules, simple rings), representations of simple rings in complete rings of linear transformations (the prelude to the Chevalley-Jacobson density theorem proved in Chapter 19), matrix rings, dual modules, dual basis lemma for projective modules, the "dual" theorem for generators of the category mod-R of all right R-modules, the trace of a module, additive categories and functors, exact sequences, left (right, half) exact functors, the left exactitude of the morphism (hom) functors $h^A : \text{mod-}R \rightsquigarrow \text{mod-}\mathbb{Z}$ and $h_A : \text{mod-}R \rightsquigarrow \text{mod-}\mathbb{Z}$, for any object A of mod-R, Baer's criterion for injectivity, enough injectives, and the existence of injective cogenerators of mod-R, fully faithful functors, idealizers, orthogonal idempotents and direct decompositions of modules, and fully invariant submodules of a module, including the radical and the socle.

Some notations introduced in this chapter are the category RINGS of rings, the category mod-R of all right R-modules, for any ring R, the endomorphism ring End M_R, for any right R-module M, and the biendomorphism[1] ring Biend M_R. An object M of mod-R is **balanced** if the canonical mapping $R \to \text{Biend}\, M_R$ is surjective. [Any generator of mod-R is balanced (Theorem of Morita, 4.1.3.) The category mod-R contains products, coproducts, and free objects. The free object on the set I is the direct sum (coproduct) $R^{(I)}$ (Proposition 4.9). The ring R_n of all $n \times n$ matrices is canonically isomorphic to the endomorphism ring of R^n, for any finite integer $n > 0$ (cf. 3.33.3). The ideal correspondence 3.1.2 is a bijection $I \mapsto C(I)$ between the set of all ideals of R and the classes of modules in mod-R closed under quotients, submodules, and products. A consequence is that an equivalence of categories mod-$R \approx$ mod-S, for rings R and S, implies an isomorphism of the ideal lattices of R and S (cf. Morita invariant, 4.31 ff.).

Additive categories are defined, and some of their elementary properties are deduced in 3.32 to 3.38 (see also Chapter 6). Left exactness of the Hom functors h_A and h^A for any object A of mod-R is demonstrated (3.39; cf. the corresponding result, 6.11, for exact additive categories).

---

[1] The introduction to Chapter 4 outlines some relationships between the various categories: mod-R, mod-End M_R, and mod-Biend M_R.

Some properties of faithfully exact functors are given (3.49), and the existence of injective cogenerators of mod-R is proved (3.53). Thus, $\mathbb{Q}/\mathbb{Z}$ is an injective cogenerator of mod-$\mathbb{Z}$, more generally, if K is the right quotient field of a principal right ideal domain R, then K/R is an injective cogenerator of mod-R (3.45). Furthermore, the character module $R' = \mathrm{Hom}_{\mathbb{Z}}(R, \mathbb{Q}/\mathbb{Z})$ of R is an injective cogenerator of mod-R (3.53). Every object M of mod-R is embeddable in a unique (up to an isomorphism) minimal injective over-module $\hat{M}$, the **injective hull** of M, and $\hat{M}$ is simultaneously a maximal essential extension of M (3.58). (The corresponding theorems for Grothendieck categories are derived in Chapter 14.)

Numerous exercises illustrate and extend theorems stated in the text. Some of these, in turn, are theorems stated in other chapters, for example, the Wedderburn-Artin Theorem and the Jacobson and Wong-Johnson density theorem for (quasi) injective modules. (Cf. Chapters 7 and 8 for the former, and Chapter 19, esp. 19.10 for the latter.)

Ring

A **ring** is an ordered triple consisting of a nonempty set R and two operations, with respect to which R is an additive group and a multiplicative semigroup, subject to the **distributive laws**:

$$(x + y)z = xz + yz$$
$$z'(x + y) = z'x + z'y$$

satisfied by all x, y, z, and $z' \in R$. Let 0 denote the zero element of the additive group, and let 1 denote the identity element of the multiplicative semigroup of R. If R consists of a single element, then $0 = 1$, and R is called the **zero ring.** If R has at least two elements, then $0 \neq 1$ because the distributive law implies that $0x = x0 = 0$ for all $x \in R$. One can verify the identities: **1.** $0x = x0 = 0$. **2.** $(-x)y = -(xy) = x(-y)$. **3.** $(-x)(-y) = xy$.

A ring is **commutative** provided that the multiplicative semigroup is commutative. A **subring** of a ring R is a subset T, which is both a subgroup of the additive group of R and a subsemigroup of the multiplicative group of R and such that either $T = \{0\}$, or else $1 \in T$. In the former case, we write $T = 0$, and in the latter case T has the same identity element as R. Thus, a subring is a ring. The **units group** of the multiplicative semigroup of a ring R is denoted units R and is called the units group of R. A **field** is a ring F with at least two elements such that every nonzero element of F is a unit. (Thus, we do not require that a field be commutative.)

Examples:

$\mathbb{N}$: the commutative semigroup under both addition and multiplication consisting of the set of natural numbers.

$\mathbb{Z} = -\mathbb{N} \cup \{0\} \cup \mathbb{N}$: the set of integers is a commutative ring with respect to the operations of addition and multiplication.

$\mathbb{R}$: the field of real numbers.

$\mathbb{Q}$: the field of rational numbers.

$\mathbb{C}$: the field of complex numbers $a + bi$, where a and b are real numbers and $i^2 = -1$, under the operations

$$(a + bi)(c + di) = (ac - bd) + (ad + bc)i,$$
$$(a + bi) + (c + di) = (a + c) + (b + d)i.$$

$\mathbb{H}$: the field of quaternions $a + bi + cj + dk$, where a, b, c, and d are real numbers and $\{\pm 1, \pm i, \pm j, \pm k\}$ is the quaternion group. Multiplication and addition are defined in a manner like the complex numbers. (To be specific, in terminology introduced in Chapter 5, $\mathbb{H}$ is the algebra over $\mathbb{R}$ generated by the set $\{1, i, j, k\}$ with the relations of the quaternion group. (See 5.75.) Similarly, $\mathbb{C}$ is the algebra over $\mathbb{R}$ generated by the set $\{1, i\}$ and the relations $\{i^2 = 1\}$.

Center

The center of a ring R, defined to be the set

$$\textbf{center } R = \{x \in R \mid xy = yx \ \forall y \in R\},$$

is a subring of R. More generally, if X is a nonempty subset of R, then

$$\textbf{centralizer } X = \{y \in R \mid xy = yx \ \forall x \in X\}$$

is a subring of R.

Homomorphism

A mapping $f : R \to R'$ of rings is a (**ring**) **homomorphism** provided that f is a homomorphism of both the additive group and multiplicative semigroup of R such that im f is a subring of R'. For additive groups R and R', the zero mapping $R \to R'$ is the mapping h such that $h(x) = 0$ for all $x \in R$, and the same terminology applies when R and R' are rings. Thus, the zero mapping $R \to R'$ of rings is a ring homomorphism, and $f(1) = 1$ for every nonzero homomorphism $f : R \to R'$.

The terms image, kernel, monomorphism, and epimorphism for a homomorphism of rings $f: R \to R'$ will apply in the same way as for the additive group of R. An **ideal** of R is defined to be the kernel of a homomorphism $f: R \to R'$ for some ring R'. If $I = \ker f$, then $\forall a, b \in I$, $r \in R$, I satisfies

(I—1) $$a - b \in I$$

(I—2) $$ra \in I \quad \text{and} \quad br \in I.$$

Conversely, let I be a nonempty subset of R satisfying (I—1) and (I—2). Then I is an additive subgroup of R, and the quotient group R/I is a ring, called the **factor**, or **quotient ring** R **modulo** I, with respect to the group operation and multiplication defined by

$$(a + I)(b + I) = ab + I.$$

Thus, the canonical group homomorphism $R \to R/I$ is a ring homomorphism, also called **canonical**, of which I is the kernel. This proves that a nonempty subset I of R is an ideal if and only if (I—1) and (I—2) hold.

The group $\mathbb{Z}_n = \mathbb{Z}/n\mathbb{Z}$ is a commutative ring consisting of n elements, where $n \in \mathbb{N}$. (Note that $n\mathbb{Z}$ is an ideal of $\mathbb{Z}$.) If R is not the zero ring, then the mapping defined by

$$\begin{cases} \mathbb{Z} \to R \\ n \mapsto n1 = 1 + 1 + \cdots + 1 \end{cases}$$

(n summands)

is a canonical ring homomorphism, and the kernel is an ideal $m\mathbb{Z}$ of $\mathbb{Z}$, for a unique $m \geq 0$, called the **characteristic** of R. Thus, a ring of characteristic zero contains a subring canonically isomorphic to the ring $\mathbb{Z}$. Usually we write n instead of $n1$, even when the characteristic is not zero.

Endomorphism Ring

3.1. Proposition and Definition. *If A is an additive abelian group, then* End A *is a ring with respect to composition and pointwise addition of homomorphisms.*

Proof. For any monoid A,

$$(f + g)h = fh + gh$$

$\forall f, g, h \in A^A$. Furthermore,

$$f(g+h) = fg + fh$$

$\forall f, g, h \in \operatorname{End} A$. Thus, End A satisfies the distributive laws, so when A is abelian, then A^A, hence End A, is an additive abelian group, so therefore End A is a ring. $\square$

3.2 *Exercises.*

3.2.1 Show that the set of all **Gaussian integers** $a + bi$, where a and b are elements of $\mathbb{Z}$, is a subring of $\mathbb{C}$. The set of all $a + bi$, where a and b are elements of $\mathbb{Q}$, is a subfield of $\mathbb{C}$. Generalize to subrings and subfields of $\mathbb{H}$.

3.2.2 If $A \neq 0$ is an abelian group, show that End A^2 is a noncommutative ring. Determine the smallest finite noncommutative ring. For any pair of integers n and m, find an injective ring homomorphism End $A^m \to$ End A^{mn}.

3.2.3 Show that $\mathbb{Z}/n\mathbb{Z}$ is a field if and only if n is a prime. Prove that any field has characteristic 0 or a prime.

3.2.4 Write $n = p_1^{e_1} \ldots p_t^{e_t}$ as a product of powers of distinct primes. Prove that $\mathbb{Z}/n\mathbb{Z}$ is isomorphic to the product $\prod_{i=1}^t \mathbb{Z}_{p^{e_i}}$ and, hence, is a product of rings of characteristics $p_1^{e_1}, \ldots, p_t^{e_t}$. Generalize to any commutative ring of characteristic n.

3.2.5 An element x in a ring R is a **zero divisor** provided that $xy = 0$ or $yx = 0$ for some nonzero $y \in R$. An **entire** ring or **domain** is one without zero divisors. Prove that any finite domain is a field. Prove that any domain has characteristic 0 or a prime.

3.2.6 A **Boolean ring** is a commutative ring R such that $a^2 = a$ for every $a \in R$. Deduce that $a + a = 0$ for every $a \in R$. Show that the following two conditions are necessary and sufficient in order that a commutative ring be boolean:

(a) If $a^n = 0$ for some $a \in R$ and some natural integer n, then $a = 0$.

(b) $(a + b)ab = 0$ for all $a, b \in R$.

3.2.7 If X is a set $\neq \varnothing$, then the set Pow X of all subsets of X is a ring under the following operations:

(1) $$A + B = A \cup B - (A \cap B)$$

(2) $$AB = A \cap B.$$

Pow X is a commutative ring (X is the ring identity and the empty set $\varnothing$ is the zero element), called the **boolean ring of subsets of** X. Every element A in this ring is idempotent.

3.2.8 If R is a ring, and if every element of R is idempotent, then R is commutative[1].

3.2.9 Let R be a boolean ring as defined in Exercise 3.2.6. Show that R is "isomorphic to" a subring of a boolean ring of subsets of some set S. (Let S denote the set of all maximal families $P = \{x_\alpha\}$ of elements x_α of R such that every finite number of elements of P has a product $\neq 0$ in R. Use Zorn's lemma to show that every $x \neq 0$ in R appears in some such P. Let $\langle x \rangle$ denote the subset of S consisting of all P with $x \in P$, and consider the mapping $x \mapsto \langle x \rangle$.)

3.2.10 A commutator element of a ring R is an element of the form $ab - ba$ with $a, b \in R$. Show that if x and y are elements of R, and if p is a prime number, then $(x + y)^p = x^p + y^p + z + pt$, where z is the sum of commutator elements and where t is an element of R. Also show that the result is not correct if p is not a prime number.

3.2.11 A not necessarily associative ring L is called a **Lie ring** in case

$$ab = -ba, \qquad a(bc) + b(ca) + c(ab) = 0$$

for all $a, b, c \in L$. Show that if A is any (associative) ring, then, by defining a new product $a \times b$ of elements a, b of A by the rule

$$a \times b = ab - ba,$$

A is a Lie ring. Show that a Lie ring L has an identity element iff $L = 0$.

3.2.12 The ring of integers can be constructed as the ring of endomorphisms of an infinite cyclic group $\mathbb{Z}$. Give a similar construction which leads to the field $\mathbb{Q}$ of rational numbers. (Consider the set T of all homomorphisms of subgroups $\neq 0$ of $\mathbb{Z}$ into $\mathbb{Z}$. For each $t \in T$, show first that there exists a unique maximal extension $[t] \in T$. Let Q be the set of all maximal elements of T and define a suitable addition in Q; the rational number a/b corresponds to the mapping $x \mapsto ax/b$ of a suitable subgroup of $\mathbb{Z}$.)

3.2.13 If e is any idempotent of a ring R, then the set $eRe = \{ere \mid r \in R\}$ is a ring under an operation induced by that of R, but is a subring iff either $e = 0$ or $e = 1$.

---

[1] A theorem of Jacobson states that if R is a ring such that for every a there corresponds an integer $n(a)$ such that $a^{n(a)} = a$, then R is commutative (see Jacobson [65, Theorem 2, p. 219]). Unlike the exercise, however, the proof of this theorem requires some sophisticated mathematics.

The Category of Rings

The objects of RINGS are rings, and the morphisms are ring homomorphisms. As noted in Chapter 1, equivalences in concrete categories are called isomorphisms. A ring homomorphism $A \to A'$ is monic if and only if it is a monomorphism, that is, injective. However, the inclusion $\mathbb{Z} \to \mathbb{Q}$ of rings, which is not an epimorphism, is an epic in RINGS.

A ring A is a category consisting of an object A; the morphisms are the elements of A. The opposite or dual category is also a ring, denoted A^{op}. A ring homomorphism $A \to A'$ defines, as in the case of semigroups, a covariant functor, and conversely. A contravariant functor $A \to A'$ induces a mapping of rings $A \to A'$, which is called an antihomomorphism. An **involution** is a duality $A \to A$.

3.3 (a) Proposition. *If R is a ring, and if K is an ideal of R, then R/K is a ring, and the canonical map $R \to R/K$ is a ring homomorphism. If $f: R \to R'$ is any ring homomorphism, then there is a unique injective homomorphism $h: R/\ker f \to R'$ such that*

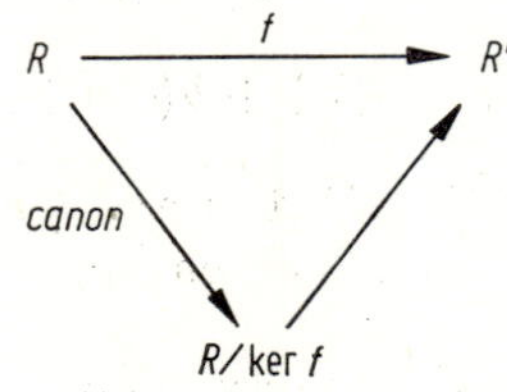

commutes. Furthermore, f induces a ring isomorphism $R/\ker f \to \operatorname{im} f$.

If $f: R \to R'$ is an antihomomorphism, then $\ker f = \{x \in R \mid f(x) = 0\}$ *is an ideal, and there is a commutative diagram*

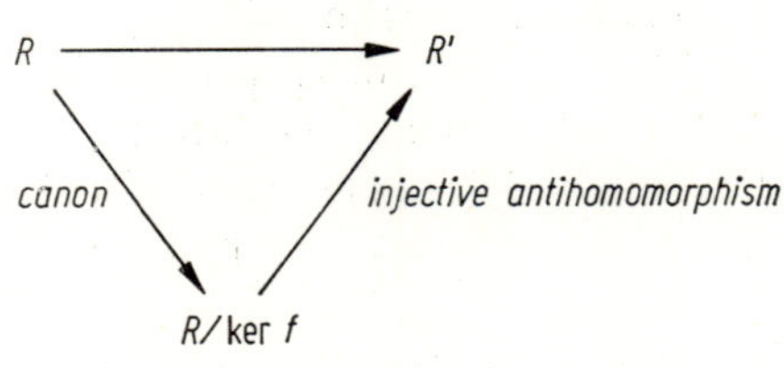

□

The cartesian product $\prod_{i \in I} X_i$ of a family of rings is a ring with respect to the operations of the direct product of the additive groups and multiplicative semigroups, respectively, of the family and is called the (**direct**) **product**. The projection mapping from the product to X_j is a ring homomorphism $\forall j \in I$, but when $|I| > 1$, the canonical injection u_j of X_j into the product is never a ring homomorphism because the image of this mapping is not a subring of the product. If 1_i is the ring identity of X_i, then $u_j(1_j) = e_j$ is an idempotent contained in the center of the product $X = \prod_{i \in I} X_i$, and $\operatorname{im} u_j = e_j X e_j \forall j \in I$. In particular, $\operatorname{im} u_j$ is an ideal X'_j of X, and $X'_j X'_k = 0 \ \forall j \neq k \in I$. Coproducts of rings are discussed in Chapter 5.

3.3 (b) *Exercises*

1. (Polynomial rings.) For any ring R, there exists a ring $R[x]$ containing R as a subring, and characterized by the following properties: There exists an element x in the center of $R[x]$ with the property that if R is any subring of any ring S, and if t is any element in the centralizer of R in S, then there exists a unique homomorphism $h: R[x] \to S$ such that $h(x) = t$, and $h(r) = r \ \forall r \in R$. When this is so, then every element $f \in R[x]$ has a unique representation $f = \sum_{i=0}^{n} a_i x^i$, with $a_i \in R$, $i = 0, \ldots, n$, and $a_n \neq 0$. Thus, if also $f = \sum_{j=0}^{m} b_j x^j$, with $b_j \in R$, $j = 0, \ldots, m$, and $b_m \neq 0$, then $n = m$, and $a_i = b_i$, $i = 0, \ldots, n$. (Then, n is called the **degree** of f.) The ring $R[x]$ is called the **polynomial ring over R in the variable, or indeterminate** x, and the elements of $R[x]$ are called **polynomials.** If $R[x']$ is also a polynomial ring over R in the variable x', then there is a ring isomorphism $g: R[x] \to R[x']$ such that $g(x) = x'$, and $g(r) = r \ \forall r \in R$.

2. For any commutative ring R, let $P(R)$ denote the ring of polynomial functions $R \to R$. Thus, an element of $P(R)$ is determined by an element $f \in R[x]$ sending $r \mapsto f(r) \ \forall r \in R$. Moreover, $P(R)$ is a subring of the product R^R, and the mapping just defined is a ring homomorphism $R[x] \to P(R)$. Determine necessary and sufficient conditions on R in order that this be an isomorphism.

3. Show that $\mathbb{C} \approx \mathbb{R}[x]/(x^2 + 1)$, where $(x^2 + 1)$ denotes the ideal of $R[x]$ consisting of multiples of $x^2 + 1$.

Module

If R is a ring, then an additive abelian group M is a **left R-module** if R operates on M in such a way that

$$(M-1) \qquad r(x + y) = rx + ry$$

$$(M-2) \qquad (r + s)x = rx + sx$$

$$(M-3) \qquad (rs)x = r(sx)$$

$$(M-4) \qquad 1x = x$$

are valid formulas $\forall r, s \in R$ and $\forall x, y \in M$. If $u: R \times M \to M$ is the operation, then (R, M, u), or more simply ${}_RM$, denotes the module. The concept of a right R-module (M, R, u) is defined by symmetry. In either case, R is called the **scalar ring**, and the elements of R are **scalars**.

3.4 **Proposition.** *Let R be a ring, and let M be an additive abelian group.*

3.4.1 *If* (R, M, u) *is a left R-module, then the mapping* $f: R \to \operatorname{End} M$ *such that* $f(r)(x) = rx \forall x \in M,\ r \in R$ *is a ring homomorphism. Conversely, if* $h: R \to \operatorname{End} M$ *is a ring homomorphism, then there is a left R-module* (R, M, v) *such that* $v(r, x) = rx = h(r)x \forall r \in R,\ x \in M$. *The symbol* $[R, M, h]$ *denotes the same module.*

3.4.2 *The statement* (3.4.1) *holds for right R-modules* (M, R, u) *replacing "left" by "right" and prefixing "homomorphism" with "anti". The symbol* $[M, R, h]$ *denotes the same module, where* $h: R \to \operatorname{End} M$ *is the ring antihomomorphism.*

Proof. Given the mapping $f: R \to \operatorname{End} M$ as defined, then $(M-1)$ holds if and only if $f(r)$ is an endomorphism of M, $(M-2)$ [resp. $M-3)$] holds if and only if f is a homomorphism of the additive (resp. multiplicative semi-) group of R, and $(M-3)$ holds if and only if $f(1)$ is the ring identity of End M. Thus, f is a ring homomorphism if and only if the operation of R on M defined by f is a left R-module. □

Each antihomomorphism of a commutative ring R is a homomorphism, so that the notions of left and right module coincide for commutative rings.

The Category mod-*R* of Right *R*-modules

The objects of the category mod-R, also denoted $\mathcal{M}_R$, consist of all right R-modules, and a morphism $f: M \to N$ is defined to be a homomorphism of the abelian groups such that

$$f(xr) = f(x)r$$

$\forall x \in M,\ r \in R$. Traditionally, $\operatorname{Hom}_R(M, N)$ is the notation for $\operatorname{Mor}_{\text{mod-}R}(M, N)$. The category operation is composition of homomorphisms. The endomorphism semigroup of an object M in mod-R is denoted $\operatorname{End} M_R$ and is a subring of the endomorphism ring End M of the abelian group M, called the **endomorphism ring of** M_R. In this notation, End $M = \operatorname{End} M_{\mathbb{Z}}$, regarding M as the right $\mathbb{Z}$-module $\mathbb{Z} \xrightarrow{\text{can}} \operatorname{End} M$. The inclusion mapping $\operatorname{End} M_R \to \operatorname{End} M$ defines M to be a left $\operatorname{End} M_R$-module.

The objects of the category R-mod consist of all left R-modules, and morphisms are group homomorphisms $f: M \to N$ such that $f(rx) = rf(x)$ for all $r \in R$, $x \in M$.

If $[M, R, g]$ (resp. $[R, M, g]$) is a right (resp. left) R-module and $f: S \to R$ is a ring homomorphism, then the right (resp. left) S-module $[M, S, gf]$ (resp. $[S, M, gf]$) is called the **pullback** S-module, and

$$\text{mod-}f \begin{cases} \text{mod-}R \rightsquigarrow \text{mod-}S & \\ [M, R, g] \mapsto [M, S, gf] & \text{(on objects)} \\ \qquad h \mapsto h & \text{(on morphisms)} \end{cases}$$

is a faithful functor, i.e., an embedding mod-$R \rightsquigarrow$ mod-S. This includes the special cases when $f: S \to R$ is injective, in particular, when S is a subring of R, and also when $f: S \to R$ is surjective, e.g., when $R = S/I$ and $S \to R$ is canonical.

If $f: S \to R$ is an antihomomorphism, then

$$\text{mod-}f \begin{cases} \text{mod-}R \rightsquigarrow S\text{-mod} & \\ [M, R, g] \mapsto [S, M, gf] & \text{(on objects)} \\ \qquad h \mapsto h & \text{(on morphisms)} \end{cases}$$

is a covariant faithful functor, and if $d^{-1}: R^{\text{op}} \to R$ is the inverse of the canonical duality, then a canonical equivalence mod-$R \rightsquigarrow R^{\text{op}}$-mod results.

3.5 Corollary. *The canonical antihomomorphism* $d: R \to R^{\text{op}}$ *induces an equivalence*

$$\text{mod-}d^{-1}: \text{mod-}R \approx R^{\text{op}}\text{-mod}.$$

Hence, there is a duality $(\text{mod-}R)^{\text{op}} \approx R^{\text{op}}$-mod. □

Homomorphisms Opposite Scalars

Henceforth, the convention of writing homomorphisms of left R-modules on the right will be adopted. Thus, given morphisms $f: M \to N$ and $g: N \to P$ in R-mod, then xf, or $(x)f$, will denote the value of f at x, and

$$(x)(fg) = (xf)g$$

denotes the value of the composition fg of f and g in R-mod. Let T denote the ring

$$T = \{f \in \text{End}\, M \mid f(rx) = rf(x) \forall r \in R, x \in M\}.$$

The inclusion $T \to \operatorname{End} M$ defines M canonically as a left T module and a right T^{op}-module. The convention of writing homomorphisms of left R-modules on the right maintains the symmetry with the situation for right R-modules in which homomorphisms are written on the side opposite to scalars. This convention is equivalent to the designation of $(\operatorname{Mor} R\text{-mod})^{op}$ for the morphisms of R-mod, and, as a consequence, for any left R-module M, $\operatorname{End}_R M = T^{op}$ is the ring opposite to T. In this case, however, the equivalence 3.5 between mod-R and R^{op}-mod does not hold as stated simply because mod-d^{-1} then sends a composition gf of morphisms $f: X \to Y$ and $g: Y \to Z$ in mod-R onto fg in R^{op}-mod, since in the latter case we write the morphisms on the right.

In order to adopt the convention of writing homomorphisms on the side opposite to scalars, and still maintain the validity of 3.5, it will be necessary to modify the definition of a covariant functor. If C and D are module categories with morphisms written on side opposite scalars, and if $T: C \rightsquigarrow D$ is a pair of mappings

$$\begin{gathered} \operatorname{obj} C \to \operatorname{obj} D \\ \operatorname{Mor} C \to \operatorname{Mor} D \end{gathered}$$

such that $T(1_A) = 1_{TA}$, for any object A of C, and such that $T(fg) = TgTf$ for all morphisms f and g of C, then T is called a **reversed covariant functor.** For example, if $f: S \to R$ is any ring antihomomorphism, then mod-f: mod-$R \rightsquigarrow$ mod-S is a reversed covariant functor when the homomorphism-opposite-scalars convention is adopted. If there are reversed functors $T: C \rightsquigarrow D$ and $S: D \rightsquigarrow C$ such that $TS \approx 1_D$ and $ST \approx 1_C$, then T is a **reversed equivalence**, and C and D are said to be equivalent categories. Thus, mod-d^{-1} in 3.5 is a reversed equivalence, and mod-$R \approx R^{op}$-mod are equivalent categories in this sense. Similarly for reversed duality.

Biendomorphism Ring

The **biendomorphism** ring $\operatorname{Biend} M_R$ of an R-module is defined to be the endomorphism ring of the module M over $S = \operatorname{End} M_R$. Thus, $\operatorname{Biend} M_R = \operatorname{End}_{\operatorname{End} M_R} M$.

$$\text{canon} \begin{cases} R \to \operatorname{Biend} M_R \\ a \mapsto a_d(M) \end{cases}$$

is a ring homomorphism, where $a_d(M)$ is the "**right multiplication** or **homothetic of** M **by** a" defined thus:

$$a_d(M) \begin{cases} M \to M \\ m \mapsto ma. \end{cases}$$

When only one module is in sight, a_d will abbreviate $a_d(M)$. Similarly $a_s(M)$ is defined when M is a left R-module. The module M is said to be **faithful** (resp. **balanced**) when the canonical mapping $R \to \text{Biend}\, M_R$ is injective (resp. surjective). Any generator, and any cogenerator, of mod-R is faithful. (Proof?) Furthermore, any generator is balanced (cf. Exercises 3.16.3, 3.19.1, and Proposition 4.1). The question of when cogenerators are balanced is taken up in Chapter 23.

3.6 *Examples.*

3.6.1 An abelian group M is a left module $[R, M, 1_R]$ over $R = \text{End}\, M$.

3.6.2 The canonical homomorphism $\mathbb{Z} \to \text{End}\, M$ defines M as a left and right $\mathbb{Z}$-module.

3.6.3 If R is a ring, then for $a \in R$ there is defined a mapping

$$a_d \begin{cases} R \to R \\ x \mapsto xa. \end{cases}$$

Clearly, $a_d \in \text{Hom}\,(R, R) = \text{End}\, R$, and the mapping

$$d = d_R \begin{cases} R \to \text{End}\, R \\ a \mapsto a_d \end{cases}$$

is a ring anti-isomorphism. Consequently, R is a right R-module $[R, R, d]$, denoted R_d. Similarly, the mappings

$$a_s \begin{cases} R \to R \\ x \mapsto ax \end{cases} \qquad s = s_R \begin{cases} R \to \text{End}\, R \\ a \mapsto a_s \end{cases}$$

define a left R-module $[R, R, s]$, denoted ${}_sR$.

If M is a right R-module, then an element $m \in M$ defines mappings

$$m_s \begin{cases} R_d \to M \\ r \mapsto mr \end{cases} \qquad \text{canon} \begin{cases} M \to \text{Hom}_R(R, M) \\ m \mapsto m_s. \end{cases}$$

The mapping on the right is an isomorphism $M \to \text{Hom}_R(R, M)$ in mod-R, called canonical. The covariant morphism functor h^R induces an autoequivalence of mod-R (cf. Equivalence of Categories, Chapter 1). This implies a canonical ring isomorphism $R \to \text{End}\, R_R = \text{Hom}_R(R, R)$. Thus, if A is an ideal of R, there are canonical ring isomorphisms $R/A \to \text{End}\, R/A_{R/A}$ and $R/A \to \text{End}_{R/A} R/A$, that is, R/A is balanced as an R-module when A is an ideal. This is not true, in general, for R/I, when I is a right ideal. (Proof?)

3.6.4 If R is a field, then a right R-module V is called a **vector space**. A morphism $V \to W$ in mod-R, is called a linear transformation (l.t.).

The ring End V_R is called a (**right**) **full linear ring** over R, and also the **full ring of l.t.'s of** V **over** R.

3.6.5 If $\{(M_i, R, u_i)\}_{i\in I}$ is a family of right R-modules, then the set product of the mapping $\{u_i, : M_i \times R \to M_i\}_{i\in I}$ is a mapping $\prod_{i\in I} M_i \times R \to \prod_{i\in I} M_i$ which when composed with the bijection $(\prod_{i\in I} M_i) \times R \to \prod_{i\in I} M_i \times R$, defines the cartesian product $\prod_{i\in I} M_i$ as a right R-module and induces a mapping $(\sum \oplus_{i\in I} M_i) \times R \to \prod_{i\in I} M_i$ which defines $\sum \oplus_{i\in I} M_i$ as a right R-module. More simply, if $f \in \prod_{i\in I} M_i$, then the module operation (in either case) is defined by the formula: $fr(i) = f(i)r \ \forall i \in I,\ r \in R$. Clearly, the family of projections $\{\prod_{i\in I} M_i \to M_i\}_{i\in I}$ is a product in mod-R, first of all, because the projections are morphisms of mod-R, and second, because the family is a product in mod-$\mathbb{Z}$. Similarly, by 2.19, *the family of injections*

$$\left\{M_i \to \bigoplus_{i\in I} M_i\right\}_{i\in I}$$

is a coproduct in mod-R. It follows that the free object functor

$$F \begin{cases} \text{SETS} \rightsquigarrow \text{mod-}R & \\ \quad I \mapsto R^{(I)} & \text{(direct sum)} \\ \quad i \mapsto f_I(i) & \end{cases}$$

exists, where the mapping $f_I : I \to R^{(I)}$ is such that for each $i \in I$ the diagram

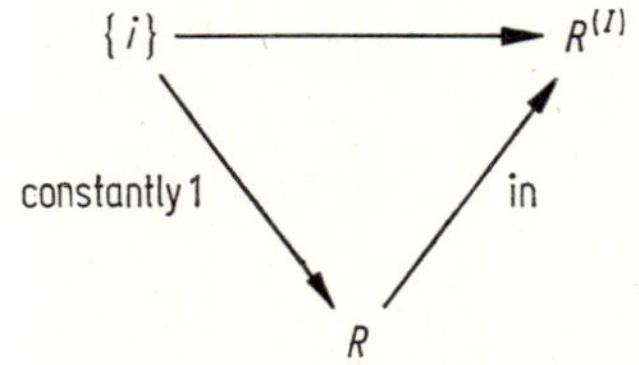

commutes, that is, $R^{(I)}$ is free on the set $\{u_i(1)\}_{i\in I}$ under the inclusion mapping, where $u_i : R \to R^{(I)}$ is the injection.

Bimodule

A (T, R)-bimodule is a module M which is both a left T-module and a right R-module such that the associative law holds: $(tx)r = t(xr) \forall t \in T$, $x \in M$, $r \in R$. Thus, M is always an (A, B)-bimodule for any subring A of End M_R and subring B of Biend M_R. The notation ${}_AM_B$ signifies an (A, B)-bimodule.

3.7 *Examples*. Let M be a right R-module.

3.7.1 (a) If N is a right R-module, and if M is a (S,R)-bimodule, then $F = \mathrm{Hom}_R(M, N)$ is a right S-module under an operation $(f, s) \mapsto fs$, $f \in F$, $s \in S$, where fs is defined by the formula: $fs(x) = f(sx) \ \forall x \in M$.

(b) Similarly, if N is a (T, R)-bimodule and M is a right R-module, then $F = \mathrm{Hom}_R(M, N)$ is a left T-module under the operation $(t, f) \to tf$, $t \in T$, $f \in F$, defined by: $tf(x) = t(f(x)) \ \forall x \in M$.

3.7.2 $\mathrm{Hom}_R(M, R)$ is a left R-module by 3.7.1 (b).

3.7.3 If R is commutative, then M is an (R, R)-bimodule, and $\mathrm{Hom}_R(M, M)$ is an (R, R)-bimodule by 3.7.1 (a) and (b).

3.7.4 Let E be any nonempty set, then the product R^E is an (R,R)-bimodule, and the direct sum $R^{(E)}$ is an (R, R)-submodule, where $(rf)(x) = rf(x)$ defines R^E as a left R-module.

3.7.5 Let $M, N \in$ mod-R, let $S = \mathrm{End}\, M_R$, and let $T = \mathrm{End}\, N_R$. Then $\mathrm{Hom}_R(M,N)$ is a (T,S)-bimodule.

Submodules

A module (N, R, p) is a **submodule** of a module (M, R, u) provided that N is a subgroup of M and that $u : M \times R \to M$ induces $p : N \times R \to N$, or equivalently, N is a subgroup of M stable under R. A **proper** submodule N is a submodule $\neq M$.

The **quotient**, or **factor**, **module** of (M, R, u) modulo a submodule N is the module $(M/N, R, t)$, where t is the unique mapping $M/N \times R \to M/N$ such that

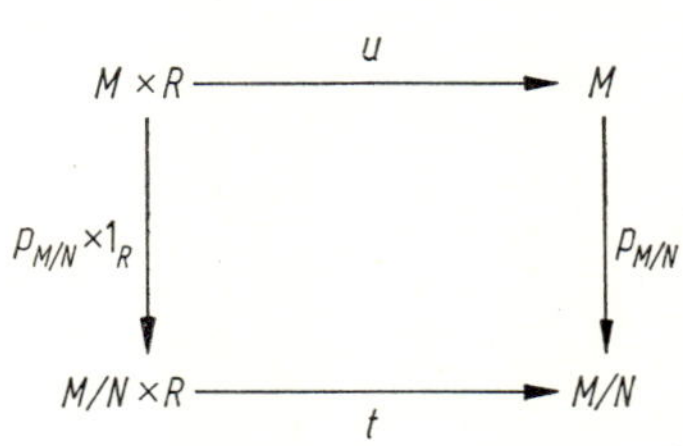

commutes. More simply, the module operation in M/N is defined by the formula $(a + N)r = ar + N \ \forall r \in R$, $a \in M$.

A submodule of R_d is called a **right ideal**, and a submodule of R_s is a **left ideal**. It follows that a subset I of R is an ideal if and only if I is a right and left ideal.

Let C be a concrete category. If B and A are objects, and $B \subseteq A$, then B is a **subobject** of A. If S is a subset, and N is a subobject, then

S **generates** N provided that N is the intersection of all subobjects of A containing S. This is denoted by $N = (S)$. A subobject M of A is **finitely**, or **countably**, or **generated by** a **elements** according as $M = (T)$ for a subset T of cardinality $|T| < \aleph_0$, or $|T| \leq \aleph_0$, or $|T| \leq a$. If A is an abelian (semi) group, then the sum $\sum_{i\in I} X_i$ of a family of (semi) groups is the sub (semi) group generated by the set union $\bigcup X_{i\in I}$ (see 2.19). When A is an R-module, and each X_i is a submodule, then the sum is a submodule. A module M is said to be **cyclic** if it is generated by a single element. A cyclic right ideal is called **principal**. If A is a right ideal of R, and M is a right R-module, then MA is defined to be the submodule generated by the set $\{ma \mid m \in M, a \in A\}$. For $x \in M$, let $xA = \{xa \mid a \in A\}$. Thus, xR is the cyclic submodule generated by x, and $MA = \sum_{m\in M} mA$. This defines the product AB of any two right ideals A and B. If C is a third ideal, then $(AB)C$ is therefore defined, and associativity of the ring operations implies that $(AB)C = A(BC)$. It follows that the set of right ideals is a semigroup with identity R. The product $A_1 \cdots A_n$ of any finite number of right ideals $A_1, \ldots, A_n$ is defined inductively $(A_1 \cdots A_{n-1}) A_n$, and

$$A_1 \cdots A_n = \left\{ \sum_{\text{finite sums}} a_{i_1} a_{i_2} \cdots a_{i_n} \mid a_{i_j} \in A_{i_j},\ j = 1, \ldots, n \right\}.$$

An ideal A is **nilpotent** (**of index** k) in case the product $A^n = A \cdots A$ (n factors) is zero (and k is the l.u.b. of all such n). Thus, the ideal $p\mathbb{Z}/p^n\mathbb{Z}$ of $\mathbb{Z}/p^n\mathbb{Z}$ is nilpotent of index n, for any integer $p > 0$. More generally, if A is any ideal of R, then A/A^n is a nilpotent ideal of R/A^n of index $\leq n$. The set of all nilpotent right ideals (of index $\leq n$) is a subsemigroup of the semigroup of all right ideals. If A and B are nilpotent of indices n and m, then $A + B$ is nilpotent of index $\leq n + m$, as the inclusion

$$(A + B)^t \subseteq A^t + B^t$$

for $t = n + m$ readily shows.

A ring R is said to be **semiprime** if 0 is the only nilpotent ideal. This is also expressible by the phrase: *R has no nilpotent ideals.* The ring R is **prime** in case a product of nonzero ideals is nonzero. Thus, a prime ring is semiprime.

A **zero divisor** in a ring is an element x such that $xy = 0$ or $yx = 0$ for some element $y \neq 0$ in the ring. A ring is called a **domain** and is said to be **entire** if it has no zero divisors other than 0. This is expressed by saying that R has no zero divisors. A field is entire, and any domain is prime. A commutative ring is entire if and only if it is prime.

An element $x \in R$ is nilpotent, if $x^n = 0$ for some $n > 0$. The **index** is the least n such that $x^n = 0$. A one-sided ideal is said to be **nil**, if every element in it is nilpotent. Thus, every nilpotent one-sided ideal is nil. The converse is not valid. (Example?) If R is commutative, then $x \in R$ is nilpotent of index n if and only if the ideal xR is nilpotent of index n. A ring may not have a maximal nilpotent ideal; however, the sum of all of the nilpotent right ideals is a nil ideal containing every nilpotent left ideal. (Proof?) The answer to the following question is still unknown: does there exist a ring R with no nil ideals (except 0) and containing a nonzero nil right ideal?

On the subject of finite generation, the next proposition is of some interest.

3.8 Proposition. *Let R be a ring, and let M be an R-module. Then M is finitely generated if and only if the set of proper submodules of M is inductive.*

Proof. Let M be finitely generated by the elements $x_1, \ldots, x_n$, and let $\{M_i \mid i \in I\}$ be a chain of submodules whose union is M. For each $j = 1, \ldots, n$, there corresponds $i_j \in I$ such that $x_j \in M_{i_j}$. Then there exists k such that $x_j \in M_{i_k}$, $j = 1, \ldots, n$, whence $M = M_{i_k}$. Conversely, well order a generating set of M of least cardinal number c, say $\{x_i \mid i < \alpha\}$, where α is the first ordinal of cardinality c, and let M_i denote the submodule generated by $\{x_j \mid j \leq i\}$. If c is infinite, then α is a limit ordinal, and M_i is proper, since M_i is generated by fewer than c elements. ▯

Free Basis

If $\{x_i\}_{i \in I}$ generates a submodule N of M, then the family is called a **basis** of N. Then $\{x_i\}_{i \in I}$ is called a **free basis** provided that the the direct sum of the inclusions $\{x_i R \to N\}_{i \in I}$ is an isomorphism $\bigoplus_{i \in I} x_i R \to N$. Then the cardinality of I is called the **basis number**. By 2.19, this is equivalent to the demand that the family $\{x_i R\}_{i \in I}$ of submodules of N be independent, and then the family $\{x_i\}_{i \in I}$ is said to be **linearly independent over** R.

3.9 Proposition.

3.9.1 *If R is a ring, and if X is an object of* mod-R, *then there exists an index set I and an epimorphism $R^{(I)} \to X$, that is, every object X of* mod-R *is isomorphic to a quotient module of a free R-module.*

3.9.2 *If $\{u_i : R \to R^{(I)}\}_{i \in I}$ are the injections into the direct sum (coproduct), then $\{u_i(1)\}_{i \in I}$ is a free basis of the free module $R^{(I)}$.*

3.9.3 *The object R is a generator of* mod-R.

Proof of 3.9.1. Follows from the Free object Theorem 2.7 which states that there is a unique morphism $R^{(X)} \to X$ such that

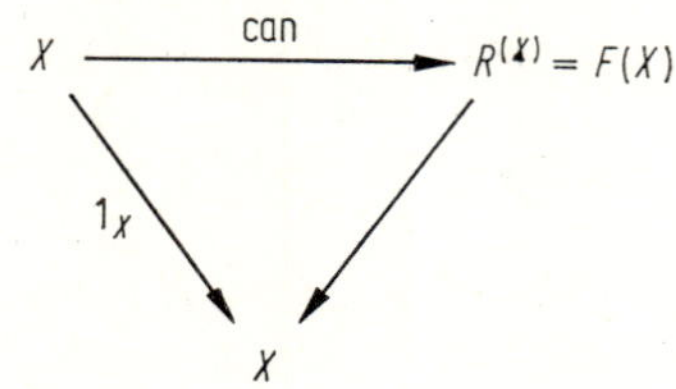

commutes. Since X obviously generates X as a module, that is, since $(X) = X$, then $R^{(X)} \to X$ is epimorphic.

Proof of 2.9.2. First of all, the set $\{u_i(1)\}_{i \in I}$ generates $R^{(I)}$ by 2.7.2 of the Free Object Functor Proposition, and second, the direct sum of the inclusions $\{u_i : R \to R^{(I)}\}_{i \in I}$ is an isomorphism by the Structure of the Direct Sum Corollary 2.20.

Proof of 3.9.3. Finally, by the equivalence $\mathrm{Hom}_R(R, \ ) \approx 1_{\text{mod-}R}$ of functors, R is a generator (cf. 3.26). □

The basis number is not, in general, independent of the choice of the free basis and hence is not an invariant of the free module $F = R^{(I)}$. It is independent, of course, when F is a free module on an infinite set I, as elementary cardinal arithmetic and the fact that the operations in R are finitary show. The basis number is also unique for a vector space V and is then called the **dimension** of V, denoted $\dim_R V$, where R is the scalar field. An **IBN-ring** is a ring R for which the basis number of every free module is unique. Any field is an IBN-ring, and so is any ring for which there exists a nonzero ring homomorphism $f : R \to S$ into an IBN-ring S(11.14). Any ring satisfying the a.c.c. on right ideals is an IBN-ring (cf. 3.10.3 and Exercises for Chapter 17).

3.10 *Exercises.*

3.10.1 An R-module M is said to be **simple** if there are precisely two submodules, namely, 0 and M.

(a) (**Schur's Lemma**). If U and V are simple R-modules, and if $f \in \mathrm{Hom}_R(U, V)$, then either $f = 0$ or else f is an isomorphism.

(b) The endomorphism ring of a simple module is a field.

(c) If U_i, $i = 1, \ldots, n$, are finitely many simple R-modules, no two of which are isomorphic, then the endomorphism ring of their direct sum is isomorphic to the ring product of the fields $D_i = \mathrm{End}_R U_i$, $i = 1, \ldots, n$.

3.10.2 Let I be any right ideal in a ring R, and define idealizer $I = \{a \in R \mid ax \in I \ \forall \ x \in I\}$. This is the smallest subring of R containing I as an ideal.

(a) Prove that

$$\begin{cases} \operatorname{End} R/I_R \to (\text{idealizer } I)/I \\ f \mapsto f(1+I) \end{cases}$$

is a ring isomorphism.

(b) (Koh) Prove that if I is a maximal annihilator right ideal, then $\operatorname{End} R/I_R$ is a domain.

3.10.3 A ring R is a field if and only if R has at least two elements and precisely two left ideals. Over a field R, every R-module is free. (Hint: Zorn's lemma). Prove that a subring of a field is an IBN-ring. Any commutative ring is IBN.

3.10.4 A principal right ideal of a ring R is one of the form xR, with $x \in R$. The ring R is a **principal right ideal ring** (PIR) provided that every right ideal is principal. A right principal ideal domain (PID) is a domain which is right PIR. Prove that over a right PID, every projective right R-module is free. A field R is a right and left PID, and so is the polynomial ring $R[x]$. Show that the polynomial ring $R[x, y]$ is not.

3.10.5 Let G denote the additive group of infinite dimensional vectors $v = (x_1, x_2, x_3, \ldots)$ with coefficients in a given field. Define four mappings of G into G as follows.

$$g\colon (x_1, x_2, x_3, \ldots) \mapsto (x_1, x_3, x_5, \ldots)$$

$$h\colon (x_1, x_2, x_3, \ldots) \mapsto (x_2, x_4, x_6, \ldots)$$

$$q\colon (x_1, x_2, x_3, \ldots) \mapsto (x_1, 0, x_2, 0, x_3, 0, \ldots)$$

$$r\colon (x_1, x_2, x_3, \ldots) \mapsto (0, x_1, 0, x_2, 0, x_3, \ldots).$$

Show that g, h, q, and r belong to the ring E of endomorphisms of G and that the following equations hold:

$$gq = hr = I, \quad hq = gr = 0, \quad qg + rh = I, \quad I = 1_G.$$

3.10.6 Let G be a nonabelian group of order 6. Find two endomorphisms whose sum is not an endomorphism.

3.10.7 Construct a ring without a maximal nilpotent ideal.

3.11 *Exercises.*

3.11.1 (a) For any A in mod-R, the group-valued functor h^A preserves products and kernels, whereas h_A converts coproducts to products and cokernels to kernels. (Note, the cokernel of $f: A \to B$, defined by dualism to the kernel, is $B/\operatorname{im} f$.)

(b) A functor mod-$R \rightsquigarrow$ mod-$\mathbb{Z}$ is **left exact** if it preserves kernels, **right exact** if it preserves cokernels, and **exact** if it is both left and right exact. Show that an object A in mod-R is projective (injective) if and only if the group-valued functor h^A (resp. h_A) is exact.

3.11.2 An R-module P is projective if and only if the following conditions, which are equivalent, are satisfied.
(a) The module P is isomorphic to a summand of a free R-module.
(b) Every exact sequence $M \to P \to 0$ splits.
Conclude that every free module is projective. Show that P is finitely generated and projective if and only if $R^n \approx P \oplus X$ for some finite integer $n > 0$ and module X.

3.11.3 If A is a divisible abelian group, and if P is any projective left R-module, then $\mathrm{Hom}(P, A)$ is an injective right R-module. (Recall, if $f: P \to A$, and if $r \in R$, then $(fr)(p) = f(rp) \forall p \in P$ defines $\mathrm{Hom}(P, A)$ as a right R-module.)

3.11.4 Show that any abelian group A embeds in a divisible abelian group.

3.11.5 Use Exercises 3.11.3 and 3.11.4 to show that every left R-module M embeds in an injective left R-module.

3.11.6 An object G of mod-R is a generator iff there exist an integer $n > 0$, and isomorphism $G^n \approx R \oplus X$ for some object X in mod-R. If R is a right PID, then every generator has a summand $\approx R$.

3.12 Ideal Proposition. *Let R be a ring. A subcategory C of* mod-R *is said to be* ***ideal*** *provided that the following conditions, which are equivalent, hold.*

3.12.1 *There exists an ideal A of R such that the canonical embedding* mod-$R/A \rightsquigarrow$ mod-R *induces a category equivalence* mod-$R/A \rightsquigarrow C$.

3.12.2 (a) *If $\{X_i\}_{i\in I}$ is any family of objects of C, and if $\{\prod_{i\in I} X_i \to X_i\}$ is the product in* mod-R, *then $\prod_{i\in I} X_i$ is an object of C.*
(b) *If X is an object of C, and*

$$0 \to X' \to X \to X'' \to 0$$

is exact in mod-R, *then X' and X'' are objects of C. A class C satisfying* 3.12.2 (a) *and* (b) *is said to be an* ***ideal class***. *The mapping*

$$C \begin{cases} \{\text{ideals of } R\} \to \{\text{ideal classes of mod-}R\} \\ \qquad A \mapsto \{M \in \text{obj mod-}R \mid \text{ann}_R M \supseteq A\} = \text{mod-}R/A \end{cases}$$

is bijective, and for any ideal class C'

$$C^{-1}(C') = \bigcap \{\text{right ideals } A \mid R/A \in \text{obj } C'\}.$$

Proof. For any ideal A, the canonical functor mod-$R/A \rightsquigarrow$ mod-R is simply the inclusion functor. Hence, assuming 3.12.1, $C =$ mod-R/A satisfies 3.12.2. Conversely, assuming 3.12.2, then evidently A is the intersection of all right ideals K such that the cyclic module R/K belongs to C. In order to see this, first note that for any $M \in \operatorname{obj}$ mod-R, that $\operatorname{ann}_R M = \bigcap_{a \in M} \operatorname{ann}_R a$. The left homothetic $a_s(R)$ induces an exact sequence

$$0 \to \operatorname{ann}_R a \to R \to aR \to 0.$$

It follows that

$$\operatorname{ann}_R M = \bigcap_{\substack{0 \to R/K \to M \\ \& \ 0 \to K \to R}\Big\} \text{exact}} K \in \operatorname{obj} \text{mod-}R,$$

that is, $\operatorname{ann}_R M$ is the intersection of all right ideals K of R such that R/K embeds in M. Hence, there is a set $\{K_j\}_{j \in J}$ of right ideals of R such that $R/K_j \in \operatorname{obj} C \ \forall j \in J$ and $A = \bigcap_{j \in J} K_j$. Let

$$0 \to K_j/A \to R/A \to R/K_j \to 0$$

be the canonical exact sequence $\forall j \in J$. Then the product $R/A \to \prod_{j \in J} R/K_j$ of the canonical projections $\{R/A \to R/K_j\}_{j \in J}$ is an embedding. Thus, $R/A \in C$ by 3.12.2(b). Then, by 3.12.2(a), C contains each direct product R/A^X, for any index set, and therefore the direct sum $R/A^{(X)}$ by (b). By the freedom of R/A in mod-R/A, any object Y of mod-R/A can be fitted into an exact sequence $R/A^{(X)} \to Y \to 0$, and then $C =$ mod-R/A by 3.12.2(b). The last statement is an immediate corollary. □

The question of what happens if 3.12.2(a) is weakened to the requirement that C contain the coproduct $\coprod_{i \in I} X_i$ of any family of objects of C is taken up in Chapter 15 under the subject of localizing subcategories. On the other hand, the question of what happens if 3.12.2(b) is strengthened to require that $X \in$ object C if and only if X' and X'' are objects of C is considered in Chapter 23 as well as in Chapter 15. (Such a class is called a **Serre** class.)

The product $C(I)C(J)$ of two classes is defined by decreeing that X belongs to the product if and only if

$$0 \to X' \to X \to X'' \to 0$$

is exact for some X' in $C(J)$ and X'' in $C(I)$. It is immediate, that $C(I)C(J) = C(IJ)$. Thus, the set of classes is a semigroup isomorphic to the semigroup of ideals of R. Furthermore, a module M of mod-R is faithful if and only if the class that M generates is mod-R, and an ideal is nilpotent if and only if $C(I)^n =$ mod-R for some $n > 0$.

The (right) **radical** of a ring R is defined to be the intersection of the maximal right ideals of the ring and is denoted rad R. Thus, rad R corresponds to the class $C(\operatorname{rad} R)$ generated by the simple right R-modules, that is, the intersection of all classes containing every simple right R-module, and therefore

$$\operatorname{rad} R = \bigcap_{\text{simple } V \in \text{obj mod-}R} \operatorname{ann}_R V$$

is an ideal of R. The radical contains every nil or nilpotent right or left ideal, and $1 + x$ is a unit for every $x \in \operatorname{rad} R$. The left radical of a ring R is defined by symmetry, and coincides with the radical of R. (See 18.5 ff.) In the literature rad R is called the **Jacobson radical** of R.

3.13 *Exercises.*

3.13.1 Prove the assertions of the last two sentences.

3.13.2 Verify $\operatorname{rad}(R/\operatorname{rad} R) = 0$.

3.13.3 Show by example that it is possible that $\operatorname{rad} R = 0$, and yet $\operatorname{rad} R/I \neq 0$ for some ideal I.

3.13.4 Prove all the theorems in Chapters 18 and 26.

3.13.5 Let A and B be ideals of R. Show that the canonical homomorphism $R/(A \cap B) \to R/A \times R/B$ is an isomorphism if and only if A and B are comaximal in the sense that $R = A + B$, and $A \cap B = AB + BA$. (The canonical homomorphism is defined as the direct product of the canonical homomorphisms $R/(A \cap B) \to R/A$ and $R/(A \cap B) \to R/B$.) Generalize this to a set of n ideals $A_1, \ldots, A_n$. [Hint: require $A_i \cap A_j = A_iA_j + A_jA_i$ and comaximality of ideals A_i and A_j when $i \neq j$.]

Simple Objects

A nonzero object A in a category C with zero is said to be **simple** provided that every morphism $A \to B$ in C is either zero or monic.

3.14 *Examples.*

3.14.1 In GROUPS, a group A is simple if and only if A has precisely two normal subgroups, namely A and the identity subgroup. The alternating group A_n on $n > 4$ is a simple group. An abelian group A is simple if and only if $A \approx \mathbb{Z}_p$, for a prime p (see Exercises for Chapter 1).

3.14.2 In RINGS, a ring A is simple if and only if A has precisely two ideals. Since A is finitely generated as an A-module, any ring A has maximal ideals by 3.8. If M is a maximal ideal, then A/M is a simple ring. Conversely, if I is an ideal, then A/I is a simple ring if and only if I is a maximal ideal. Any field is simple because it has precisely two right ideals. Furthermore, the endomorphism ring of a finitely generated module (vector space) over a field k is a simple ring (see Exercise 3.16.6).

By the ideal proposition, if mod-$A \approx$ mod-B for two rings A and B, then A is simple if and only if B is simple.

3.14.3 In mod-R, an object M is simple if and only if it has precisely two submodules. Thus, if M is a module and S a submodule, then M/S is simple if and only if S is maximal. (By 3.8, any finitely generated module M has maximal submodules, in fact, any proper submodule S is embeddable in a maximal submodule.) It follows that M is simple if and only if M is generated by any nonzero element. If $x \in M$ is nonzero, then there is an exact sequence

$$0 \to K \to R \xrightarrow{f} M \to 0,$$

where $f(a) = xa \; \forall \, a \in R$ and $K = \ker f$. It follows that K is a maximal right ideal of R (maximal in the set of all right ideals ordered by inclusion). Conversely, if K is any maximal right ideal, then R/K is a simple module. Any minimal submodule of a module (minimal in the ordered set of submodules) is simple, and conversely, a simple module is a minimal submodule of any module which contains it. If V is a simple R-module, then End V_R is a field, since the nonzero elements of End V_R are units. Thus, any simple right R-module V is a left vector space over the endomorphism ring $k = \text{End}\, V_R$, and, hence, the canonical map

$$d(V) \begin{cases} R \to \text{End}_k V = \text{Biend}\, V_R \\ r \mapsto r_d(V) \end{cases}$$

is a ring homomorphism of R into a full linear ring of l.t.'s of a left vector space. In some cases the canonical map is an embedding.

3.15 Theorem. *Any simple ring can be embedded in a complete ring of* l.t.'s *in a left (or right) vector space over a field. If V is any simple right R-module, and if R is a simple ring, then the canonical homomorphism* $R \to \text{End}_{\text{End}\, V_R} V$ *is such an embedding.*

Proof. Since R is finitely generated, as a right R-module, any ring R has maximal right ideals, and, hence, simple right R-modules, and by symmetry, simple left R-modules. But if R is a simple ring, then the kernel of the canonical map $R \to \text{End}_k V$ must be 0, where $k = \text{End}\, V_R$. (It cannot be R, since $x1 = x \; \forall \, x \in V$.) □

[The embedding referred to in the theorem is dense in the **finite topology** of $\text{End}_k V$, a basis of which consists of the subsets of $\text{End}_k V$ of the form

$$U(a, x_1, \ldots, x_n) = \{b \in \text{End}_k V \mid x_i a = x_i b, \, i = 1, \ldots, n\},$$

where $a \in \text{End}_k V$, $x_1, \ldots, x_n \in V$, and $n \in \mathbb{Z}^+$ (cf. The Chevalley-Jacobson Density Theorem 19.22).]

3.16 *Exercises.*

3.16.1 (a) Let R be a simple ring. Show that any nonzero submodule of a product R^I, in particular, any nonzero right ideal of R, is a generator in mod-R.

(b) The center of a simple ring R is a field.

3.16.2 Any generator U in mod-R, for any ring R, is finitely generated and projective over End U_R.

*3.16.3 (Morita [58]) For any ring R, a module U in mod-R is a generator if and only if U is balanced in mod-R and finitely generated projective in End U_R-mod.

3.16.4 (M. Rieffel [65]) Any nonzero right ideal of a simple ring R is balanced in mod-R.

3.16.5 Any minimal right ideal I in a semiprime ring is projective and in fact, is generated by an idempotent of R, that is, $I = eR$, for some idempotent $e \in R$.

*3.16.6 (Wedderburn-Artin). A simple ring R has a minimal right ideal if and only if R is the endomorphism ring of a finite dimensional vector space over a field k. In this case mod-$R \approx$ mod-k.

Matrix Rings

An $m \times n$ matrix $(a_{ij}) : m \times n \to A$ over a set A, for positive integers m and n, can be displayed in a rectangular array:

$$(a_{ij}) = \begin{pmatrix} a_{11} & a_{12} & \cdots & a_{1n} \\ a_{21} & a_{22} & \cdots & a_{2n} \\ \vdots & \vdots & \ddots & \vdots \\ a_{m1} & a_{m2} & \cdots & a_{mn} \end{pmatrix}$$

The **transpose** matrix is the $n \times m$ matrix

$$(a_{ji}) = (a_{ij})^t = \begin{pmatrix} a_{11} & a_{21} & \cdots & a_{m1} \\ a_{12} & a_{22} & \cdots & a_{m2} \\ \vdots & \vdots & & \vdots \\ a_{1n} & a_{2n} & \cdots & a_{nm} \end{pmatrix}.$$

If A is an additive monoid, then **matrix addition** of two $m \times n$ matrices has been defined in 2.17.4. Thus:

$$(a_{ij}) + (b_{ij}) = (c_{ij}),$$

where $c_{ij} = a_{ij} + b_{ij} \, \forall \, i, j$. The set A_{mn} of all $m \times n$ matrices over a monoid A is a monoid with respect to addition of matrices and is abelian, or a (semi) group when A is. The **zero matrix** (0) is the constantly zero function $m \times n \to A$. Furthermore, if A is also a multiplicative monoid, then the **product** $(c_{iq}) = (a_{ij})\,(b_{pq})$ of an $m \times n$ matrix (a_{ij}) and an $n \times s$ matrix (b_{pq}) is an $m \times s$ matrix, where $c_{iq} = \sum_{k=1}^{n} a_{ik} b_{kq}$. This defines an operation $A_{mn} \times A_{ns} \to A_{ms}$, called **matrix multiplication.** When A is a semigroup, the kronecker delta

$$\begin{cases} n \times n \to A \\ (i, j) \mapsto \delta_{ij} \end{cases}$$

is defined, where $\delta_{ii} = 1$ and $\delta_{ij} = 0$ for all i and $j \neq i$. If $a \in A$, then diag (a) is the $n \times n$ matrix $(a\,\delta_{ij})$. In case $0a = a0$ for all $a \in A$, then the $n \times n$ matrix diag (1) satisfies M diag $(1) = M$ and diag $(1) N = N$, for every $m \times n$ matrix M and $n \times m$ matrix N, and is called the $n \times n$ **identity matrix.**

3.17 Proposition.

3.17.1 *If A is an additive monoid, then the set A_{mn} of all $m \times n$ matrices over A is an additive monoid under matrix addition, and is abelian or a (semi) group when A is.*

3.17.2 *If A is a ring, then the set of all $n \times n$ square matrices over A is a ring with respect to matrix addition and multiplication, called the* ***full,*** *or* ***total, matrix ring over*** A ***of degree*** n, *and denoted A_n, or $M_n(A)$.*

Proof. 3.17.1 is immediate. To prove 3.17.2, let (a_{ij}), (b_{pq}), and (c_{uv}) be three $n \times n$ matrices over A. Using the associative and distributive laws, we can demonstrate the associative law for square matrices:

$$\begin{aligned} \sum_q \Big(\sum_k a_{ik} b_{kq}\Big) c_{qv} &= \sum_q \sum_k (a_{ik} b_{kq})\, c_{qv} \\ &= \sum_q \sum_k a_{ik} (b_{kq} c_{qv}) \\ &= \sum_k a_{ik} \Big(\sum_q b_{kq} c_{qv}\Big). \end{aligned}$$

The proof for the distributive law is similar. The identity matrix diag (1) is the multiplicative identity [the zero matrix (0) is the additive identity], so that A_n is a multiplicative semigroup, hence a ring. ▯

Matrix rings A_n provide simple examples of noncommutative rings. The elements $\{e_{ij}\}_{i,j \in n}$ in A_n, where e_{pq} is that matrix (e'_{ij}) such that $e'_{pq} = 1$ and $e'_{ij} = 0$ when $i \neq p$ or $j \neq q$, satisfy the formulas:

$$e_{ij} e_{pq} = \delta_{jp} e_{iq}$$

and

$$\textstyle\sum_{i=1}^{n} e_{ii} = 1.$$

Any set of elements $\{e_{ij}\}_{i,j\in n}$ in a ring R satisfying the two formulas is called a set of $n\times n$ **matrix units** in R. If R has a set of $n\times n$ matrix units, then R must be noncommutative if $n>1$, since $e_{11}e_{12}=e_{12}$, whereas $e_{12}e_{11}=0$. Thus, for $A=\mathbb{Z}_2$, the ring $M_2(\mathbb{Z}_2)$ of all 2×2 matrices over $\mathbb{Z}_2$ is noncommutative and consists of 16 elements.

3.18 Proposition. *A ring A contains a set of $n\times n$ matrix units if and only if A is isomorphic to a full matrix ring B_n of degree n over some ring B. If $M=\{e_{ij}\}_{i,j\in n}$ is a set of $n\times n$ matrix units in A, and if B is the centralizer of M in A, then every element $a\in A$ has a unique representation*

$$a=\sum_{i=1}^{n}\sum_{j=1}^{n} e_{ij}a_{ij}$$

with $a_{ij}=\sum_{k=1}^{n} e_{ki}a\,e_{jk}\in B\ \forall\, i,j\in n$, and the mapping

$$\begin{cases} A\to B_n\\ a\mapsto (a_{ij})\end{cases}$$

is a ring isomorphism. Furthermore, for each $i\in n$,

$$\begin{cases} B\to e_{ii}Ae_{ii}\\ b\mapsto e_{ii}b\,e_{ii}\end{cases}$$

is a ring isomorphism, and the ***diagonal mapping***

$$\begin{cases} B\to B_n\\ b\mapsto \operatorname{diag}(b)\end{cases}$$

is an embedding of rings.

Proof. If $a\in A$, define

$$a_{ij}=\sum_{k=1}^{n} e_{ki}a\,e_{jk},$$

$i,j=1,\ldots,n$. Then,

$$a_{ij}e_{pq}=e_{pi}a\,e_{jq}=e_{pq}a_{ij},$$

$i,j=1,\ldots,n$. Thus, $a_{ij}\in B\ \forall\, a\in A$, $i,j=1,\ldots,n$; in particular, $a_{ij}e_{ij}=e_{ii}a\,e_{jj}$. Hence, using $\sum_{i=1}^{n} e_{ii}=1$,

$$a=\sum_{i,j=1}^{n}e_{ii}a\,e_{jj}=\sum_{i,j=1}^{n}e_{ij}a_{ij}.$$

If also $a=\sum_{i,j=1}^{n} e_{ij}b_{ij}$, with $b_{ij}\in B$, $i,j=1,\ldots,n$, then

$$\begin{aligned} a_{pq}&=\sum_{k=1}^{n}e_{kp}a\,e_{qk}=\sum_{k=1}^{n}e_{kp}\Big(\sum_{i,j=1}^{n} e_{ij}b_{ij}\Big)e_{qk}\\ &=\Big(\sum_{k,j=1}^{n} e_{kj}b_{pj}e_{qk}\Big)=\sum_{k,j=1}^{n} e_{kk}b_{pq}=b_{pq},\end{aligned}$$

$p, q = 1, \ldots, n$. This proves the uniqueness part. The rest of the proof is evident. $\square$

3.19 *Exercises.*

3.19.1 Let A and B be objects of mod-R.

(a) If $M = A \oplus B$ and if $T = \text{Biend } M_R$, then A and B are T-submodules of M_T.

(b) If $M = R \oplus B$ and $S = \text{End } M_R$, then ${}_SM$ is a cyclic module generated by $(1,0)$, where 1 is the identity element of R and where (a, b) denotes an arbitrary element of M, $a \in R$, $b \in B$.

(c) Any module isomorphic to $R \oplus A$, for some right R-module A, is balanced.

(d) If $A \in$ mod-R, then for any positive integer n, A^n is balanced if and only if A is balanced.

(e) Show that G is a generator in mod-R if and only if $G^n \approx R \oplus A$ for some integer n and R-module A. Conclude that any generator in mod-R is balanced.

(f) Let R be a ring, let n be an integer > 0, and let R^n be the left R_n-module defined by the canonical ring isomorphism $R_n \to \text{End } R^n_R$. Then, the functor

$$h^{R^n} \begin{cases} \text{mod-}R \rightsquigarrow \text{mod-}R_n \\ \quad X \mapsto \text{Hom}_R(R^n, X) \approx X^n \end{cases}$$

(induced by the set-valued covariant morphism functor defined by R^n) is an equivalence and is called the (nth) **power functor.** Conclude that mod-R_n possesses every categorical property possessed by mod-R.

(g) Show that the ideals of R_n have the form I_n, where I is an ideal of R. Do this two ways: first, by computation, and second, by using (f). Show that $\text{rad } R_n = (\text{rad } R)_n$.

(h) Show that R_n satisfies the a.c.c. (resp. d.c.c.) on right ideals if and only if R does. Is this a categorical property of mod-R? Is IBN a categorical property of mod-R?

(i) Find a bijection between the right ideals of R_n and the R-submodules of R^n with ideals corresponding to (R_n, R)-bimodules [cf. (f)].

3.19.2 (a) Any module M_R is balanced over End M_R.

(b) If A is an ideal, then the cyclic module R/A is balanced.

3.19.3 Let D be an involution of a ring R. Show that D extends to *a* unique involution D' of the full matrix ring R_n such that $D'(a_{ij}) = (D a_{ij})^t$ for every matrix (a_{ij}) of R_n. Show, conversely, that every involution of R_n is obtained this way. Conclude that for commutative rings R, the transpose mapping $A \to A^t$ for matrices in R_n is an involution.

3.19.4 Consider the set A of all matrices of the form

$$\begin{pmatrix} a & b \\ -\bar{b} & \bar{a} \end{pmatrix},$$

where a and b are complex numbers and where $\bar{a}, \bar{b}$ denote the conjugate complex numbers. Show that A is a field (under the usual rules of addition and multiplication of matrices) isomorphic to the *field* $\mathbb{H}$ *of quaternions*.

*3.19.5 If F is any field, then F is a vector space over any subfield. Show that if center F contains the field $\mathbb{R}$ of real numbers, and if $\dim_{\mathbb{R}} F$ is finite, then $F \approx \mathbb{H}$, $\mathbb{C}$, or $\mathbb{R}$.

3.19.6 If B is any ring, the center of B_n is $\{\text{diag}\,(c) \mid c \in \text{center } B\}$.

3.19.7 Let D be a field.

(a) Show that D_n contains a nilpotent element of index n but does not contain an element that is nilpotent of index $>n$.

(b) Deduce that if D and H are fields such that $D_n \approx H_m$, then $n = m$. Also, $D \approx H$.

3.19.8 Let R be a ring containing matrix units $M = \{e_{ij} \mid i,j = 1,\ldots,n\}$, $n > 1$, and let $B =$ centralizer $_R M$. Verify the formulas:

(a) $x_j = x_j^{-1}$, where $x_j = 1 - e_{11} - e_{jj} + e_{1j} + e_{j1}$, $j = 1, \ldots, n$.

(b) $e_{jj} = x_j^{-1} e_{11} x_j$, $\quad j = 1, \ldots, n$.

(c) $b e_{ij} = t_{ij}(b) - e_{jj}$, $\forall\, b \in B$, where

$$t_{ij}(b) = (1 + b e_{ij}) e_{jj} (1 - b e_{ij}), \quad i, j = 1, \ldots, n.$$

(d) $e_{jj} = e_{ij} - (1 - e_{ji}) e_{ij} (1 + e_{ji}) e_{ij}$, $\quad i \neq j$.

Let $e = e_{ij}$ be any element of M, $n > 1$. Show that if S is a subring of R containing each conjugate $x^{-1}ex$ of e, x ranging over all invertible elements of R, then $S = R$. In other words, if $n > 1$, then $R = B_n$ is generated as a ring by the conjugates of any matrix unit in R.

3.19.9 A countably infinite matrix

$$(x_{ij}) = \begin{pmatrix} x_{11}, & x_{12}, & \ldots, & x_{1n}, & \ldots \\ x_{21}, & x_{22}, & \ldots, & x_{2n}, & \ldots \\ x_{31}, & x_{32}, & \ldots, & x_{3n}, & \ldots \\ \vdots & \vdots & & \vdots & \\ x_{m1}, & x_{m2}, & \ldots, & x_{mn}, & \ldots \\ \vdots & \vdots & & \vdots & \end{pmatrix}$$

with entries x_{ij} in a ring R is said to be row-finite in case each row contains only a finite number of nonzero entries. Let R_ω denote the set of all row-finite matrices over R. If (x_{ij}) and $(y_{pq}) \in R_\omega$, then $c_{ij} = \sum_{k=1}^{\infty} x_{ik} y_{kj}$ is an element of R, $\forall\, i, j$, since there exists an integer K_i such that $x_{ik} = 0 \;\forall\, k \geq K_i$.

(a) Show that R_ω is a ring under the operations:

$$(x_{ij}) + (y_{ij}) = (x_{ij} + y_{ij})$$
$$(x_{ij})\,(y_{pq}) = (c_{ij}),$$

where c_{ij} is as given.

(b) If $R^{(\omega)}$ denotes the free right R-module with a countably infinite free basis, show that the endomorphism ring of $R^{(\omega)}$ is isomorphic to R_ω.

(c) Generalize (b) to the case $R^{(I)}$, where I is any well ordered set.

*3.19.10 (Dieudonné [42]). Let K be a field, let M_n denote the units group of the multiplicative semigroup of K_n, and let C_n be the commutator subgroup of M_n. Then, $M_n/C_n \approx K^*/C$, where K^* is the multiplicative group of nonzero elements of K, and C is the commutator subgroup of K^*. (Actually, can $C = K^*$?) When K is commutative, then C is the identity subgroup, and the determinant function induces the desired isomorphism.

Dual Modules

Now $\mathrm{Hom}_R(R, M)$ consists of all maps $R \to M$, and $\mathrm{Hom}_R(M, R)$ consists of all maps $M \to R$. We have already noted that $\mathrm{Hom}_R(M, R)$ is a left R-module, which we denoted by M^*. Since $\mathrm{Hom}_R(M, R)$ is defined by duality to $\mathrm{Hom}_R(R, M)$, the natural isomorphism $M \approx \mathrm{Hom}_R(R, M)$ shows that $M^* = \mathrm{Hom}_R(M, R)$ appropriately may be called the **dual** (module) of M. Similarly, if N is any left R-module, then $\mathrm{Hom}_R(N, R)$ is a right R-module also denoted by N^*. Thus, for any right R-module M, the right R-module $(M^*)^*$ is defined and is denoted by M^{**}, called the **bidual** of M.

Now $D = \mathrm{Hom}_R(\ , R)$ is a contravariant functor mod-$R \rightsquigarrow R$-mod, so for each map $f: A \to B$ of right R-modules, there is a map $D(f) : B^* \to A^*$. Sometimes we let $(\ )^*$ denote this functor, and in this case we denote $D(f)$ by f^*. If D^2 denotes the composite functor mod-$R \rightsquigarrow$ mod-R composed of D : mod-$R \rightsquigarrow R$-mod and $D : R$-mod $\rightsquigarrow$ mod-R, we also let

$$f^{**} = D\big(D(f)\big) = (f^*)^*.$$

If $x \in M$, there is an element $x^{**} \in M^{**}$ defined by

$$(f)\,x^{**} = f(x)$$

$\forall\, f \in M^*$. Since

$$(xa + yb)^{**} = x^{**}a + y^{**}b$$

$\forall x, y \in M$, $a, b \in R$, we have an R-module homomorphism $\delta_M : M \to M^{**}$, where $\delta_M(x) = x^{**}\ \forall\, x \in M$. This map is called **canonical.** A module M is said to be **torsionless** (resp. **reflexive**) if the canonical map $M \to M^{**}$ is injective (isomorphism).

3.20 *Example. There is a natural transformation* $\delta : I \rightsquigarrow D^2$, *where* I *is the identity functor* mod-$R \rightsquigarrow$ mod-R, *defined by the diagram*:

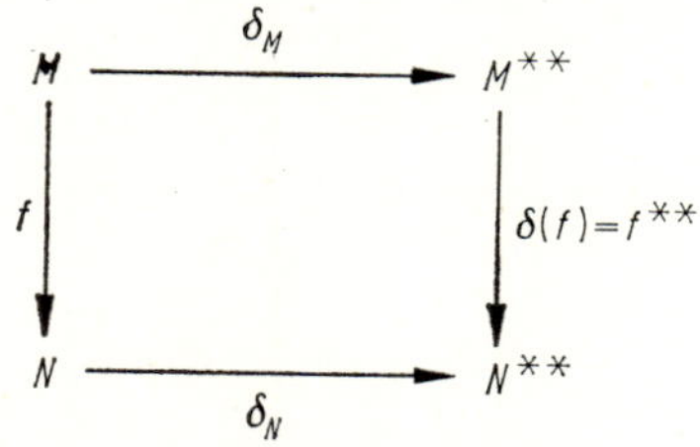

where $\delta_M(x) = x^{**}\ \forall\, x \in M$ *and* $\delta(M) = M^{**}$.

Proof. For simplicity, write all the morphisms on the left. If $\alpha \in M^{**}$, then f^{**} maps α onto αf^*:

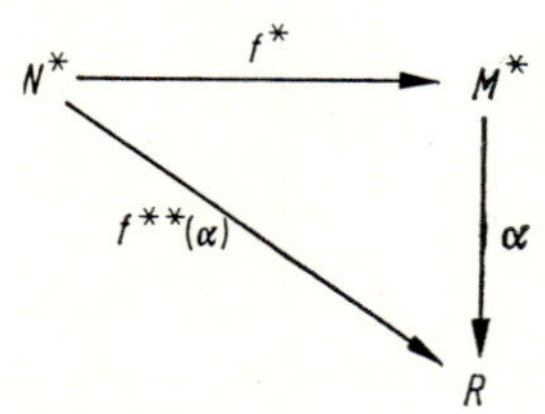

If $\alpha = x^{**}$, and if $g \in N^*$, then

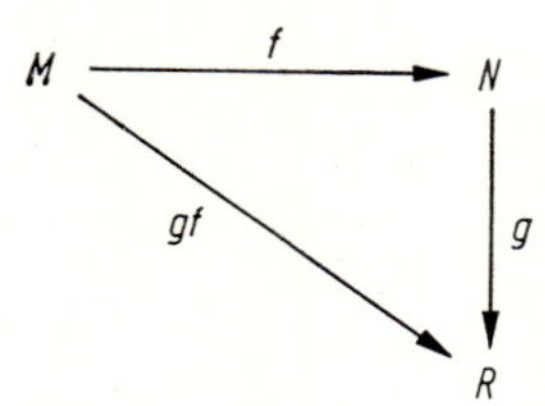

so $gf \in M^*$, and

$$x^{**} f^* : g \mapsto x^{**}(gf).$$

But

(1) $$x^{**}(gf) = gf(x)$$

$\forall\, x \in M$. This shows that f^{**} maps x^{**} onto that function in N^{**} that sends $g \in N^*$ onto $gf(x)$. But $\delta_N f(x) \in N^{**}$ does precisely the same thing:

$$\delta_N f(x) = f(x)^{**}$$

and so,

(2) $$\delta_N f(x) : g \mapsto f(x)^{**}(g) = g\big(f(x)\big) = gf(x).$$

Thus (1) and (2) show that the first diagram commutes, completing the example.

3.21 Dual Bases. *Let k be any ring, and let V be a free left k-module having a finite free basis $v_1, \ldots, v_n$. Then, the dual module $D(V)$ is a free right k-module with a free basis consisting of the n projections*

$$e_i \begin{cases} V \to k \\ \sum_{j=1}^{n} \alpha_j v_j \mapsto \alpha_i \qquad i = 1, \ldots, n \end{cases}$$

Proof. To prove that these elements form a free basis of $D(V)$, we first show they are linearly independent over k. This is true, since $f = \sum_{j=1}^{n} e_j \beta_j = 0$ implies $f(v_i) = \beta_i = 0$, $i = 1, \ldots, n$. Next, if $g \in D(V)$, then $g = \sum_{i=1}^{n} e_i g(v_i)$, since $g(v_j) = \sum_{i=1}^{n} e_i g(v_i)(v_j)$, $j = 1, \ldots, n$, that is, since $\sum_{i=1}^{n} e_i g(v_i)$ agrees with g on $v_1, \ldots, v_n$. This proves that $e_1, \ldots, e_n$ is a basis of $D(V)$. It is called the **basis of $D(V)$ dual to** $v_1, \ldots, v_n$, or **the dual basis.** (See 3.22.1 below.)

A Non-Natural Transformation. Now let k be a commutative field. It follows that $V \approx D(V)$ under a mapping

$$h \begin{cases} V \to D(V) \\ v_i \mapsto e_i, \quad i = 1, \ldots, n. \end{cases}$$

Let C denote the category of all finite dimensional vector spaces over k whose morphisms are isomorphisms. Since $h: V \to D(V)$ is defined for all $V \in \operatorname{obj} C$, and since $D(V) \in \operatorname{obj} C$, we have a transformation $k: I \to D$ mapping the covariant functor I (the identity functor) onto the contravariant functor D, where for each $V \in \operatorname{obj} C, k(V): V \to D(V)$ is the isomorphism previously defined. Now let $\bar{D}$ be the functor that assigns to each $V \in \operatorname{obj} C$ the object $D(V)$ and to each isomorphism $f: V \to V'$ the isomorphism $D(f^{-1})$. Then $k(V): V \to \bar{D}(V)$ is not natural, since if V is 1-dimensional, and if $f: V \to V$ is the isomorphism defined by scalar multiplication, $f(v_1) = \alpha v_1$, for some $\alpha \in k$, $\alpha \neq 0, 1$, then $\bar{D}(f) k(V)(v_1) = \alpha^{-1} e_1$, whereas $k(V) f(v_1) = \alpha e_1$. Thus, the following diagram is *not* commutative:

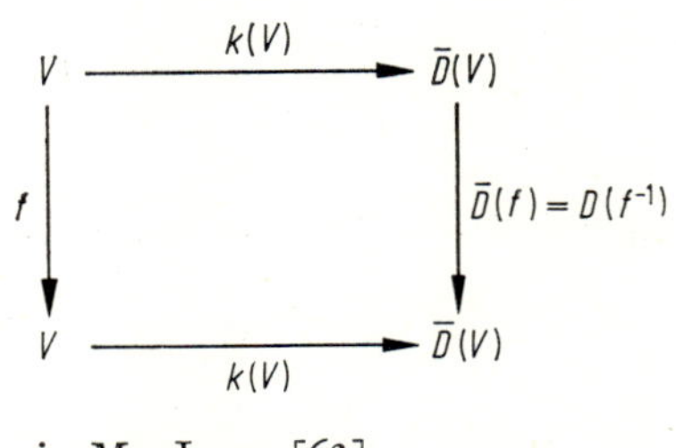

This example appears in MacLane [63].

3.22 *Exercises.*

3.22.1 Let k be a field, let C be the full subcategory of mod-k, consisting of the right vector spaces of finite dimension, and let D be the corresponding subcategory of k-mod.

(a) The k-module-valued Hom functor h_k induces a functor $T: C \rightsquigarrow D$ and, similarly, a functor $S: D \rightsquigarrow C$. Show that (T, S) is a duality pair.

(b) Next, for a field k and fixed integer $n > 0$, let C' be the full subcategory of mod-k, consisting of right vector spaces of finite dimension $\leq n$, and let E' be the right-left symmetry. A **line** in C' (resp. E') is defined to be 1-dimensional vector space. A **hyperplane** is an $(n-1)$-dimensional vector space. (A plane is 2-dimensional.) Show that h_k induces a duality $C' \rightsquigarrow E'$ mapping

$$\begin{cases} \text{lines} \to \text{hyperplanes} \\ \text{hyperplanes} \to \text{lines}. \end{cases}$$

(c) More generally, show that for any finite dimensional right vector space V, the duality functor induces a lattice anti-isomorphism between the lattices of vector subspaces of V and those $V' = \mathrm{Hom}_k(V, k)$.

3.22.2 For each right vector space V over k, the corresponding **projective space** $\overline{V}$ is the lattice of subspaces of V. To each element $X \in \overline{V}$, define the **projective dimension** $d_p X = \dim_k X - 1$. The element X is called a "point" of $\overline{V}$ if X has zero projective dimension, and a "line" if X has projective dimension 1, and so on. Thus, the lines of V become the points of $\overline{V}$, the planes of V are the "lines" of $\overline{V}$, and so forth. (The 0-subspace of V can be thought of as the empty element of $\overline{V}$.)

(a) Let V and W be vector spaces over k. A **collineation** of $\overline{V}$ is a lattice isomorphism $\overline{V} \to \overline{W}$, and a **correlation** is a lattice anti-isomorphism. Show that to each right vector space V there corresponds a left vector space W and a collineation (resp. correlation).

(b) Let θ be an automorphism of a field k, and let $f: V \to W$ be an additive group homomorphism of vector spaces over k such that

$$f(x\alpha) = f(x)\theta(\alpha)$$

$\forall x \in V,\ \alpha \in k$. Then f is called a **semilinear transformation** (see also Chapter 23), and f induces a collineation $\bar{f}: \overline{V} \to \overline{W}$, where

$$\bar{f}(X) = f(X)$$

$\forall X \in \overline{V}$.

*(c) **Fundamental theorem of projective geometry.** Any collineation $\overline{V} \to \overline{W}$ is induced by a semilinear transformation when $\dim_k V \geq 3$. (For this, and other theorems of projective geometry, see Artin [57, Chapter II, § 10, p. 85 ff.])

3.23 Dual Basis Lemma. *A right R-module P is projective if and only if there exists a family of elements* $\{x_i\}_{i \in I}$ *in P and a family* $\{f_i\}_{i \in I}$ *of elements in* $P^* = \operatorname{Hom}_R(P, R)$ *such that every* $p \in P$ *is a finite sum*

$$p = \sum_{i \in I} x_i f_i(p),$$

where $f_i(p) = 0$ *almost* $\forall\, i \in I$. *If P is projective,* $\{x_i\}_{i \in I}$ *can be any generating set.*

Proof. Let $\{x_i\}_{i \in I}$ be any generating set of P, let $F = R^{(I)}$ be a free module with a free basis $\{y_i\}_{i \in I}$, and let $h : F \to P$ be the epic mapping y_i onto x_i $\forall\, i$. Now P is projective if and only if h splits, that is, if and only if there exist a map $g : P \to F$ such that $hg = 1_P$.

Necessity. If P is projective, if $g : P \to F$ is a map such that $hg = 1_P$, and if $p \in P$, then write

$$g(p) = \sum_{i \in I} y_i r_i$$

with $r_i \in R$ depending on p ($r_i = 0$ almost $\forall\, i$). The family $\{f_i\}_{i \in I}$, where

$$f_i \begin{cases} P \to R \\ p \mapsto r_i \end{cases}$$

satisfies

$$p = hg(p) = \sum_{i \in I} x_i f_i(p).$$

Sufficiency. Let $\{x_i\}_{i \in I}$ and $\{f_i\}_{i \in I}$ be as described in the theorem. Then $\{x_i\}_{i \in I}$ generates P and there is the epic $h : F \to P$ and a map $g : P \to F$, where $p = \sum_{i \in I} x_i f_i(p) \in P$ and

$$g(p) = \sum_{i \in I} y_i f_i(p).$$

Since $hg = 1_P$, P is projective. □

Annihilators

A **pairing** is a mapping

$$\begin{cases} A \times B \to R \\ (a, b) \mapsto ab \end{cases}$$

of sets. Then A and B are **paired to** R. If sets A and B are paired to a semigroup R with zero, then the annihilator in B of an element $a \in A$

is the subset[1]

$$a^{\perp} = \mathrm{ann}_B a = \{b \in B \mid ab = 0\}.$$

If X is a subset of A, then $\mathrm{ann}_B X = \bigcap_{a \in X} \mathrm{ann}_B a$, also denoted $X^{\perp}$ ("X perp"). The notation $Xb = 0$ signifies that $b \in \mathrm{ann}_B X$. There is similar notation for $\mathrm{ann}_A Y$, for any subset Y of B, and the exponent "perp" changes sides. Thus, ${}^{\perp}Y = \mathrm{ann}_A Y$. The mapping

$$\begin{cases} \mathrm{Pow}\, A \to \mathrm{Pow}\, B \\ \quad X \mapsto X^{\perp} \end{cases}$$

is a lattice antihomomorphism, which induces a lattice anti-isomorphism of the sublattices consisting of the annihilator subsets. In other words,

$$X^{\perp} = \mathrm{ann}_B X = \mathrm{ann}_B\big(\mathrm{ann}_A(\mathrm{ann}_B X)\big) = \big({}^{\perp}(X^{\perp})\big)^{\perp}$$

for any subset X of A. The annihilator subsets form a complete sublattice of $\mathrm{Pow}\, A$ or $\mathrm{Pow}\, B$.

This defines the annihilator $\mathrm{ann}_R X$ of any subset X of a right R-module $M \times R \to R$, patently a right ideal of R, called the **right annihilator of** X. For a subset X of R_d, any right annihilator $\mathrm{ann}_R X$ of R_d is called a **right annulet**, or **annihilator right ideal.** Furthermore, for the right module M, and subset X of R, $\mathrm{ann}_M X$ is defined and is an $\mathrm{End}\, M_R$ submodule[2] of M. Left annulets and annihilator left ideals are defined symmetrically.

If $X \subseteq M_R$, and if N is a submodule, then define:

$$\mathrm{ann}_R X/N = \{r \in R \mid Xr \subseteq N\}.$$

If X is a submodule containing N, then $\mathrm{ann}_R X/N$ is just the annihilator of the factor module X/N, and therefore the notation is consistent.

If R is a subring of a ring A, then for any subset X of R we have $\mathrm{ann}_R X = R \cap \mathrm{ann}_A X$, a fact which shows that R satisfies the a.c.c. (or d.c.c.) on right annulets provided that A does.

Two Canonical Maps

Let M and G be right R-modules. The canonical mapping

$$\coprod_{f \in \mathrm{Hom}_R(G,M)} f \begin{cases} G^{(\mathrm{Hom}_R(G,M))} \to M \\ x = (\ldots, x_f, \ldots) \to \sum_{f \in \mathrm{Hom}_R(G,M)} f(x_f), \end{cases}$$

---

[1] $a^{\perp}$ is read "a perp".

[2] This is not to say that $\mathrm{ann}_M X$ is an R-submodule of M. It will be, however, if X is an ideal of R.

where $x \mapsto x_f$ is the projection $G^{(\mathrm{Hom}_R(G,M))} \to G$ determined by the element $f \in \mathrm{Hom}_R(G, M)$. The image of the canonical map is the **trace of G in M**, denoted $\mathrm{trace}_M G$. If $a \in \mathrm{End}\, M_R$, and if $f : G \to M$, then $af : G \to M$. It follows that $\mathrm{trace}_M G$ is an $(\mathrm{End}\, M_R, R)$-bimodule, and the canonical map induces an exact sequence

$$G^{(\mathrm{Hom}_R(G,M))} \to \mathrm{trace}_M G \to 0.$$

If $B = \mathrm{End}\, M_R$, and $x \in G$, then

$$O_G^M(x) = \{f(x) \mid f \in \mathrm{Hom}_R(G, M)\}$$

is a B-module, and $\mathrm{trace}_M G = \sum_{x \in G} O_G^M(x)$. For $M = R$, $\mathrm{trace}_R G$ is therefore an ideal of R, called the **trace ideal.** In this case $O_G(x)$ denotes the left ideal $O_G^R(x)$ of R, and $\mathrm{trace}_R G$ is shortened to trace G.

Dually, there is a canonical mapping

$$\prod_{f \in \mathrm{Hom}_R(M,G)} f \quad \begin{cases} M \to G^{\mathrm{Hom}_R(M,G)} \\ x \;\mapsto \langle\ , x\rangle, \end{cases}$$

where $\langle\ , x\rangle(g) = g(x)\ \forall\, g \in \mathrm{Hom}_R(M, G)$. The kernel is the annihilator $\mathrm{ann}_M \mathrm{Hom}_R(M, G)$ of $\mathrm{Hom}_R(M, G)$ in M under the pairing

$$\begin{cases} \mathrm{Hom}_R(M, G) \times M \to G \\ \qquad\qquad (g, x) \mapsto g(x). \end{cases}$$

Thus,

$$0 \to \mathrm{ann}_M \mathrm{Hom}_R(M, G) \to M \to G^{\mathrm{Hom}_R(M,G)}$$

is exact, which for $M = R$, via the canonical isomorphism $G \approx \mathrm{Hom}_R(R, G)$, yields an exact sequence

$$0 \to \mathrm{ann}_R G \to R \to G^R.$$

Therefore, the canonical map $R \to G^R$ is an embedding if and only if G is a faithful module (cf. 3.28). The dual property is the statement that the canonical map $G^{(\mathrm{Hom}_R(G,R))} \to R \to 0$ be exact. This happens if G is a generator (see the following 3.26) and for this reason, in the literature, a generator is also called a **cofaithful** module.

3.24 *Exercise.*

3.24.1 Let e be an idempotent element of a ring A. Then $\mathrm{Hom}_A(eA, A) \approx \mathrm{Hom}_A(A/(1-e)A, A) \approx Ae = \mathrm{ann}_A(1-e)A$. Also, $\mathrm{Hom}_{eAe}(eA, eAe) \approx Ae$ canonically. (If eA is a generator, then $A \approx \mathrm{End}_{eAe}\, eA$ canonically. Cf. 4.1.)

3.24.2 If $\{e_{ij}\}_{i,j\in n}$ are matrix units in the matrix ring A_n of degree n, then the right ideal $e_{ii}A_n$ of A_n is a generator of mod-A_n and is a finitely generated projective over its endomorphism ring which is canonically isomorphic to $e_{ii}Ae_{ii}$, for any $i \in n$.

*3.24.3 Let A be a ring. Then mod-$A \approx$ mod-B for a ring B if and only if there exists an integer n and an idempotent e in the total matrix ring B_n of degree n such that $B_n e B_n = B_n$ and such that the ring A is isomorphic to the ring $eB_n e$.

3.24.4 (a) Let R be any ring. If G is any right ideal, then trace G contains the ideal RG generated by G.

(b) If $G = eR$, where e is an idempotent, then trace $G = ReR$.

3.24.5 Prove all the exercises in 7.22 and 7.23.

3.24.6 If R is a field, then the matrix ring R_n is a right (left) PIR, and every one-sided ideal is generated by an idempotent. Conclude that every one-sided ideal is an annulet. Generalize.

3.25 Proposition. *If C is a category with coproducts, then for each object X of C and each object U, the coproduct of all morphisms $f: U \to X$, is a morphism*

(1) $$U^{(\mathrm{Mor}_C(U,X))} \to X$$

called **canonical**. *Furthermore, U is a generator of C if and only if the following equivalent conditions* (a) *and* (b) *hold*:

(a) *For each object X of C the canonical morphism* (1) *is epic.*

(b) *For each object X of C there is an index set I, depending on X, and an epic $U^{(I)} \to X$.*

Dually, for a cogenerator U of a category C with products.

Proof. If U is a generator, then (a) holds, and clearly (a) $\Leftrightarrow$ (b). To show that (a) implies that U is a generator, let $f: X \to Y$ and $g: X \to Y$ be morphisms. If, for every $h: U \to X$, it were true that $fh = gh$, then, necessarily, $fp = gp$ for the canonical morphism p described by (1). Since p is epic by (a), then $f = g$. Thus, $f \neq g$ would imply that $fh \neq gh$ for some $h: U \to X$ so that U is a generator. □

3.26 Proposition. *The following statements about an object G in* mod-R *are equivalent*:

3.26.1 *For each object X there exist a set I and an epimorphism $G^{(I)} \to X$.*

3.26.2 *$G^n \approx R \oplus Q$ for some integer n, and object Q.*

3.26.3 *G is a generator, that is, the set-valued functor h^G is faithful.*

3.26.4 *For each right R-module M the canonical map $G^{(F)} \to M$ is surjective, where $F = \mathrm{Hom}_R(G, M)$.*

3.26.5 *trace* $G = R$.

3.26.6 *There exist finitely many elements* $f_1, \ldots, f_n \in \mathrm{Hom}_R(G, R)$, *and* $g_1, \ldots, g_n \in G$ *such that*

$$\sum_{i=1}^n f_i(g_i) = 1.$$

Proof. (1) $\Leftrightarrow$ (3) $\Leftrightarrow$ (4) $\Leftrightarrow$ (5) by the proposition, and (5) $\Leftrightarrow$ (6) is trivial.

(1) $\Rightarrow$ (2) in the case $X = R$ because if $f: G^{(I)} \to R$ is surjective, and if J is the support of $f^{-1}(1)$, then f induces a surjection $f': G^{(J)} \to R$, hence $G^n \to R \to 0$ is exact, where n is the cardinality of J.

(2) $\Rightarrow$ (1). The statement (1) is true for $G = R$ by 3.9, and hence is true for any generator G, since there is an epimorphism $G^{(J)} \to R$, and hence, an epimorphism $G^{(I \times J)} \to R^{(I)}$ for any index set I. Thus, if $R^{(I)} \to X$ is epic, then the composition $G^{(I \times J)} \to X$ is epic. ▯

3.27 (a) Corollary. *Let R be a subring of a ring S (hence, the identity of R is the identity of S). Then the R-module S is a generator in* mod-*R if and only if R is an R-summand of S.*

Proof. Any R-module $R \oplus A$ is a generator. Conversely, assume S is a generator, and let $f_i \in \mathrm{Hom}_R(S,R)$, $x_i \in S$ be such that $\sum_1^n f_i(x_i) = 1$ (see 3.26). The correspondence

$$\varphi : y \to \sum_1^n f_i(x_i y),$$

$y \in S$, is an R-homomorphism $\varphi: S \to R$. Now $\varphi(a) = \varphi(1)a = 1a = a$ $\forall\, a \in R$. Furthermore, $\forall\, y \in S$, $\varphi(y) = y$ if and only if $y \in R$. Thus, $S = \ker \varphi \oplus R$. ▯

3.27 (b) *Example.* Let G be a group of automorphisms of a ring R such that for each $x \in R$, the set Gx of elements $\{g(x)\}_{g \in G}$ is finite, and $|Gx|$ is a unit of R. The "averaging map" sends an element x onto the element $\sum_{y \in Gx} y |Gx|^{-1}$. If A denotes the set fix $G = \{y \in R \mid g(y) = y \;\forall\, g \in G\}$, then A is a **Galois** subring of R, and the averaging map $R \to A$ splits in mod-A (also in A-mod). Conclude that A is a summand of the canonical A-module R. This holds, for example, under the assumption that G is a finite group, and either $|G|$ is a unit of R, or, more generally, there exists $x \in R$ such that $\sum_{g \in G} g(x)$ is a unit of R. (See Maschke's Theorem 13.21, and also the proof of 13.23.)

For other examples, see Exercises 3.30 (b).

Subobjects and Quotient Objects

Two morphisms $f_i : X_i \to Y$, $i = 1, 2$, in a category C with target Y are said to be **equivalent** provided that they are equivalent objects in target Y, in which case one writes $f_1 \approx f_2$. Thus, $f_1 \approx f_2 \Leftrightarrow f_1 = f_2 h$

for some equivalence $h: X_1 \to X_2$. In the case that $f_i: X_i \to Y$, $i = 1, 2$, are monics, then write $f_1 \leq f_2$ provided that f_2 allows f_1. Then, $f_1 \leq f_2$ and $f_2 \leq f_1$ if and only if $f_1 \approx f_2$. This shows that the relation $\leq$ orders the class of monics in target Y, for each object Y. If $k: K \to X$ is a monic, then K is called a **subobject**, and k is called the **embedding** or **inclusion** of K in X. The symbol $K \subseteq X$ denotes that K is a subobject of X, and i_K^X denotes the inclusion of K in X. Dually, if $q: Y \to Q$ is epic, then Q is a **quotient object.**

An object X of a category C is **Noetherian** (**Artinian**) if the class of subobjects of X satisfies the ascending (descending) chain condition, abbreviated a.c.c. (d.c.c.). Thus, if

$$X_1 \subseteq X_2 \subseteq \cdots \subseteq X_n \subseteq \cdots,$$

for any sequence of subobjects $\{X_i\}$ of X, implies that there exist an integer n and equivalences $X_i \approx X_j$ as subobjects for all $i \geq j \geq n$, then X is Noetherian. Dually for Artinian. This is equivalent to the maximum condition, namely, every nonempty class of subobjects contains a subobject which is maximal in the class of subobjects ordered by inclusion.

A ring R is **right Artinian** (**Noetherian**) if R is an Artinian (Noetherian) object of mod-R. Any field is both right and left Artinian and Noetherian, and so is the full matrix ring R_n over any field. (Proof?) These form an important class of rings, to be studied in some depth in other parts of the book and introduced here for the characterization of finitely generated faithful modules over Artinian rings stated in the next propostion.

3.28 Proposition. *Let M be a module over a ring R such that either* (a) *R is commutative and M is finitely generated or* (b) *R is right Artinian. Then M is faithful if and only if there exists an embedding $R \to M^n$ of R into a finite product of copies of M.*

Proof. If $R \to M^n$ is an embedding, then M^n, hence M, is faithful, since R is. Now assume M is faithful, let $F = \operatorname{Hom}_R(R, M)$, and assume: (*) *there exists finitely many* $\{f_i \in F \mid i = 1, \ldots, n\}$ *with* $\bigcap_{i=1}^n \ker f_i = 0$. Then, setting $F' = \{f_1, \ldots, f_n\}$, the inclusion $F' \to F$ and the canonical map $R \to M^F$ (sup. 3.25) induce a monic $R \to M^{F'} \approx M^n$. Now if R is commutative, and if $\{f_1(1), \ldots, f_n(1)\}$ generates M, where $f_i \in F$, $i = 1, \ldots, n$, then $f_1, \ldots, f_n$ satisfies (*). Also if R is Artinian, then $f_1, \ldots, f_n$ satisfies (*) for some finite n, namely, if $A = \bigcap_{i=1}^n \ker f_i$ is minimal in the set of right ideals of R which are finite intersections of kernels of elements of F, then $A \cap \ker f = A$ for all $f \in F$, that is, $0 = \bigcap_{f \in F} \ker f = A$. □

3.29 *Exercises.*

3.29.1 A ring R is right Noetherian if and only if every right ideal is finitely generated. Then, the polynomial ring $R[x]$ is right Noetherian. (Hilbert Basis Theorem 7.13.)

3.29.2 If R is right Noetherian, then any finitely generated module of mod-R is Noetherian, and the full subcategory of mod-R consisting of the finitely generated modules is Noetherian. Dually for Artinian.

*3.29.3 (Hopkins-Levitzki) Any right Artinian ring is right Noetherian.

3.29.4 If R embeds in a right R-module M^n for some integer $n > 0$, then there is a canonical morphism $B^n \to M$, where $B = \text{End}\, M_R$; if M is quasinjective (p. 172), then $B^n \to M \to 0$ is exact. Conclude from 3.28 that every quasinjective module over a right Artinian ring R is finitely generated over endomorphism ring (cf. 19.5).

3.30 (a) Proposition. *If P is a projective R-module, then*

$$(\text{trace } P)^2 = \text{trace } P$$

and

$$P(\text{trace } P) = P.$$

Furthermore, in this case, for a left ideal I of R the following conditions are equivalent: (1) $I \supseteq \text{trace } P$; (2) $(\text{trace } P)I = \text{trace } P$; (3) $PI = P$.

Proof. By the Dual Basis Lemma 3.23, there exist sets $\{x_i\}_{i\in I} \subseteq P$ and $\{f_i\}_{i\in I} \subseteq P^*$ such that every $x \in P$ can be represented as a finite sum

$$x = \sum x_i f_i(x).$$

Since $f_i(x) \in O_P(x) \subseteq \text{trace } P$, we have $P = P(\text{trace } P)$.

If I is a left ideal of R such that $PI = P$, then

$$T = \text{trace}_R P = \text{trace}_R (PI) = (\text{trace}_R P) I = TI \subseteq I.$$

Thus $(3) \Rightarrow (2)$, and $(2) \Rightarrow (1)$ is obvious since I is a left ideal. Since $PT = P$, obviously $(1) \Rightarrow (3)$. Thus (1), (2), and (3) are equivalent, whence $T^2 = T$. ▯

3.30 (b) *Exercises.* 1. If R is a commutative ring, any idempotent finitely generated ideal $I = I^2$ is generated by an idempotent element. Conclude that the annihilator of any projective module is generated by an idempotent iff its trace is finitely generated.

2. Over a commutative ring R, any finitely generated faithful projective module is a generator.

3. If a ring S is finitely generated projective as a canonical module over a commutative subring R, then R is an R-summand of S.

Antegenerators

A module M is an **antegenerator** (**antegen**, for short) in case every simple module is an epimorph of M. Dually for **antecogen**. Every generator is an antegen, and every cogenerator is an antecogen. If N is an antegen (resp. generator) and if $M \to N$ is an epimorphism, then M is an antegen (resp. generator). Dually for antecogens and cogenerators.

3.31 Proposition.

3.31.1 *A projective module G is a generator if and only if G is an antegen.*

3.31.2 *An injective module C is a cogenerator if and only if C is an antecogen.*

Proof of 3.31.1. The necessity is obvious. Conversely, let G be a projective antegen, and let $f: M \to N$ be a nonzero map of modules. Since $\ker f \neq M$, there is an $x \in M$, $x \notin \ker f$. Let $K = \ker f$ and $K' = K + xR$. Since K'/K is cyclic, there is a maximal submodule K''/K. Since $S = K'/K''$ is a simple module, there is an epimorphism $G \to S$.

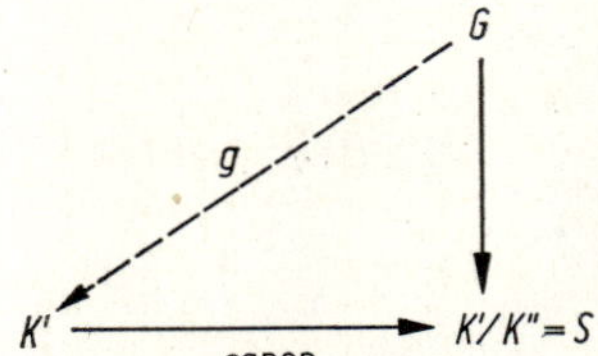

By projectivity of G, there is a map $g: G \to K'$ completing the diagram. Since $K' \subseteq M$, g induces a map $G \to M$, and $\operatorname{im} g \not\subseteq \ker f$. By 3.26.3, G is a generator. This proves 3.26.1. The proof of 3.26.2 is dual. □

Axioms for Additive Category

A category C is an **additive category** in case for each pair X, Y of objects of C, $\mathrm{Mor}_C(X, Y)$ is an abelian group subject to the following axioms:

(Ad 1) *The composition functions*

$$\mathrm{Mor}_C(Y, Z) \times \mathrm{Mor}_C(X, Y) \to \mathrm{Mor}_C(X, Z)$$

are bilinear; that is,

$$f(g_1 + g_2) = fg_1 + fg_2$$
$$(f_1 + f_2)g = f_1 g + f_2 g$$

when defined.

(Ad 2) *C has a zero object* 0.

(Ad 3) *The coproduct of any finite family of objects of C exists*[1].

The zero element 0 of the group $\mathrm{Mor}_C(X, Y)$, that is, the additive identity element, satisfies

$$0 = f0$$

for all morphisms f, since (Ad 1) implies $f0 = f(0+0) = f0 + f0$, and adding $-(f0)$ yields $0 = f0$. Also

$$f(-g) + fg = f((-g) + g) = f0 = 0$$

so that

$$f(-g) = -(fg).$$

Similarly,

$$(-f)g = -(fg) \qquad \text{and} \qquad (-f)(-g) = fg.$$

A **(co)productive** category C is one such that the (co)product of any family of morphisms exists. If the (co)products of finite families of morphisms exist, as in additive categories, then C is a **category with finite (co)products.**

Additive Functors

If C is an additive category, and if $X \in \mathrm{obj}\, C$, then $\mathrm{End}_C X$ is a ring with respect to the operations of addition in the abelian group and composition of morphisms, called the **endomorphism ring** of X. For any object Y, the group $\mathrm{Mor}_C(X, Y)$ is a left module over $\mathrm{End}_C Y$ under the induced operation

$$\mathrm{Mor}_C(Y, Y) \times \mathrm{Mor}_C(X, Y) \to \mathrm{Mor}_C(X, Y).$$

Thus, if $a \in \mathrm{End}_C Y$ and $f \in \mathrm{Mor}_C(X, Y)$, then $af \in \mathrm{Mor}_C(X, Y)$. Similarly, $\mathrm{Mor}_C(X, Y)$ is a right module over $\mathrm{End}_C X$.

If C and D are additive categories, then a functor $T: C \rightsquigarrow D$ is **additive** in case the induced morphisms $\mathrm{Mor}_C(X, Y) \rightsquigarrow \mathrm{Mor}_D(TX, TY)$ are group homomorphisms, that is,

$$T(f + g) = Tf + Tg$$

$\forall f, g \in \mathrm{Mor}_C(X, Y)$, $\forall$ objects X and Y of C. Then, if T is covariant, T induces ring homomorphisms $\mathrm{End}_C X \to \mathrm{End}_C TX$ for each object X, and ring antihomomorphisms when T is contravariant. The bilinearity

[1] In the literature, this axiom is not always assumed.

of the operation in Mor C implies (and is implied by) the additivity of the morphism functors h^X and h_X for any X. Thus, h^X factors through mod-$\mathrm{End}_C X$ and mod-$\mathbb{Z}$:

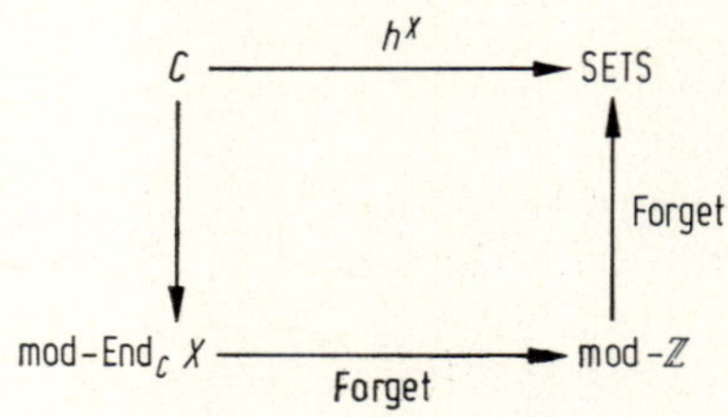

Similarly, h_X factors through $\mathrm{End}_C X$-mod.

The category of additive categories has the named objects, and morphisms are additive covariant functors. The full subcategory on the additive categories with a single object is equivalent to the category RINGS.

We let Adfunct (C, D) denote the category of additive functors $C \rightsquigarrow D$, where C and D are additive categories.

3.32 Proposition. *Let $A_1, \ldots, A_n$ be a finite collection of objects in an additive category C.*

3.32.1 *A family $\{u_i : A_i \to A \mid i = 1, \ldots, n\}$ of morphisms is a coproduct for the family $\{A_i \mid i = 1, \ldots, n\}$ if and only if there exist morphisms $\{p_i : A \to A_i\}$ such that*

$$p_i u_j = \delta_{ij}$$

and

$$\sum_{k=1}^{n} u_k p_k = 1_A .$$

3.32.2 *When 3.32.1 holds, then $\{p_i : A \to A_i\}$ is a product for the family $\{A_i \mid i = 1, \ldots, n\}$.*

3.32.3 *Finite products $\prod_{i=1}^{n} A_i$ exist in C for any finite family of objects $\{A_i \mid i = 1, \ldots, n\}$.*

3.32.4 *The dual category C^* is additive.*

3.32.5 *Every finite product is a biproduct.*

Proof of 3.32.1. If $\{u_i : A_i \to A \mid i = 1, \ldots, n\}$ is a coproduct, there exist morphisms $p_i : A \to A_i$, $i = 1, \ldots, n$, satisfying $p_i u_j = \delta_{ij}$, $i, j = 1, \ldots, n$. Then for each i,

$$\left(\sum_{k=1}^{n} u_k p_k\right) u_i = \sum_{k=1}^{n} (u_k p_k u_i) = \sum_{k=1}^{n} u_k \delta_{ki} = u_i = 1_A u_i .$$

Hence, by definition of coproduct, $\sum_{k=1}^{n} u_k p_k = 1_A$.

Conversely, if morphisms $\{p_i\}$ satisfy the given conditions, and if $f_i : A_i \to A'$ are morphisms of C, $i = 1, \ldots, n$, then the morphism $f = \sum_{k=1}^{n} f_k p_k$ satisfies

$$f u_i = \sum_{k=1}^{n} f_k p_k u_i = f_i.$$

Furthermore, if f' is another morphism such that $f' u_i = f_i$, $i = 1, \ldots, n$, then

$$f' = f' 1_A = f' \sum_{k=1}^{n} u_k p_k = \sum_{k=1}^{n} (f' u_k) p_k = \sum_{k=1}^{n} f_k p_k = f.$$

Hence, f is unique, so A is the coproduct.

In the dual category C^* if we define $f^* + g^* = (f + g)^*$ for morphisms f, g of C, then $\mathrm{Mor}_{C^*}(X^*, Y^*)$ is an abelian group $\forall X, Y \in \mathrm{obj}\, C$, and C^* satisfies (Ad 1) and (Ad 2). Assuming the notation in 3.32.1, if $\{u_i : A_i \to A\}$ is a finite coproduct in C, then $\{u_i^* : A^* \to A_i^*\}$ is a finite product in C^* such that

$$u_j^* p_i^* = \delta_{ij}$$

$$\sum_{k=1}^{n} p_k^* u_k^* = 1_{A^*}$$

for some morphisms $p_i : A \to A_i\}$. Then the proof of 3.32.1 shows that $\{p_i^* : A_i^* \to A^*\}$ is a coproduct of the family $\{A_i^* \mid i = 1, \ldots, n\}$ in C^*. (Hence C^* satisfies (Ad 3).) It follows that $\{p_i : A \to A_i\}$ is a product in C. This proves 3.32.2 and 3.32.3. Since C^* satisfies (Ad 3), C^* is additive which proves, 3.32.4. The morphism $\delta = (\delta_{ij})$ mapping the coproduct $A = \coprod A_i$ to the product $A = \prod A_i$ is just 1_A, which proves 3.32.5. □

Thus, the dual of any theorem about additive categories is a theorem about additive categories. A category equivalent to mod-A for some ring A is said to be of **module-type** and is additive. Although additive, the dual of such a category need not be of module-type. (Exercise 14.3.4.)

3.33 Proposition.

3.33.1 *Let C denote an additive category, and let $\{A_j \mid j = 1, \ldots, n\}$ and $\{B_i \mid i = 1, \ldots, m\}$ be finite families of objects. The bijection*

$$\Phi \begin{cases} G = \mathrm{Mor}_C\left(\coprod_{j=1}^{n} A_j, \prod_{i=1}^{m} B_i\right) \to \underset{(j,i) \in n \times m}{\times} \mathrm{Mor}_C(A_j, B_i) \\ \qquad f \mapsto (f_{ij}) \end{cases}$$

(stated by the Matrix Proposition 2.14) is an isomorphism of Abelian groups. (The group represented on the right is the Abelian group of all

matrices (f_{ij}) *with* $f_{ij}: A_j \to B_i$ *in* $C \forall i \in m,\ j \in n$. *In the case of* $C = \text{mod-}R$, *given a matrix* $T = (f_{ij})$, *the corresponding element* $\Phi^{-1}(T)$ *in the group G acts as follows:*

$$\Phi^{-1}(f_{ij})\,(a_1, \ldots, a_n)^t = \begin{pmatrix} f_{11} & \cdots & f_{1n} \\ \vdots & & \vdots \\ f_{m1} & \cdots & f_{mn} \end{pmatrix} \begin{pmatrix} a_1 \\ \vdots \\ a_n \end{pmatrix} = \begin{pmatrix} \sum_{j=1}^{n} f_{1j} a_j \\ \vdots \\ \sum_{j=1}^{n} f_{mj} a_j \end{pmatrix}$$

where $(a_1, \ldots, a_n) \in \prod_{i=1}^{n} A_i,\ a_i \in A_i,\ i = 1, \ldots, n.$)

3.33.2 *In particular,* $\text{Mor}_C(A^n, B^m)$ *is isomorphic to the abelian group of all* $m \times n$ *matrices* (f_{ij}) *with elements in* $\text{Mor}_C(A, B)$.

3.33.3 *Finally,* $\Phi: \text{Mor}_C A^n \to (End_C A)_n$ *is a ring isomorphism of the endomorphism ring of* A^n *with the full matrix ring of degree n over* $\text{End}_C A$. *For any ring R, this implies a canonical ring isomorphism* $R_n \to \text{End}\, R_R^n$, *that is, the endomorphism ring of the free R-module* R^n *with basis number n is canonically isomorphic to the full matrix ring* R_n *of degree n over R.* □

3.34 Proposition for an Additive Category C.

3.34.1 *If* $\{u_i : A_i \to A \mid i = 1, 2\}$ *and* $\{p_i : A \to A_i \mid i = 1, 2\}$ *are morphisms such that*

$$u_1 p_1 + u_2 p_2 = 1_A \qquad \text{and} \qquad p_i u_i = 1_{A_i}, \qquad i = 1, 2,$$

then A is the coproduct of u_1 *and* u_2.

3.34.2 *If* $u_1 : A_1 \to A$ *is a coretraction,* $p_1 u_1 = 1_{A_1}$ *for* $p_1 : A \to A_1$, *then* $e_1 = u_1 p_1$ *is idempotent, and* u_1 *is a kernel of* $1_A - e_1$.

3.34.3 *If* $e_1 \in \text{End}_C A$ *is idempotent, if* $u_1 : A_1 \to A$ *is a kernel of* e_1, *and if* $u_2 : A_2 \to A$ *is a kernel of* $1_A - e_1$, *then A is the coproduct of* u_1 *and* u_2,

$$A = \text{Ker}\, e_1 \oplus \text{Ker}(1 - e_1).$$

Proof of 3.34.1. The relations stated imply $u_2 p_1 = u_1 p_2 = 0$, so 3.34.1 follows from 3.32.

Proof of 3.34.2. Obviously, e_1 is idempotent. Next, $(1_A - e_1) u_1 = 0$, so $u_1 \leq \ker(1_A - e_1)$; however, if $(1_A - e_1) f = 0$, then $u_1(p_1 f) = e_1 f = f$. Since u_1 is monic, this shows that u_1 is a kernel of $1_A - e_1$.

Proof of 3.34.3. Since u_1 is a kernel of $1_A - e_1$, and since $(1_A - e_1) e_1 = 0$, there exists a morphism $p_1 : A \to A_1$ such that $u_1 p_1 = e_1$. Therefore, $u_1 p_1 u_1 = e_1 u_1 = u_1$, that is, $u_1(p_1 u_1 - 1_{A_1}) = 0$, so $p_1 u_1 = 1_{A_1}$ because u_1 is monic. Similarly, there exists $p_2 : A \to A_2$ such that

$$u_2 p_2 = 1_A - e_1 \qquad \text{and} \qquad p_2 u_2 = 1_{A_2}.$$

Thus, $u_1 p_1 + u_2 p_2 = e_1 + (1_A - e_1) = 1_A$, hence 3.34.3 follows from 3.34.1

3.35 Proposition and Definition. *Let C be a module type category (or more generally, an additive category which is exact in the sense of 5.24). Then a monic $u_1 : A_1 \to A$ is said to* **split** *if it satisfies the following equivalent conditions:*

3.35.1 u_1 *is a coretraction.*

3.35.2 coker u_1 *is retraction.*

3.35.3 $0 \to A_1 \overset{u_1}{\to} A \overset{p_2}{\to} A_2 \to 0$ *is split exact for some* $p_2 : A \to A_2$.

3.35.4 *For some monic* $u_2 : A_2 \to A$, *A is the coproduct of u_1 and u_2.*

Proof. (1) ⇒ (4) follows from 3.34.2 and 3.34.3. (4) ⇒ (3) is the fact, 2.13, that projections from the coproduct are retractions. (3) ⇒ (2). Given (3), u_1 is a kernel of p_2 by 3.34. Since injections into a coproduct are coretractions, 3.35.2 follows. (2) ⇒ (1). Suppose $p_2 : A \to A_2$ is a cokernel of u_1. Since every cokernel is a retraction if and only if one cokernel is, we can write $p_2 u_2 = 1_{A_2}$ for some $u_2 : A_2 \to A$. Then $e = u_2 p_2 \in \mathrm{End}_C A$ is idempotent, so 3.35.1 follows from 3.34.3, and the fact that the injections into a coproduct are coretractions. ▯

Image of a Functor

Let $T : C \rightsquigarrow D$ be a functor that is injective on objects, and let I denote class of all objects in D of the form TA, together with the morphism $T(f) : TA \to TB$, where $f : A \to B$ is a morphism. Then I is a subcategory of D, denoted im T.

3.36 Proposition. *Let $T : C \rightsquigarrow$ mod-$\mathbb{Z}$ be an additive functor. Then T is naturally equivalent to a functor that is injective on objects.*

Proof. Define $T'A$ to be the group consisting of all ordered pairs (A, x) with $x \in TA$, and define $(A, x) + (A, y) = (A, x + y)$. If $f : A \to B$ is a morphism in C, then $T'(f)(A, x) = (B, T(f)x)$. Then $h : T \to T'$, where $h_A(x) = (A, x)$, is a functor equivalence. ▯

3.37 *Exercises.* Let C be any category with zero, and let $A \prod B$ and $A \coprod B$ exist for any two objects of C. By 2.14, a morphism $f : A \coprod B \to A \prod B$ has a unique representation by a matrix

$$f = (f_{AB}),$$

where $f_{AB} = p_A f u_B$ and p_A and p_B (resp. u_A and u_B) are the projections (injections) from the product (into the coproduct).

3.37.1 Show that

$$\begin{pmatrix} 1_A & 0 \\ 0 & 1_B \end{pmatrix} : A \coprod B \to A \prod B$$

is an isomorphism. Henceforth, let $A \oplus B$ denote product and coproduct.

3.37.2 If f and g are elements of $\mathrm{Mor}_C(A, B)$, then the unique morphism h, such that

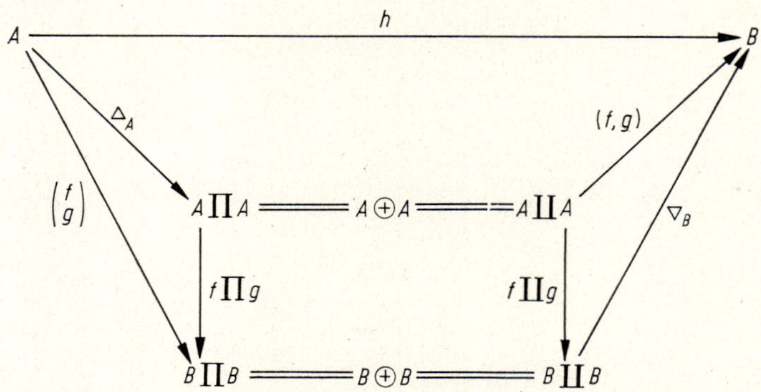

commutes, defines the sum $h = f + g$ with respect to which $\mathrm{Mor}_C(A, B)$ is an abelian semigroup (Bass [68, p. 14ff.]).

3.37.3 If A and B are objects in an additive category C, then addition $f + g$ for $f, g \in \mathrm{Mor}_C(A, B)$ is determined by any of the following compositions:

$$A \xrightarrow{\Delta} A \times A \xrightarrow{(f,g)} B$$

$$A \xrightarrow{\binom{f}{g}} B \oplus B \xrightarrow{\nabla} B$$

$$A \xrightarrow{\Delta} A \oplus A \xrightarrow{\begin{pmatrix} f & 0 \\ 0 & g \end{pmatrix}} B \oplus B \xrightarrow{\nabla} B,$$

where Δ is the diagonal and ∇ is the codiagonal morphism.

3.37.4 An object A of category C is **small** in case every morphism $A \to \coprod_{i \in I} A_i$ of A into a coproduct factors through a finite coproduct in the sense that there is a finite subset J of I and a commutative diagram

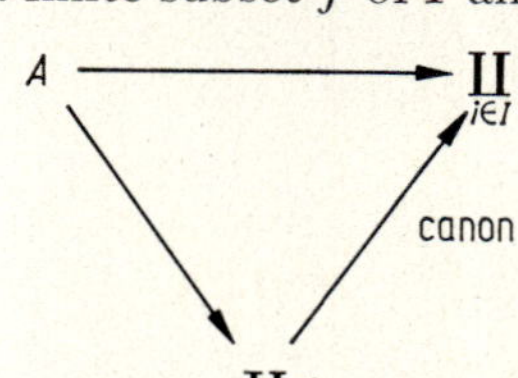

Thus, smallness of an object is a categorical property.

(a) Show that any finitely generated object of mod-R is small.

(b) Show that every small projective object P in a coproductive category C with a generator U is **finitely generated with respect to** U, that is, there exists an integer $n > 0$ and an epic $U^{(n)} \to P$. Conclude that every small projective object of mod-R is finitely generated.

3.37.5 Let C be an additive category. The following statements about an object A in C are equivalent:

(a) The object A is small.

(b) The group-valued functor h^A preserves coproducts.

(c) The functor $C \rightsquigarrow$ mod-End$_C A$ induced by h^A preserves coproducts.

3.37.6 If T : mod-$R \rightsquigarrow$ mod-S is an equivalence, then X in mod-R is finitely generated if and only if TX is finitely generated in mod-S.

3.38 Proposition for Additive Categories. *Let* $T : C \rightsquigarrow D$ *be a covariant functor of additive categories. The following conditions are equivalent:*

3.38.1 *T is additive.*

3.38.2 *T preserves finite (co)products.*

3.38.3 *T preserves (co)products $A \prod B$ $(A \coprod B)$ for each pair $A, B \in \operatorname{obj} C$.*

Proof. (1) ⇒ (2) follows immediately from the characterization of finite coproducts in additive categories. (2) ⇒ (3) is trivial.

(3) ⇒ (1). Let

$$A_i \xrightarrow{u_i} A \xrightarrow{p_j} A_j,$$

$i, j = 1, 2,$ be a (co)product in C. Then (3) implies

$$T(A_i) \xrightarrow{T(u_i)} T(A) \xrightarrow{T(p_j)} T(A_j)$$

is a (co)product in D. Let $f : A \to B$ and $g : A \to B$ be morphisms of C. Then the morphism $f + g : A \to B$ can be factored

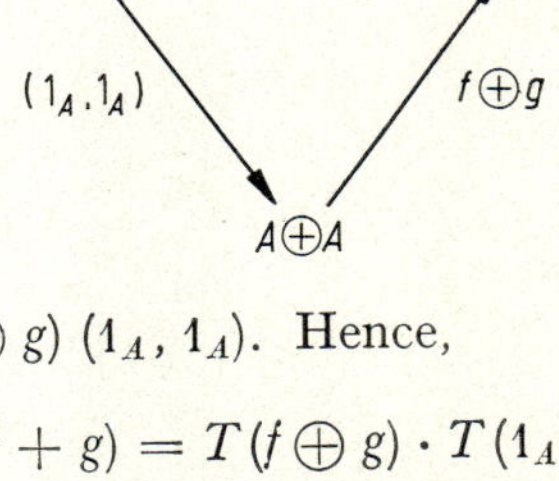

that is, $(f+g) = (f \oplus g)\,(1_A, 1_A)$. Hence,

$$\begin{aligned} T(f+g) &= T(f \oplus g) \cdot T(1_A, 1_A) \\ &= \big(T(f) \oplus T(g)\big) \cdot (1_{T(A)}, 1_{T(A)}). \end{aligned}$$

Since $T(f) + T(g)$ is that unique morphism $T(A) \to T(B)$, which has the preceding factorization, we conclude that $T(f+g) = T(f) + T(g)$, so T is additive. $\square$

Exact Sequences

A sequence of morphisms

$$\cdots \to A_1 \to A_2 \to A_3 \to A_4 \to \cdots$$

is **exact** in case, for each i,

$$\operatorname{im}(A_{i-1} \to A_i) = \ker(A_i \to A_{i+1}).$$

Otherwise, the sequence is **inexact**.

A **short exact sequence** is an exact sequence

$$0 \to A \to B \to X \to 0.$$

An exact sequence $A \to B \to X$ is **half exact**. A **left exact sequence** is an exact sequence

$$0 \to A \to B \to X$$

and, similarly for **right exact** sequences.

A covariant functor $T : C \rightsquigarrow D$ is **left** (**right**, **half**) **exact** if T preserves left (right, half) exact sequences. An exact functor is defined to be one which is both left and right exact.

3.39 Proposition. *Let A be a ring. For any object U of* mod-A, *the group valued functors h^U and h_U are left exact. Furthermore, h^U is exact if and only if U is projective object of* mod-A, *and h_U is exact if and only if U is an injective object of* mod-A. (Cf. the corresponding theorem 6.11 for exact additive categories.)

Proof. Let

$$0 \to K \xrightarrow{k} X \xrightarrow{f} Y$$

be exact in mod-A. For any object U of mod-A, the set valued functor h^U preserves monics by 2.4, so $h^U k : h^U K \to h^U X$ is injective, and obviously, $\operatorname{im} h^U k \subseteq \ker h^U f$. If $g \in \ker h^U f$, then $g : U \to X$ and $fg = 0$, so there is a canonical morphism $h : U \to K$ such that $g = kh$. Thus, $g = h^U k(h) \in \operatorname{im} h^U k$, which proves that

$$0 \to h^U K \xrightarrow{h^U k} h^U X \xrightarrow{h^U f} h^U Y$$

is exact, and so h^U is left exact. The proof for the left exactitude of h_U is dual. By 2.4, U is projective if and only if the set valued functor preserves epics, and this is tantamount to the requirement that the group-valued functor h^U preserves "cokernels", that is, right exact sequences. Dually, for the exactitude of h_U when U is injective. ▯

3.40 *Exercises.*

3.40.1 Any half exact functor $T: \text{mod-}A \rightsquigarrow \text{mod-}\mathbb{Z}$ is additive.

3.40.2 A functor $T: \text{mod-}A \rightsquigarrow \text{mod-}\mathbb{Z}$ is exact if and only if T preserves exact sequences.

Injective Modules

Projective modules are easy to collect: they constitute the summands of free modules. There can be no corresponding simple description of injective modules, and yet there is the following simple test for injectivity.

3.41 **Theorem** (Baer's Criterion). *A right R-module M is injective if and only if for each map $f: I \to M$ of a right ideal I of R into M there corresponds $m \in M$ such that $f(x) = mx \ \forall\, x \in I$.*

Proof.

Necessity. If M is injective, and if $f: I \to M$, where I is a right ideal, then f extends to a map $g: R \to M$, and then $m = g(1)$ has the desired property.

Sufficienty. Consider a module A, a submodule B, and a map $f: B \to M$. We wish to show that there is a map $g: A \to M$ extending f. Let P be a collection of pairs (A', f'), where A' is a submodule of A containing B and $f': A' \to M$ is a map extending f. P is nonempty since $(B, f) \in P$. There is a natural partial ordering of P, where $(A', f') \geq (A'', f'')$ if and only if $A' \supseteq A''$ and $f' \mid A'' = f''$. If $\{(A_i, f_i) \mid i \in I\}$ is any chain in P, and if $C = \cup A_i$, then there is a map $h: C \to M$ defined as follows: If $a \in C$, then $a \in A_i$ for some i, and set $h(a) = f_i(a)$, if also $a \in A_j$, then $f_i(a) = f_j(a)$, so $h(a)$ is independent of i. Furthermore, $(C, h) \in P$, proving that P is inductive. An application of Zorn's lemma then yields a maximal element, call it (C, h), of P. If $C = A$, the proof is complete. We shall show that $C \neq A$ leads to a contradiction.

If $x \in A$ and $x \notin C$, then $D = C + xR$ is a submodule properly containing C. Now $J = \{r \in R \mid xr \in C\}$ is a right ideal of R, and the correspondence $r \mapsto h(xr)$, defined $\forall r \in J$, is a map $\theta: J \to M$. (Note that $\theta = h \circ x_s$, where x_s is the left multiplication of J by x, so θ is a map.) By hypothesis there exists $m \in M$ such that $\theta(r) = h(xr) = mr$, $\forall r \in J$.

Next, consider

$$h' : c + xr \mapsto h(c) + mr,$$

defined $\forall c \in C$, $r \in R$. If $c_1 + xr_1 = c_2 + xr_2$, with $c_1, c_2 \in C$, $r_1, r_2 \in R$. then $x(r_1 - r_2) = c_2 - c_1 \in C$, so $r_1 - r_2 \in J$, whence

$$\begin{aligned} m(r_1 - r_2) = \theta(r_1 - r_2) &= h(x(r_1 - r_2)) = h(c_2 - c_1) \\ &= h(c_2) - h(c_1), \end{aligned}$$

that is,

$$h(c_1) + m_1 r = h(c_2) + m_2 r.$$

This demonstrates that h' is a mapping of $D = C + xR$ into M; in fact, h' is a map of D into M that extends f. Since $D \supset C$, then $(D, h') > (C, h)$, violating the maximality of (C, h) and completing the proof. ▯

Thus, M is injective if and only if for each right ideal I of R, the functor h_M induces a surjective mapping $\mathrm{Hom}_R(R, M) \to \mathrm{Hom}_R(I, M)$, which is to say that h_M is exact (see 3.39).

Regular Elements

A **right** (resp. **left**) **regular element** x of a ring is one that is not a left (resp. right) divisor of zero, that is, for which $xy = 0 \Rightarrow y = 0 \, \forall y \in R$ (resp. $yx = 0 \Rightarrow y = 0 \, \forall y \in R$). An element is **regular** if it is both left and right regular.

3.42 *Example.* Let L be the complete ring of l.t.'s in an infinite right vector space V over a field K. Let $\{x_i \mid i = 1, 2, \ldots\}$ be a basis, and let f be the linear transformation such that $f(x_{2n}) = x_n$ and $f(x_{2n-1}) = 0$, $n = 1, 2, \ldots$. Since V is projective in mod-K, there exists a linear transformation g such that $fg = 1$. Hence f is left regular since $hf = 0 \Rightarrow h = 0 \, \forall h \in L$. If f were regular, then injectivity of V in mod-K implies the existence of $g' \in L$ such that $g'f = 1$ (in which case $g' = g$), but this is impossible since f is not an automorphism of V. ▯

Divisible Groups

As stated, a right ideal I of R is principal (cyclic) if $I = xR$ for some $x \in R$. Then R is a **principal right ideal ring** in case each right ideal of R is cyclic. A **principal right ideal domain** is a principal right ideal ring that is entire. A right R-module M is **divisible** in case $M = Mx$ for each regular $x \in R$. Here $Mx = \{mx \mid m \in M\}$.

Baer's criterion for injectivity of M is the following condition: *If I is a right ideal of R and if $f: I \to M$ is a map, then there exists $m \in M$ such that $f(x) = mx \, \forall x \in I$.* If this condition is satisfied by M for a fixed right ideal I, we say that M is I-complete. In this terminology, Baer's criterion becomes: *M is injective if and only if it is I-complete for each right ideal I of R.* The first result, which follows, characterizes I-completeness of M when I is a principal right ideal. Recall that if X is a subset of either M or R, then $\operatorname{ann}_R X = \{a \in R \mid Xa = 0\} = X^{\perp}$ and if $X \subseteq R$, then

$$\operatorname{ann}_M X = \{m \in M \mid mX = 0\} = {}^{\perp}X.$$

3.43 Proposition. *If $I = xR$ is a cyclic right ideal of the ring R, then a right R-module M is I-complete if and only if ${}^{\perp}(\operatorname{ann}_R x) \subseteq Mx$.*

Proof. For brevity, let $x^{\perp}$ denote $\operatorname{ann}_R x$. First assume that M is I-complete, let $y \in {}^{\perp}(x^{\perp})$, and consider the correspondence $f: xr \mapsto yr$, defined $\forall r \in R$. If $xr = xr'$, with $r, r' \in R$, then $r - r' \in x^{\perp}$, and since $y \in {}^{\perp}(x^{\perp})$, necessarily $yr = yr'$, so f is a map of I in M. By I-completeness, there exists $m \in M$ such that $f(w) = mw \, \forall w \in I$; in particular, $y = f(x) = mx$. This proves that ${}^{\perp}(x^{\perp}) \subseteq Mx$.

Conversely, assume that ${}^{\perp}(x^{\perp}) \subseteq Mx$. If $f: I \to M$ is any map, then

$$f(x)a = f(xa) = f(0) = 0,$$

$\forall a \in x^{\perp}$, so $f(x) \in {}^{\perp}(x^{\perp})$. Consequently, $f(x) = mx$ for some $m \in M$, and then $f(w) = mw \, \forall w \in I = xR$. □

If R is an entire ring, then $x^{\perp} = 0$ for each nonzero $x \in R$, and then ${}^{\perp}(x^{\perp}) = {}^{\perp}0 = M$. Applying the proposition in this case yields the following corollary.

3.44 Corollary. *Let R be an entire ring, and let M be an R-module.*

3.44.1 *If M is injective, then M is divisible.*

3.44.2 *If M is divisible, then M is xR-complete $\forall x \in R$.* □

3.45 Corollary. *If R is a principal right ideal domain, then a right R-module is injective if and only if it is divisible.* □

The corollary is not the best possible (see Divisible Modules Over Dedekind Rings, Chapter 31, esp. 31.30). An R-module M is **torsionfree** if for every regular element $a \in R$, $xa = 0$ implies $x = 0 \, \forall x \in M$. A ring R has a (**classical**) **right quotient ring** Q provided that there is an embedding $h: R \to Q$ in RINGS such that $h(x) \in$ units Q for every regular x of R, and such that

$$Q = \{h(a)h(x)^{-1} \mid a \in R, \text{ regular } x \in R\}.$$

Henceforth, for a quotient ring Q of R, assume that $R \subseteq Q$. Thus, $Q = \{ab^{-1} \mid a \in R, \text{ regular } b \in R\}$. In order for a ring R to have a right quotient ring, it is necessary and sufficient that for each pair $a, b \in R$, and b regular, there exists $c, d \in R$ and c regular such that

$$ac = bd \qquad \textbf{(Ore condition)}.$$

(Mnemonic: $b^{-1}a = dc^{-1}$.) Any commutative ring satisfies this condition. Any domain which has a right quotient ring D is called a **right Ore domain**, and then D is necessarily a field, called the **right quotient field**. The details of the construction are given in Chapter 9.

3.46 *Exercises.*

3.46.1 (a) Any two right quotient rings $h_i : R \to Q_i$, $i = 1, 2$, are isomorphic by a map $g : Q_1 \to Q_2$ such that $gh_1 = h_2$.

3.46.1 (b) If Q is a right quotient ring of R, then Q is a left and right quotient ring of Q.

3.46.1 (c) If Q is a right quotient ring of R, then Q is a left quotient ring of R iff R has a left quotient ring.

3.46.2 (a) If R is a domain, then a torsionfree R-module is injective if and only if it is divisible. If K is the quotient field, then any torsionfree divisible module is a vector space over K. Also, K/R is not necessarily injective, although divisible.

3.46.2 (b) If R is a right Ore domain, then the right quotient field K of R is an injective R-module.

3.46.3 If R is a principal right ideal domain, more generally, any right Noetherian domain, then R is a right Ore domain. A domain is right Ore if and only if there do not exist right ideals which are infinite direct sums of nonzero submodules. (Theorem of Goldie [58].)

3.46.4 In mod-$\mathbb{Z}$, the module $\mathbb{Q}/\mathbb{Z}$ is an injective cogenerator.

3.47 **Corollary.** *If R is a principal right ideal domain, and if K is the right quotient field of R, then K/R is an injective cogenerator of* mod-R.

Proof. That R has a right quotient field K follows from the fact that $aR \cap bR = xR$, for given elements a and b of R, and some x. Thus, $a \neq 0$ and $b \neq 0$ imply that $x = ac = bd$ for elements c and d of R. Then R satisfies the Ore condition for the existence of a right quotient field K. Since K (also K/R) is divisible, then K and K/R are injective by 3.45. In order to show that K/R is an injective cogenerator, by 3.31 it suffices to show that every simple module V embeds in K/R. However,

$V \approx R/pR$, for some maximal right ideal pR of R, and then $(p^{-1}R + + R)/R$ is a submodule of K/R isomorphic to R/pR under a mapping sending the coset $p^{-1}a + R$ of K/R onto the coset $[a + pR]$ of R/pR. □

In reference to some exercises below, we remark that any product of injectives is injective (Chapter 2 Exercises 1), but the direct sum need not be.

3.48 *Exercises.*

*3.48.1 If Q is an injective module, and if a direct sum of countably many copies of Q is injective, then a direct sum of arbitrarily many copies of Q is injective. Is the dual proposition true? (Cf. Faith [66].)

*3.48.2 Show that if over a ring R each direct sum of isomorphic injective right R-modules is injective, then every direct sum of injective right R-modules is injective. Is the dual proposition true? (Cf. Faith-Walker [67].)

*3.48.3 (Cartan-Eilenberg-Bass-Papp). Show that the conditions of Exercise 3.48.2 hold if and only if R satisfies the ascending chain condition on right ideals. (Hint: If $I_1 < \cdots < I_n < \cdots$ is a chain of right ideals of R, embed the factor module R/I_n in an injective module E_n. Consider the module $E = \coprod_n^\infty E_n$, the natural mapping of $I = \bigcup_{n=1}^\infty I_n$ into E, and use Baer's criterion. In the proof of the converse note that every right ideal I is finitely generated.)

3.48.4 (Faith-Walker). Every injective projective indecomposable module is isomorphic to a summand of R (hence is cyclic).

3.48.5 Let R be a commutative entire ring, with quotient field K. For an ideal I of R prove the equivalence:

(a) I is projective.

(b) There exist finitely many elements $k_i \in K$ and $x_i \in I$ such that $k_i x_i \in R$, $i = 1, \ldots, n$, and such that $\sum_{i=1}^n k_i x_i = 1$.

(c) I is a generator.

[Hint: Use the dual basis lemma. Also use the fact that K is injective in mod-R to show that maps in $\mathrm{Hom}_R(I, R)$ are induced by multiplications by elements of K.]

3.48.6 For any left R-module M, the character module is the canonical right R-module $M' = \mathrm{Hom}(M, \mathbb{Q}/\mathbb{Z})$. For any $f: M \to N$, let $f' = h_{\mathbb{Q}/\mathbb{Z}}(f)$. Then f is monic if and only if f' is epic.

3.48.7 If P is projective in R-mod, show that the character module $P' = \mathrm{Hom}\,(P, \mathbb{Q}/\mathbb{Z})$ is injective in mod-R. Use the fact 2.12.5 that every finite abelian group is isomorphic to its own character group to conclude that $\mathbb{Z}_n$ is selfinjective, that is, injective in mod-$\mathbb{Z}_n$, for each integer $n \neq 0$. Use the character and bicharacter modules to prove that every ideal of $\mathbb{Z}_n$ is an annihilator.

Faithfully Exact Functors

A diagram

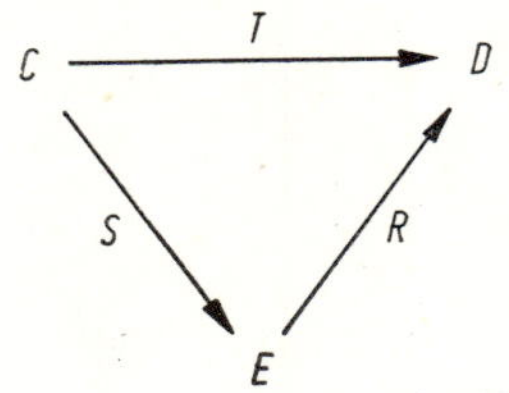

consisting of functors T, S, and R, is said to **commute** in case there is a functor equivalence $T \approx RS$. For an additive faithful functor $T : C \rightsquigarrow \text{SETS}$ there is a functor $T_{\text{mod-}\mathbb{Z}} : C \rightsquigarrow \text{mod-}\mathbb{Z}$ such that

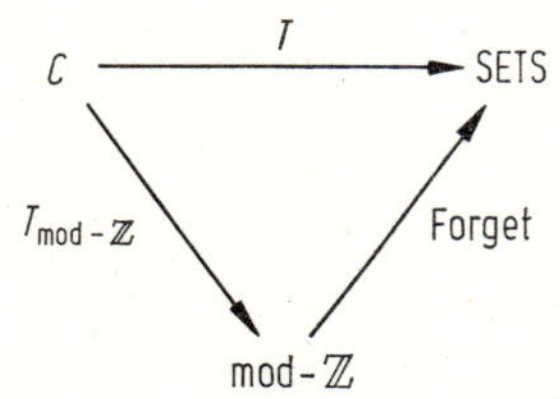

commutes, that is, T *factors through* mod-$\mathbb{Z}$. The next proposition establishes that a faithful functor $T : C \rightsquigarrow D$ reflects commutative diagrams and exact sequences when C and D are categories which are exact in the sense of 5.24. Thus, for these categories, the existence of a faithful, exact functor $T : C \rightsquigarrow D$ implies that a diagram or sequence S is commutative or exact in C if and only if the same is true of the corresponding diagram or sequence TS in D.

If U is a projective generator of an additive category C, then h^U induces an exact, faithful functor $C \rightsquigarrow \text{mod-}\mathbb{Z}$. Dually, if U is an injective cogenerator, then h_U induces a contravariant exact embedding $C \rightsquigarrow \text{mod-}\mathbb{Z}$. (Cf. the Exact Embedding Theorem 14.21.)

A **noncommutative** diagram is one containing

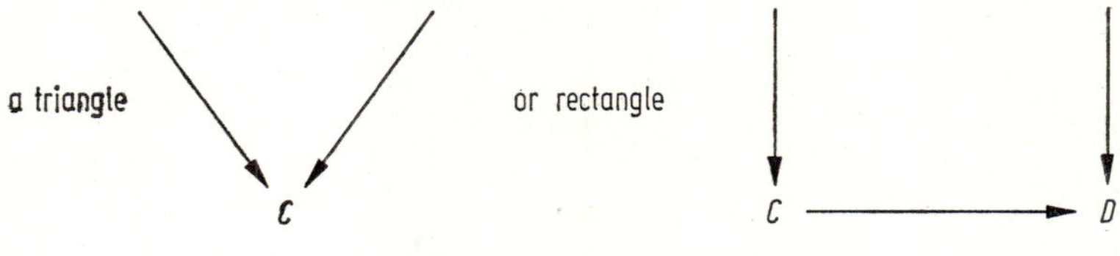

that is not commutative. **Inexact** sequences are sequences that are not exact.

3.49 Proposition. *Let C and D be categories. The following are also true for a contravariant functor T with the obvious modification on epics and monics.*

3.49.1 *A faithful functor $T: C \rightsquigarrow D$ preserves nonzero objects, nonzero morphisms, and noncommutative diagrams, that is, reflects zero objects, monics, epics, and commutative diagrams.*

3.49.2 *Let $C = \text{mod-}R$ and $D = \text{mod-}S$, for two rings R and S (or more generally, C and D may be categories which are exact in the sense of 5.24). Then a faithful functor $T: C \rightsquigarrow D$ reflects exact sequences. Thus, a faithfully exact functor $T: C \rightsquigarrow D$ reflects and preserves exact sequences.*

Proof of 3.49.1. First assume that C and D are categories with zero. A zero object is characterized by its identity morphism being the zero morphism. Thus, if $A \in \text{obj } C$, and if $T(A) \neq 0$, then $T(1_A) \neq T(0_A)$, and the fidelity of T implies $1_A \neq 0_A$.

The rest does not need the assumption of zero objects. If f, g, and h are morphisms and $f = gh$, then T preserves composites, so $T(f) = T(g)T(h)$. Thus, T preserves commutative diagrams, hence reflects noncommutative ones. If $T(f)T(h) = T(f)T(g)$, and if $T(f)$ is monic, then $T(h) = T(g)$, and the fidelity of T implies $h = g$. Thus, f is monic, so T reflects monics, dually for epics and, hence, for equivalences. The proposition for contravariant functors also holds because $T^*: D^* \to C^*$ is covariant, and all the properties considered (except that epics $\to$ monics) are preserved under $C \rightsquigarrow C^*$, where C^* is the dual category. ▯

Proof of 3.49.2 If $A \xrightarrow{f} B \neq 0$, then $A \xrightarrow{1} A \xrightarrow{f} B$ is not exact, $TA \xrightarrow{1} TA \xrightarrow{T(f)} TB$ is not exact, thus $T(f) \neq 0$. Conversely, suppose

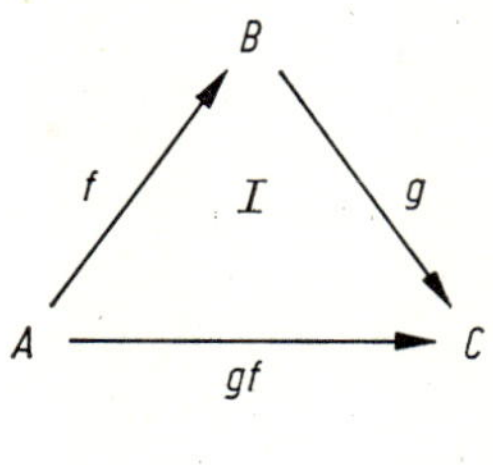

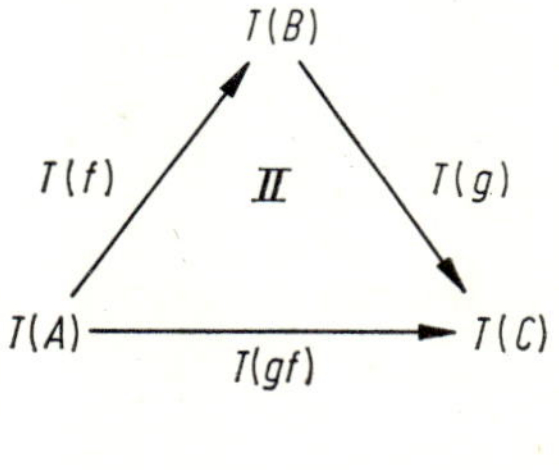

the diagram II is exact at $T(B)$. Let $K \xrightarrow{k} B$ be a kernel of g. Then $gk = 0$, and $T(gk) = T(g)T(k) = 0$. Thus, there is a morphism $T(\ker g) \to \ker(Tg)$ making the triangle I′ commutative.

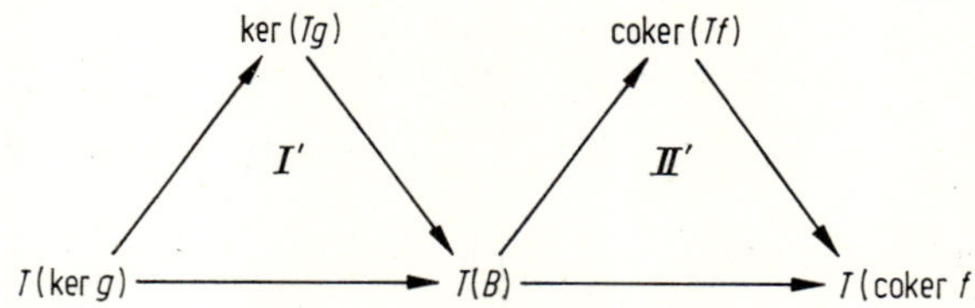

The triangle II′ is commutative by duality. Since II is exact at $T(B)$, then

$$\big(\ker(Tg) \to T(B) \to \operatorname{coker}(Tf)\big) = 0$$

so

$$\big(T(\ker g) \to T(B) \to T(\operatorname{coker} f)\big) = 0.$$

Since T is faithful, we have $\ker g \to B \to \operatorname{coker} f = 0$, which implies I is exact at B. ▯

3.50 Lemma (Eckmann-Schopf [53]). *Let R be a ring, and let G be a divisible abelian group. Then* $\operatorname{Hom}_{\mathbb{Z}}(R, G)$ *is an injective right R-module.*

Proof. Let $E = \operatorname{Hom}(R, G)$. First, if $f \in E$, $x \in R$, and $r \in R$, then $fr(x) = f(rx)$. Consider the situation

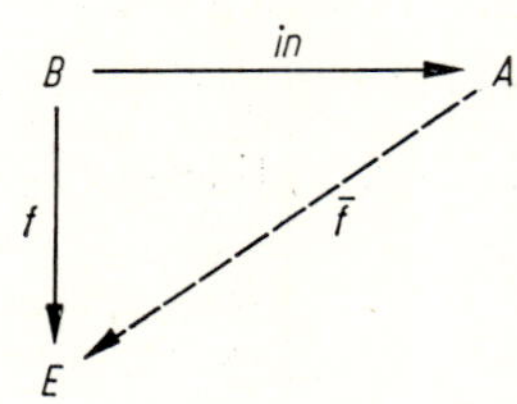

where B is a submodule of an R-module A. If $h \in E$, and $x \in R$, then $h[x]$ denotes the image of x under h. Also let π denote the $\mathbb{Z}$-homomorphism $E \to G$, that sends $h \in E$ into $h[1] \in G$. Then $\psi = \pi \circ f$ is a $\mathbb{Z}$-homomorphism of B into G. Since G is an injective $\mathbb{Z}$-module by 3.15, there is a $\mathbb{Z}$-map $\bar\psi : A \to G$ extending ψ. If $a \in A$, a_s denotes the left multiplication of R by a. Then $\bar\psi \circ a_s$ is a $\mathbb{Z}$-map of R into G; that is, $\bar\psi \circ a_s \in E \;\forall\, a \in A$. Next consider a mapping $\bar f : A \to E$, where $\bar f(a) = \bar\psi \circ a_s$, $\forall\, a \in A$. If $r \in R$, and $x \in R$, then

$$\begin{aligned}\bar f(ar)[x] &= \bar\psi \circ (ar)_s[x] = \bar\psi(ar \cdot x)\\ &= \bar\psi(a \cdot rx) = \bar\psi \circ a_s[rx] = \bar f(a)r[x],\end{aligned}$$

that is, $\bar{f}(ar) = \bar{f}(a)r \quad \forall\, r \in R, \quad a \in A,$ so $\bar{f}$ is an R-homomorphism $A \to E$. Since $\bar{\psi}$ induces ψ, it is clear that $\bar{f}$ induces f; that is, $\bar{f}$ completes the preceding diagram. □

3.51 *Exercises.*

3.51.1 If P is a projective left R-module and $G_{\mathbb{Z}}$ is divisible, show that $\mathrm{Hom}_{\mathbb{Z}}(P, G)$ is an injective right R-module. [Hint: If $P = R^{(I)}$, then $E = \mathrm{Hom}_{\mathbb{Z}}(P, G)$ is a left R-module isomorphic to F^I (direct product), where $F = \mathrm{Hom}_{\mathbb{Z}}(R, G)$]. In the general case, use a left R-module isomorphism

$$\mathrm{Hom}_{\mathbb{Z}}(P \oplus Q, G) \approx \mathrm{Hom}_{\mathbb{Z}}(P, G) \oplus \mathrm{Hom}_{\mathbb{Z}}(Q, G).$$

3.51.2 For rings A and B, let M be any (B,A)-bimodule. If M is a projective left B-module, then $\mathrm{Hom}_A(M,U)$ is an injective right B-module, for every injective right A-module U. Conclude that the functor mod-$A \rightsquigarrow$ mod-B induced by h^M preserves injective objects. Cf. Exercise 1.

The next two exercises require the notion of a left adjoint functor to the covariant morphisms functor. If M is a (B,A)-bimodule, the left adjoint to the functor h^M: mod-$A \rightsquigarrow$ mod-B is denoted by the symbol $\otimes_B M$ (cf. tensor products in Chapters 5 and 11). Thus, for any right B-module X, there are natural isomorphisms

$$\mathrm{Hom}_B(X, \mathrm{Hom}_A(M, U)) \approx \mathrm{Hom}_A(X \otimes_B M, U),$$

*3.51.3 In Exercise 3.51.2, show that the group valued contravariant morphism functor defined by the right B-module $\mathrm{Hom}_A(M,U)$ is naturally equivalent to the compose of exact functors $\otimes_B M$ and $\mathrm{Hom}_A(\ , U)$ and, therefore, is exact. This implies injectivity of $\mathrm{Hom}_A(M, U)$ in mod-B that is, h^M preserves injectives. Cf. 6.28.

3.51.4 Apply Exercise 3.51.3 to the result of 3.50, that is, for any object X of R-mod,

$$\mathrm{Hom}_R(X, \mathrm{Hom}_{\mathbb{Z}}(R, \mathbb{Q}/\mathbb{Z})) \approx \mathrm{Hom}(X \otimes_R R, \mathbb{Q}/\mathbb{Z}) \approx \mathrm{Hom}(X, \mathbb{Q}/\mathbb{Z}),$$

which shows that the contravariant morphism functor defined by $\mathrm{Hom}(R, \mathbb{Q}/\mathbb{Z})$ is naturally isomorphic to that induced by $\mathbb{Q}/\mathbb{Z}$, and implies that $\mathrm{Hom}(R, \mathbb{Q}/\mathbb{Z})$ is a cogenerator of mod-R. (Cf. 3.53.)

Character Modules

By 3.47, the group valued additive functor

$$h_{\mathbb{Q}/\mathbb{Z}} \begin{cases} \text{mod-}\mathbb{Z} \to \text{mod-}\mathbb{Z} \\ \quad A \mapsto A' = \operatorname{Hom}_{\mathbb{Z}}(A, \mathbb{Q}/\mathbb{Z}) & \text{(on objects)} \\ \quad f \mapsto f' = h_{\mathbb{Q}/\mathbb{Z}}(f) & \text{(on morphisms)} \end{cases}$$

is faithful and exact. For each abelian group A, the group A' is the **character group**, and $A'' = (A')'$ is the **bicharacter group.** Similar designations hold for the character group M' of a left R-module, for then M' is a right R-module canonically (3.7.1), called the **character module** of M.

3.52 Proposition. *The canonical mapping*

$$\pi_A \begin{cases} A \to A'' \\ a \mapsto \langle\ , a\rangle, \end{cases}$$

where

$$\langle\ , a\rangle \begin{cases} A' \to \mathbb{Q}/\mathbb{Z} \\ f \mapsto \langle f, a\rangle = f(a) \end{cases}$$

is injective. If A is a left R-module, then A'' is canonically a left R-module, and $\pi_A : A \to A''$ is a monic of R-mod.

Proof. The following three statements are equivalent:

1. $\pi_A : A \to A''$ is monic (injective).
2. For each $a \neq 0$ in A, there exists $f \neq 0$ in A' such that

   $\langle f, a\rangle = f(a) \neq 0.$

3. The canonical morphism (the dual of 3.25) $A \to \mathbb{Q}/\mathbb{Z}^{A'}$ is monic (injective).

Since $\mathbb{Q}/\mathbb{Z}$ is an injective cogenerator in mod-$\mathbb{Z}$ by 3.47, then 3. holds by the dual of 3.25, so 1. holds. If A is a left R-module, then the formula $fr(a) = f(ra)$ which holds $\forall\, a \in A$, $f \in A'$, $r \in R$ shows that $\pi_A : A \to A''$ is monic in R-mod. □

3.53 Theorem. *For any ring R, the category* mod-R *has an injective cogenerator, for example the character module of R.*

Proof. By 3.50, the character module R' of the left R module R is injective in mod-R, and by 3.31, R' is a cogenerator if and only if every object M of mod-R embeds in a product R'^I. To this end, let $R^{(I)} \to M' \to 0$

be exact in R-mod. Then $0 \to M'' \to R^{(I)\prime}$ is exact in mod-R by 3.49. Since $0 \to M \to M''$ is canonically exact by 3.52, and since $h_{\mathbb{Q}/\mathbb{Z}}$ converts coproducts into products, then $R^{(I)\prime} \approx R'^{I}$, and then there is an exact sequence $0 \to M \to R'^{I}$, which is what was to be proved. □

3.54 Corollary (Baer [40]). *For any ring R, the category* mod-*R* ***has enough injectives****, that is, for each module M of* mod-*R, there is a monic* $M \to E$ *of M into an injective module E.* □

Smallest Cogenerator

If C is a cogenerator of a category D, and if $C \to M$ is monic in D, then M is a cogenerator. A cogenerator C of D is said to be **smallest** provided that for any cogenerator M of D there exists a monic $C \to M$.

3.55 Proposition. *Let R be a ring, and let* $\{S_i \mid i \in I\}$ *be an isomorphy class of all simple R-modules. For each* $i \in I$, *let* E_i *denote the injective hull of* S_i. *Then*

3.55.1 $C = \coprod_{i\in I} E_i$ *is a smallest cogenerator in* mod-*R*.

3.55.2 *Every smallest cogenerator is isomorphic to C.*

3.55.3 *Every monic endomorphism of C is an automorphism.*

Proof of 3.55.1 and 3.55.2 Let M be a cogenerator, let $f_i : E_i \to M$ be the monics given by the dual of 3.26, and let

$$f : \coprod_{i\in I} E_i \to M$$

be the coproduct map. Since the $\{S_i\}$ are nonisomorphic, the submodules $\{f(S_i)\}$ are independent, and hence so are the submodules $\{f(E_i)\}$. This implies that f is monic as follows: If $x_i \in E_i$, and if $\sum_i x_i$ is a finite sum in $C = \coprod_{i\in I} E_i$, then $f(\sum_i x_i) = 0$ implies

$$f(\textstyle\sum_i x_i) = \sum_i f(x_i) = 0,$$

so $f(x_i) = 0$, whence $x_i = 0 \;\forall\, i \in I$. This proves that C is a smallest cogenerator.

Proof of 3.53.3 Let $h : C \to C$ be a monic. First, S_i is the unique simple submodule of E_i. We will show that h maps E_i onto E_i for every i. Suppose, for example, $h(E_1) \not\subseteq E_1$, then $h(E_1)$ has nonzero intersection with $B = \sum_{i\neq 1} \oplus E_i$. Since S_i is an essential submodule of E_i, $S = \sum_{j\neq 1} S_j$ is an essential submodule of B, so $h(E_1)$ must intersect S, hence contains some S_j, $j \neq 1$. Then E_1 contains a copy $h^{-1}S_j$ of a simple module not equal to S_1. This contradiction shows that h induces a monic endomorphism h_i of E_i for all i. The image of h_i is injective, and E_i is indecomposable, thus $\operatorname{im} h_i = E_i$, and h_i is therefore an automorphism for each i. But then h is an automorphism. ▯

3.56 Corollary. *An object E of* mod-R *is injective if and only if every monic* $E \to F$ *in* mod-R *splits.*

Proof. If $f: E \to F$ is monic in any category, and E is injective, then $h_E f: h_E(F) \to h_E(E)$ is surjective, so there exists $g: F \to E$ such that $h_E(f) g = gf = 1_E$ so that f is a coretraction. Conversely, if $f: E \to F$ is the embedding an object in an injective, then the splitting of f implies that E is injective, because if $gf = 1_E$ for some $g: F \to E$, then for a monic $h: M \to N$, and a morphism $k: M \to E$ there exists a commutative outer triangle

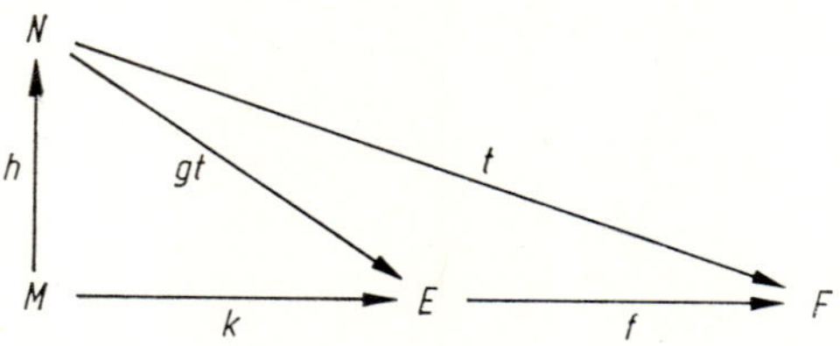

where $t: N \to F$ is a consequence of the injectivity of F. ▯

3.57 Corollary. *An object P of* mod-R *is projective if and only if every epimorphism* $P \to X$ *in* mod-R *splits.* ▯

Essential Extensions

A submodule N of a module M is **essential** in case $A \cap N \neq 0$ for every submodule $A \neq 0$. In this case M is also called an **essential extension**[1] of N. It is the same thing to require that every submodule (a) generated by an element $a \in M$ should meet N, that is, for $(a) \cap N \neq 0$, or that for each $a \in M$, there exists $r \in R$ such that $ar \neq 0$ and $ar \in N$.

For R-modules $M \supseteq N$, $(M \nabla N)_R$, or simply $M \nabla N$, denotes that N is an essential submodule of M. Trivially, if $M \nabla N$ and $N \nabla P$, then $M \nabla P$.

A map $f: A \to B$ is **essential** in case $\operatorname{im} f$ is an essential submodule.

Note that if $f: A \to B$ is a map, and if $B \nabla C$, then $A \nabla f^{-1}C$.

[1] There are those who think that then the extension should be called inessential.

3.58A Lemma. *Any row- and column-exact diagram*

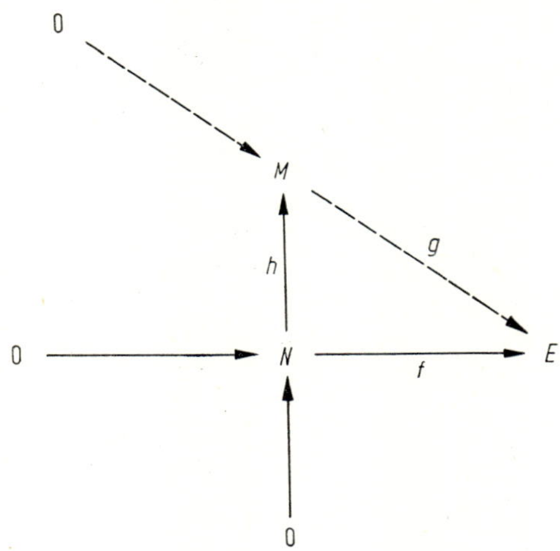

where E is injective and $N \to M$ is essential, can be completed by a diagonal exact map. Expressed otherwise, any embedding into an injective module can be extended to an embedding of any essential extension.

Proof. Injectivity of E yields the map g such that $gh = f$. Since f is monic, $\ker g \cap \operatorname{im} h = 0$, and $M \nabla \operatorname{im} h$ implies $\ker g = 0$. □

3.58B Proposition. *A module M is injective if and only if every essential monic $f: M \to A$ is an isomorphism.*

Proof. One way is given by 3.56. Conversely, let $h: M \to E$ be monic, where E is injective. Since $M \approx \operatorname{im} h$, $\operatorname{im} h$ inherits the hypothesis on M. Therefore, suppose $M = \operatorname{im} h \subseteq E$. Let S be a submodule of E given by Zorn's lemma, which is maximal in the set of those submodules K such that $K \cap M = 0$. Let $f: E \to E/S$ be canonical, and let $g: M \to E/S$ be the induced map. Since $E/S \nabla M$ and g is monic, g is an essential monic, so g is an isomorphism by hypothesis. Since $\operatorname{im} g = (M + S)/S$, this yields $E = M + S$, and $E = M \oplus S$. A summand of an injective module is injective. □

An **injective hull** $\hat{M}$ of an R-module M is defined to be a initial object in the full subcategory of source M consisting of all monics $M \to E$ such that E is an injective object of mod-R. Thus, a morphism in this category between $M \to E$ and $M \to E'$ is morphism $E \to E'$ of

mod-R such that

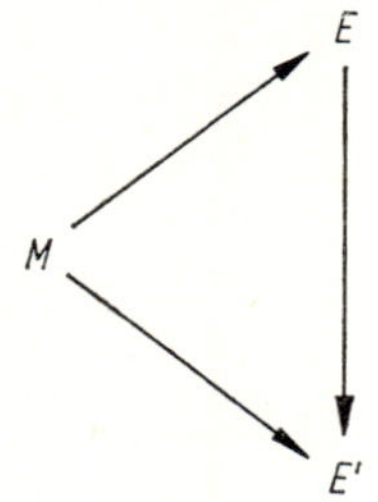

commutes, and then $E \to E'$ is necessarily monic.

A **maximal essential extension of** M is defined to be a final object in the full subcategory of source M consisting of all essential monics $M \to F$.

3.59 Theorem (Baer [40] and Eckmann-Schopf [53]). *If M is an object of* mod-R, *then the injective hull of M exists in* mod-R, *a maximal essential extension exists, and the injective hull and the maximal essential extension coincide*[1]. *Thus, $M \to E$ is an injective hull of M if and only if $M \to E$ is an maximal essential extension. Moreover, any embedding $M \to F$, where F is injective, extends to an embedding $E \to F$ of the injective hull E of M.*

Proof. Let F be an injective module, given by 3.54, which contains M. By Zorn's lemma, there is a maximal submodule among all submodules E of F such that $E \nabla M$. If $f : E \to A$ is an essential monic, then by 3.57 the monic $f^{-1} : B \to F$, where $B = \operatorname{im} f$, can be extended to a monic $g : A \to F$. Then, M is an essential submodule of $\operatorname{im} g$, and $M \subseteq E \subseteq \operatorname{im} g$, so $E = \operatorname{im} g$ by maximality of E. Thus, $B = A$ and $f^{-1} = g$, so that every essential monic $E \to A$ is an isomorphism, and E is therefore injective by the proposition. $\square$

The last sentence in the theorem follows from 3.58A, and has the following implication: *Any injective module F containing a module M contains an injective hull of M.*

Modules Isomorphic to Submodules of Each Other

Two modules A and B may be isomorphic to submodules of each other, and yet not be isomorphic. For example, let A be an entire ring that is not a principal ideal ring, and let B be a right ideal that is not principal. If x is any nonzero element of B, then the right A-module A

[1] See Notes for Chapter 3 (p. 183).

is isomorphic to the right ideal $C = xA$, and yet B is not isomorphic to A since B is not cyclic. In another example, let B denote the direct sum of infinitely many copies of the cyclic group $\mathbb{Z}_4$ of order 4, and let $A = \mathbb{Z}_2 \oplus B$. Then A and B are isomorphic to subgroups of each other, and yet A and B are not isomorphic since B has no summand isomorphic to $\mathbb{Z}_2$. Nevertheless, in each example A and B have isomorphic injective hulls. This is the substance of the next theorem and corollary. The proof is patterned after and, in fact, implies the corresponding result for sets. (Foreword Theorem 3.)

3.60 Proposition[1]. *If A and B are injective modules, and if there exist monics $A \to B$ and $B \to A$, then $A \approx B$.*

Proof. It suffices to assume that B is a submodule of A and that there is a monic $f: A \to B$. Since B is injective, B is a summand of A, so that $A = H \oplus B$. Let

$$P = H \oplus f(H) \oplus f(f(H)) \oplus \cdots.$$

Since

$$A = H \oplus B \supseteq H \oplus f(A) \supseteq H \oplus f(H) \oplus f(B) \supseteq \cdots,$$

then

$$P \cap B = f(H) \oplus f(f(H)) \oplus \cdots,$$

that is, $P \cap B = f(P)$.

Next, let Q be a maximal essential extension of $P \cap B$ in B. Since B is injective, Q is injective, so $B = Q \oplus K$ for some submodule K. Then

$$A = H \oplus B = H \oplus (Q \oplus K) = (H \oplus Q) \oplus K,$$

so that $H \oplus Q$ is injective too. The definition of P shows that $P \subseteq H + Q$, and the identity $P \cap B = f(P)$ shows that f induces a map $g: P \to Q$, which by injectivity of Q can be extended to a map $g': H \oplus Q \to Q$. The definition of Q implies that g' is an isomorphism, and then the direct sum map

$$g' \oplus 1_K : (H + Q) \oplus K \to Q \oplus K$$

is the required isomorphism $A \to B$. ▯

---

[1] The question of the validity of this theorem arose in a seminar on injective modules and quotient rings which I gave at Rutgers 1962–1963. I received two independent solutions the following week from R. Bumby and B. Osofsky. The proof here is that of Bumby [65].

Assorted Ideas

There are many, many related ideas which ought to be introduced into any study of injective modules at this point. Notably, indecomposable (injective) modules and their endomorphism rings, direct sums of injective modules (when are they injective?), fully invariant submodules, quasi-injective, or as I write, quasinjective modules, and such notions as the dual of the injective hull (called the projective cover), or the dual of a finitely generated module, and many other associated ideas. But, having covered some sixty propositions in this chapter already, we have been forced to content ourselves with a number of exercises, below and end-of-chapter, which state some of the main results I am alluding to. A fuller account of these will be given in Volume II, especially Chapters 17—22.

Briefly, a module M is **indecomposable** iff 0 and M are the only summands. Moreover, a module U is **uniform** if every submodule of U is indecomposable. A **fully invariant** submodule of a module is a submodule which every endomorphism maps into itself. A module is **quasinjective** if every homomorphism of every submodule into the module is induced by an endomorphism.

To continue, a ring R is said to be **local** if the set of non-units of R is an ideal. (One can show that R is local iff $R/\text{rad}\, R$ is a field 18.10.) Local rings are plentiful. For any commutative domain R, there is an associated quotient field K characterized up to isomorphism by the property that K is a field containing R such that $K = \{ab^{-1} \mid a, 0 \neq b \in R\}$. Then, for any prime ideal P of R, the subring RS^{-1} of K consisting of all elements as^{-1}, where $a \in R$, and s belongs to the complement $S = R - P$ of P in R, is a local ring, called **the local ring of R at the prime P**. [The construction of the quotient field K as the set of all couples (a, b), $a, 0 \neq b \in R$, with the obvious definitions for addition and multiplication proceeds analogously with the construction of $\mathbb{Q}$ out of $\mathbb{Z}$. Moreover, along similar lines, one constructs the quotient ring K of *any* commutative ring R, characterized up to isomorphism by the two properties: K is a ring containing R such that every regular element of R is a unit of K, and $K = \{ab^{-1} \mid a, b \text{ regular} \in R\}$. (See Chapter 27.) Analogously, for each prime ideal P, there exists a local ring RS^{-1}. (See Chapter 27.)]

3.61 *Exercises.*

3.61.1 A module is uniform iff it is an essential extension of every nonzero submodule.

(a) Show that M is uniform if and only if the injective hull M is indecomposable.

*(b) Show that the endomorphism ring of an indecomposable injective module M is a local ring.

(c) (Fitting's Lemma.) The endomorphism ring of any indecomposable module which is both Artinian and Noetherian is a local ring.

3.61.2 A ring A is right Noetherian if and only if every direct sum of injectives in mod-A is injective. In this case, every injective is a direct sum of indecomposable modules. (Cf. 3.48.3.)

3.61.3 If M is injective in mod-A, then every direct sum $M^{(I)}$ is injective if and only if A satisfies the a.c.c. on right ideals of the form $I = \operatorname{ann}_A X$ for some subset X of M. Then M is **$\sum$-injective.** Every $\sum$-injective module is a direct sum of indecomposable modules. Every module is $\sum$-injective if and only if every direct sum of injective modules is injective if and only if A is right Noetherian. (Cf. 3.48.2—3.)

3.61.4 Dual to the notion of essential submodule is that of a **superfluous** submodule, namely a submodule N of M such that $N + A = M \Rightarrow A = M$ for every submodule A. Then we write: $M \bigcirc N$. (This concept will be discussed again in Chapters 18 and 22.) An epimorphism $f: P \to M$ is **minimal** in case $\ker f$ is superfluous. Minimal epics are thus dual to essential monics. A **projective cover** of a module M is defined to be a minimal epic $f: P \to M$, where P is projective. When M has a projective cover it is unique in the sense dual to the uniqueness of an injective hull; and yet, unlike injective hulls, projective covers do not always exist even (especially) over $\mathbb{Z}$. (Prove this.)

*3.61.5 If A is an R-module show that the class Inj A of injective modules containing A can be ordered. (If $E, F \in \text{Inj } A$, write $E \leq F$ in case there exists a monic $f: E \to F$ that induces the identity map on A.)

*3.61.6 (a) A module M is quasinjective if and only if M is a fully invariant submodule of $E = \text{inj. hull } M$, that is, if and only if $\lambda M \subseteq M \;\forall\, \lambda \in \operatorname{End}_R E$ (Wong-Johnson [61]).

(b) For every module M there is a unique minimal essential monic $M \to Q$, where Q is quasinjective (Wong-Johnson [61]). Every minimal quasinjective extension of M is essential (Faith-Utumi [64]).

(c) Extend Theorem 3.60 to quasinjectives.

(d) If $M = A \oplus B$ is quasinjective, so are A and B.

The direct sum $R \oplus M$ is quasinjective if and only if both R and M are injective in mod-R.

(e) A direct sum of two quasinjective modules need not be quasinjective.

(f) If M is quasinjective, so is M^n for any integer $n > 0$.

(g) Let M be quasinjective, and let $S = \operatorname{End} M_R$. Then ${}^{\perp}(N^{\perp}) = N$ for each finitely generated S-submodule N (Artin-Tate (Artin [50]), Jacobson [55, p. 27] and Wong-Johnson [64]).

Idealizers

If S is a ring, and if A is a subring, then we define the **idealizer of A in S**:

$$\text{idealizer}_S\, A = \{s \in S \mid sA \subseteq A \text{ and } As \subseteq A\}.$$

This is a subring of S containing A, and A is an ideal in it. In fact, $\text{idealizer}_S\, A$ is the largest subring of S containing A as an ideal.

3.62 Proposition. *Let P be any module over a ring R, and let $S = \operatorname{End} P_R$.*

3.62.1 *If Q is a submodule of P, then the set*

$$(P:Q) = \{f \in S \mid f(P) \subseteq Q\}$$

is a subring of S, and the set

$$(Q:Q) = \{f \in S \mid f(Q) \subseteq Q\}$$

is contained in the idealizer of $(P:Q)$ in S. Furthermore, there is an exact sequence

$$0 \to (P:Q) \to (Q:Q) \xrightarrow{\varphi} \operatorname{End}\,(P/Q)_R,$$

where φ is a ring homomorphism such that if $f \in (Q:Q)$, then

$$\varphi(f): x + Q \mapsto f(x) + Q,$$

that is, $\varphi(f)(x+Q) = f(x) + Q$, where $x \in P$.

3.62.2 *If P is projective, then φ is epic, and so*

$$(Q:Q)/(P:Q) \approx \operatorname{End}\,(P/Q)_R.$$

3.62.3 *If $P = R$, then R/Q is a cyclic right R-module, and*

$$0 \to Q \to \text{idealizer}\, Q \xrightarrow{\psi} \operatorname{End} R/Q_R \to 0$$

is exact, where for each $a \in \text{idealizer}\, Q$,

$$\psi(a): x + Q \mapsto ax + Q.$$

In summary:

$$\frac{\text{idealizer } Q}{Q} \approx \text{End } (R/Q)_R.$$

Proof of 3.62.1. If $f \in (Q:Q)$, if $x, y \in P$, and if $x+Q = y+Q$, then $x-y \in Q$ and $f(x-y) = f(x)-f(y) \in Q$, so $f(x)+Q = f(y)+Q$. Thus, $x+Q \mapsto f(x)+Q$ is a mapping $P/Q \to P/Q$, which we denote by $\varphi(f)$. Clearly, $\varphi(f) \in \text{End } (P/Q)_R$, and $f \to \varphi(f)$ defines a ring homomorphism $\varphi : (Q:Q) \to \text{End } (P/Q)_R$ with $\ker \varphi = (P:Q)$. This establishes that the sequence defined in (1) is exact and shows that $(P:Q)$ is an ideal of $(Q:Q)$. Obviously, then, $(Q:Q) \subseteq \text{idealizer}_S (P:Q)$ as claimed.

Proof of 3.62.2. If P is projective, and if $k \in \text{End } (P/Q)_R$, then the diagram

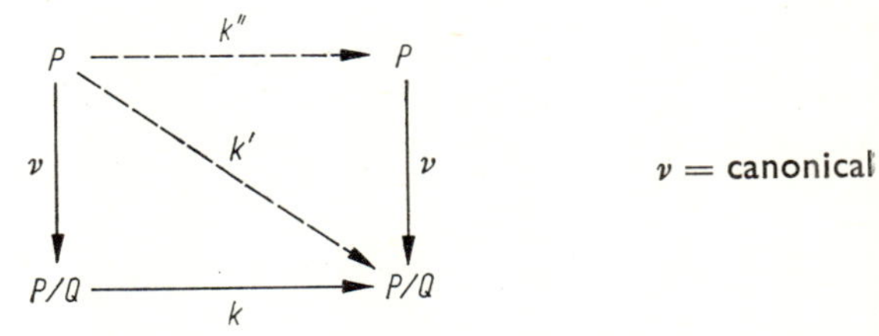

can be completed (first by k'' by projectivity of P, and then by k'). If $x \in Q$, then $v(x) = 0$, and $kv(x) = vk''(x) = 0$, hence $k''(x) \in \ker v = Q$. Thus, $k'' \in (Q:Q)$. Clearly, $\varphi(k'') = k$, so φ is epic.

Proof of 3.62.3. This is the special case of (2) when $P = R$. In this case, $S = \text{End } R_R$ is isomorphic to R under $f \mapsto f(1)$. This mapping is certainly injective, since $f(1) = g(1)$ implies $f(r) = f(1)r = g(1)r = g(r)$ $\forall r \in R$, whence $f = g$. It is surjective, since if $a \in R$, then there is a map $h \in \text{End } R_R$ defined by $h(r) = ar$ $\forall r \in R$, and clearly $h(1) = a$. Hence, if we replace S by R in (1) and (2), we obtain what we want because $(Q:Q) = \text{idealizer } Q$. □

Sum-1 Orthogonal Idempotents

The elements of a set $\{e_i \mid i \in I\}$ of idempotents in a ring S are orthogonal in case $e_i e_j = \delta_{ij} e_i$ for all $i, j \in I$. If $I = \{1, \ldots, n\}$ is a finite set and

$$1 = \sum_{i=1}^{n} e_i,$$

then the idempotents are called **sum-1 orthogonal idempotents.** A **central** element of a ring is an element in the center. Thus, a **central idempotent** is an idempotent that commutes with every element in the ring.

3.63 Proposition. *Let M be an R-module, and let* $S = \operatorname{End} M_R$.

3.63.1 *If* $\{e_i \mid i = 1, \ldots, n\}$ *are* sum-1 *orthogonal idempotents of* S, *then* $M = \sum_{i=1}^{n} \oplus e_i M$, *and* $\ker e_j = \sum_{i \neq j} e_i M$.

3.63.2 *If* $M = \sum_{i=1}^{n} \oplus M_i$, *then there exist* sum-1 *orthogonal idempotents* $e_1, \ldots, e_n \in S$ *such that* $e_i M = M_i$, *and* $\ker e_j = \sum_{i \neq j} M_i$.

3.63.3 *A submodule N of M is a summand if and only if* $N = eM$ *for some idempotent* $e \in S$. *Each element* $t \in eSe$ *induces an endomorphism* $t' \in \operatorname{End}_R eM$, *where* $t'(x) = tx \; \forall\, x \in eM$. *The mapping* $\varphi : t \mapsto t'$ *is a ring isomorphism* $eSe \to \operatorname{End}_R eM$.

Proof of 3.63.3. The mapping φ is surjective because N is a summand and is injective because

$$eSe \cap \ker \varphi \subseteq Se \cap S(1-e) = 0.$$

The proof of 3.63.1 follows by induction on n: Let $e = e_1 + \cdots + e_{n-1}$. Then $e_n = 1 - e$, and $M = eM \oplus e_n M$ by the $n = 2$ case. Since e is a sum of $n-1$ orthogonal idempotents, and since e lies in eSe, which is isomorphic to $\operatorname{End}_R eM$, the induction assumption shows that

$$\begin{aligned} eM &= e_1(eM) \oplus \cdots \oplus e_{n-1}(eM) \\ &= e_1 M \oplus \cdots \oplus e_{n-1} M, \end{aligned}$$

whence

$$M = e_1 M \oplus \cdots \oplus e_n M$$

as claimed. The $n = 2$ case is trivial. (Cf. 3.34.)

Proof of **3.63.2.** $M = M_1 \oplus M_2$, then 1_{M_1} can be extended to an element $e \in S$ such that $\ker e = M_2$. Then $M_1 = eM$ and $M_2 = (1-e)M$. This completes the $n = 2$ case, and, as in (2), an obvious induction completes the proof. □

3.64 Corollary. *If $e_1, \ldots, e_n$ are orthogonal idempotents of R, then the following statements are equivalent:*

3.64.1 $e_1, \ldots, e_n$ *are sum-1 orthogonal idempotents;*

3.64.2 $R = \sum_{i=1}^{n} \oplus e_i R$;

3.64.3 $R = \sum_{i=1}^{n} \oplus R e_i$.

Proof. (1) ⇔ (2) follows from the $M = R$ case of the proposition, and by symmetry (1) ⇔ (3).

3.65 Proposition. *Let R be a ring.*

3.65.1 *If* $e_1, \ldots, e_n$ *are* sum-1 *orthogonal central idempotents of* R, *then* $S_i = e_i R = R e_i = e_i R e_i$ *is a ring and an ideal of* R, $i = 1, \ldots, n$, *and*

$$R = S_1 \oplus \cdots \oplus S_n.$$

3.65.2 *Let*

$$R_i \xrightarrow{p_i} \prod R_i \xrightarrow{\pi_i} R_j$$

be a representation of R as a direct product of finitely many rings $R_1, \ldots, R_n$. *Then* $S_i = p_i(R_i)$ *is a ring and an ideal of* R, $i = 1, \ldots, n$. *If* e_i *is the ring identity of* S_i, *then* $e_iR = Re_i = e_iRe_i = S_i$, $i = 1, \ldots, n$, $\{e_i \mid i = 1, \ldots, n\}$ *are sum-1 orthogonal central idempotents, and* $R = S_1 \oplus \cdots \oplus S_n$.

Proof of 3.65.1. By 3.64, $R = \sum_{i=1}^{n} \oplus S_i$, where $S_i = e_iR$. Since e_i is central, $e_iRe_i = e_iR = S_i$, so S_i is a ring with identity element e_i, $i = 1, \ldots, n$. Since $RS_i = Re_iR = e_iR = S_i$, the right ideal S_i is a left ideal, whence an ideal of R.

Proof of 3.65.2. Clearly, $S_i = p_i(R_i)$, $i = 1, \ldots, n$, are ideals of $R = \prod_{i=1}^{n} R_i$ such that $S_iS_j = S_jS_i = 0$ when $i \neq j$, and $R = \sum_{i=1}^{n} \oplus S_i$. Since R_i is a ring, so is $S_i = p_i(R_i)$; let e_i be the identity element of S_i, $i = 1, \ldots, n$. Then $S_iS_j = 0$ implies $e_ie_j = 0$ whenever $i \neq j$. Thus $\{e_i \mid i = 1, \ldots, n\}$ are sum-1 orthogonal idempotents. Since $S_i = e_iR$ is an ideal, then $Re_i \subseteq S_i$, so $Re_i = S_i$ by symmetry. Since e_i commutes with everything in S_i, and since e_i annihilates everything (on both left and right) in S_j when $j \neq i$, it follows that e_i commutes with everything in R; that is, e_i lies in the center of R, $i = 1, \ldots, n$. ∎

3.66 Proposition.

3.66.1 *Let* $R = S_1 \oplus \cdots \oplus S_n$ *be a direct product of finitely many simple rings. If* A *is an ideal of* R, *then* A *is a sum of a subset of* $\{0, S_1, \ldots, S_n\}$, *and there exists a central idempotent* e *such that* $A = eR = Re = eRe$.

3.66.2 *If also* $R = S_1' \oplus \cdots \oplus S_m'$, *where* S_i' *is a simple ring,* $i = 1, \ldots, m$, *then* $n = m$, *and there is a permutation* π *of* $\{1, \ldots, n\}$ *such that* $S_i' = S_{\pi(i)}$, $i = 1, \ldots, n$.

Proof of 3.66.1. For $n = 1$, 3.66.1 is trivial. Assume this to be true for $R' = S_1 \oplus \cdots \oplus S_{n-1}$. If I is an ideal of $R = R' \oplus S_n$, and if $x \in I$, $x = a + b$, where $a \in R'$, $b \in S_n$, then $ex = ea = a$, and $e_nx = e_nb = b$, where e is the identity of R' and e_n is the identity of S_n. Thus, $a \in I \cap R'$, and $b \in I \cap S_n$. This shows that $I = (I \cap R') \oplus (I \cap S_n)$. By induction, $I \cap R'$ is a sum of a subset of $\{0, S_1, \ldots, S_{n-1}\}$, whereas $I \cap S_n = 0$ or $I \cap S_n = S_n$. Thus, I has the desired form. If $I = S_{i_1} \oplus \cdots \oplus S_{i_r}$, and if e_{i_j} is the identity of S_{i_j}, then $e = e_{i_1} + \cdots + e_{i_r}$ is a central idempotent, and $I = eR = Re = eRe$.

Proof of 3.66.2. Follows from the structure of the ideals of 3.66.1 as soon as one notices that the ideals S_j' of R are minimal ideals of R. Thus, S_j' is a sum of some of the S_i only if $S_j' = S_k$ for some k. ∎

3.67 *Exercises.*

3.67.1 Let $E_1(A)$ denote the ordered (by inclusion) set of left ideals of a ring A of the form Af, for some idempotent $f \in A$, and let $E(A)$ denote the left-right symmetry. For any right R-module M, let $S = \operatorname{End} M_R$. The mapping $N \to \operatorname{ann}_S N$ is an order anti-isomorphism of the ordered set of summands of M with the ordered set $E_1(S)$. ($I \to \operatorname{ann}_M I$ is the inverse mapping.) Thus, M is indecomposable iff S is indecomposable.

(a) R is indecomposable if and only if the only idempotents of R are 0 and 1 (and then ${}_R R$ is indecomposable).

(b) Let M be any module, and let $S = \operatorname{End} M_R$. Let e be an idempotent of S. The following four statements are equivalent: (i) the submodule eM of M is indecomposable; (ii) Se is an indecomposable left ideal of S; (iii) eS is an indecomposable right ideal of S; (iv) e is not the sum $u+v$ of nonzero orthogonal idempotents u, v. (v) eSe is indecomposable.

3.67.2 Generalize 3.63 to additive categories.

3.67.3 The following conditions on a set $\{e_i\}_{i=1}^n$ of sum-1 orthogonal idempotents in a ring R are equivalent: (1) Every e_i is central, $i = 1, \ldots, n$. (2) $e_i R(1 - e_i) = 0$, $i = 1, \ldots, n$. (3) $e_i R e_j = 0$, $j \neq i = 1, \ldots, n$. (4) $e_i R$ is an ideal, $i = 1, \ldots, n$. (5) $e_i R = R e_i$, $i = 1, \ldots, n$. (6) $(e_i R)^\perp = R(1 - e_i)$, $i = 1, \ldots, n$.

3.67.4 When the conditions of 3.67.3 hold, then $e_i R = e_i R e_i$, $i = 1, \ldots, n$, and R is canonically isomorphic to the ring product $\prod_{i=1}^n e_i R e_i$ (cf. 18.30ff.).

Fully Invariant Submodules

An R-module M is an (S,R)-bimodule, where $S = \operatorname{End} M_R$. If N is a fully invariant submodule, then each endomorphism f of M induces an endomorphism f' of N, and the correspondence $f \to f'$ is a ring homomorphism $\varphi : \operatorname{End} M_R \to \operatorname{End} N_R$. If, in addition, N is a summand of M, then φ is an epimorphism.

3.68 Proposition. *Let M be an R-module, which is a (internal) direct sum $\sum_{i \in I} \oplus M_i$, where $\{M_i\}_{i \in I}$ is a family of fully invariant submodules of M_R. Let $S = \operatorname{End} M_R$, let $S_i = \operatorname{End} (M_i)_R$, and let $p_i : S \to S_i$ be the ring epimorphism defined by the restriction of an element of S to M_i for all $i \in I$. Then the ring product $S \to \prod_{i \in I} S_i$ of the family $\{p_i : S \to S_i\}$ of ring morphisms is a ring isomorphism.* □

3.69 *Examples* of fully invariant submodules.

3.69.1 (a) The submodule of M generated by all minimal submodules of M is called the **socle of** M, denoted soc M.

(a)* [the dual of (a)] The intersection of all maximal submodules of M is defined to be the **radical of** M, denoted rad M.

If M has no minimal submodules, then $\operatorname{soc} M = 0$, and dually, $\operatorname{rad} M = M$, when M has no maximal submodules. (For example, $\mathbb{Z}$ has no minimal subgroups, and $\mathbb{Q}/\mathbb{Z}$ has no maximal subgroups.

(b) The submodule $H(N)$ generated by all submodules isomorphic to a given minimal submodule N.

(c) If A is a (left) ideal of R, then ${}^{\perp}A = \operatorname{ann}_M A$ is a fully invariant submodule of M. Any ideal A of R is a fully invariant submodule of R.

3.69.2 If $\{N_i \mid i \in I\}$ is a family of fully invariant submodules of M_R, then $\sum_{i\in I} N_i$ and $\bigcap_{i\in I} N_i$ are fully invariant submodules. (The set of fully invariant submodules is a complete lattice.)

Exercises for Chapter 3

1. (Mazur-Eilenberg) If P is a projective module that is not finitely generated, then there exists a projective module Q such that $P \oplus Q \approx Q$.

2. If I and J are projective submodules of M, then $I + J$ is projective if and only if $I \oplus J \approx (I + J) \oplus (I \cap J)$.

3. Let I be an ideal of R. If P is a projective R-module, then P/PI is a projective R/I-module.

4. Let P be a projective R-module with $S = \operatorname{End} P_R$.

(a) An element $f \in S$ is an epic endomorphism if and only if f has a right inverse in S.

(b) Give an example to show that in (a) the element f may not have a left inverse.

(c) If P satisfies the a.c.c. on submodules, or if R is commutative, then in (a) the element f is a unit.

*(d) (Vasconcelos) If M is any finitely generated module over a commutative ring R, then any epic endomorphism f is an automorphism.

5. (R. Baer [40]) Let M be contained in an injective module E.

(a) Show that there exists submodule $I(M)$ containing M satisfying the condition: Every embedding of M into an injective can be extended to an embedding of $I(M)$.

(b) Prove that $I(M)$ is a maximal essential extension of M (resp. an injective hull of M).

6. The sum of all superfluous submodules of M is a submodule denoted by superfl M. An element $x \in M$ is a **nongenerator** in case the submodule it generates is superfluous (cf. 3.61.4).

(a) Show that superfl M is the totality of nongenerators of M.

(b) Show that superfl $\mathbb{Q} = \mathbb{Q}$. Generalize.

7. Any row-exact diagram

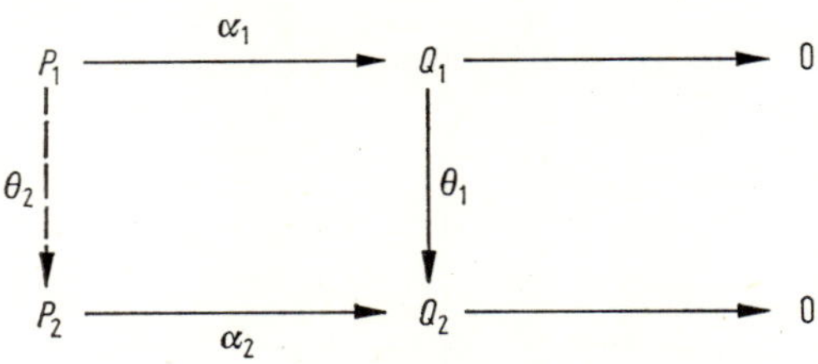

where P_1 and P_2 are projective, can be completed. Prove the following:

(a) If $P_2 \bigcirc \ker \alpha_2$, then θ_2 is an epimorphism whenever θ_1 is an epimorphism (cf. 3.61.4).

(b) If $P_2 \bigcirc \ker \alpha_2$ and $P_1 \bigcirc \ker \alpha_1$, then θ_2 is an isomorphism whenever θ_1 is an isomorphism.

8. Prove: *If P is projective, then* rad $P = P \cdot \operatorname{rad} R$. [Hint: Show that "rad" commutes with direct sums; that is, $\operatorname{rad}(\sum_i \oplus A_i) = \sum_i \oplus \operatorname{rad} A_i$.]

9. Let P_1 and P_2 be finitely generated projective right R-modules. Then $P_1/\operatorname{rad} P_1 \approx P_2/\operatorname{rad} P_2$ if and only if $P_1 \approx P_2$.

10. If P is a finitely generated projective right R-module with endomorphism ring S, then the factor ring $S/\operatorname{rad} S$ is isomorphic to $\operatorname{End}_R(P/P \cdot \operatorname{rad} R)$.

11. (**Unimodular row lemma**) Let F be a free right R-module with a free basis $x_1, \ldots, x_n$, and let $x = \sum_1^n x_i r_i \in F$, $r_i \in R, \forall i$. Let A denote the left ideal generated by $r_1, \ldots, r_n$. Prove the equivalence of the following:

(a) $A = Re$ for some idempotent $e \in R$.

(b) xR is a summand of F isomorphic to eR under a map that sends x onto e.

Conclude that if R is a **regular ring** (a ring in which every finitely generated left, or right, ideal is a summand of R), then every cyclic (resp. finitely generated) submodule of a free module is a summand.

***12.** (Kaplansky) Any projective module is a direct sum of countably generated modules.

13. Prove the equivalence of the following statements about a right ideal I:

(a) $I = eR$ for some idempotent $e \in R$.

(b) I is isomorphic to a summand of R.

(c) R/I is a projective right R-module.

14. Let I denote the set of right ideals of R of the form eR, where e is an idempotent of R. Show the equivalence of the following conditions:

(a) Every subset of I contains a maximal element.

(b) Every subset of I contains a minimal element.

(c) I does not contain an infinite independent subset.

(d) R does not contain an infinite set of orthogonal idempotents.

When this is so, show that R is a direct sum of (finitely many) indecomposable right ideals.

15. A category C of rings is **centrally closed** if for every object R of C it is true that center R is isomorphic to an object of C. Similarly, say that C is (finite) **Galois closed** if $\forall R \in C$, and (finite) groups G of automorphisms of R, it is true that fix $G \in C$. (Cf. 3.27 (b).) Discuss which categories of rings introduced in the text, here-to-fore and henceforth, are centrally or Galois closed. The full subcategory of RINGS consisting of centrally closed rings is productive.

***16.** Prove that an object E of mod-R is a cogenerator iff for any object X and subobject X_0 of X there holds $X_0 = \operatorname{ann}_X \operatorname{ann}_E X_0$. If R is a cogenerator then every right ideal is an annulet. Cf. Exercise 20.

17. Let A be a right ideal of a ring R, and let the annihilating left ideal ${}^{\perp}A$ be generated by elements $\{x_i \mid i \in I\}$.

(a) Show that the correspondence $h : a + A \mapsto (\ldots, x_i a, \ldots)$, defined for all $a \in R$, is a right R-module homomorphism $R/A \to R^I$.

(b) Prove in (a) that h is a monomorphism if and only if A is an annulet.

(c) Prove that R/A can be embedded in a direct product of copies of R if and only if A is an annulet.

(d) Prove that R/A can be embedded in a free R-module if and only if A is the annihilator of a finite subset of R.

(e) Conclude that if every right R-module is contained in a free R-module, then every right ideal of R is the annihilator of a finite subset of R.

18. Show that the following conditions on a ring R are equivalent:

(a) R satisfies the maximum condition on right annulets.

(b) R satisfies the a.c.c. on right annulets.

(c) For each right ideal A there corresponds a finitely generated right ideal $A_1 \subseteq A$ such that ${}^{\perp}A = {}^{\perp}A_1$.

19. Show that a ring R satisfies the d.c.c. on right ideals and the d.c.c. on left ideals provided that each right ideal is the annihilator of a finite subset of R and each left ideal is the annihilator of a finite subset of R. Conclude that a commutative ring R satisfies the d.c.c. on ideals in case every ideal is the annihilator of a finite subset of R. (In this case, R is selfinjective. Cf. Chapter 12 Exercises 6.)

20. (a) Prove the equivalence of the following two conditions. (i) M is torsionless. (ii) M is isomorphic to a submodule of a direct product of copies of R. Conclude that over a domain, any torsionless module is torsionfree.

(b) For any R-module M prove that M^* is torsionless.

(c) Let A be a right ideal of R. Prove that a cyclic module R/A is torsionless if and only if A is an annulet.

(d) Every right R-module is torsionsless iff R is a cogenerator of mod-R. (Cf. Exercise 16.)

21. Let R be a ring, and let A and B be right ideals.

(a) A cyclic module R/A can be injected monomorphically into a cyclic module R/B, where A and B are right ideals, if and only if there exists $x \in R$ such that $A = \text{ann}_R x/B$.

(b) Thus, there exists an injection $h: R/A \to R$ if and only if $A = x^{\perp}$ for some $x \in R$. In this case, $\text{im}\, h = xR$, and ${}^{\perp}A \supseteq Rx$.

(c) If A and B are ideals, then $R/A \approx R/B$ as R-modules if and only if $A = B$.

(d) Let $\{R_i\}_{i \in I}$ be a family of rings, and $R = \prod_{i \in I} R_i$ be the ring product. Then viewing R_i as an R-module via the (ring) projection map $R \to R_i$, $i = 1, \ldots, n$, prove that $R_i \approx R_j$ as R-modules if and only if $i = j$. Conclude that even if R_1 and R_2 are isomorphic rings, the resulting $R_1 \times R_2$-modules R_1 and R_2 are nonisomorphic. Let $R = R_1 \times R_2$, Then $R_i \approx \text{End}\,(R_i)_R$, $i = 1, 2$. Conclude that nonisomorphic modules may have isomorphic endomorphism rings.

22. Let A be a right ideal of a ring R. Show that

$$\begin{aligned} \text{Hom}_R(R/A, R) &\to R \\ f &\mapsto f(1 + A) \end{aligned}$$

is a monomorphism of left R-modules and that the image of this mapping is the left annihilator of A in R. Conclude that $\text{Hom}_R(R/A, R) \approx {}^{\perp}A$. Generalize to any submodule A of a projective right R-module P, that is

$$\text{Hom}_R(P/A, R) \approx \text{ann}_{\text{Hom}_R(P,R)} A.$$

23. If G is an abelian group, and if End G is a field of characteristic 0, then G is isomorphic to the additive group of rational numbers $\mathbb{Q}$.

24. (a) Any lattice homomorphism (resp. antihomomorphism) Pow $A \to$ Pow B is a ring homomorphism (resp. antihomomorphism) of the corresponding boolean rings.

(b) The set $E(R)$ of idempotents of a commutative ring R is a Boolean ring under $a \dotplus b = a + b$, and $a \cdot b = ab \forall a,b \in E(R)$.

25. If an R-module M satisfies the a.c.c. on submodules, then any epic endomorphism is a unit. Dualize. Conclude that if R satisfies the d.c.c. on right ideals, then a non zero divisor x is a unit.

***26.** If C is a finitely generated and projective cogenerator in mod-A, and if $A/\text{rad } A$ is semisimple, then C and A are injective in mod-A.

27. Any indecomposable projective injective object M of mod-A is isomorphic to a summand of A, that is, there exists $e = e^2 \in A$ such that $M = eA$.

***28.** (Matlis-Papp) A ring A is right Noetherian if and only if every injective object of mod-A is isomorphic to a direct sum of indecomposable modules.

29. The endomorphism ring of an indecomposable (quas-)injective module with no fully invariant submodules other than the trivial ones is a field.

30. Every torsion abelian group embeds in a direct sum of copies of $\mathbb{Q}/\mathbb{Z}$. [Hint: Reduce to the p-primary case. Then, consider the subgroup S consisting of all elements of orders $\leq p$. Use the fact that S is a vector space over the field $\mathbb{Z}/p\mathbb{Z}$, and apply 3.58.]

31. A torsion abelian group A of bounded order is a direct sum of cyclic groups. [Hint: Reduce to the p-primary case. Apply Ex. 30, and embed A in $\mathbb{Z}_{p\infty}^{(a)}$, for an appropriate cardinal a. Show that A intersects any $\mathbb{Z}_{p\infty}$ group in a cyclic group, and that A is the direct sum of its intersections with the $\mathbb{Z}_{p\infty}$ components of $\mathbb{Z}_{p\infty}^{(a)}$.]

32. Prove the Basis Theorem for finite abelian groups. (See Chapter 1, Exercise 27.)

33. (T. Szele) An abelian group H embeds in $\mathbb{Q}/\mathbb{Z}$ iff no two different subgroups are $\approx$.

Notes for Chapter 3

On the subject of history, most mathematicians stand on shaky, if hallowed, ground. Beyond folklore and one's own specialized knowledge lies a shadow world of fact, fiction, and fantasy. Yet, if the theorems themselves are the meat of the subject, surely the attributions, folklore or not, supply the spice. Only a coward, or a very wise man, would refuse to add his share to the general confusion. The often very passionate *Note Historique* appended at the end of the Bourbaki tracts in mathematics have a distinct (and recommended) Gallic flavor. In particular, the *Note Historique* for Bourbaki [58] on *Modules et Anneaux Semisimple* attributes the notions of a quotient of an algebra, ideal, and homomorphism to Molien and Cartan and outlines a history of aspects of rings and algebras not already covered in the study of Jacobson in his 1956 Colloquium Lectures for the American Mathematical Society (revised Jacobson [64]). The reader is referred to these two treatises for the major attributions.

Most books appear to be salted and peppered with attributions entirely to the taste and whims of the author. This book is no exception in this respect, but as in the other books, the intention is to supply the authors of the mathematical papers in which the work originated and not to award credit. In cases where no attribution has been made, I have been uncertain of the source, as often is the case, for example, in category theory. Here the reader may be delighted, as I was to read the stimulating account by Peter Freyd [62]. (I cannot, of course, endorse its accuracy!) Cartan-Eilenberg [56], MacLane [63], Mitchell [65], and also Curtis-Reiner [62], and Bass [68] supply some historical data on categorical and homological algebra. Also see MacLane [72].

Baer [40] introduced the idea, but not the terminology, of injective modules over an arbitrary ring (the terminology comes from Eckmann-Schopf [53]), and showed that any module type category had "enough injectives". Eckman-Schopf is generally accredited with the notion of the injective hull of a module, but actually Baer also spoke of a "smallest complete group" containing a given group. The existence of such was "assured" by theorems in his paper, but such was "only 'essentially smallest', but need not be 'actually smallest'". (The quotes are Baer's words, l.c.) Baer's complete groups over a ring R are R-modules satisfying what we have called Baer's condition, hence are injective modules. The "smallest" injective module containing a given module M is defined to be the module $I(M)$ of Exercise 5. Obviously, $I_1(M)$ and $I_2(M)$ defined with respect to two different injective modules F_1 and F_2 containing M are isomorphic by a mapping extending 1_M, and, moreover (despite what Baer said) is the smallest injective module containing M, namely $I(M) \supseteq E \supseteq M$, and E injective, implies that $I(M) = E$. Thus, no doubt, Baer had the idea the minimal injective module $I(M)$ containing M. The contribution of Eckmann and Schopf—an immense one— was to characterize $I(M)$ as a maximal essential extension. Moreover, they showed that an injective module containing M could be constructed via the embedding of M in a divisible group G.

General References

Artin, E.: The influence of J. H. M. Wedderburn on the development of modern algebra. Bull. Amer. Math. Soc. **56**, 65–72 (1950).

— Nesbitt, E., Thrall, R.: Rings with Minimum Condition. Ann Arbor: University of Michigan Press 1944.

Baer, R.: Abelian groups that are direct summands of every containing Abelian group. Bull. Amer. Math. Soc. **46**, 800–806 (1940).
Bass, H.: Algebraic K-Theory. New York: Benjamin 1968.
Bourbaki, N.: Éléments de Mathématiques, Livre II, Algèbre. Actualités Scientifiques et Industrielles, No. 1261. Paris: Hermann 1958, Chapter 8.
Brauer, R., Weiss, E.: Non-commutative Rings, Part I (Lecture Notes). Mathematics Department, Harvard University, Cambridge *circa* 1964.
Curtis, C., Reiner, I.: Representation Theory of Finite Groups and Associative Algebras. New York: Interscience Publishers, Wiley 1962.
Eckmann, B., Schopf, A.: Über injektive Moduln. Arch. Math. **4**, 75–78 (1953).
Faith, C.: Lectures on Injective Modules and Quotient Rings. Berlin/Heidelberg/New York: Springer 1967.
Herstein, I. N.: Noncommutative Rings. Carus Math. Monographs, No. 15. Amer. Math. Ass. and Wiley, New York 1968.
Jacobson, N.: The Theory of Rings, Surveys, Vol. 2. Amer. Math. Soc., Providence, 1942.
— Structure of Rings, Colloquium Publication, Vol. 37. Amer. Math. Soc., Providence 1956 (1964 revised).
Jans, J.: Rings and Homology. New York: Holt, Rinehart & Winston 1964.
Kaplansky, I.: Fields and Rings. Chicago: University of Chicago Press 1969.
Lambek, J.: Rings and Modules. New York: Blaisdell 1966.
MacLane, S.: Homology. Berlin/Göttingen/Heidelberg: Springer 1963.
— Categories for the Working Mathematician, Springer 1971.
Mitchell, B.: Theory of Categories. New York: Academic Press 1965.
Morita, K.: Duality for modules and its application to the theory of rings with minimum condition. Sci. Reports, Tokyo Kyoiku Daigaku **6**, 83–142 (1958).
Pontryagin, L.: Topological Groups. Princeton: Princeton University Press 1939.
Waerden, B. L. van der: Modern Algebra. New York: Ungar 1948.
Zariski, O., Samuel, P.: Commutative Algebra, Vol. I. Princeton/New York: Van Nostrand 1958.

Chapter 4

Correspondence Theorems for Projective Modules and the Structure of Simple Noetherian Rings

Much of Chapter 4 is devoted to the exposition of the structure theory of simple right Noetherian rings. The basic tools generalize the theorems of Morita [58] characterizing similarity of two rings A and B, that is, when there is an equivalence mod-$A \approx$ mod-B of categories. Morita's characterization 4.29 predicates the existence of a finitely generated projective module P which is a generator of the category mod-B such that A is isomorphic to $\text{End}_B P$. In the Morita situation 4.30 there is a lattice isomorphism

$$\begin{cases} \text{right } A\text{-submodules of } P \to \text{right ideals of } B \\ \qquad\qquad IP \leftrightarrow I \end{cases}$$

sending (B,A)-submodules onto ideals of B. More generally, 4.7, if U is any finitely generated projective faithful left B-module over any ring B, and $A = \operatorname{End}_B U$ then there is a lattice isomorphism

$$\begin{cases} \text{right } A\text{-submodules of } U \to (\text{right ideals of } B)\,T \\ \qquad\qquad IU \leftrightarrow I = IT \end{cases}$$

where T is the trace of U in B (Correspondence Theorem for Projective Modules.) This isomorphism sends (B, A)-submodules of U into (ideals of B)T. Thus, when $T = B$ (and only then) the two theorems coincide.

The correspondence theorem for projective modules just cited can be applied to the solution of the following problems: Characterize when the endomorphism ring $A = \operatorname{End}_B U$ of a finitely generated projective faithful module U over a ring B is (1) simple, or (2) right Noetherian, or (3) both. Thus, (1) A is simple if and only if $T = \operatorname{trace}_B U$ annihilates any ideal K of B contained in T properly, that is, $TK = 0$. In this case $(KT)^2 = 0$. If B is semiprime, then $KT = 0$, and $K = 0$, and then A is simple if and only if T is the least ideal of B 4.8.6.

Problem (2) has a similar solution. Since U is necessarily a generator of mod-A, then the correspondence theorem implies that U is right Noetherian in mod-A if and only if B satisfies the a.c.c. on right ideals of the form $I = IT$. In this case, A is right Noetherian. If, U is finitely generated in mod-A, then conversely, when A is right Noetherian, then B satisfies the a.c.c. on the right ideals $I = IT$. (Cf. 4.9).

When A is both simple and right Noetherian, then, by a theorem of Goldie [58], A contains a right ideal U such that $B = \operatorname{End} U_A$ is a right Ore domain. In this case, U is necessarily finitely generated projective over B, and so the solution to (3) can be formulated for endomorphism rings of projective modules over right Ore domains as follows:

(**A**) **Theorem.** *A ring A is simple and right Noetherian if and only if $A \approx \operatorname{End}_B U$, where U is finitely generated and projective over a right Ore domain B with just the ideals 0, $T = \operatorname{trace}_B U$, and B, and satisfying the a.c.c. on idempotent right ideals contained in T (possibly $T = B$).* (Cf. 4.22B.)

(**B**) **Proposition.** *When* (**A**) *holds, then B can be chosen such that the center C of B is a field isomorphic to the center of A, and moreover, $B = C + T$.* (Cf. 4.14 and 4.22.3A.)

(**C**) **Proposition.** *When* (**A**) *holds, then there is a right Ore domain B', a finitely generated projective left B' module U', and an isomorphism $A \approx \operatorname{End}_{B'} U'$ only if B and B' have the same right quotient field D, and B and B' are equivalent right orders of D.* (Cf. 4.37.)

The effect of (**A**), (**B**), and (**C**) is to reduce a certain amount of the structure of a simple Noetherian ring A to that of a right Ore domain B determined up to equivalence of orders in the right quotient field of B. Moreover, B *has at most one nontrivial ideal.* This is the best possible theorem in the sense that every such B which arises can be a simple ring only if A is right hereditary 4.40. However, when A has right global dimension ≤ 2, then A is similar to some simple domain, that is, some B is simple in this case (cf. 8.27). We also characterize in mod-B the condition under which A is similar to a domain. This happens if for some right ideal I of B the subring $(I:I) = \{d \in D \mid dI \subseteq I\}$ is a simple ring 4.23 B.

The relationship between the category RINGS and the categories mod-R of right R-modules, where R ranges over the objects of RINGS, is curiously circular because every module has an endomorphism ring.

Category	RINGS	mod-R	mod-End M_R	mod-Biend M_R
Objects	R			
		M		
	End M_R			
			M	
	Biend M_R			M

Some typical questions of abstract ring theory are:

1. If a ring is abstractly given, try to realize it "concretely".

(a) Is it isomorphic to a "known" ring?

(b) Is it isomorphic to the endomorphism ring of a "known" module over some "known" ring?

(b′) Can it be embedded in such a ring?

(c) Can a complete set of invariants be found? Is this a finite (countable) set?

(d) If a ring belongs to a subcategory of RINGS, what about its quotient rings and its subrings?

2. If a module M in mod-R is abstractly given, the same kind of questions can be asked, but in addition, one can ask all the questions in 1. about its endomorphism ring. Then if the answer to 1 (b) is yes, say End $M_R \approx$ End N_A, for a "known ring A and module N," how do R and A compare? And if they compare well, e.g., are isomorphic, how do M and N compare? Finally, how do R and Biend M_R (and their categories of modules) compare?

Some of the examples presently cited require the concept of an algebra and tensor product of algebras defined in the next chapter. (Hence, the examples also serve to introduce Chapter 5.)

4.1 *Examples.*

4.1.1 Every simple ring is embeddable as a dense subring of a full linear ring of l.t.'s. over a vector space. Necessary and sufficient conditions for a dense embedding of a ring into a full linear ring is the existence of a simple faithful module (Theorem of Chevalley-Jacobson). Such a ring is said to be **primitive** 19.22.

4.1.2 A ring R is isomorphic to a full linear ring over a field K if and only if R is a simple ring with minimum condition on right ideals (Theorem of Wedderburn-Artin, Chapter 8).

4.1.3 If the field K in 4.1.2 is commutative and algebraically closed, for example, if $K = \mathbb{C}$, then the ring R is a central algebra over $\mathbb{C}$ and has a finite free basis, and the constants of multiplication for the basis are known. In fact, R is a total matrix algebra over $\mathbb{C}$. For an arbitrary algebraically closed field, the field and the degree of the matrix ring are a complete set of invariants for a simple algebra of finite dimension (Theorems of Wedderburn, Chapter 13).

4.1.4 If M is a generator in mod-R, then $R \approx \text{Biend } M_R$ canonically (Theorem of Morita, 4.1 and 7.3), so every generator is a balanced module. If R is a simple ring, then every nonzero right ideal is balanced.

4.1.5 Two rings A and B are similar (that is, mod-$A \approx$ mod-B) if and only if B is isomorphic to the endomorphism ring of a finitely generated projective generator of mod-A (Theorem of Morita 4.29).

4.1.6 Dual to the notion of similarity is duality, but it is never possible to have a duality between mod-A and B-mod for any two rings C. However, there may exist subcategories $\mathcal{S}_A$ and ${}_B\mathcal{S}$ of mod-A and B-mod containing A and B respectively and such that a duality $\mathcal{S}_A \rightsquigarrow {}_B\mathcal{S}$ exists. A sufficient (and, in a sense, necessary) condition for this is the existence of a (B, A)-bimodule U which is an injective cogenerator in both categories. Then, $\text{Hom}_A(\ , U)$ and $\text{Hom}_B(\ , U)$ induce the duality pair of functors (Theorem of Morita, Chapter 23). The case $A = U = B$ characterizes Quasi-Frobenius rings (Chapter 24). These rings are right selfinjective and Artinian (Theorem of Ikeda), and every one-sided ideal is an annihilator. (Cf. Chapter 12 Exercises 6.)

4.1.7 A module M in mod-R is said to be **determined by its endomorphism ring** if and only if an isomorphism $\text{End } M_R \approx \text{End } N_R$ implies an isomorphism $M \approx N$. Any vector space is determined by its endomorphism ring (see Exercise 2 at the end of this chapter). E. A. Walker has asked a related question: For different rings A and B, when does an isomorphism $\text{End } M_A \approx \text{End } N_B$ imply a category equivalence $T: \text{mod-}A \approx \text{mod-}B$ and an equivalence $N \approx TM$ in mod-B? (Cf. 16.7 and 24.26.)

4.1.8 On the question of the relation of mod-A and mod-B, where $B = \operatorname{End} M_A$, a theorem of Popesco-Gabriel asserts in case M is a generator in mod-A that mod-A can be recovered from mod-B in two ways:

(a) mod-A is a certain full subcategory of mod-B [the embedding mod-$A \rightsquigarrow$ mod-B induced by h^M is full (Theorem of Morita and Popesco-Gabriel, 4.24)].

(b) There is an exact functor $S : \text{mod-}B \rightsquigarrow \text{mod-}A$ (a left adjoint of h^M) such that $Sh^M \approx 1_{\text{mod-}A}$ (5.70), and mod-A is an abelian category equivalent to the quotient mod-B/ker S of categories as defined by Gabriel [62] (see 15.26).

4.1.9 For a commutative ring k, the category k-ALG of all algebras over k (Chapter 5) defines an abelian semigroup with respect to tensor product $A \underset{k}{\otimes} B$ of algebras A and B. Let $[A] = [A']$ if and only if A is similar to A' as k-algebras. Then, the set of all $[A]$ is a semigroup under the operation $[A][B] = [A \underset{k}{\otimes} B]$. The **Brauer group** $Br(k)$ is defined to be the units group of this semigroup and consists of all $[A]$ such that A is an **Azumaya algebra.** (Theorems of Brauer [for k a field], Azumaya [k a local ring], and Auslander-Goldman [k arbitrary commutative]. See Chapter 32.)

4.1.10 A reduced torsionfree ring R of finite rank n is isomorphic to the endomorphism ring of a torsionfree abelian group of rank $\leq 2n$ (Corner [63]). A ring is torsionfree if the additive group is torsionfree, and a ring is **reduced** if 0 is the only divisible subgroup. The rank is the maximal number of any set of elements linearly independent over $\mathbb{Z}$. In this direction, Zassenshaus [66] proved that any ring with a finite free basis of rank n over $\mathbb{Z}$ is isomorphic to the endomorphism ring of an abelian group M of rank n.

4.1.11 Following Thrall [48], the ring A is QF-1 if every faithful module is balanced, and following Camillo [70], the ring A is balanced if every right module is balanced. Thus, A is balanced if and only if every quotient ring A/I is QF-1. Thrall proposed the problem of describing all finite dimensional QF-1 algebras over a field k. Every Quasi-Frobenius (QF) ring is QF-1, but the class of QF-1 algebras properly contains the class of QF algebras (Thrall, l.c.). (For recent results, consult Camillo [70], Dickson and Fuller [70], Fuller [71], Camillo and Fuller [72], Dlab and Ringels [72], Jans [70].)

A more general related question is when is the canonical map $A \to \operatorname{Biend} M_A$ an epic in RINGS? For example, the embedding of any ring in its classical quotient ring (e.g. the embedding of a domain in its quotient field) is a ring epic. (Any surjective ring homomorphism is, of course, an epic in RINGS.)

4.1 Proposition. *Let U be an object of* mod-A, *and let* $B = \text{End } U_A$.

4.1.1 *There is a canonical (A, B)-bimodule homomorphism*

(1) $$[\ ,\ ]\begin{cases} \text{Hom}_A(U, A) \to \text{Hom}_B(U, B) \\ \qquad f \mapsto [\ , f] \end{cases}$$

such that

(2) $$[\ , f]\begin{cases} U \to B \\ u \mapsto [u, f], \end{cases}$$

where $[u, f]$ is the endomorphism of U

(3) $$[u, f]\begin{cases} U \to U \\ v \mapsto uf(v). \end{cases}$$

Thus

(4) $$[u, f]v = uf(v)$$

$\forall u, v \in U,\ f \in \text{Hom}_A(U, A)$.

4.1.2 *Then, if* $A = \text{End}_B U$, *by symmetry, there is a canonical homomorphism*

(5) $$(\ ,\ )\begin{cases} \text{Hom}_B(U, B) \to \text{Hom}_A(U, A) \\ \qquad g \mapsto (g,\ ), \end{cases}$$

and

(6) $$v(g, u) = (v)g \cdot u.$$

(7) *In this case, $(\ ,\ )$ and $[\ ,\ ]$ are inverse isomorphisms*[1].

4.1.3 (Morita) *The following conditions on U, A, and B are equivalent:*
(a) *U is generator in* mod-A.
(b) *U is finitely generated and projective in* End U_A-mod, *and* $A \approx \text{End}_B U$ *canonically.*

4.1.4 *When U is a generator in* mod-A, *then* $[\ ,\ ]: \text{Hom}_A(U, A) \to \text{Hom}_B(U, B)$ *is an (A, B)-bimodule isomorphism, and Eq. (5) is the inverse isomorphism.*

Proof. Let $U^* = \text{Hom}_A(U, A)$ and $U' = \text{Hom}_B(U, B)$ be the canonical (A, B)-modules. If $a \in A$, $b \in B$, and $f \in U^*$, then

$$[u, afb]v = uaf(bv) = [ua, f]bv,$$

so that $[\ , afb] = a[\ , f]b$, and (1) is therefore an (A,B)-homomorphism, and by symmetry, so is (5). The identity

$$u \cdot [\ , (g,\ )](v) = [u, (g,\ )](v) = u(g, v) = (u)g \cdot v$$

[1] Morita proved this part assuming that U is a generator of mod-A.

shows that $[\ ,(g,\ )] = g \forall g \in U'$, that is, that $[\ ,(\ ,\ )] = 1_U$. When $A = \mathrm{End}_B U$, then, by symmetry, $([\ ,f],\ ) = f \forall f \in U^*$ and $([\ ,\ ],\ ) = 1_{U^*}$ This proves that $(\ ,\ )$ and $[\ ,\ ]$ are inverse isomorphisms when $A = \mathrm{End}_B U$ and $B = \mathrm{End}\ U_A$.

By 3.27, U is a generator of mod-A if and only if there exists finitely many element $f_1, \ldots, f_n$ of U^* and $u_1, \ldots, u_n$ of U such that

$$\sum_{i=1}^{n} f_i(u_i) = 1_A. \tag{8}$$

By the Dual Basis Lemma 3.23, U is finitely generated and projective in B-mod if and only if there exist finitely many elements $g_1, \ldots, g_n$ of U' and $u_1, \ldots, u_n$ of U such that

$$v = \sum_{i=1}^{n} (v) g_i u_i \tag{9}$$

$\forall v \in U$. Thus, application of (8) to $v \in U$, shows that (9) holds for $g_i = [\ ,f_i] \in U'$, $i = 1, \ldots, n$. This proves that U is a generator of mod-A only if U is finitely generated projective over B. Furthermore, if $t \in \mathrm{End}_B U$, then applying t to v in (9) yields

$$\begin{aligned} vt &= \textstyle\sum_{i=1}^{n} (v) g_i \cdot u_i t \\ &= \sum_{i=1}^{n} [v, f_i] \cdot u_i t \\ &= \sum_{i=1}^{n} v f_i(u_i t), \end{aligned} \tag{10}$$

which implies that t is the image of $a = \sum_{i=1}^{n} f_i(u_i t)$ under the canonical homomorphism

$$\begin{cases} A \to \mathrm{End}_B U \\ a \mapsto a_d, \end{cases} \tag{11}$$

that is, $vt = va \forall v \in V$, whence $t = a_d$. thus, 4.1.3 (a) implies (b) since the canonical homomorphism (11) is injective by the fact that a generator is a faithful module, and we have just shown that (11) is surjective. Conversely, 4.1.3 (b) implies (a). For assuming 4.1.3 (b), then (9) holds, and then via the homomorphism (5), there are elements $f_1, \ldots, f_n$ of U^*, namely, $f_i = (g_i,\ )$, $i = 1, \ldots, n$, such that

$$v = \sum_{i=1}^{n} (v) g_i u_i = \sum_{i=1}^{n} v(g_i, u_i) = v \sum_{i=1}^{n} f_i(u_i) \tag{12}$$

which shows that (8) holds. Then U is a generator of mod-A.

(4) Assuming U is a generator in mod-A, then $A \approx \mathrm{End}_B U$, canonically by 4.1.3 (b), and, hence, (2) holds. □

4.2 Corollary.

4.2.1 *If U is a generator in* mod-A, *then* $U^* = \mathrm{Hom}_A(U, A) \approx \mathrm{Hom}_B(U, B)$ *is a generator in* A-mod *and finitely generated projective in* mod-B, *and there are canonical isomorphisms* $B \approx \mathrm{End}_A U^*$ *and* $A \approx \mathrm{End}\, U^*_B$.

4.2.2 *If U is finitely generated and projective in* B-mod, *then U is a generator in* mod-End U_B.

Proof of 4.2.1. Let

$$\langle\ ,\ \rangle \begin{cases} U \to U^{**} \\ u \mapsto \langle\ ,u\rangle \end{cases}$$

denote the canonical (B, A) homomorphism of U into its bidual module U^{**}. Thus,

$$f\langle\ ,u\rangle = \langle f, u\rangle = f(u)$$

$\forall f \in U^*$, $u \in U$. If U is a generator in mod-A, then (8) in the proof of the theorem holds, and

$$\sum_{i=1} f_i\langle\ ,u_i\rangle = 1_A,$$

which shows that U^* is a generator in A-mod. The theorem then implies the rest.

Proof of 4.2.2. This is a corollary of the proof (3b) ⇒ (3a) of the theorem since the fact that $B = \mathrm{End}\, U_A$ was not used. ▯

4.3 Proposition and Definition. *An object U of* mod-A *is a* **progenerator** *in case U has the following equivalent properties:*

(a) *U is a generator in* mod-A *and finitely generated and projective in* mod-A.

(b) *U is a generator in* mod-A *and in* End U_A-mod.

(c) *U is finitely generated and projective in both* mod-A *and* End U_A-mod.

(d) *U is a generator in* End U_A-mod *and is finitely generated and projective over* End U_A, *and* $A \approx \mathrm{Biend}\, U_A$ *canonically.* ▯

In some places in the literature, a progenerator is called a **faithfully projective module.**

4.4 Exercise. *If B is a commutative ring, any finitely generated faithful projective module U over B is a generator of* mod-B.

The Biendomorphism Ring of a Projective Module

If U is projective and faithful over B_0, then the next proposition states that U is also projective over $B = \mathrm{Biend}_{B_0} U$, and furthermore, in a sense presently defined, B is a rational extension of B_0, and the trace of U in B_0 is a left ideal of B.

4.5 *Exercises* (Johnson, Utumi, and Findlay-Lambek).

4.5.1 If R is a ring, a right R-module M is a **rational extension** of a submodule N provided that the equivalent conditions hold:

(a) $\mathrm{Hom}_R(S/N, M) = 0$ for every module $M \supseteq S \supseteq N$.

(b) If a and b are elements of M, and if $b \neq 0$, then there exists an element $r \in R$ such that $ar \in N$ and $br \neq 0$.

4.5.2 Any rational extension is an essential extension, but the converse does not hold in general.

4.5.3 (Findlay and Lambek [58]). If M is a right R-module, with injective hull $\hat{M}$, and if $B = \mathrm{End}\,\hat{M}_R$, then

$$\overline{M} = \mathrm{ann}_{\hat{M}}\,\mathrm{ann}_B M$$

is a maximal rational extension of M. Any maximal rational extension of M is isomorphic to $\overline{M}$ under a map inducing the identity map on M. A module is **rationally closed** when $\overline{M} = M$. Note that $\overline{M}$ is always rationally closed.

4.5.4 (Johnson [51] and Utumi [56].) The maximal rational extension $\overline{R}$ of R is a ring, and R is a subring. Furthermore (Lambek [63]),

$$\begin{cases} \mathrm{Biend}\,\hat{R}_R \to \overline{R} \\ \qquad f \mapsto f(1) \end{cases}$$

is an isomorphism of rings which induces the identity mapping of R. The ring $\overline{R}$ is the (Johnson-Utumi) **maximal right quotient ring** of R. The ring $\overline{R}$ is its own maximal right quotient ring.

4.5.5 If R has a classical right quotient ring Q, then $Q \subseteq \overline{R}$, and possibly $Q \neq \overline{R}$. ▯

Proofs can be found in Lambek's book [66] or Faith's Lectures [67, Chapters 7 and 8]. However, essentially only the definition of rational extension is used in the sequel.

4.6 **Proposition.** *If U is a faithful projective left module over a ring B_0, then U is projective over $B = \mathrm{Biend}_{B_0} U$. Let G_0 and G denote the respective dual modules, and let T_0 and T denote the respective traces of U in B_0 and B. Then the canonical inclusion $B_0 \subseteq B$ induces the inclusion $G_0 \subseteq G$, and then $G = G_0 B$. Furthermore, every right ideal I of B is a rational extension of IT_0; in particular, B is contained in the maximal right quotient ring of B_0, and T_0 is a left ideal of B such that*

$$T_0^2 = T_0 = TT_0 = BT_0$$

and

$$T_0 B = T_0 T = T = T^2.$$

Thus, there is a lattice isomorphism

(1) $$\begin{cases} (\text{right ideals of } B)T \to (\text{right ideals of } B_0)T_0 \\ I = IT \mapsto IT_0 \end{cases}$$

which induces an isomorphism

(2) $$\begin{cases} T(\text{ideals of } B)T \to T_0(\text{ideals of } B_0)T_0 \\ TIT = I \mapsto IT_0 = T_0IT_0 \end{cases}$$

of multiplicative semigroups.

Proof. Let $A = \operatorname{End}_{B_0} U$, and consider the canonical (A, B)-bimodule homomorphism

$$\begin{cases} \operatorname{Hom}_{B_0}(U, B_0) \to \operatorname{Hom}_A(U, A) \\ f \mapsto (f, \ ) \end{cases}$$

such that

$$u(f, \ )v = u(f, v) = (u)f \cdot v$$

$\forall u, v \in U$. Since $B = \operatorname{End} U_A$, and since $(f, v) \in A$, then

$$(bu)fv = (bu)(f, v) = b \cdot (u)f \cdot v$$

$u, v \in U$, $b \in B$. Thus, $(bu)f = b(u)f$, which demonstrates the desired inclusion

$$G_0 = \operatorname{Hom}_{B_0}(U, B_0) \subseteq \operatorname{Hom}_B(U, B) = G.$$

By the dual basis lemma 3.23, there are a set of elements $\{f_i\}_{i\in X}$ of G_0 each having finite support and a corresponding set $\{u_i\}_{i\in X}$ of elements of U such that

$$v = \sum_{i\in X} (v) f_i u_i$$

$\forall v \in V$. Since $f_i \in \operatorname{Hom}_B(U, B)$, $\forall i \in X$, then the dual basis lemma implies that $\{f_i\}_{i\in X}$ generates the right B-module G, that is, $G = G_0 B$, and, moreover, that U is projective over B. Thus, $T = \operatorname{trace}_B U$ satisfies $TU = U$, and $T^2 = T$ by 3.30 and

$$T = \sum_{f\in G} (U)f = \sum_{f\in G} (T_0 U)f = \sum_{f\in G} T_0(U)f = T_0 T.$$

Since $TU = T_0 U = U$, then

$$T_0 = \sum_{g\in G_0} (U)g = \sum_{g\in G_0} (TU)g = \sum_{g\in G_0} T(U)g = TT_0.$$

Therefore, (1) is a bijection.

If $I = TIT$ and $J = TJT$, then $I = TI = T_0 I$ and

$$IJT_0 = IT_0JT_0 = IT_0^2JT_0 = (T_0IT_0)(T_0JT_0),$$

which proves (2).

It remains only to show that I is rational over IT_0, for any right ideal I of B. If a and b are elements of I, and $b \neq 0$, then $bU \neq 0$, so $bT_0 \neq 0$. Thus, $at \in IT_0$ and $bt \neq 0$, for some $t \in T_0$, which proves that I is rational over IT_0 in mod-B_0. In particular, B is rational over $BT_0 = BTT_0 = TT_0 = T_0$ and, therefore, rational over B_0. □

4.7 The Correspondence Theorem for Projective Modules. (Morita [58], Faith [71a, 72]). *Let U be a finitely generated faithful projective left module over B_0, let $T_0 = \text{trace}_{B_0} U$ and $A = \text{End}_{B_0} U$. If I and J are right ideals of B_0, then*

$$(I \cap J)U = IU \cap JU,$$

and the set (right ideals of B_0)T_0 *consisting of all right ideals of B_0 of the form $I = IT_0$ is a lattice such that*

$$I \vee J = I + K$$
$$I \wedge J = (I \cap J)T_0.$$

The mapping

$$(1)\quad \begin{cases} (A\text{-submodules of } U) \rightleftarrows (\text{right ideals of } B_0)T_0 \\ \qquad W \mapsto (\text{ann}_{B_0} U/W)T_0 = \bigcap_{KU=W} K = I(W) \\ \qquad KU \leftarrow K \end{cases}$$

is a lattice isomorphism, and $I(W)$ is the least right ideal of B_0 such that $I(W)U = W$. Moreover, $I(W) = \sum_{f \in G_0} (W)f$, where $G_0 = \text{Hom}_{B_0}(U, B_0)$.

The mapping (1) *induces a lattice isomorphism*

$$(2)\quad \begin{cases} (B_0, A)\text{-submodules of } U \to (\text{ideals of } B_0)T_0 \\ \qquad W \mapsto I(W). \end{cases}$$

The mapping

$$(3)\quad \begin{cases} (\text{ideals of } A) \to T_0(B_0, A)\text{-submodules of } U \\ \qquad J \mapsto UJ \end{cases}$$

is a lattice isomorphism, where, for (B_0, A)-submodules $W = T_0W$ and $V = T_0V$ of U,

$$W \wedge V = T_0(W \cap V).$$

The inverse of the mapping (3) *is induced by*

$$(4) \qquad \boldsymbol{U^*} \begin{cases} (B_0, A)\text{-submodules of } U \to \text{ideals of } A \\ \qquad\qquad W \mapsto \boldsymbol{U^*}(W), \end{cases}$$

where

$$(5) \qquad \boldsymbol{U^*}(W) = \sum_{f \in \mathrm{Hom}_A(U,A)} f(W) = \boldsymbol{U^*}(T_0 W)$$

is the image of W *under the canonical mapping* $U^{(U^*)} \to A$, where $U^* = \mathrm{Hom}_A(U, A)$.

Finally, the mapping

$$(6) \qquad S \begin{cases} \text{ideals of } A \to T_0(\text{ideals of } B_0)\, T_0 \\ \qquad J \mapsto S(J) = I(UJ) \end{cases}$$

s an isomorphism of multiplicative semigroups. Thus,

$$UJ = S(J)\, U$$

$\forall$ *ideals* J *of* A.

Proof. In view of the lattice and semigroup isomorphism (1) and (2) of 4.6, it suffices to prove the theorem under the assumption that B_0 coincides with $B = \mathrm{Biend}_{B_0} U = \mathrm{End}\, U_A$. Also let $T = T_0$. If

$$0 \to W \xrightarrow{k} U$$

is exact in mod-A, then

$$0 \to \mathrm{Hom}_A(U, W) \xrightarrow{h^U(k)} \mathrm{Hom}_A(U, U) = B$$

is exact in mod-B. If $k = 1_W^U$ is an inclusion, then

$$(7) \qquad I_0(W) = \mathrm{im}\, h^U(k) = \{b \in B / bU \subseteq W\} = \mathrm{ann}_B U/W$$

is a right ideal of B. Since U is a generator in mod-A, then by 3.26, the image of the canonical map

$$(8) \qquad U^{(\mathrm{Hom}_A(U,W))} \to W$$

is W. Thus, $I_0(W)\, U = W$. If K is a right ideal of B, and if $KU = W$, then $TU = U$ implies that $(KT)\, U = W$, that is, $I(W) = I_0(W)\, T$ is the intersection of all such right ideals K. Then $I(W)\, T = I(W)$.

If $I = IT$ and $K = KT$ are right ideals of B, then $(I \cap K)\, T$ is the sum of all right ideals of the form $Q = QT$ contained in $I \cap K$. From the definition (7),

$$(9) \qquad \begin{aligned} I_0(W_1 \cap W_2) &= I_0(W_1) \cap I_0(W_2) \\ I_0(W_1 + W_2) &= I_0(W_1) + I_0(W_2) \end{aligned}$$

for all A-submodules W_1 and W_2 of U, and then

$$(10)\qquad \begin{aligned} I(W_1 + W_2) &= I(W_1) + I(W_2) \\ I(W_1 \cap W_2) &= [I(W_1) \cap I(W_2)]T. \end{aligned}$$

The first equality in (10) is obvious from (9), since $(I + K)T = IT + KT$ for right ideals I and K of B. Next

$$\begin{aligned} I(W_1 \cap W_2) = I_0(W_1 \cap W_2)T &= [I_0(W_1) \cap I_0(W_2)]T \\ &\subseteq I_0(W_1)T \cap I_0(W_2)T, \end{aligned}$$

so that

$$I(W_1 \cap W_2) \subseteq I(W_1) \cap I(W_2),$$

and therefore,

$$(11)\qquad \begin{aligned} I(W_1 \cap W_2)U = W_1 \cap W_2 &\subseteq [I(W_1) \cap I(W_2)]U \subseteq \\ &\subseteq I(W_1)U \cap I(W_2)U = W_1 \cap W_2. \end{aligned}$$

Since $I(W) \subseteq KT$ for any right ideal K such that $KU = W$, it follows from (11) that

$$I(W_1 \cap W_2) = [I(W_1) \cap I(W_2)]T.$$

This proves that (1) is a lattice isomorphism. If W is a (B, A)-submodule, then $I(W) = I(BW) = BI(W)$ is an ideal of B, and then (2) holds.

(3) If J is any ideal of A, then

$$\boldsymbol{U}^*(UJ) = \sum_{f \in U^*} f(UJ) = \Big(\sum_{f \in U^*} f(U)\Big) J = AJ = J.$$

The isomorphism

$$\begin{cases} U^* \to U' = \operatorname{Hom}_B(U, B) \\ \quad f \mapsto [\ ,f] \end{cases}$$

given by 4.1 is such that

$$[u, f]v = uf(v).$$

Let W be a (B, A)-submodule of U, and let $J = \boldsymbol{U}^*(W)$. The submodule UJ is generated by the set

$$(12)\qquad \begin{aligned} \{ua \mid u \in U,\ a \in J\} &= \{uf(w) \mid u \in U,\ f \in U^*,\ w \in W\} \\ &= \{[u, f]w \mid u \in U,\ f \in U^*,\ w \in W\}. \end{aligned}$$

Since $[\ ,\ ] : U^* \to U'$ is an isomorphism, then the set

$$(13)\qquad \{[u, f] \mid u \in U,\ f \in U^*\} = \{(u)g \mid u \in U,\ g \in U'\}$$

generates $T = \text{trace}_B U$. Therefore, the set (13) generates both TW and UJ as (B, A)-submodules of U, that is,

$$UJ = TW. \tag{14}$$

This proves that the mapping (3) is a lattice isomorphism, namely, any (B, A)-submodule $W = TW$ can be written $W = UJ$, for a unique ideal J of A, where $J = \boldsymbol{U}^*(W)$.

(5) This follows because if $W = IU$, where $I = I(W)$, then

$$(W)U' = \sum_{f \in U'} (W)f = I\Big(\sum_{f \in U'} Uf\Big) = IT = I.$$

(6) The mapping $J \to SJ$ described in (6) is the compose of the mapping $J \to UJ$ of (3) and the mapping $W \to I(W)$ of (1). Thus, $UJ = S(J)U$, and since $TUJ = UJ$, then $TS(J)TU = UJ$, that is, $TS(J)T = S(J)$. Furthermore, $S(JK) = S(J)S(K)$, for ideals J and K of A. □

4.8 Corollary. *Let U be finitely generated faithful and projective over a ring B_0, let $A = \text{End}_{B_0} U$, and let $T_0 = \text{trace}_{B_0} U$. For example, U can be an arbitrary generator in* mod-A, *for any ring A and $B_0 = \text{End}\, U_A$.*

4.8.1 *Let P be any property of multiplicative semigroups which is preserved under homomorphisms. Then* (ideals of A) *has property P if and only if T_0*(ideals of B_0)T_0 *has property P.*

4.8.2 *A is semiprime if and only if $T_0 Q = 0$ for every nilpotent ideal Q of B_0.*

4.8.3 *A is prime if and only if $T_0 Q = 0$ for every annihilator ideal Q of B_0*

4.8.4 *If B_0 is semiprime, then B_0 is prime if and only if A is prime.*

4.8.5 *A has a finite ideal lattice if and only if the set of ideals of the form $I = T_0 I T_0$ of B_0 is finite. In particular, A is a simple ring if and only if $T_0 K = 0$ for every ideal $K \neq T_0$ contained in T_0.*

4.8.6 *If B_0 is semiprime, then A is simple if and only if T_0 is the least ideal of B_0. In this case a right ideal I of B_0 satisfies $I = IT_0$ if and only if I is an idempotent right ideal contained in T_0.*

4.8.7 *The ideals of A commute if and only if any two ideals $I = T_0 I T_0$ and $J = T_0 J T_0$ in B_0 commute.*

Proof. By 4.6, we may assume that $B_0 = B = \text{End}\, U_A$. Let $T = T_0$. The corollary is a consequence of the semigroup isomorphism

$$\text{ideals of } A \to T(\text{ideals of } B)T$$

stated in (6) of 4.6. Thus, if P is a statement that every ideal of a certain set $P(R)$ of a semigroup R is zero, then, in particular, $P(A) = 0$ implies that $P(T(\text{ideals of } B)T) = 0$. Since $TU = U$, then $HT = 0$ for a right or left ideal H of B if and only if $H = 0$. Thus, $THT = 0$ if and only if $TH = 0$. The other parts are consequences.

If B is semiprime, then $TI = 0$, for some right ideal I of B implies that $(IT)^2 = 0$, and $IT = 0$, so $I = 0$. Thus, in this case, B is prime if and only if A is prime. Furthermore, then A is simple if and only if T is the least ideal of B. In this case, if $I = I^2$ is a right ideal contained in T, then

$$I \supseteq IT \supseteq I^2 = I,$$

so $I = IT$. Conversely, $I = IT$ implies that

$$I^2 = I(TIT) = IT = I,$$

since TIT must equal T when $I \neq 0$. $\square$

4.9 Corollary. *Let U be a finitely generated faithful projective left module over a ring R, and let $T = \text{trace}_R U$.*

4.9.1 *Let L be any property of lattices preserved under homomorphisms of lattices. Then the lattice isomorphisms*

$$\begin{cases} A\text{-submodules of } U \to (\text{right ideals of } R)T \\ \qquad W \mapsto \bigcap_{KU=W} K \end{cases}$$

and

$$(R, A)\text{-submodules of } U \to (\text{ideals of } R)T$$

imply that the lattice on the left side has property L if and only if the lattice on the right side has property L[1]. *Furthermore, in the case R is semiprime and A is simple (that is, T is the least ideal of R), then in the first instance, this is equivalent to the statement that the lattice of idempotent right ideals of R contained in T has property L.*

4.9.2 *If, in addition, L is a property which is preserved under finite direct products, then the lattice of right ideals of A has property L when the lattice of A-submodules of U has property L. When U is finitely generated, the converse holds.*

Proof of 4.9.1. The first statement in (1) is obvious, and the second follows from the fact that (Goldie) dimension is such a lattice-invariant property L. The rest of (1) follows from the proof of 4.8.

Proof of 4.9.2. Since U is a generator of mod-A by 4.3, then $U^n \approx A \oplus X$ for some X in mod-A. If L is preserved under finite products, then U^n has property L, and then so does A since L is preserved

[1] Thus, in a sense defined in the following 4.19, they have the same Goldie dimension.

under the homomorphism (= projection) $U^n \to A$. Conversely, if U is finitely generated in mod-A, say $A^n \to U \to 0$ exact, and if A has property L, then A^n and then U has property L. □

4.10 Proposition. *For any left R-module M and integer* $n > 0$,

$$\operatorname{Biend}_R M = \operatorname{Biend}_R M^n$$

canonically. Thus, M is balanced in R-mod *if and only if* M^n *is.*

Proof. Let $A = \operatorname{End}_R M$. Then, by 3.33, there is a canonical isomorphism

$$\begin{cases} \operatorname{End}_R M^n \to A_n \\ \qquad f \mapsto (f_{ij}). \end{cases}$$

Since M is balanced as a right A-module, we may assume that $R = \operatorname{Biend}_R M$. There is a ring monomorphism

$$\begin{cases} R \to \operatorname{Biend}_R M^n \\ r \mapsto r' \end{cases}$$

such that

$$r'(x_1, \ldots, x_n) = (rx_1, \ldots, rx_n)$$

for all $(x_1, \ldots, x_n) \in M^n$, where $x_i \in M$, $i = 1, \ldots, n$. To show that this is an isomorphism, if $t \in \operatorname{Biend}_R M^n$, then

$$t(x_1, \ldots, x_n) = (tx_1, \ldots, tx_n)$$

since the various projections $M^n \to M$ in mod-A are induced by elements of $\operatorname{End}_R M^n = B$, and t is an element of End M^n_B. This implies that t induces an endomorphism of M in mod-A, and so, by the assumption that $R = \operatorname{Biend}_R M$, there is an element $r \in R$ such that $tx = rx \; \forall \, x \in M$. This shows that $t = r'$ and demonstrates the desired isomorphism. □

4.11 *Exercises.*

4.11.1 (a) If R is a ring, and if R is its own maximal right quotient ring, that is, if R is rationally closed (see 4.4), then any faithful projective right module over R is balanced. This is true, in particular, when R is right selfinjective.

(b) (Kato) If U is any torsionless faithful left module over any ring R, then $\operatorname{Biend}_R U$ is contained in the maximal right quotient ring of R. (Cf. 4.6.)

4.11.2 Let $U = A \oplus X$ in mod-A, and let $B = \operatorname{End} U_A$. Then U is cyclic, faithful, and projective in B-mod. Furthermore, U is balanced. However, U is a generator of B-mod if and only if X is finitely generated and projective.

4.11.3 Let $B = \operatorname{End} U_D$, where U is a right space over a field D of infinite dimension. Then U is a simple left module over B that is faithful and projective, but U is not a generator of B-mod.

4.11.4 Let R be a product of infinitely many fields. Then the right socle of R is faithful, projective, but not a generator.

4.11.5 Use $\operatorname{Biend}_R M = \operatorname{Biend}_R M^n$, for any integer n, to show, by reducing the proof to the case $U = R \oplus X$ in mod-R, that any generator is balanced (cf. 7.3).

4.11.6 Let V be a faithful and reflexive left R-module, and let $V^* = \operatorname{Hom}_R(V, R)$ be the canonical (A, R)-bimodule, where $A = \operatorname{End}_R V$.

(a) Then V is canonically isomorphic to a right ideal of A if and only if R embeds in V^*.

(b) A sufficient condition for (a) is that R be a domain.

(c) When V is finitely generated and projective over R, or when R is right self-injective, then (a) holds if and only if R embeds in V.

4.11.7 If R is a right Ore domain, then the right quotient field of R is the injective hull of R in mod-R.

4.11.8 (R. E. Johnson) If no element of R annihilates an essential right ideal, then the injective hull $\hat{R}$ of R in mod-R is a right self-injective ring such that every principal right (left) ideal is generated by an idempotent. (In the terminology of Chapter 11, $\hat{R}$ is a regular ring. Cf. Johnson quotient rings, Chapter 16.) In this case, the conditions (a), (b), and (c) are equivalent: (a) $\hat{R}$ is right Artinian and Noetherian; (b) $\hat{R}$ is left Artinian and Noetherian; (c) any family of independent right ideals of R is finite.

4.11.9 If R is an entire ring, then $\hat{R}$ is a simple right selfinjective ring (cf. Exercise 8).

4.11.10 In 4.11.9, show that every finitely generated right ideal $\neq 0$ of $\hat{R}$ is $\approx \hat{R}$, or is $= 0$.

4.12 Lemma. *If U is a (B, A)-bimodule such that the canonical maps*

$$A \to \operatorname{Biend} U_A = \operatorname{End}_B U$$
$$B \to \operatorname{Biend}_B U = \operatorname{End} U_A$$

are isomorphisms, then the canonical maps

$$\text{center } A \to \text{center } B$$
$$\text{center } B \to \text{center } A$$

are isomorphisms. In particular, if U is a generator in mod-A, *then there is a canonical isomorphism* center $A \to$ center $\operatorname{End} U_A$.

Proof. The canonical map

$$\begin{cases} A \to \mathrm{End}_B U \\ a \mapsto a_d \end{cases}$$

induces a canonical map center $A \to$ center End U_A. Assume that $B = \mathrm{End}\, U_A$ and $A = \mathrm{End}\, U_B$. Then there are canonical maps center $A \to$ center B and center $B \to$ center A which are inverses of each other, and hence these maps are isomorphisms. The general case is a consequence. Hence, if U is a generator in mod-A, and $B = \mathrm{End}\, U_A$, then by 4.1, $A \approx \mathrm{End}_B U$ canonically, and then the result just proved applies. ▯

This lemma implies that center B_n is the set of diagonal matrices diag (c), where $c \in C =$ center B, and that

$$\text{center}\,(e_{ii} B_n e_{ii}) = e_{ii}\,(\text{center}\, B_n)\, e_{ii} \approx C$$

for any idempotent matrix unit e_{ii}.

4.13 Cut Down Proposition. *Let U be projective in R-mod, let*

$$R^{(I)} \xrightarrow{p} U \to 0$$

be exact in R-mod, and let $T = \mathrm{trace}_R U$. If R_0 is any subring of R containing T, then U is projective over R_0, and p induces an exact sequence

$$R_0^{(I)} \to U \to 0.$$

Furthermore, the canonical map $\mathrm{End}_R U \to \mathrm{End}_{R_0} U$ is an isomorphism. If U is faithful over R, then center $R_0 = R_0 \cap$ center R.

Proof. Let $R_0^{(I)} \supseteq T^{(I)}$ be the canonical inclusion. Then,

$$p(R_0^{(I)}) \supseteq p(T^{(I)}) = Tp(R^{(I)}) = TU = U.$$

Since U is projective over R, then

$$R^{(I)} = \ker p \oplus W,$$

and p induces an isomorphism $\overline{p} : W \to U$ in R-mod. Then

$$R_0^{(I)} = (\ker p \cap R_0^{(I)}) \oplus W,$$

and $\overline{p} : W \to U$ is an isomorphism in R_0-mod. This shows that U is projective over R_0.

Now let $t \in T$, $u \in U$, $b \in R$, and $f \in \mathrm{End}_{R_0} U$. Then, since $bt \in T \subseteq R_0$,

$$b(tu)f = bt(u)f = (btu)f.$$

Since $TU = U$ is generated by the set $\{tu \mid t \in T, u \in U\}$, it follows that $f \in \mathrm{End}_R U$. This proves that $\mathrm{End}_R U = \mathrm{End}_{R_0} U$.

Finally, if $x \in$ center R_0, $b \in R$, and $t \in T$, then

$$x(bt) = (bt)x = (bx)t.$$

Then, $(xb - bx)T = 0$ and $(xb - bx)U = 0$, so $xb = bx$ when U is faithful. Thus, $x \in$ center R. ▯

4.14 Proposition. *Let U be a (B, A)-bimodule, with $B = \mathrm{End}\, U_A$ and $A = \mathrm{End}_B U$, such that the two equivalent conditions* (a) *and* (b) *hold:*

(a) *U is finitely generated projective over B;*

(b) *U is a generator in* mod-A.

4.14.1 *Then, the center C of B is canonically isomorphic to center A, and for any subring B_0 of B containing $T = \mathrm{trace}_B U$, in particular, for $B_0 = C + T$, it is true that U is finitely generated and projective over B_0,* $\mathrm{trace}_{B_0} U = T$, *and* $A = \mathrm{End}_{B_0} U$.

4.14.2 *Then the canonical isomorphisms of 4.7 hold:*

$$\begin{cases} \text{(right ideals of } B_0)T \to A\text{-submodules of } U \\ \qquad I = IT \mapsto IU \end{cases}$$

and

$$\begin{cases} T\text{(ideals of } B_0)T \to \text{ideals of } A \\ \qquad TIT = I \mapsto \sum_{f \in \mathrm{Hom}_A(U,A)} f(IU). \end{cases}$$

Proof. The equivalence of (a) and (b) is 4.1. Then center B = center A by 4.12, and 4.14.1 follows from 4.13. Then 4.14.2 follows from 4.6. ▯

Rings with Lateral Ideals

We define a lateral left ideal in the next lemma.

4.15 Lemma. *A nonzero left ideal V of a ring A is* **lateral** *if there holds either $Va = 0$ or ${}^\perp a \cap V = 0 \;\forall a \in A$. Then vA is a lateral right ideal $\forall v \in V$, $v \neq 0$, and* End vA_A *is a domain.*

Proof. To show that vA is a lateral right ideal, assume that $b^\perp \cap vA \neq 0$ for some $b \in A$, say $bvc = 0$ with $c \in A$. Then $bv \in {}^\perp c \cap V$. Since $vc \neq 0$, then $Vc \neq 0$, so ${}^\perp c \cap V = 0$, so $bv = 0$. Thus $bvA = 0$, proving that vA is lateral. Since V is lateral, $V^\perp = v^\perp$, and for any $x,y \in A$, $x \notin v^\perp$ and $y \notin v^\perp \Rightarrow xy \notin v^\perp$, so that

$$(\text{idealizer } v^\perp)/v^\perp$$

is a domain. Since $\operatorname{End} A/v_A^\perp =$ (idealizer $v^\perp)/v^\perp$ (see 3.61), and since $A/v^\perp \approx vA$ canonically, we have what we want. ▯

Remarks. 1. Any minimal left annulet V is lateral. Cf. 7.24.

2. If V is lateral, any nonzero left ideal $W \subseteq V$ is lateral.

3. If V is lateral, so is Va for any $a \in A$ such that $Va \neq 0$.

4. Any ring with lateral left ideals has lateral right ideals.

5. In the lemma, $\operatorname{End}_A Av$ is also a domain.

4.16 Theorem. *A simple ring A is isomorphic to the endomorphism ring of a balanced finitely generated projective left module over a domain B if and only if A has a lateral left ideal. In this case, $B = \operatorname{End} U_A$ has a least ideal.*

Proof. If A has a lateral left ideal V, then for some $x \in V$, the endomorphism ring of $U = xA$ is a domain by the lemma. Since A is simple, U is a generator of mod-A, and the result follows from 4.1. Furthermore, by 4.8, $B = \operatorname{End} U_A$ has a least ideal $T = \operatorname{trace}_B U$.

Let U be the balanced finitely generated projective left module over the domain B such that $A = \operatorname{End}_B U$. Since U is balanced, that is, since $B \approx \operatorname{End} U_A$ canonically, and since by 4.1, U is a generator of mod-A, then there is a canonical isomorphism

$$[\ ,\ ] : \operatorname{Hom}_A(U, A) = \operatorname{Hom}_B(U, B)$$

such that

$$[u, f]v = uf(v) \quad \forall\, u, v \in U, \quad f \in \operatorname{Hom}_A(U, A).$$

By 4.11.6, we can assume that $U \subseteq A$ in mod-A. Let $a \in A$ be such that $av = 0$ for some nonzero element v of U. Then the left multiplication of U by a is an element a' of $\operatorname{Hom}_A(U, A)$, and then $[u, a']\, v = 0$. Now $[u, a']$ is an element of the domain B, and v is an element of the projective B-module U which is torsionfree, so it follows that $[u, a'] = 0 \;\forall\, u \in U$, whence $a' = 0$, that is, $aU = 0$. This proves that U is lateral. ▯

4.17 Corollary. *If U is finitely generated projective and balanced over a domain B, then $A = \operatorname{End}_B U$ has lateral right and left ideals. If B is a right Ore domain, this remains true without assuming U is balanced over B.*

Proof. The proof of part of the last proposition did not require the simplicity of A. If B is a right Ore domain, then $R = \operatorname{Biend}_B U$ is contained in the maximal rational extension of B which is actually the right quotient field of B. (Proof?) Thus, R is a domain, U is finitely generated projective and balanced over R, and $A \approx \operatorname{End}_R U$ canonically, so the first result applies. ▯

4.18 Corollary. *A ring A is a simple ring with lateral left ideals if and only if A is isomorphic to the endomorphism ring of a balanced finitely generated projective module U over a domain B such that $T = \text{trace}_B U$ is a least ideal. In this case, then U is finitely generated and projective over*

$$B_0 = T + C,$$

where $C = \text{center } B$ is a field canonically isomorphic to center A. Furthermore, B_0 has at most three ideals, and $A = \text{End}_{B_0} U$ canonically.

Proof. This follows from 4.16, 4.8.6, and 4.14.1. Then, the fact that the center of a simple ring is a field implies that $C \approx \text{center } A$ is a field, and so $B_0 = C + T$ has at most three ideals. ▯

Rings with Uniform Right Ideals

We repeat a concept introduced in Exercise 3.61. An object U of mod-A is **uniform** if and only if U satisfies the equivalent conditions:

(1) Any two nonzero subobjects of U have nonzero intersection.

(2) The injective hull of U is indecomposable.

(3) Every nonzero submodule is indecomposable.

4.19 Definition and Proposition. *The following three conditions on a right R-module M are equivalent ,in which case we say that M has* ***finite (Goldie) dimension****, and write* $\dim M < \infty$.

4.19.1. *No infinite direct sum of nonzero modules can be embedded in M.*

4.19.2 *There exists a finite independent set of uniform submodules $\{U_i\}_{i=1}^n$ such that $\sum_{i=1}^n U_i$ is an essential submodule.*

4.19.3 *The injective hull of M is a direct sum of finitely many indecomposable modules $\{E_i\}_{i=1}^m$.*

Any Noetherian module has finite dimension. Any nonzero submodule of a finite dimensional module contains a nonzero uniform module.

Comment on the proof. The proposition is proved elsewhere in the text, so momentarily the proof is an exercise. Actually, it is possible to prove more than is stated, namely, $n = m$. This deeper result is not presently required, and is a consequence of the Krull-Schmidt Remak theorem discussed in Chapters 17, 18, and 21.

Any Noetherian module evidently satisfies 4.19.1, hence has finite dimension. Moreover, 4.19.2 shows that any nonzero finite dimensional module M contains a nonzero uniform module. Finally, note that any submodule of M is then also finite dimensional. ▯

4.20 Proposition. *Let U be a finitely generated projective faithful left module over B, and let $A = \mathrm{End}_B U$. Then U has finite dimension n in* mod-A *if and only if B has finite dimension n in* mod-B.

Proof. By (1) of 4.6, if W_1 and W_2 are A-submodules of U, then

$$(1) \qquad W_1 \cap W_2 = 0 \Leftrightarrow [I(W_1) \cap I(W_2)]T = 0.$$

Since $TU = U$, this is equivalent to the requirement that $I(W_1) \cap I(W_2) = 0$. We now prove the proposition for the $n = 1$ case. Thus, U is uniform in mod-A if and only if it is true that

$$(2) \qquad I \cap K = 0 \Rightarrow I = 0 \quad \text{or} \quad K = 0$$

for all right ideals $I = IT$ and $K = KT$ of B. Now B is uniform in mod-B if and only if (2) holds for all right ideals I and K of B, and, therefore, U is uniform in mod-A. Conversely, if (2) holds, and if $H \cap Q = 0$ for two right ideals of B, then $HT \cap QT = 0$. Then $HT = 0$ or $QT = 0$, whence either $H = 0$, or $Q = 0$. This proves that B is uniform in mod-B. This proves the $n = 1$ case, and the general case may be similarly proved by induction. □

4.21 Proposition[1]. *Let U be a uniform right ideal of a ring A.*

4.21.1 *Then*

$$X = \{x \in A \mid x^\perp \cap U \neq 0\}$$

is a left ideal containing the ideal XU.

4.21.2 *If A is a simple ring, or if A is a prime ring with a.c.c. on left annulets, then $X = {}^\perp U$. In either case, $B = \mathrm{End}\, U_A$ is a right Ore domain, and U is lateral.*

4.21.3 *Any simple ring with uniform right ideal has finite right* (*Goldie*) *dimension.*

Proof. First note that

$$(x + y)^\perp \supseteq x^\perp \cap y^\perp$$
$$(ax)^\perp \supseteq x^\perp$$

$\forall\, x, y \in X,\ a \in A$, which proves that X is a left ideal of A. If $x \in X$, $u \in U$, then $xu \in X$. For if $uU = 0$, then $xu \in X$, and if $uU \neq 0$, then $x^\perp \cap uU \neq 0$, and then $(xu)^\perp \cap U \neq 0$. Since each element a

[1] Everything in this proposition and proof, with the exception of the statement and proof that B is a right Ore domain when A is simple, is due to Goldie [58, pp. 593—594, Lemmas 2.2 and 2.3]. The other part essentially is due to R. E. Johnson [53] (assuming uniform left ideals) (cf. Faith and Utumi [65b]. Nevertheless, a simple ring A with uniform right ideal has a.c.c. on left [and right] annulets (Proof?), so the first case is included in the second.

of the ideal XU is a finite sum of elements of the form xu, then $a^{\perp} \cap U$ contains the finite intersection of the annihilators of such elements, hence $a^{\perp} \cap U \neq 0$. This proves that $X \supseteq XU$. Then, if A is simple, since $X \neq A$, then $XU = 0$, and therefore X is the left annihilator of U, and this implies that U is lateral.

Next assume that A is a prime ring with a.c.c. on left annulets, that is, d.c.c. on right annulets, and let M be an annulet which is minimal in the set S of nonzero right annulets of A which are finite intersections

$$a_1^{\perp} \cap \cdots \cap a_n^{\perp} \neq 0$$

of right annihilators of elements $a_1, \ldots, a_n$ in XU. Since $XU \subseteq X$, then $M \cap U \neq 0$. If N is another minimal annulet in S, then also $N \cap U \neq 0$, and then $M \cap N \neq 0$ since U is uniform. Since $M \cap N$ in an element of S, then $M = N$ is the only minimal element of S, that is, $XM = 0$, and $XUM = 0$. Since A is prime, and $M \neq 0$, then $XU = 0$, and again $X = {}^{\perp}U$.

To prove in either case that B is a domain, let $f \in B$ be such that $\ker f \neq 0$. If $u \in U$, and $uU \neq 0$, then $\ker f \cap uU \neq 0$, hence $f(u)^{\perp} \cap U \neq 0$. This shows that $f(u) \in X = {}^{\perp}U$ that is, $f(u)U = 0$ $\forall\, u \in U$. Then $(\operatorname{im} f)^2 = 0$. Since every nonzero right ideal of a prime ring is faithful, then $\operatorname{im} f = 0$. Since every nonzero endomorphism f of U in mod-A is therefore monic, then $B = \operatorname{End} U_A$ is a domain.

To conclude in either case that B is a right Ore domain, if f and g are nonzero elements of B, then $\operatorname{im} f \cap \operatorname{im} g \neq 0$, and so $(\operatorname{im} f \cap \operatorname{im} g)U \neq 0$. Write $x = f(u) = g(v)$ for elements u and v in U such that $xU \neq 0$. Then $uU \neq 0$ and $vU \neq 0$, and therefore the left multiplications of U by u and by v are endomorphisms u_s and v_s respectively such that

$$x_s = f(u)_s = fu_s = gv_s,$$

that is, B has the common right multiple (Ore) condition.

4.21.3 is left as an exercise. □

The Structure of Simple Noetherian Rings

In this section, a simple right Noetherian ring A is characterized via the endomorphism ring of a uniform right ideal of A. For this reason, we have the next proposition, which similarly characterizes simple rings with uniform right ideals. The final result in this section, 4.23 B, reduces the question of when A is similar to a domain B to a question on the structure of the right ideals of B.

4.22 A Proposition. *Let A be a ring. The following conditions are equivalent:*

4.22.1 A *A is a simple ring with a uniform right ideal U.*

4.22.2 A *A is isomorphic to the endomorphism ring of a finitely generated projective module W over a right Ore domain R with at most three ideals such that $T = \mathrm{trace}_R W$ is a least ideal.*

4.22.3 A *A is isomorphic to the endomorphism ring of a finitely generated projective module V over a right Ore domain B_0 having center C a field such that either B_0 is a simple ring, or else $B_0 = C + T_0$, where $T_0 = \mathrm{trace}_{B_0} V$, is a ring with precisely three ideals.*

Proof. (2) ⇒ (1) by 4.8.3 and 4.20.

(1) ⇒ (3) $B = \mathrm{End}\, U_A$ is a right Ore domain by 4.21, and U is finitely generated projective over B by 4.1, and then 4.14 applies. Finally, (3) ⇒ (2) is trivial. □

4.22 B Proposition. *A ring A is a simple right Noetherian (Artinian) ring if and only if A is isomorphic to the endomorphism ring of a finitely generated projective left module U over a right Ore domain B_0 satisfying* 4.22.3 A *and having the a.c.c. (d.c.c.) on idempotent right ideals contained in T_0. When this is so, then $B = \mathrm{End}\, U_A$ satisfies the a.c.c. (d.c.c.) on principal right ideals. Furthermore, A is right Artinian if and only if B_0 is a field*[1]*. In this case $B_0 = B$. Furthermore there are isomorphisms $U \approx B^n$ in B-mod, and $A \approx B_n$, the total matrix ring over B of degree n.*

Proof. If A is right Noetherian, then A contains a uniform right ideal U by 4.19, so 4.22.3 A holds, and then the a.c.c. on idempotent right ideals of B_0 contained in T_0 follows from 4.7 and 4.8.6. Next, if

$$x_1 B \subseteq \cdots \subseteq x_n B \subseteq x_{n+1} B \subseteq \cdots$$

are principal right ideals of B, then the a.c.c. on A-submodules of U implies that $x_n U = x_{n+1} U$ for some integer $n > 0$. Since B_0 and B are right Ore domains, then x_{n+1}^{-1} lies in the right quotient field K of B. Now, $xU = yU$ for two elements x and y of B implies that

$$xT = \sum_{f \in \mathrm{Hom}_B(U,B)} (xU) f = yT,$$

and so $y^{-1} x = u$ satisfies $uT = T$, and $uU = U$. Now any epimorphism u of a Noetherian module is an automorphism (Proof?), that is, u is a unit of $B = \mathrm{End}\, U_A$. Thus, $x_{n+1}^{-1} x_n$ is a unit of B which implies that $x_n B = x_{n+1} B$. This proves that B satisfies the a.c.c. on principal right ideals.

---

[1] The proof shows in the case of the d.c.c. on principal right ideals, that $B_0 = B$ is a field, and A is right Artinian. The theorem in this case is the classical Wedderburn-Artin theorem 7.9.

Conversely, assuming that 4.22.3A holds, then the a.c.c. on idempotent right ideals of B_0 contained in T_0 implies, by 4.7, 4.8.6 and 4.9.2, respectively, that U, hence A, is right Noetherian. Dually, for A right Artinian.

Finally, by 4.9, A has a minimal right ideal if and only if B_0 has a minimal right ideal, in which case B_0 is a field, whence $B = B_0$, U is a finite dimensional vector space over B_0, and $U \approx B^n$, for some integer $n > 0$. Then $A = \text{End}_B U \approx B_n$, by 3.33. This happens, in particular, whenever A is right Artinian, or equivalently, B_0 is a field. (In reference to Footnote 1, any minimal principal right ideal is a minimal right ideal.) □

Although 4.22B is a characterization of simple right Noetherian rings in terms of right Ore domains, the Ore domains which appear are in general not isomorphic and not simple. Cf. 4.40.

4.22C Corollary. *If U is finitely generated projective over any ring B, and if B is simple, then $A = \text{End}_B U$ is simple, and U is an (B, A)-bimodule progenerator.*

Proof. Simplicity of A follows from 4.8.6. Since B is simple, and $\text{Hom}_B(U, B) \neq 0$, then U is a generator of B-mod, and so U is a progenerator by 4.3. □

4.23A Proposition. *Consider the following equivalent conditions:*

(a) *U is a finitely generated projective faithful over a right Ore domain R containing a least ideal $T = \text{trace}_R U$.*

(b) *U is uniform right ideal of a simple ring A, R is a subring of $B = \text{End}\, U_A$ over which U is finitely generated and projective, and $B = \text{Biend}_R U$.*

Then we have the following:

4.23.1A *The right quotient field D of B is the right quotient field of R, and also, for any right ideal I of R, the right quotient field of the subring $(I : I) = \{d \in D \mid dI \subseteq I\}$.*

4.23.2A *Any right ideal Y of A with entire endomorphism ring is uniform, and any uniform right ideal is isomorphic to a submodule of U. Let Y be any uniform right ideal of A contained in U, and let I be a right ideal of R such that $IU = Y$ and $IT = I$, where T is the trace of U in R. Then there are canonical isomorphisms*

$$(I : I) \xrightarrow{g} \text{End}\, I_R \xrightarrow{h} \text{End}\, Y_A,$$

where

$$g(d)(x) = dx$$

and

$$h(f)(xu) = f(x)u$$

$\forall\, d \in D, x \in I, u \in U, f \in \operatorname{End} I_R$. *In particular,* End Y_A *is a right Ore domain isomorphic to the endomorphism ring of a right ideal of* R, *and* D *is the right quotient field.*

Proof. (a) $\Rightarrow$ (b) Assuming (a), then U is uniform in mod-A by 4.20, where $A = \operatorname{End}_R U$. Since R is a domain, then R embeds in the R-dual module U' of U, and then U embeds canonically in $A = \operatorname{End}_R U$ (cf. 4.11.6). Furthermore, 4.8.6 implies that A is simple.

(b) $\Rightarrow$ (a) B is a right Ore domain by 4.20. By 4.6, B is a rational extension of R, and so R is a right Ore domain with the same right quotient field D. Furthermore, the trace of U in B is the least ideal of B by 4.8, and then the trace of U in R is the least ideal of R by 4.6.

Part 4.23.1 A Let $d = ab^{-1}$ be any element of D, where $a, b \in R$. Since R is right Ore, there is an element $x \in R$ such that $x \neq 0$, and such that $ax, bx \in I$. Then, $d = (ax)(bx)^{-1}$, and ax and bx are elements of $(I : I)$. This proves that D is the right quotient field of $(I : I)$.

If Y is a right ideal such that $S = \operatorname{End} Y_A$ is a domain, then by 4.20 and 4.21, $\dim S_S = \dim Y_A < \infty$. It is an exercise (e.g. 7.18.2) to show that then $\dim S_S = 1$. Therefore, Y is a uniform right ideal. This proves that if $a \in A$, then either $aY = 0$ or $ay \neq 0$ for every nonzero $y \in Y$. Since $UY \neq 0$, this implies that Y embeds in U and $aY \neq 0$ is a submodule of U for some $a \in U$. Next assume that $Y \subseteq U$, and let $I = I(Y)$ denote the right ideal of R given by 4.7 such that $IU = Y$ and $IT = I$, where $T = \operatorname{trace}_R U$. If D denotes the right quotient field of B, then there is a ring monomorphism

$$\begin{cases} \operatorname{End} I_R \to D \\ \qquad f \mapsto f', \end{cases}$$

where the inverse mapping $f' \mapsto f$ is induced by the left multiplication of I by f'. This is true since D is the injective hull of R in mod-R (Proof?). Hence, assume that

$$Q = \operatorname{End} I_R = \{d \in D \mid dI \subseteq I\}.$$

Now, since U is finitely generated projective over R, we may assume that U is contained in a finitely generated free left R-module, say $U \subseteq R^n$. Embed R^n in D^n canonically. Then, du is an element of D^n, for any $d \in D$, and $u \in U$. However, if $q \in Q$, then

$$qY = q(IU) = (qI)U \subseteq IU = Y,$$

so that Y is a canonical left Q-module. Hence, there is a ring monomorphism

$$t \begin{cases} Q \to \operatorname{End} Y_A \\ q \mapsto \bar{q} \end{cases}$$

such that

$$\bar{q}(y) = qy$$

$q \in Q$, $y \in Y$. Moreover, t is an isomorphism, since any $f \in \text{End } Y_A$ defines an element $f_0 \in Q$ such that

$$f_0(x)u = f(xu)$$

$\forall x \in I$, $u \in U$ and the mapping

$$\begin{cases} \text{End } Y_A \to Q \\ \quad f \mapsto f_0 \end{cases}$$

is the inverse of t. □

The next proposition gives a classical response to the question of when are the domains in 4.22A simple. (Cf. Proposition 8.27.)

4.23B Proposition. *Let A be a simple ring with a uniform right ideal U, and let D denote the right quotient field of the domain $B = \text{End } U_A$.*

The following conditions are equivalent:

4.23.1B *A is isomorphic to the endomorphism ring of a finitely generated projective module W over a simple domain R.*

4.23.2B *A is isomorphic to the endomorphism ring of a finitely generated projective module W over a domain S such that the endomorphism ring of some right ideal of S is a simple ring.*

4.23.3B *Some right ideal $\neq 0$ of B has simple endomorphism ring.*

4.23.4B *The endomorphism ring of some A-submodule of U is simple.*

4.23.5B *The endomorphism ring of some uniform right ideal of A is simple.*

4.23.6B *There are a right ideal $I \neq 0$ of B and a simple subring Q of D such that $QI \subseteq I$ and D is the right quotient field of Q.*

4.23.7B *There is a progenerator V in* mod-*A such that $L = \text{End } V_A$ is a domain, and in this case L is a simple ring. (In terminology presently presented in 4.29, A is similar to a domain L, and* $\text{Hom}_A(V, \ )$ *induces a category equivalence* mod-$A \approx$ mod-L.)

Proof. (1) ⇒ (2) is trivial.

(2) ⇒ (5). First, W is canonically isomorphic to a right ideal of A by 4.11.6; in fact, W is uniform and embeds in U by 4.23.2A. Furthermore, if I is a right ideal of S such that $\text{End } I_S = R$ is simple, then 4.23A implies that IW is a uniform right ideal of A with endomorphism ring isomorphic to R.

(5) ⇒ (4) is trivial, and (4) ⇒ (3) follows from 4.23A: If Y is the A-submodule of (4), and if I is the right ideal of B such that $IU = Y$, then $\text{End } I_B = \text{End } Y_A$ is simple.

(3) ⇒ (6). If I is the right ideal of (3), then by 4.23A, $Q = (I : I)$ is isomorphic to $\text{End } I_B$, and hence Q is simple. Furthermore, by 4.23.1A, D is the right quotient field of Q.

(6) $\Rightarrow$ (7). Assuming (6), Q is contained in $R = (I : I) = \operatorname{End} I_B$. Furthermore, by 4.23 A, $Y = IU$ is a uniform right ideal, and $R = \operatorname{End} Y_A$. Now, if J is any nonzero ideal of R, since the right quotient field of Q is D, then $J \cap Q \neq 0$, and simplicity of Q implies that $J \cap Q = Q \subseteq J$. Thus, $1 \in J$, and so $J = R$. This proves that R is simple. Furthermore, since A is simple, Y is a generator of mod-A, hence by 4.1, Y is finitely generated projective over $R = \operatorname{End} Y_A$. Then simplicity of R implies that Y is a generator in R-mod, and hence Y is a progenerator of R-mod and, hence, of mod-A. This completes the proof of (6) $\Rightarrow$ (7), and (7) $\Rightarrow$ (1) is trivial. $\square$

This theorem is a "classical" answer to the question of when a ring $A = \operatorname{End}_B U$ is similar to a domain since the answer is reduced to a question on the structure of the domain B and its quotient field.

4.24 Morita and Popesco-Gabriel Full Embedding Theorem. *Let R be a ring, and let $C = $ mod-R (or in terminology of Chapter 14, C can be any Grothendieck abelian category). Then any generator U of C induces a full embedding*[1]

$$h^U : C \rightsquigarrow \text{mod-}\operatorname{End}_C U.$$

Proof. Let $A = \operatorname{End}_C U$, that is, $A = \operatorname{End} U_R$ for $C = $ mod-R, and let $T : C \rightsquigarrow$ mod-A be the indicated functor induced by h^U. The object is to find, for any morphism $t \in \operatorname{Hom}_A(TX, TX')$, a morphism $b : X \to X'$ of C such that $t = T(b)$. Let $I = TX = \operatorname{Mor}_C(U, X)$. Then $t(f) \in TX' = \operatorname{Mor}_C(U, X')\ \forall f \in I = TX$. Consider the coproduct mappings

$$g = \coprod_{f \in I} f : U^{(I)} \to X$$

and

$$g' = \coprod_{f \in I} tf : U^{(I)} \to X'.$$

Since U is a generator in C, $g : U^{(I)} \to X$ is epic. By 2.15, for each finite subset J of I, there is a row-exact commutative diagram:

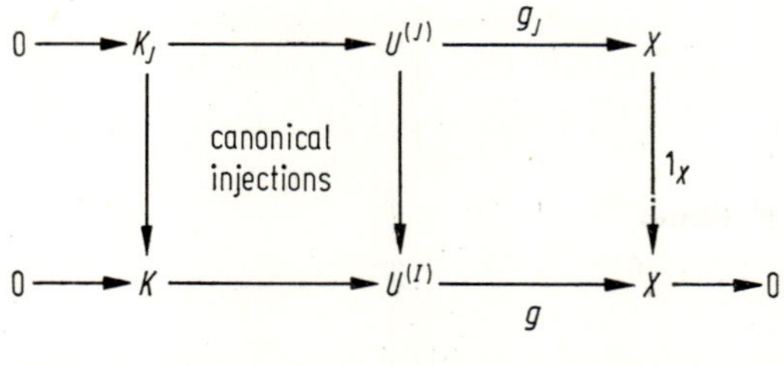

[1] In some places in the literature, an embedding is required to be an exact functor in addition to being faithful. This is not required here. The Mitchell Full Embedding Theorem (Chapter 14) is an example of an exact embedding.

In order to obtain the map $b : X \to X'$ such that $t = Tb$, it is necessary and sufficient that $\ker g' \supseteq K$, that is, that g' kills K. Now every $k \in U^{(I)}$ has finite support, hence belongs to some $U^{(J)}$, J depending on k, and therefore, every $k \in K$ belongs to some K_J. It follows that g' kills K if and only if g' kills every K_J. (If C is a Grothendieck category, then $K = \varinjlim K_J$, and $X \approx U^{(I)}/K \approx \varinjlim U^{(I)}/K_J$, by 14.4, from which the corresponding statement follows.) Since U is a generator, in order to prove that g' kills K it suffices to prove that g' kills every morphism $h : U \to U^{(I)}$ such that

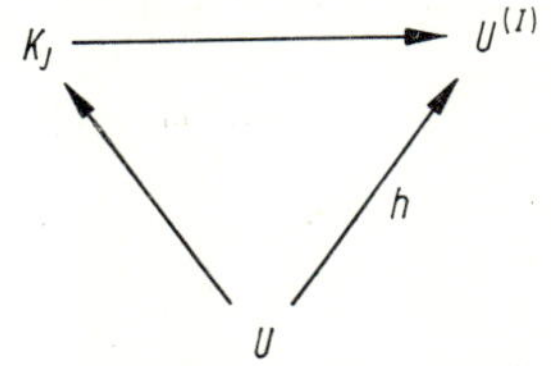

commutes.

Write $h = k_J h_1$, where $k_J : K_J \to U^{(I)}$ and $h_1 : U \to U^{(J)}$ are the compositions $K_J \to U^{(J)} \to U^{(I)}$ and $U \to K_J \to U^{(J)}$. We will show that g' kills h_1, hence h. Let u_f and p_f denote the injections and projections for the coproduct $U^{(J)}$. Then

$$0 = g_J h_1 = \coprod_{f \in J} f p_f h_1 = \coprod_{f \in J} f \alpha_f, \tag{1}$$

where $\alpha_f = p_f h_1$. Since $f \in I = TX = \mathrm{Mor}_C(U, X)$, and $t \in \mathrm{Hom}_R(TX, TX')$, then $t(f) \in \mathrm{Mor}_C(U, X')$, and $t(f\alpha) = t(f)\alpha$ for all $\alpha \in A$. Now $\alpha_f = p_f h_1 \in \mathrm{Mor}_C(U, U) = A$, and by (1), $f\alpha_f = 0$ for every $f \in I$, so

$$t(f)\,\alpha_f = t(f\alpha_f) = 0$$

$\forall f \in I$, and then

$$0 = g'_J h_1 = \coprod_{f \in J} t(f)\,\alpha_f = 0$$

($\forall$ finite subsets $J \subseteq I$.) Therefore, $g' h_1 = 0$, hence g' kills $K = \ker g$. Accordingly, there exists a morphism $b : \operatorname{coker} g \to \operatorname{coker} g' \subseteq X'$, that is, a morphism $b : X \to X'$ of C such that $g' = bg$, and then $t = T(b)$. □

An object A of a category C is said to be **small** if every morphism $f : A \to \coprod_{i \in I} A_i$ of A into a coproduct of objects of C factors through a finite coproduct, that is, there is a finite subset J of I, depending on f, and a commutative diagram

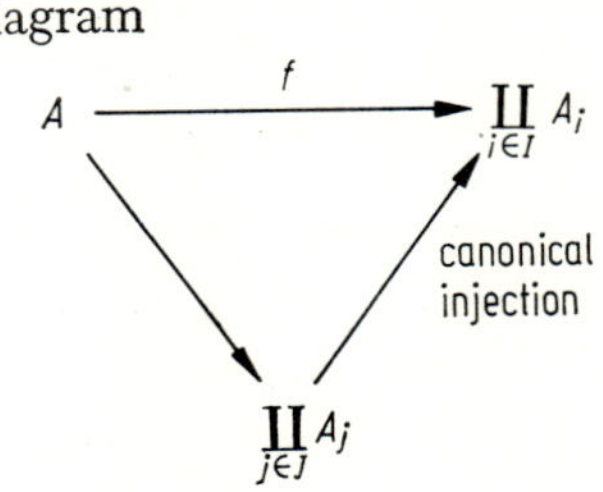

4.25 A Proposition for Additive Category. *Let* $\{A_i \mid i \in I\}$ *be a family of objects, and let* $\coprod_{i \in I} A_i$ *be a coproduct with injections* $\{u_i\}$ *and projections* $\{p_i\}$*. Let* A *be any object, and let* J *be any finite subset of* I*. Then a morphism* $f: A \to \coprod_{i \in I} A_i$ *factors through the finite coproduct* $\coprod_{j \in J} A_j$

$$A \xrightarrow{\bar{f}} \coprod_{j \in J} A_j \xrightarrow{u_{IJ}} \coprod_{i \in I} A_i$$

if and only if

$$f = \sum_{j \in J} u_j p_j f.$$

Proof. Let $\bar{u}_i$, $\bar{p}_j$ be the *ith*-injection and *jth*-projection, respectively, for the finite coproduct $\coprod_{j \in J} A_j$. Then the morphism u_{IJ} can be written: $u_{IJ} = \sum_{i \in J} u_i \bar{p}_i$. If f factors as stated, then $f = \sum_{j \in J} u_j \bar{p}_j \bar{f}$, and then, for any $k \in J$, we have

$$p_k f = \sum_{j \in J} p_k u_j \bar{p}_j \bar{f} = \delta_{kj} \bar{p}_j \bar{f} = \bar{p}_k \bar{f}.$$

Thus, $f = \sum_{j \in J} u_j p_j f$ as stated. Conversely, if f can be so written, define $\bar{f} = \sum_{j \in J} \bar{u}_j p_j f$. Then

$$u_{IJ} \bar{f} = \sum_{k \in J} u_k \bar{p}_k \sum_{j \in J} \bar{u}_j p_j f_j = f. \quad \square$$

4.25 B Proposition and Definition. *Let* C *and* D *be categories, and let* $T: C \rightsquigarrow D$ *be covariant. If* $\{A_i \mid i \in I\}$ *is a family of objects of* C *such that both* $\coprod_{i \in I} A_i$ *and* $\coprod_{i \in I} T(A_i)$ *exist, then there is a unique morphism* t *in* D *such that*

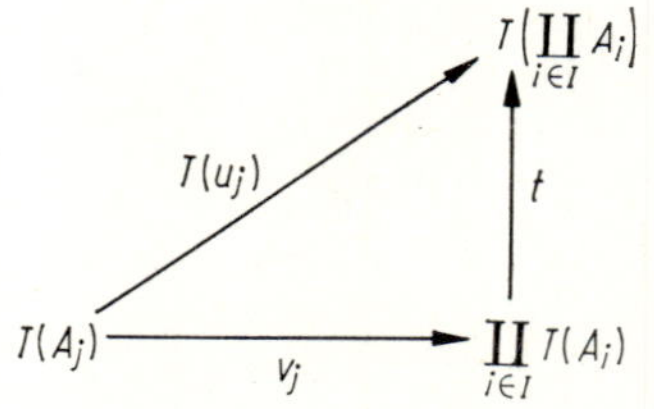

commutes for all $j \in I$*, where* $\{u_j\}$ *and* $\{v_j\}$ *are the respective canonical injections. The morphism* t *is called* **canonical.** *The functor* T *is said to* ***preserve coproducts*** *in case* t *is an equivalence for every coproduct in* C*. If* A *is any object in an additive category, then for the group-valued functor* h^A*, the canonical morphism*

$$\coprod_{i \in I} h^A(A_i) \to h^A(\coprod_{i \in I} A_i)$$

is injective (monic).

Proof. The morphism t is given by the definition of coproduct. Now assume that C is additive. For the object A, the functor $h^A: C \rightsquigarrow$ mod-$\mathbb{Z}$ is induced by $\mathrm{Mor}_C(A, \ )$. We will use the fact that in mod-$\mathbb{Z}$ coproducts are direct sums. Let $f \in \coprod_{i \in I} h^A(A_i)$. Since

$h^A(A_i) = \mathrm{Mor}_C(A, A_i)$, there exists a family $\{f_i : A \to A_i\}$ of morphisms with $f_i \neq 0$ for only finitely many $i \in I$ such that

$$f = \sum_{i \in I} v_i f_i.$$

Applying t yields

$$t(f) = \sum_{i \in I} (t v_i) f_i = \sum_{i \in I} h^A(u_i) f_i = \sum_{i \in I} u_i f_i.$$

If $f \neq 0$, then $f_k \neq 0$ for some k, so

$$p_k \sum_{i \in I} u_i f_i = f_k \neq 0.$$

Therefore, $\sum_{i \in I} u_i f_i \neq 0$, so t is monic. □

4.25 C Proposition for Additive Category with Coproducts. *An object A is small if and only if the group-valued functor h^A preserves coproducts.*

Proof. By the last result, the canonical morphism

$$t : \coprod_{i \in I} h^A(A_i) \to h^A(\coprod_{i \in I} A_i)$$

is monic, where $\coprod_{i \in I} A_i$ is any coproduct with injections $\{u_i\}$ and projections $\{p_i\}$.

Now suppose t is an isomorphism; that is, t is epic. If $g : A \to \coprod_{i \in I} A_i$ is a morphism, then $g \in h^A(\coprod_{i \in I} A_i)$, so we can write $g = \coprod_{i \in I} u_i g_i$ for some morphisms g_i, and then $p_k g = g_k \ \forall\, k \in I$. Thus, $g = \sum_{i \in I} u_i p_i g$, so by 4.25 A g factors through a finite coproduct. Consequently, A is small.

Conversely, if A is small, and if $g \in h^A(\coprod_{i \in I} A_i)$, then $g = \sum_{i \in I} u_i (p_i g)$. Thus, $t(\sum_{i \in I} v_i p_i g) = g$, so t is epic.

4.25 D Proposition. *Let R be a ring, and let A be an R-module. Then the following statements are equivalent:*

4.25.1 D h^A : mod-$R \rightsquigarrow$ mod-$\mathbb{Z}$ *preserves coproducts.*

4.25.2 D *A is small in* mod-R.

4.25.3 D *If*

$$S_1 \subseteq S_2 \subseteq \cdots \subseteq S_n \subseteq \cdots$$

is any ascending (countable) sequence of proper submodules of A, then $\bigcup_{n=1}^{\infty} S_n$ is a proper submodule of A.

Proof. (1) ⇔ (2) follows from the preceding proposition. Assume the notation of proof of this proposition. Then for any coproduct $\coprod_{i \in I} A_i$ in mod-R, the morphism $t : \coprod_{i \in I} h^A(A_i) \to h^A(\coprod_{i \in I} A_i)$ so defined is monic.

(3) ⇒ (2) If A is not small, then, by 4.25 A, there exists a coproduct $\coprod_{i \in I} A_i$ in mod-R with injections u_i and projections p_j such that for

a countable subset J of I we have $p_j f \neq 0$. (Otherwise, $f = \sum_{j \in J} u_j p_j f$ for a finite subset J.) Let $J = \bigcup_{n=1}^{\infty} J_n$, where J_n is a finite subset of J_{n+1}, and let $B_n = \coprod_{j \in J_n} A_j$, $n = 1, 2, \ldots$. Also let $B = \coprod_{j \in J} A_j$, and let $g : \coprod_{i \in I} A_i \to B$ be the projection. Then, likewise, $gf : A \to B$ cannot be factored through a finite coproduct. Let $S_n = (gf)^{-1} B_n$, $n = 1, 2, \ldots$. Since $\operatorname{Im}(gf) \nsubseteq B_n$, then $S_n \neq A$, $n \geq 1$, and yet $A = \bigcup_{n=1}^{\infty} S_n$.

(2) $\Rightarrow$ (3) Let $S_1 \subseteq S_2 \subseteq \cdots \subseteq S_n \subseteq \cdots$ be an ascending sequence of proper submodules with union A, let $f_n : A \to A/S_n$ be canonical, and let $h : A \to \prod_{n=1}^{\infty} A/S_n$ be the product map. Then $\operatorname{im} h \subseteq \coprod_{n=1}^{\infty} A/S_n$, so h defines a map $f : A \to \coprod_{n=1}^{\infty} A/S_n$. Clearly, f does not factor through any finite coproduct $\coprod_{n=1}^{k} A/S_n$, $k < \infty$, hence A is not small. $\square$

4.26 *Exercise.* Find an example of a small module which is not finitely generated.

4.27 **Corollary.** *In the notation of* 4.24, $h^U : C \rightsquigarrow$ mod-End U_R *is an equivalence if and only if* U *is a small projective generator.*

Proof. Smallness of an object is a categorical property. If h^U is an equivalence with inverse S, then $B = h^U U$ is small, hence, so is $SB = U$. Similarly, U is projective and a generator.

Conversely, if U is a small projective generator, then h^U induces a fully faithful functor $C \rightsquigarrow$ mod-B by the theorem, and by 2.2, then h^U is an equivalence if and only if h^U is representative. Hence, let M be an object of mod-B, and let

$$B^{(I)} \xrightarrow{f} B^{(J)} \to M \to 0$$

be exact in mod-B. (Such a sequence exists, e.g., $B^{(J)}$ can be chosen to be the free module on a set J of generators of M.) The fact (4.25 C) that h^U preserves coproducts when U is an arbitrary small object of C will now be used. Thus, if $T = h^U$, and if $\{u_i : U \to U^{(I)}\}_{i \in I}$ is the coproduct in C, then $\{T u_i : TU \to T(U^{(I)})\}_{i \in I}$ is a coproduct in mod-B. Since $B = TU$, this implies equivalences $t_1 : B^{(I)} \to T(U^{(I)})$ and $t_2 : B^{(J)} \to T(U^{(J)})$. Since $T = h^U$ is full, there exists $g : U^{(I)} \to U^{(J)}$ such that the diagram

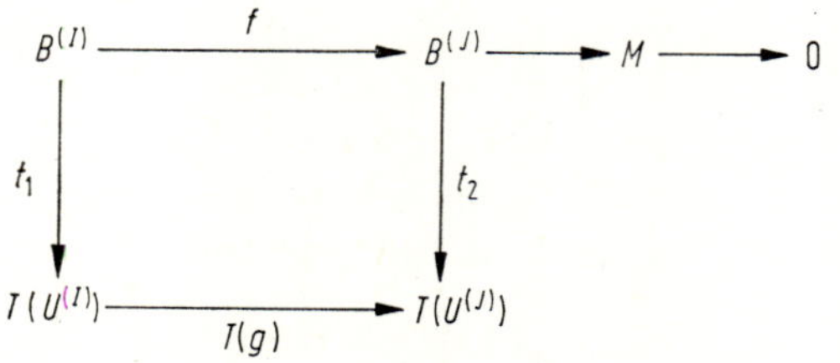

commutes. Since U is projective, $T = h^U$ is exact, hence $T(\text{coker } g) \approx \text{coker } Tg$, and then the commutative diagram implies an equivalence $M = \text{coker } f \approx T(\text{coker } g)$. This proves that $T = h^U$ is representative and, therefore, an equivalence. □

4.28 Corollary. *If R is a ring, an object U of* mod-R *is a progenerator if and only if the functor h^U :* mod-$R \rightsquigarrow$ mod-End U_R *is an equivalence.* □

4.29 Morita Theorem. *Two rings A and B are said to be* **similar,** *notationally $A \sim B$, provided that the following three equivalent conditions hold:*

4.29.1 mod-$A \approx$ mod-B.

4.29.2 A-mod $\approx$ B-mod.

4.29.3 *There exist a finitely generated projective generator (progenerator) U in* mod-A *and an isomorphism of rings $B \approx$* End U_A.

When this is so, then there is an equivalence

$$h^U = \text{Hom}_A(U, \ ) : \text{mod-}A \rightsquigarrow \text{mod-}B$$

with inverse equivalence

$$h^{U^*} = \text{Hom}_B(U^*, \ ) : \text{mod-}B \rightsquigarrow \text{mod-}A,$$

where $U^* = \text{Hom}_A(U, A)$. Thus, $h^{U^*} = (h^U)^{-1}$. Symmetrically, there is an equivalence

$$\text{Hom}_B(U, \ ) : B\text{-mod} \rightsquigarrow A\text{-mod}$$

with inverse

$$\text{Hom}_A(U', \ ) : A\text{-mod} \rightsquigarrow B\text{-mod}.$$

Any equivalence T : mod-$A \approx$ mod-B *induces, for each object X of* mod-A, *a natural isomorphism*

$$h'(X) : \text{Hom}_A(A, X) \to \text{Hom}_B(TA, TX)$$

which implies a functor equivalence $h^{TA} \approx T^{-1}$, that is, any equivalence T : mod-$A \approx$ mod-B *is representable, and T is naturally equivalent to the functor induced by the representation functor defined by the A-dual $P^* = \text{Hom}_A(P, A)$ of $P = TA$.*

Proof. Any isomorphism $f : B \to R$ of rings implies a category equivalence mod-f : mod-$R \rightsquigarrow$ mod-B. Thus, $(3) \Rightarrow (1)$ is a consequence of 4.28 and the given isomorphism End $U_A \approx B$.

Conversely, if T : mod-$A \rightsquigarrow$ mod-B is an equivalence, then for each object X of mod-A, T induces an isomorphism

$$h'(X) : \text{Hom}_A(A, X) \to \text{Hom}_B(TA, TX)$$

which is natural in X. Thus, if $h(X)$ is the compose of the natural isomorphism $X \to \mathrm{Hom}_A(A, X)$ and $h'(X)$, then there is a natural equivalence of functors

$$h : 1_{\text{mod-}A} \to h^{TA}\, T, \tag{1}$$

where $h^{TA} : \text{mod-}B \rightsquigarrow \text{mod-}A$ is induced by h^{TA}. First of all, $h(A)$ is a ring isomorphism

$$A \to \mathrm{End}\, TA_B, \tag{2}$$

which implies that $P = TA$ is an (A, B)-bimodule, and then so is $h^P Y$ for any object Y of mod-B. This explains the functor $h^P : \text{mod-}B \rightsquigarrow \text{mod-}A$ and the natural transformation (1). By symmetry, there are functor equivalences

$$h^{T^{-1}B}\, T^{-1} \approx 1_{\text{mod-}B}$$

and so

$$h^{T^{-1}B} \approx T.$$

Therefore, T is an equivalence, and h^{TA} is the inverse equivalence. By 4.28, $P = TA$ is necessarily a progenerator in mod-B and A-mod, and there is a canonical isomorphism $P^* = \mathrm{Hom}_A(P, A) \to P' = \mathrm{Hom}_B(P, B)$ by 4.1. Furthermore, P' is a progenerator in B-mod and mod-A by 4.2, so by 4.28 there are equivalences

$$\mathrm{Hom}_A(P,\ ) : A\text{-mod} \approx B\text{-mod},$$

$$\mathrm{Hom}_B(P',\ ) : B\text{-mod} \approx A\text{-mod},$$

and

$$\mathrm{Hom}_B(P,\ ) : \text{mod-}B \approx \text{mod-}A.$$

4.30 Corollary. *If U is a progenerator in* mod-A, *and if* $B = \mathrm{End}\, U_A$, *then* $T = \mathrm{trace}_B U = B$ *and in Theorem* 4.6, (1) $W \leftrightarrow I(W)$ *is an isomorphism of the lattices of A-submodules of U and right ideals of B'* (2) *mapping (B, A)-submodules of U onto ideals of B, and* (3) $J \leftrightarrow SJ$, *where* $(SJ)\,U = UJ$, *is an isomorphism of the lattices of ideals of A and ideals of B.*

Similar rings have isomorphic lattices of ideals. □

4.31 Corollary. *Let* $T : \text{mod-}A \rightsquigarrow \text{mod-}B$ *be an equivalence.*

4.31.1 *For any object Y of* mod-A, *T induces an order isomorphism*

$$[k] \mapsto [Tk]$$

of the ordered class of equivalent subobjects of Y and that of TY.

4.31.2 *T induces a bijection*

$$\begin{cases} \text{ideal classes mod-}A \to \text{ideal classes mod-}B \\ \qquad\qquad C(I) \mapsto TC(I) = C(I') \\ \qquad\qquad\quad X \mapsto TX \end{cases}$$

and a lattice isomorphism

$$\begin{cases} \text{ideals } A \to \text{ideals } B \\ \qquad I \mapsto I'. \end{cases}$$

4.31.3 *For each ideal I of A, the functor T induces an equivalence T' :* mod-$A/I \rightsquigarrow$ mod-B/I' *such that*

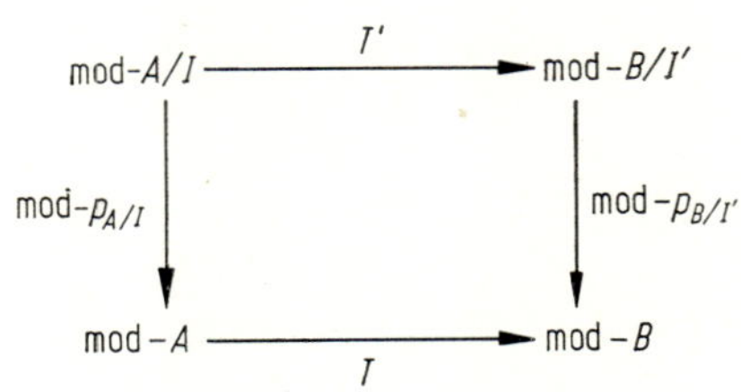

commutes, that is, $T'F = TF$ for any object or morphism F of mod-A/I. *If $T \approx h^U$ for an object U of* mod-A, *then $T' \approx h^{U/UI}$, that is, if U is a progenerator of* mod-A *with endomorphism ring canonically isomorphic to B, then U/UI is a progenerator of* mod-A/I, *with endomorphism ring canonically isomorphic to B/I'.*

Proof of 4.31.1. (Equivalent) monics is a categorical property. The fact that T is an equivalence, completes the proof.

Proof of 4.31.2. By 3.11, the set $TC(I)$ of objects of mod-B isomorphic to TX for some object $XC(I)$ is an ideal class, and the mapping $I \to I'$ such that $TC(I) = C(I')$ is bijective and order-preserving.

Proof of 4.31.3. Since T' is fully faithful and representative, then T' is an equivalence. If X is an object of mod-A/I, then $\ker f \supseteq UI$ for each morphism $f: U \to X$, and so f induces a morphism $f' : U/UI \to X$. Then there is a natural equivalence of functors

$$h' : T' \to h^{U/UI},$$

where, for all X,

$$h' \begin{cases} T'X = \operatorname{Hom}_A(U, X) \to \operatorname{Hom}_{A/I}(U/UI, X) = h^{U/UI}X \\ \qquad\qquad\qquad\qquad f \mapsto f'. \end{cases}$$

Thus, since T' is an equivalence, then U/UI is a progenerator by 4.29 and the rest follows from 4.1. □

A property P of rings is a **Morita invariant** if it is true that a ring A has property P if and only if every ring similar to A has property P. Thus, any property P of rings which for each ring A is characterizable by a categorical property of mod-A is a Morita invariant.

By the ideal proposition 3.12, and also by 4.30 and 4.31, the lattice of ideals of a ring A is a Morita invariant. A **quotient property** of rings is a property which is possessed by a ring A if and only if it is possessed by every quotient ring A/K of A.

4.32 Corollary. *A quotient Morita invariant property is a Morita invariant property.* □

A category C is **Noetherian** if every object is Noetherian, that is, the a.c.c. holds on the ordered class of equivalent subobjects of every object Y. Thus,

$$k_1 \leq k_2 \leq \cdots \leq k_n \leq \cdots$$

for any sequence of monics $\{k_i : X_i \to Y\}_{i\in\omega}$ implies that there exists an integer $m > 0$ such that $k_m \approx k_{m'} \forall m' \geq m$. A category C is **Artinian** if and only if the dual category is Noetherian, or equivalently, if and only if the d.c.c. holds on the ordered class of equivalent subobjects of every object Y of C.

4.33 *Exercises.*

4.33.1 Show that a ring A is right Noetherian (Artinian) if and only if the full subcategory of finitely generated modules is Noetherian (Artinian). Conclude that "right Noetherian (Artinian)" is a Morita invariant.

4.33.2 Show that a right Noetherian (Artinian) ring is not necessarily left Noetherian (Artinian).

*4.33.3 (Hopkins-Levitzki) Show that any right Artinian ring is right Noetherian.

4.33.4 Show that the following properties are Morita invariants:

(a) Prime.

(b) Right primitive (that is, has a faithful simple right module).

(c) Semiprime.

(d) Semiprimitive (that is, the ring has zero radical).

(e) Simple.

(f) Semisimple (that is, the ring is a direct sum of simple right modules). Show that this condition is right left symmetric and implies that each module, on either side, is semisimple.

(g) Regularity (in the sense that every right ideal is generated by an idempotent. This condition is right left symmetric).

(h) Finitude of the number of elements (but not cardinality).
(i) Cardinality if the cardinal is infinite.
(j) The ring is an injective module (self-injective).
(k) Every right ideal is projective.
(l) Every right ideal is injective.
(m) Every submodule of a projective is projective. Also the dual of this.
*(n) Every cyclic module is injective. Also the dual of this.
*(o) Every module satisfies the d.c.c. on finitely generated submodules (resp. on cyclic submodules).
*(p) Finitely generated as a ring (resp. as a module) over the center.

4.33.5 Show that the following properties are not Morita invariants:
(a) Domain.
(b) Field.
(c) Nilpotent elements $\neq 0$.
(d) Commutative.
(f) $ab = 1 \Rightarrow ba = 1$ in the ring.
(g) Every right ideal is free.
(h) Every projective module is free.
(i) Idempotents $\neq 0$ or 1.
(j) Matrix units.
(k) Every submodule of a free module is free.
*(l) The a.c.c. on right annulets $(\mathrm{acc})^{\perp}$.

4.33.6 Which of the Morita invariant properties of Exercise 4.33.4 are quotient properties?

*4.33.7 If R is a principal right ideal ring, so is every full matrix ring R_n. The converse does not hold, in general, but does hold if R is commutative. (Thus, every right ideal principal is not a Morita invariant.)

The Center of a Category

If C is a category, and if $1_C : C \to C$ is the identity functor, then the **center** of C is defined to be the set of all natural transformation $h: 1_C \to 1_C$. Thus, for each $A \in C$, there is defined a morphism $h(A): A \to A$ such that for any morphism $f: A \to B$, where $B \in C$, the diagram

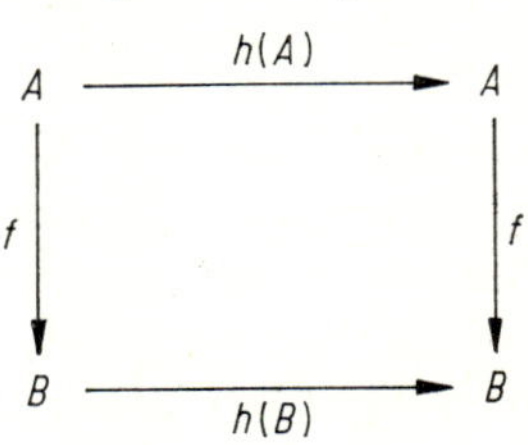

is commutative, that is, $h(A)f = fh(B)$. In particular, $h(A)$ commutes with every $f: A \to A$, and $h(A)$ is an endomorphism belonging to the center of End A_R.

4.34 Proposition. *For any ring R, there is a ring isomorphism*

$$\varphi: \text{center } R \approx \text{center mod-}R,$$

where if $x \in \text{center } R$, *if* $h = \varphi(x)$, *and if* $A \in \text{obj mod-}R$, *then* $h(A)$ *is that map* $A \to A$ *defined by the homothetic of x, that is,* $h(A)(a) = ax \forall a \in A$.

Proof. We first check that $h = \varphi(x)$ is actually a natural transformation of the identity functor. If $f: A \to B$ is a map of modules, then

$$h(B)f(a) = f(a)x = f(ax) = fh(A)(a),$$

so $h(B)f = fh(A)$ as desired. Next, if $h \in \text{center mod-}R$, then $h(R)$ maps the identity element 1 of R onto an element $x \in R$, and $h(R)(y) = yx \forall y \in R$. Now if $A \in \text{mod-}R$, then $A \approx \text{Hom}_R(R, A)$. If $a \in A$, let $f: R \to A$ be such that $f(1) = a$. Then the diagram

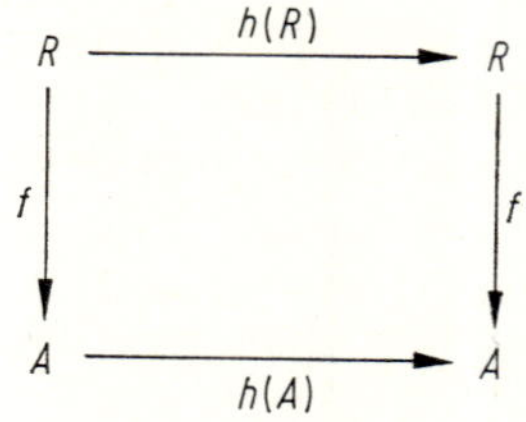

is commutative (since $h \in \text{center mod-}R$), so

$$h(A)f(y) = h(A)f(1)y = h(A)ay$$

and

$$fh(R)(y) = f(yx) = f(1)yx = ayx$$

$\forall y \in R$. Taking $y = 1$, we obtain

$$h(A)a = ax,$$

$\forall a \in A$, showing that $h = \varphi(x)$, the homothetic induced by x. Hence φ is surjective. Now $h = \varphi(x) = \varphi(y)$ implies $h(R)(1) = x = y$, so φ is $1-1$ and, hence, an isomorphism. □

The fact that

$$\text{center } R \approx \text{center } R_n$$

for any n is a consequence of the theorem and the equivalence

$$\text{mod-}R \approx \text{mod-}R_n .$$

This and the following corollary also are implied by 4.12.

4.35 Corollary. *If A is a commutative ring, and if B is similar to A, then A is isomorphic to the center of B. Thus, two commutative rings are similar if and only if they are isomorphic.* $\square$

4.36 Corollary. *If A and B are rings, and if there is an equivalence $T : \text{mod-}A \rightsquigarrow \text{mod-}B$, then T induces an isomorphism of rings* center $A \approx$ center B.

Proof. T induces an isomorphism of center mod-A and center mod-B. $\square$

A **right order** R of a ring S is a subgroup of the additive group of S which is also a submonoid of the multiplicative semigroup of S such that every element of S has the form rx^{-1} for some elements $r, x \in R$. (Note[1] that then R is a ring-1 and a subring-1 of S.) A ring R is said to have a **right quotient ring** S provided that there is a ring monomorphism $h : R \to S$ such that the image of h is a right order of S, and h maps every regular element of R onto a unit of S. If $g : R \to T$ is another quotient ring of R, then there is a unique ring isomorphism $f : S \to T$ such that $fh = g$. Thus, we may speak of the right quotient ring of R. A necessary and sufficient condition for the existence of a right quotient ring of R is the right Ore condition: *given any ordered pair (a, b), with $a, b \in R$, and b regular, there exists $a_1, b_1 \in R$, with b_1 regular, such that $ab_1 = ba_1$.* (Thus, any commutative ring has a quotient ring, as we observed in Chapter 3, *sup.* 3.46 and 3.61.) [We leave it to the reader to prove these statements, or turn to Chapter 9, where they are verified.] Henceforth, for a quotient ring $h : R \to S$, assume h is inclusion.

A ring S is said to be a **quotient ring** if it is the quotient ring of some ring R. In this case, not only is every regular element of R a unit of S, but every regular element of S is a unit. Thus, S is a quotient ring iff S is its own right and left quotient ring. [This shows that the operation of taking quotient rings is in a sense idempotent (cf. idempotent functors, Chapter 16).]

Definitions (Jacobson [42, p. 119]). *Let R_1 and R_2 be right orders in a quotient ring Q. Then*

(i) *they are* **equivalent**, $R_1 \underset{\sim}{Q} R_2$, *if there exist regular elements a_1, b_1, a_2, b_2 of Q such that* $a_1 R_1 b_1 \subseteq R_2$, $a_2 R_2 b_2 \subseteq R_1$;

(ii) *they are* **right equivalent**, $R_1 \overset{r}{\sim} R_2$, *if there exist regular elements a_1, a_2 of Q such that* $a_1 R_1 \subseteq R_2$, $a_2 R_2 \subseteq R_1$;

---

[1] See p. 383.

(iii) *they are* ***left equivalent***, $R_1 \overset{l}{\sim} R_2$, *if there exist regular elements* b_1, b_2 *of* Q *such that* $R_1 b_1 \subseteq R_2$, $R_2 b_2 \subseteq R_1$.

These are equivalence relations. Also

$$R_1 \overset{r}{\sim} R_2 \Rightarrow R_1 \underset{\sim}{Q} R_2 \quad \text{and} \quad R_1 \overset{l}{\sim} R_2 \Rightarrow R_1 \underset{\sim}{Q} R_2.$$

Lemma (Jacobson). *Let* R *be a right order in a quotient ring* Q, *and let* S *be a subring-1 of* Q.

1. *If there exist regular elements* a, b, c *and* d *of* Q *such that* $aRb \subseteq S$ *and* $cSd \subseteq R$, *then* S *is a right order in* Q, *and* $R \underset{\sim}{Q} S$.

2. *If there exist regular elements* a *and* c *of* Q *such that* $aR \subseteq S$ *and* $cS \subseteq R$, *then* S *is a right order in* Q, *and* $R \overset{r}{\sim} S$. *Similarly for left equivalence.*

Proof 1. If $q \in Q$, then

$$a^{-1} q a = x c^{-1}$$

for some $x, c \in R$, and then

$$q = (axb)(acb)^{-1},$$

so that S is a right order of Q equivalent to R. The proof of 2. is obtained by taking $b = 1$. □

4.37 Proposition. *Let* B_i *be a right Ore domain with right quotient field (quofield)* D_i, *let* U_i *be finitely generated and projective in* B_i-mod, *and let* $A_i = \text{End}_{B_i} U_i$, $i = 1, 2$. *If there is a ring isomorphism* $f: A_2 \to A_1$, *then there is an isomorphism* $g: D_2 \to D_1$ *such that* $g(B_2)$ *and* B_1 *are equivalent right orders of* D_1.

Proof. By 4.20, U_i is uniform in mod-A_i, $i = 1, 2$. By 4.23 A and its proof, U_2 is canonically isomorphic to a right ideal of A_2.

Hence, assume that U_i is a uniform right ideal of A_i, $i = 1, 2$. Then $f(U_2) = V$ is a uniform right ideal of A_1. Let $A = A_1$, $U = U_1$, $B = B_1$, and $D = D_1$. Then, by 4.23 A, V is isomorphic to an A-submodule Y of U such that $IU = Y$, for a right ideal I of B, and End V_A is isomorphic to $B' = \text{End}\, I_B$, where

$$B' = \{d \in D \mid dI \subseteq I\}$$

is a right order in D. If $x \in I$, and $x \neq 0$, then $B'x \subseteq B$ and $xB \subseteq B'$. *This proves that* B *and* B' *are equivalent right orders.* Now $R = \text{Biend}_{B_2} U_2$ is a right Ore domain containing B_2, and there is a ring isomorphism $g: R \to B'$ defined by the compositions

$$R = \text{End}\, U_{2_{A_2}} \approx \text{End}\, V_A \approx B'.$$

By 4.6, R is a rational extension of B_2 such that $RT \subseteq B_2$, where $T = \text{trace}_{B_2} U_2$. This shows that R and B_2 have the same right quofield D_2, and that R and B_2 are $\overset{l}{\sim}$ right orders, since $Rt \subseteq B_2$, for any $t \in T$, and $B_2 \subseteq R$. This also shows that B and $S = \text{Biend}_B U$ are equivalent right orders of D. Let $g: D_2 \to D$ be the unique extension of the isomorphism $R \approx B'$. Then equivalence of B_2 and R implies equivalence of the right orders $g(B_2)$ and B' and, hence, of $g(B_2)$ and B. $\square$

4.38 Proposition. *Let B_i $i = 1, 2$, be right Ore domains which have the same right quofield D. If B_1 and B_2 are similar rings, then B_1 and B_2 are equivalent right orders, there is a right ideal P of B_2, and there is a ring isomorphism*

$$B_1 \approx (P:P) = \{d \in D \mid dP \subseteq P\}.$$

Proof. By 4.29, write $B_1 = \text{End}\, P_{B_2}$ and $B_2 = \text{End}_{B_1} P$ for some (B_1, B_2)-bimodule progenerator P. Then, as in the proof of 4.23 A, there is an embedding of P in B_2 in mod-B_2, and then, since D is right self-injective, there is a ring isomorphism $B_1 \approx (P:P)$ as stated. (Cf. 4.23 A.) Then, for any $p \in P$,

$$B_1 p \subseteq P \subseteq B_2, \qquad \text{and} \qquad pB_2 \subseteq P \subseteq B_1.$$

Thus, B_1 and B_2 are equivalent right orders. $\square$

A ring R is said to be **right hereditary** provided that every right ideal is projective. Any right PID R is right hereditary since any right ideal $\neq 0$ is a free module $\approx R$. Moreover, hereditary is a Morita invariant property by Exercise 4.33.4, and, in fact right hereditary rings can be categorically characterized by the property that any submodule of a projective module is again projective. (See Chapter 31 for the proof of this, and Chapter 30 for other examples of hereditary rings. Cf. Dedekind rings.) In particular, any full matrix ring R_n over a hereditary ring R is hereditary.

4.39 A Proposition. *Let B be a right Ore domain with right quofield D, and let A be a subring of D which is a right order of D. Consider the conditions:*

4.39.1 A *A is ring similar to B, that is,* mod-$A \approx$ mod-B.

4.39.2 A *There is a right ideal I of B and a ring isomorphism* $A \approx (I:I) = \{d \in D \mid dI \subseteq I\}$.

4.39.3 A *A and B are equivalent right orders.*

Then, (1) $\Rightarrow$ (2) $\Rightarrow$ (3).

If B is a simple ring, then every ring satisfying (2) *or* (3) *is simple only if B is right hereditary and right Noetherian. Thus, when B is simple,* (1) $\Leftrightarrow$ (3) *holds only if B is right hereditary and right Noetherian. In this case,* (1) $\Leftrightarrow$ (3) *for any maximal right order A.*

Proof. (1) $\Rightarrow$ (2) $\Rightarrow$ (3) by 4.38 and its proof.

For any nonzero right ideal I of B, I is a left ideal of the ring $A = (I : I) = \{d \in D \mid dI \subseteq I\}$, and A is a right order of D equivalent to B. If B is simple, then A is ring similar to B only if A is simple. In this case, I is a generator in A-mod and in mod-B and, hence by 4.2 and 4.3, finitely generated projective in both. This proves that B is right Noetherian and right hereditary provided that every $(I : I)$ is simple. This is the case when B is simple and (3) ⇒ (1).

Conversely, assuming that B is right hereditary and right Noetherian, let A be a maximal right order equivalent to B. Then $xAy \subseteq B$ for nonzero elements $x, y \in B$. Then $xy = b \in B$, and $y = x^{-1}b$. Then, $A' = xAx^{-1}$ satisfies $A'b \subseteq B$, and hence $A'I \subseteq I$ for $I = A'bB$. Let $S = (I : I) \approx \operatorname{End} I_B$. Then, since I is a progenerator of mod-B, then by 4.29, S is similar to B. Now $A = x^{-1}Sx$ since A is a maximal right order, and therefore A is similar to B. ▯

4.39B *Exercise.* (Faith [65]) *A simple right Ore domain B is maximal in the set of right orders of its right quotient field which are left equivalent to B. Generalize.*

4.40 **Proposition.** *Let A be a simple ring with a uniform right ideal U. The following conditions are equivalent:*

4.40.1 *The endomorphism ring of every uniform right ideal V of A is simple* (*resp. similar to A*).

4.40.2 *Every uniform right ideal of A is finitely generated and projective.*

4.40.3 *A is right hereditary and right Noetherian.*

Proof. Applications of 4.22C, 4.29, and the correspondence theorem proves the equivalence of the two statements made in (1), and also (1) ⇔ (2). Clearly (3) ⇒ (2).

(1) ⇒ (3) Then $B = \operatorname{End} U_A$ is a simple ring similar to A. If I is a right ideal of B, $I \neq 0$, then, by 4.23A, $V = IU$ is a uniform right ideal of A such that $\operatorname{End} V_A$ has the same right quofield as B, and $\operatorname{End} V_A \approx \operatorname{End} I_B$. Then, by 4.39.3A, (3) holds. ▯

For other results on orders, see Chapters 7, 9, 10 and 18.

Exercises for Chapter 4

1. Prove all the theorems in Chapters 7 and 8.

2. Any simple ring A with uniform right ideal U embeds in a full matrix ring D_n, where $n = \dim A_A$, and D is the right quofield of $B = \operatorname{End} U_A$. (Cf. 4.22.) [Special case of 9.13, which implies that A has a right quotient ring $\approx D_n$.] Conclude that A satisfies the a.c.c. and the d.c.c. on right and left annulets. (See p. 142, *sup.* Two Canonical Maps; Cf. also 4.21.)

3.1. Prove that if A and B are fields, and if there exist objects V in mod-A and W in mod-B and a ring isomorphism $\text{End } V_A \to \text{End } W_B$, then there exist an equivalence $T: \text{mod-}A \approx \text{mod-}B$ and an equivalence $W \approx TV$ in mod-B. Generalize. (Cf. 16.7 and 24.26.)

3.2. If k is a field and A is a ring similar to k, then $A \approx \text{End}_k V$ for a finitely generated vector space V over k, and conversely.

4. A ring R is said to be **semisimple** if every right module is semisimple. Show that any ring R similar to a field is semisimple. Show that a ring R is semisimple if and only if R is similar to a finite product of fields. Conclude that the property of a ring being semisimple is right-left symmetrical. Show that any semisimple ring is hereditary.

5. (**Wedderburn-Artin Theorem**) (a) A ring R is semisimple if and only if R is a ring satisfying the minimum condition on right ideals and containing no nilpotent ideals $\neq 0$. Conclude from Exercise 4 that the condition is right-left symmetric.

(b) A simple ring R is semisimple if and only if R satisfies the minimum condition on right ideals (and if and only if R is similar to a field).

6. The following conditions are equivalent on a ring R.

(a) R is semisimple in mod-R.

(b) R is semisimple in R-mod.

(c) R is a finite product of full linear rings over finite dimensional vector spaces over various fields.

(d) R is a finite product of total matrix rings of various degrees over various fields.

(e) Every object of mod-R is semisimple.

(f) Every object of mod-R is injective.

(g) Every object of mod-R is projective.

7. If F is a field, then the intersection of all subfields of R is a field, called the **prime** subfield of R, which is isomorphic to either $\mathbb{Q}$ or $\mathbb{Z}_p$ for some prime p. If F is finite, then F is a vector space of finite dimension n over the prime subfield $\mathbb{Z}_p$, and therefore F has p^n elements. Show that two finite commutative fields are isomorphic if and only if they have the same number of elements (Theorem of E. H. Moore). Prove for every prime p and integer n there exists a finite field consisting of p^n elements, called the **Galois field** of p^n elements and denoted $GF(p^n)$. Prove Wedderburn's theorem which states that every finite field is necessarily commutative, hence a Galois field.

8. Prove that the multiplicative group of a finite field is cyclic and that any group of automorphisms of a finite field is cyclic of prime power order.

Notes for Chapter 4

The Original Sin and The Oedipus Complex

The mathematician-philosopher-humanist (et cetera) Bertrand Russell began his philosophical studies with two questions: 1. Is knowledge possible? 2. If the answer to 1 is yes, what can I know? I am not sure what his conclusions were, but the questions themselves are remarkable and evocative of Descartes's principles. There is a third question. 3. How can I know something? I am referring to the often disparate ways knowledge can be represented. The mathematician finds it difficult even to determine when two objects are the same — in the case of words or elements of the free group, the task proved impossible! (Unsolvability of the word problem.)

The Oedipus complex is the inability of mathematical objects to recognize their (mother) parent[1]. Is a subobject of a free object free? Is a summand of product (or coproduct) of free objects free? Many types of mathematical theorems reject the Oedipal by stating when the answers are yes. The subgroup theorem for free groups is the affirmation of the first question for groups. However, subsystems of free semigroups and algebras are Oedipal[1], and the problem is to study the incestual relations to determine axioms for the subsemigroups (algebras) which are free. (These axioms should not refer outside of the subsystem. See Cohn [62].) An affirmation of the second question is called a Krull-Schmidt, or unique decomposition, theorem, often prefixed or suffixed by Azumaya, Remak, Wedderburn, and others (Chapter 21). The unique factorization of integers into primes is the primal model.

The original sin in the algebraic garden was the failure of the unique factorization theorem for algebraic integers of an algebraic number field (= a finite dimensional field extension of $\mathbb{Q}$). Lamé gave a false proof of Fermat's last theorem based on the erroneous assumption that an algebraic integer is not Oedipal. Ideal theory was founded by Dedekind, who founded many things, in order to restore grace: factorization of ideals into a product of primes is unique in algebraic number fields (Chapters 30 and 31). But this did not satisfy Kronecker who reportedly said: "God made the integers; man made the rest". This put down a lot of Ph.D. theses coming out of the mill at the time, and also Cantor, who was way out on the transfinite (and papal?) limb of cardinal numbers.

[1] As in the Greek myth, it is also a case of the mother not recognizing the son!

Notes on Simple Rings

The fact that Ore domains appear in the structure theory of simple Noetherian rings was first noted by Goldie [58] and Lesieur-Croisot [59], generalizing Schur's lemma 3.10.1. Littlewood [33] noted that non-commutative Noetherian simple Ore domains occur in nature as the "algebra of quantum mechanics". Part of his theorem, in generalized form, is a theorem of Amitsur 7.30. New simple domains were discovered by Cozzens [70], an account of which appears in Chapter 7, esp. 7.40ff.

In the area of simple rings, Wedderburn's theorem dates to [08], the Wedderburn-Artin theorems came twenty years after (see Artin [27]); also Artin and Whaples [44], and Artin-Nesbitt-Thrall [43] — at about the same time as the density theorem (Jacobson [45]); and then there were the theorems of Goldie [58, 60] -Lesieur-Croisot [59] and Morita [58] a half-century after Wedderburn's theorem. The endomorphism ring theorem for simple rings with uniform right ideals of Faith [64] depended on the work of Goldie and R. E. Johnson [51], yet even the sharpened version of Hart [67] proved to be rather easy consequences of the Morita and Goldie-Lesieur-Croisot theorem, as Faith [67b] observed.

The correspondence theorem for projective modules, and most of the structure theory for simple rings with uniform right ideals and/or lateral right ideals, was announced by Faith [71a] and published in full in [72a]. A number of conjectures on simple rings given there are stated in Chapter 8, pp. 381—382.

Notes on Quotient Rings

Is the Ore condition a Morita invariant property? The answer is unknown in general. However, for any commutative ring R, any ring A similar to R has a quotient ring (Proof?). Also, if R has a semisimple right quotient ring, then so does any ring similar to R (cf. Chapter 9). This is only two of the many special cases for which the answer is affirmative. In these cases, $A \sim R$ implies $Q(A) \sim Q(R)$, where $Q(\ )$ denotes the quotient ring of. Even if $Q(\ )$ is a Morita invariant, does it follow that $Q(A)$ has the stated functorial property? If so, then the following holds: if S is a quotient ring, then so is S_n, for any integer n, that is, every regular element of S_n is a unit. Must this hold?

References

Artin [27], Artin, Nesbitt, and Thrall [43], Artin and Whaples [44], Auslander and Goldman [60], Bass [62a, 68], Camillo [70], Camillo and Fuller [72], Cohn [62, 66a], Cozzens [70], Dickson and Fuller [70], Dlab and Ringel [72], Faith [64, 67a, b, 71a, 72a], Faith and Utumi [65b], Findlay and Lambek [58], Gabriel [62], Goldie [58, 60], Hart [67], Jacobson [42, 45], Jans [70], Johnson [51a, 51b], Morita [58], Popesco-Gabriel [64], Shepherdson [52], Utumi [56], Wedderburn [08],

Chapter 5

Limits, Adjoints, and Algebras

A list of topics in Chapter 5 would include bifunctors, representable functors, left and right adjoint functors, adjunction bijections and morphisms, adjoint pairs, forgetful functor, and free object functor, limits, compatible families of morphisms, directed class, direct and inverse limits, kernels, cokernels, equalizers, intersections, fiber products, or push-outs, images, inverse images, epic images, complete categories, normal objects and categories, balanced categories, locally small, exact categories, diagram chasing, the Noether isomorphism theorems for exact categories, limit preserving functors, right continuous functors, Freyd's theorems, module valued adjoints, the tensor functor, flat modules, character modules, adjoints of faithful functors, Popesco-Gabriel theorem (second version), algebras over commutative rings, semigroup rings, and group rings, free algebras, commutative and non-commutative rings of polynomials.

The property of a functor $S : D \rightsquigarrow C$ being a left adjoint of a functor $T : C \rightsquigarrow D$ is defined in this chapter and is equivalent to the requirement, for each object B of D, of a natural equivalence $h^B T \approx h^{SB}$ of set valued functors. For example, the free object functor SETS $\rightsquigarrow C$ is left adjoint to the forgetful functor $C \rightsquigarrow$ SETS. A functor T has a left adjoint if and only if it is "almost" representable in the sense that the set valued functor $h^B T$ is representable, for each object B of D. Any additive functor $T : C \rightsquigarrow$ mod-R which has a left adjoint S is representable, and $T \approx h^{SR}$ (Proposition 5.52). Conversely, any representable functor $T : C \rightsquigarrow$ mod-R is left joinable if C is a locally small complete additive category with a cogenerator, for example, if $C =$ mod-A, for any ring A. (Locally small categories have at most a set of inequivalent subobjects for any object, and complete categories are defined presently.)

Freyd's theorem 5.51 characterizes functors $T : C \rightsquigarrow D$ which have a left adjoint. When C is complete, locally small, and has a cogenerator, then T has a left adjoint if and only if T is limit preserving. If, in addition, C is additive, the the three concepts coincide: limit preserving, left joinable, and representable (5.56).

Equalizers, products, intersections, and fiber products are limits of functors, and a category C is (finitely) complete if and only if C has equalizers and (finite) products. If C has a zero object, then the kernel of a morphism f is the equalizer of f and 0. An additive category has kernels if and only if it has equalizers. A functor $T : C \rightsquigarrow D$ is (finite) limit preserving if and only if T preserves equalizers and (finite) products (5.39).

If L and C are categories, then C^L denotes the category Funct (L, C) the objects of which are the functors $F : L \rightsquigarrow C$; a morphism between functors is any natural transformation. A right adjoint $\lim : C^L \rightsquigarrow C$ of the constant (diagonal) functor $c : C \rightsquigarrow C^L$ is called a limit functor. An object A of C is a limit if $A = \lim F$ for some $F : L \rightsquigarrow C$. [Thus, $A = \lim c(A)$ for any category L.] A functor $T : C \rightsquigarrow D$ is limit preserving if there is a natural equivalence of functors $\lim \overline{T} \approx T \lim$, where $\overline{T} : C^L \rightsquigarrow D^L$ is the functor induced by T sending any $F \in C^L$ onto TF. For any category C, the set-valued morphism functors h^A and h_A are limit preserving for any object A (Proposition 5.41). Any left joinable functor $T : C \rightsquigarrow D$ is limit preserving, and then the left adjoint $S : D \rightsquigarrow C$ is colimit preserving (Proposition 5.43). This implies, in particular, that any limit functor $\lim : C^L \rightsquigarrow C$ is limit preserving, and so the free object functor F : SETS $\rightsquigarrow C$ is colimit preserving. For the disjoint union $\coprod_{i \in I} S_i$ of sets in SETS, this latter result implies that the coproduct $\coprod_{i \in I} F(S_i)$ of the free objects $\{F(S_i)\}_{i \in I}$ in C is $F(\coprod_{i \in I} S_i)$ and has applications for the categories k-ALG of algebras over a commutative ring k, the category of (semi) group rings of any (semi)group G over any ring A, and the ring of (commutative or noncommutative) polynomials over any ring, all of which are taken up in this chapter.

Bifunctors

If $\{C_i\}_{i \in I}$ is a family of categories, the product $\prod_{i \in I} C_i$ category is defined on objects $\{X_i\}_{i \in I}$ which are families of objects indexed by I, with X_i an object of C_i for each $i \in I$. A morphism $\{X_i\} \to \{Y_i\}$ is an element $\{f_i\}$ of $\prod_{i \in I} \mathrm{Mor}_C(X_i, Y_i)$. If $\{g_i\}$ is another morphism, then $\{g_i\}\{f_i\}$ is defined if and only if target $f_i =$ source $g_i \forall i \in I$, and then $\{g_i\}\{f_i\} = \{g_i f_i\}$.

If C and D are categories, then $C \times D$ denotes the product category. The morphism functors define a bifunctor

$$\mathrm{Mor}_C(\ ,\ ) \begin{cases} C^{\mathrm{op}} \times C \rightsquigarrow \text{SETS} \\ (A^*, B) \mapsto \mathrm{Mor}_C(A, B) \\ (f^*, g) \mapsto M(f, g), \end{cases}$$

where, for $f : A_1 \to A_2$ and $g : B_1 \to B_2$, $M(f, g)$ is the compose of $h_{B_1}(f)$ and $h^{A_2}(g)$.

A bifunctor $F : C \times D \rightsquigarrow E$ induces, for each object Y of D and each object X of C, **partial functors**

$$F(\ , Y) \begin{cases} C \rightsquigarrow E \\ A \mapsto F(A, Y) & \text{(on objects)} \\ f \mapsto F(f, 1_Y) & \text{(on morphisms)} \end{cases}$$

and $F(X,\ ) : D \rightsquigarrow E$ (defined dually). If F and G are covariant bifunctors $C \times D \rightsquigarrow E$ and $h : F \to G$ is a natural transformation, then for each object S of C and each object Y of D, h induces natural transformations $h(\ , Y) : F(\ , Y) \to G(\ , Y)$ and $h(X,\ ) : F(X,\ ) \to G(X,\ )$ of the partial functors. The phrase "$h(X, Y)$ is natural in X and Y" expresses the same thing. Conversely, if $h'(X, Y)$ is an equivalence of $F(X, Y) \approx G(X, Y)$ in SETS for each object X of C and each object Y of D, such that $h'(X, Y)$ is natural in X and Y in this sense, then h' defines a natural transformation $F \rightsquigarrow G$ of bifunctors.

Adjoint Functors

Covariant functors $T : C \rightsquigarrow D$ and $S : D \rightsquigarrow C$ are **adjoint** provided that there is a natural equivalence

(1) $$h(\ ,\ ) : \mathrm{Mor}_C\big(S(\ ),\ \big) \to \mathrm{Mor}_D\big(\ , T(\ )\big)$$

of the set-valued bifunctors. In this case, S is **left adjoint** to T, and T is **right adjoint** of S; also S has the **adjoint** T and T has the **coadjoint** S. Thus, for each object A of C and B of D, there are bijections

(2) $$h(B, A) : \mathrm{Mor}_C(SB, A) \to \mathrm{Mor}_D(B, TA)$$

which are natural in A and B, called **adjunction** bijections.

The symbol (h, S, T, C, D), or more briefly (S, T), denotes the adjoint situation just described, that is, S is left adjoint to T. A functor $T : C \rightsquigarrow D$ is **left joinable** if T has a left adjoint S. Then (S, T) is called an **adjoint pair**. By Proposition 5.2 proved presently, if (S', T) and (S, T) are adjoint pairs, then there is a natural equivalence of functors $S' \approx S$, and one may refer to *the* left adjoint of T. In this case, by Proposition 5.67, there is a natural transformation of functors

$$\psi : ST \to 1_C$$

where

$$\psi_A = h^{-1}(TA, A)(1_{TA})$$

is the morphism $STA \to A$, defined for each object A of C. The natural transformation $ST \to 1_C$, and the morphism $\psi_A : ST(A) \to A$, are called the **adjunction morphism**. Dually, there is a natural transformation

$$\varphi : 1_D \to TS,$$

where

$$\varphi_B = h(B, SB)(1_{SB})$$

is the morphism of adjunction $B \to TS(B)$.

5.1 *Examples.*

5.1.1 **The free object functor.** This functor is left adjoint to the forgetful functor $\text{Forget}: C \rightsquigarrow \text{SETS}$. For any object Y of C and set S, let $h: S \to \text{Forget } Y$ be a mapping, and let $h': F(S) \to Y$ be the unique morphism of C such that $h' f_S = h$, where $f_S: S \to F(S)$ is canonical. The mapping

$$h(S, Y): \text{Mor}_C(FS, Y) \to \text{Mor}_{\text{SETS}}(S, \text{Forget } Y)$$

which sends a morphism $g: FS \to Y$ of C onto $g f_S = g_1$ is the inverse of the mapping $h \mapsto h'$. Going the other way $((h')_1 = h$ by definition, and $g_1' = g$ because g is the unique morphism such that $g_1' f_S = g f_S = g_1$.) Since $h(S, Y)$ is bijective and natural in S and Y, then the functors are adjoint. (Cf. 2.7, p. 95.)

5.1.2 For a (B, A) bimodule U, the contravariant functors

$$h_U : \text{mod-}A \rightleftarrows B\text{-mod} : h_U$$

are adjoint in the sense that there are bijections

$$g(X, Y): \text{Hom}_A(X, \text{Hom}_B(Y, U)) \approx \text{Hom}_B(Y, \text{Hom}_A(X, U))$$

natural in both X and Y, where

$$(y g(f))(x) = (y) f(x)$$

$\forall x \in X$, $y \in Y$ and $f \in \text{Hom}_B(Y, U)$.

5.1.3 For three sets S, A, and B, the bijections

$$g(A, B) \begin{cases} S^{A \times B} \to (S^A)^B \\ \quad f \mapsto g(f), \end{cases}$$

where $g(f) b(a) = f(a, b)$, are natural in A and B and can be written

$$g(A, B): \text{Mor}_{\text{SETS}}(A \times B, S) \approx \text{Mor}_{\text{SETS}}(B, \text{Mor}_{\text{SETS}}(A, S))$$

that is, the covariant functor h^A is right adjoint to to the functor $A \times (\ )$.

5.1.4 A functor $S: D \rightsquigarrow C$ has a right adjoint if and only if the composite functor

$$S_*^*: D^* \overset{\text{can}}{\rightsquigarrow} D \overset{S}{\rightsquigarrow} C \overset{\text{can}}{\rightsquigarrow} C^*$$

has a left adjoint. Thus, (S, T) is an adjoint pair if and only if (T_*^*, S_*^*) is an adjoint pair.

5.1.5 Let ORE DOM denote the full subcategory of RINGS on the objects which are right Ore domains. Then the forgetful functor FIELDS $\rightsquigarrow$ ORE DOM has a left adjoint which assigns to each domain its right quotient field. Generalize.

5.1.6 If U is a (B, A)-bimodule, then the mod-B valued functor

$$h^U : \text{mod-}A \rightsquigarrow \text{mod-}B$$

induced by the group valued functor h^U has a left adjoint

$$\bigotimes_B U : \text{mod-}B \rightsquigarrow \text{mod-}A\,.$$

The existence of the left adjoint of h^U is a consequence of 5.46, but a more elementary proof is The Adjoint Associativity Theorem 11.4 $\square$

5.2 Proposition and Definition. *If $T : C \rightsquigarrow D$ is a functor, then T has a left adjoint if and only if for each object Y the* **comma category** *(Y, T) has an initial object, where the objects of (Y, T) are pairs (f, X) consisting of an object X of C and a morphism $f : Y \to TX$. A morphism $(f, X) \to (g, Z)$ of (Y, T) is a morphism $h : X \to Z$ of C such that*

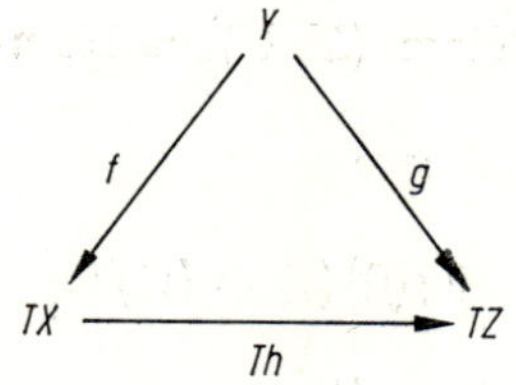

commutes.

Proof. If (f, SY) is an initial object of (Y, T), then for each object Y of D, and for each object Z of C, there is a bijection

$$(1) \qquad \begin{cases} \mathrm{Mor}_D(Y, TZ) \to \mathrm{Mor}_C(SY, Z) \\ \qquad g \mapsto g' \end{cases}$$

natural in objects Z and Y, where $g' : SY \to Z$ is the unique morphism such that $g'f = g$. If $t : Y \to Y'$ is a morphism of D, then there is a unique morphism $S(t) : SY \to SY'$ such that

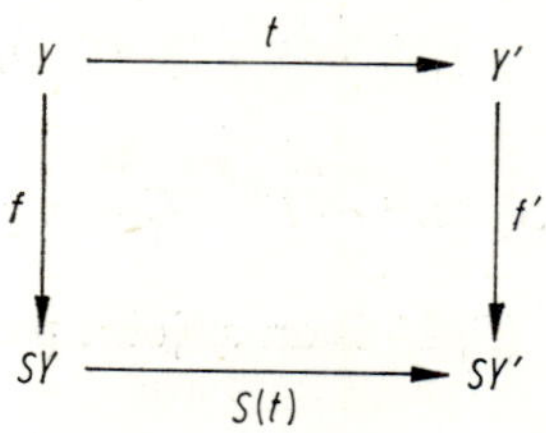

commutes, where $f' : Y' \to SY'$ is canonical. Since (f, SY) is initial in (Y, T), then necessarily $S(1_Y) = 1_{SY}$, and $S(tt') = S(t)S(t')$, for any morphisms t and t'. This proves that there is a functor $S : D \rightsquigarrow C$ which assigns SA to an morphism or object A of D. Furthermore, the bijection (1) shows that S is left adjoint to T.

Conversely, if S is a left adjoint to T, then the adjunction morphism $\varphi_Y : Y \to TSY$ defines an initial object (φ_Y, SY) of (Y, T). □

The comma category makes its next appearance at 5.50.

5.3 Proposition. *Let T and T' be functors $C \rightsquigarrow D$. Let S and S' functors $D \rightsquigarrow C$ be left adjoint to T and T', respectively. Then $T \approx T'$ (as functors) if and only if $S \approx S'$.*

Proof. Suppose we are given functor equivalences

$$h(\ ,\ ) : \mathrm{Mor}_C(S(\ ),\ ) \approx \mathrm{Mor}_D(\ , T(\ ))$$
$$h'(\ ,\ ) : \mathrm{Mor}_C(S'(\ ),\ ) \approx \mathrm{Mor}_D(\ , T'(\ )).$$

By the dual to 2.3.3, a functor equivalence $T \approx T'$ yields a functor equivalence $\mathrm{Mor}_D(\ , T(\ )) \approx \mathrm{Mor}_D(\ , T'(\ ))$ and, hence, a functor equivalence

$$k(\ ,\ ) : \mathrm{Mor}_C(S(\ ),\ ) \approx \mathrm{Mor}_C(S'(\ ),\ ).$$

This defines, for each $B \in \mathrm{obj}\, D$, a natural equivalence of functors

$$k(B,\ ) : \mathrm{Mor}_C(SB,\ ) \approx \mathrm{Mor}_C(SB',\ ).$$

By 2.3, this yields a natural equivalence of functors $S \approx S'$. The converse is obtained similarly. □

Existence of Adjoints

If we let h^A denote the morphism functor $\mathrm{Mor}(A,\ )$ [resp. $h_A = \mathrm{Mor}(\ , A)$] for any object A in a category, then we see, that if T is adjoint to S, then for each $A \in \mathrm{obj}\, C$, there is a functor equivalence

(1) $$h_A S \approx h_{TA}$$

induced by $h(\ , A)$, and for each object $B \in D$, a functor equivalence

(2) $$h^{SB} \approx h^B T.$$

induced by $h(B,\ )$.

The next proposition states that a functor which has a left adjoint is "almost" representable (cf. 5.55 and 5.56).

5.4 Proposition. *If* $T : C \rightsquigarrow D$ *is a functor, then* T *has left adjoints if and only if for each* $B \in \operatorname{obj} D$, *the functor*

$$h^B T : C \rightsquigarrow \text{SETS}$$

is representable.

Proof. If S is left adjoint to T, then by (2), there is a functor equivalence $h^B T \approx h^{SB}$. Conversely, for each $B \in \operatorname{obj} D$, choose an object SB and a natural equivalence of functors $h^{SB} \approx h^B T$ which we denote by $f(B)$. For any object A of C, $f(B)(A) : h^{SB}(A) \to h^B(T(A))$ is a bijection which is natural in the variables B and A. Thus, $h(A,B) = f(B)(A)$ defines the desired adjunction bijections.

Limits

If L and C are categories, then the **constant** (or **diagonal**) **functor** is defined

$$c = \text{const} \begin{cases} L \rightsquigarrow \text{Funct}(L, C) = C^L \\ X \mapsto c(X) = X_{\text{const}}, \end{cases}$$

where

$$c(X) = X_{\text{const}} \begin{cases} L \rightsquigarrow C & \\ M \mapsto X & \text{(on objects)} \\ f \mapsto 1_X & \text{(on morphisms).} \end{cases}$$

Clearly, a morphism $f : X \to Y$ in C determines a natural transformation $c(f) : c(X) \to c(Y)$, so that the constant functor is defined, and is fully faithful.

5.5 Proposition and Definition. *Let* L *and* C *be categories, and let* $C^L = \text{Funct}(L, C)$. *Then the constant functor* $c : C \rightsquigarrow C^L$ *is fully faithful. The right adjoint* $\lim : C^L \rightsquigarrow C$ *of* $c : C \rightsquigarrow C^L$, *if it exists, is denoted* $\lim_{L,C}$ *and called the* ***limit functor***. *For each natural equivalence* $f : F \to F'$ *of functors* $L \rightsquigarrow C$, *that is, for each morphism* f *of* C^L, $\lim f$ *is called the* ***limit morphism*** $\lim F \to \lim F'$.

Dually, a left adjoint $\operatorname{colim} : C^L \rightsquigarrow C$ *of* $c : C \rightsquigarrow C^L$ *is called the* ***colimit*** *functor.* $\square$

Compatible Families of Morphisms

Let L and C be categories, and let $F : L \rightsquigarrow C$ be a functor. A family $\{A \to F(X)\}_{X \in \operatorname{obj} L}$ of morphisms of C is said to be **compatible** with F provided that for every morphism $t : X \to Y$ of L,

the diagram

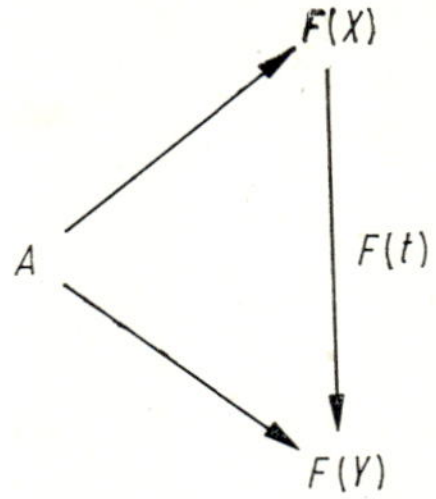

commutes [and, hence, defines a morphism $c(A) \to F$ of C^L]. The category **compat**$_L F$, of families of morphisms compatible with F, has the named objects, and a morphism between two compatible families is defined to be a morphism $A \to B$ of C such that

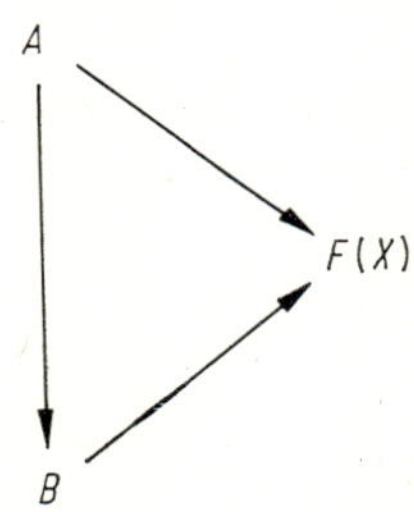

commutes for all objects X of L. When L is fixed, then compat F denotes the category. Cocompat$_L F$ is defined dually.

5.6 Proposition and Definition. *Let* $F : L \rightsquigarrow C$ *be a functor.*

5.6.1 *There is a bijection*

$$\begin{cases} \mathrm{Mor}_{C^L}\big(c(A), F\big) \to \text{compat } F \\ \qquad\qquad h \mapsto \{h(X) : A \to F(X)\}_{X \in \mathrm{ob}\, L} \end{cases}$$

for each object A *of* C.

5.6.2 *The category* compat F *has a final object if and only if there is an object* $\{f(X) : \bar{F} \to F(X)\}_{X \in \mathrm{obj}\, L}$ *of* compat F, *and a bijection*

$$\begin{cases} \mathrm{Mor}_{C^L}\big(c(A), F\big) \to \mathrm{Mor}_C(A, \bar{F}) \\ \qquad h \mapsto h' \end{cases}$$

which is natural in A. *In this case, this object is a final object of* compat F.

5.6.3 *The limit functor* $\lim : C^L \rightsquigarrow C$ *exists if and only if* compat F *has a final object, for every functor* $F : L \rightsquigarrow C$. *The functor* $C^L \rightsquigarrow C$ *which assigns* $\bar{F}$ *to* F *is right adjoint to* $c : C \rightsquigarrow C^L$, *hence naturally equivalent to* $\lim : C^L \rightsquigarrow C$.

5.6.4 *When* 5.6.2 *holds, then* $\bar{F}$ *is also denoted* $\lim F$, *or* $\underleftarrow{\lim} F$. *Then, for any morphism* $h : c(A) \to F$, *that is, for any element* $\{h(X) : A \to F(X)\}_{X \in \operatorname{obj} L}$ *of* compat F, *the unique morphism* h' *is called the* **limit,** *or canonical, morphism* $\lim h : A \to \lim F$. *Dually, in the case of colimits, the designations* $\vec{F}$ *and* colim F *hold, and for any element* $\{f(X) : F(X) \to A\}_{X \in \operatorname{obj} L}$ *of* cocompat F *there is a canonical morphism* $\operatorname{colim} f : \operatorname{colim} F \to A$.

Proof. Part 5.6.1 is clear. Assuming that $\bar{F}$ is a final object of compat F, then the mapping $h \mapsto h'$ is defined by letting $h' : A \to \bar{F}$ be the unique morphism determined by the compatible family $\{h(X) : A \to F(X)\}_{X \in \operatorname{obj} L}$. Since any morphism $f : A \to \bar{F}$ of C determines an object $\{A \to \bar{F} \to F(X)\}_{X \in \operatorname{obj} L}$ of compat F, which in turn by 5.6.1 determines a natural transformation $g : c(A) \to F$ such that $g' = f$, then the mapping $h \mapsto h'$ is bijective and natural in A. Conversely, assuming the bijection $h \mapsto h'$ stated in 5.6.2, and assuming that $\operatorname{Mor}_{C^L}(c(A), F)$ is nonempty, there is a morphism $h' : A \to \bar{F}$, and by the bijection 5.6.1, necessarily

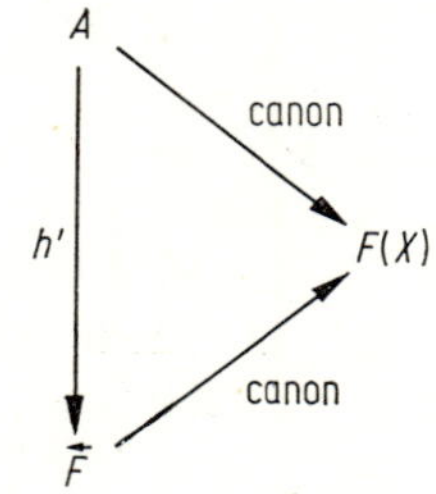

commutes $\forall X \in \operatorname{obj} L$, and h' is the unique morphism $A \to \bar{F}$ for which the diagram commutes. This proves that $\bar{F}$ is final in compat F.

Part 5.6.3. The bijections 5.6.1, which are natural in A, are also natural in F, and therefore the functor defined in 5.6.3 is a right adjoint of the constant functor as claimed. The functor equivalence folows from 5.3. □

5.7 *Exercises.*

5.7.1 Let L be a category, and let $T : C \rightsquigarrow D$ and $S : D \rightsquigarrow C$ be an adjoint pair. Let $\bar{T} : C^L \rightsquigarrow D^L$ and $\bar{S} : D^L \rightsquigarrow C^L$ be the functors induced by T and S respectively.

(a) Prove that $\bar{S}$ is a left adjoint of $\bar{T}$.

(b) Verify $h^{\bar{S}c(X)} = h^{c(SX)}$ and $h^{c(X)}T = h^{c(SX)}$ for all objects X of D.

(c) Justify $h^{h^{\lim F}} h^Y = \lim h^Y F$ for all objects Y of C and functors $F : L \rightsquigarrow C$, that is, if $S = \text{SETS}$, then $\operatorname{Mor}_{C^S}(h^{\lim F}, h^Y) = \lim h^Y F$.

(d) Justify $h^{c(X)} h^A = h^X h^{c(A)}$, for any set X and object A of C.

(e) Does Example 5.1.3 hold for categories? Let $S = \text{SETS}$, and let A and B be categories. Discuss $S^{A \times B} \rightsquigarrow (S^A)^B$, that is, discuss a functor

$$\text{Funct}\,(A \times B, S) \rightsquigarrow \text{Funct}\,\big(B, \text{Funct}\,(A, S)\big).$$

(f) Ignoring sizes, show that

$$S^{A \times B} \approx (S^B)^A \approx (S^A)^B.$$

5.7.2 If (S, T) is an adjoint pair, and if the adjunction morphism $A \to STA$ is an isomorphism for an object A of C, then the adjunction morphism $B \to TSB$ is an isomorphism for every summand B of TA. Thus, if the adjunction $1_C \to ST$ is a functor equivalence, and if every object B of D is equivalent to a summand of TA for some object A of C, then the adjunction functor morphism $TS \to 1_D$ is an equivalence, that is, $T : C \rightsquigarrow D$ is a category equivalence with inverse S.

Directed Class

An **ordered class** is a category C such that $\text{Mor}_C(A, B)$ has at most one element for each pair of objects. If $f : A \to B$ is a morphism, then we write $A \leq B$. If $A \leq B$, and $A \neq B$, then we write $A < B$. In an ordered class, always $A \leq A$. Also $A \leq B$ and $B \leq C$ implies $A \leq C$.

A **directed class** is an ordered class C such that given A and B there always exists X such that $A \leq X$ and $B \leq X$. The class is **totally ordered** when either $A \leq B$ or $B \leq A$ for all objects A and B, and both hold only if $A = B$. An **ordered subclass** is a full subcategory. An ordered class C is **inductive** in case every totally ordered subclass has an upper bound in C.

If L is a directed class, then

$$\varprojlim F, \qquad \varprojlim{}_{X \in L} F(X), \qquad \varprojlim_{X \in L} F(X)$$

are various designations for the limit of a functor $F : L \rightsquigarrow C$, and is called the **inverse limit** of F. Furthermore, $\lim : C^L \rightsquigarrow C$ is then called the **inverse limit functor.** Dually,

$$\underrightarrow{\text{colim}}\, F, \qquad \varinjlim{}_{X \in L} F(X), \qquad \varinjlim_{X \in L} F(X)$$

all denote the colimit of F, which is called the **direct limit** of F. The $\text{colim} : C^L \rightsquigarrow C$ is then called the **direct limit functor.** If this functor exists for every directed class L, then C is said to be a **category with direct limits.**

If L is a directed class, and if

$$\pi \begin{cases} L \rightsquigarrow C \\ i \mapsto A_i \text{ (on objects)} \\ (i \mapsto j) \mapsto \pi_{ji} \text{ (on morphisms)} \end{cases}$$

is a functor which is injective on objects and morphisms, then there is a directed class $\{A_i, \pi\}_{i \in L}$ on the designated objects with the two following properties:

1. There is a morphism $A_i \to A_j$ if and only if $i \leq j$, and then $\pi_{ji} : A_i \to A_j$ is the only morphism.

2. The diagram

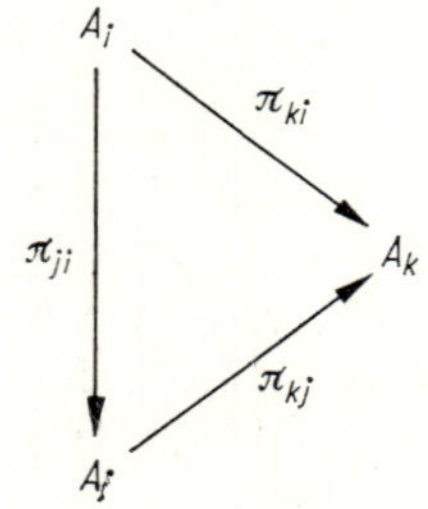

commutes for every $i \leq j \leq k$, and $\pi_{ii} = 1_{A_i}\ \forall i \in \text{obj } L$.

The category $\{A_i, \pi\}_{i \in L}$ is called a **class of objects of** C **directed by** π or **by** L. If L is small, it is called a **directed family**, or set. Conversely, if $\{A_i\}_{i \in L}$ is a subcategory of a category C, indexed by a category L so that $i \mapsto A_i$ is injective, and if $\{A_i\}_{i \in L}$ is a directed class, then (a) L is a directed class such that

$$i \leq j \Leftrightarrow \text{there is a morphism } A_i \to A_j,$$

(b) the morphisms $\pi_{ji} : A_i \to A_j$ satisfy the requirements 1. and 2., and (c) there is a functor $\pi : L \rightsquigarrow C$ such that $\{A_i, \pi\}_{i \in L}$ is directed by π.

Examples.

1. The class of subobjects of an object A in a category C is directed by inclusion. Thus, if $\{u_i : A_i \to A\}_{i \in L}$ is the class of subobjects, then the following conditions are equivalent:

(i) $A_i \leq A_j$

(ii) $u_i \leq u_j$

(iii) there exists a necessarily unique morphism $\pi_{ji} : A_i \to A_j$ such that

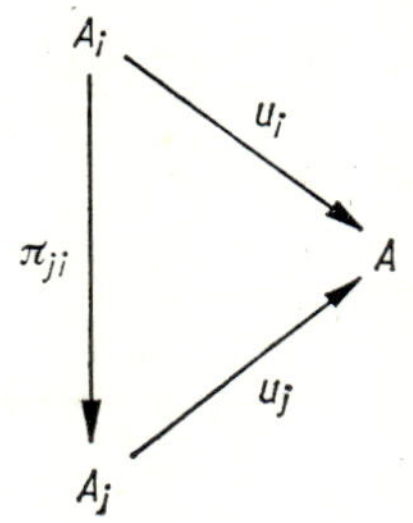

commutes. ▯

2. If L is a directed class, then a subcategory J is **cofinal** provided that for every object X of L there is an object Y of J such that $Y \geq X$. If L is a directed class, and if J is a cofinal subclass, then J is directed, and for any functor $F : L \rightsquigarrow C$, there is induced a functor $F' : J \rightsquigarrow C$ and

$$\varprojlim_{X \in L} F(X) \approx \varprojlim_{Y \in J} F'(Y)$$

is an equivalence in C. ▯

5.8 *Examples.*

5.8.1 If L is an ordered class, for any functor $F : I \rightsquigarrow L$, there is a bijection between elements $\{f(X) : A \to F(X)\}_{X \in \text{obj}\, I}$ of compat F and objects A of L contained in $F(X) \forall X \in \text{obj}\, I$. It follows, that

$$\lim F = \underset{X \in \text{obj}\, I}{\text{g.l.b.}} \{F(X)\} = \bigwedge_{X \in \text{obj}\, I} F(X) = \prod_{X \in \text{obj}\, I} F(X)$$

$$\text{colim}\, F = \underset{X \in \text{obj}\, I}{\text{l.u.b.}} \{F(X)\} = \bigvee_{X \in \text{obj}\, I} F(X) = \coprod_{X \in \text{obj}\, I} F(X).$$

Thus, limits, products, intersections, and g.l.b.'s are the same, and L is (finitely) complete if and only the g.l.b. of any (finite) family of objects of L exists. Furthermore, if L is directed and (finitely) cocomplete, then L is (finitely) complete, since

$$\text{g.l.b.} \{A_i\}_{i \in I} = \underset{X \in \text{obj}\, I}{\text{l.u.b.}} \{X \leq A_i \forall i \in I\}.$$

5.8.2 (a) For any category C and limit functor $\lim : C^L \rightsquigarrow C$, any object A of C is a limit, and $A = \lim c(A)$. Thus, it is nothing to say that an object in a category is a limit unless you say what kind of functor it is the limit of.

(b) *Final objects are limits.* If A is any final object, then $A = \lim F$, where $F : L \rightsquigarrow C$ is the empty functor defined on the empty category L.

5.8.3 *Products are limits.* Of course, products are defined to be final objects, but not necessarily in C. If $\{A \to A_i\}_{i \in I}$ is the product in a category C, then let I denote the category such that an object of I is an element and a morphism $i \to j$ exists only if $j = i$, and then $i \to i$ is the identity morphism. Then $A = \lim F$, where

$$F \begin{cases} I \rightsquigarrow C \\ i \mapsto A_i. \end{cases}$$

5.8.4 *Equalizers are limits.* If $f_i : A \to B$, $i = 1, 2$, are two morphisms of a category C, an **equalizer** is defined to be a morphism $k : K \to A$ which is final in the full subcategory of target A consisting of those morphisms $k' : K' \to A$ such that $f_1 k' = f_2 k'$. This morphism (and sometimes the object K) is denoted equalizer (f_1, f_2) and is unique up to an equivalent monic (object) in target A. If L denotes the category on two objects X and Y and two morphisms $k_i : X \to Y$, $i = 1, 2$, and if $F : L \rightsquigarrow C$ is the functor such that $FX = A$, $FY = B$, $Fk_i = f_i$, $i = 1, 2$, then equalizer (f_1, f_2) is a final object in compat F, that is, $\lim F =$ equalizer (f_1, f_2).

(a) Equalizer (f_1, f_2) *is monic.* If $k : K \to A$ is the equalizer and $q : Q \to K$ is any morphism, then $f_1 kq = f_2 kq$, so there is a unique morphism $q' : Q \to K$ such that $kq = kq'$. Thus, k is monic. By duality, coequalizer (f_1, f_2) is epic.

Let $f : A \to B$ and $g : A \to B$ be morphisms in a category C.

(b) If C is additive, then

$$\text{equalizer}\,(f, g) = \ker\,(f - g)$$
$$\text{coequalizer}\,(f, g) = \text{coker}\,(f - g)\,.$$

(c) If C is any concrete category, then equalizer (f, g) is the subobject generated by the equalizer of f and g in SETS, namely by the subset $\{x \in A \mid f(x) = g(x)\}$. For C one of (SEMI) GROUPS, RINGS, or mod-R, for any ring R, the forgetful functor maps the equalizer of f and g in C onto the equalizer of f and g in SETS.

(d) In SETS, define a relation R in B by the formula:

$$xRy \Leftrightarrow f^{-1}x \cap g^{-1}y \neq \varnothing\,.$$

Let $\bar{R}$ denote the equivalence relation in B generated by R. Then the canonical mapping

$$\begin{cases} B \to \text{Set of equivalence classes of } \bar{R} \\ x \mapsto \bar{R}_x \end{cases}$$

is the coequalizer of f and g in SETS.

(e) For the category of groups, the coequalizer of f and g is the quotient epimorphism $B \to B/H$, where H is the normal subgroup generated by the $\{f(x)g(x)^{-1} \mid x \in A\}$. Similarly, in RINGS, the coequalizer is the quotient epimorphism $B \to B/H$, where H is the ideal generated by $\{f(x) - g(x) \mid x \in A\}$.

5.8.5 Let $f_1 : A \to C$ and $f_2 : B \to C$ be morphisms of a category D, let L be the category with three objects X, Y and Z and in addition to the identity morphisms, two morphisms $k_1 : X \to Z$ and $k_2 : Y \to Z$, and let $F : L \rightsquigarrow D$ be the functor such that

$$F(X) = A, \quad F(Y) = B, \quad F(Z) = C$$
$$F(k_i) = f_i, \quad i = 1, 2.$$

Then lim F, if it exists, is called the **fiber product** of the objects A and B with respect to C, and denoted $A \prod_C B$. The objects of compat F are the class of all pairs of morphisms $P \to A$ and $P \to B$ such that

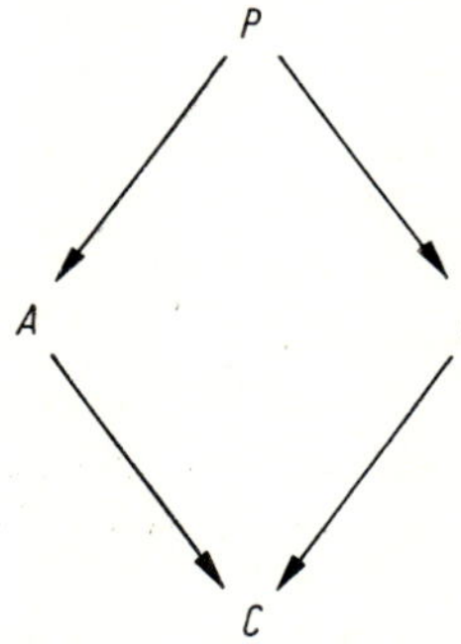

commutes. A morphism of compat F between this and a second object $P' \to A$ and $P' \to B$ of compat F is a morphism $P \to P'$ such that

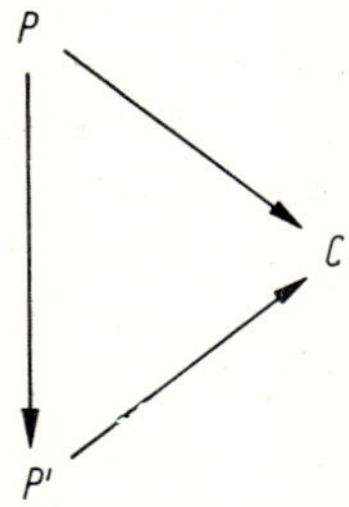

commutes. Thus, lim F is a final object P in this category, and the canonical morphism $P \to C$ is called the **fiber morphism**. The final object P and the diagram are called **pullbacks**, and the diagram is also called **cartesian**. The dual notions speak of the **fiber coproduct** $A \coprod_C B$, **pushout**, and **cocartesian**.

For any morphism $f: A \to B$ in a category with zero, the **kernel** is defined to be

$$\ker f = A \coprod_B 0 = \text{equalizer } (f, 0)$$

and the **cokernel**

$$\text{coker } f = 0 \coprod_A B = \text{coequalizer } (f, 0)$$

The canonical (fiber morphism) $\ker f \to A$ is therefore monic, and $B \to \text{coker } f$ is epic. For any monic $K \to A$, a canonical monic in the equivalence class of monics equivalent to $K \to A$ is chosen and denoted $i_K^A : K \to A$. Then, $p_{A/K} : A \to \text{coker } i_K^A$ denotes the cokernel.

Let $f: X \to Y$ be a morphism in a category C with zero, and let kill f denote the set of all morphisms g such that gf is defined and $gf = 0$. Then g **kills** f and f **is killed by** g. The dual category C^* has a zero, so we may define cokill f as the set of all morphisms k such that $k^* \in \text{kill } f^*$. Thus, cokill f is the set of all morphisms killed by f. Thus, f kills $\ker f$. (Cf. Exercise 5.10.5.)

5.8.6 A morphism $A \to B$ is said to **allow** $Y \to B$ in case there is a morphism $Y \to A$ such that

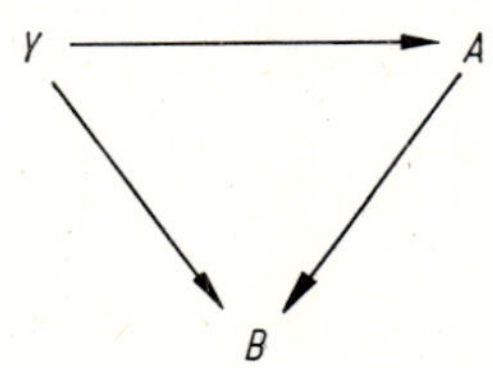

commutes.

If $\{A_i \to A\}_{i \in I}$ is a family of monics in C, and if $I' = I \coprod \{p\}$ is the ordered category such that $i \leq p$ for every $i \in I$, then the limit of the functor

$$\begin{cases} I' \to C \\ i \mapsto A_i \\ p \mapsto A \end{cases}$$

is defined to be the **intersection** $\cap A_i$ of the family. Consider the category consisting of all morphisms $A' \to A$ such that $A_i \to A$ allows $A' \to A$ $\forall i \in I$. A morphism in this category between $A' \to A$ and $B \to A$ is a morphism $A' \to B$ such that

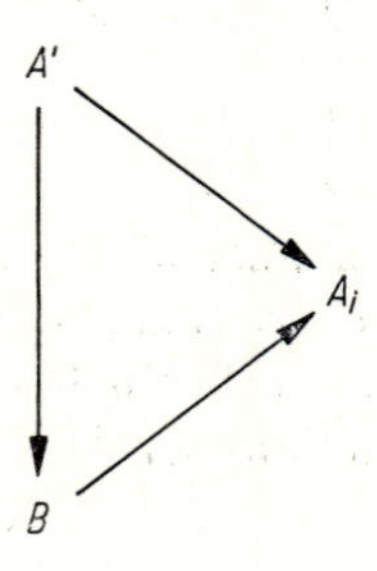

commutes for every $i \in I$. Then a final object $A' \to A$ in this category is a monic equivalent to the intersection of the family $\{A_i \to A\}_{i \in I}$.

5.8.7 The **image** of a morphism $f : A \to B$ is defined to be an initial object $\operatorname{im} f \to B$ in the category of all monics $I \to B$ that allow f.

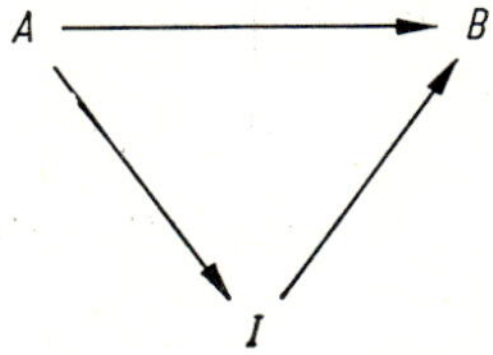

commutes. Thus, $\operatorname{im} f \to B$ is the smallest monic $I \to B$ which allows f, and any two images are equivalent monics in target B. The monic $\operatorname{im} f \to B$ and the unique morphism $A \to \operatorname{im} f$ which together compose f are both called **canonical** and are said to be induced by f. The category C is said to **have (epic) images** if every morphism $f : A \to B$ has an image (and the induced morphism $A \to \operatorname{im} f$ is always epic). If $g : A' \to A$ is monic, then $f(A')$ denotes the image of $fg : A' \to B$. Thus, $f(A) = \operatorname{im} f$.

If $f : A \to B$, and if B' is a subobject of B, then the fiber product $A \prod_B B'$ is called the **inverse** image of B', denoted $f^{-1}B'$. The fiber morphism $f' : f^{-1}B' \to B$ is said to be **induced** by f, and it **induces** a monic $\operatorname{im} f' \to B'$.

If $f : X \to Y$ is a mapping of sets, if S is a subset of Y, the inclusion mapping of which allows f, and if $g : X \to S$ is such that $f = i_S^Y g$, then $f(x) = g(x) \, \forall x \in X$, so that $\operatorname{im} f \subseteq S$. Thus, $\operatorname{im} f$ is the smallest subset of Y, the inclusion mapping of which allows f, and therefore the definition of image for categories is consistent with its prior use for SETS. Similarly for inverse images, kernel, cokernel, and intersections. □

5.9 Proposition. *If C is a category with equalizers, then for any morphism $f : A \to B$, the canonical morphism $A \to \operatorname{im} f$ is epic.*

Proof. Let $f' : A \to \operatorname{im} f$ be the canonical morphism (assuming that the image of f exists), and let $h_1 f' = h_2 f'$ for two morphisms $h_i : \operatorname{im} f \to X$, $i = 1, 2$. Then f' factors through the equalizer I of h_1 and h_2, andhence so does f. Thus, I is a subobject of $\operatorname{im} f$, hence of B, which allows f. But since $\operatorname{im} f$ is defined to be the smallest subobject of B which allows f, it follows that the inclusion of I into $\operatorname{im} f$ is an equivalence, whence $h_1 = h_2$, and f' is therefore epic. □

5.10 *Exercises.*

5.10.1 If each square of the diagram

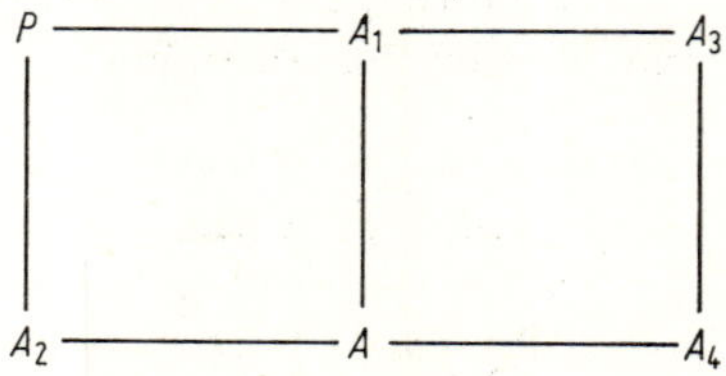

is a pullback, and if $A_3 \to A_4$ is monic, then the rectangle is a pullback.

5.10.2 If

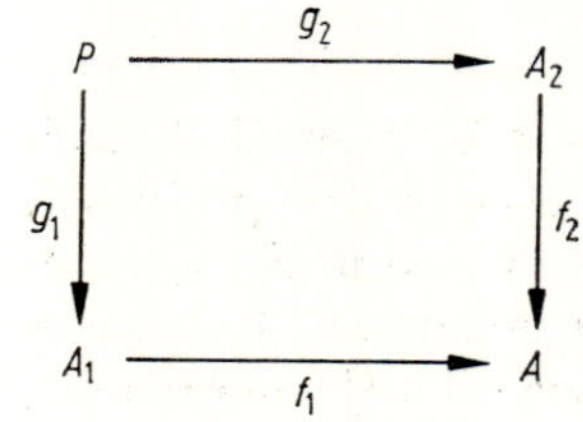

is a pullback, and if f_1 (resp. f_2) is monic, then g_2 (resp. g_1) is monic.

5.10.3 If f_1 and f_2 are monic in the diagram in 5.10.2, then the fiber product is the intersection $A_1 \cap A_2$.

5.10.4 If $A = 0$, then the diagram in 5.10.2 is cartesian if and only if $\{g_i : P \to A_i\}_{i=1,2}$ is a product.

5.10.5 In a productive category, the kernel of a product of morphisms is the product of the kernels. Thus, a product of monics is monic.

5.10.6 A morphism $f : A \to B$ is monic if and only if $A = \operatorname{im} f$.

5.10.7 Let $A_1 \subset A_2 \subset A$, $B_1 \subset B_2 \subset B$ be objects and $f : A \to B$ a morphism in a category C. If f carries a subobject A' into B', let $f(A')$ denote B'. The following relations hold whenever defined:

(a) $f(A_1) \subseteq f(A_2)$.

(b) $f^{-1}(B_1) \subseteq f^{-1}(B_2)$.

(c) $A_1 \subseteq f^{-1}(fA_1)$.

(d) $B_1 \supseteq f(f^{-1}B_1)$.

(e) $f(A_1) = f\big(f^{-1}(f(A_1))\big)$.

(f) $f^{-1}(B_1) = f^{-1}\big(f(f^{-1}(B_1))\big)$.

(g) If a monic $B' \to B$ allows f, and if $B' \cap B_1$ is defined, then $f^{-1}(B \cap B_1) = f^{-1}(B_1)$, thus, $f^{-1}(\operatorname{im} f \cap B_1) = f^{-1}B_1$.

(h) Inverse image commutes with intersections, $f^{-1}(\bigcap_{i\in I} A_i) = \bigcap_{i\in I} f^{-1} A_i$.

5.10.8 (a) *A kernel* $K \to X$ *of a morphism* $f: X \to Y$ *in a category with zero is a final object in the category* cokill $f \cap$ target X, *provided a final object exists, and is unique up to an equivalence in target* X. *Any kernel of a morphism is monic.*

(a*) *A cokernel* $Y \to Q$ *of a morphism* $f: X \to Y$ *in a category* C *with zero is an initial object in the subcategory* kill $f \cap$ source Y, *and is unique up to an equivalence in source* Y. *A cokernel is an epic.* ▯

(b) *There is a canonical morphism* coker (ker f) $\to$ ker (coker f) *such that*

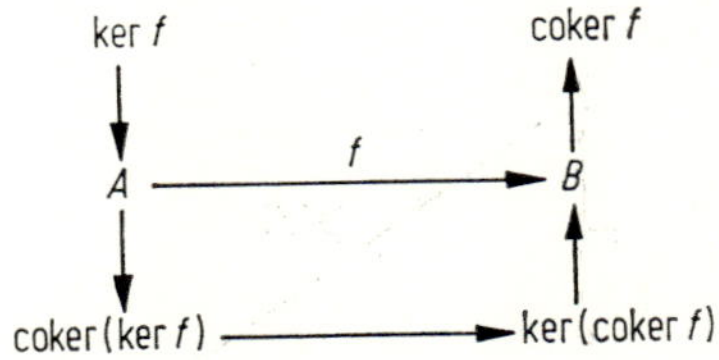

commutes (*see* 6.10).

5.10.9 If $T: I \rightsquigarrow L$ is a representative functor, and if $F: L \rightsquigarrow C$ maps the canonical morphism $(\lim T) \to T(i)$ onto a monic of C, for some $i \in I$, then $\lim FT = F \lim T$. Thus, if $\{p(i): Y \to T(i)\}_{i\in I}$ is a final object of compat T such that $F(p(i))$ is monic, for some $i \in I$, then $\{F(p(i)): FY \to FT(i)\}_{i\in I}$ is a final object of compat FT.

5.10.10 Prove, for any category C, that a coproduct is the direct limit of a functor $L \rightsquigarrow C$.

5.10.11 Prove that C is cocomplete if and only if C is a category with direct limits.

5.10.12 (Freyd) A fully faithful functor $T: C \rightsquigarrow D$ reflects limits and colimits, or equivalently, if $F: L \rightsquigarrow C$ is a functor such that $\lim_{L,D} TF$ exists, and if $TX = \lim_{L,D} TF$ for some X obj C, then $X = \lim_{L,C} F$. [Hint:

$$\begin{aligned}\mathrm{Mor}_C(Y, \lim_{L,C} F) = \mathrm{Mor}_{C^L}(c(Y), F) &= \mathrm{Mor}_{D^L}(c(TY), TF)\\ &= \mathrm{Mor}_D(TY, \lim_{L,D} TF)].\end{aligned}$$

A category C is (**finitely**) **complete** in case $\lim: C^L \rightsquigarrow C$ exists for any (finite) small category L.

5.11 Proposition. *A category* C *is* (*finitely*) *complete if and only if* C *has* (*finite*) *products and equalizers.*

Proof. By Examples 5.8.3 and 5.8.4 products and equalizers are limits. Conversely, by 5.6, it suffices to prove, for any small category L, that the category compat F, the objects of which are families of morphisms $\{A \to F(X)\}_{X \in \mathrm{obj}L}$ compatible with F, has a final object. Consider the commutative diagrams:

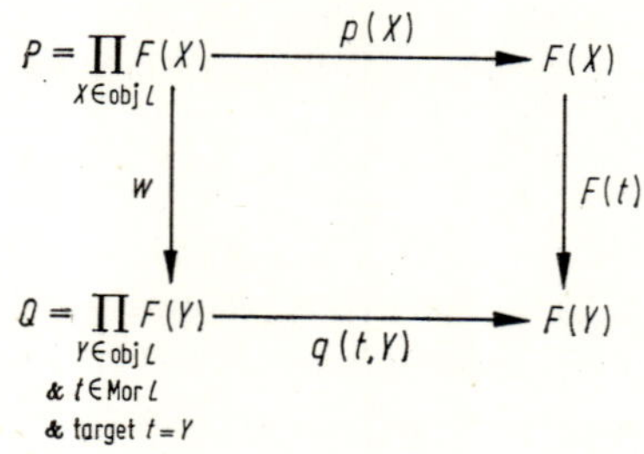

and

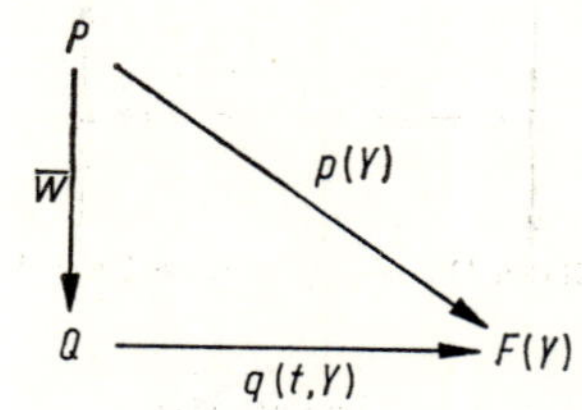

where $t: X \to Y$ is any morphism of L, $p(X): P \to F(X)$ and $q(t, Y): Q \to F(Y)$ are the canonical projections from the product, and $w: P \to Q$ and $\overline{w}: P \to Q$ are canonical morphisms. Let $k: A \to P$ denote equalizer $(w, \overline{w})$. Then the family of morphisms $\{u(X): A \to F(X)\}_{X \in \mathrm{obj}L}$ is compatible with F, where

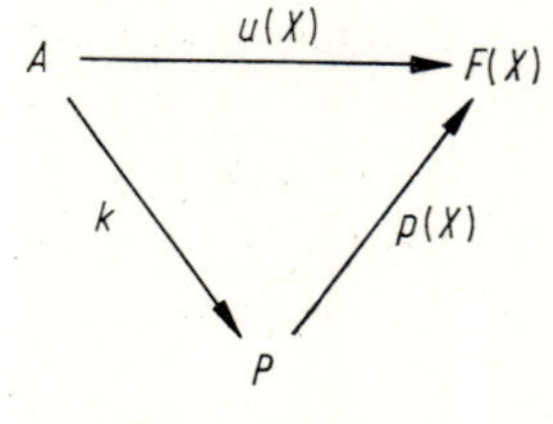

commutes. Indeed, for each $t: X \to Y$ in L,

$$F(t)\,u(X) = F(t)\,p(X)\,k = q(t, Y)\,w\,k = q(t, Y)\,\overline{w}\,k = p(Y)\,k = u(Y).$$

In order to show that $A \approx \lim F$, by the exercise, it suffices to show that A is a final object in compat F. To do this, let $\{v(X): B \to F(X)\}_{X \in \mathrm{obj}L}$ be a family of morphisms of C which is compatible with F. Then, by the definition of product, there is a unique morphism $h: B \to P$ such that $p(X)\,h = v(X)$ for any object X of L, and then, for any morphism $t: X \to Y$ in L

$$q(t, Y)\,w\,h = F(t)\,p(X)\,h = F(t)\,v(X) = v(Y)$$

and

$$q(t, Y)\overline{w}h = p(Y)h = v(Y).$$

Then, by the definition of product, $wh = \overline{w}h$. Since $k =$ equalizer$(w, \overline{w})$, there is a unique morphism $v' : B \to A$ such that $kv' = h$, and then

$$u(X)v' = p(X)kv' = p(X)h = v(X).$$

This shows that $v' : B \to A$ is a morphism of compat F, and that $\{u(X) : A \to F(X)\}_{X \in \text{obj} L}$ is the desired final object, that is, $A \approx \lim F$.

When L is finite, then the products P and Q which are required for the proof are finite, and so the parenthetical statement follows. □

5.12 Corollary. *An additive category C is (finitely) complete if and only if C has (finite) products and kernels.*

Proof. In an additive category, the kernel of the difference of two morphisms $A \to B$ is their equalizer. Thus, for additive C, the conditions of the corollary are equivalent to those of the proposition. □

5.13 Proposition. *The following statements about a category C are equivalent:*

(a) *C has finite intersections and finite products.*
(b) *C has equalizers and finite products.*
(c) *C is finitely complete.*

Proof. (a) ⇒ (b) is a consequence of the fact that given morphisms $f : A \to B$ and $g : A \to B$ in C, the diagram

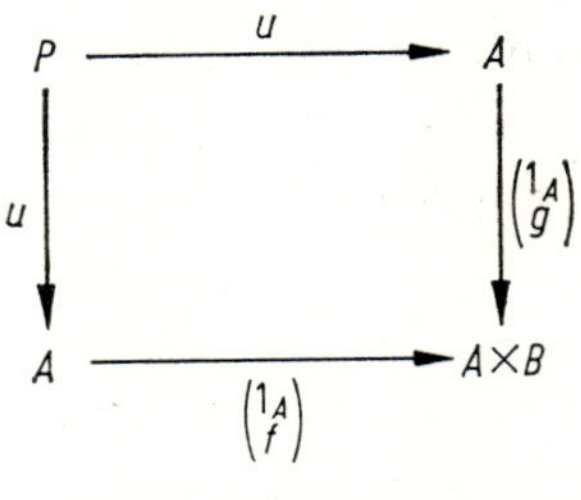

is cartesian if and only if $P \to A$ is the equalizer of the subobject defined by the vertical and horizontal morphisms $A \to A \times B$. Since then the latter two morphisms are necessarily monics, then the cartesian morphism $P \to A \times B$ is necessarily their intersection. This follows since the diagram is commutative if and only if $fu = gu$. This proves (a) ⇒ (b), and (b) ⇒ (c) is part of the theorem. Since finite products and finite intersections are finite limits, then (c) ⇒ (a). □

5.14 Corollary. *A category C is complete if and only if C has products and finite intersections.*

Proof. By the last corollary, in the presence of finite products, C has equalizers if and only if C has finite intersection. The corollary now follows from the theorem. □

5.15 Corollary. *A finitely complete category C has fiber products. Conversely, if C is a category with zero object, and if C has fiber products, then C is finitely complete.*

Proof. A fiber product is a finite limit. Conversely, the existence of fiber products imply finite intersections. Then, if C has a zero object, a fiber product

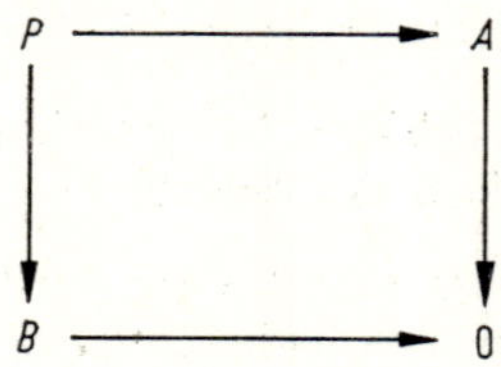

is the product of objects A and B of C. Thus, C has finite intersections and finite products, so C is finitely complete by the proposition. □

5.16 *Exercises.*

5.16.1 Show that SETS, GROUPS, mod-R (any ring R), and RINGS (also the category TOP of topological spaces) are complete.

5.16.2 Show that SETS, GROUPS, mod-R, and TOP are cocomplete.

*5.16.3 Show that RINGS is cocomplete.

5.16.4 If C is (finitely) (co)complete, then C^L is (finitely) (co)complete.

5.16.5 If C has one of the following properties, then so has C^L: Kernels, (finite) products, equalizers, additive. Extend the list as far as possible.

5.16.6 Any finitely complete category C has equalizers, and C has kernels if and only if C has a zero. Conclude that SEMIGROUPS is not finitely complete.

*5.16.7 In GROUPS, epics are surjections.

5.16.8 If $F: L \rightsquigarrow C$ is a functor, and L is small, then a limit for F is the family of compositions

$$\bigcap_{t\in \mathrm{Mor}_L(B,A)} \text{Equalizer}\,(p(A), F(t)\,p(B)) \to \prod_{X\in L} F(X) \xrightarrow{p(X)} F(X),$$

where $p(A)$ is the projection from the product for any $A \in \operatorname{obj} L$, provided the product, equalizers, and intersections exist.

5.16.9 A set $\{X_i\}_{i\in I}$ in a category C is **preinitial**, if for each object Y, there exists $i \in I$ and a morphism $X_i \to Y$. Show that a complete category C has a preinitial set if and only if C has an initial object.

Normal Categories

A subobject B of an object A is **normal** provided that B is the kernel of some morphism. A category is **normal** provided that every subobject of every object is normal. Thus, mod-$\mathbb{Z}$ is normal, but GROUPS is not.

5.17 Proposition. *If* $i_B^A : B \to A$ *is a normal subobject of* A, *then* $i_B^A = \ker(\operatorname{coker} i_B^A) = \ker p_{A/B}$.

Proof. Let $i = i_B^A$ be the kernel of a morphism $f : A \to C$. Then there exists a morphism $a : A/B \to C$ such that $ap = f$, where $p = p_{A/B}$, and it follows that f kills any monic $K \to A$ which p kills. This shows that $i_B^A = \ker p_{A/B}$. □

5.18 Corollary. *If* C *is a normal and conormal category with kernels and cokernels, then for each object* A,

$$\operatorname{coker} \begin{cases} \text{subobjects } A \to \text{quotient objects } A \\ \qquad B \mapsto A/B \end{cases}$$

is an order anti-isomorphism, and the inverse function ker *assigns to each quotient object* $A \to C$ *its kernel. That is, for each morphism* $f : X \to Y$,

$$\ker\big(\operatorname{coker}(\ker f)\big) = \ker f$$

and

$$\operatorname{coker}\big(\ker(\operatorname{coker} f)\big) = \operatorname{coker} f. \quad □$$

If $k : K \to X$ is a kernel of $f : X \to Y$, and if $h : H \to X$ is a monic killed by f, then there exists a unique morphism $g : H \to K$ such that

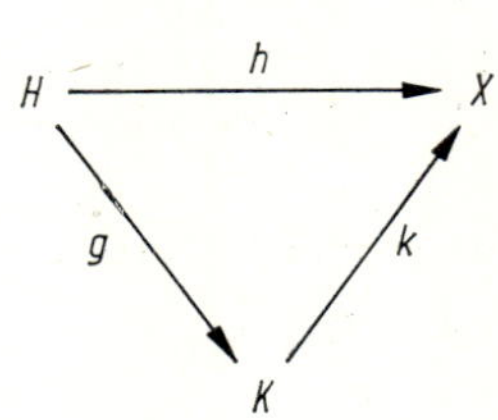

commutes. Since $kg = h$ is monic, $g : H \to K$ is necessarily a monic, that is, H is a subobject of K. Furthermore, $H \leq K$ in the ordering $\leq$ of morphisms. Thus, in two different senses, *the kernel of a morphism is the largest monic which the morphism kills.* This statement coincides with the situation in GROUPS where the kernel of a homomorphism f is the largest subgroup the inclusion mapping of which f kills. Dually, *the cokernel of a morphism* f *is the smallest epic which kills* f.

5.19 Proposition and Definition. *A category is said to be* **balanced** *in case every morphism which is both epic and monic is an equivalence.*

5.19.1 *Every normal category is balanced.*

5.19.2 *In a balanced category, a commutative diagram*

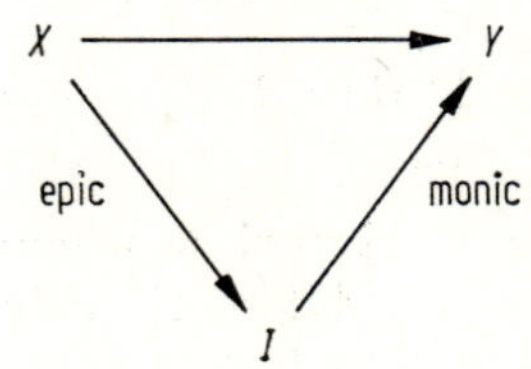

implies that $I \to Y$ is the image of $X \to Y$ provided only that $X \to Y$ has an image.

5.19.3 *In a balanced category with images and equalizers, for a commutative diagram*

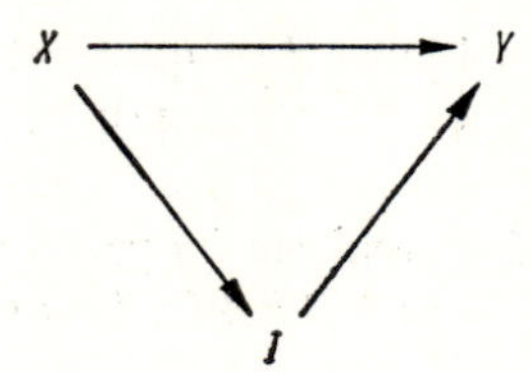

it is true that $I \to Y$ is the image of $X \to Y$ if and only if $I \to Y$ is monic and $X \to I$ is epic.

5.19.4 ***(Unique factorization)*** *In a balanced category with images, if there exists epics q, q' and monics k, k' such that a morphism f has two factorizations $f = kq = k'q'$, then $q \approx q'$ and $k \approx k'$ are equivalent epics and monics respectively.*

Proof of 5.19.1. Let $f : A \to B$ be monic and epic. Then $\operatorname{coker} f = 0$, and f is the kernel of its cokernel. But a kernel of $B \to 0$ is 1_B, so there exists $h : B \to A$ such that $fh = 1_B$. Then $fhf = 1_B f = f$, whence $hf = 1_A$, and f is therefore an equivalence.

Proof of 5.19.2. Consider a diagram

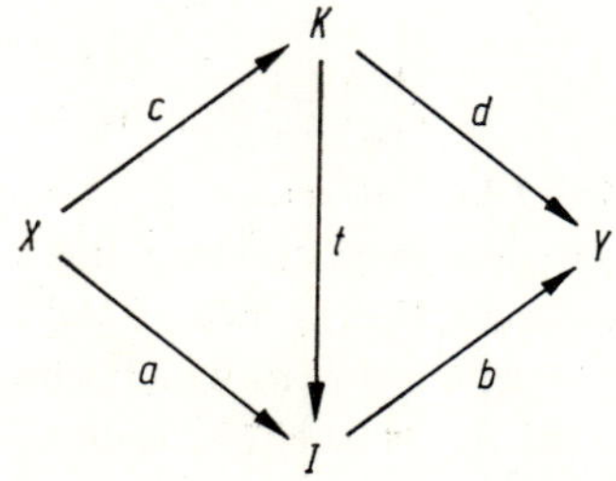

where a is epic, b is monic, d is the image of $f = ba$, and c is the canonical morphism onto image. Since b allows f, and since the image is the smallest monic which does this, the morphism t is canonically given. Then a epic implies t epic, and d monic implies t monic, and in a balanced category this implies that t is an equivalence. Therefore, $b: I \to Y$ is also an image.

Proof of 5.19.3. This follows from 5.19.2. and 5.9.

Proof of 5.19.4. This follows from 5.19.2. since both k and k' are images of f, they are equivalent monics. The equivalence t such that $kt = k'$ yields $tq' = q$, so the epics are equivalent too. ▯

5.20 Proposition. *A normal category C with kernels and cokernels has images, and for every morphism* $f: X \to Y$

$$\operatorname{im} f = \ker(\operatorname{coker} f).$$

If, in addition, C has equalizers, then the canonical morphism $X \to \operatorname{im} f$ *is epic, and whenever this is so, it is the coimage of f, that is,* $\operatorname{coim} f \approx \operatorname{im} f$, *or equivalently, the canonical morphism*

$$\operatorname{coker}(\ker f) \to \ker(\operatorname{coker} f)$$

is an equivalence.

Proof. Consider a diagram

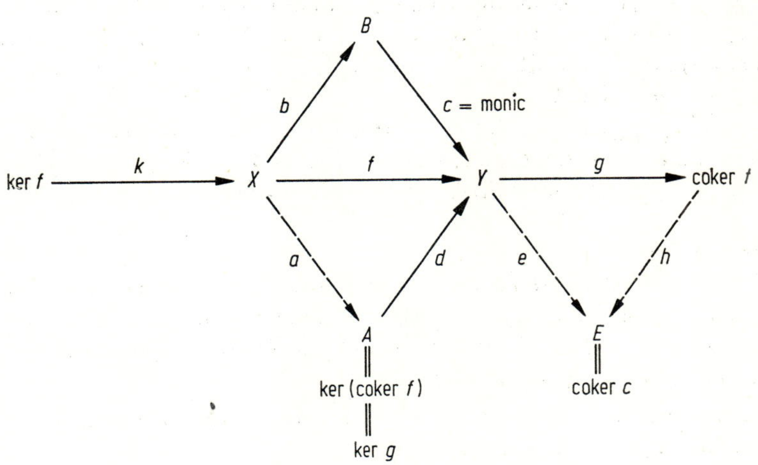

For any monic $c: B \to Y$ which allows f, let $e: Y \to E$ be the cokernel. The fact that $ec = 0$ implies that $ef = 0$, whence there exists a unique morphism h such that $hg = e$, which shows that $ed = hgd = 0$. Since the category is normal, the monic c is the kernel of its cokernel e, and hence there exists a unique morphism $t : A \to B$ such that $ct = d$. Since this is true for any monic $c : B \to Y$ which allows f, by the definition of image, it follows that $d = \ker(\operatorname{coker} f) = \operatorname{im} f$, as asserted.

The morphism $a : X \to A$ exists because $gf = 0$ and $d : A \to Y$ is the kernel of g. By 5.9, a is epic, when C has equalizers. In this case suppose that $b : X \to B$ is an epic which allows f. Then $gc = 0$, and there is a unique morphism $t : B \to A$, such that $dt = c$. Then $f = cb = dtb = da$, and the fact that d is monic yields $tb = a$. This shows that $a : X \to A$ is the largest epic which allows f, and therefore is the coimage of f. □

5.21 Corollary. *Let U be a generator in a normal, additive category C with kernels and cokernels. For each object X of C, there is a canonical exact sequence*

(1) $$U^{(\mathrm{Mor}_C(U,X))} \xrightarrow{h} X \to 0$$

provided the coproduct exists.

Proof. The canonical morphism h defined in (1) is epic by 3.25. Since the category is additive and has kernels, then it has equalizers, and by the proposition

$$\operatorname{im} h = \ker(\operatorname{coker} h) = \ker 0 = X,$$

so (1) is exact. □

A category C is **locally small** (or **well-powered**) if the equivalence class of subobjects of any object of C is a set. The category SETS is locally small by the ZF-axioms of set theory, hence, any concrete category is locally small.

A set $\{U_i\}_{i\in I}$ of objects of a category C is a **set of generators** if for every pair of distinct morphisms $f : X \to Y$ and $g : X \to Y$ there exists $i \in I$ and a morphism $h : U_i \to X$ such that $fh \neq gh$.

5.22 *Exercises.*

5.22.1 If C is a category, if $\{U_i\}_{i\in I}$ is a set of objects such that for every object A and every $i \in I$ there is a morphism $U_i \to A$, and if the coproduct $U = \coprod_{i\in I} U_i$ exists, then $\{U_i\}_{i\in I}$ is a set of generators of C if and only if U is a generator of C.

5.22.2 Any one element set is a generator of SETS, and any two-element set is a cogenerator.

5.23 Proposition. *Let C be balanced category with finite intersections and a set $\{U_i\}_{i\in I}$ of generators. Then:*

5.23.1 *If A is an object of C, a subobject $X \to A$ is characterized by those morphisms $U_i \to A$ that $X \to A$ allows.*

5.23.2 *C is locally small, that is, if A is an object of C, then the equivalence class of subobjects of A is a set.*

Proof of 5.23.1. Let $u_1 : X_1 \to A$ and $u_2 : X_2 \to A$ be different subobjects, that is, inequivalent monics. Now, if the canonical monics $v_1 : X_1 \cap X_2 \to X_1$ and $v_2 : X_1 \cap X_2 \to X_2$ were both epics, they would be equivalences, contrary to the assumption that X_1 and X_2 are not equivalent. Suppose v_1 is not epic. Then there exists an object B and distinct morphisms $f : X_1 \to B$ and $g : X_1 \to B$ such that $fv_1 = gv_1$. Hence, there exists $i \in I$ and a morphism $h : U_i \to X_1$ such that $fh \neq gh$. Then $v_1 : X_1 \cap X_2 \to X_1$ does not allow h, hence $u_2 : X_2 \to A$ does not allow $u_1 h : U_i \to A$.

Proof of 5.23.2. A set of morphisms $\{U_i \to A\}$ suffices to distinguish inequivalent subobjects of an object A. ▯

Exact Categories

5.24 Definition. A category C is defined to be **exact** if it is normal and conormal and has kernels and cokernels and if every morphism $f : X \to Y$ has a factorization

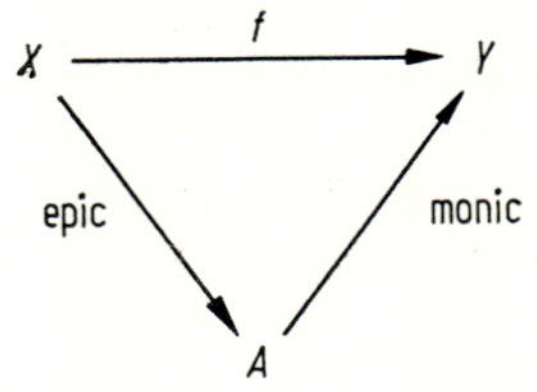

A sequence of morphisms

$$\cdots \to A_{i-1} \xrightarrow{f_{i-1}} A_i \xrightarrow{f_i} A_{i+1} \xrightarrow{f_{i+1}} A \to \cdots$$

is **exact**, if $\ker f_{i+1} = \operatorname{im} f_i$ as subobjects of $A_{i+1} \forall i$.

An exact sequence $0 \to A \to B \to X$ is said to be **left exact**. A covariant functor $T : C \rightsquigarrow D$ is said to be **left exact** if F preserves left exact sequences, that is, exactitude of the left exact sequence $0 \to A \xrightarrow{f} B \xrightarrow{g} X$ implies exactitude of $0 \to TA \xrightarrow{Tf} TB \xrightarrow{Tg} TX$. The functor T is **exact** if it is both left and right exact. Furthermore, T is said to be **half exact** if $0 \to A \xrightarrow{f} B \xrightarrow{g} X \to 0$ exact implies $TA \xrightarrow{Tf} TB \xrightarrow{Tg} TX$ exact. Clearly any left, or right, exact functor is half exact.

A contravariant functor $T : C \rightsquigarrow D$ is (left, right) exact provided that the induced functor $T^* : C^* \rightsquigarrow D$ is (left, right) exact, where $T^*(A^*) = TA$. Thus, T is left exact if $A \to B \to X \to 0$ exact implies $TA \leftarrow TB \leftarrow TX \leftarrow 0$ exact.

5.25 Proposition.

5.25.1 *An exact category has epic images, and for any morphism* $f : X \to Y$, *there is a factorization*

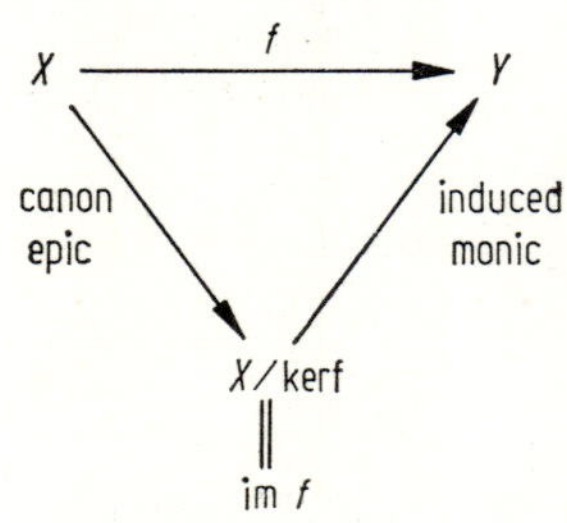

which is unique up to equivalent epics and monics, and the canonical morphism coim $f \to$ image f *is an equivalence.*

5.25.2 *A normal and conormal category with kernels and cokernels and equalizers is exact.*

5.25.3 *A normal and conormal additive category with kernels and cokernels is exact.*

Proof. By 5.20, $\operatorname{im} f = \ker(\operatorname{coker} f)$, and $\operatorname{coim} f = X/\ker f$. Furthermore, $X \to \operatorname{im} f$ is epic if and only if it is the coimage of f. Factor

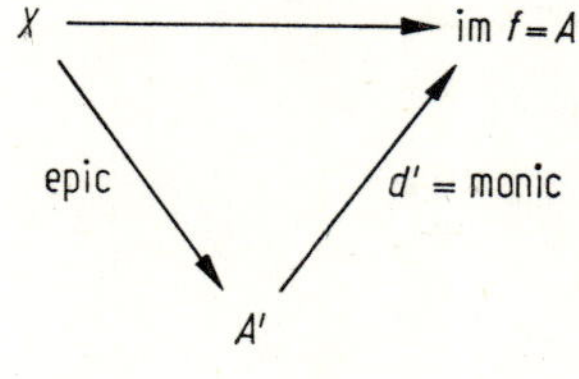

Then $dd' : A' \to Y$ is a monic which allows $f : X \to Y$, and if $C \to Y$ allows f, then $C \to Y$ allows $A \to Y$, hence $C \to Y$ allows $A' \to Y$. This shows that $A' \to Y$ is an image of f, so that $d' : A' \to A$ must be an equivalence, whence $X \to A$ is epic along with $X \to A'$. It follows that $X \to \operatorname{im} f$ is a coimage of f. Since $\operatorname{coim} f = X/(\ker f)$, the commutative triangle in the statement 5.25.1 results from the corresponding triangle induced by the morphism $X \to \operatorname{im} f$. This proves 5.25.1. When the conditions of 5.25.2 hold, then $X \to \operatorname{im} f$ is epic by 5.9, so every morphism has the epic-monic factorization. An additive category with kernels has equalizers, and therefore 5.25.2 implies 5.25.3. □

5.26 Proposition.

5.26.1 *In an exact category, the following statements hold:*

(a) $A \to B \to X$ *is exact in* $C \Leftrightarrow A^* \leftarrow B^* \leftarrow X^*$ *is exact in* C^*.

(b) $0 \to A \to B$ *is exact if and only if* $A \to B$ *is monic.*

(c) $A \to B \to 0$ *is exact if and only if* $A \to B$ *is epic.*

(d) $0 \to A \to B \to 0$ *is exact if and only if* $A \to B$ *is an equivalence.*

(e) $0 \to A \to B \to X$ *is exact if and only if* $A \to B$ *is the kernel of* $B \to X$.

(f) *A morphism* $f : A \to B$ *is monic (epic) if and only if* $\ker f = 0$ $(\operatorname{coker} f = 0)$.

5.26.2 *Let* $T : C \rightsquigarrow D$ *be a covariant functor of exact categories.* (i) *Then* T *is left exact if and only if* T *preserves kernels.* (ii) *The following conditions are equivalent:*

(a) T *is exact.*

(b) T *preserves exact sequences.*

(c) T *is kernel and cokernel preserving.*

(d) T *preserves kernels and images.*

5.26.3 *A right exact functor is exact if and only if* T *is a monofunctor.*

Proof of 5.26.1. From the definition, the dual C^* of an exact category C is exact. To prove (a), it is required to prove for $f : A \to B$ and $g : B \to X$ that $\operatorname{im} f = \ker g$ in C implies $\operatorname{im} g^* = \ker f^*$ in C^*. Now, by 5.20, $\operatorname{im} f = \ker(\operatorname{coker} f)$ and $\operatorname{coim} g = \operatorname{coker}(\ker g)$. Thus, $\operatorname{im} f = \ker g$ implies that

$$\begin{aligned}\operatorname{coim} g &= \operatorname{coker}(\operatorname{im} f)\\ &= \operatorname{coker}\big(\ker(\operatorname{coker} f)\big)\\ &= \operatorname{coker} f,\end{aligned}$$

and so $\operatorname{im} g^* = \ker f^*$. This proves (a). Next, $0 \to A \xrightarrow{f} B$ exact, implies $\ker f = \operatorname{im} 0 = 0$, so that $\operatorname{im} f = \operatorname{coker}(\ker f) = \operatorname{coker} 0$, that is, the canonical morphism $A \to \operatorname{im} f$ is an equivalence, so $f : A \to B$ is monic [this also proves (f)]. Conversely, if $f : A \to B$ is monic, then $\ker f = 0$, so $0 \to A \xrightarrow{f} B$ is exact. This proves (b), and (c) is dual. The last part (d) follows from (b) and (c) and the fact that normal categories are balanced.

To prove (e), if $f : A \to B$ is the kernel of $g : B \to X$, then f is monic, and the canonical morphism $A \to \operatorname{im} f$ is epic and monic, hence an equivalence, since C is balanced. Therefore $\operatorname{im} f = \ker g$, so

$0 \to A \xrightarrow{f} B \xrightarrow{g} C$ is exact. Conversely, if this sequence is exact, then $\ker f = \operatorname{im} 0 = 0$, so f is monic. The canonical epic $A \to \operatorname{im} f$ therefore is an equivalence, so that $\operatorname{im} f = \ker g$ implies $f = \ker g$.

Proof of 5.26.2. By 5.26.1, (a) $\Leftrightarrow$ (c), and (b) $\Rightarrow$ (a) is trivial.

(a) $\Rightarrow$ (d) If $f: A \to B$ is a morphism, then $f = \ker g$, for some morphism g. By 5.20, $\operatorname{im} f = \ker(\operatorname{coker} f)$. Since T is kernel and cokernel preserving, then

$$T(\operatorname{im} f) = \ker(T[\operatorname{coker} f]) = \ker(\operatorname{coker} Tf) = \operatorname{im} Tf,$$

that is, T preserves images.

(d) $\Rightarrow$ (b) If $A \xrightarrow{f} B \xrightarrow{g} X$ is exact, then $0 \to \operatorname{im} f \xrightarrow{f'} B \xrightarrow{g} X$, where $f' = \operatorname{im} f = \ker g$. Since T is kernel preserving, $Tf' = \ker Tg$. Since T is image preserving, then $Tf' = \operatorname{im} Tf = \ker Tg$, so T is exact.

Proof of 5.26.3. If $A \to B \to X \to 0$ is a right exact sequence, then $0 \to A \to B \to X \to 0$ is exact if and only if $0 \to A \to B$ is exact, that is, if and only if $A \to B$ is monic. Therefore, if T is right exact, and if $0 \to A \to B \to X \to 0$ is exact, then $0 \to TA \to TB \to TX \to 0$ is exact if and only if $0 \to TA \to TB$ is monic. □

5.27 *Exercises.*

5.27.1 A normal category with kernels has inverse images and finite intersections.

5.27.2 If C is exact, then so is C^L for any category L.

5.27.3 If C is exact and locally small, and if L is small, then C^L is locally small.

5.27.4 For any ring R, the category mod-R is locally and colocally small and exact additive, has a projective generator and injective cogenerator, and is complete and cocomplete.

5.27.5 Let C be a normal, additive category with coproducts, kernels, cokernels, and a generator U. Characterize obj C as a subclass X of obj C closed under coproducts and cokernels and containing a generator of C.

5.27.6 If C is exact and additive, then for each object A, the group valued functor h_A (resp. h^A) is left exact. Furthermore, h_A (resp. h^A) is exact if and only if A is an injective (projective) object of C.

5.27.7 (Puppe [67]) A category C with zero and finite products is exact and additive iff every morphism is composed of a cokernel followed by a kernel.

Sums

Let $u : A' \to A$ be a monic. A morphism $f : A \to B$ **carries** A' into B' in case there exists a monic $B' \to B$ and a morphism $A' \to B'$ such that

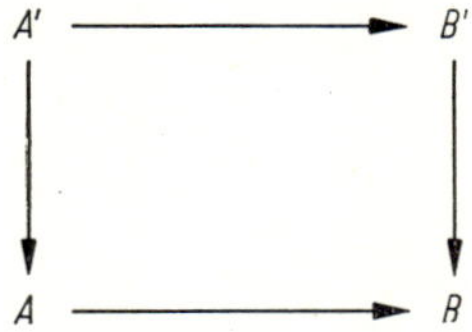

commutes. (The morphism $A' \to B'$ is necessarily unique.) Then A' is **carried into** B by f.

Let $\{u_i : A_i \to A\}$ be monics representing a family of subobjects of A. Then there is a subobject $u : A' \to A$, called the ***sum*** *of the subobjects in the family*, defined as follows: (1) Each of the monics $u_i : A_i \to A$ is smaller than u; that is, $u_i = uw_i$ for monics $w_i : A_i \to A'$. (2) Every morphism $A \to B$ that carries each A_i into B' also carries A' into B'. The sum is denoted $\sum_{i \in I} A_i$. It has the property that it precedes every subobject that is preceded by each A_i. Thus, $\sum_{i \in I} A_i$ is determined up to an equivalence[1]. A category C has **(finite) sums** if for each object of C, the sum of any (finite) family of subobjects exists.

A family of subobjects $\{A_i\}_{i \in I}$ of A is said to **generate** A provided that the canonical morphism $\sum_{i \in I} A_i \to A$ is an equivalence.

5.28 Proposition. *If C is a category with images and inverse images, then, for any family $\{A_i\}_{i \in I}$ of subobjects of an object A and any morphism $f : A \to B$, if $\sum_{i \in I} A_i$ exists, then*

$$\sum_{i \in I} f(A_i) = f\Big(\sum_{i \in I} A_i\Big).$$

Thus, morphisms commute with sums.

Proof. Since $A_i \leq \sum_{i \in I} A_i$, then $f(A_i) \leq f(\sum_{i \in I} A_i)$, so that the left side of the equality is contained in the right. If a morphism $g : B \to D$ carries each $f(A_i)$ into the same subobject U of D, then gf carries A_i into U, hence gf carries $\sum_{i \in I} A_i$ into U, so that f carries $\sum_{i \in I} A_i$ into $g^{-1}U$. Since $f(\sum_{i \in I} A_i)$ is therefore a subobject of $g^{-1}U$, then $gf(\sum_{i \in I} A_i) \leq U$, that is, $f(\sum_{i \in I} A_i)$ is carried into any subobject which each $f(A_i)$ is carried into. This proves the equality. ☐

---

[1] In the literature, the sum $\sum_{i \in I} A_i$ is sometimes denoted by $\bigcup_{i \in I} A_i$.

5.29 Proposition

5.29.1 *If C is a category with images and inverse images, and if $f: F \to c(A)$ is a morphism of C^L, for any small category L, then the image of the colimit morphism* $\operatorname{colim} F \to A$ *is the sum* $\sum_{X \in L} \operatorname{im} f(X)$ *of the images of the morphisms in the cocompatible family* $\{f(X): F(X) \to A\}_{X \in L}$.

5.29.2 *If C is a category with images, inverse images, and (finite) coproducts, then C has (finite) sums, and the sum $\sum_{i \in I} A_i$ of any (finite) family of subobjects of A is the image of the canonical morphism* $\coprod_{i \in I} A_i \to A$.

Proof. Let $I = \operatorname{im}(\operatorname{colim} f)$ be the image of the colimit morphism $\operatorname{colim} F \to A$. The morphism $I \to A$ allows $F(X) \to A$, and so there is a canonical morphism $\operatorname{im} f(X) \to I$ which allows $F(X) \to I$. It remains only to show that for any morphism $g: A \to B$, g carries I into any subobject U of B into which g carries every $\operatorname{im} f(X)$. The compose $F(X) \to \operatorname{im} f(X) \to U$ defines an element of cocompat F, so there are canonical morphisms $\operatorname{colim} F \to U$ and $\operatorname{colim} F \to B$ such that

$$\operatorname{colim} F \to B = \operatorname{colim} F \to A \xrightarrow{g} B = \operatorname{colim} F \to U \to B.$$

Therefore, $g^{-1}(U) \geq I$, and so g carries I into $g(g^{-1}(U)) \leq U$. This proves 5.29.1, and 5.29.2 follows immediately, since coproduct is a colimit by the dual of 5.8.3. ▯

5.30 Corollary. *If C is a category with images, and inverse images, for any functor $F: L \rightsquigarrow C$, and L small, the colimit of F is generated by the images of the canonical morphisms* $\{f(X): F(X) \to \operatorname{colim} F\}_{X \in L}$. *In particular, a coproduct $\coprod_{i \in I} A_i$ is generated by the images of the canonical injections $A_i \to \coprod_{i \in I} A_i$. Furthermore, if $f: F \to c(A)$ is a natural transformation of F to the constant functor $c(A): L \rightsquigarrow C$, then the image of the colimit morphism* $\operatorname{colim} F \to A$ *is the image of the coproduct of the morphisms* $\{f(X): F(X) \to A\}_{X \in L}$, *assuming the coproduct exists.*

Proof. In the special case $A = \operatorname{colim} F$, the colimit morphism $\operatorname{colim} F \to A$ is 1_A, and the image is $\operatorname{colim} F$, which by the proposition, is the sum of the images of the canonical morphisms $F(X) \to \operatorname{colim} F$. The coproduct is a colimit, and is a special case. Finally, by the proposition, the image of both the colimit, and the coproduct morphism, is the same, namely, the sum of the images of the canonical morphisms. ▯

5.31 *Exercise.*

5.31.1 (a) If C is an exact category with a small generator, then every finitely generated object of C is small.

(b) Every small projective P in a category C with generator U is finitely generated with respect to U, that is, there exists an integer $n > 0$ and an epic $U^{(n)} \to P$.

5.31.2 Prove the First and Second Noether Isomorphism Theorems (1.10 and 1.12) for exact categories.

5.31.3 (a) Let C be an exact category. (i) If A/A' is the cointersection of a family of quotient objects $\{A/A_i\}$, then A' is the sum of the family $\{A_i\}$. (ii) C has finite sums.

(b) If C is exact and additive, and if $\{u_i : A_i \to A \mid i = 1, 2\}$ is the coproduct, then $A_1 \cap A_2 = 0$, and A is the sum $A = A_1 + A_2$. Conversely, if $\{u_i : A_i \to A \mid i = 1, 2\}$ are monics with zero intersection and sum A, then A is their coproduct.

5.31.4 In an exact additive category a sequence

$$A \xrightarrow{f} B \xrightarrow{g} C$$

is exact if there exist morphisms $f' : B \to A$ and $g' : C \to B$ such that

$$ff' + g'g = 1_B.$$

5.31.5 (Freyd) A zero preserving faithful functor $T: C \rightsquigarrow D$ reflects exact sequences for exact categories C and D (cf. 5.10.12).

*5.31.6 Let (S, T) be an adjoint pair for exact additive categories. If S is exact, then T preserves injective objects. Conversely, if source T has enough injectives, then S is exact if T preserves injectives.

Some Classical Lemmas

We prove some lemmas for exact categories. Of these, the 3×3, or 9, lemma implies a Noether isomorphism theorem.

5.32 Lemma for Exact Category. *In a row exact diagram*

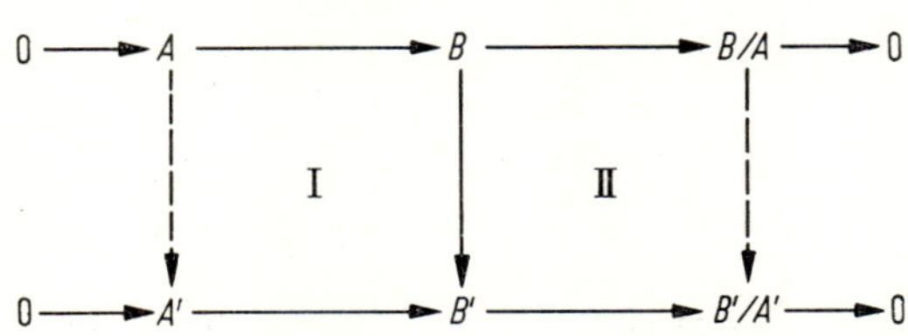

there exists a morphism $A \to A'$ that completes I if and only if there exists a morphism $B/A \to B'/A'$ that completes II.

Proof. Suppose $B/A \to B'/A'$ completes II. Then the composite morphism $A \to B \to B' \to B'/A'$ is 0. Since $A' \to B'$ is a kernel of $B' \to B'/A'$, there is a morphism $A \to A'$, which completes I. The converse is dual.

5.33 The $3 \times 3 = 9$ Lemma for Exact Category. *Any row and column exact commutative diagram*

$$\begin{array}{ccccccccc}
 & & 0 & & 0 & & 0 & & \\
 & & \downarrow & & \downarrow & & \downarrow & & \\
0 & \dashrightarrow & A_1 & \dashrightarrow & A & \dashrightarrow & A_2 & \dashrightarrow & 0 \\
 & & \downarrow & & \downarrow & & \downarrow & & \\
0 & \longrightarrow & B_1 & \longrightarrow & B & \longrightarrow & B_2 & \longrightarrow & 0 \\
 & & \downarrow & & \downarrow & & \downarrow & & \\
0 & \longrightarrow & C_1 & \longrightarrow & C & \longrightarrow & C_2 & \longrightarrow & 0 \\
 & & \downarrow & & \downarrow & & \downarrow & & \\
 & & 0 & & 0 & & 0 & &
\end{array}$$

can be completed by an exact sequence

$$0 \to A_1 \to A \to A_2 \to 0.$$

Proof. By 5.32, there exist morphisms $A_1 \to A$ and $A \to A_2$, which complete the top middle squares separately. Since $A_1 \to B_1 \to B$ is monic, so is $A_1 \to A$. We first show exactitude of

$$0 \to A_1 \to A \to B_2 \to C_2 \to 0.$$

Here $A \to B_2$ is the composite. Note that

$$A_1 \to A \to B_2 = A_1 \to B_1 \to B \to B_2 = 0.$$

Let $X \to A$ be such that $X \to A \to B \to B_2 = 0$, and choose $X \to B_1$ such that $X \to B_1 \to B = X \to A \to B$. Then

$$(X \to B_1 \to C_1 \to C) = (X \to B_1 \to B \to C) = (X \to A \to B \to C) = 0.$$

Since $C_1 \to C$ is monic, then $(X \to B_1 \to C_1) = 0$, and therefore there exists $X \to A_1$ such that $X \to A_1 \to B_1 = X \to B_1$. Therefore

$$X \to A_1 \to A \to B = X \to A_1 \to B_1 \to B = X \to B_1 \to B = X \to A \to B.$$

Since $A \to B$ is monic, this implies $X \to A_1 \to A = X \to A$, which shows that $A_1 \to A$ is a kernel of $A \to B_2$, so $0 \to A_1 \to A \to B_2$

is exact. By duality, $A \to B_2 \to C_2 \to 0$ is exact. Since $A_2 \to B_2$ is a kernel of $B_2 \to C_2$, which is the cokernel of $A \to B_2$, it follows that $A_2 \to B_2$ is an image of $A \to B_2$

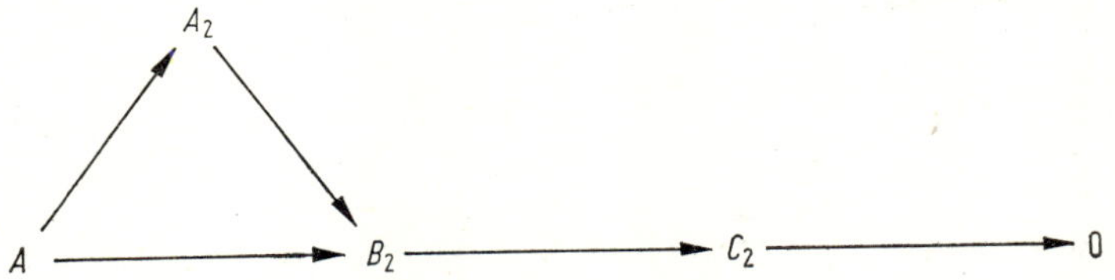

Thus, $\ker(A \to A_2) = \ker(A \to B_2) = \operatorname{im}(A_1 \to A)$, which proves exactness of $0 \to A_1 \to A \to A_2 \to 0$. □

The notation $B \subseteq A$ signifies B and A are objects, and there exists a fixed monic $B \to A$, hence an exact sequence

$$0 \to B \to A \to A/B \to 0.$$

5.34 First Noether Isomorphism Theorem for Exact Category. Let $B \subseteq A_2 \subseteq A_1$. *Then there exists a row-exact diagram*

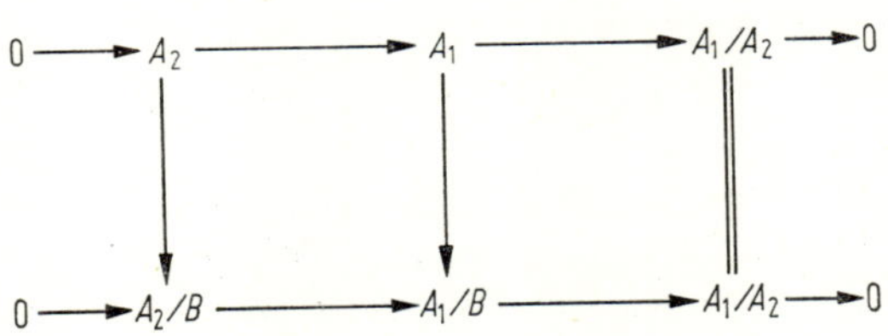

Hence, $A_2/B \subseteq A_1/B$, *and* $(A_1/B)/(A_2/B) \approx A_1/A_2$.

Proof. The proof follows from the next diagram, and an application of the proposition dual to the 3×3-lemma. □

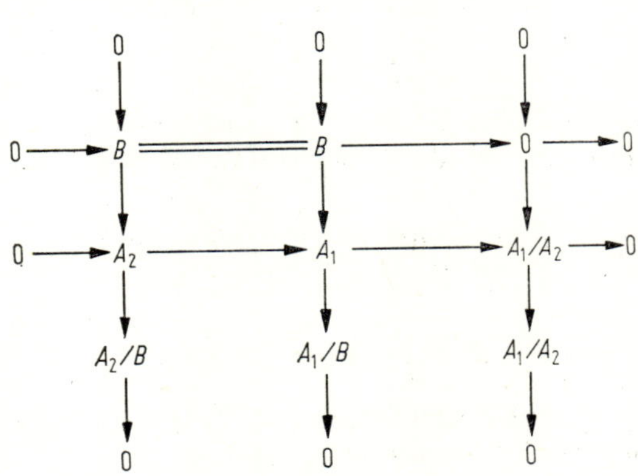

5.35 The 5-Lemma for Exact Category. *Let the following diagram be row exact and commutative.*

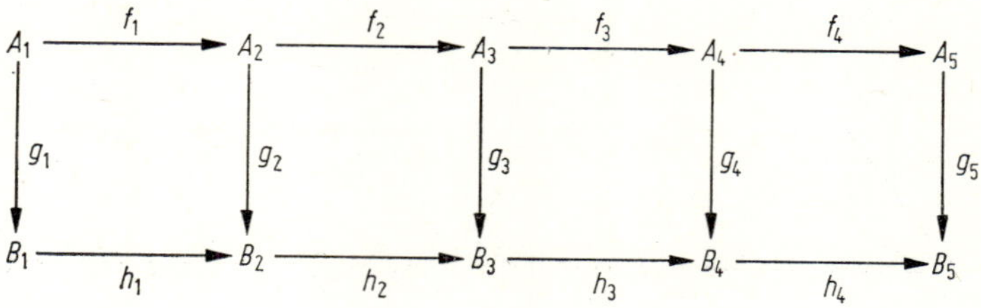

Then

5.35.1 *If g_1 is epic and g_2 and g_4 are monic, then g_3 is monic.*

5.35.2 *If g_5 is monic and g_2 and g_4 are epic, then g_3 is epic.*

5.35.3 *If g_1 is epic, g_5 is monic, and g_2 and g_4 are monic, then g_3 is an equivalence.*

Proof. We first prove this for the category mod-$\mathbb{Z}$. (i) Let $a \in \ker g_3$. Then $0 = h_3 g_3(a) = g_4 f_3(a)$. Thus, $f_3(a) = 0$ since g_4 is monic. Row exactness provides $a = f_2(a')$ for some $a' \in A_2$, and then

$$(h_2 g_2(a')) = g_3 f_2(a') = g_3(a) = 0,$$

so bottom row exactness provides $b \in B_1$ such that $g_2(a') = h_1(b)$. Now use the epic g_1 to write $b = g_1(a'')$ with $a'' \in A_1$. Then

$$g_2 f_1(a'') = h_1 g_1(a'') = h_1(b) = g_2(a').$$

The monic g_2 cancels, so $a' = f_1(a'')$, and

$$a = f_2(a') = f_2 f_1(a'') = 0(a'') = 0.$$

We now copy the proof for exact category. (i) Replace A_1 by $\operatorname{im} f_1$ and B_1 by $\operatorname{im} h_1$. Then, the induced morphism $\operatorname{im} f_1 \to \operatorname{im} h_1$ is an equivalence, so by the Noether isomorphism theorem, $\operatorname{im} f_2 \to \operatorname{im} h_2$ is monic. Hence, we can assume f_2, h_2, and g_2 are all monics. Then the hypothesis that g_4 ismonic implies g_3 is monic. (This program is suggested by Mitchell [65, p. 38, Ex. 14].) □

5.36 *Exercise for Exact Category.*

5.36.1 *If A_1 and A_2 are subobjects of an object A, then there is a row and column exact commutative diagram*

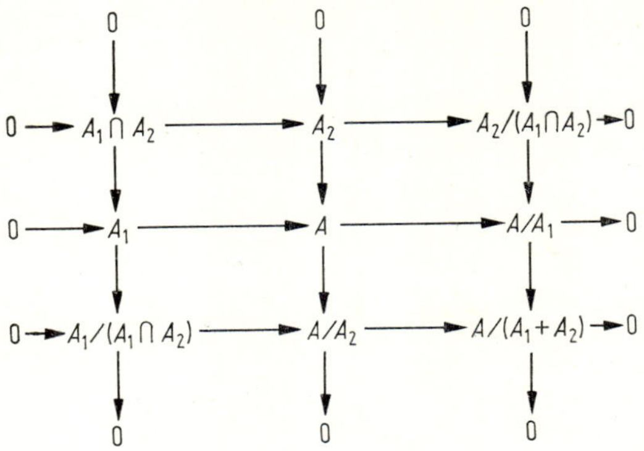

5.36.2 **Second Noether Isomorphism Theorem.** *If A_1 and A_2 are subobjects of A in an exact category, then there is a row exact commutative diagram*

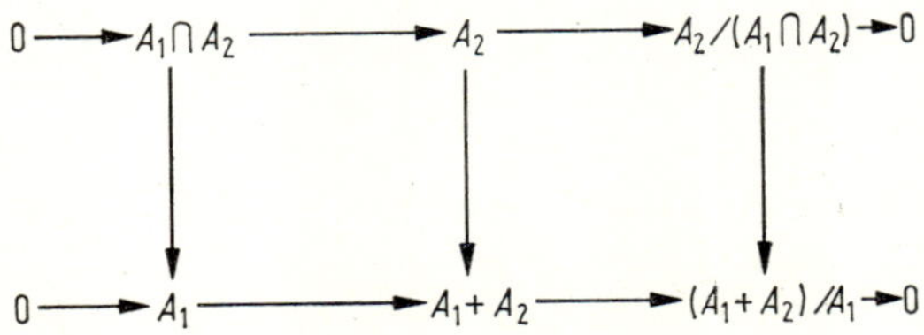

Limit Preserving

Let $F : L \rightsquigarrow C$ and $T : C \rightsquigarrow D$ be functors. Then T preserves $\lim F$ provided there is an equivalence

$$(1) \qquad t(F) : T(\lim_{L,C} F) \approx \lim_{L,D} TF$$

in D. If T preserves $\lim F$, for every $F : L \rightsquigarrow C$, and if there is a natural transformation of composite functors

$$t : T \lim_{L,C} \rightarrow \lim_{L,D} \overline{T},$$

where

$$\overline{T} \begin{cases} C^L \rightsquigarrow D^L \\ F \mapsto TF \end{cases}$$

is the functor induced by T, then T is said to *preserve* (*small*, *finite*) *limits from the* (*small*, *finite*) *category* L. In this sense, there is a commutative diagram

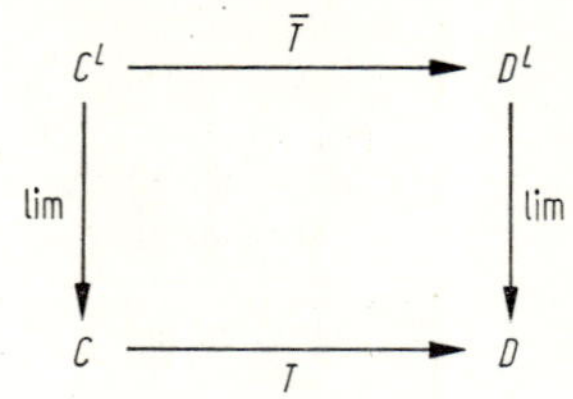

that is, $T \lim \approx \lim \overline{T}$ as functors, and for each $F : L \rightsquigarrow C$, there are equivalences (1) natural in F. *Usually, limit preservation is a term applied to functors from small categories, although a number of the propositions to follow need not be so restricted.* If T preserves limits from all (small, finite) categories L, then T is said to be **(small, finite) limit preserving.**

An **abelian** category is defined to be an exact additive category. Abelian categories, studied *in extenso* in Chapters 6 and Chapters 14 to 16, are defined here for the purpose of the next proposition. A functor is said to be **right (left) continuous** if it preserves **colimits (limits.)**

5.37 Proposition for Abelian Category. *Let C be an abelian category with coproducts and a generator. Let D be an abelian category, and let $h : S \to T$ be a natural transformation between two right continuous functors $C \rightsquigarrow D$. Suppose there exists a generator P of C such that the morphism $h(P) : S(P) \to T(P)$ is an equivalence in D. Then h is a natural equivalence.*

Proof. By 5.21, any $X \in \operatorname{obj} C$ can be fitted into an exact sequence

$$P^{(J)} \to P^{(I)} \to X \to 0,$$

where $P^{(I)}$ is a coproduct of copies of P. Since S and T are exact, the following commutative diagram has exact rows.

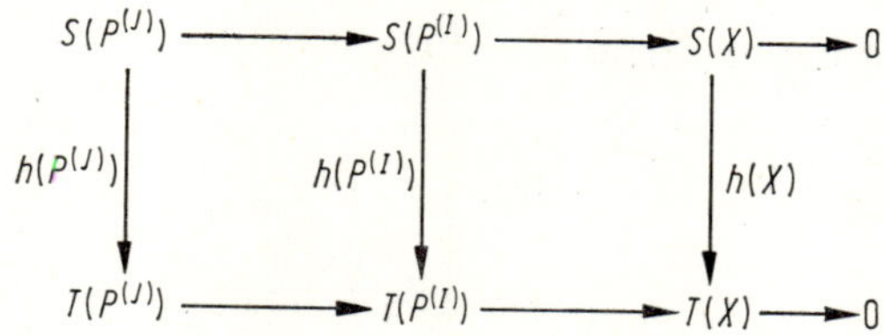

Since S and T preserve coproducts,

$$S(P^{(J)}) \approx S(P)^{(J)} \approx T(P)^{(J)} \approx T(P^{(J)}),$$

and it follows that $h(P^{(J)})$ and $h(P^{(I)})$ are equivalences. Then, by the 5-lemma, $h(X)$ is also an equivalence for every X, and, therefore, h is a natural equivalence. □

5.38 *Exercise.* A functor $T : C \rightsquigarrow D$ preserves small limits if and only if for every functor $F : L \rightsquigarrow C$, where L is small, and every family $\mathscr{F} = \{f(X) : A \to F(X)\}$ of compat F, it is true that $T\mathscr{F} = \{Tf(X) : TA \to TF(X)\}$ is a final object of compat TF whenever $\mathscr{F}$ is a final object of compat F.

5.39 **Corollary to Proposition 5.11.** *If C has products, then the following conditions on a functor $T : C \rightsquigarrow D$ are equivalent:*

5.39.1 *T is (finite) limit preserving.*

5.39.2 *T preserves (finite) products and finite intersections.*

5.39.3 *T preserves (finite) products and equalizers.*

Proof. (1) ⇒ (2) and (1) ⇒ (3) because (finite) products, (finite) intersections, and equalizers are examples of (finite) limits.

(3) ⇒ (1) Let $F : L \rightsquigarrow C$ be a functor from a small category, and, as in the proof of 5.11, write the limit of F as a final object $\{u(X) : A \to F(X)\}_{X \in \mathrm{obj} L}$, where $P = \prod_{X \in L} F(X)$; $k : A \to P$ is the equalizer of canonical morphisms $w : P \to Q$ and $\overline{w} : P \to Q$, and

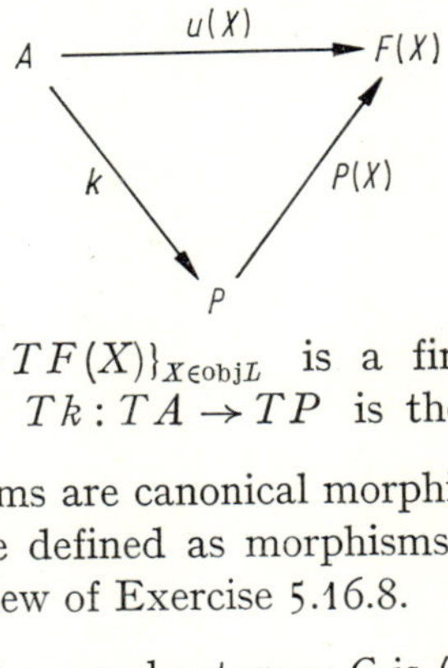

Then, $\{Tu(X) : TA \to TF(X)\}_{X \in \mathrm{obj} L}$ is a final object of D, since $TP \approx \prod_{X \in L} TF(X)$, and $Tk : TA \to TP$ is the equalizer of Tw and $T\overline{w}$. The latter morphisms are canonical morphisms $TP \to TQ$, in the same way w and $\overline{w}$ are defined as morphisms $P \to Q$. The proof of (2) ⇒ (1) is similar in view of Exercise 5.16.8. □

5.40 *Exercise.* (a) A normal category C is (finitely) complete if and only if C has (finite) products and kernels.

(b) A functor $T : C \rightsquigarrow D$ of a normal category C with (finite) products is (finite) limit preserving if and only if T preserves (finite) products and kernels.

5.41 **Proposition.** *If C is any category, and A is any object, then $h^A : C \rightsquigarrow$ SETS preserves limits, and $h_A : C \rightsquigarrow$ SETS converts colimits into limits.*

Proof. Let L be any category. It is necessary to prove a natural equivalence

$$\lim h^A \to h^A \lim$$

of functors $C^L \rightsquigarrow \text{SETS}$, or equivalently, equivalences

$$h(F) : \lim h^A F \approx h^A \lim F$$

natural in the variable $F \in \text{obj}\ C^L$. This is equivalent to showing for each F that the functors SETS $\rightsquigarrow$ SETS defined respectively by $\lim h^A F$ and $h^A \lim F$ are equivalent. To do this, since

(1) $$\text{Mor}_{\text{SETS}}(X, \lim h^A F) = \text{Mor}_{\text{SETS}^L}\big(c(X), h^A F\big)$$

and

(2) $$\text{Mor}_C(A, \lim F) = \text{Mor}_{C^L}\big(c(A), F\big)$$

by the limit adjunctions, it suffices to find bijections

$$T(X, F) = T : \text{Mor}_{\text{SETS}^L}\big(c(X), h^A F\big) \to \text{Mor}_{\text{SETS}}\big(X, \text{Mor}_{C^L}\big(c(A), F\big)\big)$$

which are natural in X and F. If $f \in \text{Mor}_{C^L}\big(c(X), h^A F\big)$, then, there is a mapping

$$T(f) : X \to \text{Mor}_{C^L}\big(c(A), F\big)$$

such that for each object M of L and element $x \in X$, the natural transformation

$$T(f)\, x : c(A) \to F$$

is such that $T(f)(x)[M] = f(M)\, x$. This defines the natural transformation, since for any morphism $a : M \to M'$ of L,

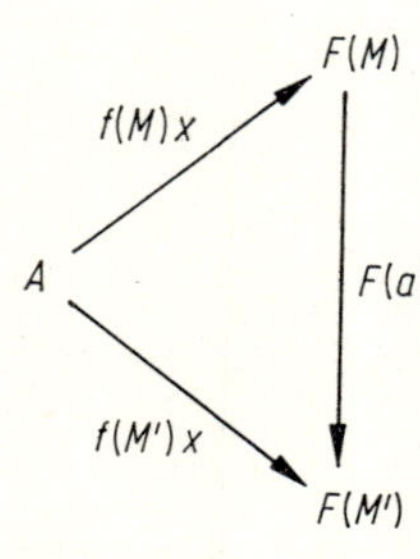

commutes. The inverse of T is the mapping which assigns to any mapping $h' : X \to \text{Mor}_{C^L}\big(c(A), F\big)$ the element $h \in \text{Mor}_{C^L}\big(c(X), h^A F\big)$ such that $h(M)\, x = h'(x)[M]$. $\square$

5.42 Corollary. *If C is any category and A is any object, then $h^A : C \rightsquigarrow \text{Sets}$ preserves products and equalizers, and $h_A : C \rightsquigarrow \text{SETS}$ converts coproducts and coequalizers to products and equalizers, respectively.*

Proof. This is an immediate consequence of the proposition, since products and equalizers are limits. It is instructive, however, to prove this directly, since by 5.39, this will provide another (and simpler) proof of the proposition.

First, let $f : X \to Y$ and $g : X \to Y$ be morphisms of C. If $k : K \to X$ is an equalizer, then by 2.4, $h^A(k) : h^A(K) \to h^A(X)$ is monic, that is, injective. The equalizer of $h^A(f)$ and $h^A(g)$ is the subset

$$\{t \in h^A(X) \mid h^A(f)t = ft = gt = h^A(g)t\}.$$

For any such t, there exists a unique morphism $t' : A \to K$ such that $kt' = t$, since $k = \text{equalizer}(f, g)$. Thus, $t = h^A(k)t'$ lies in the image of $h^A(k)$, so that $h^A(k)$ is the equalizer of $h^A f$ and $h^A g$. Dually, for h_A.

Next, let $\{p_i : P \to X_i\}_{i \in I}$ be a product in C, let $\{p'_i : P' \to h^A X_i\}_{i \in I}$ be the product in SETS, and let $g : h^A P \to P'$ be the canonical morphism such that

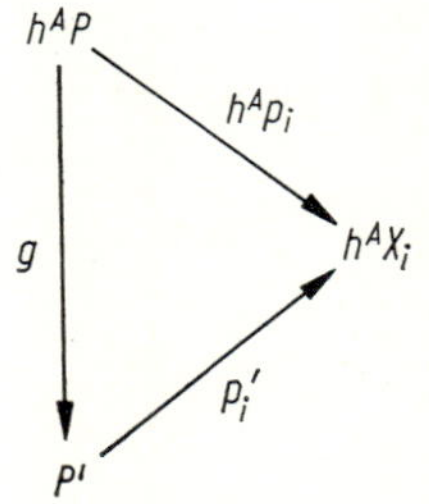

commutes. Since p_i is epic, then $h^A p_i$ is surjective by 2.4, so that given $y \in P'$, there exists $y_0 \in h^A P$ such that $h^A p_i(y_0) = p'_i(y)$ for every $i \in I$, and then $g(y_0) = y$. Thus, g is epic. If $g(a) = g(b)$, then

$$h^A p_i(a) = p_i a = p_i b = h^A p_i(b)$$

$\forall i \in I$, and therefore $a = b$. This proves that g is an equivalence, and that h^A therefore preserves products. Dually, for h_A. □

The next proposition shows that not every functor $S : D \rightsquigarrow C$ can aspire to be a left adjoint (cf. 5.55 and 5.56).

5.43 Proposition.

5.43.1 *If (S, T) is an adjoint pair, then S preserves colimits, in particular, S preserves coproducts, cokernels, coequalizers, and initial objects.*

5.43.2 *T preserves limits, in particular, T preserves products, kernels, equalizers, final objects, and intersections.*

Proof. If L is any category, and $F : L \rightsquigarrow C$ a functor, then for any object X of D,

$$\begin{aligned} h^X(T \lim F) &= h^{SX}(\lim F) \\ &= \lim h^{SX} F \\ &= \lim h^X TF \\ &= h^X(\lim TF), \end{aligned}$$

which by 2.3.3 shows that there are equivalences

$$t(F) : T \lim F \approx \lim TF$$

which are natural in F. Thus, there is a functor equivalence $T \lim \approx \lim T$. ▯

A functor is a **monofunctor** (**epifunctor**) if it preserves monics (epics).

5.44 Corollary. *If (S, T) is an adjoint pair, then S is right continuous, and T is a limit preserving monofunctor.*

Proof. The (co)limit preserving properties are given by the theorem. If f is a monic of C, then $h^{SX} f$ is a monic of D for all objects X of D, and the functor equivalence $h^X T \approx h^{SX}$ implies that $h^X Tf$ is monic in SETS for all X, which implies that Tf is monic in D (see 2.4). ▯

5.45 Corollary. *If C is complete, then for every small category L, the limit functor $\lim : C^L \rightsquigarrow C$ is a limit preserving monofunctor, and the constant functor $c : C \rightsquigarrow C^L$ is colimit preserving.*

Proof. The limit functor is right adjoint to the constant functor $C \rightsquigarrow C^L$. ▯

Adjoints of Representable Functors

The next proposition is taken from Gabriel [62].

5.46 Proposition. *Let C be an abelian category. For any object U, let $A = \operatorname{End}_C U$, and let*

$$h^U : C \rightsquigarrow \text{mod-}A$$

be the functor induced by the morphism functor. Then h^U has a left adjoint S if and only if, for any index set I, there is a coproduct $U^{(I)}$ in C.

Proof. Sufficiency. Let M be any object of mod-A, and write

(1) $$A^{(J)} \xrightarrow{f} A^{(I)} \xrightarrow{g} M \to 0$$

exact, for sets J and I. For any object U of C, we have

(2) $$h^{A^{(I)}} h^U \approx (h^A h^U)^I \approx (h^U)^I \approx h^{U^{(I)}}.$$

The first and third equivalence in (2) follows, since the contravariant set valued functor h_Y converts coproducts to products, for any $Y \in \operatorname{obj} \text{mod-}A$. The canonical equivalence $h^A \approx 1_{\text{mod-}A}$ implies the middle equivalence. The morphism $f: A^{(J)} \to A^{(I)}$ implies a canonical functor morphism

(3) $$h^{A^{(I)}} \to h^{A^{(J)}}$$

which, with (2) and the corresponding equivalence

(2′) $$h^{A^{(J)}} h^U \approx h^{U^{(J)}},$$

yields a functor morphism

(4) $$h^{U^{(I)}} \to h^{U^{(J)}}$$

which is independent of the choice of the morphism f and g in the exact sequence (1). But, by 2.4, the natural equivalence (4) is induced canonically by a morphism

(5) $$t_M : U^{(J)} \to U^{(I)}.$$

Then the functor

(6) $$S \begin{cases} \text{mod-}A \rightsquigarrow C \\ \quad M \mapsto \operatorname{coker} t_M = SM \end{cases}$$

is left adjoint to h^U. To verify this, exactitude of (1) yields exactitude in $\operatorname{Adfunct}(C, \text{mod-}\mathbb{Z})$ of the first row of

(7) $$\begin{array}{ccccccc} 0 & \longrightarrow & h^M h^U & \longrightarrow & h^{A^{(I)}} h^U & \longrightarrow & h^{A^{(J)}} h^U \\ & & \downarrow & & \downarrow & & \downarrow \\ 0 & \longrightarrow & h^{SM} & \longrightarrow & h^{U^{(I)}} & \longrightarrow & h^{U^{(J)}} \end{array}$$

where, for example, $h^M h^U$ is the compose of h^U and the group valued functor $h^M : \text{mod-}A \rightsquigarrow \text{mod-}\mathbb{Z}$. Exactitude of the bottom row of (7) is given by (5) and the fact that $SM = \operatorname{coker} t_M$. The vertical functor equivalences on the right come from (2) and (2′). Then, by an application of Lemma 5.32, there is a vertical equivalence

(8) $$h^M h^U \approx h^{SM}$$

which makes (7) commutative. This proves that S is left adjoint to h^U. (Cf. 5.4.)

Necessity. If S is left adjoint to h^U, then there is a natural equivalence of set-valued functors $h^A h^U \approx h^{SA}$. But $h^A \approx 1_{\text{mod-}A}$, hence $h^U \approx h^{SA}$ and $U \approx SA$. Since any left adjoint functor preserves coproducts by 5.43, then

$$S(A^{(I)}) \approx (SA)^{(I)} \approx U^{(I)}$$

for any index set I. ▯

A category C will be called *additive with or without finite products* if C satisfies (Ad 1) and (Ad 2) (p. 148).

5.47 Proposition. *If (h, S, T, C, D) is an adjoint situation for additive categories C and D with or without finite products, then the following conditions are equivalent:*

5.47.1 *The functor $T : C \rightsquigarrow D$ is additive.*

5.47.2 *The functor $S : D \rightsquigarrow C$ is additive.*

5.47.3 *The adjunction bijections*

$$h(B, A) : \mathrm{Mor}_C(SB, A) \to \mathrm{Mor}_D(B, TA)$$

are isomorphisms in mod-$\mathbb{Z}$ *for all* $A \in \mathrm{obj}\, C$, *and* $B \in \mathrm{obj}\, D$.

When this is so, then $h(\ ,\ )$ is a natural equivalence of the induced group-valued bifunctors

$$\mathrm{Mor}_C\big(S(\ ),\ \big) : D^{\mathrm{op}} \times C \rightsquigarrow \text{mod-}\mathbb{Z}$$

$$\mathrm{Mor}_D\big(\ , T(\ )\big) : D^{\mathrm{op}} \times C \rightsquigarrow \text{mod-}\mathbb{Z}.$$

Proof. The morphism $h_B = h(B, SB)\,(1_{SB}) : B \to TSB$, defined $\forall B \in \mathrm{obj}\, D$ satisfies

$$h(a) = T(a)\, h_B,$$

$h = h(B, A)$, for any morphism $a : SB \to A$ in C, and so, when T is additive,

$$\begin{aligned} h(a + a') &= T(a + a')\, h_B = (Ta + Ta')\, h_B \\ &= T(a)\, h_B + T(a')\, h_B = h(a) + h(a') \end{aligned}$$

when $a' : SB \to A$. This proves $(1) \Rightarrow (3)$, and $(3) \Rightarrow (1)$ follows similarly, using $g_A = h^{-1}(TA, A)\,(1_{TA})$ and

$$b g_A = h^{-1}(Tb)$$

for any morphism $b : A \to A'$ in C. Then $(2) \Leftrightarrow (3)$ by duality. ▯

5.48 Corollary. *If $T : C \rightsquigarrow D$ is a functor of additive categories, and if T has a right or left adjoint, then T is additive, and so is its adjoint.*

Proof. If T has a left adjoint S, then, by 5.43, T is limit preserving, hence T preserves products, and so T is additive by 3.38. Then S is additive by the proposition. The other possibility follows by duality.

5.49 Proposition and Definition. *A set $\{C_i\}_{i \in I}$ of objects of a category C is* ***preinitial*** *if for each object X of C there exists $i \in I$ and a morphism $C_i \to X$. If C is complete, then the limit of any preinitial set, considered as a directed set, is an initial object of C.*

Proof. Let A be the limit object of the functor described, and let $\{p_i : A \to C_i\}_{i \in I}$ be the family of canonical morphisms. First, for any morphism $f : C_i \to A$ in C, p_i is a coretraction, and $f p_i = 1_A$. This follows since

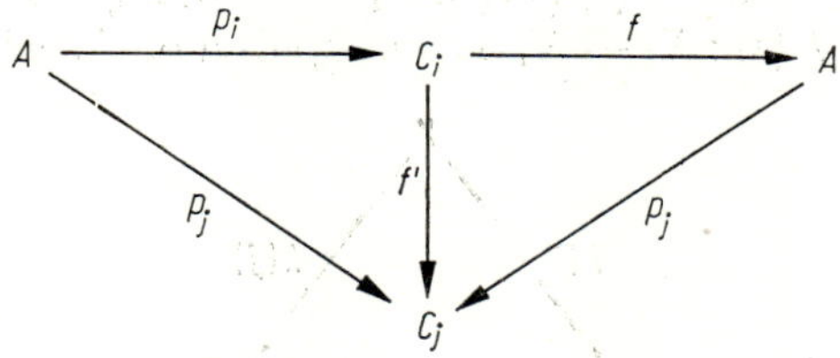

commutes for the canonically given morphism $f' : C_i \to C_j$, and 1_A is the unique morphism $g : A \to A$ such that $p_j g = p_j$ for every $j \in I$. Next, if X is any object of C, then the compose $A \to C_i \to X$ is the unique morphism $A \to X$. For suppose that f and g are two morphisms $A \to C$, and let K denote their equalizer. Then

$$A \xrightarrow{p_i} C_i \xrightarrow{\text{canon}} K \to A$$

is the identity morphism 1_A, which implies $f = g$. A therefore is an initial object of C. ▯

Existence Theorems for Adjoints

The next theorem is essentially Freyd's (cf. 5.51).

5.50 General Adjoint Functor Theorem. *If C is a complete category, then a covariant functor $T : C \rightsquigarrow D$ has a left adjoint if and only if the following conditions hold:*

5.50.1 *T preserves limits.*

5.50.2 *For each object Y of D, the comma category (Y, T) of 5.2 has a preinitial set.*

Proof. By 5.43, any left adjointable functor $T : C \rightsquigarrow D$ is limit preserving, and by 5.2, every comma category (Y, T) has an initial object.

Conversely, it suffices to prove that every (Y, T) is complete, for then (Y, T) has an initial object by the previous proposition, and then T has a left adjoint by 5.2. Let P denote

$$P \begin{cases} (Y, T) \rightsquigarrow C & \\ (f, X) \mapsto X & \text{(on objects)} \\ t \mapsto t & \text{(on morphisms).} \end{cases}$$

If $F : L \rightsquigarrow (Y, T)$ is any functor from a small category L, then $A = \lim PF$ exists, since C is complete, and $TA = \lim TPF$ since T is limit preserving. Assertion: the family $\{h(i) : Y \to TX_i\}_{i \in L}$ such that $F(i) = \big(h(i), X_i\big)$ is an element of compat TPF. For, if $a : i \to j$ is any morphism of L, then $F(a) : \big(h(i), X_i\big) \to \big(h(j), X_j\big)$ is a morphism of (Y, T), that is, $F(a) : X_i \to X_j$ is a morphism of C such that

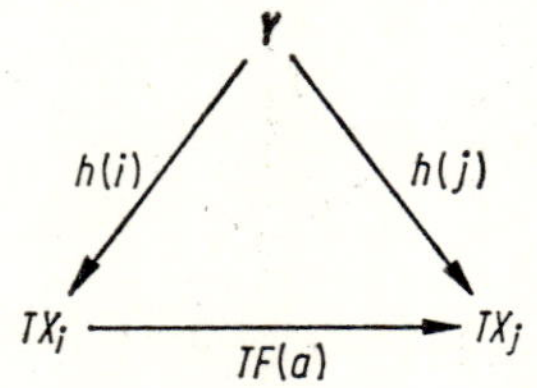

commutes. Since $TX_i = TPF(i)$, and since $TF(a) = TPF(a)$, then the assertion follows. Thus, since $TA = \lim TPF$, there is a unique morphism $t : Y \to TA$ of D such that

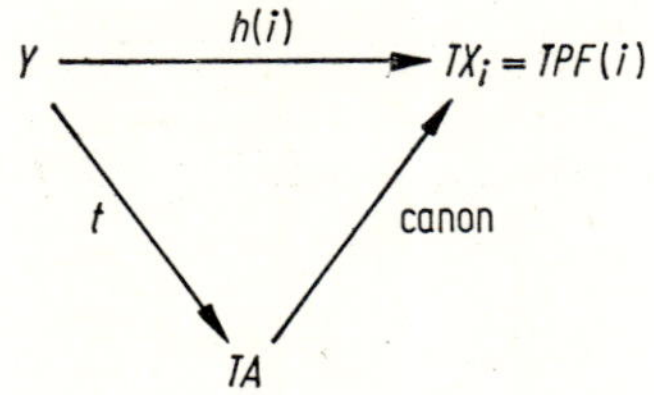

commutes. Then $(t, A) = \lim F$. For suppose that $\{g(i) : (f, X) \to (h(i), X_i)\}_{i \in L}$ is an element of compat F. Then

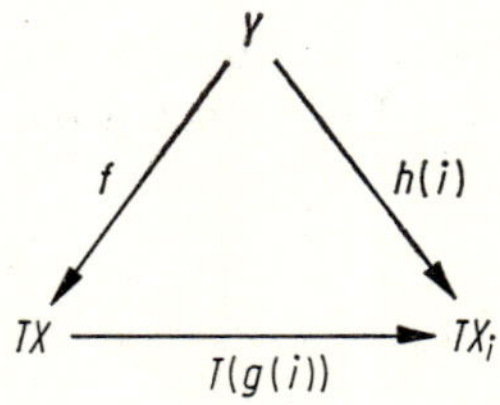

commutes, and it follows that $\{T\big(g(i)\big) : TX \to TX_i\}_{i \in L}$ is an element of compat TPF. Therefore, there is a unique morphism $b : X \to A$

such that

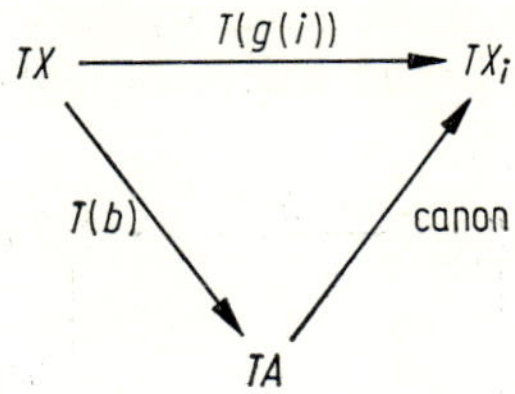

commutes for every $i \to L$. Thus, b is the unique morphism $b : (f, X) \to (t, A)$ of (Y, T) such that

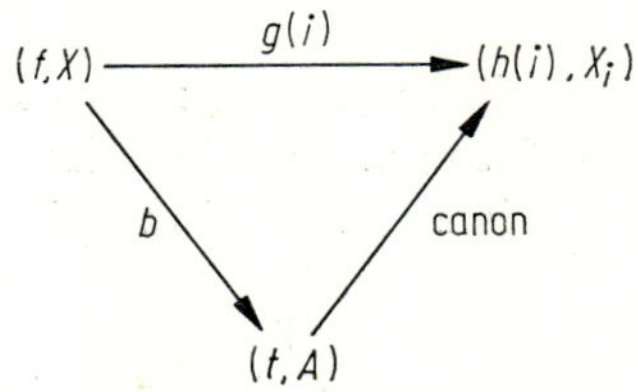

commutes, that is, $(t, A) = \lim F$. ▯

5.51 Special Adjoint Functor Theorem (Freyd [62]). *If C is a complete, locally small category with a cogenerator U, then a functor $T: C \rightsquigarrow D$ has a left adjoint if and only if T preserves limits.*

Proof. By 5.50, it is sufficient to show for each object Y of D the existence of a preinitial set in (Y, T). If (t, X) is an object of (Y, T), by completeness of C, there is a cartesian diagram

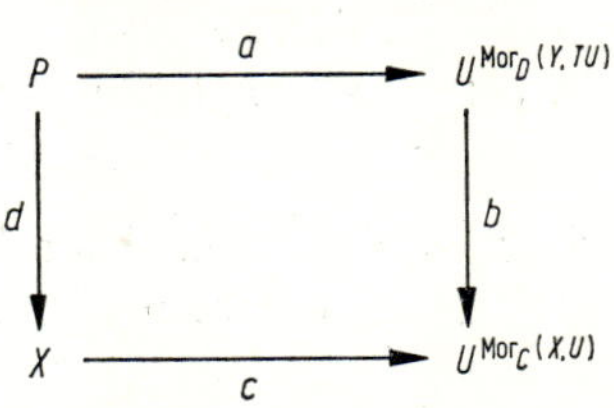

where the morphism c is canonical, and b is the unique morphism such that for any morphism $f : X \to U$ of C the diagram

(1)

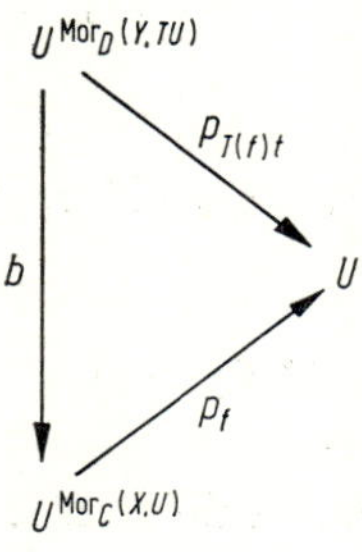

commutes, where the function p denotes the canonical projections from either product. Since U is a cogenerator, then c is monic by the dual of 3.25, and then a is monic by Exercise 5.10.2. Now

(2)

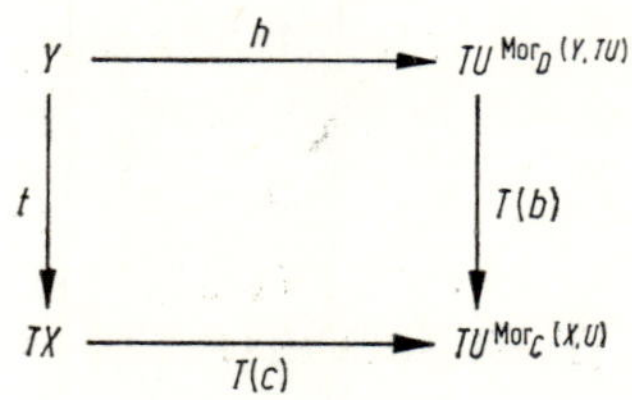

commutes, where $p_g h = f$, for any $g : Y \to TU$. Since T preserves limits, the application of T to the cartesian diagram (1) yields a cartesian diagram, and hence there is a canonical morphism $Y \to TP$ such that

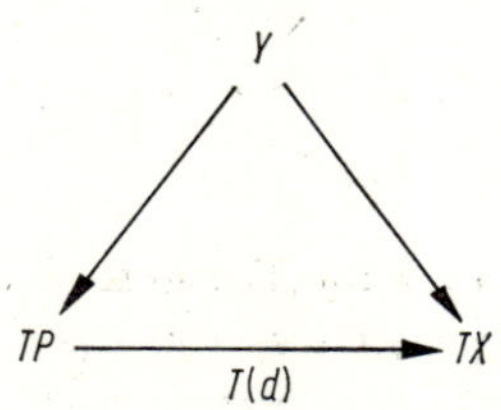

commutes. Thus, (Y, T) has a preinitial set of cardinality not exceeding the number of subobjects of $U^{\mathrm{Mor}_D(Y,TU)}$. $\square$

5.52 Corollary. *If D is a colocally small, cocomplete category with a generator, then a functor $S : D \rightsquigarrow C$ has right adjoints if and only if S is colimit preserving.*

Proof. If S has a right adjoint, then S is colimit preserving by 5.43. Conversely, suppose S is colimit preserving. Then the composite functor

$$S_*^* : D^* \overset{\text{can}}{\rightsquigarrow} D \overset{S}{\rightsquigarrow} C \overset{\text{can}}{\rightsquigarrow} C^*$$

is a limit preserving functor $D^* \rightsquigarrow C^*$, hence has a left adjoint H by The adjunction equivalences

$$\mathrm{Mor}_{D^*}(HX^*, Y^*) \approx \mathrm{Mor}_{C^*}(X^*, S_*^* Y^*)$$

yield equivalences

$$\mathrm{Mor}_D(Y, H^*X) \approx \mathrm{Mor}_C(SY, X),$$

which shows that S has a right adjoint H^*. $\square$

5.53 Corollary. *A complete locally small category C with a cogenerator is cocomplete.*

Proof. For any category C, the constant functor $c : C \rightsquigarrow C^L$ is limit preserving, since if $F : L \rightsquigarrow C$ is a functor, and if L is a final object of compat F, then $c(L)$ is a final object of compat cF. Then by Corollary 5.52, $c : C \rightsquigarrow C^L$ has a left adjoint $\operatorname{colim} : C^L \rightsquigarrow C$, and C is therefore cocomplete. ▯

Module Valued Adjoints

We now cite some applications to module-type categories.

5.54 Corollary. *Let R be a ring, and let C be a category. Then a functor $F : \text{mod-}R \rightsquigarrow C$ has right adjoints if and only if F is right continuous (colimit preserving).*

Proof. The category $D = \text{mod-}R$ satisfies the requirements of Corollary 5.52 by Exercise 5.27.4. ▯

5.55 Proposition. *If R is a ring and C is an additive category, then a functor $T : C \rightsquigarrow \text{mod-}R$ has left adjoints only if T is representable, that is, $T \approx h^A$ for some $A \in \operatorname{obj} C$.*

Proof. If $S : \text{mod-}R \rightsquigarrow C$ is left adjoint to T, then for $A = SR \in \operatorname{obj} C$ and any $X \in \operatorname{obj} C$

$$h^A X = \operatorname{Mor}_C(SR, X) \approx \operatorname{Hom}_R(R, TX) \approx TX.$$

The adjoint equivalence in the middle is an R-module isomorphism by naturality. ▯

The next two results are theorems of Watts [60] as generalized by Mitchell [65, p. 142].

5.56 Proposition. *Let C be a complete, locally small additive category with a cogenerator, let R be a ring, and let $T : C \rightsquigarrow \text{mod-}R$ be a covariant functor. Then the following properties are equivalent:*

5.56.1 *T is limit preserving.*

5.56.2 *T has a left adjoint $S : \text{mod}\ R \rightsquigarrow C$.*

6.56.3 *T is representable; that is, $T \approx h^A$ for some $A \in \operatorname{obj} C$.*

Proof. (1) $\Leftrightarrow$ (2) by 5.52, (2) $\Rightarrow$ (3) by 5.55, and (3) $\Rightarrow$ (1) by 5.42.

5.57 Corollary. *If C is a cocomplete, colocally small additive category with a generator, and if R is a ring, then a contravariant functor $T : C \rightsquigarrow \text{mod-}R$ converts colimits to limits if and only if T is representable, that is, there is a functor equivalence $T \approx h_A$ for some $A \in \operatorname{obj} C$.*

Proof. Any representable functor h_A converts colimits to limits by 5.41. Conversely, the composite functor

$$T' : C^* \rightsquigarrow C \overset{T}{\rightsquigarrow} \text{mod-}R$$

is covariant, and C^* satisfies the conditions in the preceding proposition, hence is representable. If $T' \approx h^{A^*}$, then $T \approx h_A$. ▯

5.58 Proposition.

5.58.1 *Let C and D be exact categories. Then a functor T is left (right) exact if and only if T is kernel (cokernel) preserving. In particular, any left joinable functor $T : C \rightsquigarrow D$ is left exact with right exact left adjoint.*

5.58.2 *For any object A of an exact additive category C, the group-valued functors h^A and h_A are left exact.*

5.58.3 *If C is a complete, locally small additive category with a cogenerator, then for any object A, if $B = \text{End}_C A$, then the left exact* mod-B *valued functor h^A has a right exact left adjoint.*

Proof. In an exact category C, a morphism $f : A \to B$ is monic if and only if $\ker f = 0$, hence, if and only if $0 \to A \overset{f}{\to} B$ is exact. Thus, T is left exact if T is kernel preserving. Conversely, if T is exact, and if $k : K \to A$ is the kernel of $f : A \to B$, then $0 \to K \to A \to B$ is exact, and $0 \to TK \to TA \to TB$ is exact, whence $\text{im}\, Tk = \ker Tf$, so T is kernel preserving. If T is left joinable, then it is limit, hence, kernel, preserving by 5.43, and left exact by 5.26. Similarly, h^A is limit, hence, kernel, preserving by 5.42 and 5.41, and so is left exact. This proves 5.58.2, and 5.58.3 follows from 5.56. ▯

The Tensor Functor

If U is a (B, A)-bimodule, then by 5.46 (also 5.56) the functor

$$h^U : \text{mod-}A \rightsquigarrow \text{mod-}B$$

induced by h^U has a left adjoint

$$\underset{B}{\otimes} U : \text{mod-}B \rightsquigarrow \text{mod-}A,$$

that is, there exists a functor $\underset{B}{\otimes} U$ such that, for each object X of mod-B and object Y of mod-B, there are isomorphisms

$$h(X, Y) : \text{Hom}_B\big(X \underset{B}{\otimes} U, Y\big) \approx \text{Hom}_A\big(X, \text{Hom}_B(U, Y)\big)$$

which are natural in X and Y. (Also see the Adjoint Associativity Theorem 11.4.)

Since any module is canonically a $\mathbb{Z}$-module, each object X of mod-A, and each object Y of A-mod defines group-valued functors

$$X \underset{A}{\otimes} : A\text{-mod} \rightsquigarrow \text{mod-}\mathbb{Z}$$

$$\underset{A}{\otimes} Y : \text{mod-}A \rightsquigarrow \text{mod-}\mathbb{Z}$$

which defines a bifunctor

$$(\ ) \underset{A}{\otimes} (\ ) \begin{cases} \text{mod-}A \times A\text{-mod} \rightsquigarrow \text{mod-}\mathbb{Z} & \\ \qquad (M, N) \mapsto M \underset{A}{\otimes} N & \text{(on objects)} \\ \qquad (f, g) \mapsto f \otimes g & \text{(on morphisms).} \end{cases}$$

Given $f : M \to M'$ in mod-A and $g : N \to N'$ in A-mod, then $f \otimes g : M \underset{A}{\otimes} N \to M' \underset{A}{\otimes} N'$ is the unique morphism such that

(1)

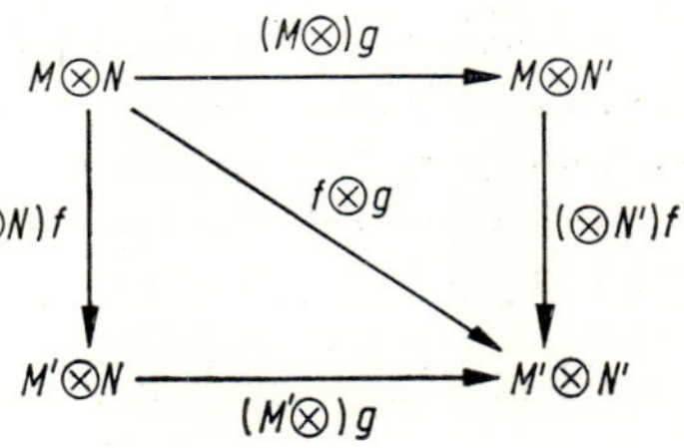

commutes. (The subscript A on the symbol $\underset{A}{\otimes}$ has been eliminated for simplicity.) As a consequence of (1), if $f' : M' \to M''$ and $g' : N' \to N''$ are morphisms of mod-A and A-mod, respectively, then

$$(f' \otimes g')(f \otimes g) = f'f \otimes g'g.$$

Commutativity of the diagram (1) can be seen as a consequence of the commutativity of the

diagram

(2)

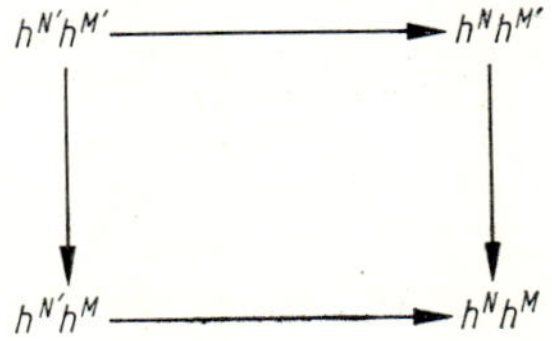

By the Yoneda Proposition 2.3, there is a bijection

(3) $$[h^{M'}, h^{M}] = \text{Hom}_A(M, M')$$

of the set $[h^{M'}, h^M]$ of natural transformations $h^{M'} \to h^M$ of functors and the set of morphisms $M \to M'$. The bijection (3) is natural in M and M'. Thus, corresponding to $f: M \to M'$, there is a natural transformation $h^{M'} \to h^M$; then the natural transformation $h^{N'} \to h^N$ corresponding to $g: N \to N'$ implies that (2) commutes, that is, for every object Y of mod-A, the diagram

$$(4) \qquad \begin{array}{ccc} h^{N'}h^{M'}Y & \longrightarrow & h^N h^{M'} Y \\ \downarrow & & \downarrow \\ h^{N'}h^{M}Y & \longrightarrow & h^N h^M Y \end{array}$$

commutes. But, by the adjunction isomorphisms

$$h^M h^N Y = \mathrm{Hom}_A(M, \mathrm{Hom}_A(N, Y)) \approx \mathrm{Hom}_A\left(M \underset{A}{\otimes} N, Y\right),$$

(4) is just the result of applying the functor h^Y to the diagram (1), for any object Y of mod-A. In particular, for any cogenerator Y of mod-A, h^Y is faithful, and so commutativity of (4) implies that of (1).

If R is a ring, and if M is an object of mod-R, then there is a canonical isomorphism $M \underset{R}{\otimes} R \to M$, since $\underset{R}{\otimes} R$ is left adjoint to $h^R \approx 1_{\text{mod-}R}$. Then, if A is any left ideal of R, the image of the canonical homomorphism

$$M \underset{R}{\otimes} A \to M \underset{R}{\otimes} R \to M$$

is MA, that is, there is a canonical exact sequence $M \underset{R}{\otimes} A \to MA \to 0$. The module M is said to be **flat** in mod-R if the group valued functor $M \underset{R}{\otimes}: R\text{-mod} \rightsquigarrow \text{mod-}\mathbb{Z}$ is exact. In this case, since for each left ideal A of R there is an exact sequence $0 \to A \to R$, then there is a canonical exact sequence

$$0 \to M \underset{R}{\otimes} A \to MA \to 0,$$

that is, $M \underset{R}{\otimes} A \approx MA$ canonically.

5.59 *Exercise.*

5.59.1 Show that an object M is flat in mod-R if and only if $M \underset{R}{\otimes} A \approx MA$ canonically for every finitely generated left ideal A of R.

5.59.2 Show that any projective object of mod-R is flat.

5.59.3 Prove all the propositions in Chapter 11.

5.59.4 If $f: A \to B$ is a ring homomorphism, show that mod-f : mod-$B \rightsquigarrow$ mod-A has a left adjoint $\underset{A}{\otimes} B$: mod-$A \rightsquigarrow$ mod-B and a right adjoint $\mathrm{Hom}_A(B, \ )$. Thus, if $T =$ mod-f, there are natural equivalences

$$\mathrm{Hom}_B\left(M \underset{A}{\otimes} B, N\right) \approx \mathrm{Hom}_A\left(M, \mathrm{Hom}_B(B, N)\right) \approx \mathrm{Hom}_A(M, TN)$$

$$\mathrm{Hom}_A(TM, N') \approx \mathrm{Hom}_A(M, N') \approx \mathrm{Hom}_A\left(M \underset{B}{\otimes} B, N'\right)$$
$$\approx \mathrm{Hom}_B\left(M, \mathrm{Hom}_A(B, N')\right)$$

for M, N in obj mod-B and N' in obj mod-A.

5.59.5 If U is a (B, A)-bimodule, and if U is flat in B-mod, then $\underset{B}{\otimes} U$: mod-$B \rightsquigarrow$ mod-A is exact. Conclude that this is the case when U is a generator in mod-A and $B = \mathrm{End}\, U_A$.

5.59.6 Let U be a generator of mod-A, and let $B = \mathrm{End}\, U_A$. Show that the functor h^U : mod-$A \rightsquigarrow$ mod-B composed with its left adjoint is $1_{\text{mod-}A}$, that is, there exist natural isomorphisms

$$X \approx \mathrm{Hom}_A(U, X) \underset{B}{\otimes} U$$

for all objects X of mod-A.

*5.59.7 A direct limit of flat modules is flat.

*5.59.8 A module is flat if and only if it is a direct limit of projective modules.

*5.59.9 A direct limit of projective modules need not be projective.

For background on the next result, consult Lambek [64a].

5.60 Theorem (Bourbaki-Lambek). *A left R-module M is flat if and only if the character module* $M' = \mathrm{Hom}_{\mathbb{Z}}(M, \mathbb{Q}/\mathbb{Z})$ *is injective in* mod-R.

Proof. The proof is sketched in Exercise 3.51.3 (see 6.28 for another way to prove this). □

5.61 Corollary. *Any projective object of R-mod is flat.*

Proof. If P is projective in R-mod, the character module $P' = h_{\mathbb{Q}/\mathbb{Z}}(P)$ is injective in mod-R (cf. 3.51, 6.28 and 11.35), and hence P is flat by the theorem. □

The proof of the theorem has the following corollary.

5.62 Corollary. *In the situation ${}_RA_S$, B_S, where ${}_RA$ is flat and B_S injective,* $\mathrm{Hom}_S(A, B)$ *is an injective right R-module.* □

A ring is (right) **selfinjective** in case R is injective in mod-R.

5.63 **Corollary.** *If A is an injective S-module that is a generator in* mod-S, *then* End A_S *is right selfinjective.*

Proof. Since A is a generator, if $R = \text{End}\, A_S$, then A is finitely generated and projective over R, hence flat over R. Now apply the corollary in the $B = A$ case. ▯

5.64.1 *Example.* Assume that S is a right self-injective ring, and let A be any injective free module. (If A is right Noetherian, then every direct sum of injective modules is injective, hence, every free S-module is then injective.) Then End A_S is right self-injective. This includes the case when S is a field and A is any right vector space, namely, any complete ring of l.t.'s of a right vector space is right self-injective. ▯

5.64.2 The category of right selfinjective rings is productive (Exercise).

5.65 *Exercises.*

5.65.1 If A is a ring, and M an object of mod-A, then

$$\text{sing}\, M = \{m \in M \mid \text{ann}_A m \text{ is an essential right ideal}\}$$

is a fully invariant A-submodule of M, called the **singular submodule** of M.

(a) If $M \nabla N$, then $\text{sing}\, M = 0$ if and only if $\text{sing}\, N = 0$.

(b) Sing A is an ideal, called the right singular ideal.

(c) If $\text{sing}\, A = 0$, then the injective hull $\hat{A}$ of A in mod-A is an (S, A)-bimodule, where $S = \text{End}\, \hat{A}_A$, canonically isomorphic to the canonical (S, A)-bimodule S. (The canonical right A-module S is defined by the mapping $A \to S$ sending $a \in A$ onto the unique element a' of S which induces the left multiplication a_s of A defines by a.)

(d) The canonical (S, A)-bimodule isomorphism $h: S \to \hat{A}$ of (c) defines $\hat{A}$ as a ring such that $h(s)h(t) = h(st)\ \forall\, s,\ t \in S$ that is, the ring A is canonically isomorphic to S.

(e) The ring $S \approx \hat{A}$ is right self-injective, that is, S is injective in mod-S.

(f) If A is a prime ring with uniform right ideal U, and $\text{sing}\, A = 0$, then $B = \text{End}\, U_A$ is a right Ore domain with right quotient field D, and the ring $\hat{A}$ is isomorphic to the full linear ring of l.t.'s on a right vector space V over D such that the canonical left $\hat{A}$-module V is isomorphic to a simple (minimal) left ideal W of $\hat{A}$ such that $W \cap A = U'$ is a uniform left ideal of A and $W = U'D$ is the vector subspace of W generated by U'.

(g) If A is a prime ring with a.c.c. on right annulets and a.c.c. on complement right ideals, then A has a uniform right ideal U such that the vector space W of (f) is finite dimensional, and then $\hat{A}$ is a classical right quotient ring of A in the sense that every regular element x of A is a unit of $\hat{A}$ (where A is included in $\hat{A}$ canonically) and

$$\hat{A} = \{ax^{-1} \mid a, \text{regular } x \in A\}.$$

(Cf. Chapters 9 and 15.)

5.65.2 A prime ring A with sing $A = 0$ contains a uniform right ideal U which is a generator of mod-A if and only if A is isomorphic to the endomorphism ring of a finitely generated projective left module V over a right Ore domain K. When this is so, the $\text{Biend}_K V$ and $\text{End } U_A$ are right Ore domains. (Cf. 4.22A.)

5.65.3 Prove a converse of 5.63: if A is a generator of mod-S, and if $\text{End } A_S$ is right selfinjective, then A (and S) are injective in mod-S.

Middle Associative Bilinear Homomorphisms

If G is an abelian group, then a mapping

$$h \begin{cases} (X, Y) \to G \\ (x, y) \mapsto h(x, y) \end{cases}$$

is **biadditive** (**bilinear**) provided that the partial mappings

$$h_x \begin{cases} Y \to G \\ y \mapsto h(x, y) \end{cases}$$

and

$$h_y \begin{cases} X \to G \\ x \mapsto h(x, y) \end{cases}$$

are group homomorphisms for all $x \in X$ and $y \in Y$. Assuming that X is a right, and Y is a left, module over a ring A, then we say that the mapping h is **middle associative** if

$$h(xa, y) = h(x, ay)$$

$\forall x \in X, y \in Y, a \in A$. The mapping h is said to be **balanced** if it is biadditive and middle associative. Let bal (X, Y) denote the subgroup of the free abelian group $F(X, Y)$ on the set (X, Y) generated by the set of all

$$\begin{cases} (x + x', y) - (x, y) - (x', y) \\ (x, y + y') - (x, y) - (x, y') \\ (xa, y) - (x, ay), \end{cases}$$

where $x \in X$, $y \in Y$, and $a \in A$. Then the compose

$$t(X, Y) : (X, Y) \xrightarrow{\text{in}} F(X, Y) \xrightarrow{\text{can}} F(X, Y)/\text{bal}(X, Y)$$

is a canonical balanced mapping. Furthermore, there is a bifunctor

$$T \begin{cases} \text{mod-}A \times A\text{-mod} \rightsquigarrow \text{mod-}\mathbb{Z} \\ \qquad (X, Y) \mapsto F(X, Y)/\text{bal}(X, Y) = T(X, Y). \end{cases}$$

If $(f, g) : (X, Y) \to (X', Y')$ is a morphism, then the unique homomorphism $F(X, Y) \to T(X', Y')$, given by the free abelian group such that

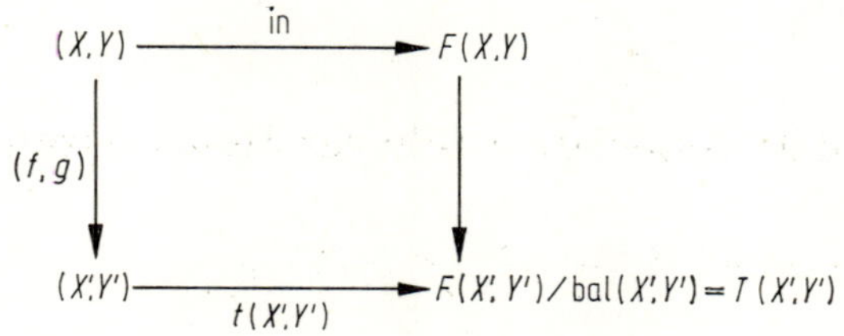

commutes, induces a mapping $T(f, g) : T(X, Y) \to T(X', Y')$ which defines T as a functor, that is, the kernel of the vertical map on the right side of the diagram contains bal(X, Y). If Y is a right B-module (as well as a left A-mod), then $T(X, Y)$ is canonically a right B-module, where

$$\sum_{(x,y)\in(X,Y)} [(x, y) + \text{bal}(X, Y)]b = \sum_{(x,y)\in(X,Y)} (x, yb) + \text{bal}(X, Y)$$

for all $b \in B$.

5.66 Adjoint Associativity. *If A and B are rings, X is an object of* mod-A, *Z is an object of* mod-B, *and Y is an (A, B)-bimodule, then there are isomorphisms*

$$\text{Hom}_B(T(X, Y), Z) \approx \text{Hom}_A(X, \text{Hom}_B(Y, Z))$$

natural in X, Y, and Z, where $T(X, Y) = F(X, Y)/\text{bal}(X, Y)$ is the canonical object of mod-B. *Thus, the partial functor*

$$T(\ , Y) \begin{cases} \text{mod-}A \rightsquigarrow \text{mod-}B \\ \qquad X \mapsto T(X, Y) \end{cases}$$

is left adjoint to

$$h^Y \begin{cases} \text{mod-}B \rightsquigarrow \text{mod-}A \\ \qquad Z \mapsto \text{hom}_B(Y, Z), \end{cases}$$

that is, there is a natural equivalence $T(\ ,Y) \approx \underset{A}{\otimes} Y$ *of right continuous functors. If* $(x, y) \in X \times Y$, *the image of* (x, y) *under the canonical mapping* $X \times Y \to T(X, Y)$ *is denoted* $x \otimes y$.

If X *is an object of* mod-A *and* Y *is an object of* A-mod, *and if* $g : X \times Y \to G$ *is a balanced mapping in* mod-$\mathbb{Z}$, *then there is a unique mapping* $h : X \underset{A}{\otimes} Y \to G$ *such that*

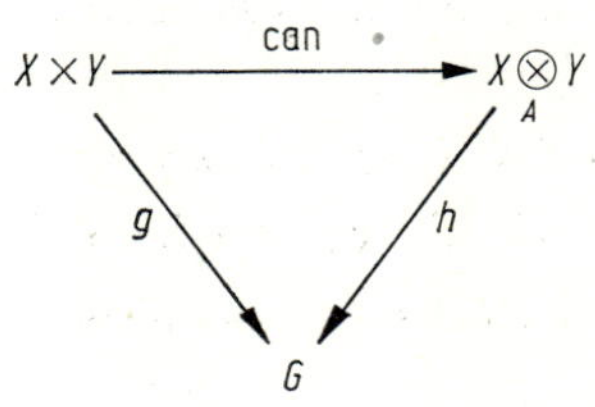

commutes. Thus, $g(x, y) = h(x \otimes y)\ \forall x \in X$ *and* $y \in Y$.

Proof. Exercise.

Adjoints of Faithful and Full Functors

5.67 Proposition. *Consider an adjoint situation* (h, S, T, C, D). *Then*

5.67.1 (a) *There is a natural transformation of functors*

$$\varphi : 1_D \to TS,$$

where for each $Y \in \text{obj}\, D$, $\varphi_Y : Y \to TSY$ *is the morphism*

$$\varphi_Y = h(Y, SY)(1_{SY}).$$

(b) *If* $X \in \text{obj}\, C$ *and* $g : Y \to TX$ *is a morphism in* D, *then* $h^{-1}g : SY \to X$ *is the unique morphism such that*

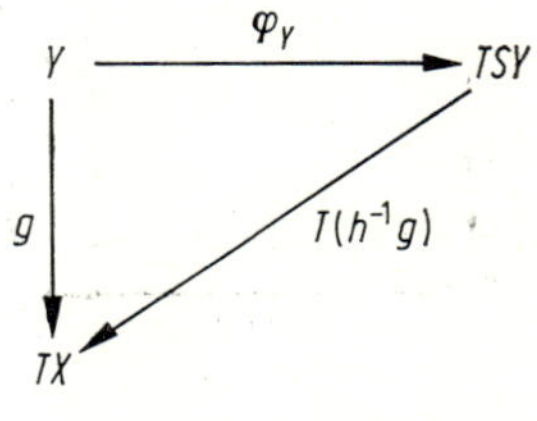

commutes. Thus, T *induces an injection* $\text{Mor}_C(SY, X) \to \text{Mor}_D(TSY, TX)$.

5.67.2 (a) *There is a natural transformation of functors*

$$\psi : ST \to 1_C,$$

where for each $X \in \text{obj}\, C$, $\psi_X : STX \to X$ *is the morphism*

$$\psi_X = h^{-1}(TX, X)(1_{TX}).$$

(b) *The following conditions are equivalent:*

(α) T *is faithful.*

(β) *If* $g : Y \to TX$ *is epic, then* $h^{-1}g$ *is epic.*

(γ) *For all* $X \in \text{obj}\, C$, $\psi_X : STX \to X$ *is epic.*

(c) *The following conditions are equivalent:*

(α) T *is full.*

(β) *For all* $X \in \text{obj}\, C$, $\psi_X : STX \to X$ *is split monic.*

Proof. If $f : X \to X'$ is a morphism in C, the commutative diagram

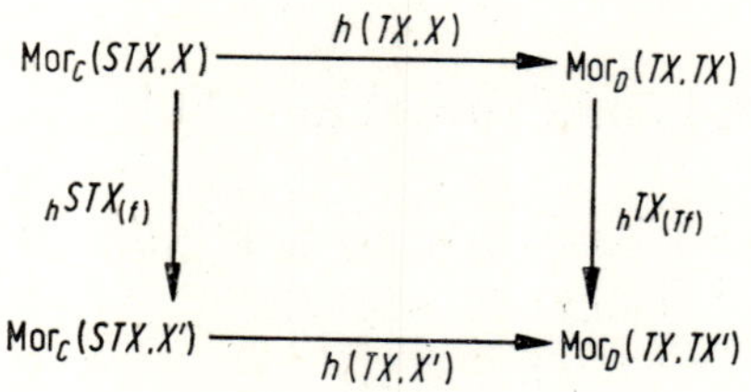

evaluated at 1_{TX} (going counterclockwise, and clockwise) yields $fh^{-1}(1_{TX}) = h^{-1}Tf$, or

(i) $$f\psi_X = h^{-1}Tf.$$

Similarly, for $g : SY \to X$, using the commutative diagram

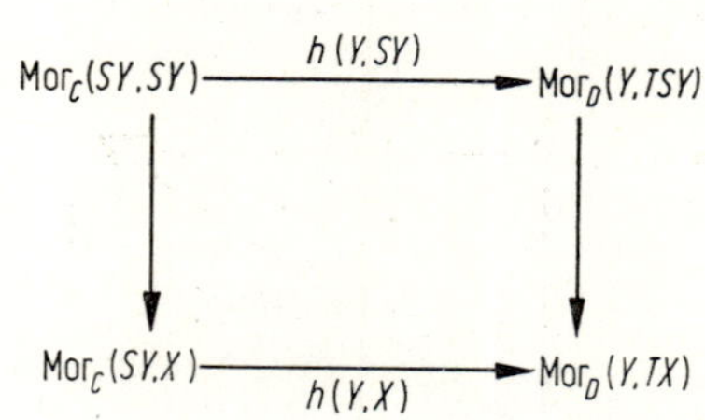

we obtain

(ii) $$h(g) = T(g)\varphi_Y.$$

The duals of (i) and (ii) are:

(i*) $\varphi_Y f = hS(f) \qquad f: Y' \to Y$

(ii*) $h^{-1}(g) = \psi_X S(g) \quad g: Y \to TX$

where $Y, Y' \in \text{obj } D$, $X, X' \in \text{obj } C$.

1. If $g: Y \to Y'$, then

(i*) $$\varphi_{Y'} g = hS(g)$$

and

(ii) $$hS(g) = TS(g)\varphi_Y$$

so

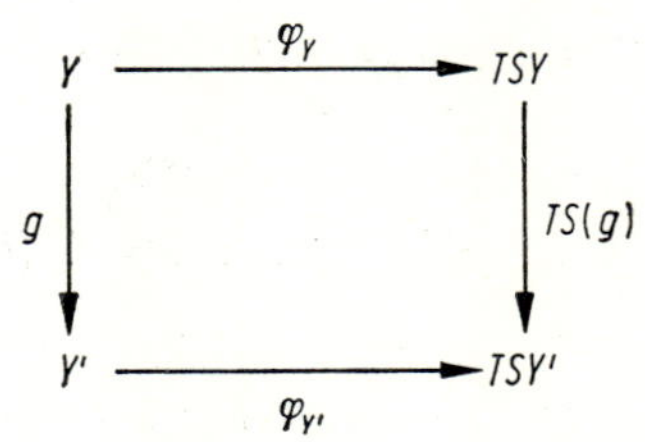

commutes proving (a). The fact that h is bijective and condition (ii) produces (b).

2(a) If $f: X \to X'$ is a morphism in C, then

(i) $$f\psi_X = h^{-1}T(f)$$

and,

(ii*) $$h^{-1}T(f) = \psi_X ST(f),$$

so $\psi_X: ST \to 1_C$ is a natural transformation.

2(b) $(\alpha) \Rightarrow (\beta)$ If $g: Y \to TX$ is epic, the second diagram in (1) shows that $T(h^{-1}g)$ is epic. A faithful functor reflects epics, so $h^{-1}g$ is epic. $(\beta) \Rightarrow (\gamma)$ follows for $Y = TX$, and $g = 1_{TX}$.

$(\gamma) \Rightarrow (\alpha)$ If $f_i: X \to X'$, $i = 1, 2$, and $T(f_1) = T(f_2)$, then naturality of ψ_X implies

$$f_1\psi_X = \psi_{X'} ST(f_1) = \psi_{X'} ST(f_2) = f_2\psi_X.$$

Since ψ_X is epic, $f_1 = f_2$, so T is faithful.

2(c) follows from (2) of the next proposition. □

5.68 Proposition. *Consider an adjoint situation* (h, S, T, C, D). *Then*

5.68.1 $$(T\psi)(\varphi T) = 1_T.$$

5.68.2 *If T is full, and if $f : X \to STX$ is such that $T(f) = \varphi_{TX} : TX \to TSTX$, then*

$$f\psi_X = 1_{STX},$$

that is, $\psi_X : STX \to X$ is a coretraction[1].

Proof. The equation (ii) for $g = \psi_X : S(TX) \to X$ yields

$$h(\psi_X) = T(\psi_X)\varphi_{TX}.$$

But $\psi_X = h^{-1}(1_{TX})$, so

$$1_{TX} = T(\psi_X)\varphi_{TX},$$

that is

$$(T\psi)(\varphi T) = 1_T.$$

Assuming T is full, then $\varphi_{TX} = T(f)$ for some $f : X \to STX$, so

$$\begin{aligned} T(1_{STX})\varphi_{TX} &= \varphi_{TX} = \varphi_{TX} T(\psi_X)\varphi_{TX} \\ &= T(f)T(\psi_X)\varphi_{TX} = T(f\psi_X)\varphi_{TX}. \end{aligned}$$

By 5.67.1 (b) the equality

$$T(1_{STX})\varphi_{TX} = T(f\psi_X)\varphi_{TX}$$

implies $1_{STX} = f\psi_X$.

5.69 Corollary. *In an adjoint situation (h, S, T, C, D); if T is fully faithful, then there is a natural equivalence of functors*

$$t : 1_C \to ST$$

where $t(X) = \psi_X^{-1}$ for all $X \in \operatorname{obj} C$.

Proof. By 5.67, $\psi_X : STX \to X$ is epic, and by 5.68.2, ψ_X is split monic, so ψ_X is an equivalence. If $f : X \to X'$, then by (i)

$$f\psi_X = h^{-1}Tf$$

and by (ii*)

$$\psi_{X'} ST(f) = h^{-1}Tf,$$

so

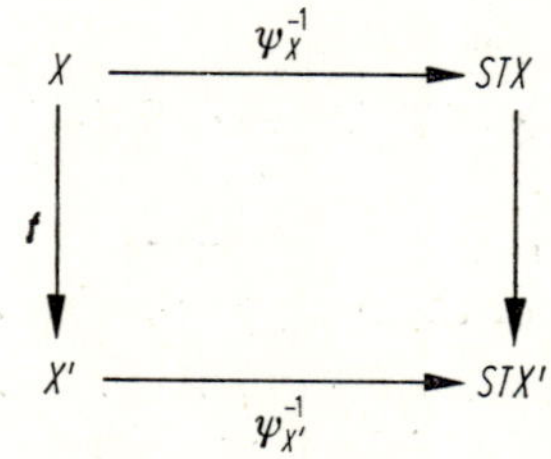

[1] This also holds when S is full (see Mitchell [65, p. 120]).

commutes. Hence, $h: 1_C \to ST$ is a natural transformation and, thus, a functor equivalence. ▯

5.70 Morita and Popesco-Gabriel Theorem (Second Version). *Let A be a ring, and let $C = \text{mod-}A$ (or C can be any Grothendieck abelian category). If U is a generator of C, and if $B = \text{End}_C U$, then the full embedding $h^U: C \rightsquigarrow \text{mod-}B$ is a coretraction, the right adjoint $\underset{B}{\otimes} U: \text{mod-}B \rightsquigarrow C$ is a retraction, and there is a natural equivalence of functors*

$$\left(\underset{B}{\otimes} U\right) h^U \approx 1_C.$$

Thus, for each object X of C, there is an equivalence

$$\text{Mor}_C(U, X) \underset{B}{\otimes} U \approx X$$

natural in X. In particular, for $C = \text{mod-}A$, there is an (A, A)-bimodule isomorphism $U^* \underset{B}{\otimes} U \approx A$.

Proof. The functor h^U is fully faithful by 4.24, and then 5.69 applies. The rest is a consequence. ▯

Remark. Since U is projective over B by 4.1, then $\otimes_B U$ is exact by 5.61. This also holds for the case C is Grothendieck (cf. 15.26), a short proof of which employing the Ext functor has been given by Takeuchi [71]. Cf. also the short proof of Lambek [71].

5.71 *Exercises.*

5.71.1 Determine necessary and sufficient conditions on an object U in order that $h^U: C \rightsquigarrow \text{mod-End}_C U$ is a full functor, where U is any cocomplete abelian category.

5.71.2 Prove all of the theorems in Chapters 7 and 8.

Algebras over Commutative Rings

Let k be a commutative ring. Considering k as an object of RINGS, there is a category Source k the objects of which are ring homomorphisms $f_A: k \to A$. Let k-ALG denote the full subcategory of Source k on the class of objects such that the image of every $f_A: k \to A$ is contained in the center of A. The compose of $f_A: k \to A$ and the canonical ring isomorphism $A \to \text{End}\, A_A$ makes A a k-module. The module operation $(a, x) \mapsto f(a)x = ax$ satisfies the formula

$$(ax)(by) = (ab)xy$$

$\forall a, b \in k$, $x, y \in A$. A morphism $g : A \to B$ of k-Alg is a ring homomorphism such that

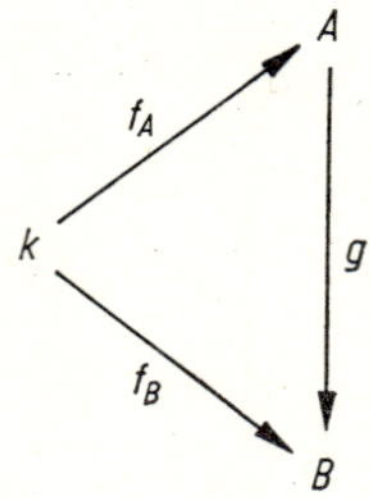

commutes. Now the image of f_A is the subring $\{a1 \mid a \in k\}$, so the condition of the diagram is simply that $g(a1) = a1\ \forall a \in k$, or, equivalently, that g induces the identity on im f_A.
Then

$$g(ax) = g(a1x) = g(a1)g(x) = a1g(x) = ag(x)$$

$\forall a \in k$, $x \in A$, that is, a morphism of k-ALG is simply a ring homomorphism $g : A \to B$ of k-algebras such that $g(ax) = ag(x)\ \ \forall a \in k$, $x \in A$. If ga denotes the composition of the homothetic $a_s : A \to A$ and g, and if ag is similarly defined, then this is simply the requirement that g commutes with homothetics defined by any $a \in k$. Thus: $ga = ag$. The k-algebra $k \to A$ is **faithful** if it is injective and is **central** if it induces an isomorphism $k \to$ center A.

If A is any algebra over k, and X is an object of mod-A, then X is canonically a k-module defined by the homomorphism $k \to A \to$ End X, and any morphism $g : X \to Y$ in mod-A satisfies $g(ax) = ag(x)$. If T : mod-$A \rightsquigarrow$ mod-B is a functor and A and B are k-algebras, then Ta is defined for all $a \in k$ by the formula

$$(Ta)f = T(af)$$

for any morphism $f : X \to Y$ in mod-A, where af is the compose of f and the left multiplication of Y by a; similarly aT is defined $\forall a \in k$. A functor T : mod-$A \rightsquigarrow$ mod-B is k-**linear** provided that $aT = Ta \forall a \in k$. Only k-linear functors are admitted in this context.

An algebra $k \to A$ is **commutative** if A is a commutative ring. The full subcategory of k-ALG consisting of all commutative k-algebras is denoted COMM k-ALG.

5.72 *Examples.*

5.72.1 If A is any ring, the canonical map $\mathbb{Z} \to A$ is an algebra over $\mathbb{Z}$.

5.72.2 If k is any commutative ring, the identity mapping $k \to k$ is an algebra over k.

5.72.3 Let M be any right A-module over a k-algebra A. As already noted, the compose $k \to A \to \text{End}\, M$ of the canonical maps $k \to A$ and $A \to \text{End}\, M$ is a k-module which induces homomorphisms $k \to \text{End}\, M_k$ and $k \to \text{End}\, M_A$. Thus, $\text{End}\, M_k$ and $\text{End}\, M_A$ are (canonical) k-algebras.

5.72.4 If M is a (B, A)-bimodule, then $\underset{B}{\otimes} M : \text{mod-}B \rightsquigarrow \text{mod-}A$ is k-linear if and only if $\alpha x = x\alpha\ \forall \alpha \in k$.

5.72.5 If M is any right A-module, and if $B = \text{End}\, M_A$ is the canonical k-algebra, then the mod-B valued functor $h^M : \text{mod-}A \rightsquigarrow \text{mod-}B$ and $\otimes_B M : \text{mod-}B \rightsquigarrow \text{mod-}A$ both are k-linear.

5.72.6 If k is any nonzero subring of center A, then the inclusion $k \to A$ is a faithful algebra, and the inclusion center $A \to A$ is a central algebra.

5.73 (Semi)Group Ring Proposition and Definition. *Let T be a multiplicative semigroup, let A be a ring, and let $f_T : T \to AT$ denote the free left A-module. For simplicity, assume that $T \subseteq AT$, and f_T is the inclusion. Thus, AT is an (A,A)-bimodule such that $at = ta \forall a \in A \forall t \in T$. Then, AT is a ring with respect to addition as a module, and multiplication defined by*

(1) $$\sum_{s \in T} a_s s \sum_{t \in T} b_t t = \sum_{r \in T} c_r r$$

where

(2) $$c_r = \sum_{st=r} a_s b_t$$

(Every element of *AT has finite support, hence the element on the right of (1) is defined.) Thus, the mapping $f_T : T \to AT$ is a semigroup homomorphism, and the mapping*

(3) $$f_A \begin{cases} A \to AT \\ a \mapsto (a, 1) \end{cases}$$

is a ring monomorphism. Henceforth assume that $A \subseteq AT$, and that f_A is the inclusion.

Let $C(AT)$ denote the category the objects of which are ordered triples (f, g, R), where R is a ring, $f : A \to R$ is a ring homomorphism, $g : T \to R$ is a semigroup homomorphism, and such that f and g commute. Thus, $f(a)g(t) = g(t)f(a)\ \forall a \in A,\ t \in T$. A morphism between this and another *object* (f', g', R') *is a ring homomorphism* $h : R \to R'$ *such that*

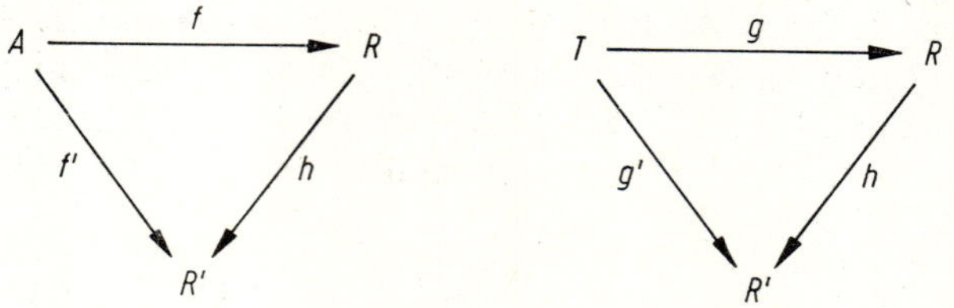

both commute. Then, the ordered triple (f_A, f_T, AT) *is an initial object of* $C(AT)$*, called the* ***semigroup ring of*** T ***over*** A *and denoted* AT*. If* A *is a commutative ring, then* AT *is an algebra with respect to the injection* f_A*, called the* ***semigroup algebra over*** A*. If* T *is a group, these designations are* ***group ring*** *and* ***algebra****, respectively.*

Proof. To verify the distributive law, write $S = U \coprod V$ for subsets U and V. Then

$$\sum_{u \in U} a_u u \sum_{t \in T} b_t t + \sum_{v \in V} a_v v \sum_{t \in T} b_t t$$

$$= \sum_{\substack{ut=k \\ u \in U \& t \in T}} a_u b_t k + \sum_{\substack{vt=q \\ v \in V \& t \in T}} a_v b_t q$$

$$= \sum_{\substack{st=r \\ s \in S \& t \in T}} a_s b_t r$$

$$= \left(\sum_{u \in U} a_u u + \sum_{v \in V} a_v v \right) \sum_{t \in T} b_t t$$

which prove distributivity on one side. The other side is a consequence of the symmetry in the definition of multiplication. Associativity:

$$\left(\sum_{s \in S} a_s s \sum_{t \in T} b_t t \right) \sum_{u \in U} c_u u$$

$$= \sum_{\substack{st=k \\ s \in S \& t \in T}} a_s b_t st \sum_{u \in U} c_u u$$

$$= \sum_{s \in S \& t \in T \& u \in U} a_s b_t c_u stu.$$

By the symmetry in the definition of multiplication, and associativity in A and T, associativity in AT results. This proves that AT is a ring. If (f, g, R) is an object of $C(AT)$, then the ring homomorphism

$f: A \to R$ defines R as a pullback left A-module, and by the Free Module Proposition 3.9, there is a morphism $h: AT \to R$ of A-mod

$$h \begin{cases} AT \to R \\ \sum_{s \in S} a_s s \mapsto \sum_{s \in S} f(a_s) g(s) \end{cases}$$

and, clearly, $h: AT \to R$ is also a ring homomorphism. Furthermore, (f_A, f_T, AT) is an initial object in $C(AT)$. ▯

Exercise.

For a finite group G, and ring A, investigate which properties of A are inherited by AG, e.g. semisimplicity, selfinjectivity etc. (Cf. 13.26.)

Let A-(SEMI)GROUP RINGS denote the category on the objects AG, where G is a (semi) group, and define a morphism $f: AS \to AG$ to be a ring homomorphism such that $f(ax) = af(x) \; \forall a \in A$. Any morphism $f: AS \to AG$ of A-(SEMI)GROUP RINGS induces homomorphisms $f_A: A \to AG$ and $f_G: S \to AG$, and (f_A, f_G, AG) is an object of $C(AS)$, as defined in the semigroup ring proposition.

5.74 Free (Semi)Group Ring Theorem.

5.74.1 *Let A be a ring. The forgetful functor* A-(SEMI)GROUP RINGS $\rightsquigarrow$ SETS *has a left adjoint*

$$\begin{cases} \text{SETS} \rightsquigarrow A\text{-(SEMI)GROUP RINGS} \\ S \mapsto A[S], \end{cases}$$

where $[S]$ denotes the free (semi)group on S, $A[S]$ is the (semi)group ring over A, and

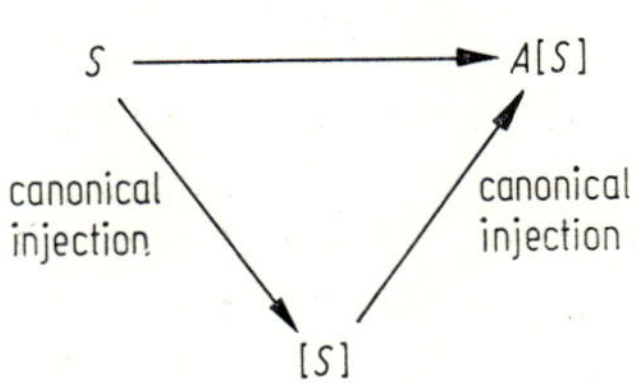

commutes. Then $S \mapsto A[S]$ is called the ***ring of noncommutative polynomials in the indeterminates, or variables****, S. When A is commutative, then $S \mapsto A[S]$ defines the free algebra functor, and $A[S]$ is called the* ***free algebra over** A **on the set** S.*

5.74.2 *Let A be any ring (commutative or noncommutative.) and let* A-ABEL (SEMI)GROUP RINGS *denote the full subcategory of* A-(SEMI)GROUP RINGS *on the objects AG, where G is an abelian (semi)group.*

The forgetful functor A-ABEL(SEMI)GROUP RINGS $\rightsquigarrow$ SETS *has a left adjoint*

$$\begin{cases} \text{SETS} \rightsquigarrow A\text{-ABEL (SEMI)GROUP RINGS} \\ \quad S \mapsto A[S]_c, \end{cases}$$

where $[S]_c$ *is the free commutative (semi)group on* S *and* $S \to A[S]_c$ *is the compose of the canonical injections* $S \to [S]_c$ *and* $[S]_c \to A[S]_c$. *When there is no possibility of confusion, then* $A[S]$ *denotes* $A[S]_c$. *This ring is called the* ***ring of polynomials in the (commuting) indeterminates, or variables,*** S.

Proof. The left adjoint to the forgetful functor has the property that for each set S and mapping $t: S \to AG$ of S into a group ring which commutes with the canonical injection $A \to AG$, there is a unique morphism $h: A[S] \to AG$ of A-(SEMI)GROUP RINGS such that

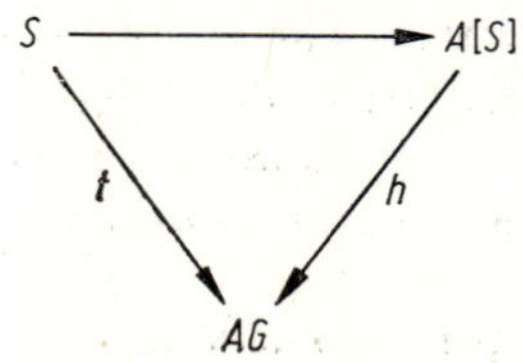

commutes, and, in fact, h is the morphism given canonically by the S-(semi)group ring $A[S]$ such that

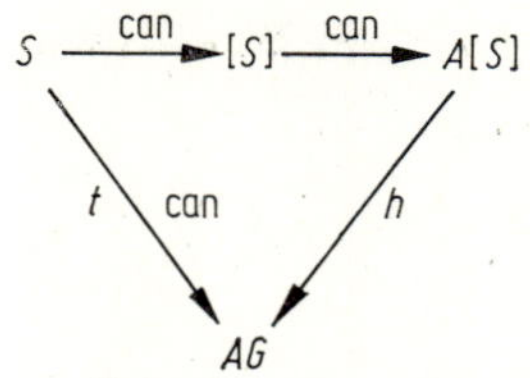

commutes. This proves existence. If A is commutative, and if $A \to R$ is an algebra, then let A' denote the image of $A \to R$. If $t: S \to R$ is a mapping, then t commutes with the canonical injection $A' \to R$, and thus there is a unique morphism $h: A[S] \to R$ such that

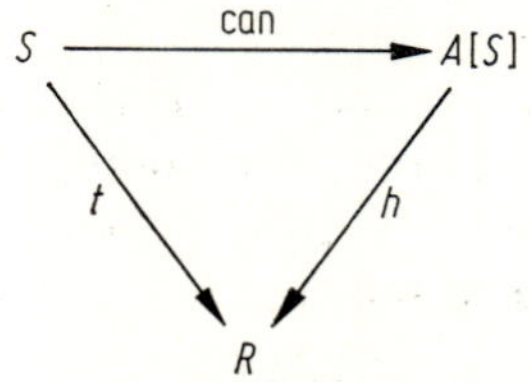

commutes, that is, $S \to A[S]$ is the free algebra on S. This completes the proof of 5.74.1, and 5.74.2 is similarly proved. $\square$

5.75 Corollary and Definition. *If A is a k-algebra, and if S is a subset of A, then the subalgebra (S) generated by S is the image of the canonical homomorphism $k[S] \to A$ from the free algebra $k[S]$ induced by the inclusion $S \to A$. If $A = (S)$, then there is an exact sequence*

$$0 \to I \to k[S] \to A \to 0,$$

where I is the kernel of $k[S] \to A$. If Δ is the ideal of $k[S]$ generated by elements $\{r_q\}_{q \in Q}$, then A is said to be the algebra with **generators** *S, and* ***relations*** *$\{r_q = 0\}$. Then $(S \mid \Delta)$ is called a* **presentation** *of A.*

Proof. Let B be the image of the canonical homomorphism $k[S] \to A$, and let $g : k[S] \to B$ be the induced map. Since $S \to k[S]$ factors through the inclusion $g^{-1}(S) \to k[S]$, by the Free Object Theorem 2.7, necessarily $g^{-1}(S) = k[S]$, and hence $B = (S)$. □

Thus, every algebra is isomorphic to a quotient ring (algebra) of a free algebra.

5.76 Theorem. *The categories* GROUPS, RINGS, *and* k-ALG, *for any commutative ring k, are biproductive.*

Proof. The forgetful functor RINGS ⇝ $\mathbb{Z}$-ALG is a category equivalence, since it is fully faithful and representative 2.2. Hence the theorem for RINGS follows from the theorem for k-ALG. The free object functor F : SETS ⇝ k-ALG (exists by 5.74) is left adjoint to the forgetful functor k-ALG ⇝ SETS, and, hence, preserves coproducts by 5.44. Let $\{t_i : S_i \to \coprod_{i \in I} S_i\}_{i \in I}$ be a disjoint union in SETS. Then $\{u_i : F(S_i) \to F(\coprod_{i \in I} S_i)\}_{i \in I}$ is a coproduct in k-ALG, where $u_i = F(t_i)\ \forall i \in I$. Set $R = F(\coprod_{i \in I} S_i)$, let S_i be a k-algebra with presentation $(S_i \mid \Delta_i)$, let Δ be the ideal of R generated by $\bigcup_{i \in I} u_i(\Delta_i)$, and let $f_i : S_i \to S = R/\Delta\ \ \forall i \in I$ be the composition

$$S_i \to F(S_i) \xrightarrow{u_i} R \xrightarrow{\text{canon}} R/\Delta = S.$$

Then $\{f_i : S_i \to S\}_{i \in I}$ is a coproduct in k-ALG, since if $\{g_i : S_i \to A\}$ is any family of morphisms in k-ALG, then there is a commutative diagram

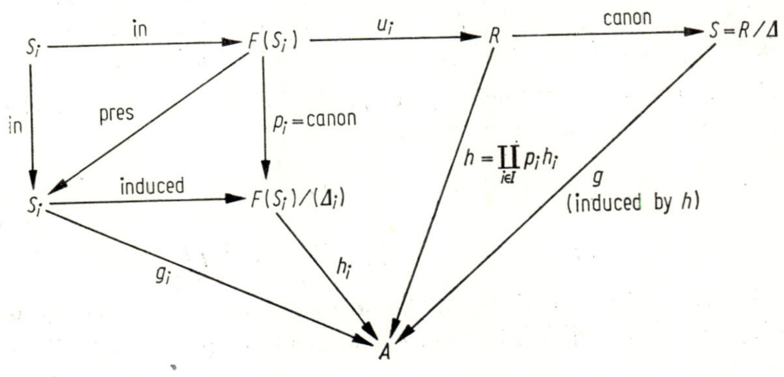

where pres denotes the presentation homomorphism $F(S_i) \to S_i$, $S_i \to F(S_i)/(\Delta_i)$ is the isomorphism induced by pres, and $h_i = g_i$ (induced)$^{-1}$. The diagram shows that ker h contains the image under u_i of Δ_i $\forall i \in I$. Since ker h therefore contains Δ, then h induces the map $g: S \to A$ such that $g f_i = g_i \forall i \in I$, and g is the unique homomorphism which does this. The proof for GROUPS is the same except for the obvious change from ideal to normal subgroup. □

5.77 Corollary. *The category of A-*(SEMI)GROUPS *is coproductive. Let* $\{u_i : S_i \to G\}_{i\in I}$ *be a coproduct in* (SEMI)GROUPS. *Then* $\{f_i : A S_i \to A G\}_{i\in I}$ *is a coproduct in A-*(SEMI)GROUP RINGS, *where* f_i *is the unique morphism of* $C(A S_i)$ *defined by the canonical injections* $A \to A S_i$ *and* $u_i : S_i \to G$. *Thus,* $\coprod_{i\in I} A S_i \to A \coprod_{i\in I} S_i$ *is an isomorphism in A-*(SEMI)GROUP RINGS.

Proof. The ring coproduct $f : \coprod_{i\in I} A S_i \to A \coprod_{i\in I} S_i$ of the family $\{f_i\}_{i\in I}$ of morphisms exists in RINGS, by 5.76, and is surjective, since $G = \coprod_{i\in I} S_i$ and A generate AG. Let $g : AG \to \coprod_{i\in I} A S_i$ be the unique morphism induced by the injection $A \to B = \coprod_{i\in I} A S_i$ and the coproduct $G \to B$ of the injections $\{S_i \to B\}_{i\in I}$ induced by the injections $\{A S_i \to B\}_{i\in I}$. Then $gf = 1$ on the images of the canonical injections $A S_i \to B$, hence $gf = 1_B$. Similarly, $fg = 1_{AG}$. □

5.78 Corollary. *If A is a ring, and X and Y disjoint sets, then there exists a unique isomorphism* $h : A[X][Y] \to A[Y][X]$ *such that*

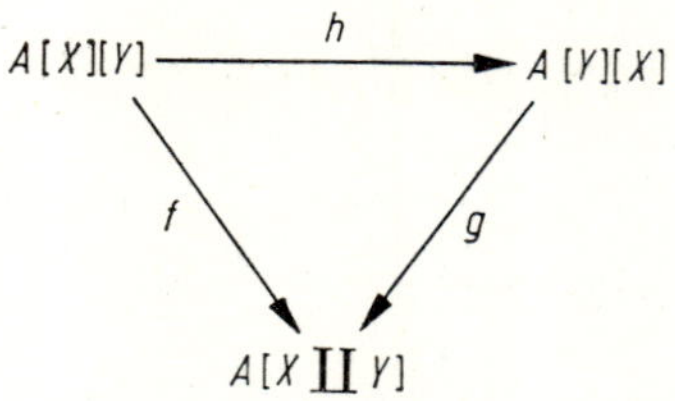

commutes, where f *is the unique morphism of* $C(A[X][Y])$ *defined by the injections* $f_1 : A[X] \to A[X \coprod Y]$ *and* $f_2 : [Y] \to A[X \coprod Y]$ *and* g *is similarly given. This also holds for commuting indeterminates.*

Proof. The homomorphism f (resp. g) is the coproduct in RINGS of f_1 and the canonical injection

$$g_1 : A[Y] \to A[X \coprod Y],$$

since

$$A[X] \coprod A[Y] \approx A[X] \coprod [Y] \approx A[X \coprod Y].$$

The first equivalence holds by 5.77, and the second holds since the free (semi)group on the disjoint union of two sets is equivalent to the coproduct of the free (semi)groups on the sets, that is, $[X] \coprod [Y] \approx [X \coprod Y]$ in (SEMI)GROUPS, which induces the latter equivalence. $\square$

For a set $S = \{y_1, \ldots, y_n\}$ and ring A, $A[y_1, \ldots, y_n]$ denotes the polynomial ring over A in the commuting indeterminates $y_1, \ldots, y_n$. Each element of the A-semigroup ring $A[x]$ has a unique representation

$$\sum_{i \in \omega} a_i x^i,$$

where $a_i \in A$ and $a_i = 0$ for all but a finite set of $i \in \omega$. Furthermore, each element in the A-group ring $A[x]$ as a unique representation

$$\sum_{i \in \mathbb{Z}} a_i x^i$$

with only finitely many $a_i \neq 0$. Thus, the A-group ring $A[x]$ is isomorphic to the A-semigroup ring $A[x, y]$. This follows since $\mathbb{Z} = \mathbb{N} \coprod (-\mathbb{N}) \coprod \{0\}$, and, in effect,

$$\begin{cases} A[x] \approx A\mathbb{Z} \approx (A\mathbb{N})\mathbb{N} \approx A[x][y] \approx A[x, y] \\ \qquad\qquad a_{ij} x^{i-j} \mapsto a_{ij} x^i y^j \end{cases}$$

is an isomorphism. To avoid confusion, **the free group ring on a single element** x is denoted $A[x, x^{-1}]$, that is, the indeterminate y is denoted by its image x^{-1} under the inverse of the isomorphism just indicated.

Exercises for Chapter 5

1. If S is a set and A is a ring, then the free semigroup ring $A[S]$ modulo the ideal COM S generated by the set

$$\{[s, t] = st - ts \mid s, t \in S\}$$

is the ring of polynomials in the commuting indeterminates S, and $S \to A[S] \to A[S]/(\text{COM } S)$ is the injection.

2. For any set S, the ring $A[S]$ of noncommutating polynomials over S is a domain if A is, and so is the polynomial ring in commuting indeterminates.

3. (P. M. Cohn [64b]) (a) For $S = \{x, y\}$ consisting of two elements and a commutative field A, show that the subalgebra B of $A[x, y]$ generated by COM S cannot be a free algebra.

(b) If I is any ideal of a free algebra $A[S]$ of rank > 1, and if I generates a subalgebra $B \neq A[S]$, then B is not a free algebra.

(c) If A is a commutative field and $A[x]$ is the polynomial ring of rank 1, then a subalgebra B of $A[x]$ is free if and only if it is (in the terminiology of Chapter 30) integrally closed in its quotient field, e.g., the subalgebra generated by x^2 and x^3 is not free.

4. (P. M. Cohn, l.c.) Let A be a field. A free A-algebra of at most countable rank can be imbedded in a free A-algebra of rank 2.

5. (Jategaonker [69]) If A is a right Ore domain which is not a left Ore domain, then the central algebra A contains a subalgebra which is a free algebra of rank 2 and, hence, by Exercise 4 contains a free algebra of uncountable rank.

***6.** (a) (Hochschild [58]) For any commutative ring k and set S, the global dimension of the semigroup ring $k[S]$ exceeds that of k by 1. When k is a field, then any right ideal of $k[S]$ is projective. (Cf. Chapter 8 for the definitions and for the case of the free commutative algebra over k.)

(b) (P. M. Cohn [64a]) A ring R is a right **fir** if R is an IBN domain, and every right ideal is a free module. For any commutative field, the free algebra $k[S]$ on any set S, and also the group algebra $k(S)$ on the free group (S) on the set S, are right and left firs.

7. The Group ring $\mathbb{Z}G$ is called the **integral group ring** of a group G. If G is finite, show that:

(a) $\mathbb{Z}G$ has no idempotents $\neq 0,1$

(b) Every projective module over $\mathbb{Z}G$ is free

(c) For any set S, the projective modules over the integral group ring of the free group on S are free. (This also holds for the free group ring over any commutative principal ideal ring. Cf. Bass [64].)

8. Let A and B be k-algebras. Then the composition

$$k \xrightarrow{\text{diag}} A \times B \xrightarrow{\text{canon}} A \underset{k}{\otimes} B$$

is a k-algebra under the group addition and the following multiplication:

$$\sum a_i \otimes \sum b_j = \sum a_i \otimes b_j.$$

The mapping

$$\begin{cases} A \to A \underset{k}{\otimes} B \\ a \mapsto a \otimes 1 \end{cases}$$

is a canonical homomorphism.

9. If A is a k-algebra and G is a group, then the mapping $A \times kG \to AG$ sending $(a, h) \mapsto ah \; \forall\, a \in A, h \in kG$ is balanced over k, and the canonical morphism $A \underset{k}{\otimes} kG \to AG$ is an isomorphism of rings.

10. Prove in COM k-ALG over a field k, that the pair of canonical homomorphisms $A \to A \underset{k}{\otimes} B$ and $B \to A \underset{k}{\otimes} B$ is a coproduct. Is this true for a general commutative ring k?

11. (Auslander-Goldman [60]) *Let A be an R-module, and* $S = \text{End } A_R$.

(a) *Then there is a commutative diagram*

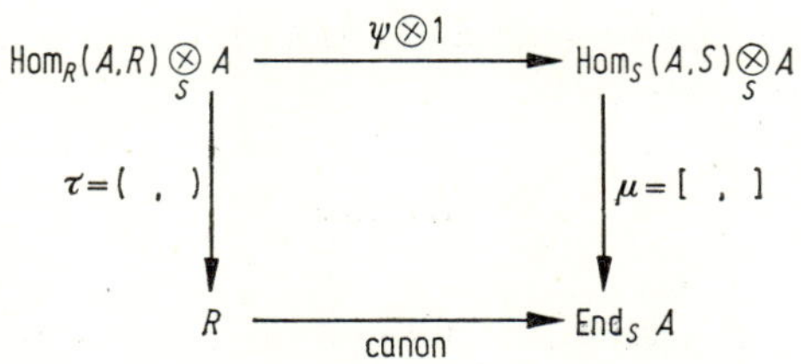

where μ *is the map* $[\ ,\ ] : \operatorname{Hom}_S(A, S) \otimes_S A \to \operatorname{End}_S A$, *that is,*

$$\mu(f \otimes a)(b) = af(b),$$

$\forall a, b \in A, \ f \in \operatorname{Hom}_S(A, S)$,

$$\psi : \operatorname{Hom}_R(A, R) \to \operatorname{Hom}_S(A, S)$$

is defined by $\psi(f)(a) = \mu(f \otimes a)$, $a \in A$; *and* τ *is the map* $(\ ,\ ) : \operatorname{Hom}_R(A, R) \otimes_S A \to R$, *that is,*

$$\tau(f \otimes a) = f(a)$$

$\forall f \in \operatorname{Hom}_R(A, R)$, $a \in A$.

(b) *The image of the map* τ *is the trace ideal,* $\operatorname{trace}_R A$, *and*

$$\operatorname{trace}_R A \cdot \operatorname{Ann}_R A = 0.$$

(c) *If* $\operatorname{trace}_R A + \operatorname{Ann}_R A = R$, *then* μ *is an epimorphism and* A *is finitely generated projective over S. In this case,*

$$0 \to \operatorname{Ann}_R A \to R \to \operatorname{End}_S A \to 0$$

is split exact.

(d) *If* A_R *is a generator, then all maps in* (a), τ, μ, ψ, and so on, *are isomorphisms.*

Notes for Chapter 5

The theory of adjoint functors was given by Kan [58], who was the first to place the tensor functor on par with the corresponding hom functor. (This perhaps also explained the mysterious absence of one or the other functors in the development of certain mathematical ideas!) Watts [60] characterized joinable module-valued functors (see 5.56—7). This also characterized continuous functors (cf. 12.9). (Cf. Eilenberg [60].) Freyd [62] provided necessary and sufficient conditions for a functor to be joinable from the left.

References

Cohn [64a, b], Eilenberg [60], Freyd [62], Gabriel [62], Gabriel-Zisman [67], Grothendieck [57], Hochschild [58], Jategaonker [69], Kan [58], Lambek [64a], MacLane [63], Mitchell [65], Morita [58], Popesco-Gabriel [64], Watts [60].

Chapter 6

Abelian Categories

An exact additive category is said to be abelian (6.1). If C is an abelian category, then so is C^L, for any (small) category L. Furthermore (Proposition 6.17), C^L inherits the following properties from C: (1) (finitely) complete; (2) (finite) products; (3) kernels or equalizers; (4) images, sums, or (finite) intersections; (5) normality with epic images; (6) additive; (7) exact; (8) abelian; (9) exact and locally small (assuming L is small). Moreover (Proposition 6.25), C^L has exact direct limits when C has (cf. Exercise 6.13). When this is true, then (Corollary 6.24) for any family $\{F(X)\}_{X\in M}$ of (left) exact functors $L \rightsquigarrow C$, the direct limit $\varinjlim_{X\in M} F(X)$ is a (left) exact functor $L \rightsquigarrow C$. This implies that a direct limit of a directed family of flat modules in mod-R is flat, for any ring R. A colimit of colimit preserving functors is always colimit preserving (6.22).

For any category C, the representation functor $H: C^{\text{op}} \rightsquigarrow \text{SETS}^C$ is limit preserving (6.23), and if C is additive, then H induces a full embedding $C^{\text{op}} \rightsquigarrow (\text{mod-}\mathbb{Z})^C$ (6.21). When, in addition, C is small, then $H(A^*) = h^A$ is a small projective object of Adfunct$(C, \text{mod-}A)$, and $G = \coprod_{A\in C} h^A$ is an embedding $L \rightsquigarrow \text{mod-}\mathbb{Z}$. Furthermore, $[G, \ ]$ is an exact embedding Adfunct$(C, \text{mod-}\mathbb{Z}) \rightsquigarrow \text{mod-}\mathbb{Z}$ (see 6.20 and 6.23). (Cf. the exact embedding theorem 14.21.)

6.1 Proposition and Definition. *A category C is said to be* **abelian** *if C satisfies any of the following conditions* I, II, III, *and* IV, *which are equivalent*:

I

(Ab 1) *Each morphism of C has a kernel and cokernel.*

(Ab 2) *C is a normal and conormal additive category [thus, every monic (epic) is the kernel (cokernel) of some morphism].*

(Ab 3) *Every morphism of C has a factorization as a composition of an epic followed by a monic. (The composition is then necessarily unique up to equivalent epics and monics.* Cf. 5.25.*)*

II

The category C is an exact, additive category.

III

The category C is normal, conormal, and additive and has fiber products and fiber coproducts.

IV

The category C is normal, conormal, and additive and has kernels and cokernels.

Proof. I with the word "additive" deleted in (Ab 2) is just the definition of an exact category (p. 255). Thus, I $\Leftrightarrow$ II.

II $\Rightarrow$ III. A category with finite products and kernels is finitely complete and, hence, has fiber products. Since and additive category has finite coproducts, then II also implies fiber coproducts, hence II $\Rightarrow$ III.

III $\Rightarrow$ IV is trivial. Assume IV. An additive category with kernels has equalizers, and then *C* is exact by (2) of 5.25. Thus, IV $\Rightarrow$ II. ▯

6.2 *Exercise.*

6.2.1 (Freyd [62]) A category *C* is abelian in case it satisfies the following axioms.

A_0. *C* has a zero object.

A_1. For every pair of objects there is a product.

A_1^*. For every pair of objects there is a coproduct.

A_2. Every morphism has a kernel.

A_2^*. Every morphism has a cokernel.

A_3. Every monic is the kernel of some morphism.

A_3^*. Every epic is the cokernel of some morphism.

6.2.2 (Puppe [67]) Consider the axioms about a category C:

K: C has kernels.

N: Every monic of C is the kernel of some morphism.

P: C has finite products.

Z: Any morphism is composed of an epic and a monic (Ab 3).

Let K^*, N^*, P^* denote the dual axioms, respectively. Show that a category C with a zero object is abelian if any one of the following conditions are satisfied:

(I) P, K, K^*, N, N^*

(I*) P^*, K, K^*, N, N^*

(II) P, N, N^*, Z

(II*) P^*, N, N^*, Z.

6.2.3 If C is a small abelian category, then Funct$(C, \text{mod-}\mathbb{Z})$ is abelian.

6.2.4 If C is any abelian category, and if F is a set of objects of C, then there is a full subcategory F' of C with the property that $F \subseteq \text{obj}\, F'$ and F' is small and abelian.

6.2.5 Let C' be a full subcategory of a category C with zero object 0, and assume that $0 \in \text{obj}\, C'$. If $F : L \rightsquigarrow C'$ is a functor, if $F' : L \rightsquigarrow C$ is the induced functor, and if $\lim_{L,C} F'$ is an object of C', then $\lim_{L,C} F' = \lim_{L,C'} F$. Thus, if C is productive, or has kernels, and if C' is closed under the taking of products or kernels of families of morphisms in C', in the sense that at least one product or kernel of any family of morphisms lies in C', then C' is productive or has kernels. Dually for colimits. Hence, if C is exact (or abelian) and if C' is closed under the taking of kernels and cokernels (and finite products), then C' is an exact (abelian) subcategory.

6.2.6 Given a square in an additive category

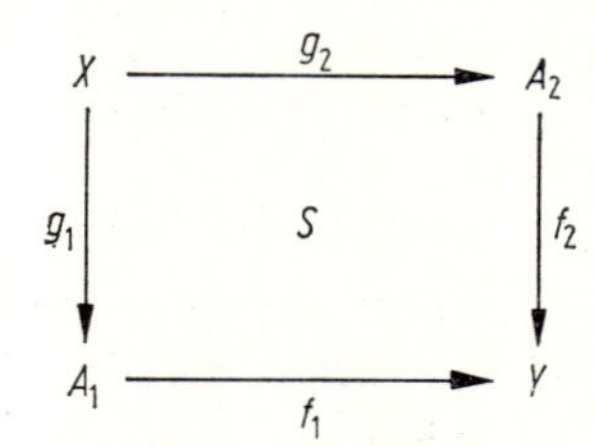

consider the composite map $X \to Y$ defined by:

$$X \xrightarrow{\text{canon}} A_1 \oplus A_2 \xrightarrow{\text{canon}} Y.$$

Deduce:

(a) $X \to Y$ is zero $\Leftrightarrow$ S commutes.

(b) $0 \to X \to A_1 \oplus A_2 \to Y$ is exact $\Leftrightarrow$ S is a fiber product.

(c) $X \to A_1 \oplus A_2 \to Y \to 0$ is exact $\Leftrightarrow$ S is fiber coproduct.

6.2.7 In an abelian category, if the square S in the preceding exercise is cartesian, and if $A_2 \to Y$ is epic, then $X \to A_1$ is epic. Dually for cocartesian, in which case $X \to A_2$ monic implies $A_1 \to Y$ is monic.

6.2.8 If C is abelian and $f: A \to B$ and $g: B \to D$ are morphisms, then

(a) $\operatorname{im} f \cap \ker g = f(\ker gf)$

(b) $\operatorname{im} f + \ker g = g^{-1}(\operatorname{im} gf)$

(c) $\operatorname{coim} g \cap \operatorname{coker} f = g^{-1}(\operatorname{coker} gf)$

(d) $\operatorname{coim} g + \operatorname{coker} f = f(\operatorname{coim} gf)$.

6.3 **Proposition.** *An abelian category C is finitely complete and finitely cocomplete. Furthermore, C is complete (cocomplete) if and only if C is productive (coproductive).*

Proof. A category is (finitely) complete if and only it has finite products and kernels, by 5.11. This, and its dual, proves the proposition. □

6.4 **Proposition.** *Let C be an abelian category. Then, the following implications hold:*

(a) *If C has a generator, then cocompleteness implies completeness.*

(b) *If C has a cogenerator, then completeness of C implies cocompleteness.*

Proof. Because of the duality between kernels and cokernels, an abelian category is locally small if and only if it is colocally small. Thus, if C has a generator, then C is locally and colocally small by 5.23. Since C^* has a cogenerator, then cocompleteness of C implies completeness of C by 5.53. This proves (a), and (b) is the dual. □

Enough Injectives

A category C has **enough injectives** in case for every object A there exists an injective object E and a monic $A \to E$. For abelian category this is equivalent to the requirement of an exact sequence $0 \to A \to E$, where E is injective.

6.5 Proposition. *If C is an abelian category with a generator U, if $\{U/K_i\}_{i\in I}$ is a family of representatives of equivalence classes of quotient objects of U, and if W is an injective object which contains U/K_i as a subobject, for every $i \in I$, then W is a cogenerator of C.*

Proof. The proof of the corresponding result 3.31 for module type categories suffices for the proposition. □

6.6 Corollary. *An abelian category C with a generator and enough injectives is cocomplete if and only if C is complete. When this is so, then C has an injective cogenerator.*

Proof. Cocompleteness implies completeness (without assuming enough injectives) by (a) of 6.4. Conversely, if C is complete, then there is an injective object W with the property of the last proposition; namely, if $0 \to U/K_i \to E_i$ is exact and E_i is injective, then W can be taken to be the product $W = \prod_{i\in I} E_i$. For, in this case, the canonical injection $E_i \to W$ induces a monic $U/K_i \to W$, for every $i \in I$, so that W is a cogenerator by the proposition. □

6.7 *Exercises.*

6.7.1 If C is an abelian category, then an object A of C is projective (injective) if and only if the group valued functor h^A (resp. h_A) is exact.

6.7.2 If C is any category, any retract of a projective object A is projective, and every epic $X \to A$ splits.

6.7.3 In any category C, a coproduct of projective objects is projective. If C has a zero, and if $\{A_i\}_{i\in I}$ is any family of objects of C, then $\coprod_{i\in I} A_i$ is projective if and only if A_i is projective for every $i \in I$.

A category C **has enough projectives** in case for each object X there corresponds a projective object P and an epic $P \to X$.

6.8 Proposition.

6.8.1 *If C is a category with enough projectives, or if C is abelian, then an object A is projective if and only if every epic $X \to A$ splits.*

6.8.2 *If C is a coproductive category with a projective generator U, then A is projective if and only if $U^{(I)} \to A$ is a retraction for some index set I.*

Proof of 6.8.1 is a consequence of Exercise 6.7.2 when C has enough projectives. If C is abelian, then given any epic $X \to Y$, and morphism $A \to Y$, the fiber product

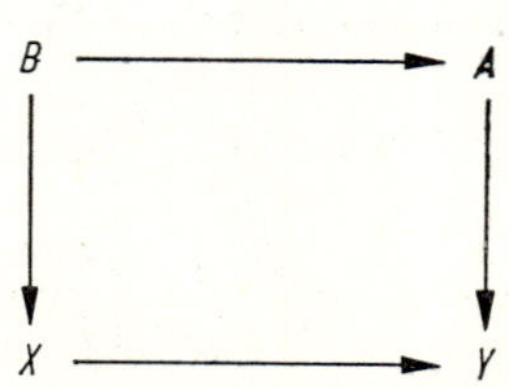

supplies a map $B \to A$ which is epic by Exercise 6.2.7. Then by Exercise 6.7.2, $(A \to B \to A) = 1_A$ for some $A \to B$, and therefore

$$A \to B \to X \to Y = A \to B \to A \to Y = A \to Y,$$

so $X \to Y$ allows $A \to Y$. ▯

Proof of 6.8.2. A category with a projective generator has enough projectives, in view of 3.25.

6.9 Proposition. *If C and D are abelian categories, any half exact functor* $T : C \rightsquigarrow D$ *is additive.*

Proof. By 3.38, a functor T of additive categories is additive if and only if T preserves finite coproducts $A \coprod B$. However, by 3.35 for exact additive (abelian) categories, any half exact functor preserves a coproduct $A \coprod B$. ▯

6.10 Proposition. *If C and D are abelian categories, then an exact functor* $T : C \rightsquigarrow D$ *is faithful if T preserves nonzero objects.*

Proof. Since C and D are additive, and since T is an additive functor by the proposition, then T is faithful if and only if T preserves nonzero morphisms. Hence, suppose that $T(f) = 0$, for some morphism $f : A \rightsquigarrow B$. Exactitude of T implies that $T(\operatorname{im} f) = \operatorname{im} Tf = 0$, so $\operatorname{im} f = 0$ by the hypothesis that T preserves nonzero objects. Then $f = 0$, so T preserves nonzero morphisms. ▯

6.11 Proposition. *The following conditions on a functor* $T : C \rightsquigarrow D$ *of abelian categories are equivalent:*

(a) *T preserves finite limits.*

(b) *T is left exact.*

(c) *T preserves finite products and kernels.*

Proof. (a) $\Leftrightarrow$ (c) by 5.39, and (c) $\Rightarrow$ (b) by 5.26. Assuming (b), then T is additive by 6.9, and so T preserves finite products by 3.38. Thus, (b) $\Rightarrow$ (c). ▯

6.12 Proposition. *Let C be a complete abelian category. The following conditions on a functor* $T : C \rightsquigarrow D$ *are equivalent:*

(a) *T is limit preserving.*

(b) *T preserves products and kernels.*

Proof. See 5.39. ▯

6.13 *Exercises.*

*6.13.1 For arbitrary categories C and D, a functor $S : D \rightsquigarrow C$ is left adjoint to a functor $T : C \rightsquigarrow D$ if and only if there exist functor morphisms $\Phi : 1_D \to TS$ and $\Psi : ST \to 1_C$ such that

(a) $$(T\Psi)(\Phi T) = 1_T,$$

or equivalently $T(\Psi_X)\Phi_{TX} = 1_{TX}$, for every object X of C. and

(b) $$\Psi_{SY} S(\Phi_Y) = 1_{SY}$$

for every object Y of D. (Cf. (1) of 5.68.)

*6.13.2 If (S, T) is an adjoint pair for abelian categories, and if S is exact, then T preserves injectives. Conversely, if $T : C \rightsquigarrow D$ preserves injectives, and if C has enough injectives, then $S : D \rightsquigarrow C$ is exact.

6.13.3 An abelian category C has **exact direct limits** if for every directed set I the colimit functor $C^I \rightsquigarrow C$ is exact.

(a) If C is a cocomplete abelian category with exact direct limits, then C satisfies the following conditions on any object A and directed family $\{A_i\}_{i\in I}$ of subobjects:

$$\left(\sum_{i\in I} A_i\right) \cap B = \sum_{i\in I} (A_i \cap B),$$

for any subobject B of A. A cocomplete category C satisfying this condition is said to be $AB5$.

(b) A cocomplete $AB5$ category has exact direct limits.

Functor Categories

6.14 Proposition. *Let L, M, and C be categories. Then, there are canonical equivalences*

(1) $$[C^L]^M \overset{h}{\rightsquigarrow} C^{M\times L} \overset{g}{\rightsquigarrow} [C^M]^L,$$

where, for example,

(2) $$h(F)(X, Y) = F(X)(Y)$$

$X \in \operatorname{obj} M$, $Y \in \operatorname{obj} L$, *for a morphism or object F of $[C^L]^M$, and the equivalence*

(3) $$gh \begin{cases} [C^L]^M \rightsquigarrow [C^M]^L \\ F \mapsto F' \end{cases} \quad \textit{(on objects \& morphisms)}$$

is such that

(4) $$F'(y)(X) = h(F)(X, Y) = F(X)(y)$$

for an object or morphism y of L and X of M. Furthermore,

$$(\lim_{M,C^L} F)(Y) = \lim_{M,C} F'(Y) \tag{5}$$

$\forall Y \in \operatorname{obj} L$ *whence*

$$\lim_{M,C^L} F = \overline{\lim}_{M,C} F', \tag{6}$$

$\forall F \in \operatorname{obj} [C^L]^M$, *is an equivalence in C. Moreover,*

$$\begin{array}{ccccccc}
C^{M\times L} & \xleftarrow{\;h\;} & [C^L]^M & \xrightarrow{\text{canon}} & [C^M]^L & \xrightarrow{\lim_{L,C^M}} & C^M \\
\downarrow{\scriptstyle \lim_{M\times L,C}} & & \downarrow{\scriptstyle \lim_{M,C^L}} & & \downarrow{\scriptstyle \overline{\lim}_{M,C}} & & \downarrow{\scriptstyle \lim_{M,C}} \\
C & \xleftarrow[\lim_{L,C}]{} & C^L & \xrightarrow[1_C]{} & C^L & \xrightarrow[\lim_{L,C}]{} & C
\end{array} \tag{7}$$

commutes, where $\overline{\lim}_{M,C}$ *is the functor induced by the limit functor* $\lim_{M,C} : C^M \rightsquigarrow C$.

Proof. It is clear that gh is an equivalence if and only if both h and g are equivalences. In either case, it is just a question of showing that, for example, gh is a functor, since the inverse is defined by $(gh)^{-1}(H)(X)(Y) = H(X, Y)$.

If $t : X \to X_1$ is any morphism of M, and if $F \in \operatorname{obj} [C^L]^M$, then $F(t) : F(X) \to F(X_1)$ is a natural transformation of functors in C^L such that for any morphism $s : Y \to Y_1$

$$\begin{array}{ccc}
F(X)(Y) & \xrightarrow{F(t)(Y)} & F(X_1)(Y) \\
\downarrow{\scriptstyle F(X)(s)} & & \downarrow{\scriptstyle F(X_1)(s)} \\
F(X)(Y_1) & \xrightarrow[F(t)(Y_1)]{} & F(X_1)(Y_1)
\end{array} \tag{8}$$

commutes. The morphism $F'(s)$ in C^M is defined by $F'(s)(X) = F(X)(s)$. Thus,

$$\begin{array}{ccc}
F'(Y)(X) & \xrightarrow{F'(Y)(t)} & F'(Y)(X_1) \\
\downarrow{\scriptstyle F'(s)(X)} & & \downarrow{\scriptstyle F'(s)(X_1)} \\
F'(Y_1)(X) & \xrightarrow[F'(Y_1)(t)]{} & F'(Y_1)(X_1)
\end{array} \tag{8'}$$

commutes, which shows that $F'(s)$ is a natural transformation for any morphism $s : Y \to Y_1$ of L, that is, $F'(s)$ is a morphism of C^M. This proves that F' is a functor, that is, an element of $[C^M]^L$ as asserted, and that gh is an equivalence with the stated inverse.

If A is an object of C^L, if $F : M \to C^L$ a functor, and if $\{A \to F(X)\}_{X \in \operatorname{obj} M}$ is an element of compat F, then the preceding diagram shows, for each $Y \in \operatorname{obj} L$, that

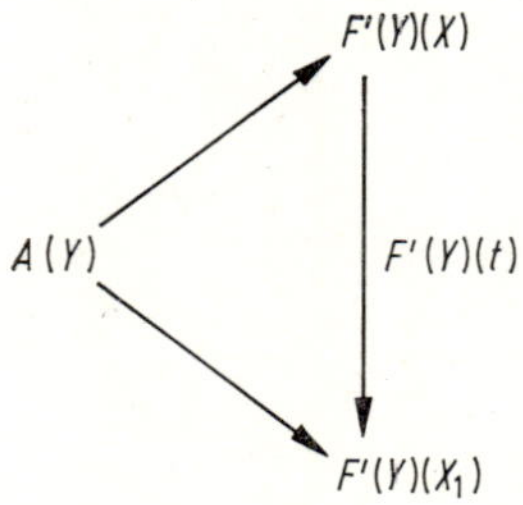

commutes for each morphism $t : X \to X_1$ of M. Therefore, for each Y, the family $\{A(Y) \to F'(Y)(X)\}_{X \in \operatorname{obj} M}$ is an element of compat F'. If A is a final object of compat F, then, $A(Y)$ is a final element of compat $F'(Y)$, for each $Y \in \operatorname{obj} L$. Conversely, if compat $F'(Y)$ has a final object $H(Y)$, for every $Y \in \operatorname{obj} L$, then there is an object H of C^L which assigns to each morphism $s : Y \to Y_1$ of L, the unique morphism $H(s) : H(Y) \to H(Y_1)$ such that

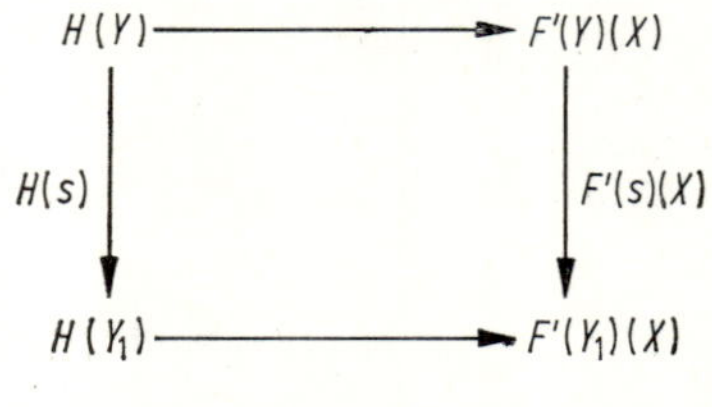

commutes for every $X \in \operatorname{obj} M$. The morphism $H(s)$ exists, because the family

$$\{H(Y) \to F'(Y)(X) \xrightarrow{F'(s)(X)} F'(Y_1)(X)\}_{X \in \operatorname{obj} M}$$

is an element of compat $F'(Y_1)$, as the diagram (8′) shows. Furthermore, for each $X \in \operatorname{obj} M$, there is a natural transformation of functors $H \to F(X)$ such that

$$\{H(Y) \to F(X)(Y) = F'(Y)(X)\}_{X \in \operatorname{obj} M}$$

is the final object of compat $F'(Y)$ for each $Y \in \operatorname{obj} L$. Thus, $\{H \to F(X)\}_{X \in \operatorname{obj} M}$ is a final object of compat F, and, hence,

$H = \lim_{M,C^L} F$. In other words,

$$(\lim_{M,C^L} F)(Y) = H(Y) = \lim_{M,C} F'(Y)$$

for each $Y \in \text{obj}\, L$. This proves that the second square of the diagram (8) commutes, since by definition $(\overline{\lim}_{M,C} F')(Y) = \lim_{M,C} F'(Y)$. The third square of the diagram commutes since $\lim_{L,C}$ is a limit preserving functor. Next, since

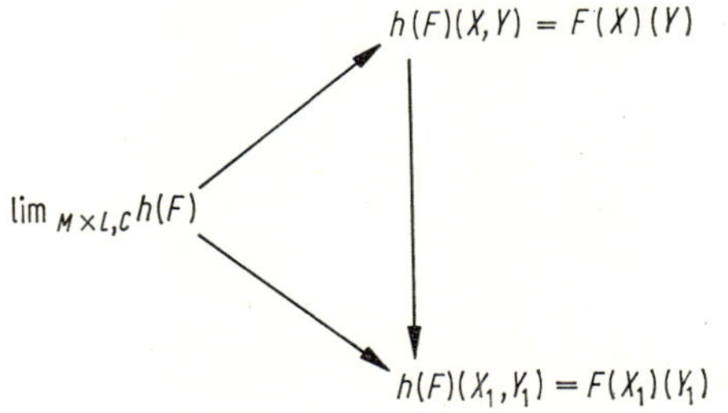

commutes for any morphism $(X, Y) \to (X_1, Y_1)$ of $M \times L$, and

$$\{\lim_{M\times L,C} h(F) \to h(F)(X, Y)\}_{(X,Y)\in M\times L}$$

is a final object of compat $h(F)$, then

$$\lim_{M\times L,C} h(F) \approx \lim_{L,C} \lim_{M,C^L} F$$

is an equivalence which is natural in F. Therefore the first square commutes. ▯

6.15 Corollary. *Let L, M, and C be categories, and for any functor $G : I \rightsquigarrow D$, adopt the notation, $\lim_{I,D} G = \lim_{i\in I} G(i)$. Then, for any functor $F : M \rightsquigarrow C^L$,*

$$\left(\lim_{X\in M} F(X)\right)(Y) = \lim_{X\in M} F(X)(Y).$$

Furthermore, if for each object Y of L, $H(Y)$ is the object of C such that

$$H(Y) = (\lim F)(Y) = \lim F'(Y) = \lim_{X\in M} F(X)(Y),$$

then the canonical morphisms $\{H(Y) \to F(X)(Y)\}_{X\in \text{obj}\, M}$ is a final object of compat $F'(Y)$, *and defines an object*

$$H = \lim_{M,C^L} F$$

of C^L.

Proof. This follows from the proposition and its proof. ▯

6.16 Proposition. *For any category L, the category C^L has any of the following properties, and their duals, provided that C has:* (1) *(finitely) complete;* (2) *(finite or fiber) products;* (3) *kernels, or equalizers;* (4) *images, sums, or (finite) intersections;* (5) *normal with epic images;* (6) *additive;* (7) *exact;* (8) *abelian;* (9) *exact and locally small (assuming that L is small). If C and L are additive, these properties also hold in Adfunct(L, C) when they hold in C.*

Proof. The formula (5) of 6.14 defines the limit $F : M \rightsquigarrow C^L$ pointwise via the limits of the corresponding functors $F'(Y) : M \rightsquigarrow C$. Thus, C^L is (finitely) complete if C is, and, moreover, C^L possesses limits of functors $F : M \rightsquigarrow C^L$ of any "type" for which C possesses limits. Thus, if C has a zero object, then the zero of C^L is the constantly zero functor. Similarly, if C has (finite) products, then C^L has (finite) products. A product $\prod_{X \in M} A(X)$ in C^L is the limit of a functor $A : M \rightsquigarrow C^L$ such that

$$\Big(\prod_{X \in M} A(X)\Big)(Y) = \prod_{X \in M} A(X)(Y) = \prod_{X \in M} A'(Y)(X)$$

$\forall\, Y \in \operatorname{obj} L$. Thus, if $\{F_i\}_{i \in I}$ is any family of objects of C^L, then:

$$\big(\prod_{i \in I} F_i\big)(Y) = \prod_{i \in I} F_i(Y).$$

The properties (2) to (4) are inherited by C^L pointwise (see p. 314), and the following proposition (6.17) establishes (5). If C is additive, then addition of natural transformations defines an operation in Mor C^L with respect to which C^L is additive. In an exact category, epics (monics) of C^L are pointwise monics by the following proposition (6.18), and then every morphism of C^L has a unique pointwise epic-monic factorization, since $\operatorname{im} f = \ker(\operatorname{coker} f)$ is defined pointwise. Therefore, C^L is exact. This proves (7), and (8) follows from (7) and (6). The next proposition suffices for (9).

If C and L are additive, then the preceding holds for Adfunct(L, C). ▯

6.17 Proposition.

6.17.1 *Let L and C be categories such that monics of C^L are pointwise monics. Then C locally small and L small imply that C^L is locally small.*

6.17.2 *If monics and epics of C^L are pointwise, and if C is balanced (or normal with epic images), then C^L is balanced (or normal with epic images).*

6.17.3 *If C is a category with kernels, and if any morphism f such that* $\ker f = 0$ *is necessarily monic, then monics of C^L are pointwise.*

6.17.4 *An exact category C satisfies* 6.17.1, 6.17.2 *and* 6.17.3.

6.17.5 *An additive category C with kernels satisfies the hypothesis and the conclusion of* 6.17.3 *and so does* Adfunct(L, C), *for any additive category L.*

Furthermore, if, in addition, C is locally small, then Adfunct(L, C) *is locally small.*

Proof of 6.17.1. If $K \to F$ is monic in C^L, then, by assumption, $K(A) \to F(A)$ is monic in C, for every A of obj L. Therefore, the class S of representatives of subobjects of F in C^L is a set, and there is an injection of S into the cartesian product of the set $\{F(A)\}_{A \in \text{obj}\, L}$.

Proof of 6.17.2. Monics and epics of C^L are pointwise, and hence if f is monic and epic in C^L, then f is a pointwise equivalence. Therefore, C^L is balanced.

If $g: Y \to Z$ is a morphism of C, and if $g': Y \to \operatorname{im} g$ is an epic image, then $\ker g = \ker g'$. Thus, if C is normal with epic images, every morphism is the kernel of an epic. Let $f: F \to G$ be a morphism of C^L, and for each $A \in \text{obj}\, L$, let $f(A) = \ker g(A)$, where $g(A): G(A) \to Q(A)$ is an epic. If $t: A \to B$ is a morphism of C, then

$$g(B) f(B) = 0 = g(B) f(B) F(t) = g(B) G(t) f(A).$$

The fact that $f(A)$ is monic implies that $g(B) G(t) = 0$, and then there is a canonical morphism $Q(t): Q(A) \to Q(B)$, such that

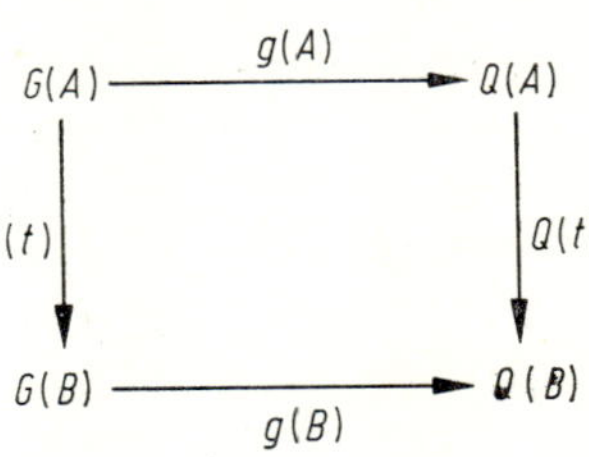

commutes. Thus, if also $t': B \to B'$, then

$$\begin{aligned} Q(t't) g(A) = g(B') G(t't) &= [g(B') G(t')] G(t) \\ &= [Q(t') g(B)] G(t) \\ &= Q(t') [g(B) G(t)] \\ &= Q(t') [Q(t) g(A)], \end{aligned}$$

and the fact that $g(A)$ is epic implies the functorial property $Q(t't) = Q(t') Q(t)$. Then, there is a functor $Q: L \rightsquigarrow C$ assigning $Q(h)$ to any object or morphism h of L, and the diagram shows that there is a natural transformation $g: G \to Q$, assigning $g(A)$ to any object A of L. Furthermore, $f: F \to G$ is the (pointwise) kernel of $g: G \to Q$. This proves normality of C^L.

Proof of 6.17.3. If f is a monic morphism of C^L, then $\ker f = 0$. Since kernels of C^L are pointwise kernels by the hypothesis that C has kernels, then $\ker f$ is pointwise zero. Therefore, f is a pointwise monic.

Proof of 6.17.4. In an exact category, by 5.24, for any morphism $f: A \to B$, f is monic if and only if $\ker f = 0$, and dually for epics.

Proof of 6.17.5. If $\ker f = 0$ in an additive category C, then $fk = fq$, for morphisms k and q, implies $f(k - q) = 0$, whence $k = q$, and therefore f is monic. Then 6.17.3, and 6.17.1 apply. The proof that C^L is locally small suffices to prove that $\mathrm{Adfunct}(L, C)$ is locally small, when L is small and additive. ▯

In order to make distinctions, the notation $[C^L]$ or $[L, C]$ also will denote $\mathrm{Funct}(L, C) = C^L$.

6.18 Corollary. *If L, M, and C are additive categories, then the full subcategory (L, C) of C^L on the additive functors $L \rightsquigarrow C$ is an additive category, also denoted by (C^L) or* $\mathrm{Adfunct}(L, C)$. *Propositions* 6.14 *and* 6.16 *hold for additive categories and additive functors, replacing $[C^M]^L$ by $((C^M))^L = \mathrm{Adfunct}(L, C^M)$, $C^{M \times L}$ by $(C^{M \times L})$ and so forth. Thus, $\lim_{L,C}$ is an additive functor, and if $F: M \rightsquigarrow C^L$ is additive, then $F': L \rightsquigarrow C^M$ and $\lim_{M,C^L} F$ are additive. In particular, (C^L) has every property listed in Proposition 6.16 that C has. Thus, (C^L) is abelian if C is. Finally, the category* $\mathrm{Sex}(L, C)$ *of left exact functors $L \rightsquigarrow C$ of abelian categories is an abelian category.*

Proof. By 3.38, any limit preserving functor of additive categories is additive, since it preserves (finite) products. Thus, $\lim_{L,C}$ is additive, for any additive categories L and C, since the limit functor is limit preserving by 5.43. Then the proofs of Propositions 6.14, 6.16, and Corollary 6.15 suffice for the proof of the corollary, except the last assertion, which is left as an exercise. (Cf. Exercises at the end of chapter.) ▯

6.19 *Exercise.*

6.19.1 Question: If C is an abelian category with enough injectives, and if L is small, does C^L necessarily have enough injectives? (Cf. $AB5$ categories, Chapter 14.)

*6.19.2 (See 6.13) A cocomplete abelian category C with exact direct limits has enough injectives. In this case, so does C^L, for any small abelian category L.

6.19.3 (a) If C is an abelian category with exact direct limits, and if L is a directed set, then $c(E)$ is an injective object of C^L whenever E is an injective object of C. Conversely, this latter condition (preservation of injectives) implies that the colimit functor $C^L \rightsquigarrow C$ is exact.

(b) Show by example that, even when C has exact direct limits, the colimit functor $C^A \rightsquigarrow C$ need not be exact for an arbitrary small category A. (Hint: Show that the functor assigning cokernels is not necessarily exact.)

6.19.4 For any ring R, show mod-$R \approx$ Adfunct (R, mod-$\mathbb{Z}$).

6.19.5 Show that as a special case of 6.18 that any limit of right R-modules is the limit considered as abelian groups supplied with the "inherited" R-module structure.

6.20 Proposition (Yoneda [54]). *For any category L and object A of L, the functor*

$$[h^A, \;] \begin{cases} \text{SETS}^L \rightsquigarrow \text{SETS} \\ \quad F \mapsto [h^A, F] \approx FA \end{cases}$$

is limit and colimit preserving. If L is small, then $G = \coprod_{A \in \text{obj}\, L} h^A$ is a generator of SETS^L *and an embedding* $G : L \rightsquigarrow \text{SETS}$.

If L is a small additive category, then h^A is a small projective object of Adfunct (L, mod-$\mathbb{Z}$), *and* $G : L \rightsquigarrow$ mod-$\mathbb{Z}$ *is an embedding, and*

$$[G, \;] : \text{Adfunct}\,(L, \text{mod-}\mathbb{Z}) \rightsquigarrow \text{mod-}\mathbb{Z}$$

is an exact embedding.

Proof. By 2.3, the Yoneda function

$$y_{A,T} \begin{cases} [h^A, F] \to FA \\ \quad t \mapsto t(A)(1_A) \end{cases}$$

is bijective and natural in both A and F, and, for any natural transformation $f : F \to G$ of functors $L \rightsquigarrow$ SETS,

(1)

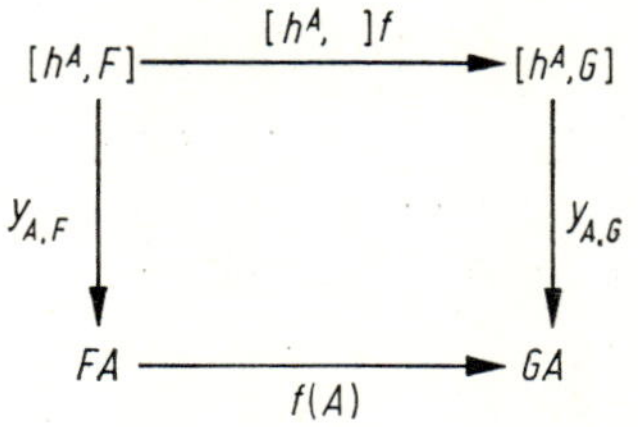

commutes (see diagram III of the proof of 2.3). It follows from this, that if $g : G \to H$ is the coequalizer of $f' : F \to G$ and $f : F \to G$ in C^L, defined pointwise, that is, if $g(A) : G(A) \to H(A)$ is the coequalizer in SETS of $f(A)$ and $f'(A)$, for every A of L, then $[h^A, \;]g : [h^A, G] \to [h^A, H]$ is the coequalizer of $[h^A, \;]f$ and $[h^A, \;]f'$ in SETS. Thus, $[h^A, \;]$ preserves coequalizers. Furthermore, if $\coprod_{i \in I} F_i$ is any coproduct in SETS^L, then

$$\Big[h^A, \coprod_{i \in I} F_i\Big] = \Big(\coprod_{i \in I} F_i\Big)(A) = \coprod_{i \in I} F_i(A) = \coprod_{i \in I} [h^A, F_i],$$

using $[h^A, T] = TA$, for any functor $T : L \rightsquigarrow$ SETS. Since $[h^A, \;]$ therefore preserves coproducts and coequalizers, then it is colimit preserving by the dual to 5.39. Furthermore, for any category C and object X, the functor $h^X : C \rightsquigarrow \text{SETS}$ is limit preserving by 5.41. This proves the first assertion. Next, if f and g are different morphisms $F \to G$, then $f(A) \neq g(A)$ for some $A \in \text{obj}\, L$, and then by the diagram (1), $[h^A, \,]f \neq [h^A, \,]g$, which implies that $ft \neq gt$ for some morphism $t : h^A \to F$ in SETS^L. When L is small, this implies that $\{h^A\}_{A \in \text{obj}\, L}$ is a family of generators of SETS^L and, therefore, the coproduct G of the family is a generator. If a and b are morphisms of L such that $G(a) = G(b)$, then $h^A(a) = h^A(b)$, for all objects A of L. Since the representation functor $L^{\text{op}} \rightsquigarrow \text{SETS}^L$ is faithful, $a = b$. Thus, $G : L \rightsquigarrow \text{SETS}$ is an embedding, inducing $G : L \rightsquigarrow \text{mod-}\mathbb{Z}$.

If L is additive, then the induced functor $h^A : L \rightsquigarrow \text{mod-}\mathbb{Z}$ is coproduct preserving, hence additive by 3.38. Since $[h^A, \,]$ is limit and colimit preserving, then h^A is a projective object of Adfunct$(L, \text{mod-}\mathbb{Z})$, and is small by 4.25 C. Since $\text{Adfunct}(L, \text{mod-}\mathbb{Z}) = (\text{mod-}\mathbb{Z})^L$ is coproductive, G is an object and a projective generator, so that

$$[G, \,] : (\text{mod-}\mathbb{Z})^L \rightsquigarrow \text{mod}\, \mathbb{Z}$$

is an exact embedding. ▯

6.21 Corollary.

6.21.1 *If L is an additive category, then there is a full embedding*

$$H \begin{cases} L^{\text{op}} \rightsquigarrow (\text{mod-}\mathbb{Z})^L \\ A^* \mapsto h^A \end{cases}$$

induced by (and called) the representation functor.

6.21.2 *If L is a small additive category, then there is an embedding $L \rightsquigarrow \text{mod-}\mathbb{Z}$ composed of the representation functor $L \rightsquigarrow (\text{mod-}\mathbb{Z})^{L^{\text{op}}}$ and the exact embedding* $(\text{mod-}\mathbb{Z})^{L^{\text{op}}} \rightsquigarrow \text{mod-}\mathbb{Z}$.

Proof. First, $H(A^*) = h^A$ is an object of Adfunct$(L, \text{mod-}\mathbb{Z})$ for any object A of an additive category L. Second, if L is small and additive, then so is L^{op}, and then the proposition applies to yield an exact embedding $(\text{mod-}\mathbb{Z})^{L^{\text{op}}} \rightsquigarrow \text{mod-}\mathbb{Z}$. Since a composition of embeddings is an embedding, the corollary follows. ▯

In order to obtain an exact embedding $H : L \rightsquigarrow \text{mod-}\mathbb{Z}$, from the embedding $G : L \rightsquigarrow \text{mod-}\mathbb{Z}$, additional hypotheses are needed. Thus, if L is a small abelian category, it will be shown that Adfunct$(L, \text{mod-}\mathbb{Z})$ has injective hulls (see Chapter 14) and that the injective hull H of G is the desired exact embedding.

A functor $F : M \rightsquigarrow C^L$ has a property P **pointwise** provided that the functor $F(X) : L \rightsquigarrow C$ has property P for every object X of M.

6.22 Corollary.

6.22.1 *If* $F: M \rightsquigarrow C^L$ *has property P pointwise, then the image* F' *of* F *under the canonical equivalence* $(C^L)^M \approx (C^M)^L$ *has property P, for any property P which is one of: (finite) limit, (finite) colimit preserving.*

Furthermore, if $F': L \rightsquigarrow C^M$ *has property P pointwise, then* F *has property P.*

6.22.2 *If* F *is pointwise limit preserving, then* $\lim_{M,C^L} F: L \rightsquigarrow C$ *is limit preserving. Dually, if* F *is pointwise colimit preserving, then* $\operatorname{colim}_{M,C^L} F: L \rightsquigarrow C$ *is colimit preserving.*

6.22.3 *If* L *and* C *are exact, and if* $F: M \rightsquigarrow C^L$ *is pointwise left (right) exact, then* F' *is left (right) exact.*

6.22.4 *If* L *and* C *are exact, and if* $F: M \rightsquigarrow C^L$ *is pointwise right exact, then* $\operatorname{colim} F$ *is right exact. Dually, if* F *is pointwise left exact, then* $\lim F$ *is left exact.*

Proof of 6.22.1. Let $T: N \rightsquigarrow L$ be a functor. Then, $Y = \lim T$ is an object of L, and, since $F(X)$ is limit preserving,

$$F'(\lim T)(X) = F(X)(\lim T) = \lim F(X)\,T$$

for every object X of L. Hence,

$$\{F'(\lim T)(X) \to F(X)\,T(Z) = F'(TZ)(X)\}_{Z \in \operatorname{obj} N}$$

is a final object of compat $F(X)\,T$ and pointwise defines a final object $\{F'(\lim T) \to (F'T)(Z)\}_{Z \in \operatorname{obj} N}$ of compat $F'T$. Thus,

$$F'(\lim T) = \lim F'T,$$

and so F' is limit preserving, that is, $F' \lim = \overline{\lim}\, F'$. If N is finite, and if F is pointwise finite limit preserving, then the proof shows that F' is finite limit preserving.

Proof of 6.22.2.

$$\begin{aligned}(\lim_{M,C^L} F)(\lim_{N,L} T) &= \lim_{M,C} F'(\lim_{N,L} T)\\ &= \lim_{M,C}(\lim_{N,C^M}[F'T])\\ &= \lim_{N,C}\left(\overline{\lim_{M,C}} \circ [F'T]\right)\\ &= \lim_{N,C}\left(\overline{\lim_{M,C} F}\right)T\\ &= \lim_{N,C}\left(\overline{\lim_{M,C^L} F}\right)T.\end{aligned}$$

This proves that $\lim_{M,C^L} F$ is limit preserving.

Proof of 6.22.3. If $0 \to Y \to Y' \to Y'' \to 0$ is exact in L, and if F is pointwise left (right) exact, then

$$0 \to F(X)(Y) \to F(X)(Y') \to F(X)(Y'') \to 0,$$

and

$$0 \to F'(Y)(X) \to F'(Y')(X) \to F'(Y'')(X) \to 0$$

is left (right) exact, for every object X of M, and this proves that

$$0 \to F'(Y) \to F'(Y') \to F'(Y'') \to 0$$

is (pointwise) left (right) exact. Thus, F' is left (right) exact.

Proof of 6.22.4. Assuming that F is pointwise right exact and the first sequence in the Proof of 6.22.3 is right exact, then the last sequence is right exact by 6.22.3. Then the fact that colim is colimit preserving implies that

$$0 \to \operatorname{colim} F'(Y) \to \operatorname{colim} F'(Y') \to \operatorname{colim} F'(Y'') \to 0$$

is right exact. But $\operatorname{colim} F'(Y) = (\operatorname{colim} F)(Y)$, by the proposition, and so the last sequence shows that colim F is right exact. ▯

6.23 Corollary. *The representation functor*

$$H \begin{cases} C^{\mathrm{op}} \rightsquigarrow \mathrm{SETS}^{C} \\ A^* \mapsto h^A \end{cases}$$

is limit preserving, for any category C.

Proof. Under the canonical equivalence $(\mathrm{SETS}^{C})^{C^{\mathrm{op}}} \approx (\mathrm{SETS}^{C^{\mathrm{op}}})^{C}$, H maps onto a functor $H' : C \rightsquigarrow \mathrm{SETS}^{C^{\mathrm{op}}}$. If A and B are objects of C, then

$$H'(A)(B^*) = H(B^*)(A) = h^B(A) = h_A D(B^*),$$

where $D : C \rightsquigarrow C^{\mathrm{op}}$ is the canonical duality. Since h_A converts colimits into limits, then $H'(A) = h_A D$ is limit preserving, for every object A of C. Then by 6.22.1, H is limit preserving. ▯

A cocomplete abelian category C is said to have **exact direct limits** provided that the following two equivalent conditions are satisfied:

1. For every directed category L, the colimit functor $C^L \rightsquigarrow C$ is an exact functor.

2. For every exact sequence $0 \to F \to G$ in C^L, where L is directed, the canonical sequence $0 \to \varinjlim F \to \varinjlim G$ is exact in C.

6.24 Corollary. *Let L and C be exact categories. If C has exact direct limits, then any pointwise (left) exact functor $F : M \rightsquigarrow C^L$ from a directed set M has a (left) exact direct limit $\varinjlim F : L \rightsquigarrow C$.*

Proof. When C has exact direct limits, then the sequence in the proof of 6.22.4 is (left) exact whenever F is pointwise (left) exact. Then, since

$$\varinjlim F'(Y) = (\varinjlim F)(Y),$$

this proves that $\varinjlim F$ is (left) exact. □

6.25 Proposition. *If C is a category with exact direct limits, then so is C^L, for any category L, and so is* Adfunct(L, C) *for any additive category L.*

Proof. Let F' denote the image of any functor F under the canonical equivalence $(C^L)^M \approx (C^M)^L$. Then $0 \to F \to G$ exact in $(C^L)^M$ implies $0 \to F' \to G'$ exact in $(C^M)^L$. Thus,

$$0 \to F'(Y) \to G'(Y)$$

is exact in C^M for any object Y of L. Since C has exact direct limits, then

$$0 \to \varinjlim F'(Y) \to \varinjlim G'(Y)$$

is exact in C, for any object Y of L. Since $\varinjlim F'(Y) = (\varinjlim F)(Y)$ for any functor F, by 6.14, it follows that $0 \to \varinjlim F \to \varinjlim G$ is exact in C^L, proving that C^L has exact direct limits. If L is additive, then the proof shows that Adfunct(L, C) has exact direct limits if C has. □

For other results on categories with exact direct limits, see the exercises at the end of chapter. The subject is taken up again in Chapter 14.

6.26 *Exercise.* Any contravariant functor $F : L \rightsquigarrow$ SETS is a colimit of representable contravariant functors. Let L^0S denote the category of contravariant functors $L \rightsquigarrow S$, where $S =$ SETS, and, for an object F of L^0S, let L/F denote the category the objects of which are morphisms $a : h_A \to F$ in L^0F, where h_A denotes the representable functor defined by an object A of L. A morphism of L/F is defined to be a morphism $f : A \to B$ in L such that

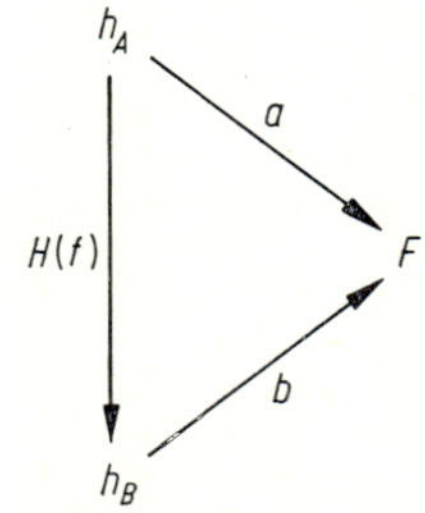

commutes, where $H(f): h_A \to h_B$ is canonical. Then

$$F = \operatorname{colim} d_F,$$

where $d_F: L/F \to L^0F$ is the functor which assigns A to an object $a: h_A \to F$ of L/F and assigns $H(f)$ to each morphism f of L/F.

6.27 *Exercise.* Let C be any cocomplete category, and let L be any category. Then any covariant functor $F: L^0S \rightsquigarrow C$ (from the category L^0S of contravariant functors $L \rightsquigarrow S$) has right adjoints if and only if F is colomit preserving. Furthermore, the functor

$$\begin{cases} C^{L^0S} \rightsquigarrow C^L \\ F \mapsto FH, \end{cases}$$

where $H: L \rightsquigarrow L^0S$ is the contravariant representation functor, induces an equivalence of the full subcategory of Funct(L^0S, C) on the colimit preserving functors and the category Funct(L, C). Thus, every functor category C^L is equivalent to the category of colimit preserving functors on a suitable functor category, assuming that C is cocomplete. (See Gabriel and Zisman [67, pp. 21—23], or Bucur and Deleanu [68, pp. 63 to 67].) [For any object A of L, $FH(A) = F(h_A)$.]

Exact Left Adjoints

6.28 **Proposition.** *Let (h, S, T, C, D) be an adjoint pair for abelian categories C and D.*

6.28.1 *If $S: D \rightsquigarrow C$ is exact, then T preserves injective objects.*

6.28.2 *If C has enough injectives* (e.g., *if C has an injective cogenerator*), *then, conversely, T preserves injective objects only if S is exact.*

Proof. If Y is any object of C, then there is a natural equivalence of contravariant set-valued functors

$$(1) \qquad h_Y S \approx h_{TY}.$$

Thus, S is exact, and Y is injective, that is, $h_Y: C \rightsquigarrow$ mod-$\mathbb{Z}$ exact implies that the group valued functor h_{TY} is exact. This proves 6.28.1. If C has an injective cogenerator Y, then (C has enough injectives) conversely, exactitude of h_{TY} implies that of S, since the faithful functor $h_Y: C \rightsquigarrow$ mod-$\mathbb{Z}$ preserves inexact sequences by 3.49.

Proof of 6.28.2 (General case). Since S is colimit, hence cokernel, preserving, it suffices to show that S is a monofunctor. Let $0 \to X \xrightarrow{k} Y$

be exact, and let $0 \to SX \xrightarrow{f} E$ be exact and E injective in C. Then, by hypothesis, TE is injective in D, and so the diagram

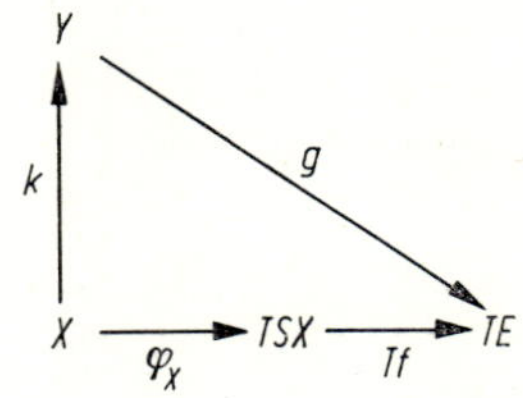

can be completed by a morphism $g: Y \to TE$. Then: $gk = T(f)\varphi_X$, and

(1) $$S(g)S(k) = S(Tf)S(\varphi_X).$$

Now suppose that $S(k)a = 0$, for some morphism a. Then $S(Tf)S(\varphi_X)a = 0$, and then commutativity of

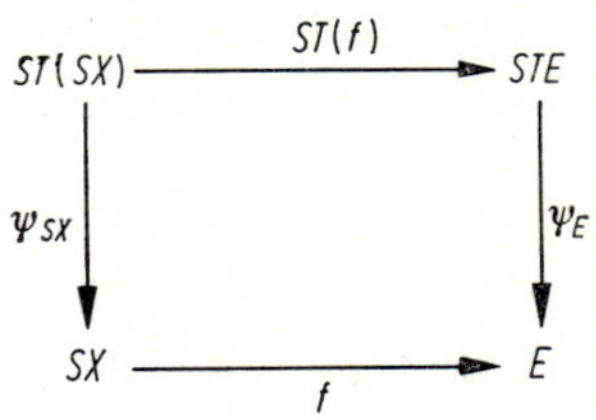

implies that $f\psi_{SX}S(\varphi_X)a = 0$. Since $\psi_{SX}S(\varphi_X) = 1_{SX}$ by the dual of (1) of 5.68, and since f is monic, then $a = 0$. This proves that $S(k)$ is monic. ▯

6.29 Corollary. *Let $f: R \to S$ be a homomorphism of rings, and let* mod-(f) : mod-$S \rightsquigarrow$ mod-R *be the induced functor* (cf. 3.4ff). *Then, every injective right S-module is injective as a canonical right R-module if and only if the canonical left R-module S is flat.* ▯

This holds, in particular, when either $S = R/A$ is the factor ring of R modulo an ideal A, or when R is a subring of S, and $f: R \to S$ is the inclusion.

Exercises for Chapter 6

1. A cocomplete abelian category C has exact direct limits if and only if the following condition is met:

For every object A of C, and directed family $\{A_i\}_{i\in I}$ of subobjects, the canonical morphism $\varinjlim_{i\in I} A_i \to A$ is monic, that is, the direct limit of a directed family of subobjects is a subobject. (Cf. 6.13.3.)

2. Show that mod-R, for any ring R, has exact direct limits.

3. If C is an abelian category with exact direct limits and a generator U, then any object A of C is the direct limit of the directed family of subobjects of A which are finitely generated with respect to U. (An object X is **finitely generated with respect to** U if X can be fitted into an exact sequence $U^n \to X \to 0$, with n an integer depending on X.)

4. If M is an object of mod-A, and if $M = \varinjlim M_i$, where $\{M_i\}_{i \in I}$ is a directed family of objects of mod-A, then there is a natural equivalence

$$\underset{A}{\otimes} M = \varinjlim \underset{A}{\otimes} M_i$$

of group valued functors. Furthermore, if M_i is flat for every $i \in I$, then M is flat, that is, a direct limit of flat modules is flat.

***5.** Let C be a small abelian category.

(a) Then any injective object of Funct$(C, \text{mod-}\mathbb{Z})$ is a right exact functor $F : C \rightsquigarrow \text{mod-}\mathbb{Z}$. If C is Noetherian, then conversely.

(b) Any essential extension in Funct$(C, \text{mod-}\mathbb{Z})$ of a monofunctor $F : C \rightsquigarrow \text{mod-}\mathbb{Z}$ is a monofunctor.

6. If C is a Noetherian small abelian category, then the representation functor $H : C^{\text{op}} \rightsquigarrow \text{Funct}(C, \text{mod-}\mathbb{Z})$ preserves injectives, hence has an exact left adjoint. [Hint: If P is projective in C, then $h^P = H(P^*)$ is injective in Funct$(C, \text{mod-}\mathbb{Z})$ by the previous exercise. Apply 6.28.]

7. (Freyd) A category B is equivalent to Adfunct$(C, \text{mod-}\mathbb{Z})$ for some small additive category C if and only if B is a cocomplete abelian category with family of generators consisting of small projective objects.

8. (Freyd) If C is a small additive category with kernels, then $T : C \rightsquigarrow \text{mod-}\mathbb{Z}$ is a small projective in Adfunct$(C, \text{mod-}\mathbb{Z})$ if and only if T is representable.

9. If C is a small abelian category, then every projective object of Adfunct$(C, \text{mod-}\mathbb{Z})$ is kernel preserving.

10. (Gabriel-Mitchell) A category C is equivalent to a module-type category iff C is an abelian category with coproducts and a small projective generator. (Cf. the Morita theorem 4.29, and also small objects 4.25.)

Notes for Chapter 6

Abelian categories were introduced by MacLane [50], Buchsbaum [55], and Grothendieck [57], to generalize the homological methods of Cartan-Eilenberg [56] (and others). However, the Freyd-Lubkin theorem

(14.21) showed that any small abelian category has an exact embedding into the category of abelian groups. Thus (to paraphrase Freyd), abelian groups lead to abelian categories, and abelian categories lead right back to abelian groups. (See Mitchell [65, Metatheorem 2.8, p. 101, and also 7.3, p. 151].) Moreover, any small abelian category has a full, exact embedding into a module-type category (Theorem of Mitchell, 15.30.) Finally, any abelian category with exact direct limits, and a generator is a quotient category (and a retract) of a module-type category (Theorem of Popesco-Gabriel 15.26; also see 5.70).

References

Buchsbaum [55], Bucur-Deleanu [68], Cartan-Eilenberg [56], Freyd [60, 62], Gabriel [62], Gabriel-Zisman [67], Grothendieck [57], Lubkin [60], Mac Lane [50], Mitchell [65], Puppe [67], Yoneda [53].

II. Structure of Noetherian Semiprime Rings

The first thing to be said about Part II is that it does not depend upon the whole of Part I. We append here a list of prerequisites (not to be taken too literally!), and recommend the rereading of the Introduction, esp. pp. XI—IV, for an overview. Also reread the first paragraph of Suggestions for Reading This Book (p. XXI).

Minimal Background

0. Foreword (Cartesian products, and injective, surjective mappings)
1. Chapter 1 (Categories)
2. Chapter 2 (Direct sum and product)
3. Chapter 3
   (a) Free module R^n (3.9)
   (b) Matrix rings R_n (3.16 and 3.17)
   (c) The canonical isomorphisms (3.38, 3.39, and 3.40)

$$\mathrm{Hom}_R(X, A \oplus B) \approx \mathrm{Hom}_R(X, A) \oplus \mathrm{Hom}_R(X, B)$$
$$\mathrm{Hom}_R(A \oplus B, X) \approx \mathrm{Hom}_R(A, X) \oplus \mathrm{Hom}_R(B, X)$$

for any three R-modules X, A, and B

(d) The canonical isomorphism $\mathrm{End}_R A^n \approx E_n$, where $E = \mathrm{End}_R A$, and where $A^n = A \oplus \cdots \oplus A$ (n-summands) (3.17)

(f) Schur's lemma [3.10.1 (a) and (b); cf. 3.14.3]

(g) Definition of projective module and characterization as a summand of a free module (3.11.2)

(h) The notion of essential extension (3.58)

(i) (Optional) Injective Hulls (3.59)

4. (Optional) Chapter 4
   (a) Correspondence theorem (4.7)
   (b) Morita theorem (4.29)

Introduction to Part II

If M is an R-module with endomorphism ring K, there is a ring homomorphism

$$\delta : R \to \mathrm{End}_K M$$

($\delta = \delta_M$), where

$$x\delta(a) = xa$$

$\forall x \in M,\ a \in R$. A **general Wedderburn theorem** is a statement that δ_M is an epimorphism. In this case, the module M is said to be **balanced.**

A ring R is **simple** in case R contains no ideals other than 0 and R. The Wedderburn-Artin theorem implies that if R is a simple ring with minimum condition on right ideals, then any minimal right ideal I of R is balanced (as an R-module). Together with Schur's lemma stating that the endomorphism ring R' of I is a field, the theorem implies that R is the endomorphism ring of a vector space. Since the minimum condition on R implies that M has finite dimension n over R', that is, that $R'^n \approx M$ (as left R'-modules), then there is a ring isomorphism $R \approx R'_n$.

An R-module M is a generator in case for each R-module N there is an index set J and an epic $M^{(J)} \to N$. (It is equivalent to say that $M^n \to R \to 0$ is exact for some n; that is, $M^n \approx R \oplus A$, for some n and R-module A.)

In 1958, Morita proved a general Wedderburn theorem: *Every generator M in* mod-R *is balanced.* Furthermore, a balanced module M is a generator if and only if M is finitely generated and projective over End M_R (7.3). The proof of this is elementary and requires very little background from Part I.

How does this apply to simple rings? *If R is simple, then every nonzero right ideal I of R is a generator hence balanced.* For RI is an ideal, therefore $RI = \sum_{a \in R} aI = R$, hence there exist $a_i \in R$, $i = 1, \ldots, n$, such that $\sum_{i=1}^{n} a_i I = R$ and, hence, a surjective mapping $I^n \to R$. This shows that general Wedderburn theorems are quite independent of the traditional chain conditions. The Wedderburn-Artin theorem is an immediate special case.

A ring satisfying the d.c.c. (a.c.c.) on right ideals is said to be right Artinian (Noetherian). Thus, simple right Artinian rings are isomorphic to matrix rings D_n over fields. The ring D_n is semisimple in the sense that every right module over it is a direct sum of simple modules. The full Wedderburn-Artin theorem (8.8) determines all semisimple rings as finite products of full matrix rings over fields, and these coincide with the right Artinian having no nilpotent ideals $\neq 0$.

The Wedderburn-Artin theorem is preceded in Chapter 7 by the corresponding result when R is a simple right Noetherian ring. In this case, a lemma states that I can be chosen so that $K = \text{End}\, I_R$ is an domain with right quotient field D, and R itself has a classical right quotient ring isomorphic to a matrix ring D_n over D for a unique integer $n > 0$. This latter result is a special case of a general theorem proved in Chapter 9, which states that a ring R has a classical right quotient ring Q that is a semisimple ring if and only if R has no nilpotent ideals $\neq 0$ and R satisfies the a.c.c. on annihilator and complement right ideals. Chapter 9 is devoted to the proof of this theorem. (Another proof is sketched in Chapter 16.)

A theorem in Chapter 10 determines all subrings R of a semilocal ring $Q = D_n$ such that Q is a classical right quotient ring of R. (Then R is said to be right order in Q.) The right orders of $Q = D_n$ are just the subrings containing K_n, where K is a right order in D (10.15 and 10.19). In case Q is a finite product of matrix rings over semilocal rings, then R contains a finite product of matrix rings over right orders in the semilocal rings.

Rings which are right orders in semilocal quotient rings D_n are characterized in Chapter 18 (e.g. 18.48), and maximal orders in the rings D_n are shown to be endomorphism rings of torsion free modules over right orders of D (10.29).

In Chapter 7, the Hilbert basis theorem states that the polynomial ring $R[x]$ (power series ring $R\langle x\rangle$) is right Noetherian if R is. The ring $A[y, D]$ of differential polynomials with respect to a derivation D of a ring A is also right Noetherian when A is (7.28). Moreover, if A is simple of characteristic 0, and if D is not an inner derivation, then $A[y, D]$ is also a simple ring. In case A is a field which is universal for the derivation D, then $A[y, D]$ is a right Noetherian V-ring in the sense that every simple module is injective (7.42). This ring has a simple injective cogenerator, but the ring itself is not semisimple. Furthermore, every submodule has a maximal submodule, and every cyclic module $\neq R$ is semisimple and injective. Every right V-ring having a semisimple classical right quotient ring is a finite product of simple right V-rings (7.36).

Over semisimple rings, every module is projective, and hence the global dimension of these rings is 0. The global dimension of the polynomial ring $R[x_1, \ldots, x_n]$ over a ring R is equal to n plus the global dimension of R. If R is semisimple, then $R[x_1, \ldots, x_n]$ has global dimension n (Hilbert Szyzgy Theorem 8.16). The Auslander theorems (8.19 to 8.21) assert

$$\text{r.gl.dim}\, R = \sup_{I \subseteq R}\{\text{proj. dim}\, R/I\},$$

and when R is not semisimple

$$\text{r.gl.dim}\, R = 1 + \sup_{I \subseteq R}\{\text{proj. dim}\, I\}.$$

Thus, R has right global dimension 1 if and only if every right ideal is projective. Simple Noetherian rings of global dimension not exceeding 2 are similar to Ore domains (8.27).

Chapter 7

General Wedderburn Theorems

The study of endomorphism rings is important both in the study of modules and in the study of abstract rings. It is important in the study of modules because many properties of a module are characterizable by properties of its endomorphism ring. Indeed, some modules over certain rings are determined by their endomorphism rings, which is a way of saying that the modules are isomorphic if and only if their endomorphism rings are isomorphic *qua* rings. This is true for vector spaces. It is important for the study of rings, since some rings, even those abstractly defined, can be embedded in a ring of endomorphisms of a relatively "nice" module, for example, a vector space.

A beautiful kind of theorem is one which states that a given algebraic structure, abstractly presented, is isomorphic to a known algebraic structure. For example, the Wedderburn-Artin theorem proved in Chapter 8 states that a ring R is isomorphic to the endomorphism ring of a vector space V of finite dimension over a field D if and only if the lattice of right ideals of R satisfies the minimum condition and the lattice of ideals of R consists of two elements. Here the information given about the ring R is extremely abstract, and yet the conclusion is contrastingly concrete. Moreover, the proof of the theorem constructs the field D and the vector space V out of parts of the ring R. To be more precise, there exists an idempotent $e \in R$ such that the ring eRe is isomorphic to the field D, and the right ideal eR is a left vector space over eRe isomorphic to V.

Balanced Modules

Let R be a ring, and let M be an R-module. We shall adopt the following notation:

$$R' = \mathscr{R}'(M) = \text{End } M_R$$
$$R'' = \mathscr{R}''(M) = \text{End}_{R'} M.$$

R'' is the **biendomorphism ring of** M, abbreviated Biend M_R.

A **homothetic** of M is a mapping $r_d : M \to M$ that for a fixed $r \in R$ assigns $x \in M$ the value xr. The subscript "d" is an abbreviation for the Latin word *dexter*; the notation r_d is standard Bourbaki, and r_d is also called "right multiplication by r". If N is a left R-module, r_s (s for the Latin *sinistra*) denotes the corresponding homothetic. The map

$R \to \text{Biend}\, M_R$ that sends $r \mapsto r_d$ is called **canonical** and is a ring homomorphism. The module M is said to be **balanced** in case the canonical map $R \to \text{Biend}\, M_R$ is an epimorphism, and **faithful** in case it is a monomorphism. The value of balanced faithful modules for representing R as the endomorphism ring of another, possibly simpler, module is clear. In this chapter, we explore this possibility in depth for simple rings.

We next provide an elementary introduction to the concept of a generator in mod-R, which does not require much from Part I. (The *cognoscenti* might skip to 7.1).

By 3.26 (or by definition in the category mod-R), a generator in mod-R is an R-module G such that every object M in mod-R is an epimorph of copies of G; that is, there exist a set I and an epimorphism $G^{(I)} \to M$. Obviously, R is a generator in mod-R. To obtain others, note that if $H \to G$ is epic, and if G is a generator, then so is H. Hence, $R \oplus A$ is a generator for any object A in mod-R, and so every free R-module is a generator.

7.1 Theorem. *Any generator in* mod-R *is balanced and faithful.*

Proof. First, consider a generator $G = R \oplus A$. If $x \in R \oplus A$, since 1 is a free basis of R, there exists $f \in \mathscr{R}'(R \oplus A)$ such that $f(1) = x$. (We view R and A embedded in $R \oplus A$ as $R \oplus 0$ and $0 \oplus A$, respectively.) Let $p : R \oplus A \to R$ be the projection, and let $t \in \mathscr{R}''(R \oplus A)$. Then

$$(1)t = (p1)t = p(1t) \in R.$$

It follows that

$$(x)t = (f1)t = f(1t) = f(1 \cdot 1t) = f(1)1t = x(1t).$$

Thus $R \to \mathscr{R}''(R \oplus A)$ is epic, which is what we wanted.

Now let G be any generator. There exist a natural number n and an isomorphism $G^n \approx R \oplus A$, for some $A \in$ mod-R. Since $R \oplus A$ is balanced, so is G^n. Moreover, since R is faithful, so is G^n, whence G. The following lemma completes the proof.

7.2 Lemma. *If G is any module, and if G^n is balanced, then G is balanced. Thus, for any module M and integer $n > 0$, there is a canonical isomorphism*

$$\text{Biend}\, M_R = \text{Biend}\, M_R^n.$$

Proof. By 3.24, $\text{End}\, G_R^n \approx R'_n$, where $R' = \mathscr{R}'(G) = \text{End}\, G_R$, so that any element $f \in \mathscr{R}'(G^n)$ can be represented by a matrix $(s_{ij}) \in R'_n$; namely, for $v = (a_1, \ldots, a_n) \in G^n$, $a_i \in G$, $i = 1, \ldots, n$, we have

$$f(v) = \begin{pmatrix} s_{11} & s_{n1} \\ \vdots & \vdots \\ s_{n1} & s_{nn} \end{pmatrix} \begin{pmatrix} a_1 \\ \vdots \\ a_n \end{pmatrix}.$$

Let $t \in \mathscr{R}''(G)$. Then the mapping

$$v = (a_1, \ldots, a_n) \mapsto (a_1 t, \ldots, a_n t),$$

denoted by $\bar{t}$, satisfies $f(v)\bar{t} = f(v\bar{t})$, and therefore $\bar{t} \in \mathscr{R}''(G^n)$. Since G^n is balanced, there exists $r \in R$ such that $v\bar{t} = vr \forall v \in G^n$. Taking $v = (a, \ldots, a)$ for arbitrary $a \in G$ yields $v\bar{t} = (at, \ldots, at) = (ar, \ldots, ar)$; that is, $at = ar$. This shows that G is balanced. ▯

7.3 Morita Proposition. *Let A be a ring. An object P is a generator in* mod-A *if and only if P is balanced in* mod-A *and finitely generated projective as a canonical module of* B-mod, *where* $B = \text{End}\, P_A$ (cf. 4.2).

Proof. By 3.26, P is a generator in mod-A if and only if $P^n \approx A \oplus X$, for some integer $n > 0$ and module X of mod-A. Since the functor $h^P : \text{mod-}A \rightsquigarrow \text{mod-}B$ preserves finite (bi) products, then (3.38ff.)

$$(1) \qquad B^n \approx \text{Hom}_A(P, P)^n \approx \text{Hom}_A(P^n, P).$$

Since the functor $h_P : \text{mod-}A \rightsquigarrow B\text{-mod}$ reverses finite products and coproducts, then

$$(2) \quad \text{Hom}_A(P^n, P) \approx \text{Hom}_A(A \oplus X, P) \approx \text{Hom}_A(A, P) \oplus \text{Hom}_A(X, P).$$

Using the canonical (B, A)-isomorphism $P \approx \text{Hom}_A(A, P)$ and setting $Y = \text{Hom}_A(X, P)$ in B-mod, then (1) and (2) yield

$$(3) \qquad B^n \approx P \oplus Y$$

in B-mod, that is, P is finitely generated and projective. The converse is obtained by reversing the steps, applying the functor h_P to (3), and using the canonical isomorphism $A \approx h_P(P)$ derived from the assumption that P is balanced in mod-A. ▯

7.4 *Exercises.*

7.4.1 Let R be a simple ring. The following statements about a right R-module M are equivalent:

(a) M is a generator in mod-R.

(b) There exists a nonzero map of M into a direct product of copies of R.

(c) $\text{Hom}_R(M, R) \neq 0$.

7.4.2 For any ring R and any R-module A, let $B = R \oplus A$, and let $S = \text{End}\, B_R$. Then ${}_S B$ is cyclic projective module, hence there exists an idempotent $e \in S$ such that ${}_S B \approx Se$ and $eSe \approx R$.

7.4.3 If R is a simple ring, and if $e \in R$ is a nonzero idempotent, then eRe is a simple ring.

7.4.4 If R is a simple ring, so is R_n for any integer $n \geq 1$.

7.4.5 A ring R is simple if and only if every nonzero right R-module is faithful.

7.4.6 If R is simple, then center R is a field.

*7.4.7 Tensor products of vector spaces and modules are discussed in Chapter 11. If A and B are simple rings, and if $K =$ center A is a subring of center B, then the tensor product $A \otimes_K B$ is a simple ring.

7.5 Theorem. *If R is a simple ring, if I is any nonzero right ideal, and if $K = \text{End}\, I_R$, then I is a finitely generated projective left K-module, and the canonical map $R \to \text{End}_K I$ is an isomorphism. The ring K is simple if and only if I is a finitely generated projective right R-module; in this case the K-module-valued functor* $\text{Hom}_R(I, \ )$ *is a category equivalence* mod-$R \approx$ mod-K.

Proof. We already know that I_R is a generator and that canonically $R \approx \text{End}_K I$. If I is finitely generated projective, for example, $F = I \oplus Q \approx R^n$, then there is an idempotent e in $R_n \approx \text{End}\, F_R$ such that $K \approx eR_ne$. By the preceding exercises, R_n, eR_ne, and hence, K are simple rings.

Conversely, suppose K is simple, and let I' denote the set of left multiplications by elements of I. Thus, if $x \in I$, then $x' \in I'$ is the homothetic: $a \mapsto xa \forall a \in I$. If $k \in K$, then $(kx)' = kx'$, thus I' is a left ideal of K, and the map $x \mapsto x'$ is a left K-module epimorphism $I \to I'$. Since K is simple, the left ideal I' is a generator of K-mod, and hence I is a generator of K-mod. By 7.3, I is a finite projective module over $R \approx \text{End}_K I$. In this case, $\text{Hom}_R(I, \ )$ is a category equivalence by Morita's theorem 4.29. □

***7.6** *Exercise* (Morita). Two rings A and B are similar if and only if there exists an integer $n > 0$, an idempotent e in the matrix ring A_n such that $A_n e A_n = A_n$, and a ring isomorphism $B \approx eA_ne$. When this is so, there is a progenerator $P \approx eA^n$, where $A_n \approx \text{End}\, A^n_A$ canonically, and $B \approx \text{End}\, P_A$.

Minimal Right Ideals

A **maximal** submodule of a nonzero module M is one which is maximal in the set of submodules $\neq M$. A **minimal** submodule is one that is minimal in the set of submodules $\neq 0$. A module M is **simple** in case $M \neq 0$ and M satisfies one of the equivalent conditions:

(a) 0 is a maximal submodule.

(b) M is a minimal submodule.

(c) Every nonzero epimorphism $M \to N$ is an isomorphism. Thus, if N is a submodule of M, then N is maximal if and only if M/N is simple.

A **maximal** (resp. minimal) right ideal I is one that is a maximal (resp. minimal) submodule of R_R. A **simple** right ideal is just a minimal right ideal.

We next prove a preliminary version of the Wedderburn-Artin theorem, which follows in 7.9.

7.7 Theorem. *Let R be a simple ring containing a simple right ideal I. Then $K = \text{End}\, I_R$ is a field, ${}_KI$ is a finite dimensional vector space, and $R \approx \text{End}_K I$. Thus, if $\dim_K I = n$, then $R \approx K_n$, the ring of all $n \times n$ matrices over K. Furthermore, there exists an idempotent $e \in R$ such that $I = eR$, and then eRe is a field $\approx K$.*

Proof. Since I_R is simple, by Schur's lemma, $K = \text{End}\, I_R$ is a field. Since R is simple, I_R is a generator, and since K is a field, ${}_KI$ is a generator. Hence, by the last theorem, I_R (resp. ${}_KI$) is finitely generated projective. Since ${}_KI$ is a vector space, and finitely generated, ${}_KI$ has a finite basis consisting of, for instance, n elements. Thus, $I \approx K^n$ as left K-modules. Since I_R is balanced, we obtain

$$R \approx \text{End}_K I \approx \text{End}_K K^n \approx K_n.$$

Since every generator is faithful, $I^2 \neq 0$, so there exist $x, y \in I$ such that $xy \neq 0$. Since I is a simple R-module, we have $I = xR$ and $I = yR$. Now I is projective, so the map $\varphi : R \to xR$, where $\varphi(a) = xa \;\forall\, a \in R$, splits 3.11.2. Write $R = \ker\varphi \oplus Q$, where Q is a right ideal isomorphic to I. This shows that $\ker\varphi$ is a maximal right ideal of R. Since $xy \neq 0$, then $xI = xyR \neq 0$, so $I \cap \ker\varphi = 0$. Since $\ker\varphi$ is a maximal right ideal, then R is the sum $\ker\varphi \oplus I$, and that sum is direct; that is, I is a summand of R, so $I = eR$, with $e = e^2 \in R$. Now for any idempotent f in any ring S, we know that $\eta : fSf \approx \text{End}(fS)_S$, where if $x \in fSf$, $\eta(x)$ is the endomorphism of fS that maps $y \in fS$ onto xy. This completes the proof. ▯

Schur's lemma 3.10 states that if V is a simple R-module, then $R' = \text{End}\, V_R$ is a field; that is, V is a left vector space over R'. If V is balanced, and faithful, then $R \approx \text{End}_{R'} V$.

7.8 Lemma. *A ring R has a balanced simple faithful right R-module if and only if R is isomorphic to a complete ring of* l.t.'s *in a left vector space.*

Proof. The necessity is trivial as just noted. Conversely, if $\varphi : R \approx \text{End}_K W$, where ${}_KW$ is a vector space, then W is naturally a right R-module and is in fact simple. Let U be any nonzero R-submodule, and let x be a nonzero element of U. If $y \in W$, then there is an element $t \in \text{End}_K W$ such that $(x)t = y$. [For example, since $W = Kx \oplus Q$, for some vector subspace Q, we can choose t such that $(kx)t = ky$ $\forall k \in K$ and $(q)t = 0 \forall q \in Q$.] If $r = \varphi^{-1}(t)$, then $y = xr \in U$, so $W \subseteq U$ and $W = U$. Therefore W_R is simple, balanced, and faithful. □

Artinian Rings and Modules

The set of submodules of an R-module M is ordered by inclusion. M is said to be **Artinian** in case M satisfies either of the following equivalent conditions (see the foreword):

(1) *The ordered set of submodules of M satisfies the minimum condition.*

(2) *The ordered set of submodules satisfies the d.c.c.*

Thus, if M is Artinian, then (1) *every nonempty collection $\{M_i\}$ of submodules contains a submodule minimal in that collection*; *and* (2) *every descending chain*

$$M_1 \supseteq M_2 \supseteq \cdots \supseteq M_n \supseteq \cdots$$

of submodules is ultimately stationary; that is, $M_n = M_{n+1} = \cdots$ *for some n.*

A ring R is (right) **Artinian** in case R_R is an Artinian module. The ring R is left Artinian if ${}_R R$ is Artinian. Artinian rings were first systematically studied by Emil Artin.

7.9 The Principal Wedderburn-Artin Theorem. *The following conditions on a ring S are equivalent:*

(a) *S is simple right Artinian.*

(b) *S is a simple ring containing a simple right ideal.*

(c) $S \approx \text{End}_K W$, *where ${}_K W$ is a left vector space of finite dimension n.*

(d) $S \approx D_n$, *where D is a field and n is an integer greater than 1.*

(e) *The right-left symmetry of* (a) to (d). *In this case, if I is any simple right ideal, every simple S-module is isomorphic to I, hence any two simple right ideals are isomorphic. If K and D are fields, and if $S \approx K_n$ and $S \approx D_m$, then $n = m$, and $K \approx D$. Finally $S \approx D_n$ is a direct sum of n simple right ideals.*

Proof. (a) ⇒ (b) is trivial. (b) ⇒ (c) follows from 7.7.

(c) ⇒ (d). If $n = \dim_K W$, then ${}_K W \approx {}_K K^n$, so

$$S \approx \text{End}_K W \approx \text{End}_K K^n = K_n.$$

(d) ⇒ (a). Let $S \approx D_n$, where D is a field. Now D is a simple ring, and a previous exercise established that D_n is a simple ring. To check, let I be an ideal containing a nonzero element $x = \sum_{i,j=1}^{n} d_{ij} e_{ij}$, where

$\mathcal{M} = \{e_{ij} \mid i, j = 1, \ldots, n\}$ are matrix units and D is the centralizer of $\mathcal{M}$ in D_n. Then $d_0 = d_{i_0 j_0} \neq 0$ for some i_0, j_0, and then

$$d_0^{-1} e_{i_0 j_0} x e_{j_0 j_0} = e_{i_0 j_0} \in I,$$

hence

$$a_{ij} e_{ii_0} e_{i_0 j_0} e_{j_0 j} = a_{ij} e_{ij} \in I$$

$\forall a_{ij} \in D$. Thus, $I \supseteq \sum_{i,j=1}^{n} D e_{ij} = S$, so S is simple. (A similar proof shows that $e_{ii}S$ is a simple right ideal, $i = 1, \ldots, n$.) Now D_n is a right (also left) vector space of dimension n^2 over D (the n^2 matrix units $\{e_{ij} \mid i, j = 1, \ldots, n\}$ are a basis), and the right ideals of D_n are D-subspaces. Since a finite dimensional vector space satisfies the minimum condition on vector subspaces, D_n satisfies the minimum condition on right ideals. Hence, so does S.

(e) is true by virtue of the symmetry in the ring D_n, where D is a field.

Finally, suppose $S \approx D_m$ and $S \approx K_n$, where D and K are both fields. Letting $W = D^m$ and $V = K^n$, the ring isomorphisms

$$S \approx D_m \approx \operatorname{End}_D W$$

$$S \approx K_n \approx \operatorname{End}_K V$$

imply that W and V are natural right S-modules, and, by the lemma preceding the theorem, W_S and V_S are simple S-modules. Let I be any simple right ideal of S, which exists by the equivalence of (a) to (d). Over a simple ring, *any* nonzero module M is faithful because $\operatorname{ann}_S M$ is an ideal; in particular, W is faithful, so $WI \neq 0$. Pick $w \in W$ so that $wI \neq 0$. Simplicity of W_S forces $wI = W$. Hence there is an epic $f: I \to W$, where $f(a) = wa \; \forall a \in I$. Since I is simple, $\ker f = 0$, so $I \approx W$. Hence, every simple module W_S is isomorphic to any simple right ideal I. Furthermore, $I_S \approx W_S$ implies

$$Q = \operatorname{End} I_S \approx D$$

and

$$\dim_Q I = m = \dim_D W.$$

This result also implies that $Q \approx K$, whence $D \approx K$, and that

$$m = \dim_Q I = \dim_K V = n.$$

Now $A = D_n$ is a direct sum of the n-simple right ideals

$$D_n = A = e_{11} A \oplus \cdots \oplus e_{nn} A.$$

(The sum is direct because $1 = \sum_{i=1}^{n} e_{ii}$ is a sum of the orthogonal idempotents $\{e_{ii} \mid i = 1, \ldots, n\}$.) Hence the same is true of $S \approx D_n$. $\square$

7.9′ *Exercises.*

7.9′.1 If $S \approx D_n$, where D is a field, and if $I_1, \ldots, I_m$ are simple right ideals of S such that $S = I_1 \oplus \cdots \oplus I_m$, then $m = n$, $I_j \approx I_1$, $j = 1, \ldots, n$, and $\text{End}\, I_R \approx D$, for any simple right ideal I.

7.9′.2 If a ring R is a sum of simple right ideals, then every right R-module is a sum of simple submodules.

7.9′.3 If $R = D_n$, where D is a field, and if e is a nonzero idempotent of R, then the ring eRe is a simple Artinian ring isomorphic to D_m.

7.9′.4 Prove the equivalence for a ring R of the following conditions:

(a) Every object in mod-R is projective.
(b) Every quotient object of R is projective.
(c) Every right ideal of R is a summand of R.
(d) Every object in mod-R is injective.
(e) Every right ideal of R is injective.
(f) Every object in mod-R is a (direct) sum of simple modules.

7.9′.5 If R and S are similar, then R satisfies the conditions of Exercise 7.9′.4 if and only if S does. This is true for $S = R_n$.

7.9′.6 Use 7.9 to prove the full Wedderburn-Artin theorem (8.8) which can be expressed as follows: If R is a right Artinian ring, and if every nonzero right ideal of R is non-nilpotent, then R is a sum of simple right ideals.

7.9′.7 Prove 7.9 as a consequence of the Morita theorem.

7.9′.8 If V is a vector space over a field D, and if I is an ideal of $\text{End}_D\, V$, then there is an ordinal number a such that $\aleph_a \leq \dim_D V$ and such that $I = \{f \in \text{End}_D\, V \mid \dim f(V) \leq \aleph_a\}$. Conclude that the ideals of any full linear ring are well ordered.

Noetherian Rings and Modules

Before citing other special cases of the general Wedderburn theorem, we introduce rings and modules first systematically studied by the mathematician Emmy Noether.

Dual to the notion of minimum condition is the maximum condition. A module M_R is **Noetherian** provided that the set of submodules of M ordered by inclusion satisfies either of the following three equivalent conditions:

(a) *maximum condition;*
(b) *a.c.c.;*
(c) *every submodule is finitely generated.*

A ring R is (right) **Noetherian** in case R_R is a Noetherian module.

The following is the proof of the equivalence. (a) $\Leftrightarrow$ (b) already has eben established (see the Foreword Proposition 12).

(a) $\Rightarrow$ (c). Let S be a submodule, and let F denote the collection of all finitely generated submodules of S. By (b), F contains a maximal element A. If x is an arbitrary element of S, then the submodule (A, x) generated by A and x is also finitely generated, so it belongs to F. Since $(A, x) \supseteq A$, we conclude that $(A, x) = A$, and then $x \in A$. This shows that $A = S$, so S is finitely generated.

(c) $\Rightarrow$ (b). Let $S_1 \subseteq \cdots \subseteq S_n \subseteq \cdots$ be a chain of submodules of M_R, and let $S = \bigcup_{i=1}^{\infty} S_n$. By the assumption (c), $S = \sum_{i=1}^{n} x_i R$, with $x_1, \ldots, x_n \in M$. By the definition of S, there exists k such that $x_1, \ldots, x_n \in S_k$. Then $S = (x_1, \ldots, x_n) \subseteq S_k \subseteq S$; that is, $S_k = S_{k+1} = \cdots = S_{k+j}, \forall j$. Thus, M satisfies the a.c.c. $\square$

A Lemma on Artinian-Noetherian Modules

The following lemma shows that the class of Noetherian modules is closed under the operations of taking epimorphic images, submodules, and finite direct sums.

7.10 Lemma.

7.10.1 *Let M be a module, and let N be a submodule. Then M is Noetherian (Artinian) if and only if both N and M/N are Noetherian (Artinian).*

7.10.2 *A finite coproduct $\coprod_{i \in n} M_i$ is Noetherian (Artinian) if and only if M_i is Noetherian (Artinian), $i = 1, \ldots, n$.*

Proof of 7.10.1. If M is Noetherian (Artinian), then any submodule N of M satisfies the same chain condition, and so does M/N because of the homomorphism theorem (1.9). Conversely, assume both M/N and N are Noetherian (Artinian), and let $\{A_n\}$ be an ascending (descending) sequence of submodules. There then exists an integer t_1 such that

$$(A_{t_1} + N)/N = (A_n + N)/N$$

for $n \geq t_1$ and an integer $t > t_1$ such that

$$A_t \cap N = A_n \cap N$$

for $n \geq t > t_1$. Let $n \geq t$, and let $x \in A_n$ (in case $\{A_n\}$ is ascending). Then, since $A_n + N = A_t + N$, we can write $x = a + b$, with $a \in A_t$, $b \in N$, and then $x - a = b \in N \cap A_n = N \cap A_t$. Thus, $x = b + a \in A_t$, so $A_n = A_t$ $\forall n \geq t$ (and similarly when $\{A_n\}$ is descending).

Proof of 7.10.2. This follows by induction on n and an application of 7.10.1 for the case $N = M_1 \oplus \cdots \oplus M_{n-1} \subseteq M$. $\square$

7.10′ *Exercises.*

7.10′.1 Recall the definition: A module M is **indecomposable** in case 0 and M are the only submodules that are summands of M. Prove that if M is any Noetherian or Artinian module, then M is a direct sum of finitely many indecomposable submodules.

7.10′.2 Let N be a submodule of a module M. Then N is **$\cap$-irreducible** (read **meet-irreducible**) in case N has the following property: If A and B are submodules of M, and if $N = A \cap B$, then either $A = N$ or $B = N$. Prove that if M is Noetherian, then each submodule of M is the intersection of finitely many $\cap$-irreducible submodules of M.

7.10′.3 Let R be a ring, and let n be a natural number. The ring R_n of all $n \times n$-matrices over R is right Noetherian (resp. Artinian) if and only if R is right Noetherian (resp. Artinian). (This can be done in at least two ways. One way is to describe the condition categorically and apply 3.27.)

7.10′.4 If R is a principal ideal entire commutative ring, and if I is an ideal $\neq 0$, then R/I is Artinian.

7.10′.5 If M_R is simple, then End M_R is a field. Does the converse hold?

7.10′.6 If R is a right Noetherian (resp. Artinian) ring, and if S is a ring epimorph of R, then S is right Noetherian (resp. Artinian).

*7.10′.7 Describe $\mathrm{End}_{\mathbb{Z}}\mathbb{Z}_{p^\infty}$. (This ring is isomorphic to the ring of p-adic integers.)

7.10′.8 Let R be a commutative principal ideal entire ring. Show that if I is any proper ideal of R, then $\bigcap_{n=1}^{\infty} I^n = 0$. (This holds more generally for commutative Noetherian entire rings. See Krull's theorem, Chapter 27.)

7.10′.9 Any subring A of a right Noetherian (Artinian) ring R satisfies the a.c.c. (d.c.c.) on right annulets.

7.10′.10 Any epimorphism $M \to M$ of a Noetherian module is an automorphism. Dually, any monomorphism $M \to M$ of an Artinian module is an automorphism.

7.10′.11 If R is either right Noetherian, or right Artinian, then for any two positive integers n and m, and isomorphism $R^n \approx R^m$ in mod-R implies $n = m$. Conclude that any two free bases of a free module R^n have the same cardinality.

Making Modules into Ideals

There is a way, which Nagata [62] calls the **principle of idealization**, to embed a given (R, S)-bimodule M into a ring A so that M becomes an ideal of A. To do this, let

$$A = \begin{pmatrix} R & M \\ 0 & S \end{pmatrix}$$

denote the set of all matrices

$$\begin{pmatrix} r & x \\ 0 & s \end{pmatrix}$$

$r \in R,\ x \in M,\ s \in S$. In A, define addition and multiplication as in ordinary matrices, e.g.,

$$\begin{pmatrix} r & x \\ 0 & s \end{pmatrix}\begin{pmatrix} r' & x' \\ 0 & s' \end{pmatrix} = \begin{pmatrix} rr' & rx' + xs' \\ 0 & ss' \end{pmatrix}.$$

Then A is a ring. For various subsets $U \subseteq R,\ V \subseteq M,\ W \subseteq S$, we let

$$\begin{pmatrix} U & V \\ 0 & W \end{pmatrix} = \left\{\begin{pmatrix} u & v \\ 0 & w \end{pmatrix} \in A \mid u \in U,\ v \in V,\ w \in W\right\}.$$

7.11 Principle of Idealization. *Let M be an (R, S)-bimodule, let*

$$A = \begin{pmatrix} R & M \\ 0 & S \end{pmatrix}$$

be the ring just constructed, and consider the mappings

$$\psi : M \to A,\ \psi : m \mapsto \begin{pmatrix} 0 & m \\ 0 & 0 \end{pmatrix},$$

$$\varphi : R \to A,\ \varphi : r \mapsto \begin{pmatrix} r & 0 \\ 0 & 0 \end{pmatrix},$$

$$\eta : S \to A,\ \eta : s \mapsto \begin{pmatrix} 0 & 0 \\ 0 & s \end{pmatrix}.$$

Then

7.11.1 *$\psi(M)$ is an ideal of A, and $\psi(M)^2 = 0$.*

7.11.2 *The mappings $\varphi : R \to A$ and $\eta : S \to A$ are ring monics*[1] *satisfying*

(a) $$\psi(rms) = \varphi(r)\,\psi(m)\,\eta(s),$$

$r \in R,\ m \in M,\ s \in S$;

(b) *the (group) direct sum map*

$$\varphi \oplus \eta \oplus \psi : R \oplus S \oplus M \to A$$

is an isomorphism;

(c) $$\varphi \oplus \eta : R \oplus S \to A$$

is a ring monic.

[1] In this instance, we are not requiring that the images contain the identity element of A.

Conversely, if A is any ring such that maps ψ, φ, and η exist satisfying the requirements 7.11.1 and 7.11.2, then there is a ring isomorphism

$$\begin{cases} A \to \begin{pmatrix} R & M \\ 0 & S \end{pmatrix} \\ \varphi(r) + \eta(s) + \psi(m) \mapsto \begin{pmatrix} r & m \\ 0 & s \end{pmatrix}. \end{cases}$$

If $f: R \to S$ is a ring homomorphism, then the subset of

$$A = \begin{pmatrix} R & M \\ 0 & S \end{pmatrix}$$

consisting of all matrices

$$\begin{pmatrix} r & m \\ 0 & f(r) \end{pmatrix},$$

where $r \in R$, $m \in M$, is a subring denoted by (A, f).

The proof is an exercise.

For subsets $U, X \subseteq R$, $V, Y \subseteq M$, $W, Z \subseteq S$, we have

$$\begin{pmatrix} U & V \\ 0 & W \end{pmatrix} \begin{pmatrix} R & M \\ 0 & S \end{pmatrix} = \begin{pmatrix} UR & UM + VS \\ 0 & WS \end{pmatrix}$$

and

$$\begin{pmatrix} R & M \\ 0 & S \end{pmatrix} \begin{pmatrix} X & Y \\ 0 & Z \end{pmatrix} = \begin{pmatrix} RX & RY + MZ \\ 0 & SZ \end{pmatrix}.$$

Thus, for a right ideal U of R, an S-submodule V of M and right ideal W of S, the following are the right ideals of A:

(1) $$\begin{pmatrix} U & UM + V \\ 0 & W \end{pmatrix}$$

including the special cases:

$$\begin{pmatrix} 0 & 0 \\ 0 & W \end{pmatrix}, \quad \begin{pmatrix} U & M \\ 0 & 0 \end{pmatrix}, \quad \begin{pmatrix} 0 & V \\ 0 & W \end{pmatrix}, \quad \begin{pmatrix} 0 & V \\ 0 & 0 \end{pmatrix}.$$

For a left ideal X of R, R-submodule Y of M, and left ideal Z of S, the left ideals of A are

(2) $$\begin{pmatrix} X & Y + MZ \\ 0 & Z \end{pmatrix}$$

including the special cases:

$$\begin{pmatrix} X & 0 \\ 0 & 0 \end{pmatrix}, \begin{pmatrix} 0 & Y \\ 0 & 0 \end{pmatrix}, \begin{pmatrix} X & Y \\ 0 & 0 \end{pmatrix}, \begin{pmatrix} 0 & M \\ 0 & Z \end{pmatrix}, \begin{pmatrix} X & M \\ 0 & Z \end{pmatrix}.$$

Hence, for an ideal U of R, an ideal W of S, and an (R, S)-submodule V of M, the ideals of A are

$$\begin{pmatrix} U & UM + V + MW \\ 0 & W \end{pmatrix} \tag{3}$$

including the special cases:

$$\begin{pmatrix} U & M \\ 0 & W \end{pmatrix}, \begin{pmatrix} 0 & V \\ 0 & 0 \end{pmatrix}, \begin{pmatrix} 0 & V \\ 0 & W \end{pmatrix}.$$

The idealization of bimodules is a useful tool in the construction of asymmetric properties of rings.

7.11′ *Examples of Non-Noetherian Rings.*

7.11′.1 *A ring A that has three left ideals and that is neither right Artinian nor Noetherian.*

Let R be a commutative field, and let $f: R \to S$ be a ring epimorphism, where S is a subfield of R such that $\dim R_S = \infty$. For example, let $R = \mathbb{Q}(x_1, \ldots, x_n, \ldots)$, the rational function field in infinitely many indeterminates; let $S = \mathbb{Q}(x_1^2, x_2^2, \ldots, x_n^2, \ldots)$; and let $f(x_i) = x_i^2$, $f(a) = a \, \forall a \in \mathbb{Q}, \forall i$. Then R is an (R, S)-bimodule, and the idealization of R, namely,

$$A = \begin{pmatrix} R & R \\ 0 & S \end{pmatrix},$$

has the property, readily deducible from the chart of ideals just listed, that

$$\begin{pmatrix} 0 & V \\ 0 & 0 \end{pmatrix}$$

is a right ideal of A, for any S-subspace V of R. Since $\dim_S R = \infty$, A is neither right Artinian nor right Noetherian; however, A has only six left ideals:

$$0 = \begin{pmatrix} 0 & 0 \\ 0 & 0 \end{pmatrix}, \begin{pmatrix} R & 0 \\ 0 & 0 \end{pmatrix}, \begin{pmatrix} 0 & R \\ 0 & 0 \end{pmatrix}, \begin{pmatrix} R & R \\ 0 & 0 \end{pmatrix}, \begin{pmatrix} 0 & R \\ 0 & S \end{pmatrix}, \begin{pmatrix} R & R \\ 0 & S \end{pmatrix} = A.$$

This follows from the general form (2) for left ideals,

$$\begin{pmatrix} \alpha R & \beta R \\ 0 & \gamma S \end{pmatrix}$$

$\alpha, \beta, \gamma = 0$ or 1. If we consider the subring (A, f) constructed via the homomorphism $f: R \to S$ (just described), then we cut down to just three left ideals

$$0 = \begin{pmatrix} 0 & 0 \\ 0 & 0 \end{pmatrix}, \quad \begin{pmatrix} 0 & R \\ 0 & 0 \end{pmatrix}, \quad \text{and } (A, f),$$

whereas the right ideals

$$\begin{pmatrix} 0 & V \\ 0 & 0 \end{pmatrix}$$

of A are also right ideals of (A, f).

7.11′.2 *A ring A that has three left ideals and is right Artinian (and Noetherian).*

Let R be a field with a subfield S such that $\dim_S R = n < \infty$, and such that there is a homomorphism $f: R \to S$. For example, let $R = \mathbb{Q}(x_1, \ldots, x_t)$, $S = \mathbb{Q}(x_1^2, \ldots, x_t^2)$, and $f: R \to S$ maps $x_i \mapsto x_i^2$, $i = 1, \ldots, t$. Let

$$A = \begin{pmatrix} R & R \\ 0 & S \end{pmatrix}$$

be the idealization. Then (A, f) has three left ideals, and every chain of right ideals has length $\leq n$.

7.11′.3 (Herstein [65]) *A right Noetherian ring A without the Krull intersection property for the radical.*

Let p be a prime number in $\mathbb{Z}$, and let $\mathbb{Z}_{(p)}$ denote the subset of $\mathbb{Q}$

$$\mathbb{Z}_{(p)} = \{ab^{-1} \in \mathbb{Q} \mid a, b \in \mathbb{Z},\ b \notin p\mathbb{Z}\}$$

Then $\mathbb{Z}_{(p)}$ is a subring of $\mathbb{Q}$, so $\mathbb{Q}$ is a $\mathbb{Z}_{(p)}$-module. The only ideals of $\mathbb{Z}_{(p)}$ occur in the chain

$$\mathbb{Z}_{(p)} \supset p\mathbb{Z}_{(p)} \supset p^2\mathbb{Z}_{(p)} \supset \cdots \supset p^n\mathbb{Z}_{(p)} \supset \cdots.$$

(Prove this.) This shows that $\mathbb{Z}_{(p)}$ is a principal ideal domain, and is Noetherian, but not Artinian.

Let

$$A = \begin{pmatrix} \mathbb{Z}_{(p)} & \mathbb{Q} \\ 0 & \mathbb{Q} \end{pmatrix}$$

be the idealization of the $(\mathbb{Z}_{(p)},\mathbb{Q})$-bimodule $\mathbb{Q}$. Using the fact that $\mathbb{Q}$ is a simple $\mathbb{Q}$-module, we find that

$$\begin{pmatrix} U & \beta\mathbb{Q} \\ 0 & \gamma\mathbb{Q} \end{pmatrix}, \tag{4}$$

$\beta, \gamma = 0$ or 1, $\beta = 1$ when $\gamma \neq 0$, constitutes all right ideals of A, as U runs over all ideals of $\mathbb{Z}_{(p)}$. Thus A is right Noetherian since $\mathbb{Z}_{(p)}$ is Noetherian; however, since

$$\begin{pmatrix} U & Q \\ 0 & 0 \end{pmatrix}$$

is an ideal of A for any ideal U of $\mathbb{Z}_{(p)}$, A is not Artinian on either side. Since

$$\begin{pmatrix} 0 & V \\ 0 & 0 \end{pmatrix}$$

is a left ideal for any $\mathbb{Z}_{(p)}$-submodule V of $\mathbb{Q}$, A is not left Noetherian.

The radical of a module M_R has been defined in an exercise (Chapter 1) as the intersection rad M of the maximal submodules of M. Thus rad R is the intersection of the maximal right ideals of R. We easily determine that

$$\text{rad}\, \mathbb{Z}_{(p)} = p\mathbb{Z}_{(p)}.$$

Consequently, for

$$A = \begin{pmatrix} \mathbb{Z}_{(p)} & \mathbb{Q} \\ 0 & \mathbb{Q} \end{pmatrix},$$

we see, by inspection of the right ideals (4), that

$$J = \text{rad}\, A = \begin{pmatrix} B & \mathbb{Q} \\ 0 & \mathbb{Q} \end{pmatrix}, \qquad B = \text{rad}\, \mathbb{Z}_{(p)}.$$

It follows that

$$J^n = \begin{pmatrix} B^n & \mathbb{Q} \\ 0 & \mathbb{Q} \end{pmatrix}$$

and

$$J_\infty = \bigcap_{n=1}^{\infty} J^n = \begin{pmatrix} 0 & \mathbb{Q} \\ 0 & \mathbb{Q} \end{pmatrix} \neq 0.$$

This example, originated by Herstein [65], is important because the intersection theorem states that

$$\bigcap_{n=1}^{\infty} (\operatorname{rad} R)^n = 0$$

for any commutative Noetherian ring R (cf. theorem of Krull 27.19). In this example, however, the ring is not left Noetherian.

7.11′.4 (Jategaonker[1] [69]). For each ordinal a, there exists a principal left ideal domain R such that $(\operatorname{rad} R)^a \neq 0$, where $I^{b+1} = I^b I$ and $I^b = \bigcap_{c<b} I^c$ is defined for any left ideal I and initial ordinal $b+1$ or limit ordinal b.

Hilbert Basis Theorem

The main purpose of this section is to enlarge our stock of Noetherian rings and modules and to make some applications.

Let R be any ring, let M be any right R-module, let $I = \{0, 1, 2, \ldots, n, \ldots\}$ be the set of non-negative integers, let $A = M^I$ (direct product), and let $B = M^{(I)}$ (direct sum). Introduce a symbol x, called an **indeterminate**, and for each $a \in M$, let ax^i represent the unique function $p \in A$, satisfying $p(i) = a$, $p(j) = 0$ if $j \neq i$. The symbol $f(x) = \sum_{i=0}^{\infty} a_i x^i$ will denote that unique function $f \in A$, satisfying $f(i) = a_i \; \forall i \in I$. The elements $\{a_i\}_{i \in I}$ are called the **coefficients of** f, and a_i is called the coefficient of x^i, $i = 0, 1, \ldots$. Let $f(x) = \sum_{i=0}^{\infty} a_i x^i \in A$. If $f(x) \neq 0$, then $f(x) \in B$ if and only if there exists an integer n, called the **degree** of $f(x)$, such that $a_n \neq 0$ and $a_m = 0 \; \forall m > n$. In this case, we write $f(x) = \sum_{i=0}^{n} a_i x^i$. *By agreement, the degree of the zero map is defined to be* $-\infty$.

The module $A = M^I$ is denoted by $M\langle x\rangle$, and the elements of $M\langle x\rangle$ are called **formal power series over** M **in the indeterminate** x. The module $B = M^{(I)}$ is denoted by $M[x]$, and the elements of $M[x]$ are called **polynomials over** M **in the indeterminate** x.

The R-module $R\langle x\rangle$ is therefore defined. If $f(x) = \sum_{i=0}^{\infty} r_i x^i$ and $g(x) = \sum_{j=0}^{\infty} s_j x^j$ are elements of $R\langle x\rangle$, then the **product** is defined to be the formal power series $\sum_{k=0}^{\infty} t_k x^k$, where $t_k = \sum_{i+j=k} r_i t_j \in R$, $k = 0, 1, 2, \ldots$, and $R\langle x\rangle$ is a ring with respect to product and sum of formal power series, called the **ring of formal power series over** R in x. If $f, g \in R\langle x\rangle$ are polynomials of degrees n and m, respectively, then fg is a

[1] These examples proved to be the first rings without the intersection property for the radical which were (1) domains or (2) principal left ideal rings, or (3) both. Also see Jategaonker's exposition [70] and other results of Jategaonker in the exercises.

polynomial of degree $\leq n + m$. Hence, $R[x]$ is a subring of $R\langle x\rangle$, called the **polynomial ring over R in the indeterminate x**. (Cf. 5.74 and 3,3'). The set of all polynomials of $R\langle x\rangle$ of degree 0 is a subring of $R\langle x\rangle$ isomorphic to R.

If $m(x) = \sum_{i=0}^{n} a_i x^i \in M[x]$, and if $p(x) = \sum_{j=0}^{q} r_j x^j \in R[x]$, then $m(x)\,p(x) = \sum_{k=0}^{n+q} b_k x^k$, where $b_k = \sum_{i+j=k} a_i r_j \in M$, $k = 0, 1, \ldots, n+q$, defines an operation $\mu : M[x] \times R[x] \to M[x]$ and $M[x]$ is an $R[x]$-module $(M[x], R[x], \mu)$.

As usual, 1 denotes the identity element of R. Then $R\langle x\rangle$ contains the element $1x$, and if we define $(1x)^0 = 1$, then it follows that $(1x)^n = 1x^n\ \forall n$. Furthermore, if M is an R-module, then each element $p(x) = \sum_{i=0}^{\infty} r_i x^i \in M\langle x\rangle$ has the unique representation $p(x) = \sum_{i=0}^{\infty} r_i (1x)^i$. Henceforth, we write x in place of $1x$. In doing so, in effect, we are then viewing the module $M\langle y\rangle$ (resp. ring $R\langle y\rangle$) of formal power series in the indeterminate $y = 1x$.

Let S be any $R[x]$-submodule of $M[x]$. If S contains a polynomial of degree n, define $L_n(S)$ to be the set of elements of M consisting of 0 and the elements $a \in M$ that appear as a coefficient of x^n of a polynomial in S having degree n. Then $L_n(S)$ is an R-submodule of M. If $m(x) = \sum_{i=0}^{n} a_i x^i \in S$ has degree n, then $m(x)x$ has degree $n+1$, and therefore $L_n(S) \subseteq L_{n+1}(S)$.

7.12 Lemma. *If S and T are $R[x]$ submodules of $M[x]$ such that $S \subseteq T$, and if $L_i(S) = L_i(T)$ for $i = 0, 1, \ldots,$ then $S = T$.*

Proof. Suppose $S \neq T$, and let $f(x) = \sum_{i=0}^{n} a_i x^i \in T$ have least degree n of all those polynomials in T that are not in S. Since $a_n \in L_n(T)$, and since $L_n(T) = L_n(S)$, there exists $g(x) \in S$ of degree n having a_n as the coefficient of x^n. Then $f(x) - g(x) = h(x)$ has degree $\leq n - 1$, so $h(x) \in S$ by the assumption on n. Then $f(x) = g(x) + h(x) \in S$, contrary to the assumption. Hence, $S = T$. □

7.13 Hilbert Basis Theorem. *Let R be a ring, and let M be a Noetherian R-module. Then the polynomial module $M[x]$ is a Noetherian $R[x]$-module.*

Proof. Let $M_1 \subseteq M_2 \subseteq \cdots \subseteq M_n \subseteq \cdots$ be an ascending sequence of $R[x]$-submodules of $M[x]$, and consider the rectangular array $\{L_i(M_j) \mid i, j = 0, 1, \ldots\}$ of R-submodules of M. For a fixed j, $L_0(M_j) \subseteq L_1(M_j) \subseteq \cdots \subseteq L_n(M_j) \subseteq \cdots$ is an ascending sequence, and since $M_j \subseteq M_m\ \forall m \geq j$, $L_i(M_j) \subseteq L_i(M_m)$, $\forall m \geq j$, we have the diagonal $L_0(M_0) \subseteq L_1(M_1) \subseteq \cdots \subseteq L_n(M_n) \subseteq \cdots$ ascending sequence. Since M is a Noetherian R-module, there exists q such that $L_q(M_q) = L_{q+1}(M_{q+1}) = \cdots$, and then $L_i(M_j) = L_q(M_q)\ \forall i \geq q$, $j \geq q$. On the other hand, the a.c.c. shows that for a fixed i there is a least integer $n(i)$ such that $L_i(M_j) = L_i(M_{n(i)})\ \forall j \geq n(i)$, and if $i \geq q$, then $n(i) = q$ will do.

Thus, the integer $n(i)$ is bounded, say by n_0, and then $L_i(M_j) = L_i(M_{n_0})$ for $i = 0, 1, \ldots$. The lemma now asserts that $M_j = M_{n_0}$, $\forall j \geq n_0$. Since $M[x]$ therefore satisfies the a.c.c. on $R[x]$-submodules, $M[x]$ is a Noetherian $R[x]$-module. ▯

7.14 Corollary. *If R is right Noetherian ring, and if M_R is a finitely generated R-module, then the polynomial module $M[x]$ is a Noetherian $R[x]$-module; in particular, the polynomial ring $R[x]$ is a right Noetherian ring.*

Proof. M is a Noetherian R-module, so the Hilbert basis theorem applies. The second statement is the $M = R$ case of the first. ▯

If $R[x]$ is the polynomial ring over R in the indeterminate x, then $R[x, y]$ will denote the polynomial ring over $R[x]$ in the indeterminate y; $R[x, y]$ is called the polynomial ring over R in the indeterminates x and y. Assuming the polynomial ring $R[x_1, \ldots, x_{n-1}]$ in $n - 1$ indeterminates is defined, let $R[x_1, \ldots, x_n]$ be a polynomial ring in the indeterminate x_n over $R[x_1, \ldots, x_{n-1}]$. This ring is called the *polynomial ring over R in n indeterminates* $x_1, \ldots, x_n$.

7.15 A Corollary. *If R is a right Noetherian ring, then so is the polynomial ring $R[x_1, \ldots, x_n]$ in n indeterminates $x_1, \ldots, x_n$.*

The proof follows from 7.14 by induction on n.

Let S be a commutative ring, and let R be a subring. If $a_1, \ldots, a_n$ are finitely many elements of S, then the subring $R[a_1, \ldots, a_n]$ generated by R and $a_1, \ldots, a_n$ is a ring epimorph of the polynomial ring $R[x_1, \ldots, x_n]$ under a mapping sending $x_i \mapsto a_i$, $i = 1, \ldots, n$, and leaving fixed the elements of R. Hence, *if R is right Noetherian, so is $R[a_1, \ldots, a_n]$.*

Conjecture of Serre: Over a polynomial ring in n variables over a field, is every projective module free? The answer is yes when $n = 1$ or 2 (Sheshadri [58]), and in any case, every projective ideal is free, but the general case remains unsolved (cf. Bass [62]).

7.15B *Exercise.*

7.15B.1 (Gilmer [68]) Prove that a polynomial $f = \sum_{i=0}^{n} f_i x^i$ of the polynomial ring $R[x]$ over a commutative ring R is a unit of $R[x]$ if and only if f_0 is a unit of R, and f_i is a nilpotent element of R, $i = 1, \ldots, n$.

*7.15B.2 (Snapper [52]) $\operatorname{rad} R[x] = N[x]$, where N is the maximal nil ideal of a commutative ring R.

7.15B.3 For any ring R, the units of $R\langle x\rangle$ are those of the form $c + xg$, where $g \in R\langle x\rangle$, and $c \in R$ is a unit of R. Conclude that $\operatorname{rad} R\langle x\rangle$ consists of all power series $r + xh$, where $h \in R\langle x\rangle$, and $r \in \operatorname{rad} R$.

7.16 Proposition. *The following conditions on a module M are equivalent:*

7.16.1 *M is uniform.*

7.16.2 *inj. hull M is indecomposable.*

Proof. Let $\hat{M}$ = inj. hull of M. If $\hat{M}$ is decomposable, then $\hat{M} = A \oplus B$ for two nonzero submodules A and B. Then $U = A \cap M \neq 0$ and $V = B \cap M \neq 0$, and $U \cap V = 0$, showing that M is not uniform. Conversely, if M contains nonzero independent submodules U and V, then we can choose respective injective hulls $\hat{U}$ and $\hat{V}$ in $\hat{M}$. Since $\hat{U}$ and $\hat{V}$ are injective, they are summands of $\hat{M}$, so $\hat{M}$ is not indecomposable. □

Ore Domains

A right ideal of R is **uniform** if it is a uniform R-module. A uniform right ideal U plays the same role in the structure of a Noetherian simple ring R as a simple right ideal does in Artinian simple ring; namely, $R = \text{End}_K U$, where U is finitely generated projective over $K = \text{End}\, U_R$. In this case, however, K is an entire ring instead of a field, as in the Artinian case. Nevertheless, K can be embedded in a field, and we now discuss how this is done.

A right **Ore domain** is an entire ring K that is a uniform right ideal in K. For example, any commutative entire ring R is Ore since $xy \in xR \cap yR$. Left Ore domains are defined symmetrically (Ore [31, 33]). Not every entire ring can be embedded in a field (Malcev [36]); among those that can are Ore domains. A field Q is a (right) **field of quotients** of an entire ring K (or a **quotient field** of K) in case K is a subring of Q such that $Q = \{ab^{-1} \mid a, b \in K,\ b \neq 0\}$. In this case, if $a, b \in K$ are nonzero, then $b^{-1}a = a_1 b_1^{-1}$, for appropriate $a_1, b_1 \in K$, and then $ab_1 = ba_1 \neq 0$; that is, $aK \cap bK \neq 0$. This proves that in order that K have a quotient field it is necessary that K be an Ore domain. Conversely, assume K is Ore, and consider the set $\mathscr{S}$ of all ordered pairs (a, b), $a, b \in K$, $b \neq 0$, and in $\mathscr{S}$ define $(a, b) \approx (c, d)$ if and only if whenever $bd_1 = db_1$ we have $ad_1 = cb_1$. One checks that this defines an equivalence relation in $\mathscr{S}$. Let a/b denote the set of pairs $(c, d) \approx (a, b)$, and let $Q = \{a/b \mid a, b \in R,\ b \neq 0\}$. In Q, we define

(1) $$a/b + c/d = (ad_1 + cb_1)/m,$$

where $m = bd_1 = db_1$. If $c \neq 0$, and if $n = bc_2 = cb_2$, then define

(2) $$(a/b)(c/d) = (ac_2/db_2) \quad \text{and} \quad (a/b)(c/d) = 0/d \quad \text{if} \quad c = 0.$$

We omit the verification that the operations (1) and (2) are independent of the choice of representatives for a/b and c/d, and we leave it to the reader to verify that Q is a field. (We discuss the entire question of quotient rings thoroughly, and in greater generality in Chapter 9.)

There is a ring monomorphism $K \to Q$ that sends $a \mapsto a/1$. [This is an immediate consequence of the formulas (1) and (2).] If we identify K with its image under this map, we see that Q is a quotient field of K since $(a/1)^{-1} = (1/a)$ $\forall a \neq 0 \in K$, and $(a/b) = (a/1)(b/1)^{-1} \forall a/b \in Q$.

(acc) ⊕

We will study the following chain condition on a module, which leads to the existence of uniform submodules. The condition is the following:

M contains no infinite set of independent submodules. This condition is abbreviated by the symbol (acc)⊕ for right R-modules, and ⊕(acc) for left. Thus, the ring R may satisfy (acc)⊕ on its right ideals, or ⊕(acc) on its left ideals. Any matrix ring D_n over a field D is a finite dimensional vector space over D, and so satisfies (acc)⊕ and ⊕(acc). Furthermore, so does any finite product of such rings.

For the definition of (Goldie) dimension in the next proposition, refer to 4.19.

7.17 Proposition.

7.17.1 *If M is a nonzero R-module satisfying* (acc)⊕, *then there exist finitely many independent uniform submodules $U_1, \ldots, U_n$ such that $U_1 + \cdots + U_n$ is an essential submodule of M. Conversely.*

7.17.2 *If M satisfies* (acc)⊕, *and if S is a nonzero submodule, then S contains a uniform submodule of M.*

7.17.3 *A module M satisfies* (acc)⊕ *iff so does every essential extension of M.*

When this is so, then $\dim M = n$.

Proof of 7.17.1. If $\mathscr{F}$ and $\mathscr{F}'$ are families of independent submodules, then we say that $\mathscr{F}'$ is a **refinement** of $\mathscr{F}$ in case every submodule in $\mathscr{F}'$ is contained in a module of $\mathscr{F}$. In this case, we say that $\mathscr{F}'$ is a **proper** refinement in case $\mathscr{F}'$ has bigger cardinal. The (acc)⊕ implies that given $\mathscr{F}$, there exists a refinement $\mathscr{F}'$ of $\mathscr{F}$ that has no proper refinements. We now show that such a family $\mathscr{F}'$ consists of uniform submodules. For if $U \in \mathscr{F}'$, and if U is not uniform, then $X \cap Y = 0$ for two nonzero submodules of U, and then the family $\mathscr{F}''$, obtained from $\mathscr{F}'$ by the deletion of U and the addition of X and Y, is a proper refinement of $\mathscr{F}'$. Hence we can assume M contains a uniform submodule U_1. If U_1 is not essential, then $U_1 \cap M_2 = 0$ for some nonzero submodule M_2.

Now M_2 contains a uniform submodule U_2. Continuing, we have a chain

$$U_1 \subset U_1 \oplus U_2 \subset \cdots \subset U_1 \oplus U_2 \oplus \cdots \oplus U_n$$

where U_i is uniform, $i = 1, \ldots, n$. The (acc)$\oplus$ shows that for some n, M is an essential extension of $\sum_{i=1}^{n} U_i$. The converse is an exercise.

Proof of 7.17.2. Any submodule of M satisfies (acc)$\oplus$, and then 7.17.1 applies. The proof of 7.17.3 is an exercise (cf. 7.18.1). □

Conceivably, if M satisfies (acc)$\oplus$, then there exist families of independent submodules of ever bigger cardinal. This is not the case. In fact, the integer n of 7.17.1 of the proposition is uniquely determined, and every family of independent submodules of M has cardinality $\leq n$. This fact can be deduced from the Krull-Schmidt-Remak theorem 21.2 (cf. Exercise 7.18.1).

7.18 *Exercises.*

7.18.1 Let M be a nonzero R-module, and let $\hat{M} =$ inj. hull M. The following conditions are equivalent:

(a) M satisfies (acc)$\oplus$.

(b) There exist finitely many independent uniform submodules $U_1, \ldots, U_n$ such that $U_1 + \cdots + U_n$ is an essential submodule of M.

(c) $\hat{M}$ is a direct sum of $m < \infty$ indecomposable modules.

(d) There exists an integer t such that M contains a family of t independent submodules, but no such family of bigger cardinality.

When this is so, then $n = m = t$.

7.18.2 (Goldie) A domain R is right Ore if and only if R satisfies (acc)$\oplus$.

7.18.3 If R is an Ore domain, so is the polynomial ring $R[x]$.

7.18.4 If K is the quotient field of a right Ore domain R, then $K =$ inj. hull R_R. More generally, show that over a right Ore domain any divisible torsionfree right module is injective.

7.18.5 If Q and Q' are quotient fields of K, then there exists a unique ring (field) isomorphism $\varphi : Q \to Q'$ such that $\varphi(a) = a \forall a \in K$.

7.18.6 If K has a right quotient field Q, then Q is a left quotient field of K if and only if K is left Ore.

***7.19** *Exercise.* Let E be an indecomposable (quas-)injective module, and let $B = \text{End } E_R$.

7.19.1 Prove that B has a unique maximal right ideal J. Conclude that $J = \text{rad } B$ is such that R/J is a field. (Then, B is a local ring. Cf. 18.10.)

7.19.2 If E has no fully invariant submodules, then B is a field.

***7.20** *Exercise.* (Azumaya's generalization [50] of the Krull-Schmidt Theorem.) A decomposition of a module M into a direct sum of indecomposable modules $\{M_i\}_{i \in I}$ such that $\text{End}(M_i)_R$ is a local ring

$\forall i \in I$ is unique in the sense that there is another such decomposition $M = \sum_{j \in J} \oplus N_j$ if and only if there is a bijection $f: I \to J$ and an automorphism of M mapping M_i onto $N_{f(i)}$ $\forall i \in I$.

Orders in Simple Artinian Rings

A ring T is a **right order** in a ring S in case T is a subring of S such that given any $x \in S$ there exist elements $a, b \in T$ such that a is a unit of S and such that $x = ba^{-1}$. For example, an Ore domain K is an order in its quotient field D, and the matrix ring K_n is an order in D_n. To see this, let $x = (d_{ij}) \in D_n$, where $d_{ij} \in D$, then $d_{ij} = k_{ij} t_{ij}^{-1}$, $k_{ij}, t_{ij} \in K$, $i, j = 1, \ldots, n$. Since K is Ore, there exists $t \neq 0$ in $\bigcap_{i,j=1}^{n} t_{ij} K$, so $t = t_{ij} g_{ij}$ for some $g_{ij} \in K$, and then $t_{ij}^{-1} t = g_{ij}$, so setting $a = \operatorname{diag} t$, we have

$$b = xa = (k_{ij} g_{ij}) \in K_n.$$

Since a is a unit in D_n, then $x = ba^{-1}$ as required. (All orders of D_n can be similarly described; cf. the Faith-Utumi theorem, Chapter 10.) More generally, if R is an order in a ring S, then R_n is an order in the matrix ring S_n.

7.21 A Proposition. *If K is a right order in D, then the canonical right K-module D is an essential extension (in fact, a rational extension 4.5) of K. If D is right Artinian, then for each integer $n > 0$, and nonzero idempotent e in the full $n \times n$ matrix ring D_n, the ring eK_ne is a right order in eD_ne.*

Proof. If $d \in D$, and $d \neq 0$, then $d = ab^{-1}$, with $a, b \in K$, and so then $a = db$ is a nonzero element of $dK \cap K$. This proves that D is essential over K. (That D is a rational extension of K is similar.)

Now assume that D is right Artinian. Then, any right regular element $d \in D$ is a unit of D. To see this, assume that k is an integer such that the descending chain $\{d^i D\}_{i=1}^{\infty}$ of right ideals is stationary after $d^k D$. Then, there exists $c \in D$ such that $d^k = d^{k+1} c$, and by right regularity of d, we must have $1 = dc$. Thus, d has the right inverse c. Now $d(cd - 1) = 0$, so that $cd = 1$, and therefore d is a unit.

In order to prove that $R = eK_ne$ is a right order of $Q = eD_ne$ we first prove (1) *that for every nonzero $q \in Q$, there corresponds $r \in R$ such that $qr \in R$, and $qr \neq 0$.* The right ideal $J = \{b \in K_n \mid qb \in K_n\}$ is an essential right ideal of K_n since, as proved above, K_n is a right order in D_n, and hence, D_n is an essential extension of K_n. Thus, qJ is a nonzero right ideal of the prime ring K_n, and hence $qJe \neq 0$. But, $q = eqe \in eD_ne$, so that there exists $r \in R$ such that $qr \neq 0$.

Next, we prove (2) *that every right regular element x of R is a unit of Q*. To do this, since Q is Artinian, it suffices to prove that x is a regular element of Q. Suppose that $xq = 0$ for some element $q \in Q$. If $q \neq 0$, then $qr \neq 0$ lies in R, for some $r \in R$, and then $xqr = 0$ violates regularity of x in R. This proves that x is right regular in Q, which, as we have shown in a right Artinian ring, is a unit.

In order to prove that R is a right order of Q, it remains only to prove that for any $q \in Q$, the set $q' = \{r \in R \mid qr \in R\}$ contains a right regular element x. [For then, x is a unit of Q, and so $q = ax^{-1}$, with $a = qx$, $x \in R$.] Since R is a subring of the right Artinian ring Q, we may choose $x \in q'$ such that the right annihilator $x^\perp$ of x in R (resp. in Q) is minimal in the set $\{y^\perp \mid y \in q'\}$. (Hint: the right annihilator in R is that in Q intersected with R.) If x is not right regular in R, then x is not a unit of Q, so $xQ \neq Q$, and therefore there exist a nonzero right ideal P of Q such that $xQ \cap P = 0$. (Proof?) Then, $I = P \cap R$ is a nonzero right ideal of R such that $xR \cap I = 0$. Since q' is an essential right ideal of R, then $J = q' \cap I \neq 0$. If $k \in J$, and if $t \in (k + x)^\perp$, then $kt = -xt \in I \cap xR$. Since $I \cap xR = 0$, then $kt = xt = 0$, so that $(k + x)^\perp \subseteq k^\perp \cap x^\perp$. The reverse inclusion is trivial, so $r_k^\perp = k^\perp \cap x^\perp$, $\forall\, r_k = k + x \,\forall\, k \in q'$. If $r_k^\perp = 0$, then r_k is the desired right regular element of q'. Otherwise, by the choice of x, we have $x^\perp = r_k^\perp = k^\perp \cap x^\perp$, $\forall\, k \in q'$, so that $q'x^\perp = 0$. Since R is a prime ring, and $q' \neq 0$, then $x^\perp = 0$, so that x is right regular as desired. □

Goldie [58, 60] and Lesieur-Croisot [59] disovered that any Noetherian prime ring is an order in an Artinian simple ring. This was historically the first finite dimensional representation theorem for a general Noetherian simple ring, and it is proved in Chapter 9. For the present we have the following corollary.

7.21 B Proposition. *If R is a simple ring with a uniform right ideal U, then R is a right order in a simple Artinian ring D_m, where D is the quotient field of the Ore domain $K = \operatorname{End} U_R$. If T is a subring of D_m containing R, then T is a simple ring, an order in D_m, and satisfies* (acc)⊕. *Moreover, there is an integer $n \geq m$, and idempotent e in the $n \times n$ matrix ring D_n, and an isomorphism*[1] *$R \approx eK_n e$.*

Proof. By 7.5, $R \approx \operatorname{End}_K U$, K is an order in a field D, and ${}_K U$ is a finitely generated projective. Suppose U is a direct summand of K^n. Then by (7.6), there exists an idempotent

$$e \in K_n \approx \operatorname{End}_K K^n$$

such that $R \approx eK_n e$. (The idempotent e is the projection idempotent $K^n \to U$.) Now K_n is an order in D_n and this implies that $eK_n e$ is an

[1] This part is Hart's theorem [67].

order in eD_ne (7.21A). By Exercise 7.10′.3, $eD_ne \approx D_m$, for some $m \leq n$. Thus $R \approx eK_ne$ is an order in a simple Artinian ring $\approx D_m$.

Next, if $D_m \supseteq T \supseteq R$, if I is an ideal in T, and if $x = ab^{-1} \in I$, then $xb = a \in I \cap R$. Now $I \cap R$ is an ideal of R, and $x \neq 0$ implies $I \cap R \neq 0$, whence $I \cap R = R$; that is, $I \supseteq R$. Then $1 \in I$, so $I = T$, proving that T is simple. Also T is trivially an order in D_m which must satisfy (acc)⊕ since D_m does (cf. 7.17.3). □

Annulets

If R is a ring with divisors of zero, then there exists a nontrivial annihilator ideal, that is, a right ideal of the form $A = \mathrm{ann}_R X$ for some subset X of R. Such a right ideal is called a (**right**) **annulet**. Thus, a right ideal A is an annulet if and only if $A = X^{\perp}$ for some subset X. Then $A = (RX)^{\perp}$, where RX is the left ideal generated by X. Let $B = {}^{\perp}A$, the left annihilator of A. Clearly, $B \supseteq (RX)$, and

$$B^{\perp} \subseteq (RX)^{\perp} = X^{\perp} = A.$$

However, $BA = 0$, so $B^{\perp} \supseteq A$. Therefore $A = B^{\perp} = ({}^{\perp}A)^{\perp}$. *Hence a right ideal A is an annulet if and only if $A = ({}^{\perp}A)^{\perp}$.*

7.22 *Exercises.*

7.22.1 Let A be a right ideal of a ring R. If $A = eR$, where $e = e^2 \in R$, then A is an annulet.

7.22.2 If $R = D_n$, where D is a field, then every right ideal of R is an annulet.

7.22.3 Every finitely generated one-sided ideal of a full linear ring is an annulet.

*7.22.4 A cyclic module R/A is contained in a product of copies of R if and only if A is a right annulet. Conclude that if R is a cogenerator in mod-R, then every right ideal of R is an annulet.

7.22.5 Show that every ideal of $\mathbb{Z}/(p^n)$ is an annulet. [Hint: The proper ideals have the form $(p^i)/(p^n)$, $i = 1, \ldots, n-1$. Use induction on i.]

Max-Min Annulets

The set of right annulets of R is ordered by inclusion, and a **minimal annulet** is one that is minimal in the ordered set of nonzero right annulets. Dually, a **maximal annulet** is one that is maximal in the set of right annulets $\neq R$. For example, if R is entire, then 0 is a maximal annulet, and R is a minimal annulet. But this is a trivial example. More typically, any right ideal A of R that is a direct summand is a right annulet, since $A = eR$ for idempotent $e \in R$, and hence $A = B^\perp$, where $B = R(1-e)$ is a left annulet. If K is a domain, the right ideal $e_{ii}R$ in a matrix $R = K_n$ over K is a minimal right annulet for any matrix unit e_{ii}. Thus, a minimal right ideal in a simple Artinian ring is a minimal annulet.

(acc)$^\perp$

The a.c.c. on right annulets of R is abbreviated (acc)$^\perp$. (See the foreword for a discussion of the a.c.c. in ordered sets.) Similarly (dcc)$^\perp$ denotes the d.c.c. on right annulets, and $^\perp$(acc) and $^\perp$(dcc) are defined by symmetry.

The following exercises are not needed for the discussion that follows, but they elucidate the basic facts about annulets and will be used elsewhere in the text.

7.23 *Exercises.*

7.23.1 The set of right annulets of R is a lattice, where $A \vee B = ({}^\perp(A+B))^\perp$ and $A \wedge B = A \cap B$.

7.23.2 The correspondence $\varphi : A \to {}^\perp A$ is a lattice anti-isomorphism between the lattice of right annulets and the lattice of left annulets. Thus R satisfies (acc)$^\perp$ if and only if R satisfies $^\perp$(dcc). Also A is a maximal annulet if and only if ${}^\perp A$ is a minimal (left) annulet. Every maximal right annulet A has the form $A = b^\perp$ for some $b \in R$.

7.23.3 If S is a subring of R, the *contraction map* $A \mapsto A \cap S$ is a surjective lattice homomorphism of the lattice of right annulets of R and that of S. Conclude that S satisfies the a.c.c. (resp. d.c.c.) on right annulets if R does. (See p. 142.)

7.23.4 If B is a minimal right annulet of R, any right subideal is lateral in the sense defined presently.

7.23.5 (Faith [67]) A ring R satisfies the a.c.c. on right annulets if and only if for each right ideal I there is a finitely generated right ideal I_1 contained in I having the same left annihilator.

7.23.6 If R is a right order in a semisimple ring S, then R satisfies (acc)$\oplus$ and (acc)$^\perp$. Thus, any simple ring R with uniform right ideal $\neq 0$ satisfies (acc)$\oplus$ and (acc)$^\perp$. Moreover, such rings are Morita invariant.

*7.23.7 If for every integer $n > 0$, the matrix ring R_n is Dedekind finite ($ab = 1 \Rightarrow ba = 1$), then the ring R is a right IBN ring (Chapter 3). Otherwise, for some n, R_n contains infinitely many orthogonal idempotents, and, in particular, satisfies neither (acc)$^\perp$ nor (acc)$\oplus$. Conclude that then R cannot be embedded into a field.

For any $x \in R$, $x^\perp$ denotes its right annihilator. A right ideal U is **lateral** in case $\forall x \in R$ the following implication holds:

$$x^\perp \cap U \neq 0 \Leftrightarrow xU = 0.$$

Any nonzero right ideal contained in a minimal right annulet is lateral. If R is simple, or a prime ring with a.c.c. on left annulets, then by (2) of 4.20, any uniform right ideal is contained in a minimal right annulet and, hence, is lateral. (See 4.15.)

7.24 Proposition.

7.24.1 *If R is a ring, and if A is a maximal right annulet of R, then the endomorphism ring of the R-module R/A is an entire ring.*

7.24.2 *Write $A = b^\perp$, where A is a maximal right annulet. Then bR is a lateral right ideal, Rb is a lateral left ideal, and* End bR_R *is an entire ring isomorphic to* End R/A_R.

Proof. To prove that $W = bR$ is lateral, let $y \in R$ be such that $y^\perp \cap bR \neq 0$; that is, $yba = 0$ for some $a \in R$ with $ba \neq 0$. If $yb \neq 0$, then $Rb \cap {}^\perp a \neq 0$. Since ${}^\perp A$ is a minimal left annulet containing Rb, and ${}^\perp A \cap {}^\perp a \neq 0$ is a nonzero left annulet, then ${}^\perp A \cap {}^\perp a = {}^\perp A$; that is, ${}^\perp a \supseteq {}^\perp A$, so $({}^\perp A)a = 0$. (Incidentally, this proves that Rb is a lateral left ideal.) Therefore $a \in ({}^\perp A)^\perp = A$, but $bA = 0$, so $ba = 0$, a contradiction. Therefore $yb = 0$, and $yW = 0$, proving that $W = bR$ is lateral. Then, End bR is entire by 4.15, and

$$0 \to b^\perp \to R \to bR \to 0$$

is canonically exact, so $R/A \approx bR$, proving 7.24. □

7.24.1 is the generalization of Koh [66] of Schur's lemma 3.10.1.

Simple Rings with Lateral Ideals

We now apply 7.24 to simple rings.

7.25 Proposition. *Let R be a simple ring with a maximal right annulet or a lateral right ideal. Then there exists a cyclic right ideal U of R such that $Q =$* End U_R *is an entire ring. Furthermore, U is a finitely generated projective Q-module, and the canonical map $R \to \mathrm{End}_Q U$ is a ring isomorphism. Thus, for some integer $n > 0$ there exist an idempotent e in the matrix ring Q_n and a ring isomorphism $R \approx eQ_ne$.* (Cf. 4.16.)

Proof. The proof is an application of the general Wedderburn theorem 7.5, 4.15, and 7.24. For the isomorphism $R \approx eQ_ne$, see the proofs of 7.5 and 7.21 B. □

7.26 *Exercises.*

7.26.1 (Hart [67]) If, in 7.25, the simple ring R has a uniform right ideal, then Q is a right Ore domain.

*7.26.2 A cyclic right R-module R/A is contained in a free R-module if and only if $X^{\perp} = A$ for some finite subset X of R.

7.26.3 If R is a ring such that, for each right ideal A, $A = X^{\perp}$ for a finite subset X and the corresponding statement for left ideals holds, then R is right and left Artinian and Noetherian.

7.26.4 If every cyclic right and left R-module is contained in a projective R-module, then R is right and left Artinian and Noetherian, and every one-sided ideal is an annulet. (Cf. Quasifrobenius rings, Chapter 24.)

Uniqueness

Simple rings are ubiquitous in noncommutative ring theory. We know that a simple Noetherian ring R is isomorphic to the ring of endomorphisms of finitely generated projective module U over an Ore domain K and that R is a right order in a simple Artinian ring D_n, where D is the field of quotients of K. The ring D_n so obtained is unique up to an isomorphism (the proof is similar to the proof of the uniqueness of a field of quotients of an Ore domain, cf. *classical quotient rings*, Chapter 9). The Wedderburn-Artin theorem (7.9) implies that two simple Artinian rings D_m and Q_n, where D and Q are fields, are isomorphic if and only if $n = m$ and $D \approx Q$. Thus the quotient fields of the various Ore domains K that arise in this way are all isomorphic. It can be shown, however, that in general K is not unique.

If U is a projective and uniform right ideal of R, and only in that case, the Ore domain K is a simple ring (7.5). Thus we are led, ineluctably, to a simple ring (not a field) without divisors of zero. It would be safe to bet that 99 per cent of the readers have not yet seen one, and yet they occur freely in nature. Littlewood [33] discussed a class of them that are related to quantum mechanics. Anyone wishing to determine the structure of all simple Noetherian rings will have to come to grips with them (as we do in the next section).

Derivations

Let A be a ring, and let $\varphi : A \to A$ be a ring monomorphism. Then a φ-derivation of A is a mapping $D : A \to A$ such that

$$D(a + b) = D(a) + D(b)$$
$$D(ab) = \varphi(a)\, D(b) + D(a)\, b$$

$\forall a, b \in A$. If $\varphi =$ identity, then D is called an **(ordinary) derivation.** For each element $x \in A$, there is a derivation

$$D_X : a \mapsto ax - xa,$$

$\forall a \in A$, called the **inner** derivation defined by x. A derivation is **outer** if it is not inner.

Examples. 1. If R is a ring, and if $A = R[x]$ is the polynomial ring, then the usual derivation D, defined by

$$D(\sum_{i=0}^n a_i x^i) = \sum_{i=0}^n i a_i x^{i-1},$$

where $a_i \in R$, $i = 1, \ldots, n$, is a derivation of A.

2. Let $A \supseteq k$ be fields such that $\dim_k A = 2$. (We do not require that $\dim A_k = 2$.) Then $A = k + kx$ for any $x \in A$, $x \notin k$. If $a \in k$, then there exist unique elements $\varphi(a)$ and $D(a) \in k$ such that

$$xa = D(a) + \varphi(a)x.$$

This defines a ring monic φ of A and a φ-derivation D. $\square$

Let D be a φ-derivation of a ring A, and let $A[y]$ be the polynomial ring. Then $A[y]$ is a ring under addition of polynomials and multiplication defined by the rule

$$ya = D(a) + \varphi(a)y \tag{1}$$

and its consequences; this ring is called the ring of **φ-differential** polynomials in D and is denoted $A[y, D, \varphi]$.

Let D be any derivation of A. (Thus, $\varphi = 1$.) Then (1) implies

$$ya = D(a) + ay \tag{2}$$

and

$$y^n a = \sum_{k=0}^{n} \binom{n}{k} D^k(a) y^{n-k}, \tag{3}$$

where $D^2 = D \circ D, \ldots, D^n = D^{n-1} \circ D = D \circ D^{n-1}$ and $\binom{n}{k}$ is the binomial coefficient. In this case, the ring is called the **ring of differential polynomials over** A **in the derivation** D and is denoted by $A[y, D]$, or $A[y,']$, where the prime represents the derivation.

7.27 Proposition. *If A is left Noetherian, with a derivation D, then the differential polynomial ring $A[y, D]$ is left Noetherian.*

Proof. Let M be any left ideal of $A[y,D]$, and, as in the proof of the Hilbert basis theorem, let $L_k(M)$ denote the left ideal consisting of zero and the leading coefficients of those elements of M of degree k. (If M

does not have elements of degree k, then $L_k(M) = 0$.) Since

$$x \sum_{k=0}^{n} a_k x^k = a_n x^{n+1} + \sum_{i=0}^{n} D a_i x_i + \sum_{i=1}^{n} a_{i-1} x^i,$$

it follows that $L_n(M) \subseteq L_{n+1}(M)$ $\forall n$. Then the verbatim proof of 7.13 (and the lemma preceding it) completes the proof. □

7.28 Theorem (Amitsur [57]) *If A is a simple ring with characteristic 0 and an outer derivation D, then the differential polynomial ring $R = A[y, D]$ is simple.*

Proof. If I is a nonzero ideal of R, then I contains an element $\neq 0$

$$(1) \qquad t = ay^n + p_t(y),$$

where $\deg p_t(y) < n$, of least degree n. Let $C(I)$ denote the set of all elements $b \in A$ such that either $b = 0$ or else there is an element of the form (1) in I with $b = a$. Obviously, $C(I)$ is a left ideal of A. If $b \in A$, then

$$(2) \qquad tb = ay^n b + p_t(y) b.$$

Now

$$(3) \qquad \begin{aligned} y^n b &= \sum_{k=0}^{n} \binom{n}{k} D^k(b) y^{n-k} \\ &= by^n + \sum_{k=1}^{n} \binom{n}{k} D^k(b) y^{n-k} \end{aligned}$$

implies that

$$(4) \qquad tb = aby^n + p_{tb}(y),$$

where $\deg p_{tb}(y) < n$, so that $ab \in C(I)$. This proves that $C(I)$ is an ideal of A. Since A is simple, and $C(I) \neq 0$, then $C(I) = A$, and $1 \in C(I)$. Thus, there is an element

$$(5) \qquad t = y^n + p_t(y),$$

where $\deg p_t(y) < n$, in I. Let b be any element of A. By applying the inner derivation D_b to t, and applying (4) and (5), there results

$$(6) \qquad \begin{aligned} D_b(t) = tb - bt &= (by^n + p_{tb}(y)) - (by^n + bp_t(y)) \\ &= p_{tb}(y) - bp_t(y) \end{aligned}$$

which is an element $D_b(t)$ of I of degree $< n$. By the definition of n, $D_b(t) = 0$ for every $b \in A$. Thus, t belongs to the centralizer of A in $A[y, D]$. The next lemma determines this to be the center. Since

$t = y^n + p_t(y)$ belongs to the center of A, and $\deg p_y(t) < n$, then by the independence of $1, y, \ldots, y^n, \ldots$ over A, necessarily, $n = 0$, and so $t = 1$ is an element of I. This proves that $I = A[y,D]$ and that $A[y,D]$ is a simple ring. ▯

7.29 Lemma. *The centralizer of A in $A[y, D]$ is contained in the center of A, in case A has characteristic 0.*

Proof. If $x = \sum_{i=0}^{m} a_i y^i$ belongs to the centralizer, and if $\deg x = m$, then

$$(7)\qquad \begin{aligned} bx &= \sum_{i=0}^{m} b a_i y^i = \sum_{i=0}^{m} a_i (y^i b) \\ &= \sum_{i=0}^{m} a_i \left(\sum_{k_i=0}^{i} \binom{i}{k_i} D^{k_i}(b) y^{i-k_i} \right). \end{aligned}$$

By comparing coefficients, if $m \geq 1$, then

$$(8)\qquad b a_m = a_m b$$

and

$$(9)\qquad b a_{m-1} = a_{m-1} b + m a_m D(b).$$

Now (8) implies that a_m belongs to the center of A, and (9) asserts that

$$(10)\qquad D(b) = bz - zb,$$

where $z = -a_{m-1}/m a_m$, so that D is the inner derivation defined by z. This contradicts the assumption that D is an outer derivation. Consequently, $m = 0$, and therefore, $x = a_0$ is an element of A, hence an element of the center of A. This completes the proof of the lemma. ▯

7.30 Proposition. *Let R be a simple ring of characteristic 0, and let $A = R[x]$ be the ordinary polynomial ring. Let $B = A[y, ']$ be the ring of differential polynomials such that $x' = 1$. Then B is a simple ring.*

Proof. Let t be a nonzero element of nonzero ideal I of B, and write

$$t = \textstyle\sum_{i=0}^{n} x^i g_i(y)$$

as a polynomial in x with coefficients $\{g_i(y)\}$ in $R[y]$, $i = 0, \ldots, n$. Define $[ty] = yt - ty$. Since commuting an element of A with y

amounts to differentiating with respect to x, we obtain

$$[\cdots[[ty]y]\cdots] = n!\, g_n(y) \in I.$$

Since A has characteristic zero, then $n!$ is a nonzero element of the field $C = \text{center}\, R$, and it follows that $g_n(y) \in I$. Suppose

$$g = g_n(y) = ry^m + \cdots \text{ (terms of degree } < m),$$

where $r \in R$, $r \neq 0$. Now

$$gx - xg = mry^{m-1} + \text{terms of degree} < m - 1.$$

After m commutations with x, we obtain $m!r \in I$, whence $r \in I$. Then $RrR = R \subseteq I$, and therefore $B \subseteq I$, so that B is simple. □

7.31 *Exercises.*

7.31.1 If A is a field, and if D is a derivation, then $R = A[y, D]$ has a left and right division algorithm; for example, if f and g are elements of R, then there exist elements q and r in A such that

$$f = gq + r,$$

$\deg r < \deg g$. This implies that R is a principal left and right ideal domain.

7.31.2 If A is not a field then $A[y, D]$ is not necessarily a principal ideal ring (for example, if $A = k[x]$ is the polynomial ring over a field k, and if $Dx = 1$).

7.31.3 Every right and left ideal of $R = k[x][y, D]$, where $Dx = 1$, is projective if the field k has characteristic 0. In this case, every maximal right (left) ideal is principal. (If k has characteristic $\neq 0$, then R has gl.dim = 2. Cf. Chapter 8, esp. 8.23.)

7.31.4 (a) If D is a derivation of A, then the inner derivation D_y of $A[y, D]$ induces D.

(b) Show that the center of $A[y, D]$ is isomorphic to the polynomial ring $T[y]$, where T is the subring of center A consisting of all a such that $D(a) = ay - ya = 0$.

(c) (Amitsur [56]) Show that units $A[y, D] \subseteq A$. Conclude that $x^{-1}Ax \subseteq A$ for all units x of $A[y, D]$, but that A is not contained in the center of $A[y, D]$ when $D \neq 0$.

7.31.5 (Jacobson [50]) (a) If R has elements x and y such that $xy = 1$, but $yx \neq 1$, then R has an infinite set of orthogonal idempotents.

(b) In a ring satisfying either the a.c.c. or d.c.c. on principal right ideals, then $xy = 1$ implies $yx = 1$, for elements x and y.

Structure of V-rings

The remainder of this chapter is devoted to the structure of semiprime right Noetherian V-rings.

7.32 A Definition and Proposition. (Villamayor) *A ring R is a* **right V-ring** *in case the following equivalent conditions are satisfied:*

(a) *Each simple right R-module is injective.*

(b) *Each right ideal is the intersection of maximal right ideals.*

(c) $\operatorname{rad} M = 0$ *for all* $M \in \text{mod-}R$.

Proof. (a) $\Rightarrow$ (c). Let S_i, $i \in I$ be a set of representatives of the distinct isomorphism classes of simple right R-modules. Then $C = \prod_{i \in I} S_i$ is an injective cogenerator in Mod-R. Hence, for $M \in \text{mod-}R$, there exists a monomorphism $0 \to M \to C^I$ (see 3.55). This readily implies that $\operatorname{rad} M = 0$.

(c) $\Rightarrow$ (b). Obvious.

(b) $\Rightarrow$ (a). Let f be a map of a right ideal I into R/M, where M is an arbitrary maximal right ideal. To show that R/M is injective, it suffices to show that f can be extended to R. $\operatorname{Ker} f = I_1$ is a maximal submodule of I. Hence, there exists $x \in I - I_1$ such that $xR + I_1 = I$. By (b), there exists a maximal right ideal M_1 of R such that $I_1 \subseteq M_1$ and $M_1 + I = R$. Now $M_1 \cap I = M_1 \cap (xR + I_1) = I_1 + (xR \cap M_1) = I_1$. Therefore, by defining $\bar{f}$ to be zero on M_1 and $\bar{f} = f$ on I, we obtain the desired extension of f to R. □

7.32 B *Exercise* (Kaplansky). A commutative ring R is a V-ring if and only if every finitely generated ideal is generated by an idempotent.

7.32 C Corollary. *If R is a ring, and if* $\operatorname{rad} R = 0$, *then R is semiprime. Thus, any right V-ring is semiprime.*

Proof. If I is a nilpotent ideal of R, and if V is a simple module, then $VI \neq V$, since $VI = V$ implies $V = VI^n = 0$, for some n. Thus, $VI = 0$. This proves that I is contained in every maximal right ideal of R, and hence $I = \operatorname{rad} R = 0$. □

A right R-module M is **semisimple** provided that M is a direct sum of simple submodules. A ring R is **semisimple** in case R is semisimple both as a right and left R-module.

7.33 A *Exercise.*

7.33 A.1 The conditions are equivalent: (a) M is semisimple (b) every submodule of M is a summand (c) Every factor module of M is semisimple.

7.33 A.2 The conditions on a ring R are equivalent: (a) R is semisimple in mod-R; (b) R is semisimple in R-mod; (c) R is a finite direct sum of simple right R-modules; (d) every right module is injective; (e) every left R-module is injective; (f) every right R-module is projective; (g) every left R-module is projective.

A ring R is **right Goldie** if R satisfies $(\mathrm{acc})^{\perp}$ and $(\mathrm{acc})\oplus$. (Any subring of a right Noetherian ring satisfies $(\mathrm{acc})^{\perp}$.)

7.33B *Exercise.* (Goldie [58, 60] and Lesieur-Croisot [60]) If a ring R is a right order in a semisimple ring Q, then R is a semiprime right Goldie ring, and every essential right ideal I contains a regular element. (See the proof of 7.21 A, and see Chapter 9 for the proof of the converse. Cf. 7.23.6.)

7.34 **Proposition.** *If A is a right V-ring, and if B is a proper ideal $\neq 0$, then $xA \not\approx A$ (xA is not isomorphic to A) for any $x \in B$.*

Proof. Let M be a maximal right ideal containing B. Then $V = A/M$ is simple, hence injective. The condition $xA \approx A$ is equivalent to the existence of elements $b, c \in A$ such that $y = xa$ has zero right annihilator $y^{\perp}$ in A, and $x = yb$. (Then $ba = 1$ and $yA = xA$.) In this case, for any $v \in V$ there is a map $xA \to V$ sending yr onto vr for any $r \in A$. Since V is injective, there exists $m \in V$ such that $v = my \in Vy$, Thus, $V = Vy$, which shows that $y \notin B$. Then $x \notin B$. $\square$

7.35 **FPS Proposition.** *A ring A is a finite product of simple rings if and only if the three conditions hold:*

FPS_1: *A has finite ideal lattice.*

FPS_2: *A has no proper ideals which are essential right ideals.*

FPS_3: *A is semiprime.*

Proof. The necessity is obvious. Conversely, assume FPS, and let B be a maximal ideal. Since B is not essential, then $B \cap I = 0$ for some nonzero right ideal I. Thus, $IB = 0$ (since $IB \subseteq I \cap B$) and then $(BI)^2 = 0$, so that semiprimeness of A implies $BI = 0$. Thus, $A_1 = \mathrm{ann}_A B$ is a nonzero ideal, and maximality of B yields a ring product $A = B \times A_1$. Since B inherits FPS, by induction on the number of ideals of A, the proof is done. $\square$

The next theorem was first proved by Faith in 1964 (published in [67]) assuming A is prime, and Ornstein [66] assuming A is right Noetherian.

7.36 A **Theorem.** *If R is a right V-ring, and if R is a right order in a semisimple ring, then R is a finite product of simple right V-rings.*

Proof. By 7.32C, any right V-ring R is semiprime (FPS_3). Assume that R is a right order in a semisimple ring Q. If Q is a direct sum of n

simple right modules, then every set of independent right ideals of R has cardinality $\leq n$. (Proof?) Furthermore, every essential right ideal contains a regular element of R. Thus, 7.34 shows that R has FPS_2. Then the bound n on the summands of R, and the proof of 7.35, show that R has FPS_1. Thus, R is a finite product of simple rings by 7.35. Clearly, each simple factor is again a V-ring □.

7.36 B *Exercise.* Prove 7.36 for right Noetherian V-rings, assuming it for the case R is prime. How does 7.36 imply the result for R right Noetherian?

Solutions of Linear Differential Equations

The rest of this chapter is devoted to the results of Cozzens [70] characterizing when differential (resp. twisted) polynomial rings are V-rings.

A (homogeneous) linear differential equation in D over k is an equation of the form

$$\left(\sum_{i=0}^{n>0} a_i D^{(i)}(x) = 0 \qquad a_i \in k,\ 0 \leq i \leq n\right)$$

$$\sum_{i=0}^{n} a_i D^{(i)}(x) = b \qquad a_i, b \in k,\ 0 \leq i \leq n.$$

For the remainder of this section, denote $k[y, D]$ by R.

7.37 **Lemma.** *Assume that every homogeneous linear differential equation in D over k has a nontrivial solution in k. If $f = \sum_{i=0}^{m} a_i y^i \in R = k[y, D]$, then there exist $a \in k$ and $g \in R$ such that $\delta(g) = m - 1$* and $af = (y+1)(g)$.

Proof. Let $g = \sum_{i=0}^{m-1} b_i y^i$. We shall determine $a, b_i \in k$, $0 \leq i \leq m-1$, such that

(1) $$af + (y+1)g = 0.$$

By expanding equation (1) and equating coefficients, we obtain

$$\begin{aligned} &a a_0 + D(b_0) + b_0 = 0 \\ &a a_1 + b_0 + D(b_1) + b_1 = 0 \\ &\vdots \\ &a a_{m-1} + b_{m-2} + D(b_{m-1}) + b_{m-1} = 0 \\ &a a_m + b_{m-1} = 0 \end{aligned}$$

Clearly, this system of equations is equivalent to an equation of the form

$$\sum_{i=0}^{m} c_i D^{(i)}(a) = 0, \tag{2}$$

$c_i \in k$, $0 \leq i \leq m$. By hypothesis, (2) has a nontrivial solution in k. □

7.38 Corollary. *If every homogeneous linear differential equation in D over k has a nontrivial solution in k, then the only irreducible elements in $R = k[y, D]$ are those of degree 1. Hence, $V_\alpha = R/(y - \alpha)R$ is a simple right R-module for all $\alpha \in k$ and conversely.* □

7.39 Lemma. *If every linear differential equation in D over k has a solution in k and every homogeneous linear differential equation in D over k has a nontrivial solution in k, then $V_\alpha = R/(y - \alpha)R$ is a divisible right R-module for all $\alpha \in k$, where $R = k[y, D]$.*

Proof. It suffices to show that

$$V_\alpha(y + \beta) = V_\alpha$$

for all $\alpha, \beta \in k$. Equivalently, given $h \in R$, $\delta(h) = 0$, there exist $f, g \in R$ such that

$$f(y + \beta) + (y + \alpha)g = h. \tag{1}$$

We shall determine $a, b \in k$ such that

$$a(y + \beta) + (y + \alpha)b = h. \tag{2}$$

Equation (2) is equivalent to an equation of the form

$$D(b) + (\alpha - \beta)b = h. \tag{3}$$

By hypothesis, there exists $b \in k$ satisfying (3). □

7.40 Theorem (Cozzens [70]) *The following conditions on the ring of differential polynomials $R = k[y, D]$ are equivalent:*

(a) *R is a right V-ring having, up to isomorphism, a unique simple right R-module.*

(b) *Every linear differential equation in D over k has a solution in k and every homogeneous linear differential equation in D over k has a nontrivial solution in k.*

Proof. (a) ⇒ (b). k is naturally a right R-module, the operation of R on k given by

$$ar = \sum_{i=0}^{n} D^{(i)}(a) r_i, \quad a \in k, \quad r = \sum_{i=0}^{n} r_i y^i \in R.$$

Clearly, as a right R-module, k is simple and, hence, injective. Let

$$\sum_{i=0}^{n} a_i D^{(i)}(x) = 0, \tag{1}$$

$a_i \in k$, $0 \leq i \leq n$, be an arbitrary homogeneous linear differential equation in D over k and let $d = \sum_{i=0}^{n} a_i y^i$.

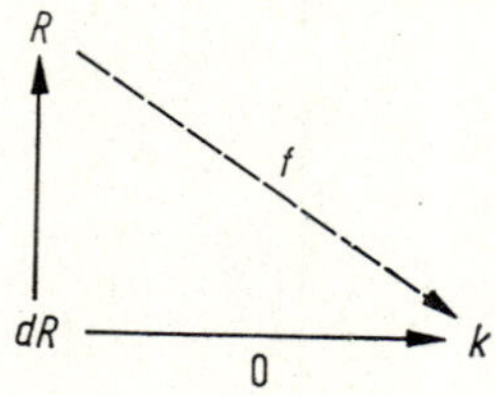

The preceding diagram can be completed with a nonzero f by injectivity of k and the fact that k is, up to isomorphism, the unique simple right R-module. $f(1)$ is a nontrivial solution for (1).

To show that linear differential equations of the form

$$\sum_{i=0}^{n} a_i D^{(i)}(x) = b,$$

$a_i, b \in k$, have a solution in k, one merely extends the map taking $d = \sum_{i=0}^{n} a_i y^i$ to b to R and observes as before, that the image of the identity of R is the desired solution.

(b) $\Rightarrow$ (a). By 7.38 and 7.39, each simple right R-module is of the form $V_\alpha = R/(y-\alpha)R$ and is divisible (i.e., $V_\alpha x = V_\alpha$ for all regular elements $x \in R$). Since divisibility is equivalent to injectivity in a principal right ideal domain 3.45, V_α is an injective right R-module for all $\alpha \in k$. Hence, R is a right V-ring.

In order to complete the proof of (b) $\Rightarrow$ (a), all that remains to be shown is that, up to isomorphism, R has a unique simple right R-module, i.e., $V_\alpha \approx V_\beta$ for all $\alpha, \beta \in k$. The hypotheses relative to homogeneous linear differential equations imply that there exist nonzero $a, b \in k$ such that $a(y-\alpha) = (y-\beta)b$. The map

$$V_\alpha \to V_\beta,$$

defined by

$$r + (y-\alpha)R \mapsto ar + (y-\beta)R,$$

is the desired isomorphism. $\square$

If k is taken to be the quotient field of $Q[X]$, Q the field of rational numbers, and D ordinary differentiation, then $k[y, D]$ is not a right V-ring; e.g., by 7.39, $D(t) = 1/x$ cannot be solved.

Universal Differential Fields

Let k be a field of characteristic 0 and D be a derivation of k. A result due to Kolchin asserts the existence of a field $U \supseteq k$ and a derivation $\overline{D}$ of U extending D such that the equation

$$p\big(x, \overline{D}(x), \ldots, \overline{D}^{(n)}(x)\big) = 0,$$

n arbitrary, has a solution $\xi \in U$ for all $p(X) \in U[X_1, \ldots, X_{n+1}] - U$. Furthermore, every homogeneous linear differential equation in $\overline{D}$ over U has a nontrivial solution in U. Such a field U is called a **universal extension** of k or a **universal differential field** (Kolchin [53]).

7.41 *Exercise.* If f is an element of degree n in the ring $k[y, D]$ of differential polynomials over a universal field, then there exist elements $a_0, a_1, \ldots, a_n$ in k such that

$$f(y) = a_0(y - a_1) \cdots (y - a_n).$$

The ring of differential polynomials over universal differential fields provides examples of Noetherian V-domains not fields.

7.42 **Theorem** (Cozzens [70]) *Let k be a universal differential field with derivation D. The ring $R = k[y, D]$ has the following properties:*

(a) *R is a principal right and left ideal domain.*

(b) *R is simple.*

(c) *R is a right V-ring.*

(d) *R is not a field.*

(e) *R has, up to isomorphism, a unique simple right R-module, hence, a simple injective cogenerator.*

Proof. (a) follows from 7.31.1. (b) is implied by 7.34, since R is a domain.

(c) and (e). Since k is a universal differential field, condition (b) of 7.40 is trivially satisfied. This, together with 7.55, implies (e), whence (c). □

Twisted Polynomials

Let $\overline{F}_2$ denote an algebraic closure of $\mathbb{Z}/2\mathbb{Z}$ and ϱ, the automorphism of $\overline{F}_2$ defined by: $z \mapsto z^2$. $\overline{F}_2[t, \varrho]$ will denote the ring of twisted polynomials in t over $\overline{F}_2$, i.e., the additive group of $\overline{F}_2[t, \varrho]$ is the additive group of the ring of polynomials in the indeterminate t with coefficients in $\overline{F}_2$, and multiplication in $\overline{F}_2[t, \varrho]$ is defined by $ta = \varrho(a)t$, for all $a \in \overline{F}_2$, and its consequences. It is well known that $\overline{F}_2[t, \varrho]$ is a principal right and left ideal domain. Furthermore, it is easy to show that the only two-sided ideals of $\overline{F}_2[t, \varrho]$ are those of the form $t^k\overline{F}_2[t, \varrho]$, k an integer ≥ 0 (Jacobson [42]).

Let $R = \overline{F}_2[t, \varrho]$, $M = \{t^k \mid k \text{ an integer } \geq 0\}$, and $R_M = \{a/m \mid a \in R,\ m \in M\}$. For a/t^k and $b/t^{k+i} \in R_M$, $a/t^k = b/t^{k+i}$ if and only if $b = t^i a$. We define addition and multiplication in R_M as follows: $a/t^k + b/t^k = (a + b)/t^k$ and $a/t^i b/t^j = \varrho^j(a)b/t^{i+j}$, where $\varrho^j(a)$ is that element of R obtained by applying ϱ^j to all the coefficients of a (Jacobson [42, p. 211]). Clearly, R_M is a simple integral domain, not a field. Moreover, R_M is a principal right and left ideal domain.

7.43 Lemma. *Given $r = \sum_{i=0}^n a_i t^i \in R$, $a_n = 1$, there exist $\alpha_i \in \overline{F}_2$, $1 \leq i \leq n$, such that $r = \prod_{i=0}^n (t - \alpha_i)$. In particular, the irreducible elements of R are those of the form $t - \alpha$, $\alpha \in \overline{F}_2$.*

Proof. Analogous to 7.37. □

One readily sees that the simple right R_M-modules are of the form R_M/pR_M, where $p = t - \alpha$, $\alpha \neq 0 \in \overline{F}_2$.

7.44 Lemma. *$R_M/(t - \alpha)R_M$ is a divisible right R_M-module for all $\alpha \in F_2 - \{0\}$.*

Proof. Analogous to 7.39. □

7.45 Theorem. *The ring R_M has the following properties:*

(a) *R_M is a principal right and left ideal domain.*

(b) *R_M is simple.*

(c) *R_M is a right V-ring.*

(d) *R_M is not a field.*

(e) *R_M has, up to isomorphism, a unique simple right R_M-module.*

Proof. Analogous to 7.42. □

The domains constructed in 7.42 and 7.45 are called **Cozzens's domains**.

7.46 *Exercise.* Show that Cozzens's domains are also left V-rings.

Exercises for Chapter 7

The first two exercises are extracted from Cozzens [70].

1. If R is a principal right and left ideal domain, then R/I is right and left Artinian, for any ideal $I \neq 0$. Let $R = k[y, D]$, where k is a universal differential field.

(a) Every cyclic module is either isomorphic to R or is a direct sum of finitely many copies of the unique simple module V.

(b) Let K denote the right quotient field of R. Then K/R is a direct sum of copies of V.

(c) Any two maximal right ideals $p_1 R$ and $p_2 R$ are such that $p_1 = u p_2$ for some unit u of R.

(d) Every element x of K has a unique expression

$$x = a + a_1/p_1 + \cdots + a_n/p_n,$$

where p_i is irreducible and $a, a_i \in R$, $i = 1, \ldots, n$.

(e) V is a simple injective cogenerator.

(f) Every module fully invariant in its injective hull is injective.

2. If mod-R has a Noetherian cogenerator, then every object of mod-R has a maximal submodule. This holds, e.g., for $R = k[y, D]$ over a universal field k. (Cf. B-rings, Chapter 22.)

3A. (Boyle [71]) A ring R is said to be **right PCI** if every cyclic module $C \not\approx R$ is injective. A ring R is **right QI** if every quasinjective module is injective. Prove that every right Noetherian right PCI ring is right hereditary. Moreover, for a left Noetherian, left hereditary ring, the following are equivalent: (1) right Noetherian right PCI; (2) right hereditary right QI; (3) right hereditary right V-ring. In this case, the ring is either semisimple, or a right Ore domain.

3B. (Koehler [70]) If every right R-module is fully invariant in its injective hull, then R is a right Noetherian V-ring.

4. (Jategaonker [68]) Let D be a ring, and let $\varrho : D \to D$ be a ring monic. In the set $D[x]$ of left polynomials, define an operation $D[x]^2 \to D[x]$ by the rule $xd = \varrho(d)x$ (and its consequences) $\forall d \in D$. [Thus: $x^n d = \varrho^n(d) x^n$ $\forall n \geq 0$.] Then, $D[x]$ is a ring, called the **twisted polynomial ring**, denoted $D[x, \varrho]$. Let D^* denote the nonzero elements of D. If D is an entire ring show that $D[x, \varrho]$ is a principal left ideal ring if and only D is a principal left ideal ring such that $\varrho(D^*) \subseteq$ units D.

5. A ring R is a generalized twisted polynomial ring (gtp) over a ring D in case for some ordinal α, R is the union $\bigcup_{\beta<\alpha} R_\beta$ of a chain of rings $\{R_\beta \mid \beta < \alpha\}$ such that

(i) $R_0 = D$,

(ii) $R_{\beta+1} = R_\beta[x_\beta, \varrho_\beta]$ is a twisted polynomial ring, and if β is a limit ordinal, then

(iii) $R_\beta = [\bigcup_{\gamma<\beta} R_\gamma][x_\beta, \varrho_\beta]$.

Next assume D is an entire ring.

(a) Show that every element $a \in R$ has unique representation as a left polynomial in finitely many of the x_β with coefficients in D.

*(b) Let $\aleph_n$ be the least cardinal such that every right ideal of R can be generated by $\aleph_n$ elements. Then

$$\text{r.gl.dim}\, R \leq (\text{w.r.gl. dim}\, R) + n + 1 .$$

(The weak dim of M is defined similarly to dim M using flat instead of projective modules.) Conclude that r.gl.dim $R = \infty$. Cf. p. 371.

(c) If R_β is a principal left ideal ring $\forall \beta < \alpha$, then R is a principal left ideal ring (and is right hereditary).

6. In the preceding exercise, assume that D is a field and that $\varrho_\beta(R_\beta) \subseteq D$ for all β. Let $|R| = \alpha$.

(a) Show that $\{1 + x_\beta\}_{\beta<\alpha}$ is a basis of a free right ideal of R.

(b) If α is a limit ordinal, then any ideal of R contains x_β for some β.

(c) Every nonzero left ideal contains a nonzero ideal of R.

(d) If I is a nonzero right ideal of R, then I contains no ideal $\neq 0$.

*(e) Conclude from (c) that R is not a left primitive ring and from (d) that R is a right primitive ring. (A ring is primitive if it has a faithful simple module, see Chapter 19.)

7. Let R be a simple ring, let I be a right ideal $\neq 0$, and let $K = \text{End}\, I_R$. Show that there is a category equivalence mod-$R \rightsquigarrow$ mod-K if and only if K is a simple ring.

8. Let A be a field, and let $D: A \to A$ be a φ-derivation of A, where φ is not an automorphism of A. For example, let $A = \mathbb{Q}(x)$, and let $\varphi: A \to A$ be defined by $\varphi: x \mapsto x^2$. Thus,

$$\varphi\left(\sum_{i=0}^n \alpha_i x^i\right) = \sum_{i=0}^n \alpha_i x^{2i}$$

$\forall \alpha_i \in \mathbb{Q}$. Let S be the corresponding ring of φ-differential polynomials.

(a) The entire ring S has a left-hand division algorithm.

(b) S is a principal left ideal ring.

(c) S is left Ore, but not right Ore. [Hint: $(y + a)S \cap y^2S = 0$.]

Notes for Chapter 7

Fairly complete references for this chapter appear in the text, and previous comments on simple rings appear in Notes for Chapter 4. Also see Cozzens and Faith [73/74].

References

Artin [27], Artin, Nesbitt, and Thrall [43], Artin and Whaples [44], Auslander-Goldman [60], Bass [62, 68], Boyle [71], Camillo [70], Camillo and Fuller [72], Cohn [62, 66a], Cozzens [70], Dickson and Fuller [72], Faith [64, 67a, b, 71a, 72a], Faith and Utumi [65b], Findlay and Lambek [58], Fuller [71], Gabriel [62], Goldie [58], Jacobson [42, 45], Jans [70], Jategaonker [68, 70a], Johnson [51a, b], Koehler [70], Morita [58], Popesco-Gabriel [64], Shepherdson [52], Utumi [56], Wedderburn [08].

Chapter 8

Semisimple Modules and Homological Dimension

A module M_R is **semisimple** in case M is a direct sum $\coprod_{i\in I} M_i$ of a family of simple submodules $\{M_i \mid i \in I\}$. A ring R is **semisimple** in case R_R is. We have seen that any simple Artinian ring is semisimple. Finite ring products of simple Artinian rings are also semisimple. The Wedderburn-Artin theorem implies the converse: Every semisimple ring is isomorphic to a finite product of full matrix rings of various degrees over various fields. These rings are also characterized by the Wedderburn-Artin theorem (8.8) as right Artinian rings with no nilpotent ideals. (The condition is right-left symmetric.)

Over semisimple rings, every module is projective. In this chapter, the projective dimension of a module is defined. The global dimension of the ring is the supremum of the projective dimensions of all modules. Semisimple rings are the rings of (right and left) global dimension zero. (Rings of global dimension 1, namely, hereditary rings, are studied in Chapter 31.) Simple Noetherian rings of global dimension ≤ 2 are similar to Ore domains (8.27).

The Hilbert-Szyzgy theorem 8.17 asserts that the polynomial ring $k[x_1, \ldots, x_n]$ over a field k has global dimension n. In general, the right global dimension equals the supremum of the projective dimensions of cyclic right R-modules (8.19) and, if R is not semisimple, equals the supremum of the projective dimensions of right ideals plus 1. (The global dimension of a free algebra $k[X]$ in any number of variables over any commutative ring k exceeds that of k by 1. Thus, $k[X]$ is hereditary when k is a field. The latter two results are not proved in the text, however.)

The following proposition asserts that in the definition of semisimple, the word "direct" can be deleted.

8.1 Proposition. *Let M be a module that is the sum $\sum_{i\in I} M_i$ of a family $\{M_i \mid i \in I\}$ of simple submodules.* Then

8.1.1 *If N is a submodule $\neq M$, then there exists a subset J of I such that $M = N \oplus (\coprod_{j\in J} M_i)$.*

8.1.2 *M is semisimple.*

Proof. Let $N \neq M$ be a submodule. If $i \in I$ is such that $M_i \cap N \neq 0$, then since M_i is simple, it follows that $M_i \cap N = M_i$; that is, $M_i \subseteq N$. Since $N \neq M$, we conclude that $M_j \cap N = 0$ for some $j \in I$. Let P denote the collection of all subsets J of I such that:

(i) The sum $\sum_{j\in J} M_j$ is direct;

(ii) $N \cap \sum_{j\in J} M_j = 0$.

Since an arbitrary collection $\{M_q \mid q \in Q\}$ is independent if and only if each subset $\{M_j \mid j \in J,\ J \subseteq Q\}$ is independent, it follows that P is inductive, and, by Zorn's lemma, P contains a maximal element J. Consider the submodule $N' = N + \sum_{j\in J} M_j$. By definition, $N' = N \oplus (\coprod_{j\in J} M_j)$. Suppose for the moment that $N' \neq M$. Then, by the preceding paragraph, there exists $q \in I$ such that $N' \cap M_q = 0$. Then, $J' = J \cup \{q\}$ belongs to P and properly contains J, a contradiction that establishes that $M = N \oplus \coprod_{j\in J} M_j$. The last statement in the proposition is the $N = 0$ case (cf. Unique decomposition theorem, Chapter 21).

8.2 Corollary. *The following conditions on a module are equivalent:*

(a) *M is semisimple.*

(b) *Every submodule of M is a summand.*

(c) *Every factor module of M is semisimple.*

Proof. (a) $\Rightarrow$ (b) follows from 8.1.1.

(b) $\Rightarrow$ (a). Assume that each submodule of M is a summand of M. Then each submodule H of M has the same property. [If G is a submodule of H, then $M = G \oplus K$ for some submodule K of M, and then $H = G \oplus (K \cap H)$.] By the proposition, M is semisimple if and only if M is a sum of simple submodules. We first show that M has at least one simple submodule. To do this, let x be any nonzero element of M, and let N be a submodule of M that is maximal in the set P of those submodules of M that do not contain x. (P is nonempty since $0 \in P$, and P is inductive, so N exists by Zorn's lemma.) Now write $M = N \oplus Q$, where Q is a submodule. Suppose for the moment that Q is not simple, and let A be a submodule, $Q \supset A \supset 0$. Then $Q = A \oplus B$, where $Q \supset B \supset 0$. Now $N + A$ and $N + B$ both properly contain N, hence both contain x. But this is a contradiction since $(N + A) \cap (N + B) = N$, and $x \notin N$. This contradiction shows that M contains a simple module Q. Next, let S denote the sum of the simple submodules of M, and write $M = S \oplus T$, where T is a submodule. By what we have shown, if $T \neq 0$, then T contains a simple submodule that would have to belong to S, which is impossible since $S \cap T = 0$. Thus, $T = 0$, and $M = S$ is semisimple as required.

(a) $\Rightarrow$ (c). If K is a submodule, then $M = K \oplus M'$ for some submodule M', and M' is semisimple by (b). Then $M/K \approx M'$ is semisimple. This proves (b) $\Rightarrow$ (c), and (c) $\Rightarrow$ (a) is trivial. $\square$

8.2′ *Exercises.*

8.2′.1 Use 8.2 to show that a direct sum of finitely many simple modules is both Artinian and Noetherian (cf. 8.10).

*8.2′.2 Show that if $M = \sum_{i=1}^{n} \oplus M_i = \sum_{j=1}^{m} \oplus M'_j$, where $\{M_i\}$ and $\{M_j\}$ are simple submodules, then $m = n$, and there exists a permutation π of $\{1, \ldots, n\}$ such that $M'_i = M_{\pi(i)}$, $i = 1, \ldots, n$ (cf. unique decomposition theorem 21.2).

The **socle** of a module M is defined to be the sum of all the simple submodules of M and is denoted socle M. By 8.2, socle M is a semisimple module containing every semisimple submodule.

8.3 **Proposition**[1]. *The socle of a module M is the intersection of all the essential submodules of M.*

Proof. If S is a simple submodule, and if A is an essential submodule, then $A \cap S \neq 0$, so $A \cap S = S$, and $S \subseteq A$. Thus, socle M is contained in every essential submodule of M.

Conversely, let Q denote the intersection of the essential submodules of M, and let N be any submodule of Q. If K is maximal in the set of submodules of M such that $N \cap K = 0$, then $H = N + K$ is an essential submodule of M. Therefore, since $H \supseteq Q$, and since $H = N \oplus K$, then $Q = N \oplus (K \cap Q)$. Since every submodule N of Q is thus a summand, then Q is zero or semisimple by the proposition. Thus, $Q \subseteq$ socle M, and $Q =$ socle M. $\square$

Prime and Semiprime

An ideal P of R is **prime** in case, given right ideals $A, B \supseteq I$, then $AB \subseteq P$ implies $A \subseteq P$ or $B \subseteq P$. (The product AB of the two right ideals A and B has been defined in Chapter 3.) The product RB is an ideal of R; hence,

$$AB = (AR)B = A(RB) \subseteq P \Leftrightarrow R(AB) = (RA)(RB) \subseteq P.$$

[1] This proposition appears in Sandomierski's thesis [64]. I have also seen it referenced as the Kasch-Sandomierski proposition.

It follows that P is prime if and only if

$$\forall \text{ ideals } A', B' \text{ of } R \ (A'B' \subseteq P \Leftrightarrow A' \subseteq P \text{ or } B' \subseteq P).$$

Thus, a maximal ideal must be prime since R is the only ideal containing it. By symmetry, an ideal P is prime if and only if the defining statement is valid with "left" substituting "right". Recall from Chapter 3 that a ring R is **prime** provided that 0 is a prime ideal; R is **semiprime** if 0 is the only nilpotent ideal. Thus, every prime ring is semiprime.

Every semisimple ring R is semiprime: If I is a right ideal, then I is a summand of R, so $I = eR$ for some idempotent e, hence $I^n = I$ for any $n \geq 1$. Thus R is semiprime.

8.4 Proposition. *If I is a simple right ideal in a ring R, then either $I^2 = 0$ or I is generated by an idempotent.*

Proof. If $I^2 \neq 0$, then there exists $b \in I$ such that $bI \neq 0$, and then simplicity of I implies $I = bI$. Since $I \neq 0$, this shows that $b^\perp$, the right annihilator of b in R, does not contain I. This fact, and the simplicity of I, implies that $b^\perp \cap I = 0$. Since $I = bI$, there exists $e \in I$ such that $b = be$, and then $be = be^2$. Since $e - e^2 \in b^\perp \cap I$, necessarily, $e = e^2$. Since $b \neq 0$, certainly, $e \neq 0$ and $eR \neq 0$. Since $e \in I$, necessarily, $eR \subseteq I$, and $I = eR$. ▯

8.5 Corollary. *If R is a semiprime ring, every simple right ideal is generated by an idempotent.*

8.6 *Exercises.*

8.6.1 Give a proof of 8.4 that uses the concept of a generator. (Since I is minimal, $D = \operatorname{End} I_R$ is a field, so I is a generator over D, and therefore I is projective over $S = \operatorname{End}_D I$. First assume R is prime. Then R is embedded in $S = \operatorname{End}_D I$, and I is cyclic. (Why?) Show that I is a summand of S.)

8.6.2 In a semiprime ring, every right ideal that is a finite sum of simple right ideals is generated by an idempotent.

8.6.3 If a prime ideal P is not an essential right ideal, then it is a right annulet.

Structure of Artinian Semiprime Rings

8.7 Theorem. *Every Artinian semiprime ring is semisimple.*

Proof. By 7.10′.1, R is a direct sum of indecomposable right ideals, $R = \coprod_{i=1}^n A_i$. Since R is Artinian, the right ideal A_1 contains a simple right ideal V_1 that by 8.5 is a summand of R, whence a summand of A_1, hence equal to A_1. Thus every A_i is simple, and R is semisimple. ▯

8.8 Wedderburn-Artin Theorem. *The following conditions on a ring $R \neq 0$ are equivalent:*

(a) *R is right Artinian and semiprime.*

(b) *R is a direct product of a finite number of simple Artinian rings.*

(c) *R is left Artinian and semiprime.*

Proof. (a) $\Rightarrow$ (b). Let M be any maximal ideal of R, and let J be minimal in the set of right ideals of R properly containing M. Then $S = R/M$ is a simple ring containing the simple right ideal J/M. By the principal Wedderburn-Artin theorem, S is Artinian (and $\approx D_n$ for a field D). If $M = 0$, the proof is complete.

R is semisimple by 8.7. Thus $R = M \oplus Q$ for some right ideal Q. Now $QM \subseteq M \cap Q$, so $QM = 0$, and therefore $(MQ)^2 = 0$. Since R is semiprime, $MQ = 0$. Write $1 = g + h$, with $g \in M$, $h \in Q$. Then

$$ga = ag = a, \qquad hb = bh = b$$

for all $a \in M$, $b \in Q$; that is, g (resp. h) is an identity element of M (resp. Q), so these ideals are rings. Furthermore, $Q \approx R/M$ is simple Artinian. Since M also is right Artinian and semiprime, we can write $M = Q_2 \oplus M_2$, where Q_2 is simple Artinian. Since R is Artinian, this procedure must stop in a finite number of steps; that is, eventually

$$R = \prod_{i=1}^{t} Q_i,$$

where Q_i is simple Artinian.

(b) $\Rightarrow$ (a). This is an immediate consequence of the fact that every simple right Artinian ring is semiprime and Artinian.

(b) $\Leftrightarrow$ (c) follows by symmetry. $\square$

We insert another proof of (a) $\Rightarrow$ (b). Since R_R is semisimple (and finitely generated), we can write

$$R_R \approx V_1^{n_1} \oplus \cdots \oplus V_k^{n_k},$$

where V_i is a simple right ideal, not isomorphic to V_j for $i \neq j$, and where $n_i > 0$. Since any epimorph of V_i is either 0 or an isomorph (Schur's lemma), we see that $\mathrm{Hom}_R(V_i, V_j) = 0 \ \forall i \neq j$. This fact, and the formula

$$\mathrm{Hom}(\prod A, \prod B) \approx \prod \mathrm{Hom}(A, B)$$

for finite products (cf. 3.33.2; also 3.38) implies

$$R \approx \mathrm{Hom}_R(R, R) \approx \prod_{i=1}^{k} \mathrm{Hom}_R(V_i^{n_i}, V_i^{n_i})$$

$$\approx \prod_{i=1}^{k} \mathrm{Hom}_R(V_i, V_i)^{n_i} = \prod_{i=1}^{k} M_{n_i}(D_i),$$

where D_i is the field $\mathrm{Hom}_R(V_i, V_i)$, and where $M_{n_i}(D_i)$ is the matrix ring, $i = 1, \ldots, n$.

8.9 Corollary. *If R is a prime Artinian ring $\neq 0$, then $R \approx D_n$, for some field D and integer $n > 0$.* □

8.10 *Exercises.*

8.10.1 If $R = \coprod_{i=1}^{t} S_i$ and $R = \coprod_{i=1}^{n} R_i$ are two representations of R as direct products of finitely many simple rings assumed to lie in R, then $n = t$, and there is a permutation π such that $S_i = R_{\pi(i)}$, $i = 1, \ldots, n$.

8.10.2 The center of a semisimple ring is a direct sum of fields.

8.11 Theorem. *The following conditions on a ring R are equivalent:*

(a) *R is semisimple in* mod-R.

(b) *R is right Artinian and semiprime.*

(c) *R is semisimple in* R-mod.

When this is so, then every right and every left R-module is semisimple.

Proof. (a) ⇒ (b). If R is semisimple, then R is a direct sum of simple right ideals, $R = \coprod_{i \in I} A_i$. Writing $1 = x_{i_1} + \cdots + x_{i_n}$, with $x_{i_j} \in A_{i_j}$, $j = 1, \ldots, n$. This shows that R is a direct sum of only finitely many of the A_i, say $R = \coprod_{i=1}^{n} A_i$, and R is then right Artinian (also Noetherian) by 7.10. Since every nonzero right ideal I of semisimple ring of R is a summand, $I = eR$, with $e = e^2 \in R$, $e \neq 0$, then I is not nilpotent.

(b) ⇒ (c). By the Wedderburn-Artin theorem, $R = \coprod_{i=1}^{n} S_i$, where S_i is a simple Artinian ring $\approx D_n$ for some field D (n and D depending on S_i), $i = 1, \ldots, n$. If $\{e_{ij} \mid i, j = 1, \ldots, n\}$ are matrix units of $A = D_n$, then

$$A = A e_{11} \oplus \cdots \oplus A e_{nn}$$

is a direct sum of left ideals that are each simple. (This is trivially verified.) Since each (simple) left ideal of S_i is a simple (left) ideal of R, R is a direct sum of simple left ideals.

By the Wedderburn-Artin theorem 8.8, we have also (b) ⇒ (a), and (c) ⇒ (b) by symmetry with (a) ⇒ (b).

8.12 *Exercise. The following propositions about a ring R are equivalent:*

8.12.1 *R is a semisimple ring.*

8.12.2 *Each right ideal of R is a summand of R_R.*

8.12.3 *Each right ideal of R is injective.*

8.12.4 *Each right R-module M is semisimple.*

8.12.5 *Each right R-module M is injective.*

8.12.6 *Each right R-module is projective.*

8.12.7 *Each cyclic right R-module is projective.*

8.12.8 *The right-left symmetry of* 8.12.2 *to* 8.12.7.

8.12.9 *R is similar to a finite product of fields.* □

Dimension of Rings and Modules

An **injective resolution** (**inj. res.**) of a module M is an infinite exact sequence

$$\text{(inj. res.)} \qquad 0 \to M \to M_0 \to M_1 \to \cdots \to M_n \to \cdots$$

such that M_n is injective, $n \geq 0$. If $M_k \neq 0$, and if $M_n = 0 \ \forall n > k$, then the inj. res. is said to have length k. Otherwise, the length is defined to be ∞.

The **injective dimension** of M, abbreviated inj. dim M, is the l.u.b. of the lengths of all possible injective resolutions. For example, M is injective if and only if inj. dim $M = 0$.

Projective resolution and projective dimension (proj. dim) are defined by dualism. The projective dimension of a module M over a ring R is also called **homological dimension**, abbreviated $\dim_R M$, or simply dim M.

The **right global dimension** of a ring R is defined to be

$$\text{r.gl. dim } R = \sup \{\dim M \mid M \in \text{mod-}R\}.$$

For example, R is semisimple if and only if r.gl. dim $R = 0$. The left global dimension is denoted l.gl. dim. One can prove that

$$\text{r.gl. dim } R = \sup \{\text{inj. dim } M \mid M \in \text{mod-}R\}$$

(see Cartan-Eilenberg [56], Northcott [60], or MacLane [63]).

8.13 A Proposition. *Every module M has an injective resolution and a projective resolution.*

Proof. We prove that inj. res. exists, and the proof for proj. res. is dual. Step one is the embedding $M \to M_0$, where $M_0 =$ inj. hull M. Assume exactitude of

$$(1) \qquad M \to M_0 \to \cdots \to M_{n-1},$$

where M_i is injective, $i = 0, \ldots, n-1$. If $M_{n-2} \to M_{n-1}$ is epic, then M has an inj. res. for which $M_k = 0$, $k \geq n$. Otherwise, $f : M_{n-2} \to M_{n-1}$ is not epic, and we embed $h : M_{n-1}/\text{im } f \to M_n$, where M_n is injective. Thus, there is an exact sequence

$$M_0 \to M_1 \to \cdots \to M_{n-1} \to M_n,$$

where $M_{n-1} \to M_n$ is fh. This completes the induction. □

8.13B Proposition. *Projective, injective, and global dimension are Morita invariants.* □

8.14A Inequality Theorem. *Let A be a right R-module, and B a submodule. Then:*

8.14A.1 *If* $\dim A > \dim B$, *then* $\dim A = \dim A/B$.

8.14A.2 *If* $\dim A < \dim B$, *then* $\dim A/B = 1 + \dim B$.

8.14A.3 *If* $\dim A = \dim B$, *then* $\dim A/B \leq 1 + \dim A$.

Proof. Exercise, or see Theorem 1.2 of Kaplansky [59], or [69a, p. 169].

8.14B Corollary. *If* $B \subseteq A$, *then* $\dim B \leq \max\{\dim A, \dim A/B\}$. *Moreover,* $\dim B = \dim A/B$ *implies* $\dim B = \dim A$.

8.14C Corollary. *Let T be a subring of a ring R, and assume that T is a summand of R as a (T, T)-bimodule. Then:*

$$\text{r.gl. dim}\, T \leq \text{r.gl. dim}\, R + \dim {}_T R.$$

Proof. Exercise, or see Kaplansky [69a].

A ring R of r.gl. dim ≤ 1 is said to be **right hereditary**. By 8.12, the semisimple rings are the rings of global dimension 0. The next result has the corollary (8.16) that

$$\text{r.gl. dim}\, R[x] = \text{r.gl. dim}\, R + 1,$$

where $R[x]$ denotes the polynomial ring. Thus, $R[x]$ is right hereditary whenever R is semisimple.

8.15A Change of Rings Theorem. *Let R be a ring, and x a central element which is not a zero divisor in R. Let R/x denote the factor ring R/(x). If A is a nonzero R/x module, and if* $\dim_{R/x} A = n$ *is finite, then* $\dim_R A = n + 1$. *Hence:*

$$\text{r.gl. dim}\, R \geq 1 + \text{r.gl. dim}\, R/x.$$

Proof. (Kaplansky [69a, p. 172]) The proof is by induction on n. Case $n = 0$. Then A is R/x-projective, and hence is a summand of a free R/x-module F. Since x is not a zero divisor, the ideal (x) is a free module, and so (proj.) $\dim_R R/x = 1$, equality holding since R/x is not a projective R module (the ideal (x) is not a summand of R). Likewise $\dim_R F = 1$, and hence $\dim_R A \leq 1$. However, A cannot be projective in mod-R, inasmuch as no nonzero projective module can be annihilated by x. Therefore, $\dim_R A = 1$.

Case $n > 0$. Map a free R/x module G onto A with kernel K. Then, $\dim_{R/x} A = n$ implies $\dim_{R/x} K = n - 1$, which by induction implies $\dim_R K = n$. Then, by the Inequality Theorem 8.14A, we get $\dim_R A = n + 1$ except when $n = 1$, in which case, we get $\dim_R A \leq 2$. Thus, the theorem is proved except when $\dim_{R/x} A = \dim_R A = 1$. In this case, we need only prove that A is R/x-projective when $\dim_{R/x} A$ and $\dim_R A$ are at most one.

Let H be a free R-module, and $0 \to T \to H \to A \to 0$ exact. Then $\dim_R A \leq 1$ implies T is R-projective. Moreover $xA = 0$ implies $T \supseteq xH$, and so $0 \to T/xH \to H/xH \to A \to 0$ is exact. Since H/xH is R/x-free, and $\dim_{R/x} A \leq 1$, then T/xH is R/x-projective, so xH/xT is a summand of T/xT. Since T is R-projective, then T/xT is R/x-projective, and hence so is $A \approx H/T \approx xH/xT$. □

There are two more "Change of Rings Theorems" which are related to 8.15A, and we state these below without proof.

8.15B Second Change of Rings Theorem. *Under the same assumptions on R and x, for any R-module A such that* $\operatorname{ann}_A x = 0$, *we have* $\dim_{R/x} A/xA \leq \dim_R A$.

8.15C Third Change of Rings Theorem. *Under the same assumptions on R, x and A, as in* 8.15B, *if R is left Noetherian, and if x belongs to the radical of R, then*

$$\dim_{R/x} A/xA = \dim_R A.$$

Since we make no use of 8.15*B* or *C*, we leave the proofs to the reader, with the warning that although the proof of 8.15B is relatively short, and straight-forward, the proof of 8.15C requires "Nakayama's lemma" given in the text in Chapter 18. (Thus, not every reader will be equipped, at this point, to prove 8.15C, at least not the way Kaplansky does it [69a, p. 178].)

The next theorem for the case of a field R is a theorem of Hilbert, and for general rings, a theorem of Eilenberg-Rosenberg-Zelinsky [57].

8.16 Hilbert Szyzgy Theorem. *For any ring $R \neq 0$, the global dimension of the polynomial ring $R[x_1, \ldots, x_n]$ in n indeterminates is:*

$$\text{r.gl. dim}\, R[x_1, \ldots, x_n] = n + \text{r.gl. dim}\, R.$$

Proof. By 8.15A, $\text{r.gl. dim}\, R[x] \geq 1 + \text{r.gl. dim}\, R[x]/x$. But, $R[x]/x \approx R$, so $\text{r.gl. dim}\, R[x] \geq 1 + \text{r.gl. dim}\, R$. The lemma below reverses this inequality, and proves the theorem for the case $n = 1$. The general result then follows by induction on n. □

8.17 Lemma. *If M is any module over a polynomial ring $R[x]$ over any ring R, then*

$$\dim_{R[x]} M \leq \dim_R M + 1 .$$

Proof. Since

$$\sum_{i \in I} \oplus M_i[x] \approx \left(\sum_{i \in I} \oplus M_i \right)[x]$$

for any family $\{M_i\}$ of R-modules, then for any R-module M, we see that $M[x]$ is $R[x]$-projective if M is R-projective. Conversely, if $M[x]$ is $R[x]$-projective, then $M[x]$ is an $R[x]$-summand of a free $R[x]$-module F, hence M is an R-summand of $M[x]$, which is an R-summand of the free R-module F, so M is R-projective; that is,

$$\dim_R M = \dim_{R[x]} M[x] .$$

If M is also an $R[x]$-module, let K denote the kernel of

$$\begin{cases} M[x] \to M \\ (m_0, \ldots, m_i, \ldots) \mapsto \sum_{i=0}^{\infty} m_i x^i . \end{cases}$$

Then there is an $R[x]$-isomorphism

$$\begin{cases} M[x] \to K \\ (m_0, m_1, \ldots, m_i, \ldots) \mapsto (m_0 x, m_1 x - m_0, \ldots, m_{i+1} x - m_i, \ldots) . \end{cases}$$

This implies

$$\dim_R M = \dim_{R[x]} M[x] = \dim_{R[x]} K \geq$$
$$\geq (\dim_{R[x]} M[x]/K) - 1 = \dim_{R[x]} M - 1 . \quad \square$$

8.18 Proposition (Auslander [55]) *Let A be a right B-module, let $\mathscr{I}$ be a nonempty well ordered set, and let $\{B_i \mid i \in \mathscr{I}\}$ be a family of submodules of A such that if $i, j \in \mathscr{I}$ and $i \leq j$, then $B_i \subseteq B_j$. If $A = \bigcup_{i \in \mathscr{I}} B_i$ and* $\dim_R (B_i / \bigcup_{j<i} B_j) \leq n$ *for all $i \in \mathscr{I}$, then* $\dim_R(A) \leq n$.

Proof. If $n = 0$, then $B_i = \bigcup_{j<i} B_j \oplus B_i'$, and $A = \sum_{i \in \mathscr{I}} \oplus B_i'$ is projective.

Now assume $n > 0$. Set $B_i' = B_i / \bigcup_{j<i} B_j$, and let F_i' be a free module mapping onto B_i' with kernel K_i'. Let $F_i = \sum_{j \leq i} \oplus F_j'$. Since

F_i' is free, there is a map $F_i' \to B_i$ such that

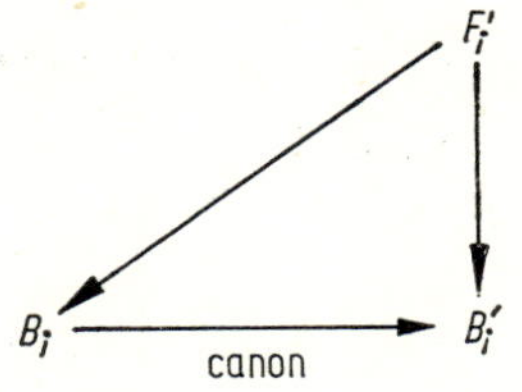

commutes, and by transfinite induction, the map $F_i' \to B_i'$ can be extended to a map $F_i \to B_i$ such that if K_i is the kernel, then

$$K_i \supseteq K_j, \qquad \text{if } i > j$$

and

$$K_i' \approx K_i / \bigcup_{j<i} K_j .$$

Thus,

$$\dim_R(K_i) = \dim_R(B_i) - 1 \leq n - 1,$$

and by the induction hypothesis, $\dim_R(\bigcup_{i \in \mathscr{I}} K_i) \leq n - 1$. Since this is the kernel of $\bigcup_{i \in \mathscr{I}} F_i \to A$, we obtain $\dim_R(A) \leq n$. □

8.19 Global Dimension Theorem (Auslander [55]). *If R is any ring, then*

$$\text{r.gl. dim } R = \sup_{I \subseteq R} \{\dim R/I\},$$

the supremum of the projective dimensions of cyclic right R-modules.

Proof. Let $d = \text{r.gl. dim } R$, and let n denote the right side of the equality. If A is an R-module, well order the elements of A, and let B_i (resp. B_i') be the submodule of A generated by all x_j such that $j \leq i$ (resp. $j < i$). Then the factor module B_i / B_i' is cyclic $\forall_i$ and, hence, has dim $\leq n$. Therefore, Auslander's proposition implies $d \leq n$. Thus $d = n$. □

8.20 Corollary. *For any ring R,*

$$\text{r.gl. dim } R = \sup \{\text{proj. dim } M \mid M \text{ fin. gen. mod-}R\}. \quad □$$

8.21 Corollary. *For any ring R, if R is not semisimple, then*

$$\text{r.gl. dim } R = 1 + \sup_{I \subseteq R} \{\dim I\}.$$

Proof. Let R/I be a cyclic module with

$$\dim R/I = \text{r.gl. dim } R = n \geq 1.$$

Then, exactitude of $0 \to I \to R \to R/I \to 0$ and projectivity of R imply that

$$1 + \dim I = \dim R/I,$$

which establishes the corollary. □

If M is an R-module, then $\mathbf{g}(M)$ denotes the smallest cardinal of any set of generators of M.

8.22 Proposition (Small [66c]) *Let k be a field of cardinality $\aleph_\varkappa$, and let $R = k[x_1, \ldots, x_n]$ be the polynomial ring, with quotient field $Q = k(x_1, \ldots, x_n)$. Let S denote a multiplicatively closed subset of R, let $M = RS^{-1}$, and let $\mathbf{g}(M) = \aleph_\mu$. Then, $\mathbf{g}(Q) = \aleph_\varkappa$, and*

$$\dim_R M \leq \mu + 1.$$

Proof. R is a unique factorization domain. For any $\alpha \in k$, $x_i - \alpha$ is a prime in R, and

$$R = k[x_1, \ldots, x_{n-1}, x_n - \alpha x_1],$$

$n \geq 2$. No R-submodule of Q generated by $< \aleph_\varkappa$ elements can contain $\{(x_1 - \alpha x_2)^{-1} \mid \alpha \in k\}$, so $\mathbf{g}(Q) = \aleph_\varkappa$.

We assume a set of generators for M is well ordered by the ordinal $\aleph_\mu$. Set $M_0 = 0$, and assume that M_i has been defined for all $i < \alpha$. Let M_α be generated by reciprocals of all denominators in any M_i for $i < \alpha$ and the first element of the set of generators not in this set, and products of these. Then M_α is generated by less than $\aleph_\mu$ elements. By transfinite induction, $M = \bigcup_{\alpha < \aleph_\mu} M_\alpha$. If $\mu > 0$, by the induction hypothesis, $\dim_R(\bigcup_{i<\alpha} M_i)$ and $\dim_R M_\alpha \leq \mu$. Hence $\dim_R(M_\alpha/\bigcup_{i<\alpha} M_i) \leq \mu + 1$, and Auslander's lemma applies.

If $\mu = 0$, then M is generated by the reciprocals of a countable set of elements in R, say $\{r_0, r_1, \ldots\}$. Set $s_0 = r_0^{-1}$, $s_{i+1} = s_i r_{i+1}^{-1}$. Then $s_i R \subseteq s_{i+1} R$, $s_i R$ is projective (in fact free), and $M = \bigcup_{i=0}^{\infty} s_i R$. Then $\dim_R(s_{i+1} R/s_i R) \leq 1$, and again by Auslander's lemma, $\dim_R M \leq 1$. No projective R-module is divisible by t_0, so M cannot be projective. Thus, $\dim_R M = 1$. □

The Hilbert theorem and the remarks at the beginning of the preceding proof show that

$$\dim_{k[x_1,\ldots,x_n]} k(x_1, \ldots, x_n) \leq \min\{\varkappa + 1, n\}.$$

Actually equality holds, but we do not prove this here. If k is a countable field, however, then obviously, $\dim_R Q = 1$ in this case. Furthermore, for any commutative local domain R with quotient field Q, a theorem of Kaplansky [66] states that $\dim_R Q = 1$ if Q is countably generated. (Cf. Small [66c] in which R can be any domain such that $\operatorname{rad} R \neq 0$, and $R/\operatorname{rad} R$ is Noetherian.)

A ring R is **right semihereditary** if every finitely generated right ideal is projective. Commutative hereditary (resp. semihereditary) entire rings are Dedekind rings (resp. Prüfer rings) studied in Chapters 30 and 31. Some of the exercises at the end of the present chapter are designed to lead the reader to investigate Chapter 31 where some, but not all, answers may be found. No further background is required for a profitable reading. (Indeed, Kaplansky's theorems on hereditary rings, which appear in Chapter 31, are presented in Chapter 1 of Cartan-Eilenberg [56].)

8.23 *Related Results and Exercises*

8.23.1 (Auslander [55]) If R is right and left Noetherian, then the right and left global dimensions are equal. [Hint: The right and left weak global dimensions are equal.]

8.23.2 (Kaplansky [58b]) The right and left global dimensions are unequal for a general ring R. Indeed, a right hereditary ring may not be left hereditary.

8.23.3 (a) For any prime p, and integer $n > 1$, gl. dim $\mathbb{Z}/p^n\mathbb{Z} = \infty$.

(b) A ring of infinite right global dimension may have an arbitrary integer $n > 0$ as left global dimension.

8.23.4 Prove the equality asserted in the remark following the proof of 8.22.

8.23.5 (Hochschild [58]) For any ring R, and set S, the (semi)group algebra $R[S]$ of the free (semi)group $[S]$ on S has

$$\text{r.gl. dim} = 1 + \text{r.gl. dim}\, R.$$

In particular, the **free ring** on S, defined to be the semigroup ring $\mathbb{Z}[S]$, has gl. dim $= 2$. Conclude that any ring T is isomorphic to the factor ring A/I of a ring A of gl. dim $= 2$.

8.23.6 The free algebra over a field is right and left hereditary.

8.23.7 The ring $T_n(R)$ of lower triangular matrices over a ring R satisfies l.gl. dim $T_n(R) = 1 +$ l.gl. dim R. Thus, $T_n(R)$ is left hereditary if and only if R is semisimple.

8.23.8 (Osofsky [67]) If R is right Noetherian, then

$$\text{r.gl. dim}\, R = \sup\{\text{inj. dim}\, R/I \mid I \subseteq R\}.$$

8.23.9 (a) (Levy [63]). If R is a right and left Noetherian hereditary semiprime ring, then it is a direct sum of right and left hereditary prime rings.

(b) (Levy, l.c.) Over such a ring R, every finitely generated right R-module is a direct sum of a torsion module and finitely many right ideals.

(c) (Webber) Any factor ring R/I is Artinian, $I \neq 0$, when R is prime.

(d) (Eisenbod-Robson) Any module over such a prime ring which is annihilated by an ideal is a direct sum of modules possessing a unique Loewy series (cf. Chapter 22), or equivalently, for any two-sided ideal $A \neq 0$, the ring R/A is serial in the sense that it is a right and left Artinian ring possessing a direct decomposition into a direct sum of right (left) ideals having a unique Jordan-Holder series (cf. Chapters 17 and 24).

8.23.10 (Rinehart [62]) (a) Let R be the algebra (ring) of differential polynomials over a field k. Then

$$t = \text{gl. dim} \bigotimes_k^n R \leq \text{gl. dim}\, R + 2(n-1),$$

and also $t \leq 2n$. If k has characteristic $\neq 0$, then $t = 2n$.

(b) The projective dimension of R as a module over the "enveloping" algebra $R \otimes_k R^{\text{op}}$ is 2.

Noetherian Rings of Global Dimension Two

A theorem of Bass, proved presently, states that a Noetherian ring R has gl. dim ≤ 2 if and only if the R-dual $X^* = \text{Hom}_R(X, R)$ of every finitely generated right R-module is projective. In reproducing Bass's proof, the concepts of torsionless and closed modules are required. Briefly, a right R-module A is torsionless provided that the canonical mapping $A \to A^{**}$ is monic:

$$\begin{cases} A \to A^{**} \\ a \mapsto \langle\ , a\rangle, \end{cases}$$

where $f\langle\ , a\rangle = f(a)$, for any $f \in A^*$; then A is reflexive if $A \to A^{**}$ is an isomorphism. (Cf. Dual Modules, pp. 137/8.) A submodule K of A is said to be **closed** if

$$K = \text{ann}_A(\text{ann}_{A^*} K).$$

For any R-module K, there is an isomorphism

$$\begin{cases} \text{ann}_{A^*} K \approx (A/K)^* \\ \quad f \mapsto \bar{f}, \end{cases}$$

where $\bar{f}(a + K) = f(a)$. Thus, there is an isomorphism

$$\text{ann}_A(\text{ann}_{A^*} K) \approx \big(A^*/(\text{ann}_{A^*} K)\big)^*.$$

If K is closed, it follows that K is isomorphic to the module on the right, that is, *every closed submodule of A is isomorphic to a dual module of a quotient of A^*.*

8.24 *Exercises.*

8.24.1 A submodule K of A is closed if and only if A/K is torsionless.

8.24.2 Every right R-module is torsionless if and only if R is a cogenerator in mod-R.

8.24.3 A submodule K of A is closed if and only if $K = \text{ann}_A H$, for some submodule H of A^*.

8.25 A Proposition. *Let R be a ring, let n be an integer > 0, and let K be a submodule of R^n in* mod-R. *Then K is closed if and only if K is isomorphic to the R-dual B^* of a left R-module B which is generated by n elements.*

Proof. If K is closed, then letting $A = R^n$, there is an isomorphism

$$K = \text{ann}_A \text{ann}_{A^*} K = (A^*/\text{ann}_{A^*} K)^*.$$

Thus, for $B = A^*/\text{ann}_{A^*} K$, then $K \approx B^*$, and since $A^* \approx R^n$ in R-mod, necessarily, B is generated by n elements.

Conversely, if $K \approx B^*$, where

$$0 \to H \to R^n \to B \to 0$$

is exact in R-mod, then, since $\text{ann}_{R^n} H \approx (R^n/H)^* \approx B^*$, K is closed by Exercise 8.24.3. □

8.25 B Proposition (Bass [60]) *If R is left and right Noetherian, then* l.gl. dim $R \leq 2$ *if and only if the R-dual of any finitely generated module is projective.*

Proof. Let M be finitely generated, and let

$$(1) \qquad 0 \to K \to R^n \to M \to 0$$

be exact. By 8.20, R has global dimension ≤ 2 if and only if the K which arise in this way all have projective dimension ≤ 1. Let

$$(2) \qquad 0 \to H \to R^m \to K \to 0$$

be exact. Then dim $K \leq 1$ if and only if dim $H = 0$. Thus, l.gl. dim $R \leq 2$ if and only if the H which arise in this way are projective. It follows from the exercises that the H which so arise are the duals of finitely generated modules. For K is torsionless, and hence by 8.24.1, H must be closed. Thus, by the isomorphism preceding the exercises, H is isomorphic to the dual of a finitely generated module. Conversely, any such dual fits into a sequence (2). □

8.26 Corollary. *If R is a right and left Noetherian ring of global dimension* ≤ 2, *then every right and left annulet of R is projective.*

Proof. If I is a right annulet, say $I = \operatorname{ann}_A X$, for a left ideal X of R, then $X \approx (R/I)^*$ is projective by Bass's theorem. □

Simple Noetherian Rings of Global Dimension ≤ 2

The Wedderburn-Artin theorem implies that a simple Artinian ring is similar to a field and, therefore, has global dimension 0. The next theorem generalizes this to simple Noetherian rings of possible global dimension 2.

8.27 Theorem[1]. *A right and left Noetherian simple ring R of global dimension* ≤ 2 *is similar to a simple right Ore domain.*

Proof. By the corollary to Bass's theorem 8.25, every right annulet of R is projective. By 4.11, 4.19, 4.29 (cf. 7.5), a simple ring R is similar to a simple right Ore domain if and only if R has a projective uniform right ideal U. Let Q be the simple right artinian ring which by Goldie's theorem 7.21 B is the classical right quotient ring of R. Let I be a minimal right ideal of Q contained in UQ. Let $A = I \cap R$. By 8.5, I is generated by an idempotent, hence is a minimal right annulet of Q, and hence, by 7.23.3, A is a right annulet of R and a minimal right annulet. Furthermore, A is uniform, since if X and Y are nonzero right ideals of R contained in A, then XQ and YQ are nonzero and contained in I. Then $XQ = YQ = I$, since I is minimal. Write

$$t = xab^{-1} = ycd^{-1}$$

for nonzero $t \in Q$, $x \in X$, $y \in Y$, $a, b, c, d \in R$. (This is possible. Cf. 9.3.) Then, writing $d^{-1}b = b_1 d_1^{-1}$ for suitable $b_1, d_1 \in R$,

$$tbd_1 = xad_1 = ycb_1$$

is a nonzero element of $X \cap Y$. This proves that the right annulet A is uniform and projective. Then $k = \operatorname{End} A_R$ is a right Ore domain (e.g. 4.21), and R is similar to k. □

[1] The author proved this theorem in 1965, and communicated it orally, and in written form, privately. Michler [69] obtained it independently in generalized form. The theorem was presented in my lectures at the Ring Theory Conference at Appalachian State U., Boone, N.C., in August 1969. Cozzens and Faith [73/74] have generalized it to the one-sided case.

Conjectures on Simple Rings

We restate Theorem 7.20 and part of 7.5.

8.28 A Theorem. *Let R be a simple ring, let U be a nonzero uniform right ideal, and let K denote the right Ore domain* End U_R. *Then U is finitely generated and projective left K-module. Furthermore, the following conditions are equivalent:*

(a) *K is a simple ring.*

(b) *U is a projective R-module.*

(c) *There is a category equivalence*

$$\text{mod-}K \approx \text{mod-}R.$$

8.28 B Conjectures.

(1) Can U be chosen so that U is a free module over K? Conjecture: No.

(2) Is $R \approx K_n$ for some integral domain K? Conjecture: No.

(3) When R is not an integral domain, must R contain nontrivial idempotents? Conjecture: No.

(4) Do there exist a simple right Ore domain K and an indecomposable finitely generated projective K-module P such that P does not embed in K? Conjecture: Yes.

(5) Can a uniform right ideal U always be chosen to be projective, or equivalently (8.28A), can one be chosen such that $K = \text{End } U_R$ is simple? Conjecture: No.

(6) Is every right Goldie simple ring (of finite global dimension) similar to an Ore domain? Conjecture: No.

(7) Answer the same questions assuming that R is a simple right and/or left Noetherian and/or a V-ring. (See Chapter 7, 7.31 ff.)

(8) Answer the same questions assuming that every quasinjective right module is injective. (Then, R is right Noetherian, and the endomorphism ring of any indecomposable injective right R-module is a field. Cf. Chapter 20.)

(9) If K is a simple right Ore domain, and if U is a finitely generated projective left K-module, then does there exist a left ideal V of K, and an isomorphism $U \approx V^n$, for some integer $n > 0$?

8.28 C *Remarks.* (1) and (2) are equivalent. Assuming (1), then U is finitely generated and free over K, say $U \approx K^n$, and then by 3.17, $R = \text{End}_K U \approx K_n$. Conversely, if $R \approx K_n$, then $R \approx \text{End}_K W$, where $W \approx K^n$ is free over K. □

(2) $\Rightarrow$ (9): If $R = \operatorname{End}_K U$, assuming $R \approx K_n$, then we can write $U = \sum_{i=1}^n Ue_{ii}$ in K-mod, for matrix units $\{e_{ij}\}_{i,j \in n}$ in R, and the K-modules Ue_{ii} are all isomorphic to $V = Ue_{11}$ via the right homothetic of U by e_{1i}. (Note, since $Re_{11} = Re_{i1}$, then $Ue_{11} = Ue_{i1}$.) This implies that $U = V^n$ in K-mod, and by 3.17, there is a ring isomorphism

$$R = \operatorname{End}_K U \approx \operatorname{End}_K V^n \approx Q_n,$$

where $Q = \operatorname{End}_K V$. By 7.21 B, R has a right quotient ring D_n, where D is the right quotient field of K, and this implies that Q is a right Ore domain with right quotient field $\approx D$. Furthermore, by 4.11.6, V embeds in Q canonically, hence $V \subseteq D$. Then, finite generation of V as a left K-module and the right Ore property of K imply the existence of a nonzero $k \in K$ such that $Vk \subseteq K$. This implies that V embeds in K, therefore, (9) holds.

(2) $\Rightarrow$ (3) trivially.

(3) $\Rightarrow$ (6). Assume (3), and let e be any nontrivial idempotent of R. Since R is simple, eR is a progenerator, and $\operatorname{End} eR_R \approx eRe$ canonically. Thus, $R \sim eRe$, so (6) holds if eRe is a domain. Otherwise, by (3), eRe has a nontrivial idempotent f, and then fRf has right Goldie dimension less than that of eRe, so by induction we may assume that fRf is similar to a domain, and then $R \sim eRe \sim fRf$ is similar to a domain.

(3) $\Leftrightarrow$ not (4). Let P be a finitely generated projective K-module. Since K is simple, then P is a progenerator of K-mod, and then $A = \operatorname{End}_K P$ is similar to K, and by 7.21B, A has a right quotient ring $\approx D_m$, for some integer $m > 0$. If P is indecomposable, then A has no nontrivial idempotents, so that (3) implies that A is a domain, so that the right quotient ring of A is a field $\approx D$. Then P embeds in K as in the proof of (2) $\Rightarrow$ (9).

Conversely, if P is an indecomposable finitely generated projective left K-module not embeddable in K, then similar arguments show that $R = \operatorname{End}_K P$ is a simple right Goldie ring without nontrivial idempotents, and, moreover, R is not a domain. (For, if R were a domain Q, then the right quotient field of Q would be D, and then the arguments above suffice to embed P in K, contrary to the assumption.)

(5) and (6) are equivalent by 8.28A.

(3) and (9) have the following relation. (3) is equivalent to the assertion that every finitely generated projective left K-module W is isomorphic to a direct sum of indecomposable left ideals of K, whereas (9) asserts these are all isomorphic. $\square$

Rings without Identity Element

A **ring-1** is a nonempty set together with two associative binary operations

$$\text{multiplication: } T \times T \to T, \qquad (a, b) \mapsto a \cdot b$$

and

$$\text{addition: } T \times T \to T, \qquad (a, b) \mapsto a + b$$

such that T is a group under addition and the operations satisfy the distributive laws

$$(a + b)c = ac + bc$$

$$a(b + c) = ab + ac.$$

$\forall a, b, c \in T$.

Thus every ring is a ring-1.

Conversely, a ring-1 T is a ring under the operations of addition and multiplication if and only if T has a multiplicative identity element.

Ring-1 homomorphisms (monomorphisms, and so on) are defined in the obvious way: a mapping $h: T \to S$ of ring is a homomorphism in case $h(ab) = h(a)\,h(b)$ and $h(a - b) = h(a) - h(b)$ $\forall a, b \in T$.

8.29A Proposition. *If T is a ring-1, let $\mathcal{T}$ denote the category whose objects are ring-1 monomorphisms*

$$h_R : T \to R,$$

where R is a ring such that

$$R = \{\text{im}\, h + n \cdot 1 \mid n \in Z\}.$$

A morphism in $\mathcal{T}$ is a ring homomorphism $f: R \to S$ completing:

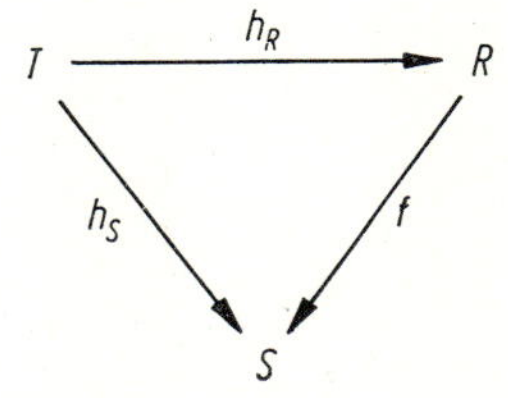

The category $\mathcal{T}$ has an initial object.

Proof. The initial object is traditionally constructed on the product set $R = T \times \mathbb{Z}$, which is a ring under the operations

$$\text{multiplication:} \quad (a, n)(b, m) = (ab + nb + ma, nm)$$

$$\text{addition:} \quad (a, n) + (b, m) = (a + b,\ n + m).$$

The identity element of R is $(0, 1)$, and $t \mapsto (t, 0)$ defines a ring-1 monic $h_R : T \to R$. Clearly, h_R belongs to our category. If also $h_S : T \to S$ is in the category, then there is a ring homomorphism

$$f : R \to S$$
$$h_R(t) + n \cdot 1 \mapsto h_S(t) + n \cdot 1 .$$

Observe that $h_R(t) + n \cdot 1 = (t, n) \in R$. Therefore, $h_R(t) + n \cdot 1 = h_R(t') + m \cdot 1$ if and only if $t = t'$ and $n = m$. This shows that $f : h_R(t) + n \cdot 1 \mapsto h_S(t) + n \cdot 1$ is indeed a mapping and is trivially a morphism of the category T. If also $f' : R \to S$ is a morphism such that $f' h_R = h_S$, then

$$\begin{aligned} f'(t, n) &= f'(t, 0) + f'(0, n) = f'(t, 0) + n \cdot 1 \\ &= f' h_R(t) + n \cdot 1 = h_S(t) + n \cdot 1 , \end{aligned}$$

that is, $f' = f$. This proves that $h_R : T \to R$ is initial. □

If T is a ring-1, not a ring, then the initial object $h_R : T \to R$ in the category $\mathcal{T}$ of 8.29A is called the **unit cover** of T. For simplicity, h_R will be considered the inclusion map, and R will be called the unit cover of T. In this case, T is an ideal of R, and R/T is a ring $\approx \mathbb{Z}$.

A (right) module (M, T, μ) over a ring-1 T is defined to be an abelian group M and a mapping

$$\mu : M \times T \to M$$
$$\mu : (m, t) \mapsto mt$$

such that axioms (1) to (3) for the definition of a module over a ring hold (see Chapter 1). A homomorphism $M \to N$ of T-modules is a group homomorphism $f : M \to N$ such that $f(xt) = f(x)t \ \forall x \in M$, $t \in T$. As in the category of rings, $\mathrm{Hom}_T(M, N)$ denotes the set of all T-homomorphisms $M \to N$. The class of all T-modules is a category $\mathcal{M}_T$ in which $\mathrm{Hom}_T(M, N)$ is the set of all morphisms $M \to N$.

8.29B Proposition. *Let T be a ring-1, and let R be the unit cover of T. Then there is a category isomorphism*

$$\varphi \begin{cases} \mathcal{M}_T \approx \mathcal{M}_R \\ (M, T, \mu) \mapsto \big(M, R, \varphi(\mu)\big), \end{cases}$$

where

$$\varphi(\mu) \begin{cases} M \times R \to M \\ \big(x, (t, n)\big) \mapsto \mu(x, t) + nx . \end{cases}$$

If $M, N \in \mathscr{M}_T$, *and if* $f \in \operatorname{Hom}_T(M, N)$, *then* $f \in \operatorname{Hom}_R(\varphi(M), \varphi(N))$, *and we let* $\varphi(f) = f$. *Then*

$$\eta \begin{cases} \mathscr{M}_R \approx \mathscr{M}_T \\ (M, R, \nu) \mapsto (M, T, \eta(\nu)), \end{cases}$$

where $\eta(\nu)$ *is the restriction of* ν *to* T.

Proof. Obviously, $\varphi\eta = \eta\varphi =$ identity on objects and morphisms. $\square$

A right ideal of a ring-1 T is a subset of T, which is a submodule under the induced operation, and similarly for left ideals of T. Then nilpotent ideals are defined as in rings.

Note that if R is a ring (with 1), then mod-R (in the usual sense) corresponds to the set of all (ring-1) R-modules which are **unital** in the sense that $x1 = x$ for every element x in the module.

8.30 Theorem. *If* T *is a ring-1 without nilpotent ideals* $\neq 0$, *and if* $T \neq 0$ *satisfies the descending condition on right ideals, then* T *has an identity element and, therefore, is a (semisimple) ring.*

The proof is left to the reader with the hint that $R = \operatorname{End} T_T$ is an Artinian semiprime ring, and $t \mapsto t_s$ is an embedding $T \to R$.

Exercises for Chapter 8

1. Let R be a ring, and let e_1 and e_2 be idempotents in R. Then

(a) The R-modules e_1R and e_2R are isomorphic if and only if there exist elements e_{12} and $e_{21} \in R$ such that

$$e_1e_{12}e_2 = e_{12}, \qquad e_2e_{21}e_1 = e_{21}$$

$$e_{12}e_{21} = e_1, \qquad e_{21}e_{12} = e_2.$$

[Hint: If φ is the isomorphism, set $e_{21} = \varphi(e_1)$, $e_{12} = \varphi^{-1}(e_2)$.]

(b) The right R-modules e_1R and e_2R are isomorphic if and only if the left R-modules Re_1 and Re_2 are isomorphic.

2. Let e be an idempotent in a semiprime ring. The following statements are equivalent: (a) eR is a minimal right ideal; (b) eRe is a field; (c) Re is a minimal left ideal.

3. The following statements about minimal right ideals I, K in a semiprime ring R are equivalent: (a) $I \approx K$; (b) $KI \neq 0$; (c) $K = xI$ for some $x \in K$.

4. If A is an ideal in a semisimple ring R, then there exists a central idempotent e (that is, $e \in$ center R) such that $A = eR = Re$.

***5.** If R is a semiprime ring, and if every right R-module is torsionless then R is semisimple.

***6.** (Brauer) If R is right Artinian, then each non-nilpotent right ideal contains an idempotent.

***7.** (Kaplansky [52]) If R satisfies the d.c.c. on principal right ideals, and if R is semiprime, then R is semisimple. (Hint: Show that $R = \text{socle}\, R$ by showing that every finitely generated right ideal contained in socle R is generated by an idempotent.) This also holds for the case of a ring-1.

8. Let $R = A_n$, where A is a ring, and let T (resp. N) denote the set of **lower (strictly lower) triangular matrices** in R. Thus

$$T = \{\sum e_{ij} d_{ij} \mid d_{ij} \in D \quad \text{and} \quad d_{ij} = 0 \;\forall j > i\}$$

$$N = \{\sum e_{ij} d_{ij} \mid d_{ij} \in D \quad \text{and} \quad d_{ij} = 0 \;\forall j \geq i\}.$$

(a) Show that T is a ring, N is a nilpotent ideal of T of index $\leq n$, and the factor ring T/N is a direct product of n copies of A.

(b) If A is semisimple, then the ring of lower triangular matrices in A_n is right hereditary. Generalize.

***9.** (Faith [66]) Let M be an R-module, and let E denote the injective hull of M. The following conditions are equivalent:

(a) A direct sum of countably many copies of E is injective.

(b) The lattice of right ideals of R that are annihilators of subsets of M satisfies the a.c.c.

(c) Any direct sum of copies of E is injective.

10. Let C be any category. If A and B are equivalent objects, then there is a category equivalence $\text{Mor}\,(A, A) \approx \text{Mor}\,(B, B)$. The converse, in general, is not true, but it is true for the category of vector spaces over a field. Prove this. (Hint: If $S = \text{End}_D V$, where V is a vector space over a field D, then the right S-module is simple and isomorphic to a right ideal I of S. Thus $K = \text{End}\, I_S$ is a field $\approx D$, and I is a left vector space over D isomorphic to V. Any two simple right ideals of S are isomorphic. Thus, if $R \approx \text{End}_Q W$, where ${}_Q W$ is a vector space over a field Q, an isomorphism $R \approx S$ implies an isomorphism $Q \approx D$ and that $\dim_Q W = \dim_D V$. This is a little more than we originally claimed.)

11. A commutative semihereditary ring has no nilpotent elements.

***12.** (Goldie [62]) If R is a principal right ideal semiprime ring, then R is right hereditary.

***13.** (Goldie [62]) If R is a principal right ideal prime ring, then $R \approx K_n$, where K is a right hereditary entire ring.

***14.** A ring R is hereditary if and only if every submodule of a projective module is projective. Conclude that every hereditary ring is semihereditary.

***15.** (Kaplansky) If R is hereditary, every projective module is isomorphic to a direct sum of right ideals of R.

***16.** (Albrecht [61]) If R is semihereditary, then every projective module is isomorphic to a direct sum of finitely generated right ideals of R.

***17.** (Bass [64]) If R is right semihereditary, then every projective left R-module is isomorphic to a direct sum of duals of finitely generated right ideals.

Notes for Chapter 8

The Wedderburn-Artin theorems on semisimple rings and modules presented in this chapter are standard, and needs no further comment, except, perhaps, the indication that it is a most powerful tool in the structure theory of Artinian nonsemisimple rings (Part V). Why is this? First, any right Artinian ring R has nilpotent radical J. This implies that any nonzero module M contains MJ as a proper submodule. But the canonical R/J-module M/MJ must be semisimple by 8.11, since R/J is a semiprime right Artinian ring. In particular, J^i/J^{i+1} is a semisimple module, for any integer i less than the index of nilpotency of J. It follows that R has a composition series (Chapter 17), the simple factors of which are by the Jordan-Hölder theorem unique up to isomorphism. Moreover, an easy consequence of this is that R is right Noetherian, that is, any right Artinian ring is right Noetherian (Theorem of Hopkins [39] and Levitzki [39]). We have just indicated an important application of the Wedderburn-Artin theorem to the structure of nonsemisimple rings. A fuller account may be found in Volume II.

As indicated by Exercise 8.23.3, there exist rings of infinite global dimension, and in fact, the example given there can be generalized: any right selfinjective right or left Noetherian ring R is either semisimple, or has gl. dim $= \infty$. [These rings are called Quasi-Frobenius (QF) rings, and coincide with the class of right and left Artinian rings in which every right or left ideal is an annulet. Cf. Chapter 12 Exercise 6.]

A theorem of Auslander and Buchsbaum [57] states that any Noetherian commutative local ring of finite global dimension is an integral domain, in fact, a unique factorization domain. Thus, any Noetherian commutative local ring with zero divisors has infinite global dimension, a result which implies 8.23.3 (a). However, there exist local commutative rings with zero divisors having finite global dimension (Osofsky [69]).

The results of Kaplansky and Small on the global dimension of a quotient field (discussed 8.22ff.) have been generalized by Osofsky [68c] (cf. also Osofsky [70] for a discussion of the global dimension of the infinite full linear ring over a field).

Mitchell [65] has an exposition of the Global Dimension theorems in the setting of abelian categories, and, moreover, proves not only the Hilbert Szyzgy theorem 8.16 for the polynomial category over an abelian category, but also proves the theorem of Hochschild stated in 8.23.5 for the "free" category over an abelian category. More recently, Mitchell has obtained both these theorems in a unified way (unpublished).

References

Albrecht [61], Artin [27], Auslander [55], Auslander-Buchsbaum [57], Bass [60a, 64], Cartan-Eilenberg [56], Eisenbud-Robson [71], Faith [66, 72a], Goldie [62], Hochschild [58], Kaplansky [52, 58a, 59, 66, 69a, 69b], Levy [63], MacLane [63], Michler [69], Mitchell [65], Northcott [60], Osofsky [68c, 69, 70], Rinehart [62], Sandomierski [64], Small [66c], Webber [70], Wedderburn [08].

Chapter 9

Noetherian Semiprime Rings

A ring S is a **(classical) right quotient ring of a subring** T if every regular element $a \in T$ has an inverse in S and

$$S = \{ab^{-1} \mid a, b \in T,\ b \text{ regular}\}.$$

Then T is an order in S (cf. 7.21). The following condition is necessary and sufficient for a ring T to possess a classical quotient ring: If $a, b \in T$, and if b is regular, then there exist $a_1, b_1 \in T$, b_1 regular, such that $ab_1 = ba_1$ (see 9.1). If T is commutative, this condition is automatic, and if T is a domain, this is the Ore condition.

It is interesting to know if a ring R (or group) can be embedded in a ring of $n \times n$ matrices over a field, in which case R is said to be **representable.** A branch of mathematics, known as representation theory, is devoted to the study of ring homomorphisms of a given ring (or group) into a matrix ring, or a ring of linear transformations, over a field. (See Curtis-Reiner [62] for a treatise on representation theory.)

A very simple but significant introduction to the subject is a theorem that states that any ring R which is a finite dimensional left vector space over a division subring F can be embedded in F_n, where $n = \dim_F R$. The embedding[1] is $a \mapsto a_d$, where a_d is the homothetic defined by $a \in R$. Such a representation is said to be **finite dimensional.**

In this connection, the Wedderburn-Artin theorem is quite strong. An Artinian prime ring is not only representable in, but is, a full matrix ring over a field. In this chapter, we show that a Noetherian prime (or simple) ring is not only representable but — what is infinitely more precise — is an order in a full matrix ring over a field. Historically the first finite dimensional representation theorem for a general Noetherian simple (or prime) ring, this theorem was proved by Goldie in 1958 for rings with ascending conditions on the right and left, and by Lesieur-Croisot in 1959 for prime rings with ascending chain conditions on the right[2]. We give the proof of Goldie [69] characterizing the rings with semisimple right quotient rings as semiprime rings with $(\mathrm{acc})^{\perp}$ and $(\mathrm{acc})\oplus$ (Theorem 9.9). A corollary is the theorem of Levitzki 9.16 stating that any nil right ideal in a right or left Noetherian ring is nilpotent. The chapter ends in a number of exercises on the maximal right (Johnson) quotient ring of a right neat ring, generalizing the classical right quotient. Moreover, a theorem of Faith and Utumi [65] determining all right orders of D_n as subrings containing F_n, for some right order F of D, is proved in the following Chapter 10.

Existence of Quotient Rings

A ring R satisfies the right **Ore condition** *in case to each pair $a, b \in R$, where b is regular, there corresponds a pair $a_1, b_1 \in R$, b_1 regular, such that*

(1) $$ab_1 = ba_1.$$

The Ore condition is obviously necessary for R to have a right quotient ring S, since then $b^{-1}a \in S$, so

(2) $$b^{-1}a = a_1 b_1^{-1}$$

for appropriate $a_1, b_1 \in R$, and (1) holds. [Thus (2) is a mnemonic for (1).] In this case, R is called a right **Ore ring.**

---

[1] An amusing pathology: R is both a left and right vector space over any subfield F, but it is possible that $n = \dim_F R < \omega$ and $\dim R_F = \omega$ (Cohn [61]). In this case, the map $a \mapsto a_s$ does not embed R in F_n, but rather in the ring F_ω of row finite matrices over F.

[2] The Chevalley-Jacobson density theorem provides a representation for any simple ring R, but it is finite dimensional only if R is Artinian. (Cf. Chapter 19.)

9.1 Proposition. *A ring R has a right quotient ring if and only if R satisfies the Ore condition.*

Proof[1]. Let G denote the set of regular elements of R. We first prove the assertion: (i) If $ab_1 = ba_1$, where $a, b, a_1 \in G$ and $b_1 \in R$, then $b_1 \in G$. To see this, let $x \in b_1^{\perp}$. Then

$$0 = b_1 x = ab_1 x = ba_1 x,$$

and $ba_1 \in G$ implies $x = 0$. Next, if $x \in {}^{\perp}b_1$, then there exist $a_2, b_2 \in R$ such that $b_2 \in G$ and $ab_2 = ba_2$, and then there exist $c \in G$, $d \in R$ such that $a_2 c = a_1 d$. Consequently,

$$ba_2 c = ab_2 c, \qquad \text{and} \qquad ba_1 d = ab_1 d$$

whence $ab_2 c = ab_1 d$. Since $a(b_2 c - b_1 d) = 0$ and $a \in G$, then $b_2 c = b_1 d$, so

$$0 = xb_1 = xb_1 d = xb_2 c.$$

Since $b_2 c \in G$, then $x = 0$, proving (i). In the product $R \times G$, define a relation $\sim$, where

$$(a, b) \sim (c, d)$$

if and only if the following implication holds: (*) $bx = dy$ for some pair $x, y \in G \Rightarrow ax = cy$. We now prove: (ii) *For a given* $(a, b) \in R \times G$, *if* (*) *holds for one ordered pair* $\langle x, y\rangle$, *it holds for any such pair* $\langle x_1, y_1\rangle \in G \times G$. We can find elements $g, g_1 \in G$ such that $yg = y_1 g_1$, and since we are assuming $bx_1 = dy_1$, and $bx = dy$, we thus have

$$bx_1 g_1 = dy_1 g_1 = dyg = bxg.$$

Consequently, $x_1 g_1 = xg$, and

$$ax_1 g_1 = axg = cyg = cy_1 g_1,$$

that is, $ax_1 = cy_1$, proving (ii). It follows from this that $\sim$ is an equivalence relation in $R \times G$. We denote the set of equivalence classes by RG^{-1}. The equivalence class of $(a, b) \in R \times G$ is denoted by a/b. Given $a/b, c/d \in RG^{-1}$, there exist $x, y \in G$ such that

$$m = bx = dy,$$

and we define

(1) $$a/b + c/d = (ax + cy)/m.$$

[1] See 16.9 for Gabriel's proof. Cf. Inverting morphisms in Gabriel-Zisman [68].

(Note that $a/b = ax/m$ and $c/d = cy/m$.) First, (1) is independent of m, since if also $m' = bx' = dy'$, with $x', y' \in G$, and if $mu = m'v$, for $u, v \in G$, then $bxu = bx'v$; that is, $xu = x'v$. Similarly, $yu = y'v$, so

$$(ax + cy)u = (ax' + cy')v,$$

whence

$$(ax + cy)/m = (ax' + cy')/m'.$$

Next, (1) is independent of the choice of representatives of a/b. For if $a/b = a'/b'$, then there exist $x', y', z \in G$ such that $m' = bx' = dy' = b'z$, and then $ax' = a'z$. Now, (1) is independent of m; that is,

$$t = a/b + c/d = (ax + cy)/m = (ax' + cy')/m'.$$

Hence $t = (a'z + cy')/m' = a'/b' + c/d$, which is what we wanted. Similarly, if $y_1 \in G$, $x_1 \in R$ are such that $n = bx_1 = cy_1$, then define

(2) $$(a/b)(c/d) = ax_1/dy_1.$$

(Note that $a/b = ax_1/n$ and $c/d = n/dy_1$.) As before, the formula (2) is independent of x_1, y_1 in an equation $bx_1 = cy_1$ and also is independent of the representatives of a/b, c/d. For example, if $a/b = a'/b'$, we choose $u, v \in G$ such that $bu = b'v$, and then $au = a'v$. Now pick $x, y \in R$, $y \in G$ such that

$$n' = bux = cvy.$$

Then, since (2) is independent of the multiple $n = bx_1 = cy_1$, we have

$$t = (a/b)(c/d) = (aux)/(dvy).$$

However, $au = a'v$, and $bu = b'v$, whence $n' = b'vx = cvy$, and

$$t = a'vx/dvy = (a'/b')(c/d),$$

which is what we wanted.

We omit the detailed computations verifying the ring axioms. Assuming this, however, the equations (1) and (2) show that the correspondence $a \to a/1$ is a ring monic $\varphi : R \to RG^{-1}$ ($a/1 = ab/b$ for all $b \in G$). Furthermore, if $a \in G$, then $a/1$ is a unit in RG^{-1}, and $a/1^{-1} = 1/a$. Finally, if $a/b \in RG^{-1}$, then $a/b = (a/1)(b/1)^{-1}$. This shows that RG^{-1} is a classical right quotient ring of im φ. It follows that R has a classical right quotient ring. □

9.2 *Exercises.*

9.2.1 If S is another classical quotient ring of R, $S = \{ab^{-1} \mid a \in R, b \in G\}$, then there exists a unique ring isomorphism $h : RG^{-1} \to S$ such that $h(a/b) = ab^{-1}$.

9.2.2 An R-module M is **torsionfree** in case, for all regular elements $a \in R$, $xa = 0$ with $x \in M$ implies $x = 0$. If M is torsionfree, and if R has a classical ring RG^{-1}, then there exists an RG^{-1}-module MG^{-1}, which is initial in the category of all maps $h: M \to N$, where N is an RG^{-1}-module and h is an R-module monomorphism such that $N = \{h(x)b^{-1} \mid x \in M,\ b \in G\}$. (Thus, there exist an R-module monomorphism $t: M \to MG^{-1}$ and a unique RG^{-1}-module homomorphism $g: MG^{-1} \to N$ such that $gt = h$. Cf. Modules of fractions, Chapter 27.)

9.2.3 Prove 9.3 (without peeking).

9.3 **Lemma.** *Let Q be a right quotient ring of a ring R. Then*

9.3.1 *If $b_1, \ldots, b_n$ are regular elements of R, there exist a regular element $c \in R$ and elements $g_i \in R$ such that $b_i^{-1} = g_i c^{-1}$, $i = 1, \ldots, n$.*

9.3.2 *If $x_1, \ldots, x_n \in Q$, there exists a regular element $c \in R$ such that $x_i c \in R$, $i = 1, \ldots, n$.*

9.3.3 *If I is a right ideal of R, then the right ideal of Q generated by I is IQ, and $IQ = \{xc^{-1} \mid x \in I,$ regular $c \in R\}$.*

9.3.4 *If $d \in R$ is regular, then dR is an essential right ideal of R.*

Proof of 9.3.1 is true for $n = 1$, since then $g_1 = b_1$, $c_1 = b_1^2$ will work. Assume the theorem for $b_1, \ldots, b_{n-1}$, and let $\bar{g}_i \in R$ and regular $\bar{d} \in R$ be such that $b_i^{-1} = \bar{g}_i \bar{d}^{-1}$, $i = 1, \ldots, n-1$. Now there exist g_n and regular $d \in R$ such that $b_n^{-1}\bar{d} = g_n d^{-1}$. Hence $c = \bar{d}d$ and $g_i = \bar{g}_i d$, $i = 1, \ldots, n-1$, are the desired elements.

Proof of 9.3.2. Write $x_i = a_i b_i^{-1}$, with $a_i, b_i \in R$, and choose c and g_i, $i = 1, \ldots, n$, in accordance with 9.3.1. Then $x_i c = a_i b_i^{-1} c = a_i g_i \in R$, $i = 1, \ldots, n$, and c is regular in R.

Proof of 9.3.3. Clearly IQ is the right ideal of Q generated by I, and $IQ \supseteq I_0 = \{xc^{-1} \mid x \in I,$ regular $c \in R\}$. A typical element $y \in IQ$ has the form $y = \sum_{i=1}^{n} y_i x_i$, with $y_i \in I$, $x_i \in Q$, $i = 1, \ldots, n$, and n an integer. By 9.3.2, there exists a regular $c \in R$ such that $x_i c = g_i \in R \ \forall\, i$.
Then $y = xc^{-1}$, where $x = \sum_1^n y_i g_i \in I$. Thus $IQ = I_0$ as asserted.

Proof of 9.3.4 is a consequence of the Ore condition. □

9.4 *Exercises.*

9.4.1 If R is right Artinian, then each right regular and each left regular element is a unit. Thus, any Artinian ring is its own quotient ring.

9.4.2 When R is commutative, or is an entire ring, 9.3.4 implies the Ore condition. To what extent is this true generally? (The author does not know the answer.)

9.4.3 The left annihilator in a ring R of any nilpotent ideal N is an essential right ideal.

Singular Ideal and Submodule

Recall the notation: $M \nabla N$ *denotes that* N *is an essential submodule.* If M is any R-module, consider the set

$$\text{sing}\, M = \{x \in M \mid R \nabla \text{ann}_R x\}.$$

If $x, y \in M$ and $a \in R$, then

(i) $$\text{ann}_R(x - y) \supseteq \text{ann}_R x \cap \text{ann}_R y$$

(ii) $$\text{ann}_R xa = \text{ann}_R(a/\text{ann}_R x).$$

Now 9.5.1 shows that $\text{ann}_R(a/I)$ is an essential right ideal of R in case I is. Hence, if $x, y \in \text{sing}\, M$, then $I = \text{ann}_R x$ is an essential right ideal, so formula (ii) shows that $xa \in \text{sing}\, M\ \forall a \in R$. Also, (i) shows that $x - y \in \text{sing}\, M\ \forall x, y,$ so $\text{sing}\, M$ is a submodule called the singular submodule of R.

The first lemma is by Johnson.

9.5 **Lemma.** *Let R be any ring, let M be a module, and let I be a submodule.*

9.5.1 *If I is an essential submodule of M, then for any $x \in M$, $\text{ann}_R x/I$ is an essential right ideal of R.*

9.5.2 *Conversely, if $\text{sing}\, M = 0$, and if $x \in M$ is such that $\text{ann}_R x/I$ is an essential right ideal, then I is essential in $I + xR$.*

Proof of 9.5.1. If A is a right ideal, and if $A \neq 0$, then $xA = 0$ implies $0 \neq A \subseteq \text{ann}\, x/I \cap A$, and $xA \neq 0$ implies $I \cap xA \neq 0$, and by choosing $y = xa \neq 0$ in this intersection, we have $a \in \text{ann}\, x/I \cap A$. Hence $\text{ann}\, x/I \cap A \neq 0$ in all cases, proving 9.5.1.

Proof of 9.5.2. If $y + xa \in I + xR$, $y \in I$, $a \in R$, we want to show that $(y + xa)R \cap I \neq 0$. Now

$$\text{ann}_R(y + xa)/I = \text{ann}_R xa/I = \text{ann}_R a/B,$$

where $B = \text{ann}_R x/I$. Since B is an essential right ideal by hypothesis,

9.5.1 implies that $\operatorname{ann}(y + xa)/I$ is an essential right ideal. Then $\operatorname{sing} M = 0$ implies that $(y + xa) \cdot \operatorname{ann}(y + xa)/I \neq 0$, and this submodule is contained in $I \cap (y + xa)R$. □

9.6 *Exercise.*

9.6.1 Sing M is a fully invariant submodule.

9.6.2 Sing R is an ideal of R.

The ideal sing R is called the (right) **singular ideal** of R.

The identity $(ra)^{n+1} = r(ar)^n a$ for elements $a, r \in R$ shows that Ra is a nil left ideal if aR is a nil right ideal. Thus, R has no nil right ideals $\neq 0$ if and only if R has no nil left ideals $\neq 0$.

9.7 **Lemma.** *Let S be a ring satisfying the maximum condition for right annulets. If S has a nonzero nil right or left ideal A, then S contains a nonzero nilpotent ideal.*

Proof. (Utumi [63a]) By the preceding remark, we can assume A is a nil left ideal $\neq 0$. Let $0 \neq a \in A$ be such that $a^\perp$ is maximal in $\{x^\perp \mid 0 \neq x \in A\}$. If $u \in S$ is such that $ua \neq 0$, then $(ua)^n = 0$ and $(ua)^{n-1} \neq 0$ for some $n > 1$. Since $[(ua)^{n-1}]^\perp \supseteq a^\perp$, and since $ua \in A$, it follows that $[(ua)^{n-1}]^\perp = a^\perp$. Since $ua \in [(ua)^{n-1}]^\perp$, then $aua = 0$. Thus, $aSa = 0$, and then $(a)^3 = 0$, where (a) is the ideal of S generated by a. □

9.8 **Lemma.** *Let R be any ring which satisfies* (acc)$^\perp$ *and* (acc)$\oplus$. *If $a \in R$ there exists $n > 0$ such that $a^n R + a^{n\perp}$ is an essential right ideal.*

Proof. There is an $n \geq 1$ such that $a^{n\perp} = a^{n+1\perp}$. Then $a^n R \cap a^{n\perp} = 0$. Let I be a right ideal and suppose that

$$I \cap (a^n R + a^{n\perp}) = 0.$$

The sum $I + a^n I + a^{2n} I + \cdots$ is direct and, because R has (acc)$\oplus$, we conclude that $I = 0$. □

9.9 **Theorem** (Goldie [58, 60], Lesieur-Croisot [59]) *A ring R has a semisimple right quotient ring if and only if R is semiprime and satisfies* (acc)$\oplus$ *and* (acc)$^\perp$.

Proof. (Goldie [69]) We first assume that R is a semiprime ring satisfying (acc)$\oplus$ and (acc)$^\perp$, and prove that an essential right ideal E contains a regular element. E is not a nil ideal 9.7; it has an element $a_1 \neq 0$ with $a_1^\perp = a_1^{2\perp}$. Either $a_1^\perp \cap E = 0$ or $a_1^\perp \cap E \neq 0$. In the latter case, choose $a_2 \in a_1^\perp \cap E$ with $a_2 \neq 0$ and $a_2^\perp = a_2^{2\perp}$. If $a_1^\perp \cap a_2^\perp \cap E \neq 0$, then the process continues.

At the general stage we have a direct sum

$$a_1 R \oplus \cdots \oplus a_k R \oplus (a_1^\perp \cap \cdots \cap a_k^\perp \cap E),$$

where $a_k \in a_1^\perp \cap \cdots \cap a_{k-1}^\perp \cap E$ and $a_k \neq 0$, $a_k^\perp = a_k^{2\perp}$.

The process has to stop, because R has (acc)$\oplus$. Let this happen at the k-th stage. Then

$$a_1^\perp \cap \cdots \cap a_k^\perp \cap E = 0 = a_1^\perp \cap \cdots \cap a_k^\perp$$

and hence

$$(a_1^2 + \cdots + a_k^2)^\perp = a_1^\perp \cap \cdots \cap a_k^\perp = 0.$$

Let Z be the singular ideal of R and $z \in Z$. Then $z^n R \oplus z^{n\perp}$ is an essential right ideal for some $n > 0$ and $z^{n\perp}$ is also essential. Hence $z^n R = 0$, Z is a nil ideal and hence is zero. Set $c = a_1^2 + \cdots + a_k^2 \in E$; as $c^\perp = 0$ we deduce that cR is essential by 9.8. Hence ${}^\perp c \subset Z$, so that ${}^\perp c = 0$ and c is a regular element. This establishes the existence of regular elements in E.

Suppose that $a, d \in R$ with d regular and set

$$E = \{x \in R \mid ax \in dR\}.$$

Then dR is essential and hence so is E, so that E contains a regular element d_1. The right Ore condition is satisfied and R has a right quotient ring Q.

Next suppose that F is an essential right ideal in Q, then $F \cap R$ is essential in R. Now $F \cap R$ has a regular element, which is a unit in Q, and hence $F = Q$. Let J be a right ideal and K a right ideal of Q such that $J \cap K = 0$ and $J \oplus K$ is essential (use Zorn's lemma); then $J \oplus K = Q$. Thus the module Q_Q is semisimple and Q is a semisimple ring.

Conversely, let R have a semisimple right quotient ring Q. Then a right ideal E of R is essential if and only if $EQ = Q$. To see this, suppose I is a non-zero right ideal of Q, then $I \cap R \neq 0$ and $I \cap R \cap E \neq 0$, taking E to be essential in R. Hence $I \cap EQ \neq 0$, which means that EQ is essential in Q; then $EQ = Q$ as Q_Q is a direct sum of simple modules. On the other hand, when $EQ = Q$ is given and I is a non-zero right ideal of R then $IQ \cap EQ \neq 0$, trivially, and hence $I \cap E \neq 0$, so that E is an essential right ideal of R. These conditions are equivalent to saying that E has a regular element, because $1 \in EQ$ and $1 = ec^{-1}$, $e \in E$, $c \in R$ and regular. On the other hand, when E has a regular element, then $EQ = Q$ and E is essential in R.

We now conclude that R is a semiprime ring, because if N is a nilpotent ideal of R, ${}^\perp N$ is essential as a right ideal 9.4.3, and has a regular element. Thus $N = 0$.

Let S be a direct sum of non-zero right ideals $\{I_\alpha \mid \alpha \in A\}$ of R which is essential as a right ideal. S has a regular element c, expressible as a finite sum

$$c = x_1 + \cdots + x_n; \qquad x_i \in I_{\alpha_i}.$$

Now cR is essential and lies in $I_{\alpha_1} + \cdots + I_{\alpha_n}$; it follows that A has only the indices $\alpha_1, \ldots, \alpha_n$, and R has (acc)$\oplus$.

Finally, the maximum condition holds for right annihilators, because R is a subring of a Noetherian ring Q (7.23). □

9.10 Corollary. *A prime ring R has a simple Artinian right quotient ring if and only if R satisfies* (acc)$\oplus$ *and* (acc)$^\perp$. □

Employing terminology introduced in Chapter 7, we say that R is **right Goldie** if R satisfies (acc)$\oplus$ and (acc)$^\perp$. Thus, by 9.9 and 9.10, a ring R has a (simple) semisimple right quotient ring iff R is a (prime) semiprime right Goldie ring.

Neat Rings

A module M is **neat** if sing $M = 0$. A ring R is **right neat** if sing $R = 0$. **Nonsingular** is a variant term for "neat."

9.11 Lemma. *Let R be a right neat ring, and let A and B be right annulets, $A \supseteq B$.*

9.11.1 *If B is essential in A, then $B = A$.*

9.11.2 *If $x, y \in R$, and if xR and yR are essential right ideals, then yxR is an essential right ideal.*

Proof of 9.11.1. If $x \in A$, then the lemma shows that $\text{ann}_R x/B$ is essential in R. Since $({}^\perp Bx)(\text{ann}_R x/B) = 0$, neatness of R implies ${}^\perp Bx = 0$, whence $x \in B = ({}^\perp B)^\perp$.

Proof of 9.11.2. Since $xR \subseteq \text{ann}\, y/yxR$, $\text{ann}\, y/yxR$ is an essential right ideal, hence by Lemma 9.5.2, yxR is essential in yR, hence in R.

9.12 Lemma.

9.12.1 *If R is a semiprime ring satisfying* (acc)$^\perp$, *then R is right neat.*

9.12.2 *Any right neat ring satisfying* (acc)$\oplus$ *also satisfies both* (acc)$^\perp$ *and* (dcc)$^\perp$.

Proof of 9.12.1 (Indirect proof) If $b \in \text{sing}\, R$ and $b \neq 0$, then $b^\perp \neq R$, hence ${}^\perp(b^\perp)$ is a left annulet $\neq 0$. Now R has d.c.c. on left annulets, therefore choose a minimal left annulet $A \subseteq {}^\perp(b^\perp)$. Since R is semiprime, $A^2 \neq 0$, and $Aa \neq 0$, with $a \in A$. Since $b^\perp$ is essential,

$b^\perp \cap aR \neq 0$, for example, $ax \in b^\perp \cap aR$, $x \in R$, $ax \neq 0$. Since $ax \in b^\perp$, then ${}^\perp(b^\perp)ax = 0$, whence $Aax = 0$, and $0 \neq Aa \subseteq {}^\perp x \cap A$. Since A is a minimal left annulet, this implies ${}^\perp x \cap A = A$, and $Ax = 0$, which contradicts $ax \neq 0$.

9.12.2 By Lemma 9.11.1, if $B \subset A$ are annulets, $B \neq A$, then A contains $B \oplus C$ for some nonzero right ideal C. In this way, from any infinite chain of right annulets, we can extract an infinite direct sum. ▯

9.13 **Corollary.** *The three conditions on a ring R are equivalent:*

(a) *R is a semiprime ring with* (acc)$\oplus$ *and* (acc)$^\perp$.

(b) *R has a semisimple right quotient ring.*

(c) *R is a right neat semiprime ring satisfying* (acc)$\oplus$.

Proof. (a) $\Leftrightarrow$ (b) is 9.9, whereas (a) $\Leftrightarrow$ (c) by 9.12. ▯

9.14 *Exercise.*

9.14.1 Let M be a module. A complement submodule N' is a submodule for which there corresponds a submodule N such that N' is maximal in the set of those submodules A such that $A \cap N = 0$. Then N' is called a **relative complement** of N in M.

(a) Show that the sum $N + N'$ of N and a relative complement N' is always an essential submodule.

(b) If A and B are independent submodules, show by Zorn's lemma that there always exists a relative complement A' of A containing B. Then if A'' is a relative complement of A' containing A, $A'' \nabla A$. A submodule C is **closed** in case C has no essential extensions in M (other than C).

(c) Show that a closed submodule is a complement, and conversely.

(d) A minimal complement submodule (if one exists) is uniform.

9.14.2 (Shock [72b]) If R has finite right Goldie dimension, then so does the polynomial ring $R[x]$, and, in fact, the dimensions are the same.

9.14.3 Let R be an order in a semisimple ring S.

(a) If B is a right ideal of R, then BS is the injective hull of B_R. If A is any R-submodule of S such that A = inj. hull of B, then $A = BS$; that is, B has a *unique* injective hull in S.

(b) The contraction map $A \to A \cap R$ is a lattice isomorphism of the lattice of all right ideals of S and the lattice of complement (closed) right ideals of R.

(c) A right ideal B of R is uniform if and only if BS is a minimal right ideal in S. Then there exists an idempotent $e \in S$ such that $BS = eS$. Furthermore, $K = \text{End}\, B_R$ is a right Ore domain with quotient field $\approx eSe$ (cf. 7.21 ff.).

Levitzki's Theorem

A (left, right) ideal N is nil if every element of N is nilpotent. In the 1930's, Köthe raised the question: If N is a nil ideal in a right Noetherian ring, is N necessarily nilpotent? Levitzki solved this problem affirmatively.

9.15 Theorem (Levitzki [45]). *If S is a right Noetherian ring, then every nil left or right ideal I of S is nilpotent.*

Proof. Let N be the maximal nilpotent ideal of S. Then the factor ring S/N contains no nilpotent ideals $\neq 0$. If I is not contained in N, $(I+N)/N$ would be a nonzero nil one-sided ideal of S/N. But S/N is right Noetherian, so $(I+N)/N$ would $\supseteq$ a nilpotent ideal by 9.7. This contradiction establishes that $I \subseteq N$, thus I is nilpotent. □

Exercises for Chapter 9

***1.** (R. E. Johnson) If R is right neat, then the injective hull $\hat{R}$ of R in mod-R is a ring canonically isomorphic to End $\hat{R}_R$, and the ring operations of $\hat{R}$ induce the module operations of $\hat{R}$ in mod-R.

a) (Johnson-Wong) The ring $\hat{R}$ is a right self injective von Neumann regular ring, and R is a canonical subring.

(b) (Johnson) The contraction map $I \to I \cap R$ induces a lattice isomorphism between the lattices $C_r(\hat{R})$ and $C_r(R)$ of closed (complement) right ideals of $\hat{R}$ and R respectively.

(c) Conclude that $\hat{R}$ is a semisimple ring if and only if R satisfies the a.c.c. (resp. d.c.c.) on complement right ideals. In this case, every regular element $d \in R$ generates an essential right ideal. (Cf. 9.4.2.)

(d) If R is a semiprime right neat ring with a.c.c. on complement right ideals, then $\hat{R}$ is the classical right quotient ring of R. Conversely, if R is a ring with a semisimple classical right quotient ring Q, then Q is the injective hull of R in mod-R.

(e) If R is right neat, then $\hat{R}$ is isomorphic to a product of full linear rings of transformations on right vector spaces if and only if every complement right ideal of R contains a minimal complement right ideal.

(f) (Faith-Utumi [65]) If I is any right ideal of R, then $\hat{I} = eS$ for some $e = e^2 \in S = \hat{R}$, and eSe is a right selfinjective regular ring. Furthermore,

$$K = \{x \in eSe \mid xI \subseteq I\} \approx \text{End } I_R$$

is a subring of eSe canonically isomorphic to End I_R, and eSe is the injective hull of K in mod-K. Thus, since $eSe \approx \text{End } eS_S$ canonically, and since $eS = e\hat{R}$ is the injective hull $\hat{I}$ of I in mod-R, then the formula results:

$$\text{End}\,\widehat{I_R} = \text{End}\,\hat{I}_{\hat{R}}.$$

(g) (Faith-Utumi, l. c.) Assume the notation of (f). Then, if either $\hat{R}$ is an essential extension of R in R-mod or e is a primitive idempotent (that is, eS is a minimal right ideal of $S = \hat{R}$, or, what is the same, if I is a uniform right ideal of R), and if $Q = eSe \cap R \neq 0$, then $eSe = \hat{Q}$.

***2.** Let R be a nil ring-1. Then R is nilpotent if R satisfies one of:

(a) (Herstein-Small and Levitzki) R satisfies the a.c.c. on left and right annulets.

(b) (Lanski) R satisfies (acc)$^{\perp}$ and (acc)$\oplus$.

3. A ring R has a right quotient ring S if and only if for every module M of mod-R the set of torsion elements (= the set of elements of M annihilated by some regular element of R) is a submodule of M. (Gentile-Levy) In this case, a module M is an R-submodule of an S-module if and only if M is torsionfree, and then $MS \approx M \otimes_R S$ under the mapping sending $ms \mapsto m \otimes s$. Furthermore, if also N is an R-submodule of an S-module, any R-homomorphism $f : M \to N$ has a unique extension $\bar{f} : MS \to NS$ to an S-homomorphism $\bar{f}$. Thus, if $N \otimes_R S$ is injective in mod-S, then it is injective in mod-R, and is the injective hull of N. In particular, S is the injective hull of mod-R, whenever S is right selfinjective. (Cf. Levy [63]; also 6.28.)

4. (Levy [63]) A right hereditary ring R with a semisimple right quotient ring is right Noetherian. Moreover, R is a finite product of hereditary prime rings $\{R_i\}_{i=1}^{n}$, where R_i is a minimal annihilator ideal.

5. (Levy [63]) If every finitely generated torsionfree right R-module is embedable in a free R-module, then R is said to be right TF. If R has a two-sided quotient ring, S, and if R is right TF, then so is S. A semiprime ring R with right quotient ring S is right TF only if S is semisimple and the left quotient ring of R. (A quotient ring S is right TF iff every finitely generated module embeds in a free module. A right and left TF ring S is QF. Cf. Faith and Walker [67] and also Faith [73/74b].)

6. Let R be any semiprime ring satisfying (acc)$\oplus$, and let A be any right ideal. If $a \in A$ is such that $a^{\perp}$ is minimal in $\{x^{\perp} \mid x \in A\}$, then aR is essential in A.

7. Let R be a semiprime ring.

(a) If R satisfies the a.c.c. on right annulets, and if $a \in R$ is such that aR is an essential right ideal, then a is regular; that is, $a^{\perp} = {}^{\perp}a = 0$.

(b) If R satisfies (acc)$\oplus$, and if $a \in R$ is such that $a^{\perp} = 0$, then aR is an essential right ideal.

(c) If R satisfies both (acc)$\oplus$ and the a.c.c. on right annulets, then every essential right ideal contains a regular element.

8. Prove that the property of being a right Ore ring is Morita invariant for semiprime right Goldie ring.

9. If R is semiprime right Goldie, then so is eRe, for any idempotent $e \neq 0$. (Cf. proof of 7.21 A.)

Notes for Chapter 9

Prior to the work of Goldie, fundamental and very general theorems on quotient rings and concommitant embedding theorems had been obtained by Johnson [51]. With the publication of Goldie's theorems, mathematicians saw how Johnson's theorems would apply to yield new proofs. The author gave an account of this viewpoint in lectures at Penn State in 1961 and Rutgers in 1962, 1963, and published in 1967. A number of others are listed here: Gabriel [62], Lesieur-Croisot [59], Freyd [62], Johnson [65], Procesi and Herstein (in Appendix B to Jacobson [64]), Procesi and Small [65]. The proofs of Zelmanowitz [69] and Goldie [69] (presented here) contain elements of all of these.

In 1960, Goldie gave a proof of Levitzki's theorem based on his structure theory for Noetherian semiprime rings. Next, Herstein [62] published a simplification of Goldie's proof. Utumi's proof [63] (reproduced here) may be the last word on the subject. This history has a moral. It is never easy to find the optimal proof — the hypotheses are likely to be stronger than necessary. Actually, Levitzki did look at the right objects, namely annulets, and the theorem of Levitzki [64] (edited by Amitsur) states that any nil ideal is nilpotent if the ring satisfies the a.c.c. on right and left annulets, a result proved independently by Herstein and Small [64]. (Cf. Exercise 2 which also states the theorem of Lanski.) Also, see other nil and nilpotency theorems of Fitting and Shock in Chapters 17 and 18.)

Elizarov [69] has published a survey of quotient rings, including those which have Artinian (also perfect or Quasi-Frobenius) quotient rings (e.g. Jans [67], Jategaonker [70], Gupta [68], Gupta and Saha [67], Mewborn and Winton [69], Robson [67a], Small [66b], and Talintyre [63, 66].) We discuss rings with semilocal right quotient rings in Chapters 10 and 18.

In another direction Small [66a] proved that right and left Noetherian hereditary rings have right quotient rings (see References for related papers on hereditary Noetherian rings). Chatters [72] proved these rings are direct sums of Artinian rings and of prime rings. (Cf. Faith [73/74a].)

Remaining unsolved is the question: Are right Ore rings Morita invariant? If so, do similar right Ore rings have similar right quotient rings? (Cf. Notes on Quotient Rings, p. 229.)

References

Chatters [70, 72], Curtis-Reiner [62], Elizarov [69], Eisenbud and Griffith [71], Eisenbud and Robson [70], Faith [67a], Faith and Utumi [65b], Freyd [62], Gabriel [62], Gabriel and Zisman [68], Goldie [58, 60, 69], Gupta [68], Gupta and Saha [67], Herstein [62], Herstein and Small [64], Jacobson [64], Jans [67], Jategaonker [70, 72], Johnson [51a, 51b], Johnson and Wong [59], Lanski [69], Lesieur and Croisot [59], Levitzki [45, 64], Levy [63], Mewborn and Winton [69], Michler [69d], Ore [31], Procesi [64], Procesi and Small [65], Robson [67a, 68], Shock [71b], Small [66a, 66b, 68], Talintyre [63, 66], Utumi [63], Webber [70], Zelmanowitz [69].

Chapter 10

Orders in Semilocal Matrix Rings

A ring R is **semilocal** if $R/\text{rad}\, R$ is semisimple. [Since $R/\text{rad}\, R$ is semiprime, by the W-A Theorem 8.8, this is equivalent to the requirement that $R/\text{rad}\, R$ is right (or left) Artinian.] The main theorem in this chapter is the determination of right orders in semilocal matrix rings D_n. A **right order** R of a ring S is a subring-1 such that

$$S = \{ac^{-1} \mid a, \text{ regular } c \in R\}.$$

The theorem of Faith and Utumi [65] implies that every order R in a matrix ring $S = D_n$ contains a order of the form F_n, for some set M of $n \times n$ matrix units, where F is a subring of $D =$ centralizer M. Moreover, when D_n is a semilocal quotient ring, by 10.15 and 10.19, then F is a right order in D, and

$$D_n = \{af^{-1} \mid a \in F_n,\ f \in F\}.$$

Thus, the "denominators" can be chosen to be scalars $f \in F$ (cf. 10.10). When R is also a left order, then this holds independently of the choice of M.

A corollary is a theorem of Goldie which states that every prime principal right ideal ring R (which by Theorem 9.10 is a right order in a total matrix ring D_n over a field D) is isomorphic to a total matrix ring F_n over a right Ore domain (with right quotient field D). In this case, F is right hereditary, but need not be a principal right ideal ring.

In the last part of the chapter, maximal right orders in semilocal quotient rings D_n are determined a la Faith [64] as endomorphism rings of torsionfree modules over right orders in D. (This extends Robson's generalization [67] for D right Artinian.)

The characterization of right orders in semilocal rings is taken up in Chapter 18, esp. 18.47.

For convenience, we repeat some concepts from Chapter 4 (supra 4.37) and Chapter 7 (supra 7.21).

10.1 Definitions (Jacobson [42]) Let R_1 and R_2 be right orders in a quotient ring Q. Then

(i) they are **equivalent**, $R_1 \overset{Q}{\sim} R_2$, if there exist regular elements a_1, b_1, a_2, b_2 of Q such that $a_1 R_1 b_1 \subseteq R_2$, $a_2 R_2 b_2 \subseteq R_1$;

(ii) they are **right equivalent**, $R \overset{r}{\sim} R_2$, if there exist regular elements a_1, a_2 of Q such that $a_1 R_1 \subseteq R_2$, $a_2 R_2 \subseteq R_1$;

(iii) they are **left equivalent**, $R_1 \overset{l}{\sim} R_2$, if there exist regular elements b_1, b_2 of Q such that $R_1 b_1 \subseteq R_2$, $R_2 b_2 \subseteq R_1$.

These are equivalence relations. Also, $R_1 \overset{r}{\sim} R_2 \Rightarrow R_1 \overset{Q}{\sim} R_2$ and $R_1 \overset{l}{\sim} R_2 \Rightarrow R_1 \overset{Q}{\sim} R_2$.

10.2 Lemma (Jacobson) *Let R be a right order in a quotient ring Q, and let S be a* subring-1 *of Q.*

10.2.1 *If there exist regular elements a, b, c, d of Q such that* $aRb \subseteq S$, $cSd \subseteq R$, *then S is a right order in Q, and* $R \overset{Q}{\sim} S$.

10.2.2 *If there exist regular elements a, c of Q such that* $aR \subseteq S$, $cS \subseteq R$, *then S is a right order in Q, and* $R \overset{r}{\sim} S$. *Similarly for left equivalence.*

Proof of 10.2.1. If $q \in Q$, then

$$a^{-1}qa = xc^{-1}$$

for some $x, c \in R$, and then

$$q = (axb)(acb)^{-1},$$

so that S is a right order of Q similar to R.

Proof of 10.2.2 is obtained by taking $b = 1$. □

10.3 Definition. Q is a **right quotient ring (quoring)** of R. This designation may hold in the case R is a subring-1 of Q such that:

(Q1) R has regular elements.

(Q2) Every regular element of R is a unit of Q.

(Q3) If $q \in Q$, then $q = xc^{-1}$, with x, regular $c \in R$.

Is S is a right quotient ring of a ring R, then R is a right order of S, and, conversely, if S is right Artinian (see 10.7).

10.4 Proposition. *A ring-1 R has a right quotient ring Q if and only if R has a regular element and the right Ore condition holds, namely, for every pair (a, c), with $a, c \in R$, and c regular, there is a pair (a_1, c_1) of elements of R, with c_1 regular, such that $ac_1 = ca_1$.*

Proof. Cf. 9.1.

10.5 Proposition. *Any two right quotient rings of a ring R are isomorphic under a map which induces 1_R.*

Proof. Let D and Q be two quotient rings of R, and let $x \cdot y$ be the operation in Q. Now, the Ore condition is that

$$ac^{-1} = bd^{-1} \Leftrightarrow ad_1 = bc_1,$$

with d_1, regular $c_1 \in R$ (where $d^{-1}c = c_1 d_1^{-1}$). This proves that

$$ac^{-1} = bd^{-1} \text{ in } D \Leftrightarrow a \cdot c^{-1} = b \cdot d^{-1} \text{ in } Q$$

and that $ac^{-1} \mapsto a \cdot c^{-1}$ yields the desired isomorphism. □

A ring is said to be (von Neumann) **regular** provided that every principal right ideal is generated by an idempotent, and then every principal left ideal is generated by an idempotent (see 11.24).

A ring R is **Dedekind finite** provided that the implication $xy = 1 \Rightarrow yx = 1$ is valid in R.

10.6 *Exercises.*

10.6.1 The following conditions are equivalent:

(a) R is semisimple.

(b) R is regular and right Noetherian.

(c) R is regular and right Artinian.

(d) R satisfies the d.c.c. on principal right ideals, and $\operatorname{rad} R = 0$.

10.6.2 Any semilocal ring is Dedekind finite.

10.6.3 A full linear ring $L = \operatorname{End}_D V$, where V is a left vector space over D, is left selfinjective and regular. Moreover, L is Dedekind finite if and only if V is finite dimensional.

10.6.4 Any product of regular rings is regular. Any matrix ring D_n over a regular ring D is regular.

*10.6.5 (Utumi [65]) Every right and left selfinjective ring R is Dedekind finite (cf. Chapter 19).

*10.6.6 (Roos [68]) A left selfinjective regular ring R is Dedekind finite if and only if R is right selfinjective.

10.7 Definition and Proposition.

10.7.I *A ring S is said to be a* **quoring, or quotient ring,** *provided that S satisfies the equivalent conditions:*

(a) *S is the right quotient ring of S.*

(b) *S is the right quotient ring of some ring-1 R.*

(c) *Every regular element of S is a unit of S.*

(d) *S is the left quotient ring of S.*

In this case S is called the **right quoring** *of R.*

10.7.II *A ring S is a quoring provided that S satisfies any of the conditions:*

(1) *S satisfies the* d.c.c. *on principal right ideals of the form* $x^n S$, *where x is any regular element.*

(2) *S is right selfinjective, and Dedekind finite.*

(3) *S is regular.*

Proof. The equivalence of (a), (c), and (d) is trivial, and so is the implication (a) $\Rightarrow$ (b). Assuming (b), if $q = xc^{-1}$ is a regular element of Q, with $x, c \in R$, then x is a regular element of R, and so $q^{-1} = cx^{-1}$ is a unit of Q. Thus, (b) $\Rightarrow$ (c).

(1) If $x^{n+1}S = x^n S$, then $x^n = x^{n+1}q$, for some $q \in S$. Then $x^n(1 - xq) = 0$, so the regularity of x (even right regularity) implies $(1 - xq) = 0$. Then $x(1 - qx) = 0$, and so $qx = 1 = xq$, and x is a unit with inverse q. This proves (1).

(2) If S is injective, and if x is (right) regular, then $yx = 1$ for some $y \in S$. Then, (2) implies that $xy = 1$, and therefore S is a quoring.

(3) If $x \in S$, then $xS = eS$ for some $e = e^2 \in S$, and so $(1 - e)x = 0$. Regularity of x then implies that $e = 1$, that is, $xS = S$. Thus, (3) $\Rightarrow$ (1), so S is a quoring. □

10.8 *Exercise.* If D is a ring satisfying 10.7.II (1), (2), or (3), then so does the matrix ring D_n.

10.9 **Proposition.** *Consider the following conditions on an element x of a right order R of a quoring S.*

(a) *x is right regular in R.*

(b) *x is right regular in S.*

(c) *x is regular in R.*

(d) *x is regular in S.*

(e) *x is a unit of S.*

Then,

$$(e) \Leftrightarrow (d) \Rightarrow (c) \Rightarrow (b) \Leftrightarrow (a).$$

Thus, conditions (a) to (e) *are equivalent if and only if* (b) $\Leftrightarrow$ (d). *In this case, S is a right quoring of R, for example, when S satisfies the* d.c.c. *on principal right ideals, or when S is Dedekind finite and left self-injective.*

Proof. (e) $\Rightarrow$ (d) $\Rightarrow$ (c) $\Rightarrow$ (a), and (b) $\Rightarrow$ (a) are trivial. To prove (a) $\Rightarrow$ (b), let $xy = 0$, for some $y \in S$. Then $yc = a \in R$, for some regular element c of R, and $xa = 0$. Then, (a) implies that $a = yc = 0$, and $y = 0$. This proves the first chain of implications. If (b) $\Rightarrow$ (d), then (a) $\Rightarrow$ (e), and the cycle of implications is complete. When (1) or (2) of 10.7.II of the last proposition hold, then (b) $\Rightarrow$ (d). ▯

10.10 Proposition. *Let F be a subring-1 of a ring D. Then every element of the matrix ring D_n has the form*

$$fc^{-1},$$

$f \in F_n$, $c \in F$, *if and only if F is a right order of D. Furthermore, if D_n is a quoring, then so is D, and, conversely, if D is a ring satisfying* 10.7.II (1), (2), *or* (3).

Proof. If F is a right order of D, and if

$$x = \sum_{i,j=1}^{n} e_{ij} d_{ij},$$

$d_{ij} \in D \forall i, j$, is an element of D_n, then by 9.3, there exist elements $f_{ij}, c \in F$ such that $d_{ij} = f_{ij} c^{-1}$, and then $x = fc^{-1}$, where $f = \sum_{i,j=1}^{n} e_{ij} f_{ij} \in F_n$.

Conversely, if $d = xc^{-1} \in D$, with $x \in F_n$ and $c \in F$, then $x = dc \in F_n \cap D = F$, and so F is a right order of D.

Finally, if D_n is a quoring, and if $d \in D$ is regular in D, then d is regular in D_n, so d^{-1} exists in D_n. But, then d^{-1} belongs to the centralizer of the matrix units $\{e_{ij}\}$, and so $d^{-1} \in D$. Thus, D is a quotient ring.

The converse follows from 10.7 and 10.8. ▯

10.11 Proposition and Definition. *A right ideal I of a ring-1 R is* **closed** *if the implication holds:*

$$xc \in I \Rightarrow x \in I$$

$\forall x \in I$, *regular* $c \in R$. *Every right annulet of a ring-1 is closed, and every right ideal of a quoring is closed. Furthermore, if Q is a quoring, and if R is a subring-1 such that every regular element of R is a unit of Q, then the contraction mapping*

$$\begin{cases} \text{right ideals of } Q \to \text{closed right ideals of } R \\ \qquad\qquad I \mapsto I \cap R \end{cases}$$

is a lattice homomorphism. If, in addition, R is a right order of Q, then the contraction mapping $I \mapsto I \cap R$ *is a lattice isomorphism, and the inverse maps*

$$J \mapsto JQ = \{xc^{-1} \mid x \in J, \text{ regular } c \in R\}$$

for each closed right ideal J of R. If J is a closed right ideal, then JQ is an ideal of Q if and only if J is an ideal of R which is closed on the left.

Proof. If I is a right ideal of Q, if $x \in R$, and if $xc \in I$ for some regular element $c \in R$, then $x \in I \cap R$. Thus, $I \cap R$ is closed in R. If R is a right order of Q, and if $y \in I$, then $x = yc \in I \cap R$ for some regular element $c \in R$, and so $y = xc^{-1}$. This shows that $I = (I \cap R)Q$ consists of all such elements $y = xc^{-1}$. Furthermore, if J is any right ideal of R, then

$$J' = \{xc^{-1} \mid x \in J, \text{ regular } c \in R\}$$

is a right ideal of Q, and

$$J' = JQ = (JQ \cap R)Q.$$

Therefore, if $y \in JQ \cap R$, then $y = xc^{-1}$, with $x \in J$, $c \in R$. If J is a closed right ideal of R, then $x = yc \in J$ implies that $y \in J$, and so $JQ \cap R = J$. This proves the stated lattice isomorphism.

Next, let J be an ideal of R which is closed on the left, let c be a regular element of R, and let $x \in J$. Then

$$y = c^{-1}x = x_1 c_1^{-1}$$

for $x_1, c_1 \in R$, and then $xc_1 = cx_1 \in J$. Since J is closed on the left, then $x_1 \in J$, and hence every element $y = c^{-1}x$, with $x \in J$, $c \in R$, belongs to JQ. It follows that JQ is an ideal of Q. Conversely, $J = JQ \cap R$ implies that J is an ideal of R whenever JQ is an ideal of Q. Morever, the contraction of any left ideal of Q is a closed left ideal of R. ▯

10.12 Definition. *Let C denote of class a quorings which is closed under similarity. The property of being a right order in a member of C will be said to be a Morita invariant provided that for every right order R of a member S of C it is true that every ring similar to the subring* R_1 *of S generated by R has a class C right quoring. (Here* $R_1 = R + \mathbb{Z}1$ *equals R in case R is a ring, that is, in case* $1 \in R$*.)*

10.13 *Exercises.*

10.13.1 The class of right orders in semisimple rings is Morita invariant, in particular, if $R = F_n$ is a right order in a full matrix ring $S = D_n$, where D is semisimple, then F is a right order in D.

10.13.2 If R is a commutative quotient ring, then the $n \times n$ matrix ring is a quotient ring. (Hint: If R is a commutative ring, and if $x \in R_n$, then x is regular in R_n if and only if det x is regular in R.)

10.14 Proposition.

10.14.1 *The following conditions on a ring R are equivalent:*

(a) *R is semiprime right Goldie.*

(b) *R has a semisimple right quoring S.*

(c) *R is semiprime, and for every finitely generated projective right R-module P, the endomorphism ring* End $\hat{P}_R$ *of the injective hull $\hat{P}$ of P in* mod-R *is semisimple.*

10.14.2 *Semiprime right Goldie rings are Morita invariant.*

Proof. If $T : C \rightsquigarrow D$ is a category equivalence, then T induces a bijection $\mathrm{Mor}_C(X, Y) \to \mathrm{Mor}_D(TX, TY)$, for any pair of objects X and Y of C, and for abelian categories, this is an isomorphism of abelian groups. Furthermore, in this case, T induces a ring isomorphism $\mathrm{End}_C X \approx \mathrm{End}_D TX$, for any object X. Thus, 10.14.2 follows from 10.14.1 (c).

Proof of 10.14.1. (a) $\Leftrightarrow$ (b) by 9.9. Assuming (a) and (b), then sing $R = 0$ by 9.14, and so, by 4.5, $\hat{R}$ is the maximal rational extension $\bar{R}$ of R. Now the right quoring S of R is a rational extension of R and, hence, embeds canonically in $\bar{R}$. Since S is semisimple, then S is injective in mod-S, hence S is a summand of $\bar{R}$ in mod-S. However, $\bar{R}$ is an essential extension of R in mod-R, and so $\bar{R} = S = \hat{R}$ is a semisimple ring. This proves that (a) $\Rightarrow$ (c) for the case $P = R$, and then (c) follows trivially.

Conversely, assume (c). Then $B = \mathrm{End}\,\hat{R}_R$ is semisimple. Since $\hat{R}$ contains R, then $\hat{R}$ is a cyclic canonical B-module, and then $A = \mathrm{End}_B\hat{R}$ is semisimple. Since $\bar{R} = \mathrm{Biend}\,\hat{R}_R$ canonically by 4.5, then $\bar{R} = A$ is semisimple. Since a semisimple ring satisfies (acc)$\oplus$, and since $\bar{R}$ is an essential extension of R, then R satisfies (acc)$\oplus$. Moreover, any subring of a ring satisfying (acc)$^\perp$ has the same property, and so (c) $\Rightarrow$ (a) by 9.9. ▯

10.15 Faith-Utumi Theorem [65]. *Let R be any subring-1 of a matrix ring $S = D_n$, where $D = \mathrm{centralizer}_S M$ of a set $M = \{e_{ij}\}_{i,j=1}^n$ of matrix units. Let*

$$A = A_M = \{a \in R \mid aM \subseteq R\}$$

and

$$B = B_M = \{b \in R \mid Mb \subseteq R\}.$$

10.15.1 *Then*

$$BA = (BA \cap D)_n = \sum_{i,j=1}^{n} (BA \cap D) e_{ij}$$

is a subring-1 of R, AB is an ideal of R, and $(BA)^2 \subseteq AB$. Furthermore, $F = BA \cap D$ and $U = B \cap D = A \cap D$ are ideals of $P = R \cap D$ such that $F \supseteq U^2$. Thus, $R \supseteq BA = F_n \supseteq U_n^2$.

10.15.2 *If D_n is a right quoring of R, then M can be so chosen so that A contains a unit a of S, B contains a unit b of S, and $BA = F_n$ is a right order of D_n equivalent to R and to P_n. If R is also a left order of D_n, then any choice of matrix units M has these properties.*

10.15.3 *In case 10.15.2, if D is a semisimple*[1] *ring, then D is the right quoring of F, and*

$$D_n = \{af^{-1} \mid a \in F_n, \text{ regular } f \in F\}.$$

Proof of 10.15.1. Since $AP \subseteq A$, $PB \subseteq B$, and $P \subseteq D$, then $PA \subseteq F$, and $FP \subseteq F$, so that F is an ideal of P. Since B is a right ideal of R, then $U = B \cap D$ is a right ideal of $P = R \cap D$, and $U = A \cap D$ is a left ideal of P. Furthermore,

$$U \supseteq F = BA \cap D \supseteq (B \cap D)(A \cap D) = U^2.$$

By 3.18, every element $t \in D_n$ has a unique representation $t = \sum_{i,j=1}^{n} t_{ij} e_{ij}$, where

$$t_{ij} = \sum_{k=1}^{n} e_{ik} t e_{kj} \in D.$$

Thus, if $t = ba \in BA$, $b \in B$, $a \in A$, then $t_{ij} \in BA \cap D = F$ since $MB \subseteq B$ and $AM \subseteq A$. This proves that $BA = F_n$. Moreover, $(BA)^2 = BABA \subset AB$, since A is a left, and B is a right ideal.

Proof of 10.15.2. Now assume that D_n is a right quoring of R. First, let M' be any set of $n \times n$ matrix units of D_n, with centralizer D'. By 9.3, there is a common denominator a in R of the elements in M', that is, a regular element $a \in R$ such that $M'a \subseteq R$. Then,

$$M = a^{-1} M' a$$

is a set of matrix units of S, and

$$D = a^{-1} D' a = \text{centralizer}_S M$$

[1] This holds more generally for any semilocal ring D (10.19). The original theorem of Faith-Utumi [65] was stated for a field D. (Most of 10.15.1 and some of 10.15.2 is taken from Faith [65].)

is a ring isomorphic to D'. Furthermore, a is a unit of S contained in

$$A = \{x \in R \mid xM \subseteq R\}.$$

Moreover

$$B = \{y \in R \mid My \subseteq R\}$$

also contains a unit b of S. We now show that any set M of matrix units of S, for which A and B just defined each contains a unit of S, has the stated properties. Since every set M of matrix units satisfies this requirement in the case R is also a left order in S, then every set of matrix units has the stated properties in this case.

Since AB is an ideal of R, and since AB contains a regular element $t = ab \in R$, then D_n is a right quoring of AB, and every $y \in D_n$ has the form $y = xt(ct)^{-1}$, for elements x, regular $c \in R$. Since $bSb^{-1} = S$, where $S = D_n$, every element $y \in S$ has the form $y = by_1b^{-1}$, for some $y_1 = xr^{-1} \in S$, where x, regular $r \in R$. Thus,

$$y = b(xr^{-1})b^{-1} = (bxa)(bra)^{-1}$$

with $bxa \in BA$ and $bra \in BA$. This proves that D_n is a right quoring of BA. Now $bRa \subseteq BA$, and $BA \subseteq R$, so that BA is a right order of D_n equivalent to R. Moreover, since F is an ideal of P by 10.15.1, then F_n is an ideal of P_n containing a regular element t. Thus, $tP_n \subseteq F_n$, so F_n is equivalent to P_n.

Proof of 10.15.3. Assume that D_n is a semisimple right quoring of F_n. Since semiprime right Goldie rings are Morita invariant, by 10.14, then F is semiprime right Goldie and, hence, F has a semisimple right quoring Q. (If F is a subring-1 of D, then the subring F' generated by F is semiprime right Goldie, with semisimple right quoring Q, and then Q is the right quoring of F.) Since every regular element $f \in F$ is regular in F_n, hence $f^{-1} \in D_n$, and in fact, $f^{-1} \in D$, then Q can be embedded in D by a mapping which induces 1_F. Then, F_n is a right order in Q_n, and by 10.10,

$$Q_n = \{ac^{-1} \mid a \in F_n,\ c \in F\}.$$

However, Q_n is semisimple, and hence every right regular element of Q_n is a unit, so by 10.9, Q_n is a right quoring of F_n. By 10.5, this proves $Q_n = D_n$, so that $Q = D$ is the right quoring of F. ▯

10.16 Proposition. *Let S be a right quotient ring of R. If I is an ideal of S such that every right regular element of $\bar{S} = S/I$ is a unit of $\bar{S}$ (e.g., if*

S/I has the d.c.c. *on principal right ideals; cf. also* 10.7*), then $\bar{S}$ is canonically the right quotient ring of $\bar{R} = R/(I \cap R)$, and $(I \cap R)S = I$.*

Proof. If $y \in S$, then $y = ac^{-1}$, with $a, c \in R$, and then $\bar{y} = \bar{a}\bar{c}^{-1}$, with $\bar{a}, \bar{c} \in \bar{R}$, where the bar notation indicates the images under the canonical map $S \to \bar{S} = S/I$. Thus, $\bar{R} = (R + I)/I \approx R/(I \cap R)$ is a right order in S, and by 10.9, the assumption on $\bar{S}$ implies that $\bar{S}$ is the right quotient ring of $\bar{R}$.

A ring S is **semilocal** provided that $S/\text{rad}\, S$ is semisimple.

10.17 Corollary. *If R has a semilocal right quotient ring S, then $\bar{R} = R/(R \cap \text{rad}\, S)$ is a right Goldie ring with right quotient ring $\bar{S} = S/\text{rad}\, S$.*

Proof. Since S is semisimple (Artinian) modulo $I = \text{rad}\, S$, then the proposition applies. Since then $\bar{R}$ has semisimple right quoring S/I, then $\bar{R}$ is right Goldie 10.14. □

10.18 Lemma[1]. *Let $R = F_n$ have a semilocal right quoring $S = D_n$, and choose a set M of $n \times n$ matrix units of S in accordance with* 10.15 *so that $D = \text{centralizer}_S M$ contains F. For any $x \in S$, let $x = (x_{ij})$ denote the matrix representation with $x_{ij} \in D$, $i, j = 1, \ldots, n$. Then, for any regular element $t \in F_n$ there exists a regular element $a \in F_n$ such that the diagonal elements x_{ii} of the matrix $x = ta$ are regular elements of F, and the off-diagonal elements x_{ij} are contained in $F \cap \text{rad}\, D$, $i = 1, \ldots, n$, $i \neq j = 1, \ldots, n$.*

Proof. By 10.17, $\bar{S} = S/\text{rad}\, S$ is the right quoring of $\bar{R} = R/(R \cap \text{rad}\, S)$. Then, by 10.15, every element of $\bar{S} = \bar{D}_n$ has the form $\bar{a}\bar{c}^{-1}$, where $a \in F_n$ and $c \in F$ (that is, $\bar{a} \in \bar{R}$ and $\bar{c} \in \bar{F}$). In particular, t^{-1} for any regular $t \in F_n$ has this form, and so

$$tac^{-1} = 1 + (w_{ij}),$$

where $(w_{ij}) \in (\text{rad}\, D)_n$. Then, $x = ta \in F_n$ is such that

$$x_{ii} = (1 + w_{ii})c \in F$$

is regular (since $1 + w_{ii}$ and c are units of D), and

$$x_{ij} = w_{ij}c \in F \cap \text{rad}\, D$$

as asserted. □

[1] This lemma was proved by Faith and Utumi [65] for the case D is a field, and the extension to the case D is semisimple is immediate. Robson [67, 3.3, p. 607] extended this to the case D is right Artinian.

10.19 Proposition[1]. *If R has a semilocal right quoring $S = D_n$, then there is a set M of $n \times n$ matrix units of S such that R contains a right order F_n of D_n, and F is a right order of $D =$* centralizer$_S M$. *Moreover,*

$$D_n = \{ac^{-1} \mid a \in F_n,\ c \in F\}.$$

Proof. Everything follows from 10.15, except that F is a right order in D in general. By 10.15, F_n is a right order of D_n equivalent to R, and so every element d of D_n has the form $d = yt^{-1}$, with y, regular $t \in F_n$; in particular, any element $d \in D$ can be so written. Apply 10.18 to write $x = ta \in F_n$, with $x_{ii} \in F$ invertible in D, $i = 1, \ldots, n$. Since

$$ya = dta = dx = d(x_{ij}) = (dx_{ij}) \in F_n,$$

then $f_{ij} = dx_{ij} \in F_n$ and $d = f_{11}x_{11}^{-1}$, with $x_{11} \in F$. This proves that D is a right quoring of F. □

10.20 Corollary. *If T is a right order in a quotient ring S which is a finite product of matrix rings, say $S = \prod_{i=1}^{n} S_i$, where $S_i = M_{n_i}(D_i)$ is a total matrix ring of degree n_i over a semilocal ring D_i, then $T_i = S_i \cap T$ is a right order in S_i, so there exists a right order F_i of D_i such that T contains a finite product $\prod_{i=1}^{n} M_{n_i}(F_i)$.*

Proof. One easily verifies that T_i is a right order of S_i (using the centrality in S of the identity element of S_i), and then the proposition can be applied to T_i to obtain the corollary. □

Principal Ideal Rings

If R is a right order in D_n, the theorem of Faith and Utumi states that there exists an order F in D and an inclusion $F_n \subseteq R \subseteq D_n$. Even though F_n is an order in D_n, the ring structure of R does not depend entirely on F_n. Under a stronger hypothesis, however, it does.

10.21 Proposition (Goldie [62]) *If R is a prime ring and a principal right ideal ring, then there exist a right Ore domain F, an integer $n > 0$, and an isomorphism $R \approx F_n$.*

Proof. (Faith-Utumi [64a]). Using the notation of 10.15, we can write $B = cR$, for some $c \in R$. Since B contains a regular element, c is a regular element. Since $e_{ij}B \subseteq B$, that is, $e_{ij}cR \subseteq cR$, then

$$c^{-1}e_{ij}c = f_{ij} \in R$$

[1] This is the Faith and Utumi theorem for semilocal rings, and the comment in the footnote to Lemma 10.18 applies here. Also see Notes for Chapter 10.

$(i, j = 1, \ldots, n)$. Now $M = \{f_{ij} \mid i, j = 1, \ldots, n\}$ is a complete set of matrix units in S, with associated field $H = c^{-1}Dc$. Since $M \subseteq R$, then by 3.18, $R = \sum_{i,j=1}^{n} f_{ij}K$, where $K = H \cap R$. Since R is an order in H_n, then K is an order in H; that is, K is an Ore domain 10.19. □

F need not be a principal ideal ring (cf. Swan [61]). Nevertheless, F is right Noetherian and hereditary. (Proof?)

Supplementary Remarks. 1. Let T be a right order in the ring $S = D_n$, $n > 1$. If f is any idempotent of S such that fS is a minimal right ideal, then $f = ac^{-1}$, with $a, c \in R$. Thus $e = c^{-1}fc$ is idempotent and $0 \neq ce \in Se \cap R$. Since $Se \cap R \neq 0$, it follows from primeness of T that $(eS \cap R)(Se \cap R) \neq 0$, so that $eSe \cap R \neq 0$. Then the proof of 10.15 shows that $F = eSe \cap T$ is a right Ore domain, and $D \approx eSe$ is its right quotient field; we thus obtain that $S \approx D_n$, where D contains a right Ore domain K, which is contained in R. This illustrates the precise nature of 10.15.2, which states much more. □

2. Next we show that, in general, not every set M of matrix units has the property 10.15.2 without the hypothesis that T is also a left order in S.

Let K be a right Ore domain with right quotient field D that is not a left Ore domain. Let x, y be nonzero elements of K such that $Kx \cap Ky = 0$, and let

$$T = \begin{pmatrix} Kx, & Ky \\ Kx, & Ky \end{pmatrix}.$$

[T is the set of all 2×2 matrices $\left(\begin{smallmatrix} a b \\ c d \end{smallmatrix}\right)$ with $a, c \in Kx$, and $b, d \in Ky$.] Since K is right Ore, if $A = \left(\begin{smallmatrix} a b \\ c d \end{smallmatrix}\right)$ is an arbitrary element of D_2, there exists $0 \neq q \in K$ such that $aq, bq, cq, dq \in K$, and then

$$B = \begin{pmatrix} a & b \\ c & d \end{pmatrix} \begin{pmatrix} qx & 0 \\ 0 & qy \end{pmatrix} \in T.$$

Thus, $A = BC^{-1}$, with $B, C = \left(\begin{smallmatrix} qx & 0 \\ 0 & qy \end{smallmatrix}\right) \in T$. Hence, T is a right order in D_2.

As in the theorem, identify D with the subring of D_2 consisting of all scalar matrices $\left(\begin{smallmatrix} k & 0 \\ 0 & k \end{smallmatrix}\right)$ with $k \in D$. Now assume for the moment that T contains a subring F_2, where F is an entire ring $\subseteq D$. The contradiction is immediately evident since the form of T,

$$T = \begin{pmatrix} Kx, & Ky \\ Kx, & Ky \end{pmatrix},$$

where $Kx \cap Ky = 0$, precludes the possibility of its containing a nonzero scalar matrix $\left(\begin{smallmatrix} d & 0 \\ 0 & d \end{smallmatrix}\right)$ with $d \in K$. □

3. A ring R, as in 10.6, need not contain nontrivial idempotents. Perhaps the simplest example is as follows: Let $S = \mathbb{Q}_2$ be the ring of

all 2×2 matrices over the rational number field $\mathbb{Q}$, and let R be the subring consisting of all matrices $\begin{pmatrix} a & b \\ c & d \end{pmatrix}$, where b, c are even integers and a, d are integers that are either both even or both odd. Then $R = (2\mathbb{Z})_2 + \mathbb{Z}$ is not an entire ring, and R is a right order in the ring $\mathbb{Q}_2$, but R does not contain idempotents $\neq 0, 1$. □

10.22 *Exercises.*

10.22.1 (Goldie) If $R = F_n$ is a principal right ideal ring, where F is a domain, then F is right hereditary, and every right ideal of F is generated by n (or fewer) elements.

10.22.2 For any ring F, if every F-submodule of F^n is generated by $\leq n$ elements then $R = F_n$ is a principal right ideal ring, and conversely.

*10.22.3 (Asano-Goldie) If R is a principal ideal ring, then R is a direct product of finitely many rings each of which is either a prime ring or an Artinian ring. An Artinian principal ideal ring is a direct product of finitely many rings, each of which is a full matrix ring A_n over a local ring A, for various rings A and various n. (Cf. local rings, Chapter 18.)

10.22.4 Generalize 10.21 to the case where R has a semilocal right quoring D_n. If R is a principal right ideal ring, show that $R = F_n$, where F is a right order of D. (Note that 10.22.3 implies than any right and left principal ideal ring R does have a semilocal right quoring.)

Maximal Orders in Semilocal Matrix Rings

In this section, maximal right orders in semilocal matrix rings D_n are determined to be endomorphism rings of torsionfree modules over right orders in D. This theorem was first proved in Faith [64] for a simple right Goldie ring R which proved to be a maximal right order in D_n, where D is a field. With the Faith-Utumi theorem 10.19 for semilocal rings at hand, the adaption of the proof of Faith [64] by Robson [67] to right Artinian rings D is applicable here to the case D is semilocal.

10.23 Lemma. *Let R and S be right orders in a quotient ring Q, with $R \subseteq S$, $R \sim S$. Then there exist right orders T and T' in Q such that*

$$R \subseteq T \subseteq S, \qquad R \overset{l}{\sim} T \overset{r}{\sim} S,$$

$$R \subseteq T' \subseteq S, \qquad R \overset{r}{\sim} T' \overset{l}{\sim} S.$$

Proof. There exist regular elements x and y of Q such that $xSy \subseteq R$. Say $x = ab^{-1}$, $y = cd^{-1}$, $a, b, c, d \in R$. Now $d \in R \Rightarrow Rd \subseteq R$, and

$$b \in R \Rightarrow b \in S \Rightarrow bS \subseteq S \Rightarrow S \subseteq b^{-1}S.$$

Thus

$$aSc \subseteq ab^{-1}Sc \subseteq Rd \subseteq R.$$

Now let T be the subring of S generated by aS and R. Then $T = R + aS + RaS$. It is easy to see that $aS \subseteq T$ and $Tc \subseteq R$. Also $R \subseteq T \subseteq S$. By 10.2, $R \overset{l}{\sim} T \overset{r}{\sim} S$. Similarly one can construct T'. ▯

10.24 Definitions. A **maximal $\sim$ right order** is a right order maximal amongst the right orders which are equivalent to it. **Maximal $\overset{r}{\sim}$** and **maximal $\overset{l}{\sim}$** right orders are defined likewise.

The connection between these three definitions is shown by the following result.

10.25 Theorem. *The following conditions on a right order R in a quotient ring Q are equivalent:*

(i) *R is a maximal $\sim$ right order;*

(ii) *R is both a maximal $\overset{r}{\sim}$ and a maximal $\overset{l}{\sim}$ right order.*

Proof. (i) $\Rightarrow$ (ii). This follows from the definitions.

(ii) $\Rightarrow$ (i). Say $S \supseteq R$, $S \sim R$. Then there exists T such that $R \subseteq T \subseteq S$, $R \overset{l}{\sim} T \overset{r}{\sim} S$, by 10.23. But R is a maximal $\overset{l}{\sim}$ right order; so $T = R$ and $S \overset{r}{\sim} R$. But R is a maximal $\overset{r}{\sim}$ right order, so $S = R$. Thus R is a maximal $\sim$ right order. ▯

10.26 Definitions. A ring R with an identity is said to be an **ipri-ring** if every (two-sided) ideal of R is a principal right ideal. **Ipli-rings** are defined analogously.

It is immediate that a simple ring with an identity is both an ipri-ring and an ipli-ring. It is also obvious that a **pri-ring** (a ring in which every right ideal is principal) is an ipri-ring.

10.27 Theorem. *Let R be an ipri- (or ipli-) ring with a semilocal right quotient ring Q. Then R is a maximal $\overset{r}{\sim}$ (or $\overset{l}{\sim}$) right order in Q.*

Proof. Let $S \overset{r}{\sim} R$ be a right order containing R, and let $c \in Q$ be a unit such that $cS \subseteq R$. Then $R(RcS) = RcS$ and $(RcS)R = RcSR = RcS$, so that RcS is an ideal of R, and $RcS = dR$, for some $d \in R$. Since $1 \in R$, then $c \in dR$, and $c = da$, for some $a \in R$. Since c is a unit of Q, and Q is semilocal, this implies that d is a unit of Q. If $q \in Q$ is such that $dRq \subseteq dR$, then $dq = db$, for some $b \in R$, and hence $q \in R$. This proves that

$$R = \{q \in Q \mid dRq \subseteq dR\}.$$

Then, since $dRS = (RcS)S = RcS = dR$, this proves that $R = S$, and so R is a maximal $\overset{r}{\sim}$ order.

A similar proof deals with the case in which R is an ipli-ring. ∎

10.28 Corollary (Robson [67]) *A semiprime pri-ring is a maximal $\overset{l}{\sim}$ right order in its right quotient ring.*

(Faith [64]) *A simple right Goldie ring with an identity is a maximal $\sim$ right order in its right quotient ring.* Cf. 4.39.B. ∎

We are now in a position to generalize to semilocal rings the aforementioned theorem of Faith [64] and Robson [67a].

10.29 Theorem. *Let R be a maximal $\overset{l}{\sim}$ right order in a semilocal quotient ring Q, where Q is a complete $n \times n$ matrix ring. Then R is isomorphic to the endomorphism ring of a torsionfree F'-module K over a ring F', where F' is a right order in a semilocal ring D, and $Q \cong D_n$.*

Proof. (Cf. Faith [64] and Robson [67]) Choose a complete set M of matrix units of Q as specified in 10.19. Then $R \supseteq F_n$, where F is a right order in D, and $Q \approx D_n$. Let $e_{11} = e$. Then Qe is a right D-module, and $Q \cong \Omega = \mathrm{Hom}_D(Qe, Qe)$ under the map φ given by $\varphi(q)x = qx$, for all x in Qe, and all q in Q.

Let F' be the subring of D generated by F and the identity. Then F' is a right order in D. Furthermore, if $Qe \cap R = K$, then K is a unital torsionfree right F'-module. For $K \subseteq Qe$, F' is a right order in D, and Qe is clearly a unital torsionfree right D-module.

Now $(Qe \cap R)D \supseteq F_n eD = F_n De = Qe$. But evidently $(Qe \cap R)D \subseteq Qe$. Hence $KD = Qe$. But D is the right quotient ring of F', so

$$Qe = \{xd^{-1} \mid x \in K,\ d \text{ regular in } F'\}.$$

Thus, any element γ of $\Gamma = \mathrm{Hom}_{F'}(K, K)$ has a unique extension in Ω. The natural isomorphism $Q \cong \Omega$ implies that

$$\Gamma \cong T = \{q \in Q \mid qK \subseteq K\}.$$

Clearly $T \supseteq R$. So, if it can be shown that $T \overset{l}{\sim} R$, the result will follow. Let $K = K_1 = Qe_{11} \cap R$, and let $K_i = Qe_{ii} \cap R$. Then $K_i \supseteq F_n e_{ii}$. Let f be a regular element in F. Then $K_1 K_i \supseteq K_1 e_{1i} f = K_i'$, say. Since K_i is a left ideal, $K_i' \subseteq K_i$. Now $TK_1 \subseteq K_1$, by definition of T. So $TK_i' \subseteq K_i'$. But

$$f = \sum_i fe_{ii} = \sum_i fe_{i1} \cdot e_{1i} \in \sum_i K_1 e_{1i}.$$

Therefore $f^2 \in \sum_i K_i'$. Also, $T \sum_i K_i' \subseteq \sum_i K_i' \subseteq R$. Thus $Tf^2 \subseteq R$. However, f^2 is a regular element of Q. Hence $T \overset{l}{\sim} R$, and so $T = R \cong \Gamma$. ∎

10.30 Corollary (Faith [64]) *Let R be a prime right Goldie ring, and let R be a maximal $\sim$ right order in its right quotient ring, Q. Then R is isomorpic to the endomorphism ring of a unital torsionfree module of finite Goldie dimension over an integral domain.*

Proof. This follows directly from 10.29, since Q is a simple Artinian ring. In the notation of 10.29, the fact that K has finite dimension over F' (i.e., contains no infinite direct sum of submodules) follows since Qe has finite dimension over D. ▯

Regarding 10.30, if R is simple, then one can say much more 4.22A.

10.31 Corollary (Faith [65]) *Let R be a right order of $S = D_n$, where D is a field, and assume that R is a simple ring with identity element.*

10.31.1 *If z is an element of S such that $Rz \subseteq zR$, then z is invertible in S and $z, z^{-1} \in R$.*

10.31.2 *R contains the center of S, that is, center R = center S.*

Proof of 10.31.1. Since R is a right order of S, $I = zR \cap R \neq 0$. Thus I is a nonzero right ideal of R, and the relation $Rz \subseteq zR$ implies that I is an ideal of R. Since R is simple, $I = R$, so $zR \supseteq R$. It follows that z is not a right zero divisor in S and since S is right Artinian, we conclude that $z^{-1} \in S$. Since $R \supseteq z^{-1}R$, then $z^{-1} \in R$. Now simplicity of R implies that $Rz^{-1}R$, the ideal of R generated by z^{-1}, equals R. Since $z^{-1}R \subseteq Rz^{-1}$, we obtain that

$$R = Rz^{-1}R \subseteq Rz^{-1} \subseteq R,$$

that is, that $Rz^{-1} = R$. Thus, $(z^{-1})^{-1} = z \in R$, proving 10.31.1.

Proof of 10.31.2 is an immediate consequence. ▯

Notes for Chapter 10

More complete references have been given in Chapter 10 than has been possible in some other chapters, but something I omitted is the simplification of Lambek [66] of the proof of the Faith-Utumi theorem. (This did much to clear the air, I think, for the generalizations appearing in Chapter 10.) Jacobson also gave a sharper proof of this theorem in Appendix B of his revised colloquium publication [64].

A consequence of the theorem is that any prime right Goldie ring R must contain nilpotent elements if it is not a domain. This follows since $R \supseteq F_n$, and $n > 1$ when R is not a domain. However, *any* prime ring not a domain must contain nilpotent elements! (Oral communication from Robson—the proof is trivial!)

We have commented on Robson's contributions [67a, b], some of which are generalized in the text from Artinian to semilocal rings. (Actually, in 10.27, Q can be any Dedekind finite ring.)

The question of the uniqueness of the ring F of 10.19 has been raised (orally by L. Levy), and the relationship between R and F is not clear. Along the lines of the theorems on orders in Chapter 4, esp. 4.37ff., one might prove that any two domains F and K such that F_n and K_n are $\sim$ right orders in D_n, where D is a field, are $\sim$ right orders of D. Moreover, R and F might be connected, at least when R is simple, by a correspondence theorem similar to 4.7. Thus, in this case F should contain a least ideal T, and then R should be right Noetherian iff F satisfies the a.c.c. on idempotent right ideals.

References

Faith [64, 65], Faith-Utumi [65a], Goldie [62], Jacobson [42, 64], Lambek [66], Robson [67], Roos [68], Utumi [65].

III. Tensor Algebra

If M is a (B, A)-bimodule, then $\mathrm{Hom}_A(M, X)$ is canonically a right B-module, for any object X of mod-A, and the resulting functor

$$\mathrm{Hom}_A(M, \ )\begin{cases} \text{mod-}A \rightsquigarrow \text{mod-}B \\ \quad X \mapsto \mathrm{Hom}_A(M, X) \end{cases}$$

is also denoted $\mathrm{Hom}_A(M, \ )$, or simply h^M. (Dually for $\mathrm{Hom}_A(\ , M) = h_M$.) This functor is left continuous in the sense that it preserves products and kernels 5.44, and the left adjoint

$$\underset{B}{\otimes} M \begin{cases} \text{mod-}B \rightsquigarrow \text{mod-}A \\ \quad Y \mapsto Y \underset{B}{\otimes} M \end{cases}$$

is right continuous, that is, preserves coproducts and cokernels. (Cf. Adjoint associativity 11.4.) The freedom of the tensor product of free modules is a consequence 11.10.

A module M is flat in mod-B if and only if $\underset{B}{\otimes} M : \text{mod-}B \rightsquigarrow \text{mod-}\mathbb{Z}$ is exact. The fundamental properties of flat modules are taken up in Chapter 11. Thus, a module is flat if and only if M is the direct limit of projective modules (see 11.32 and 6.24). Hence, every projective module is flat 11.22. Furthermore, every finitely related flat module is projective. Over a Noetherian ring, this implies that every finitely generated flat module is projective 11.31. Regular rings are characterized 11.24 by the property that every right module is flat. (This property is right-left symmetric.) As is shown in Chapter 19, the injective hull R of any right neat ring R (that is, a ring with zero right singular ideal) is a right self-injective regular ring, and more generally, the endomorphism ring of any quasinjective non-singular right R-module M is regular and right selfinjective. (Any right full linear ring is such an example.)

The main content of Chapter 12 is an exposition of the Morita theorems using tensor products in a form suitable for the descriptions of the structure of the Picard group coming at the end of the chapter, and the Brauer group in Chapter 32. Thus, $\mathrm{Pic}_k A$ is the group of isomorphism classes (P) of invertible (A, A)-bimodules under the operation $(P)(Q) = \left(P \underset{k}{\otimes} Q\right)$, where A is any algebra over a commutative ring k.

There is an isomorphism 12.15

$$\begin{cases} \operatorname{Pic}_k A \to \operatorname{Pic}(A\text{-mod}) \\ \quad (P) \mapsto P \underset{A}{\otimes} \\ \quad TA \leftarrow T \end{cases}$$

where Pic (mod-A) is the group of k-linear autoequivalences of mod-A. If A is commutative, then there is an isomorphism $\operatorname{Aut}_k A \approx \operatorname{Pic}_k A/\operatorname{Pic}_A A$ of the group of k-algebra automorphisms 12.18.

Chapter 13 is an exposition of some properties of finite dimensional algebras over a field k. These include the Wedderburn-Artin theorem 13.5 for algebraically closed k, the nilpotence of algebras with nil basis 13.28, and the theorem of Maschke on the semisimplicity of group algebras of characteristic not dividing the order of the group 13.21. The Brauer group and the Skolem-Noether theorem are indicated in exercises, and proved more generally for separable algebras over commutative rings in Chapter 32.

Chapter 11

Tensor Products and Flat Modules

Let R be a ring, M a right and N a left R-module. As usual, $M \times N$ denotes cartesian product. If G is an abelian group, and $g : M \times N \to G$ is a mapping of sets, then for each $y \in N$ there is a mapping $g_y : M \to G$, defined by the formula $g_y(a) = g(a, y) \; \forall a \in M$. Symmetrically, if $x \in M$, then $g_x : N \to G$ is defined by the formula $g_x(b) = g(x, b)$ $\forall b \in M$. A mapping $g : M \times N \to G$ is **bilinear** in case $g_y : M \to G$ and $g_x : N \to G$ are group homomorphisms $\forall x \in M$, $y \in N$. A **balanced** map $g : M \times N \to G$ is a bilinear map such that

$$g(xr, y) = g(x, ry)$$

$\forall x \in M$, $y \in N$, $r \in R$.

Consider the category $T(M, R, N)$, consisting of balanced mappings $M \times N \to G$, where G is an abelian group. An initial object in this category is uniquely determined up to a unique isomorphism, and is denoted $M \underset{R}{\otimes} N$, or $M \otimes_R N$, read M **tensor** N (over R), also called the **tensor product** of M and N (over R). When the context is clear, we simplify this to $M \otimes N$, however the latter symbol ordinarily represents $M \otimes_{\mathbb{Z}} N$.

11.1 ***Existence of*** $M \otimes_R N$. *If* $M \in$ mod-R *and* $N \in R$-mod, *then there exists an abelian group* $M \otimes_R N$ *(unique up to isomorphism) and a balanced mapping* $t: M \times N \to M \otimes_R N$ *such that for each balanced mapping* $f: M \times N \to G$ *into an abelian group there is a unique homomorphism* h *completing the diagram*

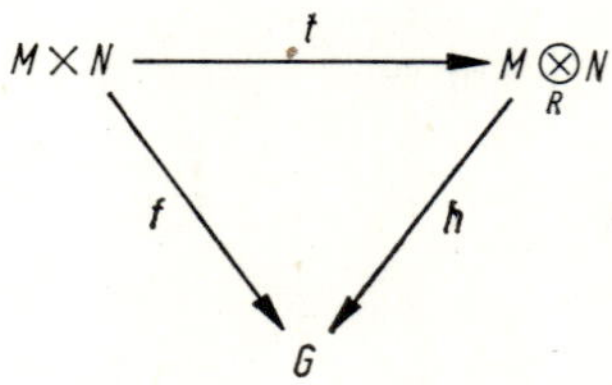

(tensor)

Proof. Let F be the free abelian group on the elements of $M \times N$. Thus F is a $\mathbb{Z}$-module for which there is defined an injective mapping $p: M \times N \to F$ such that $p(M \times N)$ is a basis of F. *For simplicity assume* $p =$ *identity.* By the mapping property of free groups (2.10), if $f: M \times N \to H$ is balanced, there is a unique homomorphism completing the diagram

(1)

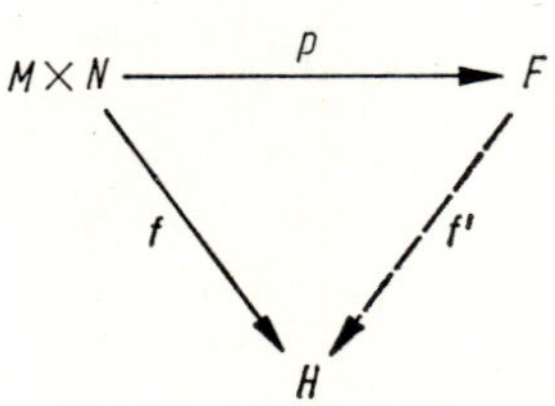

The intersection W of the kernels of all homomorphisms $f': F \to H$ obtained this way is the submodule of F generated by all elements of the form

$$\left.\begin{array}{l}(x + y', y) - (x, y) - (x', y)\\ (x, y + y') - (x, y) - (x, y')\\ (x, ry) - (xr, y)\end{array}\right\} x, x' \in M,\ \ y, y' \in N,\ \ r \in R.$$

Let π be the canonical map $F \to F/W$, and denote F/W by $M \otimes N$. Then $\tau = \pi p$ is a balanced map $M \times N \to M \otimes N$. If $f: M \times N \to H$

is any balanced map, then $\ker f' \supseteq W$, so the diagram

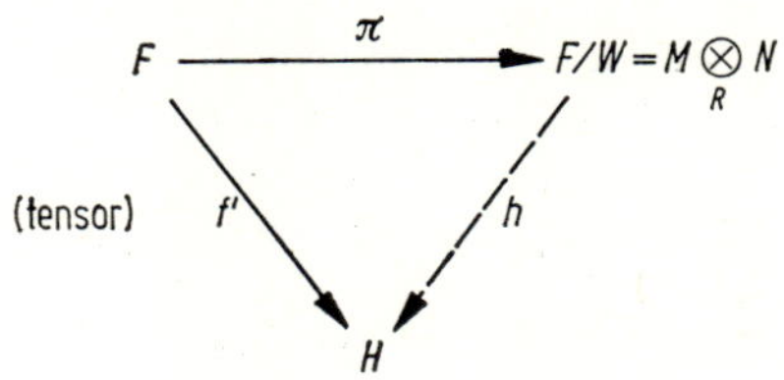

can be completed by a unique homomorphism h, and then $f = f'p = h\pi p = h\tau$ completes the tensor diagram. To show that h is the unique homomorphism $M \otimes N \to H$ that will do this, assume that k is any other homomorphism such that $f = k\tau$. Then

$$f = k\tau = (k\pi)p = (h\pi)p = f'p.$$

By the uniqueness of the map f' completing the diagram (1), we deduce $f' = k\pi$; that is, $h\pi = k\pi$. By the uniqueness of any map $h : M \otimes N \to H$ satisfying $h\pi = f'$, we see that $h = k$. $\square$

The image $\tau(x, y)$ of $(x, y) \in M \times N$ is denoted by $x \otimes y$, read x **tensor** y, and we have the identities:

$$\begin{aligned} (x + x') \otimes y &= (x \otimes y) + (x' \otimes y) \\ x \otimes (y + y') &= (x \otimes y) + (x \otimes y') \\ x \otimes (ry) &= (xr) \otimes y. \end{aligned}$$

The group $M \otimes N$ is generated by the set $\{x \otimes y \mid x \in M,\ y \in N\}$; that is, $M \otimes N$ consists of all finite sums of elements of this form. Consequently, if $\{x_i \mid i \in I\}$ generates M, and if $\{y_j \mid j \in J\}$ generates N, then $\{x_i \otimes y_j \mid i \in I,\ j \in J\}$ generates $M \otimes N$.

11.2 Proposition. *Let A be a left ideal of R, and let M be a right R-module. There is a unique homomorphism*

$$g_A : M \underset{R}{\otimes} A \to M$$

such that

$$g_A : m \otimes a \mapsto ma \qquad m \in M,\ a \in A.$$

This homomorphism is called ***canonical,*** *and* $\operatorname{im} g_A$ *is the subgroup MA, consisting of all finite sums $\sum_i m_i a_i$ with $m_i \in M$, and $a_i \in A$. The induced homomorphism $M \underset{R}{\otimes} A \to MA$ is called canonical. The canonical map $M \underset{R}{\otimes} R \to M$ is an isomorphism.*

Proof. The mapping $M \times A \to M$ sending $(m, a) \mapsto ma$ is balanced, therefore g_A exists by 11.1. Since $M \otimes A$ consists of all finite sums of the form $\sum_i m_i \otimes a_i$, then $\operatorname{im} g_A = MA$ as stated. If $\sum_i m_i \otimes a_i$ lies in the kernel of the canonical map $M \otimes R \to M$, then $\sum_i m_i a_i = 0$, so $\sum_i m_i \otimes a_i = \sum_i m_i a_i \otimes 1 = 0$, and g_R is therefore monic. Since $\operatorname{im} g_R = MR = M$, g_R is an isomorphism. □

11.3 Proposition. *Let M, M' be right, and N, N' be left modules over R, and let f* (resp. g) *be an R-module homomorphism of M into M'* (resp. N *into* N').

11.3.1 *Then there is a group homomorphism, denoted by $f \otimes g$, of $M \underset{R}{\otimes} N$ into $M' \underset{R}{\otimes} N'$ such that $(f \otimes g)(x \otimes y) = f(x) \otimes g(y)$ $\forall x \in M$, $y \in N$.*

11.3.2 *If f and g are epimorphisms, then $f \otimes g$ is an epimorphism and $\ker(f \otimes g)$ is the subgroup of $M \underset{R}{\otimes} N$ generated by the union of the sets:*

$$\{u \otimes y \mid u \in \ker f,\ y \in N\}$$

$$\{x \otimes v \mid v \in \ker g,\ x \in M\}.$$

11.3.3 *If f and g are isomorphisms, then $f \otimes g$ is an isomorphism.*

Proof of 11.3.1. Since the mapping $k : M \times N \to M' \otimes N'$ defined by $k : (x, y) \to f(x) \otimes g(y)$ is balanced, by the definition of $M \underset{R}{\otimes} N$ there exists a unique homomorphism $h : M \otimes N \to M' \otimes N'$, satisfying $h(x \otimes y) = f(x) \otimes g(y)$ $\forall x \otimes y \in M \otimes N$; we let $f \otimes g$ denote the map h.

Proof of 11.3.2. If f and g are epimorphisms, and if $x' \otimes y' \in M' \otimes N'$, $x' \in M'$, $y' \in N'$, then there exist $x \in M$, $y \in N$ such that $f(x) = x'$, $g(y) = y'$, and then $(f \otimes g)(x \otimes y) = x' \otimes y'$. Since $M \otimes N$ is generated by the elements $\{x \otimes y \mid x \in M,\ y \in N\}$, $f \otimes g$ is epic.

Trivially, $\ker(f \otimes g)$ contains the subgroup H generated by the two sets defined in 11.3.2. Let ν be the canonical map $\nu : M \otimes N \to (M \otimes N)/H$. If $x, x_1 \in M$, $y, y_1 \in N$ are such that $u = x_1 - x \in \ker f$, $v = y_1 - y \in \ker g$, then

$$x_1 \otimes y_1 = x \otimes y + (u \otimes y + x \otimes v + u \otimes v).$$

Since the right summand (in parenthesis) lies in H, then $\nu(x_1 \otimes y_1) = \nu(x \otimes y)$, and there is a balanced mapping $p : M' \times N' \to (M \otimes N)/H$ defined by $p(f(x), g(y)) = \nu(x \otimes y)$. Thus there exists a homomorphism $h : M' \otimes N' \to (M \otimes N)/H$ such that $h(x' \otimes y') = p(x', y')$. Since $h(f(x), g(y)) = \nu(x \otimes y)$, $\forall x \in M$, $y \in N$, then $h \circ (f \otimes g) = \nu$. If

$w \in \ker f \otimes g$, then

$$\nu(w) = h(f \otimes g)(w) = h(0) = 0$$

and then $w \in H = \ker \nu$. Thus, $\ker(f \otimes g) = H$, as asserted.

Proof of 11.3.3 is an immediate consequence. □

Tensor as Functor

If A, M, M', M'' are right R-modules, if B, N, N', N'' are left R-modules, and if

$$M \xrightarrow{f} M' \xrightarrow{f'} M'', \quad N \xrightarrow{g} N' \xrightarrow{g'} N''$$

are maps, then $f'f \otimes g'g$ maps $M \underset{R}{\otimes} N \to M'' \underset{R}{\otimes} N''$, and

$$f'f \otimes g'g = (f' \otimes g')(f \otimes g).$$

We now define a functor:

For each $A \in \text{mod-}R$, there is a functor

$$\begin{cases} A \underset{R}{\otimes} : R\text{-mod} \rightsquigarrow \text{ABELGRPS} & \\ A \underset{R}{\otimes} : N \mapsto A \otimes_R N & \text{on objects} \\ A \underset{R}{\otimes} : g \mapsto 1_A \otimes g & \text{on morphisms.} \end{cases}$$

Similarly, if $B \in R\text{-mod}$, there is a functor

$$\begin{cases} \underset{R}{\otimes} B : \text{mod-}R \rightsquigarrow \text{ABELGRPS} & \\ \underset{R}{\otimes} B : M \mapsto M \otimes B & \text{(on objects)} \\ \underset{R}{\otimes} B : f \mapsto f \otimes 1_B & \text{(on morphisms).} \end{cases}$$

The functors $A \underset{R}{\otimes}$ and $\underset{R}{\otimes} B$ are each covariant because

$$f'f \otimes 1 = (f' \otimes 1)(f \otimes 1)$$

and

$$1 \otimes g'g = (1 \otimes g')(1 \otimes g).$$

If R and S are rings, and if A, B, and C are modules in the situation A_R, ${}_RB_S$, C_S, then

(1) $$\text{Hom}_S(B, C) \in \text{mod-}R$$

(2) $$A \underset{R}{\otimes} B \in \text{mod-}S.$$

If $f \in \mathrm{Hom}_S(B, C)$, $r \in R$, then fr is defined to be fr_s, where r_s is the left homothetic $b \mapsto rb$. Also $(a \otimes b)t = a \otimes bt$ for all $a \in A$, $b \in B$, $t \in S$.

11.4 Adjoint Associativity. *Let R and S be rings, and let A, B, and C be modules in the context A_R, ${}_RB_S$, and C_S. Then there is a natural isomorphism of abelian groups*

$$\mathrm{Hom}_S(A \underset{R}{\otimes} B, C) \underset{\varphi}{\overset{\eta}{\rightleftarrows}} \mathrm{Hom}_R(A, \mathrm{Hom}_S(B, C))$$

defined by the formula

$$[\eta f](a)(b) = f(a \otimes b)$$

$\forall a \in A$, $b \in B$, $f \in \mathrm{Hom}_S(A \underset{R}{\otimes} B, C)$. *The inverse isomorphism φ is defined by the formula*

$$\varphi(g)(a \otimes b) = [g(a)](b)$$

where $g \in \mathrm{Hom}_R(A, \mathrm{Hom}_S(B,C))$. (Cf. 11.19.4.)

Proof. (MacLane [63]) For each $a \in A$, and each $f \in \mathrm{Hom}_S(A \otimes B, C)$, the function $(\eta f)(a)$ that sends $b \in B$ onto $f(a \otimes b)$ is actually an S-mapping $B \to C$ because, if $s \in S$, then

$$[\eta f(a)](bs) = f(a \otimes (bs)) = f(a \otimes b)s = [\eta f(a)](b)s.$$

Similarly, $\eta(f_1 + f_2) = \eta f_1 + \eta f_2$, $\forall f_1, f_2$. Thus, η is a group homomorphism.

To show that η is an isomorphism, we shall construct an inverse map φ. If $g: A \to \mathrm{Hom}_S(B, C)$ is an R-map, and if $a \in A$, $b \in B$, then $g(a) \in \mathrm{Hom}_S(B, C)$, $[g(a)](b) \in C$, and $\forall r \in R$

$$[g(ar)](b) = [g(a)r](b) = g(a)(rb).$$

The mapping $A \times B \to C$ that sends $(a, b) \mapsto [g(a)](b)$ is obviously bilinear, hence balanced (over R), thus by 11.1 there exists a unique homomorphism $\varphi(g): A \underset{R}{\otimes} B \to C$ such that

$$\varphi(g)(a \otimes b) = [g(a)](b).$$

If $s \in S$, then

$$\varphi(g)[(a \otimes b)s] = \varphi(g)[(a \otimes bs)] = [g(a)](bs) = [g(a)](b)s$$
$$= \varphi(g)[a \otimes b]s$$

and $\varphi(g) \in \operatorname{Hom}_S(A \underset{R}{\otimes} B, C)\ \forall g$. If also $h: A \to \operatorname{Hom}_S(B, C)$, then

$$\begin{aligned}\varphi(g+h)(a \otimes b) &= [(g+h)(a)](b) = [g(a)]b + [h(a)]b \\ &= \varphi(g)(a \otimes b) + \varphi(h)(a \otimes b)\end{aligned}$$

that is, $\varphi(g+h) = \varphi(g) + \varphi(h)$, therefore the mapping $g \to \varphi(g)$ is a homomorphism

$$\varphi: \operatorname{Hom}_R\big(A, \operatorname{Hom}_S(B, C)\big) \to \operatorname{Hom}_S\big(A \underset{R}{\otimes} B, C\big).$$

We need only check $\eta\varphi = \varphi\eta =$ identity. If g is as previously noted, then $\varphi(g): A \otimes B \to C$, and

$$[\eta\varphi(g)](a)(b) = \varphi(g)(a \otimes b) = g(a)(b)$$

thus $\eta\varphi = 1$; i.e., $\eta\varphi(g) = g\ \forall g$. If $f: A \underset{R}{\otimes} B \to C$, then $\eta f: A \to \operatorname{Hom}_S(B, C)$, and

$$[\varphi\eta f](a \otimes b) = [\eta f(a)](b) = f(a \otimes b)$$

so $\varphi\eta = 1$. Thus η is an isomorphism.

Since $\operatorname{Hom}_R(X, Y)$ is covariant in X, contravariant in Y, and since $X \underset{R}{\otimes} Y$ is covariant in X and Y, we see that $T_1 = \operatorname{Hom}_R\big(A, \operatorname{Hom}_S(B, C)\big)$ and $T_2 = \operatorname{Hom}_S(A \underset{R}{\otimes} B, C)$ are contravariant in A and B, and covariant in C.

We check that φ and η are natural transformations as follows: Let $\alpha: A' \to A$, $\beta: B' \to B$, $\gamma: C \to C'$; we shall show that the following diagram commutes.

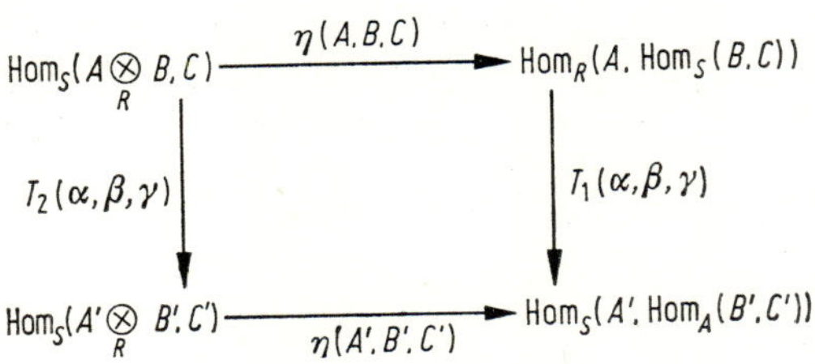

Now, if $a' \in A'$, $b' \in B'$, $c' \in C'$, and if $f \in \operatorname{Hom}_S\big(A \underset{R}{\otimes} B, C\big)$, then $T_1\eta(f)$ is defined by:

$$\big((T_1\eta(f))(a')\big)(b') = \gamma[\eta(f)(\alpha a')](\beta b') = \gamma[f(\alpha a' \otimes \beta b')]$$

$\eta T_2(f)$ is defined by:

$$\big[\big(\eta(T_2 f)\big)(a')\big](b') = T_2 f(a' \otimes b') = \gamma[f(\alpha a' \otimes \beta b')].$$

This proves commutativity of the diagram, and completes the proof. $\square$

11.5 Definition.

11.5.1 If R and S are rings, then for any (R, S)-bimodule M, the functor

$$(1) \qquad \otimes_R M : \text{mod-}R \rightsquigarrow \text{mod-}S$$

is said to be a **left adjoint** of the functor

$$(2) \qquad \mathrm{Hom}_S(M, \ ) : \text{mod-}S \rightsquigarrow \text{mod-}R.$$

(Cf. Adjoint functors, Chapter 5.)

11.5.2 A functor $T : C \rightsquigarrow D$ of abelian categories is **left** (resp. **right**) **continuous** if it preserves products and kernels (resp. coproducts and cokernels).

Continuity was previously discussed in Chapter 5.

11.6 *Exercise.* Assume the terminology of 11.5.

11.6.1 (A) Prove that the group-valued functor $\mathrm{Hom}_S(M, \ )$ is left continuous.

(B) Use adjoint associativity (11.4) to prove that the adjoint functor $\otimes_R M$ is right continuous.

(When *finished* with (*B*), cf. 5.43.)

11.6.2 Use the fact that the group-valued functor $\mathrm{Hom}_S(M, \ )$ is additive to prove that the group-valued functor $\otimes_R M$ is additive. (Cf. Chapter 5, especialy 5.43, 5.48, and 5.58.)

11.6.3 Show that M is a flat left R-module if and only if the group valued functor $\mathrm{Hom}_S(M, \ )$ preserves injective modules. (Cf. 6.28.)

Example 11.7.1 shows that, in general, neither of the tensor functors are left exact.

11.7 *Examples.*

11.7.1 Let R be a ring, and let A be an ideal such that $A^2 \neq A \neq R$. (For example, if $R = \mathbb{Z}$, then any nonzero ideal A satisfies $A \neq A^2$.) If $r + A \in R/A$, $r \in R$, and $a \in A$, then $a \otimes' (r + A) \in A \otimes R/A$, where $\otimes'$ denotes tensor in $A \otimes R/A$. Since

$$(1_A \otimes 1_{R/A})\big(a \otimes' (r + A)\big) = a \otimes (r + A) = 1 \otimes (ar + A) = 1 \otimes 0 = 0$$

then $\ker(1_A \otimes 1_{R/A}) = A \otimes_R R/A$. By 11.3, the kernel K of $A \otimes R \to A \otimes R/A$ is generated by $\{a \otimes b \mid a \in A, \ b \in R\}$. The map $\varphi : A \otimes R \to A$ such that $\varphi(a \otimes r) = ar$ maps K onto A^2. Since $A \neq A^2$, then $K \neq A \otimes R$, which proves that $A \otimes R/A \neq 0$, and hence

$$0 \to A \otimes R/A \to R \otimes R/A$$

is not exact (although $0 \to A \to R$ is).

11.7.2 Let A be any abelian group, m a natural number, $\mathbb{Z}_m = \mathbb{Z}/m\mathbb{Z}$ the cyclic group of order m, and $mA = \{ma \mid a \in A\}$. *Then there is an isomorphism* $h : A/mA \to \mathbb{Z}_m \otimes A$ (tensored over $\mathbb{Z}$) *given by the formula*: $h(a + mA) = u \otimes a$, where u is any generator of $\mathbb{Z}_m$, for example, $u = 1 + m\mathbb{Z}$. If $a, a' \in A$, and if $a - a' \in mA$, that is, $a - a' = ma''$, with $a'' \in A$, then

$$u \otimes (a - a') = u \otimes ma'' = mu \otimes a'' = 0 \otimes a'' = 0,$$

so

$$h(a + mA) = u \otimes a = u \otimes a' = h(a' + mA).$$

This proves h is a homomorphism of $A/mA \to \mathbb{Z}_m \otimes A$. An arbitrary element of $\mathbb{Z}_m \otimes A$ has the form $ku \otimes a$ for some natural number k, and some $a \in A$. There is a homomorphism $g : \mathbb{Z}_m \otimes A \to A/mA$ given by the formula: $g(ku \otimes a) = ka + mA$. To see this, consider the mapping $t : \mathbb{Z}_m \times A \to A/mA$, defined by the formula: $t(ku, a) = ka + mA$. If k' is a natural number such that $k'u = ku$, then m divides $k' - k$, and so $(k' - k)a \in mA$; that is,

$$t(k'u, a) = k'a + mA = ka + mA = t(ku, a).$$

Therefore, t is a mapping, in fact, a balanced mapping of $\mathbb{Z}_m \times A$ into A/mA. By the universal mapping theorem, there is a homomorphism $g : \mathbb{Z}_m \otimes A$ into A/mA, satisfying $g(ku \otimes a) = ka + mA$. Now

$$hg(ku \otimes a) = h(ka + mA) = u \otimes ka = ku \otimes a$$

$$gh(ka + mA) = g(u \otimes ka) = g(ku \otimes a) = ka + mA$$

so hg and gh are identity maps. Therefore h is an isomorphism (and $h^{-1} = g$).

11.7.3 If A is any divisible group, e.g., if $A = \mathbb{Q}$, then $mA = A$, and therefore $\mathbb{Z}_m \otimes A \approx A/A = 0$. On the other hand, if A is any group of order m, for example, $A = \mathbb{Z}_m$, then $\mathbb{Z}_m \otimes A \approx A$.

11.8 *Exercises.*

11.8.1 If m and n are elements of $\mathbb{Z}$, and $d = (m, n)$ is the greatest common divisor, then $(\mathbb{Z}/m\mathbb{Z}) \otimes (\mathbb{Z}/n\mathbb{Z}) \approx \mathbb{Z}/d\mathbb{Z}$. Thus, $(m, n) = 1$ implies that $\mathbb{Z}_m \otimes \mathbb{Z}_n = 0$.

11.8.2 If R and S are rings, and A, B, C are modules in the situation A_R, ${}_RB_S$, ${}_SC$, then there is a natural isomorphism

$$(A \otimes_R B) \otimes_S C \approx A \otimes_R (B \otimes_S C).$$

11.8.3 If R is a commutative ring, and A and B are R-modules, then there is an isomorphism $A \otimes_R B \approx B \otimes_R A$, mapping $a \otimes b \mapsto b \otimes a$, where $a \in A$, $b \in B$.

Tensored Bimodules

An (S, R)-**homomorphism** (or map) of an (S, R)-bimodule M is a group homomorphism h of M such that

$$h(sxr) = sh(x)r \ \forall x \in M, \ s \in S, \ r \in R.$$

11.9 Proposition. *Let M, M' be (S, R)-bimodules, let N, N' be left R-modules, let $f: M \to M'$ be an (S, R)-homomorphism, and let $g: N \to N'$ be an R-homomorphism. Then $f \otimes g$ is an S-homomorphism of the left S-modules $M \otimes_R N$ and $M' \otimes_R N'$.*

Proof. We already have shown that $M \otimes_R N$ and $M' \otimes_R N'$ are left S-modules. If $s \in S$, if $x \in M$, $y \in N$, and if $u = x \otimes y$, then

$$su = sx \otimes y$$

and

$$\begin{aligned}(f \otimes g)(su) &= f(sx) \otimes g(y) = sf(x) \otimes g(y) \\ &= s(f \otimes g)(x \otimes y) = s(f \otimes g)(u).\end{aligned}$$

Since $M \otimes_R N$ is generated by all elements $u = x \otimes y$ as x ranges over M, and y ranges over N, then

$$(f \otimes g)(sw) = s(f \otimes g)(w)$$

$\forall w \in M \otimes_R N$, $\forall s \in S$. ▯

11.10 Corollary. *Let R be a ring, and let I and J be nonempty sets. Then the composition*

$$R^{(I)} \otimes_R R^{(J)} \to (R^{(I)} \otimes_R R)^{(J)} \to (R^{(I)})^{(J)} \to R^{(I \times J)}$$

is an (R, R)-bimodule isomorphism. (A tensor product of two free modules is a free module.)

Proof. Use the canonical isomorphism $R \otimes_R M \approx M$, the fact (11.6) that $\otimes_R$ preserves coproducts, and 11.9. ▯

11.11 Corollary. *If I is a set, if E is a left R-module, and if $\{x_i \mid i \in I\}$ is a free basis of a free right R-module $R^{(I)}$, then each element a in $R^{(I)} \otimes_R E$ has a unique representation as a finite sum*

$$a = \sum_{i \in I'} x_i \otimes y_i$$

where I' is a finite subset of I, and $y_i \in E \; \forall i \in I$.

Proof. This follows quite easily from the fact 11.6.1 that any tensor functor preserves coproducts. Thus, $R^{(I)} \underset{R}{\otimes} E \approx E^{(I)}$ canonically. $\square$

Change of Rings

An R-module $[M, R, f]$ is an abelian group M together with a ring antihomomorphism $f: R \to \operatorname{End} M_{\mathbb{Z}}$ (3.4). Thus, if $h: S \to R$ is a ring homomorphism, we have **pullback S-module** $(M, S, fh]$. Notation: (M_S). Similarly, a left R-module N becomes a left S-module $({}_SN)$.

11.12 Pullback Lemma. *For a ring homomorphism $h: S \to R$, and R-modules M_R, ${}_RN$, and ${}_RA$, there are homomorphisms:*

$$\begin{cases} h_{\#} : (M_S) \otimes_S ({}_SN) \to M \otimes_R N \\ \qquad x \otimes_S y \mapsto x \otimes_R y \end{cases}$$

$$h^{\#} : \operatorname{Hom}_R(A, N) \to \operatorname{Hom}_S\big(({}_SA), ({}_SN)\big).$$

If h is epic, then $h_{\#}$ and $h^{\#}$ are isomorphisms.

Proof. Consider the mappings

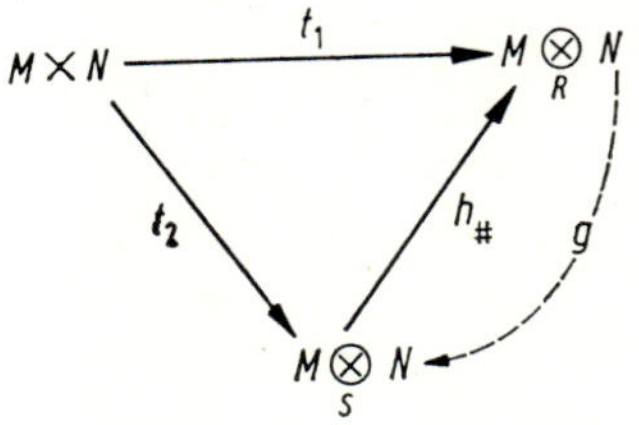

where

$$t_1(x, y) = x \otimes_R y$$

$$t_2(x, y) = x \otimes_S y.$$

If $s \in S$, then

$$(xs) \otimes_R y = \big(x h(s)\big) \otimes_R y = x \otimes_R \big(h(s) y\big) = x \otimes_R (sy).$$

Thus, t_1 is a balanced mapping over S, and $h_\#$ therefore exists by 11.1. Similarly, if $r \in R$, then $r = h(s)$ for some $s \in S$, and

$$t_2(xr, y) = t_2(xs, y) = t_2(x, sy).$$

Therefore, t_2 is balanced over R, and g exists by 11.1. Since $h_\# g = g h_\# =$ identity, $h_\#$ is an isomorphism.

Actually $\mathrm{Hom}_R(A, N) = \mathrm{Hom}_S(A, N)$, when h is epic, thus for $h^\#$ we can take the identity map.

11.13 Corollary. *If $h: S \to R$ is a ring homomorphism, and I any set, then R is a (pullback) left S-module and*

$$S^{(I)} \underset{S}{\otimes} R$$

is a free right R-module. If $\{x_i \mid i \in I\}$ is a basis for the free S-module $S^{(I)}$, then $\{x_i \otimes 1 \mid i \in I\}$ is a basis for $S^{(I)} \underset{S}{\otimes} R$.

Proof. This follows from 11.11.

Recall that a free R-module has an **invariant basis number** in case any two bases have the same cardinality (Chapter 3). An IBN-ring is one over which every free right R-module has an invariant basis number. This is right-left symmetric.

11.14 Theorem. *Let $f: S \to R$ be a nonzero ring homomorphism. If R is an* IBN-*ring, so is S.*

Proof. Let $F \approx S^{(I)}$ be a free right S-module. By the corollary, $F \otimes_S R$ is a free right R-module $\approx R^{(I)}$. Since R is an IBN-ring, every basis of F has cardinality $|I|$, and S is IBN. ▯

Any commutative ring S (also any local ring) satisfies the hypotheses, since S modulo a maximal ideal is a field. In fact, since every Noetherian ring is an IBN-ring, if S modulo any ideal is a Noetherian ring, then S is IBN (cf. Leavitt [51] and Cohn [66b]).

11.15 Corollary. *Let R and S be rings, let P be a projective R-module, and let Q be an (R, S)-bimodule that is a projective S-module. Then $P \otimes_R Q$ is a projective S-module.*

Proof. An S-module A is projective if and only if the functor $\mathrm{Hom}_S(A, \ )$ is exact; that is, $B \to C \to 0$ exact implies $\mathrm{Hom}_S(A, B) \to \mathrm{Hom}_S(A, C) \to 0$ exact. Since P is a projective R-module, $\mathrm{Hom}_R(P, \ )$ is exact, and

$$\mathrm{Hom}_R\big(P, \mathrm{Hom}_S(Q, B)\big) \to \mathrm{Hom}_R\big(P, \mathrm{Hom}_S(Q, C)\big) \to 0$$

is exact. Applying adjoint associativity, the sequence

$$\mathrm{Hom}_S(P \otimes_R Q, B) \to \mathrm{Hom}_S(P \otimes_R Q, C) \to 0$$

is exact, so the functor $\mathrm{Hom}_S(P \otimes_R Q, \ )$ is exact, whence $P \otimes_R Q$ is S-projective. $\square$

The following consequence is the basis of the Brauer group to be discussed in Chapter 32.

11.16 Corollary. *If R is a commutative ring, then the tensor product of two projective R-modules is a projective R-module.* $\square$

If A is a right R-module, its dual module $A^* = \mathrm{Hom}_R(A, R)$ is a canonical left R-module. If B is a left R-module, denote its dual by $B^* = \mathrm{Hom}_R(B, R)$.

11.17 Corollary. *If A is a right R-module, and B is an (R, R)-bimodule, then there is a natural isomorphism* $\mathrm{Hom}_R(A, B^*) \approx (A \underset{R}{\otimes} B)^*$.

Proof. Apply 11.4 to the situation A_R, ${}_RB^*_R$, R_R. $\square$

11.18 Corollary. *If D is any abelian group, and A is a right R-module, then $A' = \mathrm{Hom}_{\mathbb{Z}}(A, D)$ is a left R-module. If B is any left R-module, there is a natural isomorphism*

$$\mathrm{Hom}_R(A, B') \approx (A \otimes_R B)'.$$

Proof. Apply 11.4 to the situation A_R, ${}_RB_{\mathbb{Z}}$, $D_{\mathbb{Z}}$. $\square$

If P and A are R-modules there is a homomorphism

$$\mu : A \underset{R}{\otimes} P^* \to \mathrm{Hom}_R(P, A),$$

where $\mu(a \otimes f)(p) = af(p)\ \forall f \in P^*,\ a \in A,\ p \in P$ (see the following exercises).

11.19 *Exercises.*

11.19.1 If P_R is finitely generated and projective, then $\mu : A \otimes_R P^* \to \mathrm{Hom}_R(P, A)$ is an isomorphism, where $P^* = \mathrm{Hom}_R(P, R)$. Let $S = \mathrm{End}_R P$. Then there is an (S, S)-bimodule isomorphism

$$P \otimes_R P^* \to \mathrm{End}_R P.$$

11.19.2 If P is finitely generated and projective over a commutative ring R, then there is an natural isomorphism $P^* \underset{R}{\otimes} A^* \approx \left(P \underset{R}{\otimes} A\right)^*$ $\forall$ R-modules A.

11.19.3 In the situation P_R, ${}_SM_R$, ${}_SN$, there is a natural transformation

$$\Phi : P \otimes_R \mathrm{Hom}_S(M, N) \to \mathrm{Hom}_S\big(\mathrm{Hom}_R(P, M), N\big)$$

where

$$\Phi(p \otimes f)(g) = fgp$$

$\forall p \in P$, $f \in \operatorname{Hom}_S(M, N)$, $g \in \operatorname{Hom}_R(P, M)$. (a) If P is finitely generated projective over R, Φ is a natural isomorphism. (Hint: this is trivial for $P = R$.) (b) If $S = \operatorname{End} P_R$, this yields an isomorphism

$$P \otimes_R \operatorname{Hom}_S(P, S) \approx S.$$

[Note: $\operatorname{Hom}_S(P, S)$ is the S-dual of P.]

11.19.4 Consider the situation A_R, ${}_SB_R$, and ${}_SC$, for rings R and S, and modules A, B, C (cf. 11.4). Then, there exists a natural isomorphism:

$$h : \operatorname{Hom}_R(A, \operatorname{Hom}_S(B,C)) \to \operatorname{Hom}_S(B \otimes A, C)$$

such that for each $f: A \to \operatorname{Hom}_S(B,C)$, we have

$$h[f](b \otimes a) = f(a)[b].$$

Characterizations of Flat Modules

As defined in Chapter 5, a left R-module M is flat in case the group valued functor $\underset{R}{\otimes} M$ is exact. Presented in this paragraph are a number of characterizations of flatness of a module.

11.20 Proposition. *The conditions* (a) *and* (b) *on a right R-module M are equivalent:*

(a) *M is flat.*

(b) *For each finitely generated left ideal A of R, the canonical map*

$$g \begin{cases} M \otimes_R A \to MA \\ m \otimes a \mapsto ma \qquad (m \in M,\ a \in A) \end{cases}$$

is an isomorphism. When (a) *or* (b) *holds, then $M \otimes_R A \to MA$ is an isomorphism for each left ideal A.*

Proof. (a) $\Rightarrow$ (b). The canonical map $M \otimes R \xrightarrow{f} M$ is an isomorphism. Since M is flat, $M \otimes A \xrightarrow{g} M \otimes R$ is monic, hence $M \otimes A \xrightarrow{fg} M$ is monic. Since $\operatorname{im} fg = MA$, the canonical map g is an isomorphism. This is true for any left ideal A.

(b) $\Rightarrow$ (a). Let A be any left ideal, and suppose $a_i \in A$, $m_i \in M$, $i = 1, \ldots, t$ are such that $\sum_{i=1}^t m_i a_i = 0$. If B denotes the left ideal $B = \sum_{i=1}^t R a_i$, since B is finitely generated, the canonical map $M \otimes B \to MB$ is by hypothesis an isomorphism. Thus $\sum_{i=1}^t m_i \otimes a_i = 0$ in $M \otimes_R B$. In view of the canonical homomorphism $M \otimes_R B \to M \otimes_R A$, then $\sum_{i=1}^t m_i \otimes a_i = 0$ in $M \otimes_R A$, so $M \otimes_R A \to MA$ is an isomorphism for all left ideals. This implies that $M \otimes_R A \to M$ is monic, whence $M' \to (M \otimes_R A)'$ is epic. Since $(M \otimes_R A)' \approx \mathrm{Hom}_R(A, M')$ by the result corresponding to 11.18 given by 11.19.4 (instead of 11.4)

$$\mathrm{Hom}_R(R, M') \to \mathrm{Hom}_R(A, M')$$

is epic. Thus, Baer's criterion 3.41 for injectivity implies that M' is an injective left R-module, and, by 5.60, M is a flat R-module. □

11.21 Corollary. *Let P be a flat R-module, and K a submodule. Then P/K is flat if and only if*

$$(1) \qquad KA = PA \cap K$$

for each finitely generated left ideal A of R. When this is so, then (1) *holds for each left ideal A.*

Proof. Since $\otimes_R A$ is right exact, the sequence

$$K \otimes_R A \to P \otimes_R A \to P/K \otimes_R A \to 0$$

is exact. Since P is flat, the canonical map $P \otimes_R A \to PA$ is an isomorphism that maps $K \otimes_R A$ onto KA. Thus, $P/K \otimes_R A \approx PA/KA$. Therefore, by the proposition, P/K is flat if and only if

$$(2) \qquad (P/K)A \approx PA/KA$$

for all (finitely generated) left ideals A. We shall show that this isomorphism is equivalent to (1). By the homomorphism theorem (1.14), we have

$$(3) \qquad (P/K)A = (PA + K)/K \approx PA/PA \cap K.$$

Thus (2) is equivalent to

$$(4) \qquad PA/KA \approx PA/PA \cap K.$$

Since $PA \cap K \supseteq KA$, and since the isomorphisms (2), (3), and (4) are canonical, (2) is equivalent to (1). Since flatness of P/K is equivalent to (2), we have the desired result. □

11.22 Corollary. *Any projective R-module is flat.* (Cf. 5.61.) □

11.23 Corollary. *If P is a flat R-module, and K a submodule, then P/K is flat if and only if $Pa \cap K \subseteq Ka$ $\forall a \in R$. (Then $Pa \cap K = Ka$.)*

Proof. $PA \cap K \supseteq KA$ for any left ideal of R, so (1) of the proposition holds if and only if (1′) $PA \cap K \subseteq KA$. Now $Pa \cap K \subseteq Ka$ $\forall a \in A$ implies (1′). Conversely, (1′) for finitely generated left ideals implies $PRa \cap K \subseteq KRa$, or $Pa \cap K \subseteq Ka\ \forall a \in A$, completing the proof. ▯

The rings in the following definition were introduced by von Neumann [31], and characterized via the flatness of modules by Harada [56] and Auslander [57].

11.24 Theorem and Definition. *A ring R is said to be* **regular** *in case the following equivalent statements hold:*

(a) *Every right R-module is flat.*

(a′) *Every left R-module is flat.*

(b) *For each $a \in R$ there corresponds $x \in R$ such that $a = axa$.*

(c) *Every cyclic right ideal is generated by an idempotent.*

(c′) *Every cyclic left ideal is generated by an idempotent.*

(d) *Every finitely generated right ideal is generated by an idempotent.*

(e + *i*) *The left-right symmetry of* (*i*), $i =$ a, b, c, d.

Proof. Assume every R-module is flat. If $a \in R$, then R/aR is flat, and by the corollary for the case $A = Ra$, $P = R$, and $K = aR$, we have $Ra \cap aR \subseteq aRa$. Thus, $a = axa$ for some $x \in R$, whence $e = ax$ is an idempotent that generates aR. Thus, (a) $\Rightarrow$ (b). Conversely, assume (b), let P be a free module, K a submodule, and $a \in R$. Then $Ra = Re$ for some idempotent $e \in R$, so $Ma = Me$ for any R-module M. Since e is idempotent $Pe \cap K \subseteq Ke$, so $Pa \cap K \subseteq Ka$, which by the corollary proves that P/K is flat. Since any R-module M is isomorphic to P/K for a suitable free R-module P and submodule K, every R-module is flat.

(b) $\Leftrightarrow$ (c). If $a = axa$, then $e = ax$ is an idempotent that generates the cyclic right ideal generated by a. The converse is similar.

(c) $\Rightarrow$ (d) is an exercise (also established by 19.7 to follow), and (d) $\Rightarrow$ (c) is trivial. Then, (e + i) is a consequence of the right-left symmetry of (b). ▯

11.25 *Exercise.*

*11.25.1 A summand of a flat module is flat, and any direct limit of flat modules is flat. In particular, a direct sum of flat modules is flat. Every flat module is a direct limit of projective modules.

11.25.2 Any flat module is torsionfree.

*11.25.3 Let R be a commutative ring. Then an R-module M is flat iff (in a sense discussed in Chapter 27) M is "locally flat" (that is, iff for every prime ideal P, the "local module" $M_P = M \underset{R}{\otimes} R_P$, where R_P is the local ring at P, is flat). Conclude that the local ring R_P is flat, for each prime P.

11.25.4 Let R be a commutative hereditary domain. Then, an R-module M is flat iff M is torsionfree. Thus, any overring of R is flat. (An overring of a ring is a subring of the quotient ring containing the ring. See Exercises for Chapter 11.)

11.25.5 Over a commutative ring R, a tensor product of two flat modules is flat.

11.25.6 (a) Let P be a flat right R-module, and let K be a submodule. If P/K is flat, then $Pa \cap K = 0$ $\forall a \in \mathrm{ann}_R K$.

(b) If I is a right ideal of R, and if R/I is flat, then $I^{\perp} \cap I = 0$.

11.26 Proposition. *Let R be a ring, and $0 \to K \to F \to A \to 0$ be an exact sequence of right R-modules, where F is free with a free basis $\{x_\alpha\}$. If $u = x_{\alpha_1} a_1 + \cdots + x_{\alpha_r} a_r$ is an element of F, define I_u to be the left ideal in R generated by $a_1, \ldots, a_r$. Then the following statements are equivalent:*

(a) *A is flat.*

(b) *If $u \in K$, then $u \in KI_u$.*

Proof. (a) $\Rightarrow$ (b): If A is flat and $u \in K$, then $u \in K \cap FI_u = KI_u$, by 11.21.

(b) $\Rightarrow$ (a): Let I be any left ideal in R, and $u \in K \cap FI$. Then clearly $I_u \subseteq I$, and so $u \in KI_u \subseteq KI$. This is true for all $u \in K \cap FI$, and so it follows that $K \cap FI = KI$. Hence, applying 11.21, A is flat, completing the proof. $\square$

The next characterization of flat modules is attributed to Villamayor (unpublished) by Chase [60].

11.27 Proposition. *Let R be a ring, and $0 \to K \to F \to A \to 0$ be an exact sequence of right R-modules, where F is free. Then the following statements are equivalent:*

(a) *A is flat.*

(b) *Given any $u \in K$, there exists a homomorphism $\theta : F \to K$ such that $\theta(u) = u$.*

(c) *Given any $u_1, \ldots, u_n$ in K, there exists a homomorphism $\theta : F \to K$ such that $\theta(u_i) = u_i$ for $i = 1, \ldots, n$.*

Proof. (a) $\Rightarrow$ (b): Let $u \in K$. Define I_u as in 11.26, that is, $u = x_{\alpha_1} a_1 + \cdots + x_{\alpha_r} a_r$ (where $\{x_\alpha\}$ is a free basis of F) and $I_u = \sum_{i=1}^{r} Ra_i$. Since A is flat, 11.26 implies that $u \in KI_u$, and so $u = v_1 a_1 + \cdots + v_r a_r$, with $v_i \in K$. Define the homomorphism $\theta : F \to K$ by $\theta(x_{\alpha_i}) = v_i$ for $i = 1, \ldots, r$, and $\theta(x_\alpha) = 0$ if $\alpha \neq \alpha_1, \ldots, \alpha_r$. Then clearly θ has the desired property.

(b) $\Rightarrow$ (a): Given $u = x_{\alpha_1} a_1 + \cdots + x_{\alpha_r} a_r$ in K, let $\theta : F \to K$ be a homomorphism such that $\theta(u) = u$. Then $u = \theta(x_{\alpha_1}) a_1 + \cdots + \theta(x_{\alpha_r}) a_r$, and is thus in KI_u. It then follows from 11.26 that A is flat.

(b) $\Rightarrow$ (c): Let $u_1, \ldots, u_n \in K$. If $n = 1$, then the existence of θ follows from (b). Proceed by induction on n; assume that $n > 1$ and (c) holds for $k < n$. Let $\theta_n : F \to K$ be a homomorphism such that $\theta_n(u_n) = u_n$. Let $v_i = u_i - \theta_n(u_i)$ for $i = 1, \ldots, n-1$, then by the induction assumption, there exists a homomorphism $\theta' : F \to K$ such that $\theta'(v_i) = v_i$ for all $i = 1, \ldots, n-1$. Define the homomorphism $\theta : F \to K$ by $\theta = 1 - (1 - \theta')(1 - \theta_n)$; then θ has the desired properties.

(c) $\Rightarrow$ (b) is trivial. $\square$

Finitely Presented Flat Modules are Projective

In this paragraph it is demonstrated that a flat module $M = F/K$, where F is free, and K is finitely generated, is necessarily projective (see 11.30). This implies that finitely generated flat modules over Noetherian rings are projective (11.31).

A module M over R is **finitely related**, or **finitely presented**, if there is an exact sequence

$$0 \to K \to R^n \to M \to 0 \tag{1}$$

where n is an integer > 0, and K is finitely generated. This is equivalent to the requirement that there exist integers n and m such that

$$R^m \to R^n \to M \to 0 \tag{2}$$

is exact. The following lemma implies that when a module M is finitely related, then any exact sequence (1) implies that K is finitely generated.

11.28 Lemma (Schanuel) *Let*

$$0 \to P_1 \to P_0 \xrightarrow{f} M \to 0$$

$$0 \to P_1' \to P_0' \xrightarrow{f} M \to 0$$

be exact, and assume that P_0 and P_0' are projective. Then there is an isomorphism

$$P_0 \oplus P_1' \approx P_0' \oplus P_1$$

Proof. There is a row and column exact commutative diagram

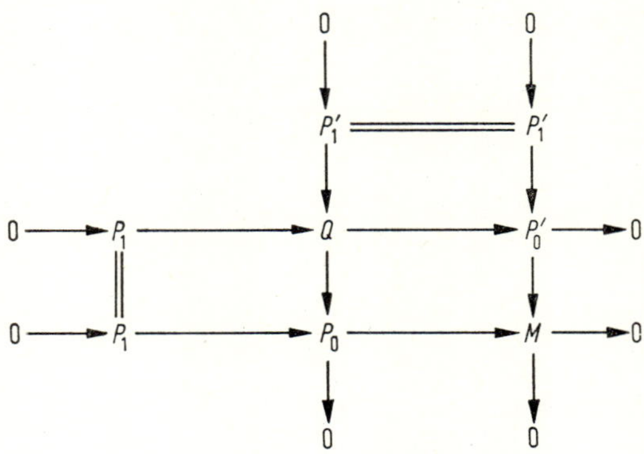

where $Q \to M$ is the fiber product of f and f'. Since P_0 is projective, then $Q \to P_0$ splits, and so $Q \approx P_0 \oplus P'_1$. Similarly, $Q \approx P'_0 \oplus P$. ▯

11.29 *Exercise.* If

$$0 \to P_n \to \cdots \to P_0 \to M \to 0$$

$$0 \to P'_n \to \cdots \to P'_0 \to M \to 0$$

are exact, and $\{P_i\}_{i=0}^{n-1}$ and $\{P'_i\}_{i=0}^{n-1}$ are projective, then there is an isomorphism

$$P_0 \oplus P'_1 \oplus P_2 \oplus \cdots \approx P'_0 \oplus P_1 \oplus P'_2 \oplus \cdots$$

11.30 **Corollary.** *Let R be any ring. If M is flat, if F is free, and*

$$0 \to K \to F \to M \to 0$$

is exact, and if K is finitely generated, then M is projective. Thus, any finitely related flat module is projective.

Proof. If $M \approx F/K$ is flat, and K is finitely generated, then 11.27 (c) implies that there is a map $u : F \to K$ extending the identity map $K \to K$. It follows that $F = K \oplus X$, where $X = \{u(x) - x \mid x \in F\}$, and so $M \approx X$ is projective along with F. ▯

11.31 **Corollary.** *If R is right Noetherian, then any finitely generated flat right R-module is projective.*

Proof. Over a right Noetherian ring, any finitely generated module is finitely presented, so 11.30 applies. ▯

11.32 Proposition (Lazard [64]). *Over any ring R, any flat module is a direct limit of projective modules.*

Proof. Let $M = F/K$ be flat, where F is projective. If $\{K_i\}_{i\in I}$ is the directed family of finitely generated submodules of K, then by Exercises for Chapter 6, $K = \varinjlim_{i\in I} K_i$, and then $F/K = \varinjlim_{i\in I} F/K_i$. (Cf. Chapter 14.) Now, by 11.27(c), and the proof of 11.30, there is a cofinal set of finitely generated projective summands $\{P_j\}_{j\in J}$ contained in the family $\{K_i\}_{i\in I}$. Thus, $K = \varinjlim_{j\in J} P_j$, and therefore $M = F/K = \varinjlim_{j\in J} F/P_j$ is a direct limit of projectives. □

The next several propositions and proofs are taken from Chase [60].

11.33 Proposition. *Let R be a ring, and A be a right R-module. Then the following statements are equivalent:*

(a) *A is flat.*

(b) *If $a_1\lambda_1 + \cdots + a_r\lambda_r = 0$, where $a_k \in A$ and $\lambda_k \in R$, then there exist $b_1, \ldots, b_n \in A$ and $\{\mu_{ik}\} \subseteq R$ $(i = 1, \ldots, n;\ k = 1, \ldots, r)$ such that $a_k = \sum_{i=1}^n b_i\mu_{ik}$ and $\sum_{k=1}^r \mu_{ik}\lambda_k = 0$.*

(c) *If $\sum_{k=1}^r a_k\lambda_{kj} = 0$, where $a_k \in A$ and $\lambda_{kj} \in R$ $(j = 1, \ldots, s)$ then there exist $b_1, \ldots, b_n \in A$ and $\{\mu_{ik}\} \subseteq R$ $(i = 1, \ldots, n;\ k = 1, \ldots, r)$ such that $a_k = \sum_{i=1}^n b_i\mu_{ik}$ and $\sum_{k=1}^r \mu_{ik}\lambda_{kj} = 0$.*

Proof. (a) ⇒ (c): Let $f: F \to A$ be an epimorphism, where F is a free right R-module, and let $K = \ker(f)$. Select $x_1, \ldots, x_r$ in F such that $f(x_k) = a_k$, and set $u_j = \sum_{k=1}^r x_k\lambda_{kj}$ for $j = 1, \ldots, s$. Then $f(u_j) = \sum_{k=1}^r a_k\lambda_{kj} = 0$. Since A is flat, we have from 11.29 that there exists a homomorphism $\theta: F \to K$ such that $\theta(u_j) = u_j$. Write $x_k - \theta(x_k) = \sum_{i=1}^n z_i\mu_{ik}$, where $z_1, \ldots, z_n$ are part of a basis of F. Set $b_i = f(z_i)$; then $a_k = f(x_k) = f\{x_k - \theta(x_k)\} = \sum_{i=1}^n b_i\mu_{ik}$. Finally, $0 = u_j - \theta(u_j) = \sum_{k=1}^r \{x_k - \theta(x_k)\}\lambda_{kj} = \sum_{k=1}^r (\sum_{i=1}^n z_i\mu_{ik})\lambda_{kj} = \sum_{i=1}^n z_i(\sum_{k=1}^r \mu_{ik}\lambda_{kj})$. Since $z_1, \ldots, z_n$ is part of a basis of F, it then follows that $\sum_{k=1}^r \mu_{ik}\lambda_{kj} = 0$.

(c) ⇒ (b): Trivial.

(b) ⇒ (a): Let $f: F \to A$ be an epimorphism, where F is a free right R-module with basis $\{x_\alpha\}$. Let $K = \ker(f)$, and suppose $u = x_{\alpha_1}\lambda_1 + \cdots + x_{\alpha_r}\lambda_r$ is in K. Let $a_k = f(x_{\alpha_k})$; then $a_1\lambda_1 + \cdots + a_r\lambda_r = f(u) = 0$. By hypothesis, there exist $b_1, \ldots, b_n$ in A and $\{\mu_{ik}\} \subseteq R\,(i \leq n,\ k \leq r)$ such that $a_k = \sum_{i=1}^n b_i\mu_{ik}$ and $\sum_{k=1}^r \mu_{ik}\lambda_k = 0$. Select $z_i \in F$ such that $f(z_i) = b_i$. Define a homomorphism $\theta: F \to F$ by $\theta(x_{\alpha_k}) = x_{\alpha_k} - \sum_{i=1}^n z_i\mu_{ik}$ for $k \leq r$ and $\theta(x_\alpha) = 0$ for $\alpha \neq \alpha_1, \ldots, \alpha_r$. Then $f\{\theta(x_{\alpha_k})\} = a_k - \sum_{i=1}^n b_i\mu_{ik} = 0$, and so $\theta(F) \subseteq K$. Finally, we have that

$$\theta(u) = u - \sum_{k=1}^{r}\left(\sum_{i=1}^{n} z_i u_{ik}\right)\lambda_k = u - \sum_{i=1}^{n} z_i\left(\sum_{k=1}^{r}\mu_{ik}\lambda_k\right) = u.$$

It then follows from 11.29 that A is flat. □

A functorial proof of the next proposition (employing the functor Tor) has been given by Colby and Rutter [71, p. 248]. (Cf. Lenzing [69].)

11.34 Proposition. *For any ring R the following statements are equivalent in which case R is said to be* ***left coherent:***

(a) *The direct product of any family of flat right R-modules is flat.*

(b) *The direct product of any family of copies of R is flat as a right R-module.*

(c) *Any finitely generated submodule of a free left R-module is finitely related.*

(d) *Any finitely generated left ideal in R is finitely related.*

Proof. (a) ⇒ (b): Trivial, since R is a flat right R-module.

(b) ⇒ (c): Let G be a free left R-module, and L be a finitely generated submodule of G. Clearly we may assume that G is finitely generated; hence, for some integer $s > 0$, identify G with the left R-module of all s-tuples $(\lambda_1, \ldots, \lambda_s)$ of elements of R. Let $u_1, \ldots, u_r$ generate L, where $u_k = (\lambda_{k1}, \ldots, \lambda_{ks})$. Let F be a free left R-module with basis $x_1, \ldots, x_r$, and define an epimorphism $f: F \to L$ by $f(x_k) = u_k$. Set $K = \ker(f)$, and for each $\alpha \in K$ let $R^{(\alpha)}$ be a copy of R. Define $A = \prod_{\alpha\in K} R^{(\alpha)}$, a right R-module. If $\alpha = a_1(\alpha)x_1 + \cdots + a_r(\alpha)x_r$ is in K, then $a_1(\alpha)u_1 + \cdots + a_r(\alpha)u_r = f(\alpha) = 0$, and so $\sum_{k=1}^{r} a_k(\alpha)\lambda_{kj} = 0$ for all $j \leq s$. Thus, setting $a_k = \{a_k(\alpha)\} \in A$ for $k = 1, \ldots, r$, we get that $\sum_{k=1}^{r} a_k\lambda_{kj} = 0$ for $j \leq s$.

By hypothesis, A is a flat right R-module; hence there exist $b_1, \ldots, b_n \in A$ and $\{\mu_{ik}\} \subseteq R$ $(i \leq n,\ k \leq r)$ satisfying the conditions of Proposition 2.3(c). Set $z_i = \sum_{k=1}^{r}\mu_{ik}x_k \in F$ for $i \leq n$; then $f(z_i) = \sum_{k=1}^{r}\mu_{ik}u_k = 0$, since $\sum_{k=1}^{r}\mu_{ik}\lambda_{kj} = 0$ for all $j \leq s$. Hence $z_1, \ldots, z_n \in K$. Write $b_i = \{b_i(\alpha)\}$, where $b_i(\alpha) \in R$; then, since $a_k = \sum_{i=1}^{n} b_i\mu_{ik}$ for all $k \leq r$, we obtain easily that $a_k(\alpha) = \sum_{i=1}^{n} b_i(\alpha)\mu_{ik}$ for $k \leq r$ and $\alpha \in K$. It then follows that $\alpha = \sum_{k=1}^{r} a_k(\alpha)x_k = \sum_{i=1}^{n} b_i(\alpha)z_i$. Hence $z_1, \ldots, z_n$ generate K, completing the proof that L is finitely related.

(c) ⇒ (d): Trivial.

(d) ⇒ (b): Let $\{R^{(\alpha)}\}$ be any family of copies of R, $A = \prod_\alpha R^{(\alpha)}$, and suppose that $a_1\lambda_1 + \cdots + a_r\lambda_r = 0$, where $a_k = \{a_k(\alpha)\} \in A$ and $\lambda_k \in R$, $k \leq r$. Let I be the left ideal in R generated by $\lambda_1, \ldots, \lambda_r$, let F be a free left R-module with free basis $x_1, \ldots, x_r$, and $f: F \to I$ be the epimorphism defined by $f(x_k) = \lambda_k$. Let K be the kernel of f; by hypothesis, K is finitely generated. Let $z_1, \ldots, z_n$ be a set of generators of K, and write $z_i = \sum_{k=1}^r \mu_{ik} x_k$. Setting $u(\alpha) = a_1(\alpha)x_1 + \cdots + a_r(\alpha)x_r$ we have that $f(u(\alpha)) = a_1(\alpha)\lambda_1 + \cdots + a_r(\alpha)\lambda_r = 0$, and so there exist $b_i(\alpha) \in R$ such that $u(\alpha) = \sum_{i=1}^n b_i(\alpha)z_i = \sum_{k=1}^r (\sum_{i=1}^n b_i(\alpha)\mu_{ik}) x_k$. Since $x_1, \ldots, x_r$ is a basis of F, it follows that $a_k(\alpha) = \sum_{i=1}^n b_i(\alpha)\mu_{ik}$ for all α, and so $a_k = \sum_{i=1}^n b_i \mu_{ik}$ for $k = 1, \ldots, r$. Finally, $\sum_{k=1}^r \mu_{ik}\lambda_k = f(z_i) = 0$ for each $i \leq n$, since $z_i \in K$. It then follows from 11.33 that A is flat.

The proof of (c) ⇒ (a) is an exercise, or refer to Chase [60, p. 461]. □

We include this Chapter with an application of a theorem from Chapter 6.

11.35 Proposition. (Corollary to 6.28.)

11.35.1 *If M is a (B, A)-bimodule, then the following are equivalent:*

(a) $\mathrm{Hom}_A(M, \ )$: mod-$A \rightsquigarrow$ mod-B *preserves injectives.*

(b) *If Q is the smallest injective cogenerator of* mod-A, *then* $\mathrm{Hom}_A(M, Q)$ *is injective in* mod-B.

(c) *The left adjoint of the functor* (a) *is exact.*

(d) *M is a flat B-module.*

11.35.2 *If I is an ideal of a ring A, then every injective right A/I module is injective in* mod-A *if and only if A/I is flat in A*-mod.

11.35.3 *If B is a subring of A, then every injective right A-module is an injective right (canonical) B-module iff A is flat as a (canonical) left B-module.* □

This yields another proof of 5.60 which states:

A left B-module M is flat if and only if the character module $\mathrm{Hom}(M, \mathbb{Q}/\mathbb{Z})$ *of M is injective as the canonical right B-module.*

Exercises for Chapter 11

1. The classes of rings are left coherent:

(a) Left Noetherian.

(b) Left semihereditary.

(c) The ring of polynomials, and the ring of formal power series, in any number of variables over any field.

(d) Any factor ring R/I of a left coherent ring R provided that I is finitely generated as a left ideal.

e) End F_R, where F is free over a left coherent ring R.

2. (Chase [60]) The statements are equivalent for any ring R.

(a) R is left coherent.

(b) If I is a finitely generated left ideal, then so is $(I : a) = \{r \in R \mid ra \in I\}$, for every $a \in R$.

(c) The intersection of two finitely generated left ideals is finitely generated, and so is $(0 : a)$, for any $a \in R$.

When these conditions hold, then $(I : C) = \{r \in R \mid rC \subseteq I\}$ is finitely generated for any finitely generated left ideal I, and finite subset C of R.

***3.** Let $f: R \to S$ be a morphism of RINGS. The f.a.e.: (1) $R \to S$ is epic; (2) The canonical homomorphism $S \otimes_R S \to S$ is an isomorphism; (3) The inclusion mod-$S \rightsquigarrow$ mod-R is a full embedding.

***4.** (Richman [65]) A commutative integral domain is semihereditary (i.e. Prüfer, see Chapter 30) iff every overring is flat. (Cf. 11.25.4.) In this case every overring is semihereditary.

***5.** (Gentile [60], Sandomierski [60]) Let R be a right semihereditary right neat ring. Then, the maximal right quotient ring $\hat{R}$ of R is left flat over R. In this case, any finitely generated flat submodule of a free right R-module is projective.

6. (Gentile [60]) If R is a right hereditary domain, then any torsion-free left R-module is flat. Generalize to (right neat) right hereditary rings. (Cf. 11.25.4.)

7. (Gentile [60]) If R is a right Ore domain, then the right quotient field K of R is left flat over R. Furthermore, then K is right flat over R iff K is also the left quotient field of R. Generalize to right Goldie semiprime rings.

***8.** (Bass [60]) Every flat right R-module is projective iff R is a ring with the d.c.c. on principal left ideals.

9. A ring is said to be semiprimary if $R/\text{rad } R$ is semisimple, and rad R is nilpotent. (See Chapter 18.) Any semiprimary ring satisfies the d.c.c. on finitely generated right ideals. (Dually, for finitely generated left ideals.) Thus, over semiprimary rings, in particular, over Artinian rings, any flat module is projective.

***10.** If R is a ring such that $R/\text{rad } R$ is semisimple, and idempotents of $R/\text{rad } R$ "lift" (cf. Chapter 22), then any finitely generated flat module is projective.

***11.** (Endo [62]) Over a commutative domain, any finitely generated flat module is projective. (Cf. Exercise 15.)

12. A ring R is a right IBN ring if every matrix ring R_n over R is Dedekind finite, $n = 1, 2, \ldots$ A sufficient condition for this is that R_n satisfies the a.c.c. on principal right ideals. (If every right ideal of R is free, then this condition is also necessary. Cf. Cohn [67].) [Hint for the proof: in any ring S the relations $ab = 1$ and $ba \neq 1$ implies the existence of infinitely many orthogonal idempotents $e_{ij} = b^{i-1}(1 - ba)a^{j-1}$ (Jacobson [50]).] [Hint: $e_{ij}e_{kq} = \delta_{jk}e_{iq}$.]

13. Let $R = \operatorname{End}_D V$, for an infinite dimensional vector space V over a field D. Write $V = W_1 \oplus W_2$, for subspaces $W_1 \approx W_2 \approx V$. Thus, $V \approx V^2$, and hence, conclude that $R = \operatorname{End}_D V \approx R_2 = \operatorname{End}_D V^2$. Now R_2 is a free R-module with a free basis consisting of four elements. Conclude that $R \approx R^4$, so R is neither right nor left IBN.

14. If R is any domain, not a right Ore domain, then the maximal right quotient ring $\hat{R}$ of R is such that every finitely generated nonzero right ideal is $\approx \hat{R}$. Show that $\hat{R} \approx \hat{R}^2$ in mod-$\hat{R}$, so $\hat{R}$ is not right IBN.

***15.** (Vasconcelos [69]) A finitely generated ideal I of a commutative ring R is projective iff I is flat and the annihilator of I is finitely generated. (More generally, a finitely generated flat module M is projective iff the "invariant factors" of M are finitely generated.)

Notes for Chapter 11

We discussed adjoint functors in Chapter 5, and Kan's theorem stating that $\otimes_R M$ is left adjoint to $\operatorname{Hom}_R(M, \ )$. Previously, Whitney [44] had defined the tensor product of abelian groups. Moreover, the old-fashioned name for tensor product is Kronecker product: the "Kronecker" product of two vector spaces of dimensions m and n is simply the vector space of dimension mn.

The weak global dimension theorem was proved by Auslander [55]: any ring has the same weak dimension on either side. (Also, over a right Noetherian ring, the w.gl. dim R = gl. dim R. See Chapter 8.) Moreover, rings of w.gl. dim = 0 coincide with the class of regular rings 11.24 (Auslander [57] and Harada [56]).

Bass [60] characterized when flat modules are projective. (See Exercise 8, and Chapter 22.) Lazard [64] showed that every flat is the direct limit of a directed set of finitely generated projective (in fact, free) modules 11.32. Actually, Bass [60] characterizes when the direct limit of projective modules is projective, or when a module has the "same weak as projective dimension". Chase [60] characterized when direct products of flat modules are flat 11.34, and, moreover, when a product of projectives is projective. (See Chapter 20.)

References

Auslander [57], Bass [60], Bourbaki [61], Chase [60], Cohn [66b, 67], Colby and Rutter [71], Gentile [60], Harada [56], Jacobson [50], Kan [58], Lazard [64], Leavitt [51], Lenzing [69], MacLane [63], Richmond [65], Rutter [71], Sandomierski [67, 68], Storrer [69a], Vasoncelos [69, 70a, b], Whitney [44].

Chapter 12

Morita Theorems and the Picard Group

The Morita theorem stated and proved in Chapter 4 is taken up again in this chapter in expanded and more general form, to be used in the determination of the Picard group Pic(mod-A) of all k-linear auto-equivalences of mod-A for an arbitrary algebra A, and also (in Exercises for Chapter 12, and in Chapter 32) of the Brauer group of a commutative ring k.

By the Morita theorem 4.29, any equivalence $T:\text{mod-}A \approx \text{mod-}B$ is representable by the A-dual module $P^* = \text{Hom}_A(P, A)$, where $P = TA$. Thus, there is a natural equivalence of functors

$$T \approx h^{P^*} \ \&\ T^{-1} \approx h^P.$$

Moreover, by adjoint associativity 11.4

$$\text{Hom}_A(P^*,\ ) \approx \underset{A}{\otimes} P$$

and

$$\text{Hom}_B(P,\ ) \approx \underset{B}{\otimes} P^*$$

are naturally equivalent. This implies natural isomorphisms

$$P \underset{B}{\otimes} P^* \approx A \qquad \text{and} \qquad P^* \underset{A}{\otimes} P \approx B$$

of (A, A) and (B, B) bimodules respectively. Then P and P^* are said to be **invertible** (A, B) and (B, A) bimodules, respectively. For any k-algebra A, the group $\text{Pic}_k A$ of classes (P) of invertible (A, A) — bimodules is isomorphic to Pic(A-mod) under the mapping $(P) \mapsto P_A$ (see 12.15).

In Chapter 5, we proved a theorem 5.70 of Popesco and Gabriel: *If U is a generator in $C = \text{mod-}A$, (or more generally, in a Grothendieck category C) and if $B = \text{End}_C U$, then the* mod-B*-valued functor $T = \text{Mor}_C(U,\ )$ possesses a left adjoint $G = \underset{B}{\otimes} U$, and*

$$GT \approx 1_C.$$

(Actually, G is an exact functor 15.26.)

This theorem implies the Morita theorem just stated. Furthermore, the category C is equivalent to a quotient category mod-$B/\text{Ker}\, G$, as described in Chapter 15. Although some of the theorems in the chapter can be obtained as corollaries to theorems just noted and proved in other parts of the text, the proofs of the Morita theorems given in this chapter require no new apparatus beyond the classical theory of tensor products (Chapter 11), whereas the greater generality of the theorem of Gabriel and Popesco has to be paid for by much preparation on categories.

Projectives and Generators over Commutative Rings

When is a projective R-module P a generator? When R is commutative, and P is either finitely generated, or R is Noetherian, we find 12.2 that P is a generator if and only if P is faithful. This result provides a wide class of generators. Also tensor products of generators are generators 12.4.

12.1 Lemma. *Let R be a commutative ring, and let M be a finitely generated R-module, and A any ideal of R such that $MA = M$. Then $M(1-a) = 0$ for some $a \in A$.*

Proof. If $x_1, \ldots, x_n$ generate M, then

$$x_i = \sum_{j=1}^{n} x_j a_{ij}$$

for some $a_{ij} \in A$, $i, j = 1, \ldots, n$, whence

$$\sum_{j=1}^{n} x_i (\delta_{ij} - a_{ij}) = 0$$

$\forall i$. Then $x_i \det(\delta_{ij} - a_{ij}) = 0$, $i = 1, \ldots, n$, so $M \cdot \det(\delta_{ij} - a_{ij}) = 0$. Clearly, $\det(\delta_{ij} - a_{ij}) = 1 - a$ for some $a \in A$. (See 27.17 for another proof of this.) ▯

12.2 Proposition. *Let R be a commutative ring, and let P be a projective R-module. If P is finitely generated, or if R is Noetherian, then* trace P *is generated by an idempotent e, and* ann P *is generated by $(1-e)$. Thus*

$$R = \operatorname{trace} P \oplus \operatorname{ann} P.$$

In this case, P is a generator of mod-R *if and only if P is faithful.*

Proof. If P is finitely generated over R, then so is $P^* = \operatorname{Hom}_R(P, R)$, and the ideal trace P generated by all $\{f(x) \mid x \in P,\ f \in P^*\}$ is finitely generated. Thus, in either case, the ideal $T = \operatorname{trace} P$ is finitely generated. By 3.30, $PT = P$ and $T^2 = T$. Taking $M = T$ in the lemma, we have $T(1-e) = 0$ for some $e \in T$. Thus, $e(1-e) = 0$, so $e = e^2$, and $T = Te$. This proves that $T = eR$. Now $P(1-e) = PT(1-e) = 0$, so ann $P \supseteq (1-e)R$, and clearly ann $P = (1-e)R$. This proves $R = \operatorname{trace} P \oplus \operatorname{ann} P$. Also, P is a generator if and only if trace $P = R$, hence if and only if ann $P = 0$. This proves the proposition. ▯

12.3 Corollary. *Let R be a commutative ring.*

12.3.1 *A module P is a progenerator if and only if P is finitely generated, projective, and faithful.*

12.3.2 *If R does not contain idempotents other than* 0 *or* 1, *then every projective R-module is a generator.*

12.3.3 *If R is Noetherian, then any faithful projective module is a generator of* mod-R. □

12.4 Proposition. *Let P and Q be modules over a commutative ring R. Then $P \otimes_R Q$ is finitely generated projective and faithful over R (resp. a progenerator) if and only if P and Q are.*

Proof (Bass). By the corollary, over a commutative ring a module is a progenerator if and only if it is finitely generated projective and faithful. If P and Q are progenerators, then $P^n \approx R \oplus A$, $Q^m \approx R \oplus B$, for suitable integers n and m, and R-modules A and B. Then, by additivity of adjoint functors 5.48, the module

$$P^{nm} \underset{R}{\otimes} Q^{nm} \approx (P \underset{R}{\otimes} Q)^{nm}$$

contains R as a summand in mod-R, Then, by 3.26, $P \underset{R}{\otimes} Q$ is a generator. Since $P \underset{R}{\otimes} Q$ is projective 11.16, we conclude that $P \underset{R}{\otimes} Q$ is a progenerator.

Conversely, suppose $P \underset{R}{\otimes} Q = G$ is a progenerator. Then P is faithful since G is. Now $G = P \underset{R}{\otimes} Q$ is finitely generated, for instance, by $\{x_i \otimes q_i \mid i = 1, \ldots, p\}$, and then $G = \sum_{i=1}^{p} (x_i \otimes Q)$. Let F be a free R-module on p letters. There exists an epimorphism $f: F \to P$. Then $f \otimes 1_Q : F \otimes Q \to P \otimes Q = G$ is epic, and splits since G is projective. Therefore, $f \otimes 1_{Q \otimes P} : F \otimes Q \otimes P \to P \otimes Q \otimes P$ is surjective and splits. Thus,

$$P \otimes_R G \approx P \otimes Q \otimes P$$

is isomorphic to a summand of

$$F \otimes_R G \approx F \otimes Q \otimes P.$$

Since $F \otimes_R G$ is a progenerator by the first part of the theorem, we deduce that $P \otimes_R G$ is finitely generated projective. Since G is a generator, $G^n \approx R \oplus A$ for a suitable integer n and module A. Then $(P \otimes_R G)^n$ is still projective, and

$$(P \otimes_R G)^n \approx P^n \otimes G^n \approx (P^n \otimes_R R) \oplus (P^n \otimes_R A) \approx P^n \oplus (P^n \otimes_R A).$$

Thus, P^n is isomorphic to a summand of the projective R-module $(P \otimes_R G)^n$, so P^n, whence P is projective. Thus, we have proved that P is finitely generated projective and faithful; that is, P is a progenerator. ▯

12.5 Corollary. *If P and Q are modules over a commutative ring R such that $P \otimes_R Q \approx R^n$ for some integer $n > 0$, then P and Q are both finitely generated projective faithful R-modules (progenerators).* ▯

Actually, for any finitely generated faithful projective P, there exists Q as in 12.5. (See Bass [67, p. 37].)

Definition of (,) and [,]

The tensor product $P^* \otimes_S P$ is naturally an (R, R)-bimodule, where if $r, r' \in R$, $x \in P$, $y \in P^*$, $r(y \otimes x)r' = (ry) \otimes (xr')$. There is a natural (R, R)-module map $P^* \otimes_S P \to R$ denoted by (,), and defined by

$$(\, , \,) : \sum u_i \otimes x_i \to \sum u_i(x_i).$$

[Here $u_i \in P^*$, $x_i \in P$, and $u_i(x_i)$ is the value of u_i at x_i, and for this reason, (,) is called the evaluation map.] If r, $r' \in R$, and $\sum_1^n u_i \otimes x_i \in P^* \otimes_S P$, then

$$(\, , \,) : \textstyle\sum_1^n (r' u_i \otimes x_i r) \to r'\left(\sum_1^n u_i(x_i)\right)r$$

so (,) is an (R, R)-map as claimed.

If $u \in P^*$, $x \in P$, then there is an element $[x, u] \in S = \operatorname{End} P_R$ such that

(1) $$[x, u](y) = x \cdot u(y) = x(u, y)$$

$\forall y \in P$. Since $u(y) \in R$, $x \cdot u(y) \in P$, so the correspondence $y \to x \cdot u(y)$ is indeed an element $[x, u]$ of $S = \operatorname{End} P_R$. Furthermore, one verifies immediately that the mapping

$$[\, , \,] : P \otimes_R P^* \to S$$

$$\textstyle\sum_1^n x_i \otimes u_i \mapsto \sum_1^n [x_i, u_i]$$

(where $x_i \in P$, $u_i \in P^*$, $i = 1, \ldots, n$) is an (S, S)-homomorphism of the (S, S)-bimodule $P \otimes_R P^*$ into S. Then, as verified in the proof of 4.1, there is an identity

(2) $$g[x, f] = g(x)f = (g, x)f$$

for all g, $f \in P^*$, and $x \in P$.

Equivalence Data

A set of pre-equivalence data (A, B, P, Q, f, g) consists of k-algebras A and B, bimodules ${}_AP_B$ and ${}_BQ_A$, and bimodule homomorphisms

$$f : P \otimes_B Q \to A$$

$$g : Q \otimes_A P \to B$$

such that by setting $f(p \otimes q) = pq$ and $g(q \otimes p) = qp$, we have the associative law

(1′) $$(pq)p' = p(qp'),$$

(2′) $$(qp)q' = q(pq')$$

$\forall p, p' \in P,\ q, q' \in Q$. If f and g are isomorphisms, then (A, B, P, Q, f, g) is called a **set of equivalence data.**

12.6 Proposition. *Let P be an R-module, and let* $P^* = \mathrm{Hom}_R(P, R)$. *Then*

$$\big(\mathrm{End}\, P_R, R, P, P^*, [\ ,\ ], (\ ,\ )\big)$$

is a set of pre-equivalence data.

Proof. Here $f = [\ ,\ ]$, $g = (\ ,\ )$, $Q = P^*$, $pq = [p, q]$, and $qp = (q, p) = q(p)$. Then $(i)'$ is the identity (i), $i = 1$ and 2, defined in the previous section. □

12.7 Proposition. *Let (A, B, P, Q, f, g) be a set of pre-equivalence data. If f is surjective, then*

12.7.1 *f is an isomorphism.*

12.7.2 *P and Q are generators as A-modules.*

12.7.3 *P and Q are finitely generated and projective B-modules.*

12.7.4 *g induces bimodule isomorphisms*

$$P \approx \mathrm{Hom}_B(Q, B), \qquad Q \approx \mathrm{Hom}_B(P, B).$$

12.7.5 *The k-algebra homothetic maps*

$$A \to \mathrm{End}\, P_B$$

$$A \to \mathrm{End}_B Q$$

are isomorphisms.

Proof. Since f is surjective, we can write

(*) $$\sum_{i=1}^{n} p_i q_i = 1 \in A$$

Proof of 12.7.1. If $\sum p'_j \otimes q'_j \in \ker f$, then

$$\sum_j p'_j \otimes q'_j = \sum_{j,i} (p'_j \otimes q'_j) p_i q_i = \sum p'_j \otimes ((q'_j p_i) q_i)$$
$$= \sum_{j,i} (p'_j (q'_j p_i)) \otimes q_i = \sum (p'_j q'_j)(p_i \otimes q_i)$$
$$= \sum_j (p'_j q'_j)(\sum_i p_i q_i) = 0, \text{ since } \sum p'_j q'_j = 0.$$

Proof of 12.7.2. Let $h_i : P \to A$ be an A-map defined by $h_i(p) = p q_i$, $i = 1, \ldots, n$. Then (*) implies that the coproduct map $(h_1 \oplus \cdots \oplus h_n)$ of $P^n \to A$ is surjective, so ${}_AP$ is a generator by 3.26, and similar proof for Q_A.

Proof of 12.7.3 follows from 4.1.2, or 7.3.

Proof of 12.7.4. The map g induces an (A, B)-bimodule homomorphism

$$h : P \to \mathrm{Hom}_B(Q, B)$$
$$h(p)(q) = qp.$$

(Note, we are abandoning our convention of writing maps opposite scalars).

If $h(p) = 0$, then

$$p = \sum_{i=1}^n (p_i q_i) p = \sum_{i=1}^n p_i (q_i p) = \sum_{i=1}^n p_i 0 = 0.$$

If $f : Q \to B$ is a B-map, then

$$f(q) = f\Big(\sum_i q(p_i q_i)\Big) = f(\sum_i (q p_i) q_i) = \sum_i (q p_i) f(q_i) = \sum_i q(p_i f(q_i))$$

so

$$f = h\Big(\sum_i p_i f(q_i)\Big)$$

that is, f is surjective, whence an isomorphism. Similarly, $Q \approx \mathrm{Hom}_B(P, B)$

Proof of 12.7.5. Similar to the proof of 12.7.4. □

12.8 Corollary. *Let P be an R-module, $S = \mathrm{End}\, P_R$ and $P^* = \mathrm{Hom}_R(P, R)$.*

(a) *The (R, R)-homomorphism $(\ ,\ ) : P^* \otimes_S P \to R$ is surjective if and only if P_R is a mod-R generator, in which case, $[\ ,\ ]$ is an isomorphism, $P^*_S \approx \mathrm{Hom}_S(P, S)$, $R \approx \mathrm{End}_S P$, and ${}_RP^*$ is a generator.*

(b) *The* (S, S)*-homomorphism* $[\ ,\ ] : P \otimes_R P^* \to S$ *is surjective if and only if* P_R *is finitely generated and projective, in which case,* $[\ ,\ ]$ *is an isomorphism,* $S \approx \operatorname{End}_R P^*$, *and* ${}_SP$ *is a generator for* S-mod.

(c) $(S, R, P, P^*, [\ ,\ ], (\ ,\ ))$ *is a set of equivalence data if and only if* P_R *is a finitely generated projective generator (progenerator).*

Proof. (c) follows from (a), (b), and 12.6. Also the proposition, together with 12.6, yields the implication $\Rightarrow$ in both (a) and (b), and the stated ring isomorphisms. (Cf. also 4.1 and 7.3.)

(a) If P_R is a generator, then $\operatorname{trace}_R P = R$ by 3.26, but $\operatorname{trace}_R P$ is the image of $(\ ,\ )$, so $(\ ,\ )$ is surjective.

(b) By the dual basis lemma 3.23 P is finitely generated and projective if and only if there exist finitely many elements $x_1, \dots, x_n \in P$, and corresponding elements $u_1, \dots, u_n \in P^*$ such that

$$x = \textstyle\sum_1^n x_i u_i(x) \qquad \forall x \in P$$

or, equivalently,

$$\textstyle\sum_1^n [x_i, u_i] = 1$$

the identity of S. This latter condition holds, for some choice of $x_i \in P$, $u_i \in P^*$, if and only if $[\ ,\ ]$ is an epimorphism. ▯

12.9 Proposition (Eilenberg-Watts) *Let* A *and* B *be* k*-algebras, and* ${}_AM_B$ *and* ${}_AN_B$ *bimodules.*

12.9.1 *If* $f : M \to N$ *is an* (A, B)*-bimodule homomorphism, then there is a natural transformation* $\otimes_A f : \otimes_A M \to \otimes_A N$ *of functors, where* $(\otimes_A f)(P) = 1_P \otimes_A f$ *for each* $P \in \operatorname{obj}$ mod-A.

12.9.2 *If* $f : M \to N$ *is an* (A, B)*-bimodule isomorphism, then* $\otimes_A f$ *is a natural equivalence of* k*-linear functors.*

12.9.3 *Two functors* $\otimes_A M$ *and* $\otimes_A N$ *from* mod-A *to* mod-B *are naturally equivalent if and only if* ${}_AM_B \approx {}_AN_B$ *as bimodules.*

12.9.4 *The correspondence* $M \mapsto \otimes_A M$ *induces a bijection between the set of isomorphism classes of* (A, B)*-bimodules and the set of natural equivalences between right continuous* k*-linear functors* mod-$A \rightsquigarrow$ mod-B. *If* C *is another* k*-algebra, then in the situation* ${}_AM_B$, ${}_BN_C$, *the functor*

$$\otimes_A (M \otimes_B N) : \text{mod-}A \rightsquigarrow \text{mod-}C$$

is naturally equivalent to the composite functor

$$(\otimes_B N) \circ (\otimes_A M) : \text{mod-}A \rightsquigarrow \text{mod-}C.$$

Proof. 12.9.1, 2, and 3 follow from Yoneda's theorem 2.3 (cf. 6.20 and 6.21), or they may be verified directly. By 5.53 and 5.54, any right continuous functor S from mod-A to mod-B is left adjoint to a representable functor h^{SA}. By 5.3, two such, S_1 and S_2, are equivalent iff $h^{S_1A} \approx h^{S_2A}$. Then, $S_1A \approx S_2A$ are equivalent right B-modules by Yoneda 2.3, and this isomorphism is patently an (A,B)-bimodule morphism. So far, none of this speaks of k-linear functors, but this part follows naturally. (Cf. 5.72.4 and 5.) This proves 4, except for the last statement which follows from:

$$\begin{aligned}(\otimes_B N)\circ(\otimes_A M)(P) &= (\otimes_B N)(P\otimes_A M)\\ &= (P\otimes_A M)\otimes_B N\\ &\approx P\otimes_A(M\otimes_B N)\\ &= \otimes_A(M\otimes_B N)(P). \quad \square\end{aligned}$$

12.10 Morita I. *Let* (A, B, P, Q, f, g) *be a set of equivalence data,* e.g., *let* P_B *be a progenerator,* $A = \operatorname{End} P_B$, $f = [\ ,\ ]$, $g = (\ ,\ )$. (See 12.8.) *Then*

12.10.1
$$\otimes_B Q : \text{mod-}B \rightsquigarrow \text{mod-}A$$
$$\otimes_A P : \text{mod-}A \rightsquigarrow \text{mod-}B$$

are inverse equivalences of categories, and so are

$$Q\otimes_A : A\text{-mod} \rightsquigarrow B\text{-mod}$$
$$P\otimes_B : B\text{-mod} \rightsquigarrow A\text{-mod}.$$

12.10.2 *P and Q are each progenerators, both as A and as B-modules.*

12.10.3 *f and g induce bimodule isomorphisms*

$$P \approx \operatorname{Hom}_B(Q, B) \approx \operatorname{Hom}_A(Q, A)$$
$$Q \approx \operatorname{Hom}_A(P, A) \approx \operatorname{Hom}_B(P, B).$$

12.10.4 *The k-algebra homomorphisms*

$$\operatorname{End} P_B \leftarrow A \rightarrow \operatorname{End}_B Q$$
$$\operatorname{End}_A P \leftarrow B \rightarrow \operatorname{End} Q_A$$

induced by the respective canonical maps are all isomorphisms.

12.10.5 *The bimodule endomorphism rings of A, B, P, and Q are all isomorphic to the centers of A, B,* mod-A, *and* mod-B.

12.10.6 *The lattice of right ideals of A is isomorphic to the lattice of B-submodules of P under $C \mapsto CP \approx C \otimes_A P$, with ideals corresponding to (A, B)-submodules of P. Similar statements about the other lattice isomorphisms follow by symmetry. In particular, A and B have isomorphic lattices of ideals.* (Cf. 4.7.)

Proof of 12.10.1. Since $Q \approx \mathrm{Hom}_B(P, B)$, and $P \approx \mathrm{Hom}_A(Q, A)$ by 12.7, by 12.8 we have bimodule isomorphisms

$$Q \otimes_A P \approx B \qquad \text{and} \qquad P \otimes_B Q \approx A$$

thus 12.9 implies an equivalence of functors

$$(\otimes_A P) \circ (\otimes_B Q) \approx \otimes_B (Q \otimes_A P) \approx \otimes_B B \approx 1_{\text{mod-}B}.$$

Similarly,

$$(\otimes_B Q) \circ (\otimes_A P) \approx 1_{\text{mod-}A}$$

so $\otimes_B Q$ and $\otimes_A P$ are inverse equivalences.

Next, proofs of 12.10.2, 12.10.3, and 12.10.4 follow immediately from 12.7.

Proof of 12.10.5. We have isomorphisms

$$\text{Center } A \approx \mathrm{Hom}_{(A,A)}(A, A) \xrightarrow{\otimes_A P} \mathrm{Hom}_{(A,B)}(P, P)$$

$$\text{Center } B \approx \mathrm{Hom}_{(B,B)}(B, B) \xrightarrow{P \otimes_B} \mathrm{Hom}_{(A,B)}(P, P)$$

and similarly for Q. Then 12.10.5 follows from these isomorphisms, and from 4.34.

Proof of 12.10.6. Since ${}_AP$ is projective, the canonical map $C \otimes_A P \to CP$ is an isomorphism (11.20). Since $\otimes_A P : \text{mod-}A \rightsquigarrow \text{mod-}B$ is an equivalence, the mapping $C \mapsto CP$ is the desired lattice isomorphism, with fully invariant (A, A)-submodules, that is, ideals of A, corresponding to fully invariant B-submodules of P, which by 12.10.4 are the (A, B)-submodules of P. $\square$

12.11 Statement. For equivalence data $\big(S, R, P, P^*, [\ ,\ ], (\ ,\ )\big)$, the proposition yields the equivalence

$$\otimes_R P^* : \text{mod-}R \rightsquigarrow \text{mod-}S.$$

There is a natural equivalence of functors $\otimes_R P^* \approx \mathrm{Hom}_R(P,\ )$, so the latter functor is also an equivalence with inverse $\mathrm{Hom}_S(P^*,\ )$.

Morita II

This theorem determines that all category equivalences mod-$A \rightsquigarrow$ mod-B come from sets of equivalence data. The concept of k-linear equivalences is introduced supra 5.72 (cf. 5.72.4 and 5).

12.12 Morita II. *Let A and B be k-algebras, and let*

$$G : \text{mod-}A \rightsquigarrow \text{mod-}B$$
$$H : \text{mod-}B \rightsquigarrow \text{mod-}A$$

be inverse k-linear equivalences. Set $P = GA$, $Q = HB$. Then P is a (A,B)-bimodule and Q is a (B,A)-bimodule. Furthermore:

12.12.1 *There are natural equivalences of functors*

$$G \approx \otimes_A P \quad \text{and} \quad H \approx \otimes_B Q .$$

12.12.2 *There are bimodule isomorphisms*

$$f : P \otimes_B Q \to A , \quad \text{and} \quad g : Q \otimes_A P \to B .$$

12.12.3 *f and g may be chosen so that the following diagrams commute.*

$$\begin{array}{ccc} P \otimes_B Q \otimes_A P & \xrightarrow{f \otimes 1_P} & A \otimes_A P \\ \Big\downarrow{\scriptstyle 1_P \otimes g} & & \Big\downarrow{\scriptstyle \text{canon}} \\ P \otimes_B B & \xrightarrow[\text{canon}]{} & P \end{array} \qquad \begin{array}{ccc} Q \otimes_A P \otimes_B Q & \xrightarrow{g \otimes 1_Q} & B \otimes_B Q \\ \Big\downarrow{\scriptstyle 1_Q \otimes f} & & \Big\downarrow{\scriptstyle \text{canon}} \\ Q \otimes_A A & \xrightarrow[\text{canon}]{} & Q \end{array}$$

Proof of 12.12.1 and 2. Exercise, or (if necessary) refer to the proof of 4.29.

Proof of 12.12.3. All the maps designated in the diagram are bimodule isomorphisms. If $a : A \otimes_A P \to P$ and $b : P \otimes_B B \to P$ are the canonical maps, then there exists an (A, B)-automorphism u of P such that

$$b(1 \otimes g) = ua(f \otimes 1) .$$

Therefore $u \in \operatorname{End} P_B \approx A$, so u can be supposed as an element of A, whence a unit in A. Since u is also an A-homomorphism of P, and ${}_AP$ is faithful, then $u \in \operatorname{center} A$. Since $ua = a(u_s \otimes 1_P)$, if we replace f by uf, the first square commutes. Next write

$$f(p \otimes q) = pq , \quad \text{and} \quad g(q \otimes p) = qp$$

$\forall p \in P$, $q \in Q$. Then $(pq)p' = p(qp')$ $\forall p, p' \in P$, $q \in Q$. We need to prove that $(qp)q' = q(pq')$ $\forall q' \in Q$. Now

$$\begin{aligned}((qp)q')p' &= (qp)(q'p') && (g \text{ is a left } B\text{-map}) \\ &= q(p(q'p')) && (g \text{ is a right } B\text{-map}) \\ &= q((pq')p') && (\text{by assumption}) \\ &= (q(pq'))p' && (q \otimes ap = qa \otimes p \; \forall a \in A).\end{aligned}$$

Hence, if $d = (qp)q' - q(pq')$, then $dp' = 0$ $\forall p' \in P$. Let $h: A \to Q$ be defined by $h(a) = da$. Then $h \otimes 1_P : A \otimes_A P \to Q \otimes_A P$ followed by g is zero. Since g is an isomorphism, $h \otimes 1_P = 0$. Since $\otimes_A P$ is fully faithful, then $h = 0$, whence $d = 0$. ▯

12.13 Definition and Proposition. *A bimodule ${}_AM_B$ is* ***invertible*** *in case it satisfies the following equivalent conditions.*

(a) *The functor* $\otimes_A M$: mod-A $\rightsquigarrow$ mod-B *is an equivalence.*

(b) *There exists a bimodule* ${}_BN_A$, *and bimodule isomorphisms* $M \otimes_B N \approx A$ *and* $N \otimes_A M \approx B$.

(c) *The functor* $M \otimes_B$: mod-B $\rightsquigarrow$ mod-A *is an equivalence.*

Proof. (a) $\Rightarrow$ (b). If $H = \otimes_A M$ is an equivalence, then there exists a functor G : mod-$B \rightsquigarrow$ mod-A such that $HG \approx 1_{\text{mod-}A}$ and $GH \approx 1_{\text{mod-}B}$. The last proposition implies $H = \otimes_B N$ for some bimodule ${}_BN_A$. Now any functor Q : mod-$A \rightsquigarrow$ mod-A, which is naturally equivalent to $1_{\text{mod-}A}$, is by the last proposition naturally equivalent to $\otimes_A A$. Thus,

$$HG \approx \otimes_A(M \otimes_B N) \approx \otimes_A A$$

so by (a) of the last proposition, we have a bimodule isomorphism $M \otimes_B N \approx A$. Similarly, $N \otimes_A M \approx B$. This proves (*a*) $\Rightarrow$ (b), and (*b*) $\Rightarrow$ (a) follows by reversing the steps. Since (b) is right-left symmetric, the proof is complete. ▯

12.14 Morita III. *If A and B are k-algebras, then the correspondence $P \mapsto \otimes_A P$ is a bijection between the set of isomorphism classes of invertible (A, B)-bimodules and the set of natural equivalences of k-linear category equivalences from* mod-A *to* mod-B.

Proof. This follows immediately from the preceding definition, Morita II, and 12.9. So far, this says nothing about k-linear functors, but inasmuch as $B = \text{End } P_A$, the functors involved are automatically k-linear. (See 5.72.4/5.) ▯

Autoequivalences and the Picard Group

If $\mathscr{A}$ is a k-linear abelian category, and k a commutative ring, then

$$\mathrm{Pic}_k\, \mathscr{A}$$

denotes the group of isomorphism classes of k-linear equivalences[1] $T : \mathscr{A} \rightsquigarrow \mathscr{A}$. Thus, if also $S : \mathscr{A} \rightsquigarrow \mathscr{A}$, then $S = T$ in $\mathrm{Pic}_k\, \mathscr{A}$ if and only if there exists a natural equivalence $h : S \to T$ of functors. The group operation is composition of functors.

If A is a k-algebra, define

$$\mathrm{Pic}_k A$$

to be the group of isomorphism classes (P) of invertible (A, A)-bimodules under the operation

$$(P)\,(Q) = (P \otimes_A Q).$$

It follows from 12.4 and 12.14 that $\mathrm{Pic}_k A$ is a group under this operation, and $(P)^{-1} = \big(\mathrm{Hom}_A(P, A)\big)$. [In $\mathrm{Hom}_A(P, A)$, either the left or the right A-module structure can be used.]

The next proposition follows from 12.9 and 12.14.

12.15 Proposition. *Let A be a k-algebra. The functors*

$$\begin{cases} (P) \mapsto P \otimes_A \\ \mathrm{Pic}_k A \rightsquigarrow \mathrm{Pic}\,(A\text{-mod}) \end{cases}$$

and

$$\begin{cases} \mathrm{Pic}\,(A\text{-mod}) \rightsquigarrow \mathrm{Pic}_k A \\ T \mapsto TA \end{cases}$$

are inverse equivalences. □

Let A be an algebra over k. The algebra isomorphisms $A \to A$ form a group under composition of maps. This group is denoted $\mathrm{Aut}_k A$, and the elements are called ***automorphisms.*** An automorphism f is ***inner*** provided that there exists a unit $x \in A$ such that $f(a) = x^{-1} a x$ for all $a \in A$. Then f is called the inner **automorphism effected (or defined) by** x. The set of inner automorphisms is a normal subgroup of $\mathrm{Aut}_k A$, which is denoted by $\mathrm{InAut}_k A$.

Let P be an invertible (A, A)-bimodule. If $\alpha, \beta, \in \mathrm{Aut}_k A$ are k-algebra automorphisms, then

$$_{\alpha}P_{\beta}$$

---

[1] *Supra* 5.72. Ignoring sizes, the center of an Abelian category $\mathscr{A}$ is a ring C. If C is *a* k-algebra, then $\mathrm{Mor}_{\mathscr{A}}(X, Y)$ becomes a k-module canonically, for any objects X and Y. Then, a k-linear equivalence is defined so the homomorphism $(X, Y) \to (TX, TY)$ induced by T is a k-linear transformation.

denotes the (A, A)-bimodule defined by

$$a \cdot p \cdot b = \alpha(a)\, p\beta(b)$$

$\forall a, b \in A,\ p \in P$. Thus, ${}_1P_1 = P$.

If $f : P \to Q$ is a left A-isomorphism where $(P), (Q) \in \mathrm{Pic}_k A$, and if $a \in A$, then the mapping

$$\begin{cases} P \to P \\ p \mapsto ((p)fa)f^{-1} \end{cases}$$

is an element of $\mathrm{Hom}_A({}_AP, {}_AP) = A$, which we denote by $\alpha(a)$. Thus

$$(p\alpha(a))f = (p)fa \qquad \forall p \in P.$$

The mapping $a \mapsto \alpha(a)$ is an algebra automorphism $\alpha \in \mathrm{Aut}_k A$, and $f : {}_1P_\alpha \to Q$ is a bimodule isomorphism.

12.16 Lemma. *If* $\alpha, \beta, \gamma \in \mathrm{Aut}_k A$, *then there exist the following bimodule isomorphisms*:

12.16.1 ${}_\alpha A_\beta \approx {}_{\gamma\alpha}A_{\gamma\beta}$

12.16.2 ${}_1A_\alpha \otimes_A {}_1A_\beta \approx {}_1A_{\alpha\beta}$

12.16.3 ${}_1A_\alpha \approx {}_1A_1 \Leftrightarrow \alpha \in \mathrm{InAut}_k A$.

12.16.4 *If* $(P) \in \mathrm{Pic}_k A$, *and* ${}_AP \approx {}_AA$, *then* $P \approx {}_1A_\alpha$ *for some* $\alpha \in \mathrm{Aut}_k A$.

Proof of 12.16.1. The map

$$\begin{cases} {}_\alpha A_\beta \to {}_{\gamma\alpha}A_{\gamma\beta} \\ x \mapsto \gamma(x) \end{cases}$$

is the desired (A, A)-bimodule isomorphism.

Proof of 12.16.2 By 12.16.1,

$${}_1A_\alpha \otimes_A {}_1A_\beta \approx {}_{\alpha^{-1}}A_1 \otimes_A {}_1A_\beta \approx {}_{\alpha^{-1}}A_\beta \approx {}_1A_{\alpha\beta}.$$

Proof of 12.16.3. If $f : {}_1A_\alpha \to {}_1A_1$ is a bimodule isomorphism, then as a left A-module isomorphism, $f(x) = xu$, where $f(1) = u$ is a unit of A. Also

$$\alpha(a)u = \alpha(a)f(1) = f(\alpha(a \cdot 1)) = f(1 \cdot a) = ua$$

so $\alpha(a) = uau^{-1}\ \forall a \in A$, and $\alpha \in \mathrm{InAut}_k A$. Conversely, if $\alpha(a) = uau^{-1}$, then there is a bimodule isomorphism $f : {}_1A_\alpha \to {}_1A_1$ that sends $x \mapsto xu\ \forall x \in A$.

Proof of 12.16.4. This was proved in the paragraph immediately preceding the lemma. ∎

12.17 Proposition. *Let Q be a left A-progenerator, and let $B = \mathrm{End}_A Q$ be considered as a k-algebra. Then there is an exact sequence*

$$(1) \qquad 1 \to \mathrm{InAut}_k B \to \mathrm{Aut}_k B \xrightarrow{\varphi_Q} \mathrm{Pic}_k A$$

where

$$\mathrm{im}\, \varphi_Q = \{(P) \in \mathrm{Pic}_k A \mid P \otimes_A Q \approx Q \text{ as left } A\text{-modules}\}.$$

Proof. Suppose first that $Q = A$. Then $B = A$, and we define $\varphi_A(\alpha) = {}_1A_\alpha \ \forall \alpha \in \mathrm{Aut}_k A$. By the lemma, $\varphi_A : \mathrm{Aut}_k A \to \mathrm{Pic}_k A$ is a map with kernel $\mathrm{InAut}_k A$, and $\mathrm{im}\, \varphi_A$ is as described.

In the general case, by Morita I (12.10), the functor $T = \mathrm{Hom}_A(Q, \ ) \approx Q^* \otimes_A$ is an equivalence A-mod $\rightsquigarrow$ B-mod, with $B = TQ$, where $Q^* = \mathrm{Hom}_A(Q, A)$. The equivalence T induces an isomorphism $\mathrm{Pic}_k(A\text{-mod}) \to \mathrm{Pic}_k(B\text{-mod})$, which by 12.15 yields an isomorphism

$$\begin{cases} h : \mathrm{Pic}_k A \to \mathrm{Pic}_k B \\ \quad h : (P) \mapsto (Q^* \otimes_A P \otimes {}_A Q)\,. \end{cases}$$

Let $\varphi_Q : \mathrm{Aut}_k B \to \mathrm{Pic}_k A$ be the composite

$$\mathrm{Aut}_k B \xrightarrow{\varphi_B} \mathrm{Pic}_k B \xrightarrow{h^{-1}} \mathrm{Pic}_k A\,.$$

The exactness of the sequence (1) follows from the exactness for the special case $Q = A$. Also, if $(P) \in \mathrm{Pic}_k A$, then $P \otimes_A Q \approx Q$ as left A-modules if and only if $Q^* \otimes_A P \otimes_A Q \approx Q^* \otimes_A Q$ as left B-modules. Since $Q^* \otimes_A Q \approx B$ as (B, B)-modules by 12.14, the last statement also follows from the special case. ▯

12.18 Proposition. *If A is a k-algebra with center C, then there is an exact sequence*

$$0 \to \mathrm{Pic}_C A \to \mathrm{Pic}_k A \to \mathrm{Aut}_k C\,.$$

If A is commutative, then

$$0 \to \mathrm{Pic}_A A \to \mathrm{Pic}_k A \to \mathrm{Aut}_k A \to 1$$

is split exact.

Proof. Let P be an invertible (A, A)-bimodule, and let $t \in C$. Since $p \mapsto pt$ is a bimodule endomorphism of P, there exists $\alpha_P(t) \in C$ such that $\alpha_P(t)p = pt \ \forall p \in P$. Since $\lambda p = p\lambda \ \forall \lambda \in k$, $\alpha_P(t\lambda) = \alpha_P(t)\lambda \ \forall \lambda \in k$. Clearly, the mapping

$$\begin{cases} \alpha_P : C \to C \\ \alpha_P : t \to \alpha_P(t) \end{cases}$$

is an algebra endomorphism.

If $p \otimes q \in P \otimes_A Q$, and $t \in C$, then

$$(p \otimes q)t = p \otimes (qt) = p \otimes \alpha_Q(t)q = p\alpha_Q(t) \otimes q = \alpha_P\alpha_Q(t)(p \otimes q)$$

whence

$$\alpha_P\alpha_Q = \alpha_{P\otimes_A Q}.$$

Since $\alpha_A = 1_C$, this implies

$$1_C = \alpha_A = \alpha_{P\otimes_A P^*} = \alpha_P\alpha_{P^*} \qquad (= \alpha_{P^*}\alpha_P)$$

which proves that α_P is a unit in $\text{End}_k A$; that is, $\alpha_P \in \text{Aut}_k A$. We next observe that $\alpha_P = \alpha_{P'}$ $\forall P' \in (P)$; that is, if $p \mapsto p'$ is an (A, A)-bimodule isomorphism $P \to P'$, and if $t \in C$, then

$$p't = (pt)' = (\alpha_P(t)p)' = \alpha_P(t)p'$$

shows that $\alpha_{P'}(t) = \alpha_P(t)$. This shows that

$$\begin{cases} \text{Pic}_k A \to \text{Aut}_k C \\ \quad (P) \mapsto \alpha_P \end{cases}$$

is a mapping, which by $\alpha_P\alpha_Q = \alpha_{P\otimes_A Q}$ is a homomorphism. If (P) is an element of the kernel of this homomorphism, that is, if $\alpha_P = 1_C$, then $tp = pt\,\forall t \in C, p \in P$, whence $(P) \in \text{Pic}_C A$. Conversely, if $P \in \text{Pic}_C A$, then $P \in \text{Pic}_k A$ and $\alpha_P = 1_C$. Thus, the first sequence in the proposition is exact. If A is commutative, then the mapping $\alpha \mapsto {}_1A_\alpha$ of 21.16 gives the required splitting $\text{Aut}_k A \to \text{Pic}_k A$. □

In reference to the examples which follow, a **(Prüfer) Dedekind domain** is a commutative (semi) hereditary domain.

12.19 *Examples.*

12.19.1 If k is a Prüfer domain (cf. Chapter 30), $\text{Pic}_k k$ is isomorphic to the ideal class group of k. (The ideal class group is the group of isomorphism classes of invertible ideals of k, and is isomorphic to the factor group Invert k modulo principal fractional ideals.) The reason for this is that if $(P) \in \text{Pic}_k k$, then $k = \text{End}\, P_k$, so P is an indecomposable finitely generated projective k-module. Since k is Prüfer, hence semihereditary, every projective module is isomorphic to a direct sum of ideals of k. Thus, $(P) = (P')$ for some $P' \in \text{Invert}\, k$.

12.19.2 Let A be the ring of (algebraic) integers in an algebraic number field K. (An algebraic number field is a finite field extension of $\mathbb{Q}$, see Chapter 30.) Let $G = G(K/\mathbb{Q})$ denote the group of automorphisms of K, leaving fixed the elements of $\mathbb{Q}$. Every automorphism f of G induces an automorphism f' of A, which leaves fixed the elements of Z. Thus, $f' \in \text{Aut}_{\mathbb{Z}} A$, and conversely, every $g \in \text{Aut}_{\mathbb{Z}} A$ has a unique extension $h \in G(K/\mathbb{Q})$. In short, $G(K/\mathbb{Q}) \approx \text{Aut}_{\mathbb{Z}} A$, under $f \mapsto f'$.

Since A is a Dedekind domain, the first example asserts that $\mathrm{Pic}_A A$ is the ideal class group of A. By the proposition, $\mathrm{Pic}_{\mathbb{Z}} A \approx \mathrm{Aut}_{\mathbb{Z}} A \oplus \mathrm{Pic}_A A$. Now $\mathrm{Pic}_A A$ is also the group of autoequivalences of the category A-mod, and is finite (finitude of the class number).

Exercises for Chapter 12

1. A ring R is a cogenerator in mod-R if and only if every right R-module is torsionless. In this case, every right ideal of R is a right annulet.

***2.** If R is a commutative entire ring, then a right ideal $I \neq 0$ is projective if and only if I is a generator. In this case, I is necessarily finitely generated, and the canonical map $R \to \mathrm{End}\, I_R$ is an isomorphism (cf. invertible ideals in Chapter 31). Conclude that a Dedekind domain is Noetherian.

3. Determine all generators in various rings, e.g., a simple ring. Show that if R is a ring in which every nonzero right ideal is a generator, then R is a prime ring. (Then if R is commutative, every right ideal of R is projective; that is, R is hereditary. See 30.17.)

***4.** Prove the statement following 12.5.

***5.** (Morita [58], Azumaya [59]) Dualize the Morita theorems for category equivalence mod-$A \rightsquigarrow$ mod-B. First note that there is no duality mod-$A \rightsquigarrow B$-mod, for any rings A and B. Next, find necessary and sufficient conditions on Abelian subcategories $\mathscr{S}_A$ of mod-A and ${}_B\mathscr{S}$ of B-mod, with A an object of $\mathscr{S}_A$ and B an object of ${}_B\mathscr{S}$ such that a duality $D: \mathscr{S}_A \rightsquigarrow {}_B\mathscr{S}$ exists. Show that then any exact sequence $0 \to X \to Y \to Z \to 0$ in mod-A is such that $Y \in \mathscr{S}_A$ iff $X \in \mathscr{S}_A$ and $Z \in \mathscr{S}_A$. Moreover, show that $U = D(A)$ is an (B,A)-bimodule such that the functor mod-$A \rightsquigarrow B$-mod defined by the contravariant morphism functor h_U induces the duality D, and similarly for D^{-1}. Show that a necessary and sufficient condition on a (B,A)-bimodule U in order that U induce such a duality is that U be an injective cogenerator both in mod-A and B-mod and $B \approx \mathrm{End}\, U_A$, $A \approx \mathrm{End}_B U$. Such a duality is then called a U-**duality**. Show that a necessary condition on A and B is that both rings are semilocal rings in which idempotents lift modulo the radical. [Or equivalently, by Chapter 22, every finitely generated module (on either side, and for either ring) has a projective cover.] Conclude that not every ring A can possess a duality in the sense described. Moreover, if A is right Artinian, then show that B is left Artinian, and that U is finitely generated in both categories. Conclude that a necessary and sufficient condition for an A-duality is that A be an injective cogenerator in mod-A and A-mod. For a right Artinian ring A, a necessary and sufficient condition for an A-duality is that every right, and every left, ideal be an annulet. [Then A is said to be **Quasi-Frobenius** (QF).]

***6.** (Ikeda [52] and Eilenberg-Nakayama [55]) A necessary and sufficient condition for a ring A to be QF is that A be right selfinjective and right or left Noetherian. (Cf. Faith [66a, p. 182].)

***7.** (Connell [63]) Any group ring AG of a finite group G over a QF ring A is QF.

***8.** If A is a finite dimensional algebra over a field k, and if $U = \mathrm{Hom}_k(A,k)$, then A possesses a U-duality from the category of finitely generated right, to the category of finitely generated left, A-modules.

***9.** (Azumaya [59], Morita [58]) Over an Artinian commutative ring A, every indecomposable injective A-module is finitely generated. Conclude that A has a finitely generated injective cogenerator U, and hence that A possesses a U-duality.

***10.** (Morita [58]) Over an Artinian commutative ring A, there is a bijection between the equivalence class of naturally equivalent dualities mod-$A \rightsquigarrow$ mod-A, and the set of automorphisms of A of orders ≤ 2. If U and W are right A-modules defining a U- and W-duality respectively, then there exists an automorphism d of A of order ≤ 2, and an additive group isomorphism $f: U \to W$ such that $f(ua) = f(u)\,d(a)\ \forall\, u \in U,\ a \in A$. (Such a mapping is called a **semilinear transformation**, see Chapter 23, supra 23.34.)

Notes for Chapter 12

As indicated in the introduction, most of this chapter is taken from Bass [67, 68], who acknowledges Chase and Schanuel [67, p. 48] in the exposition ot the Morita theorems. The Picard group will make its next formal appearance in Chapter 32 (specifically 32.40 and theorems of Skolem-Noether-Rosenberg-Zelinsky). [However, as noted in 12.19.1, for any semihereditary domain A, $\mathrm{Pic}_A A$ is isomorphic to the ideal class group of A, so the Picard group also appears (informally) in Chapter 30.]

The fundamental results on a generator of mod-A come from Azumaya [51, 66], Auslander-Goldman [60], and Morita [58]. For the exercises on duality and QF-rings, consult Morita [58] and Azumaya [59]. (More detailed references will be given in Chapters 23 and 24.)

Eilenberg [60] and Watts [60] characterized right continuous functors as tensors 12.9.

References

Auslander-Goldman [60], Azumaya [51, 59], Bass [62, 67, 68], Connell [63], Eilenberg [60], Eilenberg-Nakayama [55], Faith [66a], Ikeda [52], Morita [58], Watts [60].

Chapter 13

Algebras over Fields

This chapter is a brief introduction into the structure of algebras, mostly finite dimensional, over any field k. The main contents are the Wedderburn theorems for a finite dimensional algebras A over an algebraically closed field k. If A has no nilpotent ideals $\neq 0$, then A is a finite product of total matrix algebras over k. In this case, the set $d(A)$ of degrees of the total matrix algebras is a complete set of invariants of A. Thus, 13.7 two finite dimensional semiprime algebras A and B over k are isomorphic if and only if $d(A) = d(B)$.

The Wedderburn factor theorem (13.18) determines the structure of an arbitrary finite dimensional algebra A over an algebraically closed or perfect field k as follows: If N is the maximal nilpotent ideal (thus, $N = \text{rad}\, A$), then there is a semisimple subalgebra S isomorphic to A/N such that $A = S + N$ and $S \cap N = 0$. Thus, every such algebra A is a direct sum over k of a semisimple and a nilpotent algebra-1. Essentially, this reduces the structure of A to that of nilpotent algebras-1, and to the effect on N of the k-linear transformations induced by multiplication of N by elements of S.

Included in this chapter are (1): Maschke's theorem 13.21 on the semisimplicity of finite group algebras kG over fields k of characteristic not dividing $|G|$; and (2) Wedderburn's theorem on the nilpotency of algebras A having a finite basis consisting of nilpotent elements. (A related theorem of Nagata and Higman[1], not included in the text, states that any nil algebra whose elements satisfy the identity $x^n = 0$ is nilpotent of index $\leq 2^n - 1$ when the characteristic is 0 or bigger than n.) The Skolem-Noether theorem, and the Brauer group of a field, are treated in exercises, and proved in Chapter 32 in the more general context of separable algebras.

The Regular Representations of an Algebra

If k is any commutative ring, and if A is an algebra over k, then the canonical homomorphism

$$\begin{cases} A \to \text{End}_k A \\ a \mapsto a_d \qquad\qquad (x)\, a_d = xa\ \forall x \in A \end{cases}$$

[1] Higgins's short proof of this theorem is given in Jacobson [64, Appendix C, p. 274].

is called the **right regular representation** of A. (Here, the algebra is regarded as a left vector space over k, and, in keeping with the convention stated in Chapter 3, endomorphisms are written on the side opposite scalars.) Dually, there is an anti-embedding $A \to \operatorname{End}_k A$ via $a \mapsto a_s$, where $(x)a_s = ax$, called the **left regular representation of** A. If A is a free module over k, say $A \approx k^m$, then A is said to have a **(free)** k**-basis.** Then the right regular representation embeds A into the k-algebra k_m of $m \times m$ row-finite matrices over k, since there is a canonical isomorphism $k_m \approx \operatorname{End}_k k^m$. This is the case whenever k is a field, since then A is a vector space over k. Then, when $\dim_k A = m < \infty$, the first regular representation embeds A into the matrix algebra k_m.

Scalar Extensions

If k is a subring of a commutative ring K, and if A is a k-algebra, then $A_K = A \otimes_k K$ is a K-algebra called the **scalar extension of** A **by** K. If A is flat over k, then

$$0 \to A \underset{k}{\otimes} k \to A \underset{k}{\otimes} K$$

is exact, so there is an embedding of $A \approx A \underset{k}{\otimes} k$ in A_K, justifying the term extension in this case. Furthermore, if K is flat over k, and if B is a k-submodule of A, then B_K is canonically isomorphic to a K-submodule of A_K.

13.1 Corollary. *If A and B are faithful algebras over k, and if A is a free k-module with basis $\{x_i \mid i \in I\}$, then every element $x \in A \otimes_k B$ has a unique representation*

$$x = \textstyle\sum_{i\in I} x_i \otimes b_i \qquad (\textit{finite sum})$$

with $b_i \in B\ \forall i \in I$. In this case the canonical map $a \mapsto a \otimes 1$ is an algebra monic $A \to A \underset{k}{\otimes} B$.

Proof. The first statement follows from 11.13. Second, $a \mapsto a \otimes 1$ is an algebra homomorphism. If $a = \sum_i x_i \alpha_i \in A$, and if $a \otimes 1 = 0$, then $\sum_i x_i \otimes \alpha_i 1 = 0$, so $\alpha_i 1 = 0$ in B, whence $\alpha_i = 0\ \forall i \in I$, and then $a = 0$. □

13.2 *Exercises.*

13.2.1 If A is a finite dimensional algebra over a (commutative) field k, then A is a finite dimensional vector space over k, so there exists a vector space basis $\{u_i\}_{i=1}^n$. Each $a \in A$ can be uniquely expressed in the form

$$a = \textstyle\sum_{i=1}^n \alpha_i u_i, \qquad (\alpha_i \in k,\ i = 1, \ldots, n).$$

In particular, there exist n^3 scalars $\gamma_{ijg} \in k$ such that

$$u_i u_j = \sum_{g=1}^{n} \gamma_{ijg} u_g, \qquad (i, j = 1, \ldots, n).$$

Let K be any field containing k, and consider a vector space B over K of dimension equal to n. Choosing a basis $\{w_i\}_{i=1}^{n}$ of B, show that B is a K-algebra under multiplication defined by

$$w_i w_j = \sum_{g=1}^{n} \gamma_{ijg} w_g \qquad (i, j = 1, \ldots, n).$$

(Since this depends only on the n^3 scalars γ_{ijg}, multiplication is associative in B since it is in A.) Show that $B \approx A \otimes_k K$ as K-algebras.

Generalize to algebras over a commutative field k.

13.2.2 Let $A \underset{k}{\otimes} K$ be a scalar extension of the k-algebra A, and let K be a flat k-module.

(a) If B is a (right, left) ideal of A, then B_K is a (right, left) ideal of A_K.

(b) If K is free over k, then B is a nilpotent (right, left) ideal of A, and B_K is a nilpotent ideal of A_K.

(c) A is semiprime if A_K is. The converse fails (see the following example.)

13.3 *Example*. Let $F = GF(2)$, the finite field consisting of 2 elements, let $A = F(x)$ be a transcendental field extension, and let $k = F(x^2)$. Then A is an algebra over k with a free basis 1, x, and x is a zero of the polynomial $f(X) = X^2 - \beta$, where $\beta = x^2 \in k$. Let K be a field extension of $k, \varphi : A \approx K$ an isomorphism with $\varphi \mid k = \text{id.}$, and $A_K = A \otimes_k K$. If we identify A with its canonical image under $A \to A_K$, then A_K has a basis 1, x over K. Let $\theta = \varphi(x)$. Then $x - \theta \cdot 1 \in A_K$, and

$$(x - \theta 1)^2 = x^2 - \theta^2 \cdot 1.$$

Since

$$\theta^2 = (\varphi(x))^2 = \varphi(x^2) = \varphi(\beta) = \beta$$

we have

$$f(x) = x^2 - \beta 1 = (x - \theta \cdot 1)^2 = 0$$

that is, $x - \theta 1$ is nilpotent of index 2. Since A_K is commutative, $(x - \theta 1)$ generates a nilpotent ideal of index 2. Thus, A_K is not semiprime, although A is a field. □

Algebras over Algebraically Closed Fields

If k is a commutative field, and if $p(x) = \sum_{i=0}^{n} \alpha_i x^i$ is a polynomial in x with coefficients in k, that is, if $p(x) \in k[x]$, then for each element $\beta \in k$, the element $\sum_{i=0}^{n} \alpha_i \beta^i$ lies in k, and is denoted by

$p(\beta)$; β is a zero of $p(x)$ in case $p(\beta) = 0$. The field k is ***algebraically closed*** when to each polynomial $p(x) \in k[x]$ there corresponds $\beta \in k$ such that $p(\beta) = 0$ (for example, the field $\mathbb{C}$ of complex numbers). In this case, by induction, there exist $\beta_0, \beta_1, \ldots, \beta_n \in k$ such that $p(x) = \beta_0 \prod_{i=1}^{n} (x - \beta_i)$ factors into linear factors in $k[x]$.

If D is an algebra over a field k, and if D is a (not necessarily commutative) field, then D is a **division algebra over** k.

13.4 Proposition. *If D is a finite dimensional division algebra over k, and if k is algebraically closed, then $D \approx k$ canonically.*

Proof. Let n denote the dimension of D over k. If $a \in D$, then $1, a, \ldots, a^n$ are linearly dependent over k, so there exist $\alpha_i \in k$ not all zero such that $\sum_{i=0}^{n} \alpha_i a^i = 0$. Thus, if L denotes the subfield of D generated by k and a, then the polynomial $p(x) = \sum_{i=0}^{n} \alpha_i x^i \in k[x]$ has a zero $a \in L$. Hence let $g(x)$ be the polynomial of least degree $t \leq n$ such that $g(a) = 0$ in L. Since k is algebraically closed, $g(x) = (x - \alpha) \times q(x)$, with $\alpha \in k$, $q(x) \in k[x]$. Then in L, $g(a) = (a - \alpha)\, q(a) = 0$, and $q(a) \neq 0$ by the definition of $g(x)$. Since D is a field, necessarily, $(a - \alpha) 1 = 0$; that is, $a = \alpha 1 \in k1$. This shows that D coincides with the field $k \cdot 1$, which is canonically isomorphic to k under the correspondence $\beta 1 \to \beta$ defined $\forall \beta \in k$. □

13.5 Corollary. *Let R be any finite dimensional algebra over an algebraically closed field k. If R has no nilpotent ideals, then R is a direct product of finitely many total matrix algebras over k.*

Proof. Since R is a semiprime Artinian ring, by the Wedderburn-Artin theorem (8.8), R is a ring direct product $R_1 \oplus \cdots \oplus R_t$, where each R_i is a simple Artinian ring. Since we may assume that R_i is an ideal of R, it follows that R_i is a subalgebra of R, and hence an algebra over k, $i = 1, \ldots, t$. Thus, we may assume that R is itself simple. Then $R \approx D_n$ for some division algebra D over k. Since R, and hence D, is finite dimensional, the proposition implies that $D \approx k$. Since R is therefore isomorphic to the total matrix algebra k_n, the proof is complete. □

13.6 Corollary. *If A is a finite dimensional and simple algebra over a field k, and if K is an algebraically closed field containing k, then the algebra $A \underset{k}{\otimes} K$ is isomorphic to a total matrix algebra K_n.*

Proof. By 13.5, it suffices to prove that A_K is simple. (Exercise.) □

Any simple Artinian ring R is isomorphic to D_n, where D is a field, and D and n are uniquely determined (8.9). In a direct product decomposition $R = \prod_{i=1}^{t} R_i$ of a finite dimensional semisimple algebra over a fixed algebraically closed field k, where each algebra R_i is a simple matrix algebra k_{n_i} of degree n_i, the algebras $R_1, \ldots, R_t$ are also uniquely determined, and so is the set $d(R) = \{n_1, \ldots, n_t\}$.

13.7 Theorem (Wedderburn). *The set $d(R)$ is a complete set of invariants of a finite dimensional semisimple algebra R over an algebraically closed field k.* $\square$

Expressed otherwise, if S is another such algebra, then $d(S) = d(R)$ if and only if $S \approx R$.

Simple Algebras

An algebra A over k is simple in case A is a simple ring. The center of a simple ring is a field. (Proof?)

13.8 Proposition. *Let k be a field, let A and B be algebras over k, and A central. If A and B are simple, then $A \underset{k}{\otimes} B$ is simple, and*

$$\text{center } A \underset{k}{\otimes} B = 1 \otimes \text{center } B \approx \text{center } B.$$

Proof. Let I be a nonzero ideal of $R = A \otimes B$ (all tensors over k). If $x \in R$, then

$$x = a_1 \otimes b_1 + \cdots + a_n \otimes b_n \tag{1}$$

and $a_i \in A$, $b_j \in B$. If $x \neq 0$, then the length of x is defined to be the smallest number of nonzero a_i in any representation (1) of x. Choose x to be an element of shortest length. If $a', a'' \in A$, then

$$(a' \otimes 1)\, x\, (a'' \otimes 1) = a' a_1 a'' \otimes b_1 + \cdots + a' a_n a'' \otimes b_n \tag{2}$$

lies in I. Since A is simple, $A = A a_1 A$, so 1 is a finite sum of elements of the form $a'' a_1 a'$. Hence, adding up elements of the form (2), we have

$$0 \neq x_1 = 1 \otimes b_1 + \cdots + c_n \otimes b_n \in I. \tag{3}$$

Then, for an arbitrary $a \in A$, x_1 lies in I and has shorter length than x.

$$(a \otimes 1) x_1 - x_1 (a \otimes 1) = (c_2 a - a c_2) \otimes b_2 + \cdots + (c_n a - a c_n) \otimes b_n.$$

Thus, by the choice of x, $(a \otimes 1) x_1 - x_1 (a \otimes 1) = 0$, whence $c_i a = a c_i$ for each i and every $a \in A$. Thus, $c_i \in \text{center } A = k$, so that $c_i \otimes b_i = 1 \otimes c_i b_i$, $i = 2, \ldots, n$, and from (3) we get

$$x_1 = 1 \otimes (b_1 + c_2 b_2 + \cdots + c_n b_n). \tag{4}$$

Since $x_1 \neq 0$, $y = b_1 + c_2 b_2 + \cdots + c_n b_n \neq 0$. Since $x_1 = 1 \otimes y \in I$, and $y \neq 0$, by symmetry with (3) we get $y \otimes 1 \in I$, so $A \otimes B = I$. This proves simplicity.

Next let $u = \sum_{i=1}^{n} a_i \otimes b_i \in \text{center } A \otimes B$, where $u \neq 0$, and the b_i are linearly independent over k. By the same reasoning as just stated, $a_i \in k = \text{center } A$, and $u = 1 \otimes y$, where $y = c_1 b_1 + \cdots + c_n b_n$. If $b \in B$, then $1 \otimes (by - yb) = 0$, so $by = yb$, whence $y \in \text{center } B$. Thus,

$$\text{center } A \underset{k}{\otimes} B = \{1 \otimes y \mid y \in \text{center } B\}$$

as asserted. □

13.9 Corollary. *Every scalar extension $A \otimes_k K$ of a central simple algebra A over k by a field K is central simple.* □

Burnside's Theorem

We now cite a theorem of Burnside which can be proved using 13.4, and the Wedderburn-Artin theorems and (to add a new twist to an old theorem) a theorem of Morita.

13.10 A Theorem. *Let V be a finite dimensional vector space over an algebraically closed field k, let $L = \text{End}_k V$, and let R be a subalgebra of the canonical k-algebra L. Then, V is a canonical R-module which is simple iff $L = R$.*

Proof. Every R-endomorphism of V is a k-endomorphism, and hence belongs to L, that is, $K = \text{End } V_R \subseteq L$. Since $\dim_k L = n^2$, where $n = \dim_k V$, and since K is a field by Schur's lemma, then 13.4 implies that $K = k$. If V is a simple R-module, then faithfulness of V implies that R has no nilpotent ideal $\neq 0$, and that R is a prime ring. Thus, the Wedderburn-Artin theorem for finite dimensional algebras implies that R is simple. Then, every simple right R-module embeds in R, and hence V is a generator of mod-R (7.4.1). Then, Morita's theorem 7.3 implies that $R \approx \text{End}_k V = L$ canonically, so that $R = L$. Conversely, V is a simple right L-module (Proof?). □

13.10B *Exercise.*

13.10B.1 *Let A and B be algebras over a field k. If I is an ideal of A, then the image I' of the canonical map $I \otimes_k B \to A \otimes_k B$ is an ideal of $A \otimes_k B$. If A is central, and B simple, and if J is an ideal of $A \otimes_k B$, then $J = I'$ for some ideal I of A.*

13.10.2 *In the category of commutative rings, $R \otimes_{\mathbb{Z}} S$ is a coproduct $R \coprod S$.*

13.10B.3 *If k is a commutative ring and if A and B are k-algebras, then*

$$A_n \underset{k}{\otimes} B_m \approx (A \underset{k}{\otimes} B)_{nm}.$$

13.10B.4 *Prove* 13.10B.3 *as a special case of the more general k-algebra isomorphism*

$$\varphi : (\mathrm{End}_A M) \otimes_k' (\mathrm{End}_B N) \to \mathrm{End}_{A \underset{k}{\otimes} B}(M \otimes_k N)$$

where

$$\varphi(f \otimes' g) = f \otimes g \quad (\textit{tensor product of maps})$$

$\forall f \in \mathrm{End}_k M$, $g \in \mathrm{End}_k N$, *and M and N are projective A and B-modules, respectively*. (Cf. 32.20.)

Separable Field Extensions

Let L be an algebra over a field k. An element $a \in L$ is **algebraic** over k provided that there exists a nonzero polynomial $f(x) = \sum_{i=0}^{n} \alpha_i x^i \in k[x]$, where $\alpha_i \in k$, $i = 0, \ldots, n$, such that

$$f(a) = \sum_{i=0}^{n} \alpha_i a^i = 0.$$

The set of all $g(x) \in k[x]$ *such that* $g(a) = 0$ *is an ideal* $Z(a)$ *of* $k[x]$. Since $k[x]$ is a principal ideal ring, there exists a polynomial $f_a(x)$ such that $Z(a) = f_a(x)\, k[x]$. For uniqueness, choose $f_a(x)$ such that if degree $f_a(x) = n$, then the coefficient of x^n is 1. Then $f_a(x)$ is **monic**, is called the **minimal polynomial of** a **over** k.

A polynomial $p(x) \in k[x]$ is **irreducible** if any factorization $p(x) = a(x)\, b(x)$ into a product of polynomials $a(x), b(x) \in k[x]$ implies that $a(x)$ or $b(x)$ is a constant, that is, an element of k. The minimal polynomial $f_a(x)$ of an element $a \in k$ is irreducible.

The algebra L/k is **algebraic** if every element of L is algebraic over k. One easily sees that if $\dim_k L$ is finite, then L/k is algebraic, for if $a \in L$, then $1, a, \ldots, a^n$ are linearly dependent over k for any $n \geq \dim_k L$, so $\sum_{i=0}^{n} \alpha_i a^i = 0$ with $\alpha_i \in k$ not all zero, and therefore $\sum_{i=0}^{n} \alpha_i x^i \in Z(a)$. The algebra L/k is **transcendental** if it is not algebraic.

If $k[x]$ denotes the polynomial ring over k, then a element $a \in L$ is not algebraic. (= transcendental) over k iff the canonical k-algebra homomorphism 5.74.1 $k[x] \to k[a[$ sending $x \mapsto a$ is an isomorphism.

13.11 *Exercises.*

13.11.1 Let $a \in k$. Then $h(x) \in Z(a)$ generates $Z(a)$ if and only if $h(x)$ is a polynomial of least degree in $Z(a)$.

13.11.2 If $h(x) \in k[x]$, then $h(x)$ is irreducible if and only if $h(x)$ generates a maximal ideal of $k[x]$. Then, $k[x]/(h(x))$ is a field.

13.11.3 The minimal polynomial $f_a(x)$ of an element $a \in k$ is irreducible. Let L denote the field $k[x]/(f_a(x))$. The canonical map $k[x] \to L$ maps $k \subseteq k[x]$ onto a field $\bar{k} \subseteq L$, and maps each $g(x) \in k[x]$ onto $\bar{g}(\bar{x})$, and thus $\bar{f}_a(\bar{x}) = 0$. Conclude by induction that there exists a field E containing k in which $f_a(x)$ factors into linear factors. Such a field of least degree is called a **splitting field** of $f_a(x)$ over k. Show that any two splitting fields of $f_a(x)$ over k are isomorphic under a mapping which induces the identity mapping of k.

13.11.4 Let L/k be a field extension, let $a \in L$, and let $k(a)$ (resp. $k[a]$) be the field (resp. ring) generated by a and k. The following statements are equivalent:

(a) a is algebraic over k.

(b) $\dim_k k(a) < \infty$.

(c) $k(a) = k[a]$.

*13.11.5 (Cf. Hilbert Nullstellensatz, 30.49). Generalize 4 as follows: If $a_1, \ldots, a_n \in L$, then the subring $k[a_1, \ldots, a_n]$ is a field if and only if L/k is algebraic.

*13.11.6 (Amitsur [56] If D is a transcendental division algebra over k, then $\dim_k D \geq |k|$ [Hint: Consider the linear independence over k of the elements $(t-a)^{-1}$, where $a \in k$.]

*13.11.7 (Amitsur [56]) Let $k(X)$ denote the division subalgebra and let $k[X]$ denote the subalgebra, generated by a subset X of a division algebra D over a field k, and assume that k is an uncountable field of cardinality $> |X|$. Then, if $D = k[X]$, then D is an algebraic algebra. (Cf. 13.11.5.)

13.11.8 (a) If D is any central division algebra over k, then the polynomial ring $D[x]$ is a right and left PID such that every ideal has the form $(p(x))$ for a central polynomial $p(x)$, that is, for $p(x) \in k[x]$. (Cf. Jacobson [42].)

(b) Any right and left PID has the property that any factor ring R/I, for an ideal $I \neq 0$, R, is QF. (Cf. Exercise 6, Chapter 12.) In particular, R/I is (right and left) Artinian. Conclude that every prime ideal is maximal. If K is the right quotient field of R (existence?), and if P is a maximal ideal of R, then there is a simple ring $RS^{-1} = \{as^{-1} \mid a \in R, s \in S = R - P\}$ contained in K, called the localization of R at P (also denoted R_P.)

*13.11.9 (cf. Jacobson [64], pp. 241 and 248.) Let D be a central division algebra over a field k. Prove the equivalence of the conditions (a)—(d):

(a) If $L = k(x)$ is the quotient field of the polynomial algebra $k[x]$, then $D \otimes_k k(x)$ is the right quotient field of $D[x] \approx D \otimes_k k[x]$.

(b) Every $d(x) \in D[x]$ is the factor of a central polynomial. [Hint: write $d(x)^{-1} = f(x)\, a(x)^{-1}$, with $f(x) \in D[x]$ and $a(x) \in k[x]$. This can be done if $D \otimes_k k(x)$ is the right quotient field of $D[x]$.]

(c) Every nonzero right ideal of $D[x]$ contains an ideal generated by a nonzero polynomial $a(x) \in k[x]$.

(d) $D[x]$ has no faithful simple modules (in the terminology of Chapter 19 supra 19.22, $D[x]$ is not a primitive ring).

*13.11.10 (Amitsur [56]) Let D be a central algebraic division algebra over k and assume one of the following: (A) $[D:k] < \infty$; (B) D/k is locally finite dimensional in the sense that any finite subset generates a finite dimensional subalgebra; (C) For any integer $n > 0$, the $n \times n$ matrix algebra D_n is algebraic over k; (D) k is uncountable. Then, D satisfies the equivalent conditions (a)—(d) of 13.11.9.

13.11.11 If D is a transcendental central division algebra over k, then the polynomial ring $D[x]$ is a primitive ring; thus $D \underset{k}{\otimes} k(x)$ is a (right and left) PID, and, moreover a simple ring not a field. (Indeed, these properties hold whenever D is a division algebra and $D[x]$ is primitive.)

*13.11.12 (Amitsur, l.c.) If A is an algebra over a field k, and if $\dim_k A < |k|$, then rad A is an algebraic subalgebra, in fact, rad A is a nil ideal.

*13.11.13 (Amitsur, l.c.) The radical of a finitely generated algebra A over an uncountable field k is a nil ideal.

*13.11.14 Let k be an algebraically closed field, and R the polynomial ring in n algebraically independent variables $x_1, \ldots, x_n$. For any ideal I of R, define the zero set $Z(I)$ in k^n in the obvious way, and dually for any subset X of k^n, define the ideal $I(X)$ to be the largest ideal of R such that $Z(I(X)) = X$. The **Hilbert Nullstellensatz** states, for any ideal $I \neq R$, that there exists an integer m (depending on I) such that $f^m \in I \; \forall f \in I(Z(I))$. Prove this for the $n = 1$ case.

13.12 Exercise and Definition. An **algebraic closure** of a field k is a field L containing k satisfying the equivalent conditions:

13.12.1 L is maximal in the ordered set Alg(k) of all algebraic field extensions of k.

13.12.2 L is minimal in the ordered set of all algebraically closed fields containing k.

13.12.3 L is algebraic over k, and L is algebraically closed.

13.12.4 L is algebraic over k, and every non-constant $f(x) \in k[x]$ has a zero in L.

Prove: If L and K are two algebraic closures of k, then there is an isomorphism $L \approx K$ which induces 1_k. □

Let $\bar{k}$ denote the algebraic closure of k. Then every polynomial $f(x) \in k[x]$ factors linearly in $\bar{k}$,

$$f(x) = \alpha_0 \prod_{i=1}^{n} (x - \alpha_i)$$

where $\alpha_i \in \bar{k}$, $i = 0, \ldots, n$. The polynomial $f(x)$ is **separable** over k provided that the elements $\alpha_1, \alpha_2, \ldots, \alpha_n$ are distinct. Since 13.12 any two algebraic closures of k are isomorphic over k, this definition does not depend upon $\bar{k}$. If C/k is a field extension, an element $a \in C$ is **separable** if a is algebraic over k and the minimal polynomial of a over k is separable over k. The field extension C/k is **separable** provided that every element of C is separable over k. **Inseparable** means not separable. To distinguish from separability of algebras, we designate the concept just defined to be **field-separability**, a modification which will be abandoned when the appropriate theorem (13.16 and 32.34) tells us it can be.

Field Separability Criteria

There is a simple test for separability. Ideals A and B of a ring R are **relatively prime**, or **comaximal**, in case $R = A + B$. (Cf. 3.13.5.)

13.13 Separability Test. *Let C/k be an algebraic field extension.*

13.13.1 *An element $a \in C$ is separable over k if and only if the minimal polynomial $f_a(x) = \sum_{i=0}^{n} \alpha_i x^i$ and its derivative $f'_a(x) = \sum_{i=1}^{n} i\alpha_i x^{i-1}$ are relatively prime in $k[x]$.*

13.13.2 *An element a is inseparable if and only if k has prime characteristic p and $f_a(x) = g(x^p)$ for some polynomial $g(x) \in k[x]$.* □

13.14 *Exercises.*

13.14.1 Let $C \supseteq k$ be fields, and C/k finite dimensional. Then $C = k(a)$ for a single element $a \in C$ if and only if there are only finitely many fields D between C and k, $C \supseteq D \supseteq k$. Conclude that if $C = k(a)$ is finite dimensional, and if D is a field between C and k, then $D = k(b)$ for some $b \in C$.

*13.14.2 If C/k is a finite dimensional field extension, and if C/k is separable, then $C = k(a)$ for some $a \in C$.

13.15 Proposition. *Let k be a field, and let $C = k(a)$ be an algebraic field extension generated by a single element a. The following conditions are equivalent.*

(a) *C/k is field-separable.*

(b) *The element a is separable over k.*

(c) *The tensor algebra $C \otimes_k L$ has no nilpotent elements $\neq 0$ for any field extension $L \supseteq k$.*

(d) *For any field* $L \supseteq k$, $C \otimes_k L$ *is a (ring) product of finitely many copies of* L.

The statement (i) *when* L *is any algebraic closed field* $\supseteq k$, $i = c$ *or* d.

When this is so, then D/k *is field-separable for any field* $C \supseteq D \supseteq k$.

Proof. (a) $\Rightarrow$ (b) is trivial. (b) $\Rightarrow$ (c). Since C/k is algebraic, then $C = \sum_{i=0}^{n-1} k a^i$ is a direct sum, where n is the degree of the minimal polynomial $f_a(x)$ of a. Consequently, $\{a^i \otimes 1 \mid i = 0, \ldots, n-1\}$ is a basis of the L-algebra $C \otimes_k L$; that is, there exists an exact sequence

$$0 \to f_a(x) \cdot L[x] \to L[x] \to C \otimes_k L \to 0$$

under which $x \mapsto a \otimes 1$. If $y \in C \otimes_k L$, then $y = G(a)$, where $G[x] \in L[x]$ is irreducible. If y is nilpotent, say $y^t = 0$, then $G(x)^t \in f_a(x) \cdot L$. Since $f_a(x)$ is separable, $f_a(x)$ does not divide the square of any irreducible polynomial over L. Therefore, $f_a(x)$ divides $G(x)$, and $y = G(a) = 0$. Thus, any nilpotent element is zero.

(c) $\Rightarrow$ (a). Deny (a). If $b \in C$ is not separable, then k has characteristic $p \neq 0$, and $f_b(x) = g(x^p)$ for some polynomial $g(x) \in k[x]$ (see 13.13). If L is any field extension containing k and the p-th roots of the coefficients of $g(x)$ (thus, L may be the algebraic closure of k), then $\alpha = \beta^p$ for some $\beta \in L$, whenever α is a coefficient of $g(x)$. This implies that $f_b(x) = (h(x))^p$ for $h(x) \in L[x]$. Since $\deg h < \deg f_b$, then $h(b) \in C \otimes_k L$ is nonzero, and

$$f_b(b) = h(b)^p = 0.$$

Hence, falsity of (a) implies that of (c).

(c) $\Leftrightarrow$ (d). Since C is finite dimensional over k, the algebra $A = C \otimes_k L$ is finite dimensional over L. Therefore, since A is Artinian, the Wedderburn-Artin theorem 8.8 applies; namely, A is a product of finitely many fields if and only if A has no nilpotent ideals. Since A is commutative, this is equivalent to the demand that A have no nilpotent elements.

This proves the equivalence of (a) — (d). The equivalence of (c) or (d) to the corresponding statement with L any algebraically closed field $\supseteq k$ follows, since any field K containing k embeds in such a field. Thus, if (c) holds for $L = \overline{K}$, then (c) holds for $L = K$, since $C \otimes_k K \subseteq C \otimes_k \overline{K}$ has no nilpotent elements $\neq 0$. Similarly, if (d) holds for $L = \overline{K}$, then (c) holds for $L = \overline{K}$ and K, whence (d) holds for $L = K$ by the proof of (c) $\Rightarrow$ (d). We now prove the last statement. Since

$$0 \to D \otimes_k L \to C \otimes_k L$$

is exact, $D \otimes_k L \subseteq C \otimes_k L$; hence given (b), $C \otimes_k L$, whence $D \otimes_k L$, has no nilpotent elements. Since $D = k(b)$ for some $b \in D$ (see 13.14) then D/k is separable by (c) $\Rightarrow$ (a). ▯

13.16 Separability Criterion. *Let A be a finite dimensional algebra over a field k. Then, A is said to be separable over k, or, A/k is separable, provided that the following conditions, which are equivalent, hold:*

13.16.1 *$L \underset{k}{\otimes} A$ is semisimple for all field extensions L of k.*

13.16.2 *For some algebraically closed field $L \supseteq k$, the L-algebra $L \underset{k}{\otimes} A$ is a product of total matrix algebras over L.*

13.16.3 *A is semisimple, and* center *A is a product of finitely many field-separable extensions of k.*

In this case, 13.16.2 *holds for $L = \bar{k}$, the algebraic closure of k, and also for some finite separable field extension L contained in $\bar{k}$.*

Proof. Exercise, or see the proof of 32.34. ▯

Note that 13.16.3 implies that any separable finite dimensional field extension A/k is a separable algebra, and conversely.

A field is said to be **perfect** if every (finite) algebraic extension is separable.

13.17 *Exercise.* A field k is perfect in case iff either (a) k has characteristic 0, or (b) k has characteristic $p > 0$ and k contains all p-th roots of elements of k, or equivalently, every element of k is a p-th power. (This holds for any algebraic extension of a finite field.)

13.18 Wedderburn Factor Theorem. *If A is an algebra of finite dimension over a field k, and if $A/\text{rad}\, A$ is separable, then there is a semisimple subalgebra S of A canonically isomorphic to $A/\text{rad}\, A$ such that $A = S + \text{rad}\, A$ and $S \cap \text{rad}\, A = 0$. The algebra S is the* ***semisimple factor,*** *or* ***part,*** *of A.*

Proof. The proof can be resolved into two possibilities for $N = \text{rad}\, A$, namely, (1) $N^2 \neq 0$, or (2) $N^2 = 0$. In any case, the truth of the theorem can be assumed for all algebras of dimension $< n = \dim_k A$, and hence in case (1) that there exists a subalgebra S_1 containing N^2 such that

$$\bar{A} = \bar{S}_1 + \bar{N} \;\&\; \bar{S}_1 \cap \bar{N} = 0$$

where the bars denote images under the canonical map $A \to \bar{A} = A/N^2$. Then, $A = S_1 + N$, and $S_1 \cap N = N^2$. Since the radical of S_1 is N^2, and since $\dim_k S_1 < n = \dim_k A$, then by the induction hypothesis, there is a subalgebra S of S_1 such that $S_1 = S + N^2$, and $S \cap N^2 = 0$. But, then $A = S + N$, and $S \cap N = 0$ is the desired decomposition of A, since $S \approx A/N$ is semisimple.

We prove (2) *only in the case A/N is a product of total matrix algebras over k.* Thus will the theorem hold for the scalar extension $\bar{A}_P = \bar{A} \underset{k}{\otimes} P$, where P is a finite separable extension of k contained in the algebraic closure $\bar{k}$ (and for $P = \bar{k}$), for then, by (13.16), A_P is a product of total matrix algebras over P. The reduction to $\bar{A}$ of the desired decomposition for $\bar{A}_P$ is then left for an exercise, or the reader may follow Jacobson [42, p. 117] (as we have been doing).

Hence, suppose that $\bar{A} = \prod_{i=1}^t M_{n_i}(k)$ is a product of total matrix algebras over k of degrees n_i, $i = 1, \ldots, t$. It can be shown that orthogonal idempotents of $\bar{A} = A/\text{rad}\, A$ can be "lifted" to orthogonal idempotents of A, in particular, there exist orthogonal idempotents $\{e_i\}_{i=1}^t$ of A such that $\bar{e}_i$ is the identity of $M_{n_i}(k)$, $i = 1, \ldots, t$ (e. g. 18.21). Then $B_i = e_i A e_i$ is an algebra over k with identity element e_i, and $\text{rad}\, B_i = e_i N e_i$, and $B_i/e_i N e_i = M_{n_i}(k)$, $i = 1, \ldots, t$. If $t = 1$, then matrix units of $M_n(k)$ can be "lifted" to matrix units $\{f_{ij}\}_{i,j=1}^n$ of $A = B_1$, and then $k_n = \sum_{i,j=1}^n f_{ij} k$ is a subalgebra of A which is the semisimple part, since clearly $k_n \cap N = 0$, where $k = k1$ is the scalar subalgebra. For the case $t > 1$, if S_i denotes the semisimple part of B_i, $i = 1, \ldots, t$, then the sum $S = S_1 + \cdots + S_t$ has dimension $m = \sum_{i=1}^t n_i$ over k, and $S \cap N = 0$. Since $\dim_k A/N = m$, it follows that $A = S + N$, and so S is the semisimple part of A. □

13.19 Corollary. *Over a perfect field k, every finite dimensional algebra A has a semisimple part.* □

Exercises on Galois Theory

An algebraic extension field L/k is **normal** if every $f(x) \in k[x]$ which has one zero in L factors into linear factors in $L[x]$. This may be paraphrased: every polynomial over k which has one zero in L has every zero in L. Thus, if L and the set $Z(f)$ zeros of $f(x)$ are considered as canonical subsets of $\bar{k}$, then either $Z(f) \cap L = \varnothing$, or else $Z(f) \subseteq L$. A field extension L/k is **finite** if $\dim_k L$ is finite.

13.20 *Exercise.*

13.20.1 Every finite field extension L/k is contained in a finite normal extension $N(L/k)$ of least degree, and any two such are isomorphic by a map which induces 1_k. Then, $N(L/k)$ is the **normal closure** of L/k. A finite extension L/k is normal iff L is the splitting field of a polynomial over k.

13.20.2 The normal closure of a finite separable extension is a finite separable extension.

13.20.3 An extension L/k is finite, separable and normal if and only if L/k is the splitting field of a separable polynomial over k.

13.20.4 An extension L/k is **Galois** if there is a group G of automorphisms of L such that

$$\text{fix}\, G = \{x \in L \mid g(x) = x \,\forall g \in G\}$$

equals k. Then $\dim_k L$ is finite if and only if $|G|$ is finite, in which case $\dim_k L = |G|$.

13.20.5 Any finite extension of a finite field is Galois.

13.20.6 *Fundamental Theorem of Galois.* A finite extension L/k is Galois, say $k = \text{fix}\, G$, if and only if L/k is finite separable and normal. Then, the mapping

$$\begin{cases} \text{subgroups of } G \to \text{subfields of } L \supseteq k \\ S \mapsto \text{fix}\, S \\ G(L/M) \leftarrow\!\shortmid M \end{cases}$$

is a bijection. Then, $G(L/M)$ is called the **Galois group** of L over M, and M is the corresponding **Galois subfield.** Moreover, S is a normal subgroup of G if and only if fix S is a normal extension of k. In the latter case, fix S/k is Galois, every $g \in G(L/k)$ induces an automorphism $\bar{q} \in G(\text{fix}\, S/k)$, and there is a canonical exact sequence

$$(1) \to G(L/\text{fix}\, S) \to G(L/k) \to G(\text{fix}\, S/k) \to (1).$$

13.20.7 If G is a finite group of automorphisms of L, then $G = G(L/\text{fix}\, G)$.

13.20.8 If L/k is a finite separable extension, then $L = k[a]$, for some $a \in L$. (Hint: by the Fundamental theorem there are finitely many intermediate extensions M such that $L \supseteq M \supseteq k$.)

The Trace of a Matrix

Let $\{e_{ij}\}_{i,j=1}^n$ be matrix units in a total matrix algebra k_n over a field k. Then every element $a \in k_n$ has a unique representation

$$a = \sum_{i,j=1}^{n} \alpha_{ij} e_{ij} \to (\alpha_{ij})$$

with $\alpha_{ij} \in k$, $i, j = 1, \ldots, n$. The trace of a matrix a is defined to be the sum $Tr(a)$ of the diagonal elements in its matrix representation

$$Tr(a) = \sum_{i=1}^{n} \alpha_{ii} \tag{1}$$

Then:

$$Tr(a+b) = Tr(a) + Tr(b) \tag{2}$$

and

$$Tr(ab) = Tr(ba) \qquad \forall a, b \in k_n. \tag{3}$$

(These formulas hold, in fact, for any commutative ring k.)

$Tr(a)$ is independent of the choice of matrix units in k_n; that is, if also $\{f_{ij}\ i, j = 1, \ldots, n\}$ are matrix units in k_n, and if

$$a = \textstyle\sum_{i,j=1}^n \beta_{ij} f_{ij} \qquad (\beta_{ij} \in k)$$

then $\sum_{i=1}^n \beta_{ii} = \sum_{i=1}^n \alpha_{ii}$. There are many ways to see this. One way is to show that if

$$f(x) = x^t - \alpha_1 x^{t-1} - \cdots - \alpha_t$$

is the minimal polynomial of a over k, then $Tr(a) = \alpha_1$. (The element a satisfies some polynomial since $1, a, a^2, \ldots, a^n, \ldots$ generate a vector subspace of finite dimension less than n^2.) In particular, if a is nilpotent, say $a^n = 0$, then the minimal polynomial $f(x)$ of a divides the polynomial x^t, and it follows that $f(x) = x^t$, and so $Tr(a) = 0$.

Another way to prove that the trace is independent of the choice of the matrix units is to show that there is a unit $y \in k_n$ such that $y e_{ij} y^{-1} = f_{ij}$, $i, j = 1, \ldots, n$. (This is done in Chapters 21 and 22.) Then:

$$a = \sum_{i,j=1}^{n} \alpha_{ij} e_{ij} = \sum_{i,j=1}^{n} \beta_{ij} f_{ij} = y \left(\sum_{i,j=1}^{n} \beta_{ij} e_{ij} \right) y^{-1}$$

and the fact that $Tr(ay) = Tr(ya)$, that is, $Tr(y^{-1}ay) = Tr(a)$, implies that $\sum_{i=1}^n \alpha_{ii} = \sum_{i=1}^n \beta_{ii}$.

Maschke's Theorem

Let G be a finite group of order n, and consider the left regular representation of the group algebra

$$kG \to \mathrm{End}_k kG \approx k_n$$

$$a \mapsto a_s \leftrightarrow (\alpha_{ij}) = \bar{a}$$

Write $G = \{g_1 = 1, g_2, \ldots, g_n\}$. Then, $\bar{a} = (\alpha_{ij})$ is the matrix such that

$$a g_j = \sum_{i=1}^{n} \alpha_{ij} g_i \qquad (j = 1, \ldots, n)$$

Then, $\bar{g}_i$ is simply a **permutation matrix**, $i = 1, \ldots, n$, that is, *a matrix such that* 1 *appears in each row once and only once, the rest of the entries in that row are zero, and no column has more than a single* 1. This corresponds to the following facts: (i) $g_i g_j$ is a basis element (that is, an element of G), and (ii) $g_i G = G$, $i, j = 1, \ldots, n$. If the matrix $\bar{g}_i$ has 1 on its diagonal, say in the (p, p)-position, $g_i g_p = g_p$, and then necessarily $g_i = 1$, that is, $i = 1$; consequently, $Tr(\bar{g}_i) = 0$, $i > 1$, and $Tr(\bar{g}_1) = n$. This simple fact on matrices will be used in the proof of Maschke's theorem which originated in [98].

13.21 Theorem (Maschke) *Let $A = kG$ be the group algebra of a finite group G of order n over a field k. Then A is a semisimple algebra if and only if the characteristic π of k does not divide n.*

Proof. First assume that π divides n, and let $h = \sum_{i=1}^{n} g_i$. Since $g_i h = h g_i = h$, $i = 1, \ldots, n$, we deduce two facts: (a) h commutes with *every* element of A; (b) $h^2 = nh$. Since π divides n, (b) implies that $h^2 = 0$. But (a) implies that $hA = Ah$, so $(hA)^2 = h^2 A = 0$, and hA is therefore a nilpotent ideal of A. Since A is a finite dimensional algebra over k, we conclude from the Wedderburn-Artin theorem (8.8) that A is not semisimple.

Conversely, if A is not semisimple, then by the Wedderburn-Artin theorem, A has a nonzero nilpotent ideal N. Let $x = \sum_{i=1}^{n} g_i \alpha_i$ be a nonzero element of N. Then $\gamma = \alpha_i \neq 0$ for some i, and then $y = x\gamma^{-1} g_i^{-1}$ is an element of N of the form

$$y = 1 + g_2 \beta_2 + \cdots + g_n \beta_n$$

with $\beta_i \in K$, $i = 2, \ldots, n$. Since N is nilpotent, so is y. Already we have determined that any nilpotent matrix has zero trace, so $Tr(y) = 0$. But, since the trace function is additive,

$$Tr(y) = Tr(1) + Tr(g_2)\beta_2 + \cdots + Tr(g_n)\beta_n.$$

By the remarks preceding the theorem we know that

$$Tr(g_2) = \cdots = Tr(g_n) = 0,$$

and $Tr(1) = n$. Thus, $Tr(y) = n = 0$, so π divides n, completing the proof. □

Maschke's theorem implies that for the field $\mathbb{C}$ of complex numbers (or any field of characteristic zero), the group algebra $\mathbb{C}G$ is semisimple. The question of which semisimple algebras over $\mathbb{C}$ are group algebras is unsolved, and appears to be extremely difficult (see Brauer's discussion [63, pp. 135–137]).

The proof just given employs the Wedderburn-Artin theorem and some matrix theory. It is an example of what happens in "systems" mathematics: Build up a mathematical theory (in our case the Wedderburn-Artin theorem), and then consistently use it to solve related problems. This is a very powerful principle in mathematics, as many questions have natural formulations within a system that would not be formulated otherwise. In many ways it is an efficient principle inasmuch as a wide class of problems may yield to a system with little or no extra effort. But such a principle also has deficiencies, which may be listed: (1) Every problem has its own axiom structure (the hypotheses), which may lead to proofs that are markedly simpler than those derived from a system; (2) attempts to solve the problem from first principles may be discouraging, resulting in a coarsening of mathematical intuition. The next theorem, or its proof, provides a fundamental proof of Maschke's theorem. This remarkably different proof is instructive.

We first introduce a lemma on the trace function of a finite group, and then the concept of a split-split algebra.

13.22 Lemma and Definition. *Let k be a commutative ring, let G be a finite group, and let E, A, B, A_1, and B_1 be kG-modules. The mapping*

$$\begin{cases} Tr_G : B \to E \\ \qquad a \mapsto \sum_{g \in G} g(a) \end{cases}$$

is a kG-map *such that*

$$(1) \qquad h\, Tr_G(x) = Tr_G(hx) \qquad \forall x \in E,\ h \in G.$$

Furthermore,

$$(2) \qquad Tr_G(f_1 f f_2) = f_1 Tr_G(f) f_2$$

for any k-map f, *and* $k(G)$-maps f_1 *and* f_2:

$$A_1 \xrightarrow{f_2} A \xrightarrow{f} B \xrightarrow{f_1} B_1 .$$

Proof. (1) is obvious, since $hG = Gh = G$ for all $h \in G$; that is, multiplication by h permutes the elements of G. (2) $\mathrm{Hom}_k(A, B)$ is an evident kG-module, so (1) is defined. Moreover

$$\begin{aligned} Tr_G(f_1 f f_2) &= \sum_{g \in G} g(f_1 f f_2) \\ &= \sum_{g \in G} (g f_1)(g f)(g f_2) \\ &= \sum_{g \in G} f_1 (g f) f_2 = f_1 Tr_G(f) f_2 . \quad \square \end{aligned}$$

A k-algebra A is said to be a k-**split-split** algebra provided that for each exact sequence

$$(1) \qquad 0 \to P \to Q \to T \to 0$$

of A-modules it is true that (1) is split exact as A-modules if and only if (1) is split exact as k-modules.[1]

Since A-split always implies k-split, the substance of this condition is that k-split must imply A-split. It is equivalent to show that if an A-submodule of Q is a k-summand, then it is an A-summand.

13.23 Theorem. *Let G be a group of finite order n, and let k be a commutative ring such that $n = n \cdot 1$ is a unit in k. Then the group algebra $A = kG$ is a k-split-split algebra* (cf. 32.6).

Proof. Let M be an A-module, and let N be an A-submodule such that N is a summand of the left k-module M, and write $M = N \oplus N'$, where N' is a k-submodule of M. Let $p : M \to N$ be the projection parallel to N'. Then

$$px = x$$

and

$$pgx = gx = gpx \qquad \forall g \in G, \ x \in M$$

so p is a kG-map such that

$$g^{-1}pgx = x$$

$$(\textstyle\sum_{g \in G} g^{-1}pg)x = nx$$

where n is the order of G. By the lemma,

$$np = \sum_{g \in G} g^{-1}pgp = p\,Tr_G(p) = Tr_G(p).$$

Hence $(n^{-1}Tr_G) : N \to M$ is a kG-map such that $n^{-1}Tr_G \circ p = 1_N$, proving that the kG-map $p : M \to N$ splits, so N is a kG-summand.

13.24 *Exercise.* Apply 13.23 to prove Maschke's theorem.

Orders in Algebras

13.25 Definition. Let A be an algebra with a finite free basis over a commutative quotient ring k, and let R be a right order of k. Then an R-order of A is an R-subalgebra-1 B which is finitely generated in mod-R such that the equivalent conditions hold:

---

[1] In the literature, such an algebra is said to be "semisimple", or to have zero relative global dimension. (Cf. Hochschild [56, 58].)

13.25.1 B contains a free k-basis of A

13.25.2 $Bk = A$

Then,

$$A = \{b\alpha^{-1} \mid b \in B, \text{ regular } \alpha \in R\}.$$

Proof. Exercise.

13.26 *Example*.

13.26.1 If B is an algebra with a finite free basis over R, then B is an R-order in the k-algebra $B_k = B \underset{R}{\otimes} k$.

13.26.2 The group algebra RG of a finite group is an R-order in the k-algebra kG.

13.26.3 If B is an R-order in a k-algebra over a field k, then, since $\dim_k A$ is finite, A is semilocal, in fact, Artinian. Thus, the results of Chapter 10 (cf. also Chapter 25) apply to these orders. Thus, Maschke's theorem implies that a finite group algebra RG is an order in a semisimple algebra kG when R is an order in k, and the characteristic of k does not divide the order of G. In any case, RG is an order in a Quasi-Frobenius (QF) algebra. (See 13.27.7.) There is a vast literature on R-orders. See, for example, Jacobson [42], Deuring [48], Auslander-Goldman [60a], and Curtis-Reiner [62, esp. p. 515 *ff*.].

13.27 *Exercises*.

13.27.1 (a) If G is a finite group, the (integral) group ring $\mathbb{Z}G$ contains no idempotents $\neq 0,1$. (b) Every projective $\mathbb{Z}G$-module is free.

13.27.2 If G is a finite group, if R is commutative, and if V is a simple RG-module, then V is finite dimensional over End V_{RG}.

13.27.3 If K is a field of characteristic 0, if G is a finite group, and if $d(KG) = \{n_1, \ldots, n_d\}$ are the Wedderburn invariants (11.42) for the semisimple algebra KG, then d is the number of conjugacy classes of G.

13.27.4 (Connell [63]). Let G be a finite group. If A is a right self-injective ring, then the group ring AG is right self-injective.

13.27.5 (Connell). If a group ring AG over a ring A is right selfinjective, then A is right selfinjective, and G is a torsion group; that is, every element of G has finite order.

13.27.6 (Connell). Let G be a group, and A, a ring. Then AG is a right Artinian ring if and only if A is right Artinian and G is a finite group.

13.27.7 If G is a finite group, then the group ring AG over the ring A is quasifrobenius if an only if A is quasifrobenius QF. (Cf. Chapter 12, Exercise 6.)

13.27.8 Find necessary and sufficient conditions on R and G so that RG is a prime ring [resp. perfect ring (see Chapter 22)]. (Cf. Connell [63]; Woods [71].)

13.27.9 Let G and H be finite groups. For any commutative ring R, there is a canonical isomorphism

$$R(G \times H) \approx RG \underset{R}{\otimes} RH$$

of R-algebras, where $G \times H$ is the group product.

13.27.10 (Auslander [58]) A group is **locally finite** if every finite subset generates a finite subgroup. (Thus, an abelian group is locally finite if and only if it is a torsion group.) The group algebra kG of a locally finite group G is a regular ring if and only if k is a regular ring uniquely divisible by the order of each element of G. In order for kG to be regular, without assuming local finiteness of G, it is necessary that G be a torsion group, and k have the stated divisibility property.

Nilpotent Bases

In this section the notion of scalar extension, together with the Wedderburn-Artin theorem for algebras, is applied to deduce a theorem of Wedderburn on algebras having a nilpotent basis. (They are nilpotent.) Like the first proof of Maschke's theorem, the proof shows how the theory of matrices is pertinent to the structure of algebras of finite dimensions.

Let K be an algebraically closed field, and let a be a matrix in K_n, the total matrix algebra of degree n over K. Thus

$$a = \sum_{i,j=1}^{n} \alpha_{ij} e_{ij} \qquad (\alpha_{ij} \in k)$$

13.28 Theorem (Wedderburn [37]) *If R is a finite dimensional algebra over a field k, and if a subalgebra B has a basis $x_1, \ldots, x_n$ consisting of nilpotent elements, then B is nilpotent.*

Proof. The set $A = \{\alpha + b \mid \alpha \in k,\ b \in B\}$ is a subalgebra. Since $1 \in A$, A is an algebra, and B is an ideal of A such that A/B is a field $\approx k$. Let $N = \operatorname{rad} A$. Since A is Artinian, N is nilpotent, A/N is semisimple, and so N is the intersection of the maximal ideals of A. Now B is nilpotent if and only if $B \subseteq N$. If $B \not\subseteq N$, then B lies outside a maximal ideal M. Since B is an ideal, then $B + M = A$. Then the simple algebra A/M has a basis consisting of $[x_1 + M], \ldots, [x_t + M]$ (cosets), where $t \leq n$; that is, there exists a simple algebra A which has a nilpotent basis. (We wish to show that this is a contradiction.) We can suppose that A is central. Then if K is an algebraically closed field $\supseteq k$, then $B = A \otimes_k K$ is a simple central algebra over K 13.9, with a nilpotent basis. Let $b_1, \ldots, b_g$ be a nilpotent basis of B. By Wedderburn's theorem 13.6, B is a total matrix algebra $\approx K_n$. If $a = \sum_{i=1}^{g} \alpha_i b_i$, $\alpha_i \in K$, $i = 1, \ldots, n$, then $Tr(a) = \sum_{i=1}^{g} \alpha_i Tr(b_i)$. Since b_i is nilpotent, and K is algebraically closed, by a remark preceding 13.21, $Tr(b_i) = 0$, $i = 1, \ldots, g$, and therefore $Tr(a) = 0$. This is true for all $a \in K_n$. Since K_n contains nilpotent elements, K_n cannot be a field, so $n > 1$. If $\{e_{ij}\}_{i,j=1}^{n}$ is any set of $n \times n$ matrix units of K_n, then $Tr(e_{11}) = 1$, which is the desired contradiction. □

Exercises for Chapter 13

1. If A is a finite dimensional central simple algebra over C, then $\dim_C A = n^2$, where n is an integer. (Hint: Let K be an algebraically closed field containing C. Then $A_K = A \otimes_C K \approx K_n$ by 11.41.)

2. If A is a finite dimensional central simple algebra over C, and if $n = \dim_C A$, then $A^{\text{op}} \otimes_C A \approx C_n$. [Hint: Let $R = A^{\text{op}} \otimes_C A$. Then A is a right R-module, where $c(a^0 \otimes b) = acb \; \forall a, b, c \in A$. Let $f \in \text{End}\, A_R$. Show that $f \mapsto f(1)$ is an isomorphism $\text{End}\; A_R \approx C$. Since R is simple algebra, the map $\varphi : R \to A$ is a splitting R-epimorphism, where $\varphi\,(a^0 \otimes b) = ab$. This map splits because R is a semisimple algebra, so every R-module is projective. Then A_R is a generator (why?), hence balanced; so that $R \approx \text{End}_C A$. Cf. the Brauer group, Exercises 4(b) and 10(b).]

3. Use Exercise 2 to show that a tensor product of central division algebras need not be a field (in fact, need not be entire).

4. *Brauer group*. Two finite dimensional central algebras A and B over K are **similar** $A \sim B$, if there exist total matrix algebras K_m and K_n, and an isomorphism

$$A \otimes K_m \approx B \otimes K_n .$$

(a) Show that $A \sim B$ is an equivalence relation (Hint: $K_p \otimes K_q \approx K_{pq}$.) Prove that this definition coincides with its previous use, namely that $A \sim B$ if and only if mod-$A \approx$ mod-B.

(b) Let $[A]$ denote the equivalence class of a finite dimensional simple algebra A. Let $Br(K)$ denote the set of equivalence classes of such algebras over K. By Wedderburn's theorem, $[A] = [D]$ for some division algebra D. (If D and E are division algebras such that $[D] = [E]$ then $D \approx E$.) Show

$$[A][B] = [A \otimes_K B]$$

is a binary operation on $Br(K)$ under which $Br(K)$ is an abelian group, called the **Brauer group of** K (cf. Chapter 32). By Exercise 2, $[A]^{-1} = [A^{\text{op}}]$.

*(c) Show that $Br(K)$ is a torsion group for every field K. If $\dim_K A = n$, show that $[A]^n =$ identity; that is, if $\otimes_K^n A$ denotes the tensor operation n times, then $\otimes_K^n A \approx K_m$ for some m.

5. A **splitting field** for a central simple algebra A over K is a field $F \supseteq K$ such that $A \otimes_K F$ is total matrix over F, that is, such that $[A \otimes_K F]$ is the group identity in $B_r(F)$. A **maximal subfield** of A is a commutative subfield F maximal in the set of all commutative subfields of A. (Necessarily, F contains center A.)

(a) Let D be a finite dimensional central division algebra over K. Then any maximal subfield of D is a splitting field. Moreover, D has a separable maximal subfield, hence a separable splitting field.

(b) Any simple central algebra of finite dimension has a separable splitting field.

The material taken up in Exercises 6 to 10 *is the subject of Chapter* 32.

6. The **enveloping algebra** of an algebra A over k is defined to be $A^e = A \otimes_k A^{\text{op}}$, for any commutative ring k. Then A is a canonical right A^e-module such that

$$x(a \otimes b^0) = bxa \qquad \forall a, b, x \in A.$$

Then, A is a **separable** algebra if A is A^e-projective.

(a) (Hochshild) Show that A is separable if and only if A^e is a k-split-split algebra. In this case, A is a k-split-split algebra.

(b) If k is a field, and A is finite dimensional, show that the condition (a) characterizes separable algebras.

7. An algebra A over a commutative ring k is said to be an **Azumaya algebra** provided that A is finitely generated and central over k, and separable. In this case there is a canonical isomorphism $A^e \approx \text{End}_k A$, and a k-linear equivalence mod-$A^e \approx$ mod-k. Conversely.

8. An algebra A is Azumaya if and only if for every maximal ideal Q of k, the canonical k/Q algebra A/QA is central simple, and A is finitely generated projective and faithful over k.

9. An algebra A over k is **ideal** when the *ideal mapping* $Q \mapsto QA$ between the ideals of k and those of A is bijective. A is an Azumaya algebra iff the enveloping algebra A^e is an ideal algebra, and A_k is finitely generated. Every Azumaya algebra is ideal. (Cf. Rao [72].)

10. (a) Two Azumaya algebras A and B over k are **similar** provided that there is a k-linear equivalence mod-$A \approx$ mod-B. This is equivalent to an isomorphism of k-algebras

$$A \underset{k}{\otimes} \text{End}_k P \approx B \underset{k}{\otimes} \text{End}_k M$$

for some choice of two finitely generated projective faithful k-modules P and M.

(b) The Brauer group for commutative rings (Azumaya [51], Auslander and Goldman [60]). Let (A) denote the similarity class of an Azumaya algebra A over k. The set $Br(k)$ of similarity classes is an abelian group with respect to the operation

$$(A)(B) = (A \otimes B).$$

Show that $Br(k)$ is a torsion group (Grothendieck [65]; cf. Knus and Ojanguren [72].)

11. A central algebra A over k is said to be **centralizing** provided that the canonical mapping $A \underset{k}{\otimes} (\text{centralizer}_B A) \to B$ is an isomorphism for every central algebra B over k which contains A. Every Azumaya algebra, in particular, every central simple finite dimensional algebra, is centralizing. Conversely, if A is finitely generated and projective over k, and if A is centralizing, then A is an Azumaya algebra.

Notes for Chapter 13

The theorems on this chapter are classical, and a number of extensions of them are treated in Exercises and in Chapter 32, so further "notes" may seem uncalled for. Nevertheless, we take this opportunity to discuss some related theorems in the literature.

A simple central algebra A of finite dimension over k is a **crossed product** provided that there is a separable normal maximal subfield N. In this case, each element g_i of the Galois group $G = \{g_1, \ldots, g_n\}$ of N/k is induced by an inner automorphism I_{u_i} of A, $i = 1, \ldots, n$, and $\{u_1, \ldots, u_n\}$ is a vector space basis of A over N. The algebra A is **cyclic** if A is a crossed product with G a cyclic group. Cyclic algebras were discovered by L. E. Dickson (see, for example, his book [23] entitled "Algebras and their Arithmetics") and generalized to crossed products by Brauer and Noether (cf. Deuring [48, p. 53, pp. 55, 56, and *Vorwort*]). A **rational algebra** is an algebra of finite dimension over an algebraic number field k (that is, over a finite field extension k of $\mathbb{Q}$). Every rational simple central algebra is cyclic (Theorem of Brauer-Noether-Hasse; cf. Deuring [48, p. 118]). The first non-cyclic algebras were discovered by Albert (in a 1938 Bull. Amer. Math. Soc. paper), and the first "uncrossed" algebra by Amitsur (see Amitsur [72/73]).

In [68], Roiter proved the **Brauer-Thrall conjecture**: Let A be a finite dimensional algebra over a field k, and assume that there exists an integer N such that every indecomposable right A-module M has $\dim_k M < N$. Then, the isomorphy class of finitely generated indecomposable right A-modules is finite. [Furthermore, because of the duality between finitely generated right, and finitely generated left, modules (discussed in Chapter 12 Exercise 8), the isomorphy class of finitely generated indecomposable left modules is finite.] Actually, although Roiter proves these results for finite dimensional algebras he *states* that they hold for Artinian rings, where the bound N is placed on the number of elements required to generate any indecomposable module[1]. (Cf. Chapter 20 bis. For the Brauer-Thrall conjecture for group algebras, consult Janusz [69].)

The definition we have given of an R-order B in an algebra A over a field k is equivalent to that of Auslander-Goldman [60a] in case R is an integrally closed domain, in which case B is contained in a maximal order under the assumption that A is central simple. (In [68], Bass proves this under slightly different assumptions. See his Theorem 8.5, p. 154.) In case R is a Dedekind domain, any maximal order in a central simple algebra A is a hereditary Noetherian ring (see, for example, Auslander-Goldman [60a, p. 8, Theorem 2.9] and Bass [68, p. 159; Theorem 8.7] for this, and also a partial converse. Cf. also Deuring [48] and Jacobson [42]).

Formanek [72] proved that if B is a right order in a finite dimensional simple algebra A, then the center of B is an order in the center of A. Moreover, if B is a right order in a field A, and if G is a finite group of automorphisms of B canonically extended to a finite group of automorphisms of A, then the fixed subring of B under G is a right order in the fixed subring of A under G (Faith [72c]).

The semisimplicity of a finite group algebra kG over a field k of characteristic 0 has been generalized by Amitsur [56] (and others) to the statement that kG has zero radical for any uncountable field of characteristic 0 (see Chapter 26). The problem for countable fields is still open. Formanek [72b] proved that kS is semisimple over any field k, $S = \bigcup_{n\in\omega} S_n$; and Formanek and Snider [72] proved the existence of primitive group algebras, using the characterization of Connell [63] of when a group ring $R[G]$ is prime. (This happens iff R is prime and G has no finite normal subgroups.)

References

Amitsur [56], [72/73], Auslander [58], Auslander-Goldman [60a], Azumaya [51], Brauer [63], Connell [63], Curtis-Reiner [62], Deuring [48], Dickson [23], Faith [72c], Formanek [72a, 72b], Formanek and Snider [72], Grothendieck [65], Hochschild [56, 58], Jacobson [42, 64], Janusz [69], Knus and Ojanguren [72], Maschke [98], Rao [72], Roiter [68], Wedderburn [08, 37], Woods [71].

[1] Te proof, however, does not appear to adept itself to the Artinian case without certain unproven supplementary lemmas.

IV. Structure of Abelian Categories

An abelian category C with coproducts is cocomplete 6.3, and is called an $AB3$ category. The sum of any family of subobjects of an object exist, and if, in addition, sum distributes over intersection in the sense that

$$\sum_{i\in I} A_i \cap B = \sum_{i\in I} (A_i \cap B)$$

where $\{A_i\}_{i\in I}$ is any directed family of subobjects of any object A of C and B is any subobject of A, then C is said to be an $AB5$ category (Grothendieck [57]). (An $AB5$ category in which sum distributes over arbitrary intersections is called an $AB6$ category. See Chapter 15 Exercises.) An $AB3$ category is $AB5$ if and only if it has exact direct limits 14.6, or equivalently, the direct limit of any directed family of subobjects is a subobject (and in this case the direct limit is the sum).

A Grothendieck category is an $AB5$ category with a generator. In such a category, every object has an injective hull 14.17. This is needed to prove the main result of Chapter 14, the (Freyd, Lubkin) embedding theorem 14.21, which states that every small abelian category C has an exact, covariant embedding $C \rightsquigarrow \text{mod-}\mathbb{Z}$. The (Mitchell) full embedding theorem 15.30 states that a small abelian category C has a full exact embedding $C \rightsquigarrow \text{mod-}A$, for a suitable ring A. (This also provides an exact embedding $C \rightsquigarrow \text{mod-}\mathbb{Z}$.)

Quotient categories are introduced in Chapter 15. A full subcategory $\mathscr{S}$ of an abelian category $\mathscr{A}$ is called a **Serre class** if for every exact sequence $0 \to X \to Y \to Z \to 0$ in $\mathscr{A}$ it is true that $Y \in \mathscr{S} \Leftrightarrow X \,\&\, Z \in \mathscr{S}$. Then, there is a category $\mathscr{A}/\mathscr{S}$ on the objects of $\mathscr{A}$, and morphisms are defined in $\mathscr{A}/\mathscr{S}$ by:

$$\operatorname{Mor}_{\mathscr{A}/\mathscr{S}}(X, Y) = \varinjlim_{Y', X/X' \in \mathscr{S}} \operatorname{Mor}_{\mathscr{A}}(X', Y/Y')$$

The category $\mathscr{A}/\mathscr{S}$ is called a **quotient category**, and is abelian 15.7. The Serre subcategory $\mathscr{S}$ is a **localizing** subcategory provided that the canonical functor $T: \mathscr{A} \rightsquigarrow \mathscr{A}/\mathscr{S}$ has a right adjoint $S: \mathscr{A}/\mathscr{S} \rightsquigarrow \mathscr{A}$ (called the **section** functor), and the composition $\mathscr{A} \rightsquigarrow \mathscr{A}$ is called the **localizing** functor. If $\mathscr{A}$ is Grothendieck, a necessary and sufficient condition for this is that $\mathscr{S}$ should be a subcategory with arbitrary coproducts, or equivalently, direct limits taken in $\mathscr{A}$ of directed families of subobjects of $\mathscr{S}$ should belong to $\mathscr{S}$. In this case, for each object M of $\mathscr{A}$, there is an exact sequence

$$0 \to X \to M \xrightarrow{u} N \to Y \to 0$$

where X, $Y \in \mathscr{S}$, and N is $\mathscr{S}$-closed in the sense 15.15. If C denotes the full subcategory of $\mathscr{S}$-closed objects, then the localizing functor induces an equivalence

$$\psi(N) : N \to STN,$$

and there is an equivalence of categories $\mathscr{A}/\mathscr{S} \approx C$. The morphism $u : M \to N$ is determined uniquely up to an equivalence of $\mathscr{A}$, and is called the $\mathscr{S}$-envelope of M, denoted $u_{\mathscr{S}} : M \to M_{\mathscr{S}}$. If $\mathscr{A} = \text{mod-}A$, for some ring A, then $A_{\mathscr{S}}$ is a ring, and in classical situations 16.8 when the localizing functor is exact, this functor is naturally isomorphic to $\underset{A}{\otimes} A_{\mathscr{S}}$, and there is an equivalence of categories

$$\text{mod-}A/\mathscr{S} \approx \text{mod-}A_{\mathscr{S}}.$$

This situation occurs for the localization theorem 16.15 generalizing to Grothendieck categories localization theorems for commutative rings (Chapters 27 and 28). This method also applies to partial right quotient rings 16.9 generalizing the right quotient rings of Chapters 9 and 10, and to the Johnson maximal right quotient rings of right neat rings 16.12. (Cf. also Chapter 19.)

A classical aspect of module theory is the relating to a ring A intrinsically some extrinsic property of mod-A. Thus, 8.12 every module is semisimple if and only if A is semisimple or 8.21,

$$\text{r.gl. dim } A = \sup \{\dim I \mid I \subseteq R\} + 1$$

when A is not semisimple; or the characterisation 11.24 of rings with every module flat, as regular rings in which every principal right ideal is generated by an idempotent. These examples can be multiplied at will, for in a sense, this is ring (and module) theory.

In this vein, a natural question raised by localizing subcategories of mod-A is the intrinsic relation to mod-A. This is solved by Gabriel's theorem 16.3 which establishes a mapping

$$C \mapsto \mathscr{F}(C) = \{I \subseteq A \mid A/I \in \text{obj } C\}$$

of the class of localizing subcategories of mod-A to the class consisting of sets of all right ideals of A. The set $\mathscr{F}(C)$ is idempotent and topologizing

(= idemtop) in the sense of 16.3, and conversely, any idemtop set F of right ideals of A corresponds to the localizing subcategory $\mathcal{S}(F)$ of mod-A consisting of all modules M such that for every $x \in M$ the right ideal $\operatorname{ann}_A x \in F$. Moreover, as stated above for Grothendieck categories the quotient category mod-$A/\mathcal{S}$ is equivalent to the full subcategory C of mod-A consisting of $\mathcal{S}$-closed modules, or in the terminology of torsion theories taken up in Chapter 16, the subcategory consisting of torsionfree divisible modules. The same correspondence can be used to characterize Serre classes, and classes of closed full subcategories of mod-A 16.2. In these cases, then the $\mathcal{S}(F)$-envelope of a module M is denoted M_F instead of $M_{\mathcal{S}(F)}$, and in particular, A_F denotes $A_{\mathcal{S}(F)}$.

Chapter 15 contains the Popesco-Gabriel theorem 15.26 which characterizes Grothendieck categories as quotient categories of module type categories.

Chapter 16 concludes with a theorem 16.20 of E. A. and C. L. Walker stating that mod-A is determined in a very strong sense by the endomorphism ring of any proper generator. This generalizes the known theorems for principal right ideal domains, and quasi-Frobenius rings. (Cf. Chapter 24.)

Chapter 14

Grothendieck Categories

Grothendieck [57] introduced the notation for abelian categories which follows:

*AB*3. A coproductive, hence cocomplete, abelian category.

*AB*4. An *AB*3 category in which every coproduct $\coprod_{i\in I} A_i \to \coprod_{i\in I} B_i$ of monics $\{A_i \to B_i\}_{i\in I}$ is monic.

*AB*5. An *AB*3 category such that for any directed family $\{A_i\}_{i\in I}$ of subobjects of any object X, and any subobject B,

$$\Big(\sum_{i\in I} A_i\Big) \cap B = \sum_{i\in I} (A_i \cap B).$$

ABX^*, the dual of ABX, $X = 3, 4, 5$.

For any totally ordered family $\{A_i\}_{i\in I}$ of subobjects of any object X, the sum is called **union**, and is denoted by $\bigcup_{i\in I} A_i$. If the family is finite, and consists of $A_1 \le \cdots \le A_n$, then $\bigcup_{i\in I} A_i = A_n$.

14.1 Proposition and Definition. *An $AB5$ category with a generator is called a* **Grothendieck category**. *Any Grothendieck category is locally small.*[1]

Proof. A cocomplete abelian category with a generator is locally small 5.23. □

14.2 Theorem. *If C is an abelian category, and if C is any of $AB3$, $AB4$, $AB5$, locally small or Grothendieck, then so is C^L, for any small category L. If L is also additive, the same is true of the category (L, C) of additive functors $L \rightsquigarrow C$.*

Proof. $AB3$ is inherited by C^L by 6.18.8 and 6.18.2*. Then $AB4$ follows from the fact that coproducts in C^L are defined pointwise, and from the fact 6.17 that in an exact category so are monics. If C is $AB5$, and if $\{H_i\}_{i\in I}$ is a family of objects of C^L, then for any object K of C^L, and object A of C,

$$\Big(K \cap \sum_{i\in I} H_i\Big)A = KA \cap \Big(\sum_{i\in I} H_i\Big)A = KA \cap \sum_{i\in I} H_i A$$
$$= \sum_{i\in I} (KA \cap H_i A) = \Big(\sum_{i\in I} (K \cap H_i)\Big)A.$$

Therefore, C^L is $AB5$. If C is locally small, then so is C^L, in view of 6.16.9. The same results hold for (L, C), when L is small and additive. (Cf. 6.17 and 6.18.) Moreover, in the latter case, (L, C) has a generator 6.20. Thus, if C is $AB5$, then (L, C) must be Grothendieck. (Cf. 6.25.) □

14.3 *Exercise.*

14.3.1 Let $T: C \rightsquigarrow D$ be a fully faithful functor of abelian categories which has an exact left adjoint $S: D \rightsquigarrow C$. Then C inherits any of the following conditions placed on D: $AB3$, $AB4$, $AB5$, and $AB3^*$. If further, T is exact, then the conditions $AB4^*$ and $AB5^*$ are inherited by C.

14.3.2 Let C be an abelian category with enough injectives, and let D be abelian. Then a fully faithful functor $T: C \rightsquigarrow D$ which preserves injectives satisfies the requirements of the previous exercise, that is, any left adjoint of T is exact.

14.3.3 (Mitchell [65, p. 84]) If $\{A_i, \pi\}_{i\in L}$ is a directed family of objects in an $AB5$ category, and if $\{f_i: A \to A_i\}_{i\in L}$ is a final object of co-compat π (thus, $A = \varinjlim \pi$), then

$$\ker f_n = \sum_{i\le n} \ker f_i .$$

[1] Locally small categories are defined *supra* 5.22; in the literature, such a category is also said to be well-powered (e.g. Freyd [64].)

14.3.4 If C is $AB5$ and $AB5^*$, then $C = 0$.

14.4 Proposition. *Let C be an $AB3$ category, and let $\{u_i : A_i \to A\}_{i\in I}$ be a family of monics directed by inclusion. Then $\{A/A_i\}_{i\in I}$ is a directed family, and*

$$\varinjlim A/A_i = A/\textstyle\sum_{i\in I} A_i.$$

Proof. The fact that $\{u_i : A_i \to A\}_{i\in I}$ is directed by inclusion means that I becomes a directed set when ordered by $i \leq j$ if and only if $u_i \leq u_j$. If $p_i : A \to A/A_i$ is the cokernel of $u_i : A_i \to A$ for each $i \in I$, then by the first Noether isomorphism theorem 5.34,

$$0 \to A_j/A_i \to A/A_i \to A/A_j \to 0$$

is exact whenever $i \leq j$, so $\{A/A_i\}_{i\in I}$ is a directed family. Since the direct limit functor preserves cokernels 5.43, from

$$0 \to A_i \to A \to A/A_i \to 0$$

exact, we get

$$\varinjlim A_i \xrightarrow{u} A \to \varinjlim A/A_i \to 0$$

exact, where $u = \varinjlim u_i$. By 5.29, $\sum_{i\in I} A_i$ is the image of u, hence

$$0 \to \sum_{i\in I} A_i \to A \to \varinjlim A/A_i \to 0$$

is exact, which is what we wanted. ▯

An abelian category C is said to have **exact direct limits** *provided that for each directed set L, the following equivalent conditions hold:*

(1) *The colimit functor $C^L \rightsquigarrow C$ is exact.*

(2) *If $0 \to F \to G$ is exact in C^L, then $0 \to \varinjlim F \to \varinjlim G$ is exact in C.*

14.5 Proposition. *If C is an abelian category with enough injectives, then C has exact direct limits if and only if for every directed set L the constant functor $C \rightsquigarrow C^L$ preserves injectives.*

Proof. By 6.28 a left adjoinable functor $T : C \rightsquigarrow D$ has an exact left adjoint if and only if T preserves injectives, provided that C has enough injectives. ▯

14.6 Proposition. *The following conditions on an $AB3$ category C are equivalent:*

(a) *C is $AB5$.*

(b) *C has exact direct limits.*

(c) *Inverse images commute with sums. Thus, for each morphism* $f: B \to A$, *and directed family* $\{A_i\}_{i \in L}$ *of subobjects of* A, *it is true that*

$$f^{-1}\Big(\sum_{i \in L} A_i\Big) = \sum_{i \in L} f^{-1}(A_i).$$

(d) *For any directed set L, if A is an object of C, and if* $0 \to F \to c(A)$ *is exact in* C^L, *then* $0 \to \varinjlim F \to A$ *is exact in C.*

(e) *If* $\{F(i) : A_i \to A\}_{i \in L}$ *is any directed family of subobjects of an object A of C, then*

$$\varinjlim_{i \in L} A_i = \textstyle\sum_{i \in L} A_i.$$

(f) *The direct limit of any directed family of subobjects of any object A of C is a subobject of A.*

Proof. For the most part, the proof uses the arguments of Mitchell [65, pp. 82–85].

(c) $\Rightarrow$ (a). If f denotes the inclusion of a subobject B in an object A, and if $\{A_i\}_{i \in I}$ is a family of subobjects of A, then $B \cap A_i = f^{-1}A_i$, so that (c) implies

$$\sum_{i \in I} (B \cap A_i) = B \cap \sum_{i \in I} A_i = f^{-1}\big(\sum_{i \in I} A_i\big).$$

(b) $\Rightarrow$ (d) $\Leftrightarrow$ (e) $\Leftrightarrow$ (f). By 5.29, in an $AB3$ category, the image of the canonical morphism $\varinjlim_{i \in L} A_i \to A$, where $\{A_i\}_{i \in L}$ is a directed family of subobjects of A, is the sum $\sum_{i \in L} A_i$. This establishes (d) $\Leftrightarrow$ (e). Moreover, (f) is merely a restatment of (d), whereas (d) is a special case of (b). Thus, (b) $\Rightarrow$ (d) $\Leftrightarrow$ (e) $\Leftrightarrow$ (f).

(b) $\Rightarrow$ (c). Since inverse image is a fiber product (see Example 5.8.7 immediately preceding 5.9) and since exactitude of the direct limit functor implies finite limit preservation, it follows that

$$f^{-1}(\varinjlim_{i \in L} A_i) = \varinjlim_{i \in L} f^{-1}(A_i).$$

Now, (b) $\Rightarrow$ (e), and an application of (e) yields (c). (a) & (e) $\Rightarrow$ (c). Let $J = \operatorname{im} f$. If $X \to A$ is any monic, then

$$0 \to f^{-1}X \to B \to J/X \cap J \to 0$$

is exact. Thus, for every $i \in I$,

$$0 \to f^{-1}A_i \to B \to J/J \cap A_i \to 0$$

is exact, so, by (e)

$$0 \to \varinjlim_{i \in I} f^{-1}A_i \to B \to \varinjlim_{i \in I} J/J \cap A_i \to 0$$

is exact; that is, by 5.29 and 14.4,

$$0 \to \sum_{i\in I} f^{-1} A_i \to B \to J / \sum_{i\in I} (J \cap A_i) \to 0$$

is exact. By the second Noether isomorphism theorem 5.36 for exact categories,

$$\begin{aligned}\sum_{i\in I} f^{-1} A_i &= f^{-1} \sum_{i\in I} (J \cap A_i) = f^{-1} J \cap f^{-1}\big(\sum_{i\in I} A_i\big) \\ &= f^{-1}\big(\operatorname{im} f \cap \sum_{i\in I} A_i\big) = f^{-1} \sum_{i\in I} A_i.\end{aligned}$$

(a) & (c) $\Rightarrow$ (b). Let $h: F \to G$ be a monic in C^L, where L is directed, let $A = \varinjlim F$, $B = \varinjlim G$, and let $u: A \to B$ be the limit morphism. There are canonical morphisms $\{a_i : F(i) \to A\}_{i\in L}$, $\{b_i : G(i) \to B\}_{i\in L}$, and if $A'_i = \operatorname{im} a_i \ \forall i \in L$, then by 5.29, $A = \sum_{i\in L} A'_i$. In order to prove exactitude of the direct limit functor, it is necessary and sufficient to prove left exactitude, since 5.43 the direct limit functor is colimit preserving. Thus, it is enough to show that the limit morphism $u = \lim h$ is monic. We will show that the assumption that $K = \ker u \neq 0$ leads to a contradiction. First, then

$$A'_k \cap K \neq 0 \quad \text{and} \quad M = a_k^{-1}(A'_k \cap K) \neq 0$$

for some $k \in L$. Furthermore,

$$b_k\big(h(k)(M)\big) = u\big(a_k(M)\big) \subseteq u(A'_k \cap K) \subseteq u(K) = 0$$

that is, $h(k)(M)$ is contained in $\sum_{k\le n} L_{kn} = \ker b_k$, where $L_{kn} = \ker G(f_{kn})$, and $f_{kn} : k \to n$ (see Exercise 14.3.3). Since C is $AB5$, then

$$h(k)(M) = \sum_{k\le n} L_{kn} \cap h(k)(M).$$

Since sums commute with inverse images by (c), this implies

$$\begin{aligned}M = h(k)^{-1} h(k)(M) &= h(k)^{-1}\Big(\sum_{k\le n} L_{kn} \cap h(k)(M)\Big) \\ &= \sum_{k\le n} h(k)^{-1}\big(L_{kn} \cap h(k)(M)\big).\end{aligned}$$

Since $h(i)$ is monic, $\forall i \in L$, then

$$G(f_{kn})\Big(h(k)^{-1}\big(L_{kn} \cap h(k)(M)\big)\Big) = 0$$

$\forall n \ge k$, and so $a_k\Big(h(k)^{-1}\big(L_{kn} \cap h(k)(M)\big)\Big) = 0$ for all $n \ge k$. Since morphisms commute with sums (see 5.28), the representation of M as a sum in the display above shows that $a_k(M) = 0$. But a_k is monic, so $M = 0$, a contradiction. Thus, $K = 0$, so u is monic.

(a) $\Rightarrow$ (e) Let $\{f_i : A_i \to A\}$ be a directed family of monics in C, let $f : L \to A$ be the limit of the family of morphism $\{f_i\}_{i \in L}$, and let $\{g_i : A_i \to L\}_{i \in L}$ be the canonical morphisms. By 5.29, L is the sum $\sum_{i \in L} g_i(A_i)$ of the images of the family $\{g_i\}_{i \in L}$, and $\operatorname{im} f = \sum_{i \in L} A_i$. If $\ker f \neq 0$, then $\ker f \cap g_i(A_i) \neq 0$ for some $i \in L$, and then $g_i^{-1}(\ker f) \neq 0$. Thus, $\ker f_i \neq 0$, which contradicts the assumption that f_i is monic.

(e) $\Rightarrow$ (a) Assume that a direct limit of a directed family of subobjects $\{A_i\}$ is a subobject. Then, for a subobject B, the family $\{A_i + B\}$ is also directed, and the sum $\sum_i (A_i + B) = (\sum A_i) + B$ is the direct limit. Since direct limit preserves cokernels,

$$0 \to B \to A_i + B \to (A_i + B)/B \to 0$$

induces an exact sequence

$$B \to (\sum A_i) + B \to \varinjlim (A_i + B)/B \to 0$$

that is,

$$\varinjlim (A_i + B)/B = \big((\sum A_i) + B\big)/B.$$

In the same way,

$$0 \to A_i \cap B \to A_i \to (A_i + B)/B \to 0$$

induces an exact direct limit

$$0 \to \sum_i (A_i \cap B) \to \sum A_i \to \big((\sum A_i) + B\big)/B \to 0.$$

But also

$$0 \to (\sum A_i) \cap B \to \sum A_i \to \big((\sum A_i) + B\big)/B \to 0$$

is exact (First Noether Isomorphism Theorem 5.34), so that the $AB5$ condition $\sum_i (A_i \cap B) = (\sum_i A_i) \cap B$ holds. □

14.7 Corollary for Grothendieck Category with Generator. *Every object A is the direct limit of the directed family of finitely generated subobjects of A.*

Proof. Since the category C has a generator U, then by 14.1 C is locally small, so the class of subobjects of an object A is a set. Since the sum $A_1 + A_2$ of two subobjects is the image of the coproduct morphism $A_1 \oplus A_2 \to A$, it follows that $A_1 + A_2$ is finitely generated whenever A_1 and A_2 are. The family $\{A_i\}_{i \in I}$ of finitely generated subobjects is therefore a directed system, and the direct limit morphism $f : L \to A$ is monic by the proposition. If f is not an equivalence, then there is a morphism $h : U \to A$, which f does not allow. Since $\operatorname{im} h$ is finitely generated, $\operatorname{im} h$ precedes f, that is, f allows h, which is a contradiction. □

14.8 *Exercise.* A category C is **locally Noetherian** if C is a Grothendieck category and possesses a set of generators $\{U_i\}_{i\in I}$ such that U_i is a Noetherian object of C $\forall i \in I$.

If C is a Grothendieck category, and M is a Noetherian object of C, and if $P = \varinjlim P_i$ is an object which is the direct limit of a directed family $\{P_i\}_{i\in I}$ of subobjects of P, then the canonical morphism

$$\varinjlim_{i\in I} \mathrm{Mor}_C(M, P_i) \to \mathrm{Mor}_C\left(M, \varinjlim_{i\in I} P_i\right)$$

is an equivalence. Deduce that in a locally Noetherian category, a coproduct of injective objects is injective.

Injective Hulls

In this section, C will denote an abelian category. If A is an object of C, then an **extension** of A is a monic $A \to E$. The extension $A \to E$ is: **proper** if it is not an equivalence; **trivial** if $A \to E$ splits; and **essential** provided that $A' \cap A \neq 0$ for every monic $A' \to E$.

14.9 **Proposition.** *An extension $A \to E$ is essential if and only if a morphism $E \to B$ is monic whenever $A \to E \to B$ is monic.*

Proof. If $\ker(E \to B) \neq 0$, then $0 \neq \ker(E \to B) \cap A = \ker(A \to B)$.

Conversely, suppose $A \cap A' = 0$ for some nonzero monic $A' \to E$. Clearly, $A \to E \to E/A'$ is monic, but $E \to E/A'$ is not. ▯

14.10 **Proposition.** *An object A in C is injective if and only if A has only trivial extensions (every monic splits).* ▯

This is the dual to 5.5.

14.11 **Theorem.** *In a locally small $AB5$ category C, an object A is injective if and only if A has no proper essential extensions.*

Proof. Every extension of an injective object is trivial, hence not essential if proper.

Assume that A has no proper essential extensions, and let $A \to E$ be any extension. Since C is $AB5$, the ordered set S of subobjects A' of E satisfying $A' \cap A = 0$ is inductive; that is,

$$(\textstyle\sum_{i\in I} A_i) \cap A = \sum_{i\in I} (A_i \cap A) = 0$$

for every chain $\{A_i \mid i \in I\}$ in S. By Zorn's lemma, there exists a maximal element A' in S, a subobject of E. Then $A \to E \to E/A'$ is an essential monic, and is an equivalence only if A is a retract of E. Cf. 3.59. ▯

14.12 Baer's Criterion. *Let R be a ring, and let A be in* mod-R. *Then A is injective if and only if, for each right ideal I of R, $h_A(R) \to h_A(I)$ is epic.*

Proof (Freyd) If $a \in A$, let $R \xrightarrow{a} A$ be the map $1 \mapsto a$. We will show that the hypothesis implies that A has no proper essential extensions. Let $A \subset B$, let $b \in B$, $b \notin A$, and consider the pullback

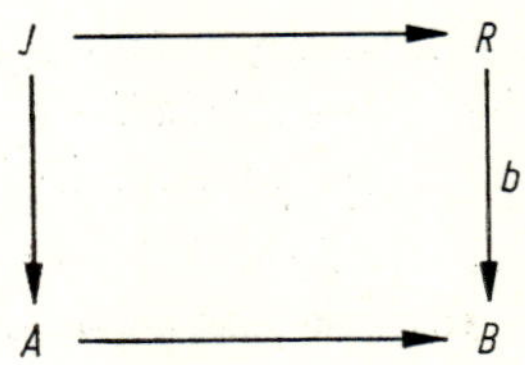

By 5.10.2, $J \to R$ is monic, so by hypothesis there exists $a \in A$ such that $J \to R \xrightarrow{a} A = J \to A$. Then $b - a \neq 0$, and $(b-a)R \cap A = 0$. Thus, B is not an essential extension of A. ▯

14.13 Lemma. *If $A \to E$ and $E \to B$ are extensions, then $(A \to B) = (A \to E \to B)$ is essential if and only if $A \to E$ and $E \to B$ are both essential.*

The proof is left to the reader. ▯

The **injective hull** of an object A is an essential extension $A \to E$, where E is injective.

14.14 Proposition. *If $u_i : A \to E_i$, $i = 1, 2$ are injective hulls of A, then there exists a unique equivalence $f : E_1 \to E_2$ such that*

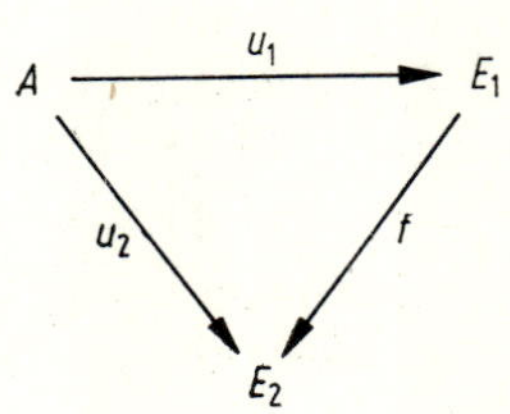

commutes.

Proof. $f : E_1 \to E_2$ exists by injectivity of E_2. Since $fu_1 = u_2$ is monic, and u_1 essential, f is monic by 14.9. Since u_2 is essential, f is essential by the preceding lemma. By 14.11, the extension $f : E_1 \to E_2$ is not proper; that is, f is an equivalence. ▯

14.15 Proposition. *In an* $AB5$ *category* C, *if* $\{u_i : A_i \to A\}$ *is a directed family of subobjects of an object* A, *and if* $\{B \to A_i\}$ *is an essential extension for each* i, *then* $B \to \sum_{i \in I} A_i$ *is an essential extension.*

Proof. Here $B \to \sum A_i = B \to A_j \to \sum A_i$ for any j. If X is a nonzero subobject of $\sum A_i$, then by the $AB5$ axiom $X = \sum (A_i \cap X)$, so $A_i \cap X \neq 0$ for some i. But $B \to A_i$ essential implies $B \cap (A_i \cap X) \neq 0$, so $B \cap X \neq 0$. ▯

14.16 Corollary. *If* C *is an* $AB5$ *category, then an object* A *has an injective hull if and only if* A *has an injective extension* X.

Proof. X is an extension of every essential extension of A. Let E denote a functor such that $EY = Y$ if $Y \in \operatorname{obj} C$ is injective, and otherwise EY is a proper essential extension of Y. For each ordinal α, define

$$E_{\alpha+1}A = E(E_\alpha A)$$

and if α is a limit ordinal, then

$$E_\alpha A = \varinjlim_{\beta<\alpha} E_\beta A = \bigcup\nolimits_{\beta<\alpha} E_\beta A .$$

Hence by the last result, $E_\alpha A$ is an essential extension of A for all A. If α is an ordinal of cardinal bigger than the cardinality of the set of subobjects of X, then necessarily $E_\gamma A = E_\alpha A$ for all $\gamma \geq \alpha$. Since $E_\alpha A$ has no proper essential extensions, it follows that it is an injective hull of A. ▯

14.17 Theorem. *Let* C *be a Grothendieck category. Then,* C *has enough injectives, an injective cogenerator, and every object has an injective hull.*

Comments on the proof. (1) By 14.16, it suffices to show that C has enough injectives.

(2) Mitchell [65, pp. 89—90] gives a proof by considering the embedding $h^U : C \rightsquigarrow \text{mod-}\mathbb{Z}$, where U is a generator of C, and using the fact that every object of mod-$\mathbb{Z}$ has an injective hull.

(3) By the Popesco-Gabriel theorem 4.29 and 5.70, h^U induces a full embedding $T : C \rightsquigarrow \text{mod-}A$, where $A = \operatorname{End}_C U$. Moreover, 5.70, T is a coretraction, that is, T has a left adjoint S and $ST \approx 1_C$. It can be shown 15.24 that there is a quotient category mod-$A/\ker S$ which has injective hulls, and by 15.26, there is an equivalence $C \approx \text{mod-}A/\ker S$.

With these remarks, the proof is left for an exercise. ▯

14.18 Proposition. *If* C *is a small abelian category, and if* T *is an injective object of* Adfunct $(C, \text{mod-}\mathbb{Z})$, *then* T *is a right exact functor.*

Proof. If $A' \to A \to A'' \to 0$ is exact in C, then by 6.23, $0 \to h^{A''} \to h^A \to h^{A'}$ is exact in $F = \text{Funct}\,(C, \text{mod-}\mathbb{Z})$. By injectivity of T in F, $\text{Mor}_F(\ ,T)$ is an exact functor, and so

$$\text{Mor}_F(h^{A'}, T) \to \text{Mor}_F(h^A, T) \to \text{Mor}_F(h^{A''}, T) \to 0$$

is exact. The natural isomorphism $\text{Mor}_F(h^B, T) \approx TB$ of 2.3 yields $TA' \to TA \to TA'' \to 0$ exact, so T is right exact. □

14.19 *Exercise.* (Gabriel [62, p. 357]) The converse of 14.18 holds when C is a Noetherian category.

14.20 Lemma. *Let C be a small abelian category, and let $M \to E$ be an essential extension in* Adfunct $(C, \text{mod-}\mathbb{Z})$. *If M is a monofunctor, so is E.*

Proof (Freyd) Suppose $f: A' \to A$ is monic in C, but $E(f)$ is not. If $x \in EA'$, is nonzero, and $E(f)x = 0$, define a subobject F of E by

$$FB = \{E(t)x \mid t \in \text{Mor}_C(A', B)\}.$$

Do this for every B in C. Then, for $g: B' \to B$, $E(g)FB' \subseteq FB$, and thus $F(g)$ is defined by restriction of $E(g)$ to FB'. Since FB is a subgroup of EB, F is a group-valued functor (and is the image of the functor morphism $\eta: h^{A'} \to E$, where $\eta(1_{A'}) = x$). Since $F \neq 0$, there exists $B \in \text{obj}\, C$ such that, $FB \cap EB \neq 0$, that is, there is a morphism $t: A' \to B$ with $0 \neq E(t)x \in FB$. If

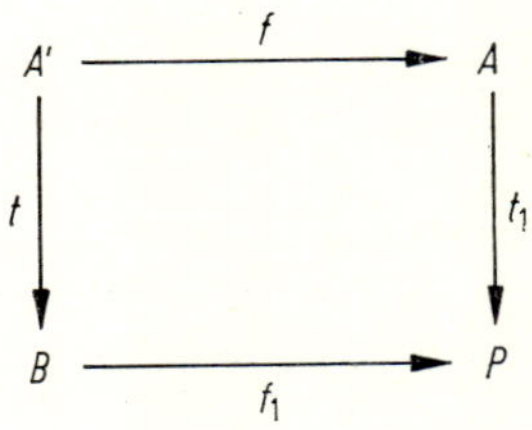

is cocartesian, then 5.10.2 implies that f_1 is monic, since f is. Because $E(f)x = 0$, we have

$$0 = E(t_1)E(f)x = E(f_1)E(t)x = M(f_1)E(t)x.$$

Since M is a monofunctor, $M(f_1)$ is monic, so $E(t)x = 0$, a contradiction. □

14.21 The Exact Embedding Theorem (Freyd [60], Lubkin [60]) *If C is a small abelian category, then there is an exact covariant embedding $C \rightsquigarrow$* mod-$\mathbb{Z}$.

Proof. By 14.2, Adfunct (C, mod-$\mathbb{Z}$) is Grothendieck, and, therefore, every object of Adfunct (C, mod-$\mathbb{Z}$) has an injective hull by 14.17. Now, by 6.20, Adfunct (C, mod-$\mathbb{Z}$) has a projective generator G such that $G : C \rightsquigarrow$ mod-$\mathbb{Z}$ is an embedding. Then, if H is an injective hull of G, then H is right exact by 14.18, and a monofunctor by 14.20. Therefore, H is the desired exact embedding of C in mod-$\mathbb{Z}$. $\square$

***14.22** *Exercise*. 14.22.1 An abelian category C with a generator and cogenerator has an exact embedding $C \rightsquigarrow$ mod-$\mathbb{Z}$.

14.22.1 Every small additive category A is equivalent to a full subcategory C of mod-R, for some ring R. Moreover, the inclusion $C \rightsquigarrow$ mod-R preserves limits, and C is a generating set of mod-R consisting of projective modules. (Cf. 6.20, p. 313.)

Coreflective Categories

Let C' be a subcategory of C. If $A \in \text{obj } C$, a **coreflection** of A in C' is a morphism

$$r(A) : A \to RA$$

such that $RA \in \text{obj } C'$ and for every object A' of C', and morphism $A \to A'$ of C, there exists a unique morphism $RA \to A'$ such that

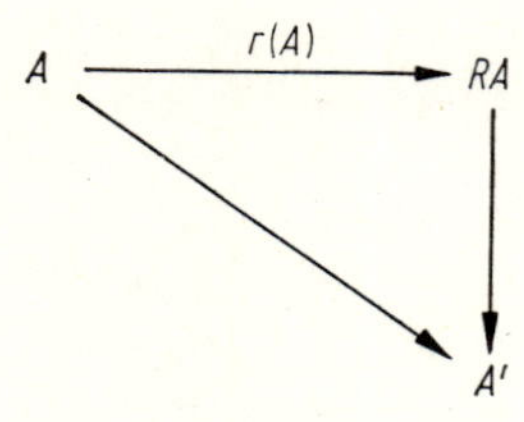

commutes, or equivalently, there is a bijection

$$\text{Mor}_{C'}(RA, A') = \text{Mor}_C(A, 1^C_{C'} A')$$

for every $A' \in \text{obj } C'$, where $1^C_{C'} : C' \rightsquigarrow C$ is the inclusion functor. The category C' is **coreflective** provided that the following two equivalent conditions are fulfilled:

R_1. Every object A of C has a coreflection RA in C'.

R_2. The inclusion functor $C' \rightsquigarrow C$ has a left adjoint $R : C \rightsquigarrow C'$, called the **coreflector** of C' in C.

When this is so, then for the functor $R : C \rightsquigarrow C'$, and object A of C, the object RA is the **coreflection** of A in C'. If C' is full and coreflective, then it is **fully coreflective**.[1]

By duality we say a subcategory C' is **reflective** if the inclusion functor has a right adjoint.

Exercises for Chapter 14

1. If C' is fully coreflective in C, and if $F : L \rightsquigarrow C'$ is a functor, and if $F' : L \rightsquigarrow C$ is the induced functor, then

$$\lim_{L,C'} F = R(\lim_{L,C} F') .$$

***2.** A fully coreflective subcategory C' of a category C is abelian (and $AB5$) if C is abelian (and $AB5$), provided that the coreflector $R : C \rightsquigarrow C'$ is exact. ($R : C \rightsquigarrow C'$ is always colimit preserving, for any category C.)

***3.** If C is small, and abelian, then every object F of Adfunct (C, mod-$\mathbb{Z}$) has a coreflector in the full subcategory Sex (C, mod-$\mathbb{Z}$) consisting of left exact functors.

***4.** If C is small, and abelian, then the full subcategory Sex (C, mod-$\mathbb{Z}$) of Adfunct (C, mod-$\mathbb{Z}$), consisting of all left exact functors $C \rightsquigarrow$ mod-$\mathbb{Z}$, is coreflective, and the coreflector is an exact colimit preserving functor. Furthermore, Sex (C, mod-$\mathbb{Z}$) is an $AB5$ category with a generator and injective cogenerator.

***5.** If C is a small abelian category, then the full subcategory of Adfunct (C, mod-$\mathbb{Z}$) consisting of all the right exact functors is reflective.

Notes for Chapter 14

The introduction to Mitchell's book, the notes to MacLane [63, 72], and the afterword of Freyd [64], constitute a brief history of abelian categories. Gabriel [62, p. 363] credits the embedding theorem 14.21 to "students of Eilenberg", and in my off-print, Gabriel wrote in "P. Freyd and S. Lubkin". The theorem in Bass [68, p. 22] is ascribed to Freyd, Grothendieck, and Lubkin, but no references are given.

References

Bucur and Deleanu [68], Freyd [60, 64], Gabriel [62], Grothendieck [57], Lubkin [60], MacLane [63, 72], Popesco and Gabriel [64], Mitchell [65], Takeuchi [71].

[1] In this we adopt the notation of Mitchell [65]. However elsewhere (e.g. MacLane [72, p. 89]) reflective is employed instead of coreflective.

Chapter 15

Quotient Categories and Localizing Functors

Let $\mathscr{A}$ be an abelian category. A **Serre class** or **subcategory** of $\mathscr{A}$ is a nonempty full subcategory $\mathscr{S}$ such that for every exact sequence

$$0 \to A \to B \to C \to 0$$

in $\mathscr{A}$ it is true that

$$B \in \mathscr{S} \Leftrightarrow A \in \mathscr{S} \,\&\, C \in \mathscr{S}.$$

Thus, $B \in \mathscr{S}$ iff every sub- and quotient-object of B is in $\mathscr{S}$. The intersection of any number of Serre classes is a Serre class.

15.1 *Examples.*

15.1.1 The subclass of mod-A consisting of Noetherian (resp. Artinian) right A-modules is a Serre class.

15.1.2 If $F : \mathscr{A} \rightsquigarrow \mathscr{B}$ is a functor, then the **kernel** of F is defined to be the full subcategory with objects:

$$\operatorname{Ker} F = \mathscr{S} = \{A \in \operatorname{obj} \mathscr{A} \mid FA = 0\} = \{A \in \operatorname{obj} \mathscr{A} \mid F(1_A) = 0\}.$$

If F is exact, then the kernel of F is a Serre class closed under coproducts.

15.1.3 If $\mathscr{F}$ is a family of injective objects in $\mathscr{A}$, then there is a Serre class

$$\mathscr{S} = \{A \in \mathscr{A} \mid h_E(A) = 0 \;\forall E \in \mathscr{F}\} = \bigcap_{E \in \mathscr{F}} \ker h_E.$$

This class is closed under coproducts.

The Serre class in Example 15.1.3 is the intersection of the Serre classes defined for each of the exact functors h_E. If $\{A_i \mid i \in I\}$ is a family of objects in $\mathscr{S}$, then, since h_E converts coproducts into products (13.8)*, we have

$$h_E(\textstyle\coprod_i A_i) \approx \prod_i h_E(A_i) = 0$$

so $\coprod_i A_i \in \mathscr{S}$. Hence, if, in Example 15.1.3, $\mathscr{A}$ has arbitrary coproducts, so does $\mathscr{S}$.

15.1.4 The subcategories of mod-$\mathbb{Z}$ listed are Serre classes: (a) all torsion groups, (b) all finitely generated groups, (c) all finite groups, (d) all p-groups.

15.1′ *Exercises.*

15.1′.1 A nonempty subcategory $\mathscr{S}$ of an abelian category $\mathscr{A}$ is a Serre subcategory only if

$$A \to B \to C$$

exact in $\mathscr{A}$, and A, C objects of $\mathscr{S}$ implies $B \in \mathscr{S}$.

15.1′.2 If $\mathscr{S}$ is a Serre subcategory of $\mathscr{A}$, then a morphism $f: A \to B$ is said to be $\mathscr{S}$-monic (resp. $\mathscr{S}$-epic; $\mathscr{S}$-equivalence) if and only if $\ker f \in \mathscr{S}$ (resp. $\operatorname{coker} f \in \mathscr{S}$; f is both $\mathscr{S}$-monic and $\mathscr{S}$-epic). If $A \xrightarrow{f} B \xrightarrow{g} C$ is exact in $\mathscr{A}$, then prove:

(a) f and g $\mathscr{S}$-blank implies that gf is (fill in the blank).

(b) If gf is $\mathscr{S}$-blank, then f is.

(c) If f is a $\mathscr{S}$-equivalence, then g is $\mathscr{S}$-blank if and only if gf is.

(d) If g is a $\mathscr{S}$-equivalence, then f is $\mathscr{S}$-blank if and only if gf is.

15.2 **Proposition.** *If $\mathscr{A}$ is an abelian category, then any Serre subcategory is abelian.*

Proof. If $A_i \in \mathscr{S}$, $i = 1, 2$, where $\mathscr{S}$ is a Serre subcategory, then there is an exact sequence

$$0 \to A_1 \xrightarrow{u_1} \prod_{i=1}^{2} A_i \xrightarrow{p_2} A_2 \to 0$$

which shows that $\prod_{i=1}^{2} A_i \in \mathscr{S}$. By induction, $\mathscr{S}$ has finite products (coproducts), so $\mathscr{S}$ is additive. The axioms for abelian category are satisfied in $\mathscr{S}$ because kernels and cokernels of morphisms of $\mathscr{S}$ exist in $\mathscr{A}$, and belong to $\mathscr{S}$. From this ($ab1$), ($ab2$), and ($ab3$) of 6.1 follow. ▯

The notion of a Serre class $\mathscr{S}$ of $\mathscr{A}$ will be used to define a **quotient category** $\mathscr{A}/\mathscr{S}$ (after Grothendieck [57]). The objects of $\mathscr{A}/\mathscr{S}$ are the objects of $\mathscr{A}$, and morphisms of $\mathscr{A}/\mathscr{S}$ are defined by 15.3.

Abbreviate $\operatorname{Mor}_{\mathscr{A}}(A, B)$ by ${}_{\mathscr{A}}(A, B)$, or (A, B). If A, B, A', B' are objects of $\mathscr{A}$, and $A' \subseteq A$, $B' \subseteq B$, the canonical epic $p_{B/B'}: B \to B/B'$, and monic $i^A_{A'}: A' \to A$, define a homomorphism

$$(i^A_{A'}, p_{B/B'}): (A, B) \to (A', B/B') \tag{1}$$

which sends $f: A \to B$ onto $p_{B/B'} \circ f_0$, where $f_0: A' \to B$ is induced by f.

Let I denote the class of ordered pairs $\langle A', B' \rangle$, where $A' \subseteq A$, $B' \subseteq B$, and $A/A' \in \mathscr{S}$, $B' \in \mathscr{S}$. Order I by

$$\langle A', B' \rangle \leq \langle A'', B'' \rangle \Leftrightarrow A' \supseteq A'' \;\&\; B' \subseteq B''.$$

The class I is directed by this order since $\langle A' \cap A'', B' + B''\rangle$ belongs to I, and follows both $\langle A', B'\rangle$ and $\langle A'', B''\rangle$ in the ordering. Furthermore, the homomorphisms (1) order the class of groups

$$\{(A', B/B') \mid A/A' \in \mathscr{S}, \ B' \in \mathscr{S}\} \tag{2}$$

which is directed by the class I just defined; that is, $(A' \cap A'', B/(B' + B''))$ follows both $(A', B/B')$ and $(A'', B/B'')$ in this ordering of (2). Thus:

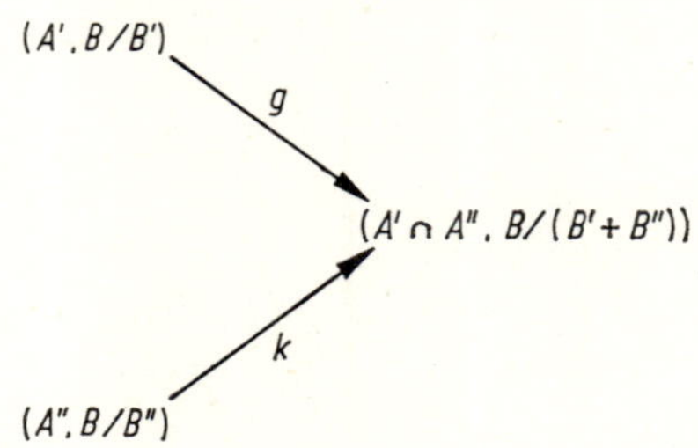

where g is defined as in (1) by the projection $B/B' \to B/B' \,/\, (B' + B'')/B' \approx B/(B' + B'')$, and the injection $A' \cap A'' \to A'$, and similarly for k. This shows that the class (2) ordered by the homomorphisms (1) is directed by I. We now proceed to the following definition.

15.3 Definition.

$$\operatorname{Mor}_{\mathscr{A}/\mathscr{S}}(A, B) = {}_{\mathscr{A}/\mathscr{S}}(A, B) = \varinjlim_{\substack{A/A' \in \mathscr{S} \\ B' \in \mathscr{S}}} (A', B/B').$$

If $f' : A' \to B/B'$ and $f'' : A'' \to B/B''$ are morphisms of $\mathscr{A}$, where $\langle A', B'\rangle$ and $\langle A'', B''\rangle$ are elements of I, then we say $f' \equiv f'' \pmod{\mathscr{S}}$ if and only if there exists $\langle A''', B'''\rangle$ in I following both $\langle A', B'\rangle$ and $\langle A'', B''\rangle$ such that f' and f'' induce the same morphism $A''' \to B/B'''$. Clearly this defines an equivalence relation in Mor $\mathscr{A}$. Let G denote the set of all equivalence classes $\{f\}$ under this relation. If $f_i : A_i \to B/B_i$, where $\langle A_i, B_i\rangle \in I$, $i = 1, 2$, then define addition in G by

$$\{f_1\} + \{f_2\} = \{\bar{f}_1 + \bar{f}_2\}$$

where $\bar{f}_i$ denotes the morphism $A_1 \cap A_2 \to B/(B_1 + B_2)$ induced by f_i, $i = 1, 2$. This definition is independent of the representatives f_i in $\{f_i\}$, $i = 1, 2$, and G is an abelian group under this operation. If $\langle A_1, B_1\rangle \in I$, then we have a mapping

$$\begin{cases} \operatorname{Mor}(A_1, B/B_1) \to G \\ \qquad\qquad\quad f \mapsto \{f\} \end{cases}$$

which is compatible with our directed system, and G is the direct limit; that is, there is an isomorphism

$$\operatorname{Mor}_{\mathscr{A}/\mathscr{S}}(A, B) = \varinjlim_{\langle A', B'\rangle \in I} \operatorname{Mor}_{\mathscr{A}}(A', B/B') \approx G.$$

Let L denote the isomorphism $\operatorname{Mor}_{\mathscr{A}/\mathscr{S}}(A, B) \to G$ just defined. If $t \in \operatorname{Mor}_{\mathscr{A}/\mathscr{S}}(A, B)$, and if $L(t) = \{f\}$, we say that t is **represented by** a morphism $f : A' \to B/B'$, where $A/A' \in \mathscr{S}$, and $B' \in \mathscr{S}$. Other terminologies for this are: *t comes from*, or *t is the image in the direct limit of*, a morphism f just described.

To define composition of maps in $\mathscr{A}/\mathscr{S}$, let $\bar{f} \in {}_{\mathscr{A}/\mathscr{S}}(A, B)$, $\bar{g} \in {}_{\mathscr{A}/\mathscr{S}}(B, C)$. Then $\bar{f}$ comes from an $f \in {}_{\mathscr{A}}(A', B/B')$, and $\bar{g}$ comes from a $g \in (B'', C/C')$, with A/A', B', B/B'', $C' \in \mathscr{S}$. If $A'' = f^{-1}\big((B'' + B')/B'\big)$, then $A/A'' \in \mathscr{S}$, and f induces a map $f' \in {}_{\mathscr{A}}\big(A'', (B'' + B')/B'\big)$. If $C''/C' = g(B'' \cap B')$, then $C'' \in \mathscr{S}$, and g induces a map $g' \in {}_{\mathscr{A}}(B''/(B'' \cap B'), C/C'')$. Now let h be the composition of the maps

$$A'' \xrightarrow{f'} (B'' + B')/B' \approx B''/(B'' \cap B') \xrightarrow{g'} C/C''.$$

Since A/A'' and C'' are in $\mathscr{S}$, h determines an element $\bar{h} \in {}_{\mathscr{A}/\mathscr{S}}(A, C)$. The element $\bar{h}$ so determined is unique, and $\bar{g} \circ \bar{f} = \bar{h}$ defines a bilinear composition, making $\mathscr{A}/\mathscr{S}$ an additive category. This sketches part of a proof of the following lemma.

15.4 Lemma. *If $\mathscr{A}$ is an abelian category, and if $\mathscr{S}$ is a Serre class, then $\mathscr{A}/\mathscr{S}$ is an additive category, and the functor $T : \mathscr{A} \rightsquigarrow \mathscr{A}/\mathscr{S}$ that maps $A \to A$ for every object $A \in \mathscr{A}$, and that maps $f : A \to B$ onto the limit Tf in the direct limit $\varinjlim {}_{\mathscr{A}}(A', B/B')$, is an additive functor, called* ***canonical.***

Proof. To show that $\mathscr{A}/\mathscr{S}$ is additive, it remains only to show that finite coproducts exist. This however follows from Lemma 15.6 which shows that T preserves exact sequences. Any exact functor preserves finite coproducts, and so T is additive (see 3.38 and 3.40.1), and $\mathscr{A}/\mathscr{S}$ is additive. □

Actually, $\mathscr{A}/\mathscr{S}$ is an abelian category 15.7.

15.5 Lemma. *If $f : A \to B$ is a morphism of $\mathscr{A}$, then*

(a) $Tf = 0 \Leftrightarrow \operatorname{im} f \in \mathscr{S}$;

(b) Tf *is monic* $\Leftrightarrow \ker f \in \mathscr{S}$;

(c) Tf *is epic* $\Leftrightarrow \operatorname{coker} f \in \mathscr{S}$.

Proof. (a) If $\operatorname{im} f \in \mathscr{S}$, then the image of f in ${}_{\mathscr{A}}(A, B/\operatorname{im} f)$ is zero, hence the image of f in the direct limit $\varinjlim {}_{\mathscr{A}}(A', B/B')$ is zero; that is,

$Tf = 0$. Conversely, if $Tf = 0$, then there exist $A' \subset A$, $B' \subset B$ such that f induces the zero morphism $A' \to B/B'$, whence f carries A' into $f(A') \subseteq B'$. Then $f(A') \in \mathscr{S}$, and exactitude of

$$0 \to f(A') \to \operatorname{im} f \to A/(A' + \operatorname{Ker} f) \to 0$$

implies $\operatorname{im} f \in \mathscr{S}$.

(b) If Tf is monic, and $i : \ker f \to A$ is canonical, then $f \circ i = 0$, and $Tf \circ Ti = 0$, so $Ti = 0$, whence $\ker f = \operatorname{im} i \in \mathscr{S}$. Conversely, let $\bar{f} : TC \to TA$ be a nonzero morphism in $\mathscr{A}/\mathscr{S}$, which is the image of $f' : C' \to A/A'$, where $C/C', A' \in \mathscr{S}$. Replacing A' by $A' + \ker f$, we can suppose $\ker f \subseteq A'$. Then f induces a monic $f'' : A/A' \to B/f(A')$. Since $\bar{f} \neq 0$, $\operatorname{im} f$ and $\operatorname{im}(f'' \circ f')$ lie outside of $\mathscr{S}$, hence $Tf \circ \bar{f} \neq 0$. Thus Tf is monic.

(c) is proved dually to (b). □

15.6 Lemma. *Let*

$$0 \to K \xrightarrow{i} A \xrightarrow{f} B \xrightarrow{q} Q \to 0$$

be exact in $\mathscr{A}$.

15.6.1 *Then*

$$0 \to TK \xrightarrow{Ti} TA \xrightarrow{Tf} TB \xrightarrow{Tq} TQ \to 0$$

is exact in $\mathscr{A}/\mathscr{S}$; *that is,* $T : \mathscr{A} \to \mathscr{A}/\mathscr{S}$ *is an exact functor.*

15.6.2 Tf *is an equivalence in* $\mathscr{A}/\mathscr{S} \Leftrightarrow \ker f \in \mathscr{S}$ *and* $\operatorname{coker} f \in \mathscr{S}$.

Proof of 15.6.1. Ti is monic by the preceding lemma. If $\bar{k} : TX \to TA$ is a morphism such that $Tf \circ \bar{k} = 0$, then we must show that

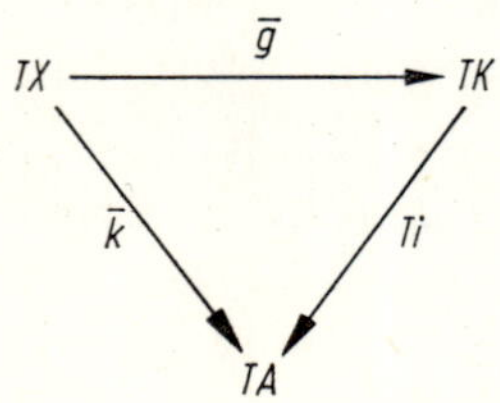

commutes for a suitable morphism $\bar{g}$. To do this, note that $\bar{k}$ is the image of a morphism $k : X' \to A/A'$ for some $X/X', A' \in \mathscr{S}$. The diagram

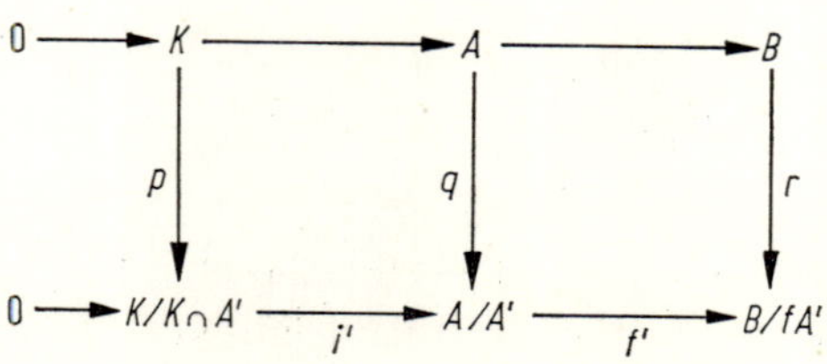

commutes, where p, q, and r are the canonical morphisms. Since $Tf \circ \bar{k} = 0$, then $f' \circ k(X') \in \mathscr{S}$. Let $X'' = k^{-1}(\operatorname{im} i')$. Then X'/X'' and $X/X'' \in \mathscr{S}$, and k restricted to X'' is the composition of $g: X'' \to K/K \cap A'$ and i'. If $\bar{g}$ is the image of g in ${}_{\mathscr{A}/\mathscr{S}}(X, K)$, then $Ti \circ \bar{g} = \bar{k}$. The cokernel part of 15.6.2 is dual.

Proof of 15.6.2. If Tf is an equivalence, then $\ker f \in \mathscr{S}$ and $\operatorname{coker} f \in \mathscr{S}$ by 15.5. Conversely, suppose this is so. Let $q: A \to \operatorname{coim} f$, $j: \operatorname{im} f \to B$, and $h: \operatorname{coim} f \to \operatorname{im} f$ be the canonical morphisms. The identity morphism $\operatorname{coim} f \to \operatorname{coim} f$ belongs to ${}_{\mathscr{A}}(\operatorname{coim} f, A/i(K))$, and its image in ${}_{\mathscr{A}/\mathscr{S}}(\operatorname{coim} f, A)$ is an inverse of Tq. Similarly, Tj is an equivalence, hence so is $Tf = Tj \circ Th \circ Tq$. ▯

15.7 Theorem. *If $\mathscr{S}$ is a Serre class of an abelian category $\mathscr{A}$, then the quotient category $\mathscr{A}/\mathscr{S}$ is abelian, and the canonical functor $\mathscr{A} \rightsquigarrow \mathscr{A}/\mathscr{S}$ is exact.*

Proof. Let $\bar{f}: TA \to TB$ be any morphism in $\mathscr{A}/\mathscr{S}$. Then $\bar{f}$ is represented by some $f \in {}_{\mathscr{A}}(A', B/B')$, where $A/A', B' \in \mathscr{S}$.

The commutative diagram

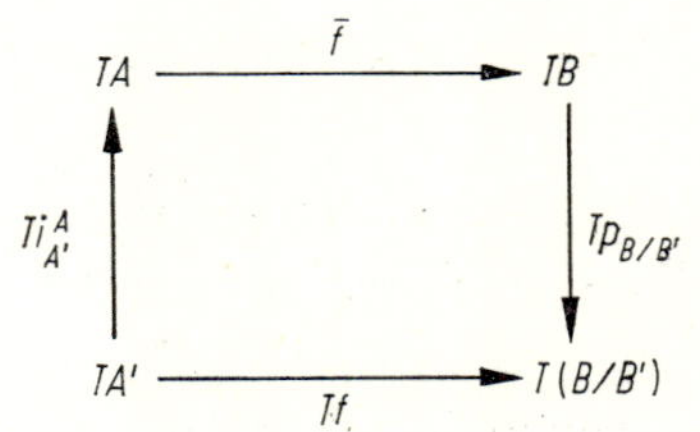

and the fact that $Ti_{A'}^{A}$ and $Tp_{B/B'}$ are equivalences, prove that $\bar{f}$ has a kernel (cokernel) if and only if Tf does. But Tf has a kernel and cokernel by Lemma 15.6. This proves $(Ab1)$ and, moreover, 15.6.1 similarly establishes $(Ab2)$ and $(Ab3)$. ▯

15.8 Corollary. *Let $\mathscr{S}$ be a Serre class of $\mathscr{A}$, and consider*

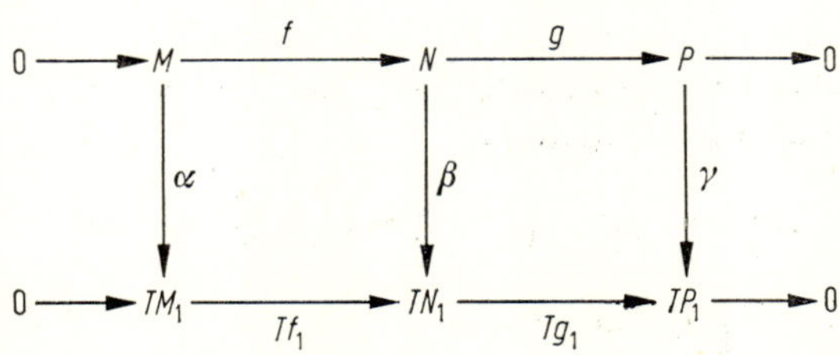

where $M, N, P \in$ obj $\mathscr{A}/\mathscr{S}$, and $M_1, N_1, P_1 \in$ obj $\mathscr{A}$. If the top row is exact, then there exists an exact sequence $0 \to M_1 \xrightarrow{f_1} N_1 \xrightarrow{g_1} P_1 \to 0$ of $\mathscr{A}$, and equivalences *α, β, and γ in $\mathscr{A}/\mathscr{S}$ such that:*

(a) *the diagram commutes;*

(b) *the bottom row is exact.*

Proof. The morphism f is the image in the direct limit of an element $f' \in_{\mathscr{A}} (M', N/N')$, where M/M', $N' \in \mathscr{S}$. Since Tf' is monic, then $\ker f' \in \mathscr{S}$ by a preceding lemma. Let $q : M' \to \operatorname{coim} f'$ be the canonical morphism, let $M_1 = \operatorname{coim} f'$, let $N_1 = N/N'$, let f_1 be the morphism induced by f', let $P_1 = \operatorname{coker} f_1$, let $g_1 : N_1 \to P_1$ be canonical, let $\alpha = Tq \circ (T i^M_{M'})^{-1}$, let $\beta = T p_{N/N'}$, and let γ be the morphism induced by $(Tg_1) \circ \beta$. □

15.9 Corollary. *Let $F : \mathscr{A} \rightsquigarrow \mathscr{B}$ be a functor of abelian categories such that $\{X \in \operatorname{obj} \mathscr{A} \mid FX = 0\}$ contains a Serre class $\mathscr{S}$. Then there exists a unique functor $H : \mathscr{A}/\mathscr{S} \to \mathscr{B}$ such that*

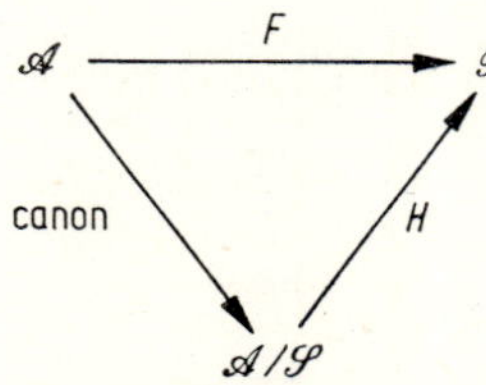

commutes, that is, such that there exists a natural equivalence of functors $H \circ T \approx F$, where $T : \mathscr{A} \rightsquigarrow \mathscr{A}/\mathscr{S}$ is canonical. If H' is another such functor, then $H \approx H'$.

Proof. Consider a commutative diagram

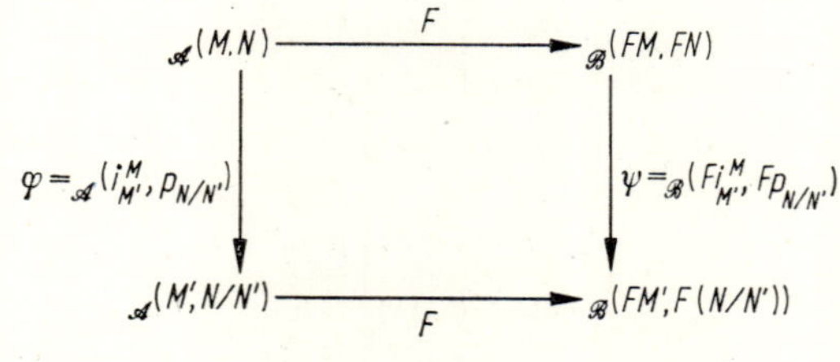

where φ is defined by (1) following 15.2, and M/M', $N' \in \mathscr{S}$. Then ψ is bijective, and we have a mapping

$$\psi^{-1} \circ F : \varinjlim {}_{\mathscr{A}}(M', N/N') \to {}_{\mathscr{B}}(FM, FN).$$

Thus,

$$\begin{cases} H : \mathscr{A}/\mathscr{S} \rightsquigarrow \mathscr{B} \\ H : M \mapsto FM \\ H : {}_{\mathscr{A}/\mathscr{S}}(M, N) \mapsto {}_{\mathscr{B}}(FM, FN) \end{cases}$$

is the desired functor. *If also* $H' \circ T \approx F$, *then* $H(TX) \approx H'(TX)$, *that is* $H(X) \approx H'(X)$, *for every object* X *of* $\mathscr{A}/\mathscr{S}$, *so* $H \approx H'$. ∎

15.10 Corollary. *Let* $\mathscr{A}$ *and* $\mathscr{B}$ *be abelian categories, let* $\mathscr{S}$ *be a Serre class in* $\mathscr{A}$, *and let* $H : \mathscr{A}/\mathscr{S} \to \mathscr{B}$ *be a functor. Then H is exact if and only if the composition*

$$\mathscr{A} \underset{\text{canon}}{\rightsquigarrow} \mathscr{A}/\mathscr{S} \underset{H}{\rightsquigarrow} \mathscr{B}$$

is an exact functor.

Proof. Since the canonical functor $T : \mathscr{A} \rightsquigarrow \mathscr{A}/\mathscr{S}$ is exact, then $H \circ T$ is exact if H is exact. Conversely, assume $H \circ T$ is exact. If $0 \to M \xrightarrow{f} N \xrightarrow{g} P \to 0$ is exact in $\mathscr{A}/\mathscr{S}$, and if the notations of the first corollary are made, then

$$0 \to HTM_1 \xrightarrow{HTf_1} HTN_1 \xrightarrow{HTg_1} HTP_1 \to 0$$

is exact, and therefore

$$0 \to HM \xrightarrow{Hf} HN \xrightarrow{Hg} HP \to 0$$

is exact. ∎

The Section Functor

Let $\mathscr{S}$ be a Serre class of an abelian category $\mathscr{A}$, and assume that the canonical functor $T : \mathscr{A} \rightsquigarrow \mathscr{A}/\mathscr{S}$ has a right adjoint $\mathscr{A}/\mathscr{S} \rightsquigarrow \mathscr{A}$. Any two adjoints to T are equivalent functors 5.3. Choose a representative right adjoint functor $S : \mathscr{A}/\mathscr{S} \rightsquigarrow \mathscr{A}$, and call it the **section functor**. A Serre subcategory $\mathscr{S}$ is an abelian category, and $\mathscr{S}$ is called a **localizing subcategory** of $\mathscr{A}$ whenever the section functor $S : \mathscr{A}/\mathscr{S} \rightsquigarrow \mathscr{A}$ exists. Then, the composite functor $ST : \mathscr{A} \rightsquigarrow \mathscr{A}$ is called the **localizing functor.**

15.11 Theorem. *Let* $\mathscr{A}$ *be a Grothendieck category, and let* $\mathscr{S}$ *be a Serre subcategory of* $\mathscr{A}$. *Then, the following conditions are equivalent:*

(a) $\mathscr{S}$ *is a localizing subcategory of* $\mathscr{A}$.

(b) *The full subcategory* $\mathscr{S}$ *of* $\mathscr{A}$ *has arbitrary coproducts.*

(c) *The category* $\mathscr{S}$ *is cocomplete.*

(d) *The direct limit in* $\mathscr{A}$ *of a directed family of objects in* $\mathscr{S}$ *belongs to* $\mathscr{S}$.

(e) *The canonical functor* $T : \mathscr{A} \rightsquigarrow \mathscr{A}/\mathscr{S}$ *is right continuous.*

Proof. An abelian category is cocomplete if and only if it has coproducts 6.3, so (b) ⇔ (c).

(b) $\Leftrightarrow$ (d). Since $\mathscr{S}$ contains, along with any $X \in \text{obj}\,\mathscr{S}$, any object in $\mathscr{A}$ equivalent to X, it follows that (b) holds if and only if the coproduct in $\mathscr{A}$ of any family of objects of $\mathscr{S}$ belong to $\mathscr{S}$. Since a coproduct is a direct limit 5.8, and since a direct limit is a cokernel of a coproduct 5.11, we deduce (b) $\Leftrightarrow$ (d).

(c) $\Leftrightarrow$ (e). Assuming (e), we have

$$\textstyle\coprod_i T A_i \approx T(\coprod_i A_i) \in \text{obj}\,\mathscr{S}$$

for any collection of objects $\{T A_i\}$ of $\mathscr{S}$, so $\mathscr{S}$ has (arbitrary) coproducts. Conversely, if $\coprod_i T A_i \in \text{obj}\,\mathscr{S}$, and if

$$f : \textstyle\coprod_i T A_i \to T(\coprod_i A_i)$$

is the canonical morphism in $\mathscr{A}$, then, since cokernel and kernel belong to $\mathscr{S}$, by 15.5, $T(f)$ is an isomorphism in $\mathscr{A}/\mathscr{S}$. Thus, T is right continuous.

Let T be a functor $\mathscr{A} \rightsquigarrow \mathscr{B}$ of abelian categories. If $\mathscr{A}$ has a generator, is cocomplete, and colocally small, then T has an adjoint if and only if T is right continuous 5.52. Under the hypotheses of the theorem, by 14.17, $\mathscr{A}$ has an (injective) cogenerator, hence is colocally small, so (a) $\Leftrightarrow$ (e). ▯

15.12 Proposition. *Let M be an object of $\mathscr{A}$, and let $\mathscr{S}$ be a Serre class of $\mathscr{A}$. Then M is said to be* $\mathscr{S}$**-closed** *provided that M satisfies the following equivalent conditions:*

(a) *If $f : P \to Q$ is a morphism such that* $\ker f$ *and* $\operatorname{coker} f$ *belong to $\mathscr{S}$, then $h_M(f) : h_M(Q) \to h_M(P)$ is a bijection.*

(b) *Every subobject of M belonging to $\mathscr{S}$ is zero, and for each $P \in \text{obj}\,\mathscr{S}$, every exact sequence*

$$0 \to M \xrightarrow{f} N \to P \to 0$$

splits.

(c) *For every $P \in \text{obj}\,\mathscr{A}$, the canonical functor $T : \mathscr{A} \to \mathscr{A}/\mathscr{S}$ induces a bijection*

$$T(P, M) : h_M(P) \to h_{TM}(TP).$$

Proof. (a) $\Rightarrow$ (b). If $K \subseteq M$, and $K \in \text{obj}\,\mathscr{S}$, then take $P = M$, and $f : M \to M/K$ canonical in (a), and write $1_M = v \circ f = h_M(f)$, where $v : M/K \to M$. Thus, f is monic, so $K = 0$.

For the second part, by (a), there exists $g : N \to M$ such that $gf = 1_M$, so f splits.

(b) $\Rightarrow$ (c). By (b), if $P,\ M \in \text{obj}\,\mathscr{A}$, we can write

$$_{\mathscr{A}/\mathscr{S}}(TP, TM) = \varinjlim_{P/P' \in \mathscr{S}} {}_{\mathscr{A}}(P', M)$$

so every element in ${}_{\mathscr{A}/\mathscr{S}}(TP, TM)$ is the image of a morphism $f : P' \to M$. By 5.10, the pushout

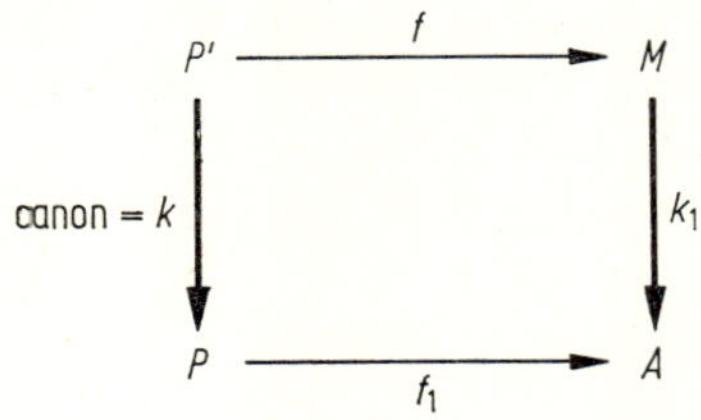

yields a monic $k_1 : M \to A$ such that $\operatorname{coker} k_1 \approx P/P'$, so by (b), there exists a morphism $v : A \to M$ such that $v \circ k_1 = 1_M$. Then

$$f = 1_M f = v \circ (k_1 f) = v \circ (f_1 k) = (v \circ f_1) k$$

so T induces a surjection $h_M(P) \to h_{TM}(TP)$. If $g : P \to M$ is nonzero, then $\operatorname{im} g \notin \operatorname{obj} \mathscr{S}$ by (b), and hence $Tg : TP \to TM$ is nonzero. Thus, T induces a bijection.

(c) $\Rightarrow$ (a). Given $f : P \to Q$, the diagram

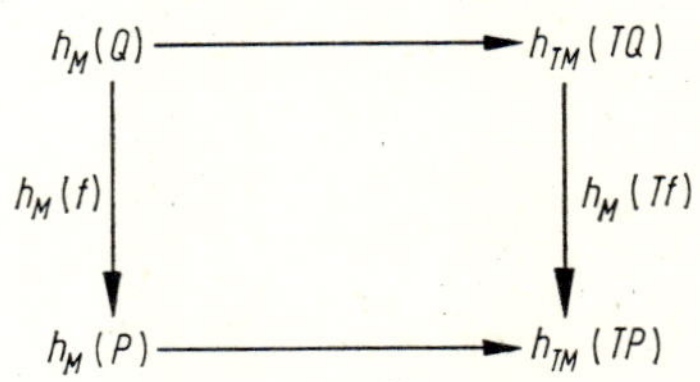

commutes and the horizontal maps are bijections. If $\ker f$ and $\operatorname{coker} f$ belong to $\mathscr{S}$, then Tf, whence $h_M(f)$, is an equivalence. ▯

15.13 Corollary. *If $\mathscr{S}$ is a localizing subcategory of $\mathscr{A}$, and if $S : \mathscr{A}/\mathscr{S} \rightsquigarrow \mathscr{A}$ denotes the section functor, then every object of $\mathscr{A}$ of the form SY is $\mathscr{S}$-closed.*

Proof. Let $f : P \to Q$ have kernel and cokernel in $\mathscr{S}$. The diagram

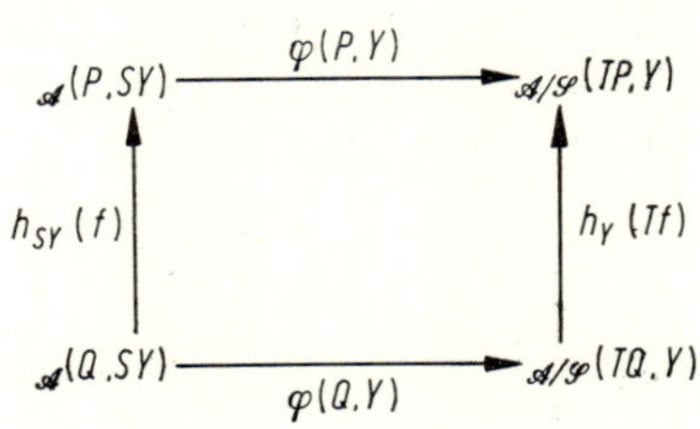

commutes, and the horizontal maps are isomorphisms. Since Tf is an isomorphism, then $h_Y(Tf)$ is bijective, and so is $h_{SY}(f)$. Thus 15.12(a) holds. ▯

Since S is adjoint to T, by 5.67, we have the natural isomorphism

$$\varphi = \varphi_{(X,Y)} : {}_{\mathscr{A}}(X, SY) \to {}_{\mathscr{A}/\mathscr{S}}(TX, Y)$$

with inverse

$$\psi = \psi_{(X,Y)} : {}_{\mathscr{A}/\mathscr{S}}(TX, Y) \to {}_{\mathscr{A}}(X, SY)$$

and natural transformations of functors

$$\Phi : T \circ S \to 1_{\mathscr{A}/\mathscr{S}}$$

$$\Psi : 1_{\mathscr{A}} \to S \circ T.$$

15.14 Proposition. *Let $\mathscr{S}$ be a localizing subcategory of $\mathscr{A}$, and let $S : \mathscr{A}/\mathscr{S} \rightsquigarrow \mathscr{A}$ be the section functor. Then:*

15.14.1 *The natural transformation*

$$\Phi : TS \to 1_{\mathscr{A}/\mathscr{S}}$$

is an equivalence.

15.14.2 *$\mathscr{S}$ contains* $\ker \Psi(M)$ *and* $\operatorname{coker} \Psi(M)$ *for all* $M \in \mathscr{A}$.

Proof of 15.14.1. If $M \in \mathscr{A}$ and $N \in \mathscr{A}/\mathscr{S}$, then

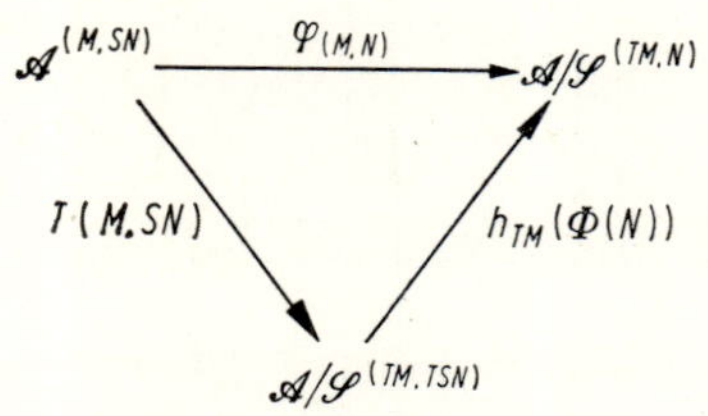

commutes. By the corollary, SN is $\mathscr{S}$-closed, so by the proposition, $T(M, SN)$ is a bijection. Since $\varphi_{(M,N)}$ is an isomorphism, it follows that $h_{TM}(\Phi(N))$ is a bijection for all M. Since every object of $\mathscr{A}/\mathscr{S}$ is equivalent to an object of the form TM, then $h_{TM} \approx 1_{\mathscr{A}/\mathscr{S}}$, so $\Phi(N)$ is an equivalence.

Proof of 15.14.2. By 5.67.2(a) the composition of $T\Psi(M)$ and $\Phi(TM)$ is the identity morphism of TM. Since $\Phi(TM)$ is an isomorphism, it follows that $T\Psi(M)$ is an isomorphism,

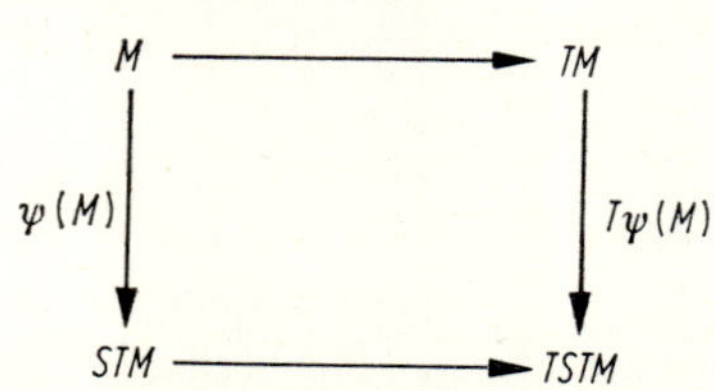

and that $\ker \Psi(M) \in \ker T = \mathscr{S}$. Similarly, $\operatorname{coker} \Psi(M) \in \mathscr{S}$. $\square$

15.15 Corollary. *If $\mathscr{S}$ is a localizing subcategory of $\mathscr{A}$, then an object M of $\mathscr{A}$ is $\mathscr{S}$-closed if and only if*

$$\Psi(M) : M \to STM$$

is an equivalence. Moreover, for any object M, $\ker \Psi(M)$ *is the largest subobject of M belong to $\mathscr{S}$, denoted $\mathscr{S}(M)$, and $\Psi(M/\mathscr{S}(M))$ is an equivalence.*

Proof. By a preceding corollary, any object in the image of S is $\mathscr{S}$-closed. Thus, $M \approx STM$ implies M is $\mathscr{S}$-closed. Conversely, since $\ker \Psi(M)$ belongs to $\mathscr{S}$ for any M, if M is $\mathscr{S}$-closed, necessarily $\ker \Psi(M) = 0$. Dually, $STM \to \operatorname{coker} \Psi(M)$ splits by 15.12.2. Since every subobject of STM belonging to $\mathscr{S}$ is zero, $\operatorname{coker} \Psi(M) = 0$, and $\Psi(M)$ is an equivalence.

From the functorial property of Ψ, it follows for any object M of $\mathscr{A}$ that $\Psi(M/\mathscr{S}(M))$ is an equivalence. Thus, $M/\mathscr{S}(M)$ is $\mathscr{S}$-closed, and hence $M/\mathscr{S}(M)$ has no subobjects belonging to $\mathscr{S}$ (15.12). This implies that $\mathscr{S}(M)$ contains every subobject of M in $\mathscr{S}$. □

15.16 Corollary. *If $\mathscr{S}$ is a localizing subcategory of $\mathscr{A}$, and if $\{U_i \mid i \in I\}$ is a family of generators of $\mathscr{A}$, then $\{TU_i \mid i \in I\}$ is family of generators of $\mathscr{A}/\mathscr{S}$.*

Proof. Exercise.

15.17 *Exercise.*

15.17.1 If $\mathscr{S}$ is a localizing subcategory of $\mathscr{A}$, and if C denotes the full subcategory of $\mathscr{A}$ consisting of the class of $\mathscr{S}$-closed objects of $\mathscr{A}$, then C is a coreflective subcategory, and the inclusion functor $C \rightsquigarrow \mathscr{A}$ has a left adjoint which assigns to each object M, an $\mathscr{S}$-closed object $\approx M/\mathscr{S}(M)$. Conclude that C is an abelian category (but not necessarily an abelian subcategory of $\mathscr{A}$.)

15.17.2 Let C be a small abelian category. A covariant additive functor $F : C \rightsquigarrow$ mod-$\mathbb{Z}$ is **weakly effaceable** if for all objects A of C, and each $x \in F(A)$, there is a monic $0 \to A \xrightarrow{f} B$ such that $F(f)x = 0$.

The full subcategory $\mathscr{S}$ of Adfunct (C, mod-$\mathbb{Z}$) consisting of all weakly effaceable functors is a localizing subcategory. Moreover, the objects of the full subcategory of Adfunct (C, mod-$\mathbb{Z}$) consisting of $\mathscr{S}$-closed objects consists of the left exact functors $C \rightsquigarrow$ mod-$\mathbb{Z}$. (Cf. Exercises Chapter 14, especially 4.)

*15.17.3 (Mitchell) Let C be small abelian. There is an exact fully faithful functor $C \rightsquigarrow$ mod-A, for some ring A. (15.30. Cf. Swan [68, pp. 14—22].)

15.18 Proposition. *Let*

$$T_1 : \mathscr{A} \to \mathscr{B}$$

be an exact functor of abelian categories, and let

$$S_1 : \mathscr{B} \to \mathscr{A}$$

be an adjoint to T_1 *in the situation* $(\varphi, T_1, S_1, \mathscr{A}, \mathscr{B})$. *Assume that the natural transformation*

$$\Phi_1 : T_1 \circ S_1 \to 1_{\mathscr{B}}$$

defined by φ *is an equivalence. Then* $\ker T_1$ *is a localizing subcategory of* $\mathscr{A}$, *and there exists a category equivalence* $H : \mathscr{A}/\ker T_1 \approx \mathscr{B}$ *such that*

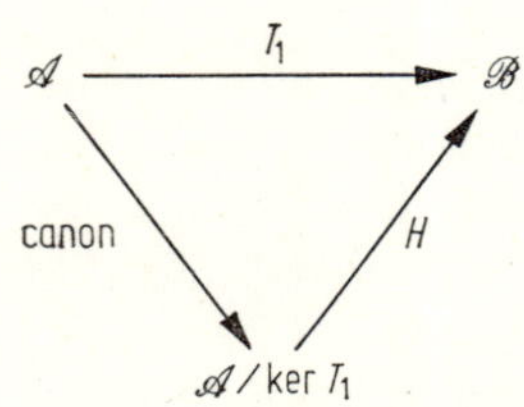

commutes. (Then H is said to be the equivalence induced by T_1.)

Proof. Since T_1 is exact, $\ker T_1 = \{X \in \text{obj}\, \mathscr{A} \mid T_1 X = 0\}$ is a Serre class. Let $J : \mathscr{A} \to \mathscr{A}/\ker T_1$ be the canonical functor, and $H : \mathscr{A}/\ker T_1$ be the unique functor (15.10) such that $T_1 = H \circ J$. By 15.5, if f is a morphism of $\mathscr{A}$, $T_1 f$ is an equivalence if and only if both $J(\ker f)$ and $J(\text{coker}\, f)$ are zero, or equivalently, if J is an equivalence. Apply this to $\Psi_1(M) : M \to S_1 \circ T_1(M)$, where $M \in \text{obj}\, \mathscr{A}$. Then

$$T_1 \Psi_1(M) \circ \Phi_1(T_1 M) = 1_{T_1 M}.$$

Since $\Phi_1(T_1 M)$ is an equivalence, we see that $T_1 \Psi_1(M)$ is an equivalence, and by the preceding remark, so is $J\Psi_1(M)$. Since every object M of $\mathscr{A}/\ker T_1$ has the form JM, the morphism

$$J\Psi_1(M) : JM \to JS_1 T_1(M) = JS_1 HJ(M)$$

defines a natural equivalence of functors

$$\mathscr{A}/\ker T_1 \approx (JS_1) \circ H.$$

Since $\Phi : T_1 S_1 \approx 1_{\mathscr{B}}$ gives a natural equivalence $H \circ (JS_1) \approx 1_{\mathscr{B}}$, then H is an equivalence of categories $\mathscr{A}/\ker T_1 \approx \mathscr{B}$ (with inverse JS_1).

It remains only to show that ker T_1 is localizing. We do this assuming that $\mathscr{A}$ is a Grothendieck category, and leave the general case as an exercise. Since T_1 has an adjoint, T_1 preserves coproducts 5.43. Since $T_1 = H \circ J$, and H is an equivalence, it follows that J preserves coproducts, so ker T_1 is localizing (15.11). ▯

Remark. $S_1 \circ H$ is adjoint to J.

15.19A Proposition. *Let $\mathscr{S}$ be a Serre class of the abelian category $\mathscr{A}$. Then the following are equivalent:*

(a) *$\mathscr{S}$ is a localizing subcategory of $\mathscr{A}$.*

(b) *Every object M of $\mathscr{A}$ contains a subobject $\mathscr{S}(M)$ that is largest in the class of subobjects of M which belong to $\mathscr{S}$, and if $\mathscr{S}(M) = 0$, then there is a monic $M \to N$, where N is $\mathscr{S}$-closed.*

(c) *For each object M of $\mathscr{A}$, there is an exact sequence*

$$0 \to A' \to M \xrightarrow{u} N \to A''' \to 0$$

*where $A' \in \mathscr{S}$ & $A''' \in \mathscr{S}$ and N is $\mathscr{S}$-closed. The morphism u is then called an **$\mathscr{S}$-envelope** of M.*

Proof. (a) ⇒ (b). By 15.14, $\mathscr{S}(M) = \ker \psi(M)$ is the largest subobject of M that belongs to $\mathscr{S}$. If $\ker \psi(M) = 0$, then $\psi(M): M \to STM$ is a monic, and STM is $\mathscr{S}$-closed.

(b) ⇒ (c). Let M' be a maximal subobject of M belonging to $\mathscr{S}$, and let $i: M/M' \to F$ be a monic, where F is $\mathscr{S}$-closed. Let N be the inverse image in F of the maximal subobject of coker i that belongs to $\mathscr{S}$. Then N is $\mathscr{S}$-closed by 15.12(a). Since $i(M/M') \subseteq N$, then i induces a morphism $j: M/M' \to N$, and

$$j \circ p_{M/M'}: M \to N$$

is an $\mathscr{S}$-envelope of M.

(c) ⇒ (a). By definition, $\mathscr{S}$ is localizing provided that the canonical functor $T: \mathscr{A} \rightsquigarrow \mathscr{A}/\mathscr{S}$ has an adjoint $S: \mathscr{A}/\mathscr{S} \rightsquigarrow \mathscr{A}$. If $N \in \operatorname{obj} \mathscr{A}/\mathscr{S}$, let $u(N): N \to SN$ be an $\mathscr{S}$-envelope. Then the mapping

$${}_{\mathscr{A}/\mathscr{S}}(TM, N) = {}_{\mathscr{A}/\mathscr{S}}(TM, TN) \xrightarrow{\operatorname{Mor}(TM,\, Tu(N))} {}_{\mathscr{A}/\mathscr{S}}(TM, TSN)$$

is bijective by 15.12, and

$${}_{\mathscr{A}}(M, SN) \xrightarrow{T(M,\, SN)} {}_{\mathscr{A}/\mathscr{S}}(TM, TSN)$$

is bijective by 15.12(c), and the composition of these two mappings gives a natural isomorphism

$$\operatorname{Mor}_{\mathscr{A}/\mathscr{S}}(T(\ ), N) \approx \operatorname{Mor}_{\mathscr{A}}(\ , SN).$$

Since $h_N \circ T = \mathrm{Mor}_{\mathscr{A}/\mathscr{S}}(T(\ ), N)$ is therefore representable, then T has an adjoint by the dual of 5.4. □

15.19B Proposition. *If $\mathscr{S}$ is a localizing subcategory of an abelian category $\mathscr{A}$, then the canonical functor $T : \mathscr{A} \rightsquigarrow \mathscr{A}/\mathscr{S}$ induces an equivalence $T' : C \rightsquigarrow \mathscr{A}/\mathscr{S}$ of the full subcategory C consisting of all $\mathscr{S}$-closed objects, and the inverse equivalence S' is induced by the section functor $S : \mathscr{A}/\mathscr{S} \rightsquigarrow \mathscr{A}$.*

Proof. Since every object SY is $\mathscr{S}$-closed by 15.13, then S induces a functor S' as stated. Moreover, the equivalence $TS \approx 1_{\mathscr{A}/\mathscr{S}}$ stated in 15.14.1 yields $T'S \approx 1_{\mathscr{A}/\mathscr{S}}$. Since by 15.15 $\Psi(M) : M \to ST(M)$ is an equivalence for every $\mathscr{S}$-closed object M, then Ψ induces a natural equivalence $\Psi' : 1_C \to ST'$ as needed. □

It follows that any two $\mathscr{S}$-envelopes $u_i : M \to N_i$ $i = 1, 2,$ of an object M of $\mathscr{A}$ are equivalent; for from 15.15, we have $ST(N) \approx N$ for any $\mathscr{S}$-closed object N, and so, using 15.14.2, we have

$$ST(M) \approx ST(N_i) \approx N_i,$$

$i = 1, 2,$ and so $N_1 \approx N_2$.

The following corollary introduces some terminology which will be used in the concept of a torsion theory discussed in Chapter 16.

15.19C Corollary and Definition. *If $\mathscr{S}$ is a localizing subcategory of the abelian category $\mathscr{A}$, then an object M is said to be* ***$\mathscr{S}$-torsion****, or simply* ***torsion*** *(resp.* ***$\mathscr{S}$-torsionfree****, or* ***torsionfree****) provided that $M = \mathscr{S}(M)$ (resp. $\mathscr{S}(M) = 0$). Let M be torsionfree. Then a monic $u : M \to N$ is an $\mathscr{S}$-envelope iff N is $\mathscr{S}$-closed and N/M is torsion. In this case we have: (1) u is an essential monic; (2) If E is an essential extension of N, then E and E/N are torsionfree. Conversely, if (1) and (2) hold, and N/M is torsion, then u is an $\mathscr{S}$-envelope. Moreover, assuming that the injective hull $\hat{M}$ exists, then the $\mathscr{S}$-envelope of any torsionfree object M is the largest subobject D of the injective hull such that D/M is torsion. Thus, a torsionfree object M is $\mathscr{S}$-closed iff $\hat{M}/M$ is torsionfree.*

Proof. If $u : M \to N$ is an $\mathscr{S}$-envelope, and if K is a subobject of N such that $K \cap M = 0$, then K embeds canonically in N/M which by the definition 15.19A is an object of $\mathscr{S}$. Thus, K belongs to $\mathscr{S}$. But, $\mathscr{S}(N) = 0$, so $K = 0$. This proves that u is essential. Similarly, any essential extension E of N is torsionfree. Next, if $p : E \to E/N$ is canonical, and if $P = \mathscr{S}(E/N)$, then for a suitable subobject H of E, there is an exact sequence $0 \to N \to H \to P \to 0$, which splits since N is $\mathscr{S}$-closed and $P \in \mathscr{S}$. Since N is an essential submodule of H, this happens only if $N = H$ and $P = 0$. This proves (2). Conversely, since N/M is torsion, we need only show that N is $\mathscr{S}$-closed. Since N is torsionfree, by 15.12 we need only show that any sequence $0 \to N \to Q \to P \to 0$, with $P \in \mathscr{S}$, splits. However, if K is maximal in the set of subobjects of Q such that $K \cap N = 0$, then the canonical monic $N \to Q/K$ is essential. Thus, $[N+K]/K$ is an essential subobject of Q/K. Assuming (2), then the quotient object Q/K modulo $[N+K]/K$ must be torsionfree, that is, $Q/N+K$ is torsionfree. However, $Q/N+K$ is a quotient object of $P \approx Q/N$. Since $P \in \mathscr{S}$, then $Q/N+K$ is torsion, and so $= 0$, that is, $Q = N+K = N \oplus K$, which yields the desired splitting. This proves that N is $\mathscr{S}$-closed, and that u is an $\mathscr{S}$-envelope.

This proves the converse. (2) implies that no essential extension of N is torsion modulo N. This proves the necessity of the next to last assertion. The sufficiency follows from the fact that D/M is torsion and D satisfies (1) and (2). The last assertion follows quite clearly from this. ▯

15.20 *Exercise. If $\mathscr{S}$ is a localizing subcategory of $\mathscr{A}$, then the following conditions are equivalent:*

15.20.1 *The localizing functor is exact.*

15.20.2 *If $N \subseteq M$, and if N and M are $\mathscr{S}$-closed, then M/N is $\mathscr{S}$-closed.*

Recall the definition: a monic $f: M \to N$ is essential in case $f(M) \cap P \neq 0$, or equivalently, provided that $f^{-1}P \neq 0$ for every nonzero subobject P of N.

15.21 **Proposition.** *Let $\mathscr{S}$ be a localizing subcategory of $\mathscr{A}$. Let $f: M \to N$ be a morphism, and let M' (resp. N') be a maximal subobject of M (resp. N) belonging to $\mathscr{S}$, and let $f': M/M' \to N/N'$ denote the morphism induced by f. Then, $Tf: TM \to TN$ is an essential extension if and only if $f': M/M' \to N/N'$ is an essential extension.*

Proof. By 15.5, $Tp_{M/M'}: TM \to T(M/M')$ and $Tp_{N/N'}$ are equivalences; so Tf is an essential monic if and only if Tf' is. Hence, it suffices to prove the proposition when $M' = N' = 0$. If f is an essential monic, for any nonzero subobject TP of $N = TN$, $f^{-1}P \neq 0$ and $f^{-1}P \notin \mathscr{S}$. Then $(Tf)^{-1}TP \approx T(f^{-1}P)$ is nonzero, and so Tf is an essential monic. Conversely, if Tf is an essential monic, and $P \neq 0$ a subobject of N, then $TP \neq 0$ because $N' = 0$ implies $P \notin \mathscr{S}$. Thus $f^{-1}P \neq 0$. ▯

15.22 **Proposition.** *Let $\mathscr{S}$ be a Serre class of an abelian category $\mathscr{A}$, and let $u: M \to E$ be an injective hull in $\mathscr{A}$ of an object M of $\mathscr{A}$. If M has no nonzero subobjects belonging to $\mathscr{S}$, then E is $\mathscr{S}$-closed and $Tu: TM \to TE$ is an injective hull in $\mathscr{A}/\mathscr{S}$ of TM, where $T: \mathscr{A} \rightsquigarrow \mathscr{A}/\mathscr{S}$ is canonical.*

Proof. If $N \in \mathscr{S}$ is a subobject of E, then $M \cap N \in \mathscr{S}$, and $M \cap N \subseteq M$, hence $M \cap N = 0$. Since u is essential, necessarily, $N = 0$, and so E is $\mathscr{S}$-closed by 15.21. Furthermore, $N' = M' = 0$ in 15.21, so Tu is an essential monic. To prove that TE is injective, let $f' : TE \to TM$ be a monic in $\mathscr{A}/\mathscr{S}$, the image of a morphism $f'' : E' \to M/M'$, where E/E', M' belong to $\mathscr{S}$. Since Tf'' is monic, $\ker f'' \in \mathscr{S}$, hence $\ker f'' = 0$. Let $i : E' \to E$ be the inclusion. Since E is injective, there exists $g: M/M' \to E$ such that $i = gf$. Then

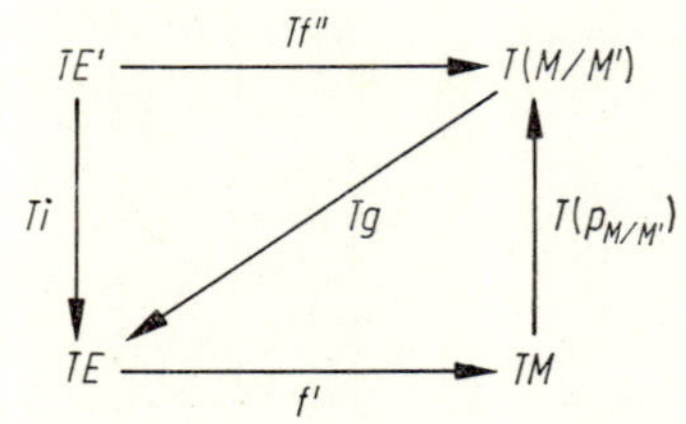

commutes. Since Ti and $T(p_{M/M'})$ are equivalences, f' splits, and so TE is injective. □

15.23 Corollary. *Let $\mathscr{S}$ be a Serre class of an abelian category $\mathscr{A}$ having injective envelopes. Then the following conditions are equivalent:*

15.23.1 *$\mathscr{S}$ is a localizing subcategory of $\mathscr{A}$.*

15.23.2 *Every object M of $\mathscr{A}$ contains a largest subobject belonging to $\mathscr{S}$.*

Proof. Apply 15.22 and 15.19 □

15.24 Corollary. *Let $\mathscr{A}$ be an abelian category with injective hulls, and let $\mathscr{S}$ be a localizing subcategory. Then:*

15.24.1 *$\mathscr{A}/\mathscr{S}$ is a category with injective hulls, and every injective object of $\mathscr{A}/\mathscr{S}$ has the form TE, where E is injective in $\mathscr{A}$. Moreover, the section functor $\mathscr{A}/\mathscr{S} \rightsquigarrow \mathscr{A}$ preserves injectives.*

15.24.2 *If E is injective in $\mathscr{A}$, and if E_1 is the injective hull of a maximal subobject M_1 of E belonging to $\mathscr{S}$, then $E \approx E_1 \oplus SE_2$, where E_2 is an injective object of $\mathscr{A}/\mathscr{S}$. Furthermore, E_1 and E_2 are unique up to equivalences in $\mathscr{A}$.*

15.24.3 *If for every object X of $\mathscr{S}$, it is true that the injective hull of X belongs to $\mathscr{S}$, then the canonical functor $T : \mathscr{A} \rightsquigarrow \mathscr{A}/\mathscr{S}$ preserves injective objects; that is, TE is injective in $\mathscr{A}/\mathscr{S}$ whenever E is injective in $\mathscr{A}$.*

Proof of 15.24.1. By 15.22, every object X of $\mathscr{A}/\mathscr{S}$ is equivalent to an object TM, where $M \in \operatorname{obj} \mathscr{A}$ is such that $\mathscr{S}(M) = 0$. By 15.22, if $u : M \to E$ is an injective hull in $\mathscr{A}$, then $T(u) : TM \to TE$

is an injective hull in $\mathscr{A}/\mathscr{S}$. Since $X \approx TM$, $\mathscr{A}/\mathscr{S}$ has injective hulls. If X is injective in $\mathscr{A}/\mathscr{S}$, then $X \approx TM \approx TE$. Then, $SX \approx STE \approx E$ by 15.15, so S preserves injectives. (Cf. 6.28.)

Proof of 15.24.2. Since E is injective, and $M_1 \to E_1$ is essential, the monic $M_1 \to E$ can be extended to a monic $E_1 \to E$. Since $E_1 \to E$ splits, $E \approx E_1 \oplus F$, where $\mathscr{S}(F) = 0$. By 15.22, $E_2 = TF$ is injective in $\mathscr{A}/\mathscr{S}$, and by 15.15, $F \approx STF = SE_2$.

Proof of 15.24.3. The hypothesis implies that $E_1 \in \operatorname{obj} \mathscr{S}$, whence $TE_1 = 0$, and

$$TE \approx TSE_2 \approx TF \approx E_2$$

is injective in $\mathscr{S}$. ▯

15.25 Corollary. *Let $\mathscr{A}$ be a Grothendieck category. If $\mathscr{F}$ is a family of injective objects in $\mathscr{A}$, then*

$$\mathscr{S}(\mathscr{F}) = \{X \in \mathscr{A} \mid h_E(X) = 0 \,\forall E \in \mathscr{F}\}$$

is a localizing subcategory of $\mathscr{A}$. Conversely, if $\mathscr{S}$ is a localizing subcategory of $\mathscr{A}$, then

$$\mathscr{S} = \mathscr{S}(\mathscr{F}(\mathscr{S}))$$

where $\mathscr{F}(\mathscr{S})$ is the family of $\mathscr{S}$-closed injective objects of $\mathscr{A}$.

Proof. The first statement follows from 15.11, and the fact (15.1) that $\mathscr{S}(\mathscr{F})$ has coproducts .Conversely, if $\mathscr{S}$ is localizing, then $\mathscr{A}/\mathscr{S}$ has injective hulls (because $\mathscr{A}$ does). In this case, an object M of $\mathscr{A}$ belongs to $\mathscr{S}$ if and only if $\operatorname{Mor}_{\mathscr{A}}(M, E) = 0$ for all $E \in \mathscr{F}(\mathscr{S})$. ▯

Characterization of Grothendieck Categories

15.26 Theorem (Popesco-Gabriel [64]) (Third Version). *Let C be a Grothendieck category. Let $U \in \operatorname{obj} C$, let $A = \operatorname{End}_C U$, let $T = h^U$ denote the* mod-A *valued functor induced by U, and let S denote the functor left adjoint to T. The following conditions are equivalent:*

(a) *U is a generator of C.*

(b) *T is fully faithful.*

(c) *The natural transformation of functors*

$$\Psi : ST \to 1_C$$

given by the adjoint situation (η, T, S, C, mod-A*) is an equivalence, and S is an exact functor.*

(d) *S is exact; that is, $\mathscr{S} = \operatorname{Ker} S$ is a localizing subcategory of* mod-A, *and S induces an equivalence $H : \text{mod-}A/\mathscr{S} \rightsquigarrow C$ such that*

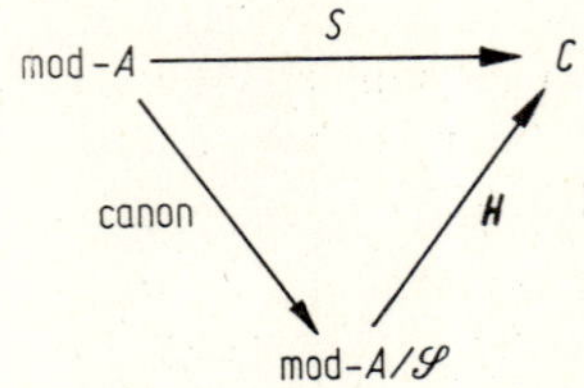

commutes. (Thus, if $J : \text{mod-}A \rightsquigarrow \text{mod-}A/\mathscr{S}$ is canonical, then $H \circ J \approx S$.)

Proof. The implications (c) ⇒ (b) ⇒ (a) are obvious, and (c) ⇒ (d) follows from 15.19. We now prove that (a) ⇒ (c).

(a) ⇒ (c). By 5.69 and 5.70 the adjoint situation yields a natural equivalence $ST \approx 1_C$. Hence, it remains only to prove that S is exact. This is done by Popesco-Gabriel in a long sequence of lemmas. The details of the proofs of these lemmas have been written by Popesco for Bucur and Deleanu [68, Lemmas 6.26—6.30, pp. 148—151].

We give instead a short proof of Takeuchi requiring properties of the so-called left derived functors (making their first and only appearance in the text!). (Also see Lambek's short proof [71].)

15.26.1 **Lemma.** *If $Y \subseteq TX$, then $SY \subseteq X \approx STX$ is monic.*

Proof. Since $TX = \operatorname{Mor}_C(U, X)$, there is a canonical morphism $U^{(Y)} \to X$. The image I of this morphism is such that $Y \subset TI \subset TX$. Applying the method of the proof of the statement in 4.24 that $h^U : C \rightsquigarrow \text{mod-}A$ is a full functor, we see that the composition

$$\operatorname{Mor}_C(I, Z) \xrightarrow{T} \operatorname{Hom}_A(TI, TZ) \to \operatorname{Hom}_A(Y, TZ)$$

is a bijection for any object Z of C. Hence, I can be identified with SY. Then, $I \subseteq X$ means that $SY \to X$ is monic. □

15.26.2 **Lemma.** *If Y is a subobject of a free object F of* mod-A, *then $SY \to SF$ is monic.*

Proof. There exists a directed family $\{F_a\}$ of finitely generated free submodules of F, such that $F = \sum_a F_a$. Then, $Y = \sum_a (Y \cap F_a)$. Since, to each a there is an integer n_a such that

$$F_a \approx A^{n_a} = (TU)^{n_a} \approx T(U^{n_a})$$

then Lemma 1 shows that $S(Y \cap F_a) \to SF_a$ is monic. Since S commutes with direct limits, and since C has exact direct limits, then

$$SY = \varinjlim S(Y \cap F_a) \leq \varinjlim SF_a = SF.$$

15.26.3 **Lemma.** *If* $0 \leftarrow Y \leftarrow F_0 \leftarrow F_1 \leftarrow \cdots$ *is a projective resolution and* F_i *free,* $i \geq 0$, *then* $0 \leftarrow SY \leftarrow SF_0 \leftarrow SF_1 \cdots$ *is exact.*

Proof. S is right exact (being a left adjoint to h^U), and then Lemma 2 gives what we want. ▯

This completes the proof of 15.26, since the last lemma shows that the n-th left derived functor of S is zero for $n > 0$. (See, for example, MacLane [63, pp. 392, 393).) ▯

15.27 Corollary. *A Grothendieck category is a category with injective hulls, and an injective cogenerator* 14.17.

Proof. Apply 15.24 to 15.26. ▯

15.28 Corollary (Gabriel [62], Mitchell [64]). *A category C is equivalent to a category* mod-*A*, *for some ring A, if and only if C is an AB*3 *category with a small projective generator U. In this case,* $h^U : C \rightsquigarrow$ mod-$\text{End}_C U$ *is an equivalence, and C is necessarily AB*5. ▯

15.29 Corollary (Popesco-Gabriel). *A category C is equivalent to an AB*-5 *category with a generator if and only if there exists a ring A, a localizing subcategory* $\mathcal{S}$ *of* mod-*A and an equivalence* $C \approx$ mod-$A/\mathcal{S}$. *Moreover, C is equivalent to the full subcategory of* mod-*A consisting of* $\mathcal{S}$*-closed modules.* ▯

15.30 The Full, Exact Embedding Theorem (Mitchell [64]) *Every small abelian category C has a full, exact covariant embedding* $C \rightsquigarrow$ mod-*A*, *for some ring A.*

Proof (in which it is assumed that C has a full exact contravariant embedding $T : C \rightsquigarrow C_1$ *in a Grothendieck category* C_1 *with a generator U such that for every object X of C, TX is a quotient object of U.* Cf. Popesco in Bucur-Deleanu [68, p. 153], and Mitchell [65, p. 151]). Then, C_1 contains an injective cogenerator V such that every TX is a subobject of V. Then, the opposite category C_1^{op} contains a projective generator $G = V^*$. If A is the C_1^{op} endomorphism ring of G, then, by 15.28, there is an equivalence $C_1^{\text{op}} \approx$ mod-A. Then, the composition

$$C \rightsquigarrow C_1 \overset{\text{canon}}{\rightsquigarrow} C_1^{\text{op}} \approx \text{mod-}A$$

is a full, exact embedding. ▯

15.31 Corollary. *If* $\mathcal{A}$ *is a Grothendieck category, and if* $\mathcal{S}$ *is a localizing subcategory, then* $\mathcal{A}/\mathcal{S}$ *is a Grothendieck category.*

Proof. Exercise. Cf. Gabriel [62, p. 378]. ▯

AB6 Categories

An $AB3$ category C is said to be $AB6$ provided that for every family $\{\{C_{ab}\}_{b\in I_a}\}_{a\in J}$ of directed families of subobjects of objects of C the canonical morphism

$$\sum_{b\in I_a} \bigcap_{a\in J} C_{ab} \to \bigcap_{a\in J} \sum_{b\in I_a} C_{ab}$$

is an isomorphism, assuming the existence of the intersection.

Exercises for Chapter 15

1. (Roos [65]; cf. Roos [67]) The conditions on an abelian category C are equivalent.

1.1. C is an $AB4^*$ and an $AB6$ category with a generator.

1.2. There is a ring A, and a localizing category B of mod-A equivalent to mod-A_0, for some ring A_0, and an equivalence $C \approx (\text{mod-}A)/B$.

1.3. There is a generator U of C such that the kernel of the exact left adjoint

$$S : \text{mod-}A \rightsquigarrow C \qquad \text{where } A = \text{End}_C U$$

to the canonical functor $C \rightsquigarrow$ mod-A defined by $\text{Mor}_C(U, \cdot)$ is equivalent to mod-A_0, for some ring A_0.

1.4. C is equivalent to a category of the form mod (A, a), where a is an idempotent ideal of a ring A, and mod (A, a) denotes the full subcategory of mod-A consisting of all objects M for which the canonical homomorphism

$$M \approx \text{Hom}_A(A, M) \to \text{Hom}_A(a, M)$$

is an isomorphism.

2. For any ring R, let $\mathscr{S}$ denote the full subcategory of mod-R consisting of all modules M such that R is a rational extension of the annihilator of any element of M. Show that $\mathscr{S}$ is a localizing category. If $\overline{M}$ is the image of any object M of mod-R under the localizing functor, show that $\overline{M}$ is the maximal rational extension of M (in the sense of 4.5), and that $\overline{R}$ is a ring. Verify 4.5, and Chapter 9, Exercise 1. Also specialize to the case of right Goldie semiprime rings, and right Ore domains.

3. (Popesco-Gabriel) Is C is a Grothendieck category with generator U, then there exists a coreflective subcategory C_1 of mod-A, with exact coreflector, and an equivalence $C \approx C_1$. A right A-module N is said to be *negligible* if for every object X in C and morphism $f: A \to N$, the functor h_{TX} maps the monic $K \to A$ onto a bijection $h_{TX}(f): h_{TX}(A) \to h_{TX}(K)$ where $K = \ker f$. Let B denote the full subcategory of negligible modules. Then, B is a localizing subcategory of mod-A, and C is equivalent to the subcategory C_1 of B-closed modules.

Notes for Chapter 15

Notes for this chapter have been included in the introduction to Part IV. I could not find the explicit statement of 15.19B in Gabriel [62], although the proof and most of the chapter come from this source. Proposition 15.19B has been stated, however, for module type categories by Popesco-Gabriel [64] (see Exercise 3, and also Proposition 16.8C). The characterization of $\mathscr{S}$-closed objects in 15.19C was suggested by the corresponding result for module type categories (cf. Lambek [71, p. 10]).

References

Bucur and Deleanu [68], Gabriel [62], Gabriel and Popesco [64], Lambek [71], Mitchell [64], Takeuchi [71].

Chapter 16

Torsion Theories, Radicals, and Idempotent, Topologizing, and Multiplicative Sets

This chapter contains applications of the localizing functors, and quotient categories defined in the preceding chapter. The application 16.9 is to multiplicative subsets S of a ring R, and the corresponding (partial) ring of quotients R_S with respect to S. Theorem 16.12 yields the Johnson maximal right quotient ring $\hat{R}$ of a right neat ring R (cf. Chapter 19). When R is semiprime, and satisfies the (acc)$^\perp$ and (acc)$\oplus$, then $\hat{R}$ is the classical right quoring.

Proposition 16.3 establishes a bijection between the class of all localizing subcategories of mod-R, and a class consisting of sets of right ideals of R which are idempotent and topologizing in a sense presently defined. (If R is right neat, then the set of essential right ideals of R is such a set.) A torsion theory for a Grothendieck category $\mathscr{A}$ is an ordered pair (B, C) of subcategories such that, letting $(\ ,\ ) = \mathrm{Mor}_{\mathscr{A}}(\ ,\ )$

$$X \in B \text{ iff } (X, \hat{Y}) = 0 \ \forall\ Y \in C, \text{ where } \hat{Y} \text{ is the injective hull}$$

$$Y \in C \text{ iff } (X, \hat{Y}) = 0 \ \forall\ X \in B.$$

Then the objects of B are said to be torsion and those of C torsionfree. There is a bijection between torsion theories (B, C) and localizing subcategories of $\mathscr{A}$ (16.8B).

Let $\mathscr{F}$ be a set of right ideals of a ring A. Then $\mathscr{F}$ is said to be **topologizing** provided that the following conditions hold:

F.1 A belongs to $\mathscr{F}$, and every right ideal containing a member of $\mathscr{F}$ belongs to $\mathscr{F}$.

F.2 The intersection of any two members of $\mathscr{F}$ belongs to $\mathscr{F}$.

F.3 If $I \in \mathscr{F}$ and $a \in A$ then $\mathrm{ann}_A(a + I)I \in \mathscr{F}$

We abbreviate, by $(a/I)^{\perp}$ or $\mathrm{ann}_A a/I$, the right ideal

$$\mathrm{ann}_A(a + I)/I = \{b \in A \mid ab \in I\}.$$

(In the literature, this is also denoted $a^{-1}I$.) For future reference, we state:

F.4 If $J \in \mathscr{F}$ and if $\mathrm{ann}_A a/I \in \mathscr{F}\ \forall\ a \in J$ then $I \in \mathscr{F}$.

An A-module M is said to be $\mathscr{F}$-**torsion** (Gabriel: $\mathscr{F}$-**negligible**) provided that $\mathscr{F}$ contains the annihilator of every element of M. Thus, M is $\mathscr{F}$-torsion if and only if for every $x \in M$ there corresponds $I \in \mathscr{F}$ such that $xI = 0$.

A full subcategory C of an abelian category $\mathscr{A}$ is said to be **closed** if the following conditions are satisfied:

C.1 If $Y \in \mathrm{obj}\, C$, then every subobject and every quotient object of Y belongs to C.

C.2 C contains the coproduct of any pair of objects of C.

C.3 C contains the direct limit in $\mathscr{A}$ of any directed family of objects of C.

16.1 Exercise and Definition.

16.1.1 If $M \in \mathrm{obj}\,\mathscr{A}$, and if C is a closed subcategory, then M contains a subobject CM, which contains every subobject of C contained in M, and $f(CM) \subseteq CN$ for all morphisms $f: M \to N$ of $\mathscr{A}$. Conclude that the functor $T: \mathscr{A} \rightsquigarrow C$ mapping each object $M \mapsto CM$ is adjoint to the embedding functor $C \rightsquigarrow \mathscr{A}$, and that T is left exact.

16.1.2 Show that a topologizing set F of right ideals of a ring A is a basis of neighborhoods of 0 for a topology on A.

16.1.3 Show that F.4 implies F.2 and F.3, assuming $A \in F$.

16.1.4 Show that the axioms for closed subcategory are redundant.

If G is a set of right ideals of mod-A, then $\mathscr{S}(G)$ denotes full subcategory of mod-A consisting of G-torsion modules.

16.2 Lemma. *The mapping $\mathscr{S}$ that associates to each topologizing set G of right ideals of* mod-A *the set $\mathscr{S}(G)$ of G-torsion modules is a bijection between the class of all topologizing sets of right ideals and the class of all closed full subcategories of* mod-A. *The inverse mapping $\mathscr{F}$ associates to a subcategory C of* mod-A, *the set*

$$\mathscr{F}(C) = \{I \mid I \subseteq A \ \& \ A/I \in \text{obj } C\}.$$

Proof. Let C be a closed subcategory of mod-A, and let $F = \mathscr{F}(C)$. If $A \supseteq I \supseteq J$, and $A/J \in C$, then exactitude of

(1) $$0 \to I/J \to A/J \to A/I \to 0$$

yields F.1. Next, A/I and A/J in C imply an exact sequence

$$0 \to I \cap J \to A \to A/I \times A/J$$

(the product of the obvious projections), and hence there is an exact sequence

(2) $$0 \to A/(I \cap J) \to A/I \times A/J$$

which yields F.2. To prove that F.3 holds, let $A/I \in C$. Then, if $J = \text{ann}_A a/I$, there is a monic

(3) $$\begin{cases} A/J \to A/I \\ b + J \mapsto ab + I. \end{cases}$$

Then C.1 implies $A/J \in C$, and $J \in F$, proving F.3.

Conversely, assume that G is topologizing, and let $C = \mathscr{S}(G)$. Let

(4) $$0 \to X \to Y \to Z \to 0$$

be exact in mod-A, and let $Y \in C$. Then X is also G-torsion. If $a \in Y$, then $\text{ann}_A a/X \supseteq \text{ann}_A a$, and then F.1 implies $\text{ann}_A a/X \in G$. This proves that $Z \approx Y/X$ is G-torsion, so C.1 holds. Let $Y = \coprod_{t \in T} X_t$ be a coproduct of objects of C. Each $y \in Y$ is a finite sum $y = \sum_t x_t$, with $x_t \in X_t$, and $\text{ann}_A y$ contains the finite intersection $I = \bigcap_t \text{ann}_A x_t$.

Since $\mathrm{ann}_A x_t \in G$ for every t, then F.2 implies $I \in G$, and then F.1 implies $\mathrm{ann}_A y \in G$, proving that $Y \in C$. This established C.2. Since any direct limit in mod-A is a quotient object of a coproduct, then this proves C.3.

That $\mathscr{F} = \mathscr{S}^{-1}$ is obvious. □

If C and D are closed subcategories of $\mathscr{A}$, then the **product** $C \cdot D$ is defined to be the full subcategory of $\mathscr{A}$ consisting of those objects X of $\mathscr{A}$ such that X/DX belongs to C. The operation $(C, D) \to C \cdot D$ defined on the class of closed subcategories is associative. Then, C^n is defined for any n in the usual way for associative systems.

16.3 Proposition and Definition (Gabriel [62]) *Let A be a ring.*

16.3.1 *If F and G are topologizing sets of right ideals of* mod-A, *then the product* $\mathscr{S}(F) \cdot \mathscr{S}(G)$ *of the closed subcategories* $\mathscr{S}(F)$ *and* $\mathscr{S}(G)$ *is a closed subcategory. We denote by* $F \cdot G$ *the topologizing set corresponding to* $\mathscr{S}(F) \cdot \mathscr{S}(G)$. *Thus,*

$$\mathscr{S}(F \cdot G) = \mathscr{S}(F) \cdot \mathscr{S}(G).$$

16.3.2 *If I is a right ideal Of A, then*

$$I \in F \cdot G \Leftrightarrow \exists J \in F \text{ such that } \mathrm{ann}_A a/I \in G \; \forall a \in J.$$

The set F is said to **idempotent** *in case* $F \circ F = F$, *or, equivalently, in case the following condition holds:*

F.4 $J \in F \;\&\; \mathrm{ann}_A a/I \in F \; \forall a \in J \Rightarrow I \in F$.

16.3.3 *The mapping* $\mathscr{S}$ *is a bijection on the class of all idempotent topologizing sets of right ideals, and the class of all localizing subcategories.*

Proof of 16.3.1. Let

$$0 \to X \to Y \to Z \to 0 \tag{1'}$$

be exact in mod-A. If $Y \in \mathscr{S}(F) \cdot \mathscr{S}(G)$, then $Y/\mathscr{S}(G)Y \in \mathscr{S}(F)$, where $\mathscr{S}(G)Y$ is the largest subobject of Y belonging to $\mathscr{S}(G)$ (see Exercise). Now $\mathscr{S}(G)X = \mathscr{S}(G)Y \cap X$, and

$$(X + \mathscr{S}(G)Y)/\mathscr{S}(G)Y \approx X/X \cap \mathscr{S}(G)Y = X/\mathscr{S}(G)X.$$

Therefore, C.1 implies that $X/\mathscr{S}(G)X \in \mathscr{S}(F)$; that is, $X \in \mathscr{S}(F) \cdot \mathscr{S}(G)$. The proof that $Z \in \mathscr{S}(F) \cdot \mathscr{S}(G)$ is similar. Hence, C.1 holds for $\mathscr{S}(F) \cdot \mathscr{S}(G)$. Let $\{X_t \mid t \in T\}$ be a family of objects in $\mathscr{S}(F) \cdot \mathscr{S}(G)$. Then let $Y = \coprod_{t\in T} X_t$, and let $Z = \coprod_{t\in T} X_t/\mathscr{S}(G)X_t \in \mathscr{S}(F)$. Now

$\mathscr{S}(G)Y \approx \coprod_{t\in T}\mathscr{S}(G)X_t$, and the obvious embedding

$$\coprod_{t\in T}X_t/\coprod_{t\in T}\mathscr{S}(G)X_t \to \coprod_{t\in T}(X_t/\mathscr{S}(G)X_t) = Z$$

induces an embedding $Y/\mathscr{S}(G)Y \to Z$. Since $\mathscr{S}(F)$ is closed, we have $Y/\mathscr{S}(G)Y \in \mathscr{S}(F)$, so $Y \in \mathscr{S}(F)\cdot\mathscr{S}(G)$. This proves that $\mathscr{S}(F)\cdot\mathscr{S}(G)$ is closed.

Proof of 16.3.2. Let $I \in F\cdot G$. Then $A/I \in \mathscr{S}(F)\cdot\mathscr{S}(G)$. and setting $D = \mathscr{S}(G)$, then $A/I/D(A/I) \in \mathscr{S}(F)$. Now $D(A/I)$ is the largest submodule J/I of A/I belonging to D. Since $J/I \in D$, then by F.3, $\operatorname{ann}_A a/I \in G\,\forall a \in J$. Conversely, let the right side of the arrow in 16.3.2 hold. Then $(J+I)/I \in \mathscr{S}(G) = D$ since every $\operatorname{ann}_A a/I \in G\,\forall a \in (J+I)/I$. Hence, $D(A/I) \supseteq (J+I)/I$, so $A/I/D(A/I)$ is a quotient object of $(A/I)/(J+I)/I \approx A/J+I$. Thus, $A/I/D(A/I) \approx A/Q$, where $Q \supseteq J+I$. Since $J \in F$, then $Q \in F$ by F.1, and hence $A/I/D(A/I) \in \mathscr{S}(F)$. This proves that $A/I \in \mathscr{S}(F)\cdot\mathscr{S}(G)$, or that $I \in F\cdot G$.

Proof of 16.3.3. By 16.2, we must prove that F is idempotent if and only if $\mathscr{S}(F)$ is localizing. Assume that F is idempotent (and topologizing), then C.1 holds. Next, assume that (1′) is exact, let $y \in Y$, and let $I = \operatorname{ann}_A y$. Now $J = \operatorname{ann}_A y/X \in F$. Since $yJ \subseteq X$, then $\operatorname{ann}_A a/I \in F\,\forall a \in J$, and so F.4 implies that $Y \in \mathscr{S}(F)$. Thus $C = \mathscr{S}(F)$ is a Serre class. Since C is closed, C has arbitrary coproducts (see the proof of 16.2), so C is localizing by 15.11. The converse is similar. □

16.4 Definition. An **idemtop** in mod-A is an idempotent topologizing set of right ideals of mod-A. For each $M \in$ mod-A, choose an $\mathscr{S}(F)$-envelope $u_M : M \to M_F$ [cf. 15.19. Thus, M_F is $\mathscr{S}(F)$-closed and both $\ker u_M$ and coker u_M belong to $\mathscr{S}(F)$.] For each $m \in M_F$, the mapping

$$\begin{cases} m_s : A \to M_F \\ \qquad a \mapsto ma \end{cases}$$

has a unique extension $\overline{m}_s : A_F \to M_F$. The map

$$\begin{cases} \varphi_M : M_F \times A_F \to M_F \\ \varphi_M : (m, x) \mapsto \overline{m}_s(x) \end{cases}$$

is bilinear. Furthermore, A_F is a ring under the composition $\varphi_A : A_F \times A_F \to A_F$, and then M_F is a right A_F-module under φ_M. Finally, if I is a right ideal of A, then I_F can be assumed to be a right ideal of A_F.

16.5 Definition and Lemma. *Let F be idemtop in* mod-A, *let* $\mathscr{S}(F)$ *denote the corresponding localizing subcategory, let* $T: \text{mod-}A \rightsquigarrow \text{mod-}A/\mathscr{S}(F)$ *be the canonical functor, and let* $S': \text{mod-}A/\mathscr{S}(F) \rightsquigarrow \text{mod-}A_F$ *be the functor induced by the localizing functor. If* ϱ *is the functor that assigns to each* A_F*-module the underlying* A*-module, then* S' *is such that* $\varrho \circ S'$ *is right adjoint to* T.

For each $M \in \text{mod-}A$, *and* $N \in \text{mod-}A_F$,

$$\varrho(N, M_F): \text{Hom}_{A_F}(N, M_F) \to \text{Hom}_A(\varrho N, \varrho M_F)$$

is bijective.

Proof. Everything is clear up to the last sentence. Every A_F-map $N \to M_F$ is an A-map, so $\varrho(N, M_F)$ is injective. If $f: N \to M_F$ is an A-map, then for a fixed $x \in N$, the mapping $g: a \mapsto f(xa) \; \forall a \in A_F$ is an A-map, and so is $h: a \mapsto f(x)a$. Since $gu_A = hu_A$, then $g = h$ 15.12. This proves that $\varrho(N, M_F)$ is surjective. □

16.6 Proposition. *Assume the same hypothesis as the lemma. The set* F' *of right ideals* I *of* A_F *such that* $\varrho(A_F/I)$ *is* F*-torsion is idemtop in* mod-A_F; *that is,* $\mathscr{S}(F')$ *is a localizing subcategory of* mod-A_F. *Let* $T': \text{mod-}A_F \rightsquigarrow \text{mod-}A_F/\mathscr{S}(F')$ *be the canonical functor.*

16.6.1 *The functor* $T \circ \varrho$ *induces an equivalence*

$$\text{mod-}A_F/\mathscr{S}(F') \rightsquigarrow \text{mod-}A/\mathscr{S}(F).$$

16.6.2 *An* A_F*-module* N *is* $\mathscr{S}(F')$*-closed* [*resp.* $\mathscr{S}(F')$*-closed and injective*] *if and only if* ϱN *is* $\mathscr{S}(F)$*-closed* [*resp.* $\mathscr{S}(F)$*-closed and injective*].

Proof. Let $B = \text{mod}\, A/\mathscr{S}(F)$. If $M \in \text{obj}\, B$, $N \in \text{mod-}A_F$, then adjointness of $\varrho \circ S'$ and T implies

$$\text{Mor}_B(T\varrho N, M) \approx \text{Hom}_A(\varrho N, \varrho S'M)$$

which, by the lemma, yields

$$\text{Mor}_B(T\varrho N, M) = \text{Hom}_{A_F}(N, S'M).$$

Thus, $T\varrho$ is left adjoint to S', so the category equivalence comes from 15.18. Now 16.6.2 follows from 16.6.1 and 15.15. □

16.7 A Proposition. *If* F *is idemtop in* mod-A, *then the localizing functor defined by* F, *or, equivalently, by the localizing subcategory* $\mathscr{S}(F)$, *is naturally isomorphic to the functor defined by:*

$$M \to \varinjlim_{I \in F} \text{Hom}_A(I, M/\mathscr{S}(F)M)$$

Proof. Identify $\varrho(M_F) \approx \operatorname{Hom}_A(A, M_F)$. By the preceding proof,

$$\operatorname{Hom}_A(A, M_F) \approx \operatorname{Hom}_{\text{mod-}A/\mathscr{S}(F)}(A, M) = \varinjlim_{I \in F} \operatorname{Hom}_A\big(I, M/\mathscr{S}(F)M\big).$$

The direct limit can be chosen so that (2) yields (1) as an isomorphism of A-modules. □

16.7B Corollary. *Let F be idemtop in* mod-A *that contains a cofinal subset G consisting of finitely generated right ideals, and assume that the localizing functor defined by F is exact. Then the localizing functor is naturally isomorphic to $\otimes_A A_F$, and induces a category equivalence*

$$(\text{mod-}A)/\mathscr{S}(F) \rightsquigarrow \text{mod-}A_F.$$

Proof. The following functors preserves coproducts:

(1) $M \mapsto \mathscr{S}(F)M$

(2) $M \mapsto M/\mathscr{S}(F)M$

(3) $\operatorname{Hom}_A(N, \ )$ $\forall$ finitely generated $N \in$ mod-A. (See 4.25 D.)

Thus, the localizing functor induces a functor

$$\begin{cases} H : \text{mod-}A \rightsquigarrow \text{mod-}A_F \\ H : M \mapsto \varinjlim_{I \in G} \operatorname{Hom}_A\big(I, M/\mathscr{S}(F)M\big) \end{cases}$$

which commutes with coproducts, and is right continuous. Using 5.54, and 5.56, (also 12.9.4) we get $H \approx \otimes_A A_F$.

For the last part, let $\mathscr{B} = \text{mod-}A/\mathscr{S}(F)$, let $T : \text{mod-}A \rightsquigarrow \mathscr{B}$ be canonical, and let S be a right adjoint of T. Since

$$\operatorname{Mor}_{\mathscr{B}}(A_F, \ ) \approx \operatorname{Hom}_A\big(A, S(\ )\big)$$

exactitude of S implies projectivity of A_F in $\mathscr{B}$. Since S commutes with coproducts, so does $\operatorname{Mor}_{\mathscr{B}}(A_F, \ )$, and therefore A_F is a small object in $\mathscr{B}$ 4.25 D. Since $\operatorname{End}_{\mathscr{B}}(A_F) \approx A_F$, by 4.29, $\operatorname{Mor}_{\mathscr{B}}(A_F, \ )$ induces a category equivalence $\mathscr{B} \rightsquigarrow \text{mod-}A_F$. □

Torsion Theories

We now give a brief exposition (and generalization) of the torsion theories of S. E. Dickson [66], and others, following Lambek's exposition in [71] for modules.

A **radical** on, or of, an abelian category $\mathscr{A}$ is a function T defined on the objects of $\mathscr{A}$ such that the following conditions hold:

R.1 $T(M) \subseteq M$,

R.2 $f: M \to N$ implies $f(T(M)) \subseteq T(N)$,

R.3 $T(M/T(M)) = 0$.

(Actually, R.2 implies R.1 and R.3.) The radical T is **idempotent** provided that T is an idempotent operation: $T(T(M)) = T(M)\ \forall$ objects M of A. Appropriately, for any ring R, the function which assigns to each object M of mod-R its radical, namely the intersection $\operatorname{rad} M$ of all maximal submodules of M, is a radical on mod-R, but it is rarely idempotent. (It is, nevertheless, for an important class of rings: V-rings 7.32A, but in this case it is the zero function.) However, if C is any closed subcategory of an abelian category $\mathscr{A}$, then there is an idempotent radical which assigns to any object M the subobject CM (defined in 16.1.1.). Examples of these are the subcategory of semisimple modules (over any ring), or the subcategory of torsion modules over commutative domains. (The complete analogy with torsion modules is brought out in 16.8B.)

Abbreviate $\operatorname{Mor}_{\mathscr{A}}(X, Y)$ by (X, Y), and for any subclass B of $\mathscr{A}$, define:

$$B^{\perp} = \{Y \in \operatorname{obj} \mathscr{A} \quad | \quad (X, Y) = 0\ \forall\ X \in B\}$$

and define ${}^{\perp}C$ by symmetry, for any subclass C. Furthermore, assuming enough injectives, define

$$B^{R} = \{Y \in \operatorname{obj} \mathscr{A} \quad | \quad (X, \hat{Y}) = 0\ \forall\ X \in B\},$$

where $\hat{Y}$ denotes the injective hull of Y. For any subclass C of $\mathscr{A}$, let:

$$C^{L} = \{X \in \operatorname{obj} \mathscr{A} \quad | \quad (X, \hat{Y}) = 0\ \forall\ Y \in C\}.$$

A subclass B of $\mathscr{A}$ is said to be a **semi-Serre class** or subcategory[1] if one-half of the condition for a Serre class holds, namely for $0 \to X \to M \to Y \to 0$ exact in $\mathscr{A}$, we have $M \in B$ whenever X and Y are in B.

16.8A Proposition. *If B and C are full subcategories of a Grothendieck category $\mathscr{A}$, then the following are equivalent conditions:*

PTF.1 $B = {}^{\perp}C$ *and* $C = B^{\perp}$.

PTF.2 *B is a semi-Serre class containing coproducts, and quotient objects, of any family of objects in B, and $C = B^{\perp}$.*

[1] In Lambek [71], the term "closed under group extensions" is used.

PTF.3 *C is a semi-Serre class containing products, and subobjects, of any family of objects in C, and $B = {}^{\perp}C$.*

PTF.4 *There is an idempotent radical on $\mathscr{A}$ such that*

$$B = \{X \mid T(X) = X\} \quad \textit{and} \quad C = \{Y \mid T(Y) = 0\}.$$

When this is so, then C is a coreflective and B is a reflective subcategory of $\mathscr{A}$, and the left adjoint to the inclusion functor $C \rightsquigarrow \mathscr{A}$ sends any object X to $X/T(X)$.

Proof. $1 \Rightarrow 3$ is a consequence of the fact 5.41 that h^X is limit preserving, for any X, and dually for $1 \Rightarrow 2$. To show $2 \Rightarrow 4$, let $T(M)$ be the sum of all subobjects of M which are in B, and dually for $3 \Rightarrow 4$. Assuming 4, then any morphism $f: X \to Y$, with $X \in B$ and $Y \in C$, must satisfy $f(T(X)) = f(X) = 0$. Thus, ${}^{\perp}C \supseteq B$ and $B^{\perp} \supseteq C$. If $X \in {}^{\perp}C$, then since $X/TX \in C$, we get $(X, X/TX) = 0$ only if $X = TX \in B$. Thus, $B = {}^{\perp}C$ and dually for $C = B^{\perp}$, so 1 holds. This proves the equivalence of PTF.1—4. Moreover, since $(TX, Y) = 0$, left exactness of $(\ , Y)$ implies

$$(X/T(X), Y) = (X, Y)$$

for any pair $X \in \mathscr{A}$ and $Y \in C$, so the inclusion functor $C \rightsquigarrow \mathscr{A}$ has the stated left adjoint. Thus, C is coreflective (*supra* Chapter 14 Exercises; cf. also 5.51). □

The ordered pair (B,C) of the proposition is called a **pretorsion** theory. In addition to those already stated, well-known examples are (torsion, torsionfree) in the category of abelian groups, and (divisible, reduced) in the category of divisible abelian groups, where a group is "reduced" if it has no divisible subgroups $\neq 0$. It is clear that any subclass A of $\mathscr{A}$ defines a pretorsion theory $({}^{\perp}A, ({}^{\perp}A)^{\perp})$, called the *pretorsion theory generated by A*.

16.8B Proposition and Definition. *An ordered pair (B, C) of full subcategories of a Grothendieck category $\mathscr{A}$ is said to be a* **torsion theory** *provided that the following equivalent conditions hold:*

TF.1 *$B = C^L$ and $C = B^R$.*

TF.2 *B is a localizing subcategory, and $C = B^{\perp}$.*

TF.3 *C is a semi-Serre class, containing the product and injective hulls of any family of objects of C, and $B = {}^{\perp}C$.*

TF.4 *There exists a radical T on $\mathscr{A}$ such that $M \subseteq N$ implies that $M \cap T(N) = T(M)$ and, moreover, $B = \{M \mid T(M) = M\}$ and $C = \{M \mid T(M) = 0\}$.*

Moreover, in case $\mathscr{A} = \text{mod-}R$, *for some ring* R, *these conditions are equivalent to:*

TF.5 *There exists an idemtop* F *of* R *such that* $B = \mathscr{S}(F)$.

Proof. We make some preliminary remarks:

(1) TF.1 $\Rightarrow B = {}^{\perp}C$.

(2) TF.2 = PTF.2 + (2′) B is closed under subobjects.

(3) TF.3 = PTF.3 + (3′) C is closed under injective hulls.

(4) TF.4 = PTF.4 + (4′) $T(M) = T(N) \cap M$ $\forall$ objects N and subobjects M.

(5) TF.5 $\Leftrightarrow$ TF.2 by 16.3.3.

(1) holds because TF.1 implies that $C = B^R$ is closed under injective hulls. The equivalence of PT.2—4 readily provides that of TF.2—4, that is, of (2)—(4). For example, (2) $\Rightarrow$ (4) since, given PT.4, we have from 15.19 that for any object Y, $T(Y)$ is the largest subobject of Y belonging to B. Thus, $M \subseteq N$ implies that $T(M) \subseteq T(N)$ by R.2 (and so $T(M) \subseteq M \cap T(N)$) whereas the reverse inclusion comes from the fact that $M \cap T(N)$ is a subobject of $T(N)$, and (2′) implies that $M \cap T(N)$ lies in B, hence in $T(M)$.

The other implications have about the same degree of difficulty. Let us prove (2) $\Rightarrow$ (1). Then, we may assume TF.4. If $Y \in C$, then $\hat{Y} \in C$, since $T(\hat{Y}) \cap Y = T(Y) = 0$, so $T(\hat{Y}) = 0$. Since C is thereby closed under injective hulls (incidentally, this gives us TF.4 $\Rightarrow$ TF.3), then $B^R \supseteq C$, and the reverse inclusion follows from left exactness of h^X, that is, $h^X(\hat{Y}) = 0$ implies $h^X(Y) = 0$. Thus, $B^R = C$. Then, as we have noted, $C^L = {}^{\perp}C$. Now $B = {}^{\perp}C$ by PTF.3, so we get $B = C^L$.

By 16.8A, C is coreflective, and since the inclusion $C \rightsquigarrow \mathscr{A}$ preserves injectives, the coreflector of C is exact 6.28. □

In 16.8B, as in 15.19C, the objects of B are torsion, those of C are torsionfree, and $T(M)$ is the **torsion subobject** of M. A radical T satisfying TF.4 is called a **torsion**, or **hereditary**, **radical**. Furthermore, a module M is said to be **divisible** (with respect to the torsion theory) provided that $\hat{M}/M$ is torsionfree, which by 15.19C is equivalent to the condition that M is $\mathscr{S}$-closed. Henceforth, then, for mod-R, we adopt the terminology "divisible" which is more suggestive than $\mathscr{S}$-closed. (Note, however, for the torsion theory generated by 0, every object is "divisible".) Furthermore, we say **divisible hull** instead of $\mathscr{S}$-envelope.

We restate 15.29 in this terminology.

16.8 C Proposition. *A category* C *is equivalent to the full subcategory of torsionfree divisible modules relative to some torsion theory for some module type category iff* C *is a Grothendieck category.* □

If the corresponding idemtop F for the torsion theory consists of the right ideals I which are **dense** in the sense that R is a rational extension of I, then the divisible hull is the maximal rational extension of M. If, on the other hand, F contains every essential right ideal, then a torsionfree module M is divisible iff injective, inasmuch as $M = \hat{M}$. (See Chapter 16 Exercises.)

Multiplicative Sets

Let S be a subset of a ring A. Then S is **multiplicative** if $ab \in S$ $\forall a \in S,\ b \in S$. The set

$$F_S = \{I \subseteq A \mid \forall a \in A\ \exists s \in S (as \in I)\}$$

of right ideals of A is idemtop. Thus, an A-module M is F_S-torsion if and only if every $m \in M$ is annihilated by an element $s \in S$. If $I \in F_S$, then $I \cap S \neq 0$; that is, I meets S. (In 16.9, the converse condition is denoted S.1.)

Definition. Let $\varphi : A \to B$ be a ring homomorphism, and let S be a multiplicative set of A. Then B is called the **right quotient ring with respect to** S, notation (B, φ), if the following conditions hold:

Q.1 Ker φ is F_S-torsion, or equivalently

$$\varphi(a) = 0 \Rightarrow \exists s \in S (as = 0).$$

Q.2 $\varphi(S) \subseteq$ units B.

Q.3 $B = \{\varphi(a)\,\varphi(s)^{-1} \mid a \in A,\ s \in S\}$.

Thus, if S denotes the set of regular elements of A, then B is the classical quotient ring of A discussed in Chapter 9. For any $M \in$ mod-A, let M_S denote M_{F_S} (see 16.4).

16.9 Proposition. *Let A be a ring, and S a multiplicative subset. Then the following three conditions are equivalent:*

(a) *A has a right quotient ring with respect to S.*

(b) *S satisfies the following two conditions:*

S.1 $\forall s \in S,\ \forall a \in A,\ \exists t \in S,\ \exists b \in A\ \ (at = sb)$;

S.2 $a \in A,\ s \in S\ \&\ sa = 0 \Rightarrow \exists t \in S\ \ (at = 0)$.

(c) *If $u_A : A \to A_S$ is the canonical ring homomorphism (recall that $A_S = A_{F_S}$), then $u_A(S) \subseteq$ units A_S.*

When these conditions hold, and if (B, φ) is a right quotient right with respect to S, then there exists a unique ring isomorphism $\psi : B \to A_S$ such that

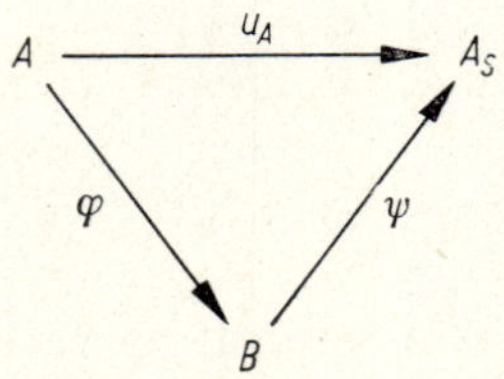

commutes. In this case, the localizing functor $M \mapsto M_S$ is exact and naturally equivalent to $\otimes_A A_S$.

Proof. (a) $\Rightarrow$ (b). To verify S.2, let $sa = 0$. Then $\varphi(s)\varphi(a) = 0$, so $\varphi(a) = 0$, and then $at = 0$ for some $t \in S$ by Q.1.

For S.1, let us write

$$\varphi(s)^{-1}\varphi(a) = \varphi(c)\varphi(r)^{-1}$$

for suitable $r \in S$, $c \in A$. Then $\varphi(a)\varphi(r) = \varphi(s)\varphi(c)$, so $ar - sc \in \operatorname{Ker}\varphi$, and hence, by Q.1, $\exists u \in S$ with $aru - scu = 0$. Then $t = ru$, $b = cu$ satisfies S.1.

(b) $\Rightarrow$ (c). We first show that the localizing functor is exact. Let

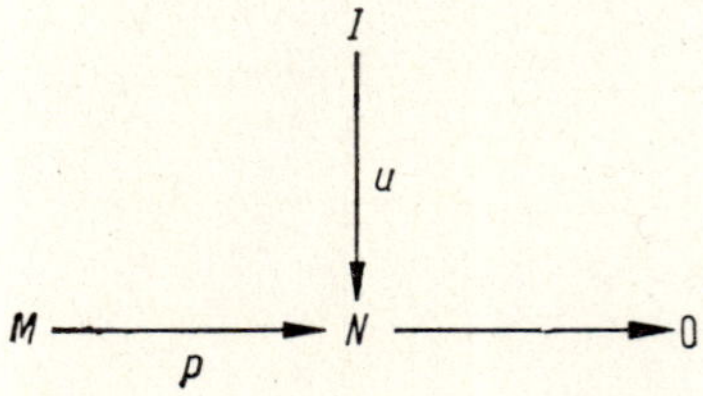

be a diagram such that $I \in F_S$, and M and $\ker p$ are F_S-closed. It suffices to show that I contains a right ideal $J \in F_S$, such that $u|J = p \circ v$, where $v : J \to M$. If $s \in I \cap S$, then $u(s) = p(m)$ with $m \in M$, and $sa = 0 \Rightarrow ma = 0$. Then $J = sA$, $v(sa) = ma$ fits the requirements.

If $t \in S$, and $t_s : a \mapsto sa$ is the homothetic, then coker t_s is F_S-torsion, so $t_s : A_A \to A_A$ is epic; that is, $u_A(t)$ has a right inverse. Let t' be such that $u_A(t)t' = 1$. Then

$$u_A(t)\,(t' u_A(t) - 1) = 0.$$

Since $u_A(t)$ is not a zero divisor in A_S, then $t' u_A(t) = 1$, so $u_A(t)$ is a unit.

(c) $\Rightarrow$ (a). (A_S, u_A) is a right quotient ring with respect to S, and is unique in the sense of commutativity of the diagram. $\square$

16.10 *Exercises.*

16.10.1 Show that S.1 and S.2 hold whenever $S \subseteq$ center A.

16.10.2 When A is commutative, and F is idemtop, then A_F is commutative, and

$$P \mapsto P_F$$

is a bijection of the set of prime ideals P of A not in F, and those of A_F not in F' (cf. 27.4).

Let M and N be right A-modules. Recall from Chapter 4 that $M \nabla N$ denotes that N is an essential submodule of M. Furthermore,

$$\operatorname{sing} A = \{a \in A \mid A \nabla \operatorname{ann}_A a\}$$

is an ideal of A (9.6). Moreover (*loc. cit.*), A is *right neat* if $\operatorname{sing} A = 0$.

16.11 *Exercise.* If $\operatorname{sing} A = 0$, then $F = \{I \subseteq A \mid A \nabla I\}$ is idemtop.

16.12 **Theorem and Definition.** *Let A be a right neat ring, and let F be the idemtop set $\{I \subseteq A \mid A \nabla I\}$. Then A_F is a regular ring, is right self-injective, and is the injective hull of the right A-module A. The ring A_F is called the (Johnson)* ***maximal right quotient ring of*** *A, and is denoted by $\hat{A}$.*

Proof (Gabriel [62]). (a) We first prove that A_F is injective in mod-A. Let $f: I \to A_F$ be an A-map, where I is a right ideal of A, and let J be a relative complement right ideal of A. Then, $I + J \approx I \oplus J$, so f can be extended to $\bar{f}: I + J \to A_F$. Since $I + J \in F$, then $\bar{f}$ can be extended to $A \to A_F$, and so A_F is injective. Since $A_F \nabla A$, A_F is the injective hull of A.

To prove that A_F is regular, we first prove (b) that every $\mathscr{S}F$-closed A-module $M \subseteq A_F$ is an A_F-summand of A_F. If the A-module N is a relative complement of $M \cap A$ in A, then $(M \cap A) + N \in F$, and $A_F/(M + N)$ is therefore F-torsion. Then, the injection $M + N \to A_F$ is an $\mathscr{S}F$-envelope of $M + N$. Thus

$$A_F = (M \oplus N)_F = M_F \oplus N_F = M \oplus N_F.$$

We now prove that A_F is regular; that is, every cyclic right ideal $a A_F$ is generated by an idempotent. The mapping

$$\begin{cases} A_F \to \operatorname{End} A_F \\ f \mapsto f_s \end{cases}$$

(where f_s is the left homothet of A_F determined by $f \in A_F$) is a ring isomorphism. Since A_F is F-closed, ker f_s is also, and is a summand of A_F by (b). The A_F-module M complementary to ker f_s is injective by (a), and f_s induces an isomorphism of M on a summand of A_F. It follows that aA_F is a summand of A, hence generated by an idempotent. This proves that A_F is regular. Furthermore, A_F is an injective A_F-module by 16.6.2. ▯

16.13 *Exercises.*

16.13.1 If A is an entire ring, or if A is a simple ring, then sing $A = 0$, and $\hat{A}$ is a simple ring.

16.13.2 If sing $A = 0$, then sing $\hat{A} = 0$.

16.13.3 If sing $A = 0$, then $\hat{A}$ is a semisimple ring if and only if A satisfies (acc)$\oplus$.

16.13.4 If A is semiprime, and if A satisfies (acc)$^\perp$, then sing $A = 0$.

16.14 **Theorem** (cf. 9.9) *If A is a semiprime ring satisfying* (acc)$\oplus$ *and* (acc)$^\perp$, *then $\hat{A}$ is a semisimple ring, and is the right quotient ring of A.*

Proof (Gabriel [62]). $\hat{A}$ is a semisimple ring by the exercises. We let the reader verify, with the following hints, that (A_F, u_A) is a right quotient ring of A with respect to the set S of nonzero divisors of A. If $t \in S$, then $t_s : A \to A$ is an injective map $A \to A_F$, and extends to an automorphism of A_F, proving that t is invertible in A_F. By 16.12, we need only show that a right ideal I is essential if and only if I meets S. This computation is left as an exercise, or the reader might refer to the proof of 9.9. ▯

16.15 **Proposition.** *Let $\mathscr{A}$ be an abelian category with a set of generators and exact direct limits, let $C =$ center $\mathscr{A}$ (= the ring consisting of all natural transformations $1_{\mathscr{A}} \to 1_{\mathscr{A}}$), let P denote a multiplicative subset of C, and let $\mathscr{A}_P$ denote the full subcategory of $\mathscr{A}$ whose objects are those M in $\mathscr{A}$ such that M is the upperbound of the subobjects N of M for which there exists $s \in P$ such that $s(N) = 0$.*

(a) *Then $\mathscr{A}_P$ is a localizing subcategory $\mathscr{A}$.*

(b) *An object M of $\mathscr{A}$ is $\mathscr{A}_P$-closed if and only if $s(M)$ is an automorphism of M for every $s \in P$.*

Proof. (a) is obvious. (See 15.11.)

(b) Let M be $\mathscr{A}_P$-closed. Then the definition tells us that $s(M)$ is monic, so im $s(M)$ is $\mathscr{A}_P$-closed, and by 15.19, im $s(M) \to M$ splits. But then coker $s(M)$ belongs to $\mathscr{A}_P$, so coker $s(M) = 0$. Thus, im $s(M) = M$, and $s(M)$ is an automorphism.

Conversely, if $s(M)$ is an automorphism for all s, then every subobject of M in $\mathscr{A}_P$ is 0. Let $j_M : M \to M_P$ be an $\mathscr{A}_P$-envelope of M, and suppose N is a subobject of M_P containing M such that im $s(N) \subseteq M$ for some $s \in P$. If $i : M \to N$ is the inclusion, then $1_M = s(M)^{-1} \circ s(N) \circ i$, so $i : M \to N$ splits. Therefore, $M = N$, whence $M = M_P$. □

16.6 Corollary. *The localizing functor* $M \rightsquigarrow M_P$ *is exact.*

Proof. Refer to 15.20. □

Endomorphisms Rings of Proper Generators

16.17 Definition (C. L. and E. A. Walker [65]) A generator U of an abelian category C is said to be **proper** provided that for every subobject V of U, the mapping induced by h_U

$$\mathrm{Mor}_C(V, U) \to \mathrm{Mor}_C(U, U)$$

is bijective if and only if $V = U$.

16.18 *Example.* If G is a generator, and if C is a cogenerator, then $U = G \coprod C$ is a proper generator, since, for every subobject $V \neq U$, there is a morphism $f : U \to U$ so that f composed with the canonical morphism $q : U \to U/V$ is nonzero. Thus, if $k : V \to U$ is the inclusion, then $f \neq kg$, for any $g : U \to V$.

16.19 *Exercise.*

16.19.1 Any principal right ideal domain A is a proper generator of mod-A.

*16.19.2 (Walker (l.c. p. 419) In mod-$\mathbb{Z}/p^n\mathbb{Z}$, where p is a prime, a generator G is proper provided either that G has a divisible subgroup $\neq 0$, or G is a direct sum of cyclic subgroups.

16.20 Theorem (Walker, C. L. and E. A. [65]) *Let A and R be rings, let U be a proper generator of* mod-A, *and let V be a proper generator of* mod-R. *If there is a ring isomorphism* $\varphi : \mathrm{End}\, U_A \to \mathrm{End}\, V_R$, *then there is a category equivalence* $F :$ mod-$A \rightsquigarrow$ mod-R *such that* $FU = V$, *and F induces* φ.

Proof (Walker) Let $B = \mathrm{End}\, U_A$. By 16.17, and 15.19, if T denotes $\mathrm{Hom}_A(U, \ )$, then

$$\underset{B}{\otimes} U : \text{mod-}B \rightsquigarrow \text{mod-}A$$

is the left adjoint functor and there exists a commutative diagram

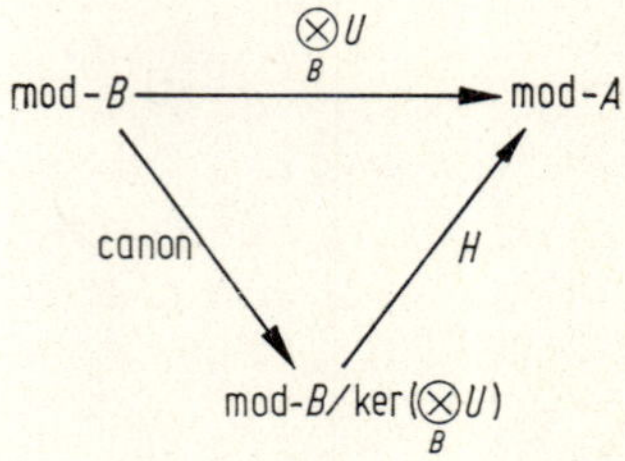

where H is a category equivalence. Let $\mathscr{F}_U = \mathscr{F}\left(\operatorname{Ker}\left(\underset{B}{\otimes} U\right)\right)$ denote the idemtop set of right ideals of B determined by the localizing subcategory $\operatorname{Ker}\left(\underset{B}{\otimes} U\right)$. Thus, by 16.3,

$$I \in \mathscr{F}_U \Leftrightarrow B/I \underset{B}{\otimes} U = 0.$$

By adjoint associativity

$$\operatorname{Hom}_A\left(B/I \underset{B}{\otimes} U, U\right) \approx \operatorname{Hom}_B(B/I, \operatorname{Hom}_A(U, U)) \approx \operatorname{Hom}_B(B/I, B).$$

Thus, we get the equivalence of the following conditions on a right ideal I of B:

(a) $I \in \mathscr{F}_U$;

(b) $\operatorname{Hom}_A\left(B/I \underset{B}{\otimes} U, U\right) = 0$;

(c) $\operatorname{Hom}_B(B/I, B) = 0$.

Thus, $\mathscr{F} = \mathscr{F}\left(\operatorname{Ker} \underset{B}{\otimes} U\right)$ is determined by $B = \operatorname{End} U_A$. Identifying B with $\operatorname{End} V_R$, we see that $\mathscr{F}_U = \mathscr{F}_V$. Consequently,

$$K = \operatorname{Ker}\left(\underset{B}{\otimes} U\right) = \mathscr{S}(\mathscr{F}_U) = \mathscr{S}(\mathscr{F}_V) = \operatorname{Ker}\left(\underset{B}{\otimes} V\right)$$

and

$$\text{mod-}A \approx \text{mod-}B/K \approx \text{mod-}R.$$

The first equivalence is H^{-1}, and $\underset{B}{\otimes} V$ induces the second equivalence. Then $F = \left(\underset{B}{\otimes} V\right) \circ H^{-1}$ is the desired equivalence. □

Exercises for Chapter 16

Most of the exercises which follow have been culled from Lambek [71] and [72].

1. Torsion theories can be ordered by $(B,C) \leq (B',C')$ if $B \subseteq B'$, or equivalently, $C' \subseteq C$.

1.1 Given any class A of modules, the largest torsion theory in which all objects of A are torsionfree is $(A^L, A^{LR} = A^\perp)$. Dually, (A^{RL}, A^R) is the smallest torsion theory in which all the objects of A are torsion modules.

1.2 The class of dense right ideals of a ring is an idemtop F, and the corresponding torsion theory 16.8B is the largest for which the object R is torsionfree. The $\mathscr{S}(F)$-envelope of any module M is the maximal rational extension of M.

1.3 Show that any torsion theory is the largest torsion theory relative to which a certain right R-module M is torsionfree. If I is the injective hull of M, then a module X is torsion iff $\operatorname{Hom}_R(X,I) = 0$.

1.4 Any torsion theory has the form (A^{RL}, A^R) where A is the set of cyclic torsion modules. (Any torsion theory is generated by cyclic modules.)

1.5 Any torsion theory has the form (A^L, A^{LR}), where A is the set of injective hulls of cyclic modules R/F, such that R/F is torsionfree.

2.1 There is a bijection between torsion theories and full subcategories of mod-R with exact coreflectors: to every torsion theory corresponds the category of torsionfree divisible modules.

2.2 Let A denote the full subcategory of torsionfree divisible modules for a torsion theory (B,C) of mod-R, and let Q denote the coreflector of the inclusion $A \rightsquigarrow$ mod-R (composed of the left adjoints of the inclusions $A \rightsquigarrow C$ and $C \rightsquigarrow$ mod-R). Then, the object function of Q is such that, letting $D(X)$ denote the divisible hull of any module S,

$$Q(M) = D\big(M/T(M)\big)$$

and is called the **module of quotients of** M. Moreover, $Q(R)$ is a ring, called the **ring of quotients of** R relative to the torsion theory (B,C).

2.3 Let F denote the idemtop corresponding to B. For each right torsionfree R-module M, define (as in 15.3):

$$L(M) = \varinjlim_{I \in F} \operatorname{Hom}_R(I, M).$$

Then, L is a functor $C \rightsquigarrow$ mod-R, and for each right R-module M,

$$Q(M) \approx L\big(M/T(M)\big).$$

Show that the inclusion functor $A \rightsquigarrow C$ has a left adjoint D which assigns to each torsionfree module its divisible hull.

2.4 (Gabriel [62, p. 376], Goldman [69]) Let A be the category of torsion free divisible R-modules as in Exercise 2.2. The inclusion

$A \rightsquigarrow$ mod-R is exact iff A is closed under cokernels iff every factor module X/X' of an object X of A is divisible. (A sufficient condition for this is for R to be right hereditary.)

3.1 (Lambek [72]) Let $g: R \to Q$ be the canonical homomorphism of R into the ring $Q = Q(R)$ of 2.2. Also, let $h: R \to S$ be the canonical homomorphism of R into the biendomorphism ring S of the injective module I which is such that (B,C) is the largest torsion theory relative to which I is torsionfree (see 1.3). Show that S contains Q, and that $\ker g = \ker h$. If, moreover, I is finitely generated over End I_R (finendo), then $Q = S$. (For a discussion of results comparing Q and S, consult Lambek [71, p. 52].)

3.2 A right ideal K is the torsion submodule of R for some torsion theory of mod-R iff K is an ideal and $K = \operatorname{ann}_R \widehat{R/K}$ (equivalently: $\widehat{R/K}$ is an R/K-module). In this case, K is the kernel of the canonical homomorphism $h: R \to Q(R)$.

3.3 (Silver [67]) A ring morphism $h: R \to S$ is epic in RINGS iff mod-h : mod-$S \rightsquigarrow$ mod-R is a full functor if $S \otimes_R S \approx S$ canonically.

3.4 (Lambek [71, p. 45]) If $h: R \to S$ is a flat epimorphism of rings (meaning that h is a ring epic, that is, $S \otimes_R S \approx S$ canonically, and S is a flat left R-module), then h is a ring of quotients of R with respect to the largest torsion theory (B, C) such that the inclusion mod-h : mod-$S \rightsquigarrow$ mod-R maps every S-module into a torsion module.

3.5 A ring homomorphism $h: R \to S$ is a ring of quotients of R (relative to some torsion theory; see 2.2) iff there exists an injective right R-module I with biendomorphism ring S such that I is cyclic over End I_R and h is the canonical homomorphism $R \to S$.

3.6 Is S is a faithful ring of quotients of R (meaning that $\ker h = 0$ in 3.5), then S is the biendomorphism ring of $\hat{S} \oplus \hat{S}/S$.

4. An ideal P of a commutative ring R is prime provided that the complement $R - P$ is a multiplicative set. In this case, the quotient ring with respect to P is denoted R_P, when $P \neq 0$. (a) Show that R_P is a local ring (the local ring at P). (b) If K is the full quotient ring of R, then R_P is isomorphic to the subring $\{x \in K \mid \exists b \in R - P \,\&\, xb \in R\}$.

Notes for Chapter 16

Most of this chapter was extracted from Gabriel [62]. The exposition of torsion theories was modeled after Lambek's [71], and his introduction to the subject I assign as required reading (see pp. V and VI, and the attributions, p. 1). Radicals on categories I found in a paper by Amitsur [54, pp. 119—122] in the setting of "bicategories" postulated in MacLane's paper [50] ("Duality in groups"). (Kurosch and others also published on this subject.)

Far from exhausting topics treated in Gabriel's thesis [62], instead I have presented (in Chapters 15 and 16) only the concepts most fundamental to a reading of the entire work, which I strongly urge upon my readers. (Similarly, the work of Morita [58] provides a seemingly inexhaustable source of inspiration and ideas. Also see Walker and Walker [72].)

References

The reader is referred to the bibliography of Lambek [71] for additional references.

Amitsur [54], Dickson [66], Gabriel [62], Johnson [51a], Lambek [66, 71], MacLane [50], Morita [58, 70], Utumi [56], Walker and Walker [65, 72].

Bibliography

In the text, a reference to Morita [58] refers to a paper of K. Morita written in 1958. When more than one paper is written in the same year, we distinguish them as in Formanek [72a] and [72b].

The following list consists of papers either cited in the text, or closely related to papers cited in the text, or pointing in directions away from the text. Artin [50], Elizarov [69], and Swan [59] are respective examples.

[65] Abian, A.: The Theory of Sets and Transfinite Arithmetic. Philadelphia: Saunders 1965.

[39] Albert, A. A.: Structure of Algebras, Colloquium Publication, Vol. 24. Amer. Math. Soc., Providence 1939.

[61] Albrecht, F.: On projective modules over semi-hereditary rings. Proc. Amer. Math. Soc. **12**, 638–639 (1961).

[54] Amitsur, S. A.: A general theory of radicals, II. Radicals in rings and bicategories. Amer. J. Math. **76**, 100–125 (1954).

[56a] Amitsur, S. A.: Algebras over infinite fields. Proc. Amer. Math. Soc. **7**, 35–48 (1956).

[56b] Amitsur, S. A.: Radicals of polynomial rings. Canad. J. Math. **8**, 355–361 (1956).

[57a] Amitsur, S. A.: Derivations in simple rings. Proc. Lond. Math. Soc. (3) **7** 87–112 (1957).

[68] Amitsur, S. A.: Rings with involution. Israel J. Math. **6**, 99–106 (1968).

[72] Amitsur, S. A.: On central division algebras. Israel J. Math. **12**, 408–420 (1972).

[27a] Artin, E.: Zur Theorie der hyperkomplexen Zahlen. Abh. Math. Sem. Univ. Hamburg **5**, 251–260 (1927).

[27b] Artin, E.: Zur Arithmetik hyperkomplexer Zahlen. Abh. Math. Sem. Univ. Hamburg **5**, 261–289 (1927).

[50] Artin, E.: The influence of J.H.M. Wedderburn on the development of modern algebra. Bull. Amer. Math. Soc. **56**, 65–72 (1950).

[55] Artin, E.: Galois Theory. South Bend: Notre Dame University 1955.

[57] Artin, E.: Geometric Algebra. New York: Interscience Publishers 1957.

[43] Artin, E., Whaples, G.: The theory of simple rings. Amer. J. Math. **65**, 87–107 (1943).

[44] Artin, E., Nesbitt, E., Thrall, R.: Rings with Minimum Condition. Ann Arbor: University of Michigan Press 1944.

[55] Auslander, M.: On the dimension of modules and algebras, III. Nagoya Math. J. **9**, 67–77 (1955).

[57] Auslander, M.: On regular group rings. Proc. Amer. Math. Soc. **8**, 658–664 (1957).

[57] Auslander, M., Buchsbaum, D.: Homological dimension in local rings. Trans. Amer. Math. Soc. **85**, 390–405 (1957).

[66b] Auslander, M.: Coherent functors. In: Proceedings of the Conference on Categorical Algebra (La Jolla 1965). Berlin/Heidelberg/New York: Springer 1966, pp. 189—231.

[60a] Auslander, M., Goldman, O.: Maximal orders. Trans. Amer. Math. Soc. **97**, 1—24 (1960).

[60b] Auslander, M., Goldman, O.: The Brauer group of a commutative ring. Trans. Amer. Math. Soc. **97**, 367—409 (1960).

[51] Azumaya, G.: On maximally central algebras. Nagoya Math. J. 2, 119—150 (1951).

[59] Azumaya, G.: A duality theory for injective modules. (Theory of quasi-Frobenius modules.) Amer. J. Math. **81**, 249—278 (1959).

[66] Azumaya, G.: Completely faithful modules and self-injective rings. Nagoya Math. J. **27**, 697—708 (1966).

[40] Baer, R.: Abelian groups that are direct summands of every containing abelian group. Bull. Amer. Math. Soc. **46**, 800—806 (1940).

[66] Balcerzyk, S.: On projective dimension of direct limits of modules. Bull. Acad. Polon. Sci., Sér. Sci. Math. Astron. Phys. **14**, 241—244 (1966).

Bar-Hillel, Y., Fraenkel, A. A. (see Fraenkel and Bar-Hillel).

[60] Bass, H.: Finitistic dimension and a homological generalization of semi-primary rings. Trans. Amer. Math. Soc. **95**, 466—488 (1960).

[62a] Bass, H.: The Morita Theorems. (Lecture Notes.) Department of Mathematics, University of Oregon, Eugene 1962.

[67] Bass, H.: Lectures on Algebraic K-Theory. Tata Institute for Advanced Study, Colaba 1967.

[68] Bass, H.: Algebraic K-Theory. New York: Benjamin 1968.

[64] Bergman, G. M.: A ring primitive on the right but not the left. Proc. Amer. Math. Soc. **15**, 473—475 (1964).

[67] Birkhoff, G.: Lattice Theory, Colloquium Publication, Vol. 25 (revised). Amer. Math. Soc., Providence 1967.

[67] Birkhoff, G. D., MacLane, S.: Algebra. New York: Macmillan 1967.

[60] Bourbaki, N.: Eléments de Mathématique. Théorie des Ensembles, Chaps. 1 and 2. (A.S.I. No. 1212), Paris: Hermann 1960.

[61] Bourbaki, N.: Eléments de Mathématique. Algebra Commutative, Chaps. 1 (Modules Plat) and 2 (Localisation). (A.S.I. No. 1290), Paris: Hermann 1961.

[63] Bourbaki, N.: Eléments de Mathématique. Théorie des Ensembles, Chap. 3. (A.S.I. No. 1243), Paris: Hermann 1963.

[71] Boyle, A. K.: Ph. D. Thesis, 1971. Rutgers University, New Brunswick.

[63] Brauer, R.: Representations of finite groups, Lectures on Modern Mathematics, Vol. I (T. L. Saaty, Ed.). New York: Wiley 1953, pp. 133—175.

[64] Brauer, R., Weiss, E.: Non-commutative Rings, Part I (Lecture Notes). Mathematics Department, Harvard University, Cambridge 1964.

[55] Buchsbaum, D. A.: Exact categories and duality. Trans. Amer. Math. Soc. **80**, 1—34 (1955).

Buchsbaum, D., Auslander, M. (see Auslander and Buchsbaum).

[68] Bucur, I., Deleanu, A.: Introduction to the Theory of Categories and Functors. London/New York/Sydney: Wiley-Interscience 1968.

[65] Bumby, R. T.: Modules which are isomorphic to submodules of each other. Arch. math. **16**, 184—185 (1965).

[70] Camillo, V. P.: Balanced rings and a problem of Thrall. Trans. Amer. Math. Soc. **149**, 143—153 (1970).

[72] Camillo, V., Fuller, K. R.: Balanced and QF — 1 algebras. Proc. Amer. Math. Soc. **23**, 373—378 (1972).

[42] Cantor, G.: Contributions to the Founding of the Theory of Transfinite Numbers. Dover/New York 1942.

[56] Cartan, H., Eilenberg, S.: Homological Algebra. Princeton: Princeton University Press 1956.

[60] Chase, S. U.: Direct products of modules. Trans. Amer. Math. Soc. **97**, 457—473 (1960).

[62a] Chase, S. U.: On direct products and sums of modules. Pac. J. Math. **12**, 847—854 (1962).

[62b] Chase, S. U.: A remark on direct products of modules. Proc. Amer. Math. Soc. **13**, 214—216 (1962).

[65] Chase, S. U., Faith, C.: Quotient rings and direct products of full linear rings. Math. Zeit. **88**, 250—264 (1965).

[70] Chatters, A. W.: The restricted minimum condition in Noetherian hereditary rings. J. London Math. Soc. **4**, 83—87 (1971).

[72] Chatters, A. W.: A decomposition theorem for Noetherian hereditary rings. Bull. London Math. Soc. (to appear).

[55] Chevalley, C.: The Construction and Study of Certain Important Algebras. The Mathematical Society of Japan, Tokyo 1955.

[56] Chevalley, C.: Fundamental Concepts of Algebra. New York: Academic Press 1956.

[61] Clifford, A. H., Preston, G. B.: Algebraic Theory of Semigroups, Vol. I. Surveys of the Amer. Math. Soc., Vol. 7. Providence 1961.

[66] Cohen, P. J.: Set Theory and the Continuum Hypothesis. New York/Amsterdam: Benjamin 1966.

[62] Cohn, P. M.: On subsemigroups of free semigroups. Proc. Amer. Math. Soc. **13**, 347—351 (1962).

[64a] Cohn, P. M.: Free ideal rings. J. Algebra **1**, 47—69 (1964).

[64b] Cohn, P. M.: Subalgebras of free associative algebras. Proc. Lond. Math. Soc. (3) **14**, 618—632 (1964).

[65] Cohn, P. M.: Universal Algebra. New York: Harper Row 1965.

[66a] Cohn, P. M.: Morita Equivalence and Duality. University of London, Bookstore, Queen Mary College, Mile End Road, London 1966.

[66b] Cohn, P. M.: Some remarks on the invariant basis property. Topology **5**, 215—228 (1966).

[69] Cohn, P. M.: Free associative algebras. Bull. Lond. Math. Soc. **1**, 1—39 (1969).

[71] Cohn, P. M.: Free Rings and Their Relations. New York: Academic Press 1971.

[71] Colby, R. R., Rutter, E. A., Jr.: Π-Flat and Π-Projective modules. Arch. Math. **22**, 246—251 (1971).

[63] Connell, I.: On the group ring. Canad. J. Math. **15**, 650—685 (1963).

[63] Corner, A. L. S.: Every countable reduced torsion-free ring is an endomorphism ring. Proc. Lond. Math. Soc. (3) **13**, 687—710 (1963).

[65] Courter, R. C.: The dimension of a maximal commutative subalgebra of K_n. Duke Math. J. **32**, 225—232 (1965).

[70] Cozzens, J. H.: Homological properties of the ring of differential polynomials. Bull. Amer. Math. Soc. **76**, 75—79 (1970).

[72] Cozzens, J. H.: Simple principal left ideal domains. J. Alg. **23**, 66—75 (1972).

[73/74] Cozzens, J. H., Faith, C.: Simple rings with ascending chain conditions: A survey.

Croisot, R., Lesieur, L. (see Lesieur and Croisot).

[62] Curtis, C. W., Reiner, I.: Representation Theory of Finite Groups and Associative Algebras. New York: Interscience Publishers 1962.

[71] Dade, E. C.: Deux groupes finis distincts ayant la même algebre de groupe sur tout corps. Math. Z. **119**, 345—348 (1971).

[48] Deuring, M.: Algebren. New York: Chelsea 1948.
Deleanu, A., Bucur, I. (see Bucur and Deleanu).
[23] Dickson, L. E.: Algebras and their Arithmetics. University of Chicago, 1923, and Stechert & Co. (reprint) 1938.
[70] Dickson, S. E., Fuller, K. R.: Commutative QF – 1 artinian rings are QF. Proc. Amer. Math. Soc. **24**, 667–670 (1970).
[42] Dieudonné, J. A.: Les determinants sur un corps non-commutatif. Bull. Soc. Math. France **70**, 27–45 (1942).
Dieudonné, J. A., Grothendieck, A. (see Grothendieck and Dieudonné).
[72] Dlab, V., Ringel, C. M.: Balanced rings. J. Algebra **22**, 480–501 (1972).
[72] Dlab, V., Ringel, C. M.: Balanced rings. Proc. Tulane University Symposium in Ring Theory, New Orleans. Lecture Notes in Mathematics, No. 246. Berlin/Heidelberg/New York: Springer, pp. 73–143. (1972).
[53] Eckmann, B., Schopf, A.: Über injektive Moduln. Archiv Math. **4**, 75–78 (1953).
[56] Eilenberg, S.: Homological dimension and syzygies. Ann. Math. **64**, 328–336 (1956).
[60] Eilenberg, S.: Abstract description of some basic functors. J. Indian Math. Soc. **24**, 221–234 (1960).
[42] Eilenberg, S., MacLane, S.: Natural isomorphisms in group theory. Proc. Nat. Acad. Sci. U.S.A. **28**, 537–543 (1942).
[45] Eilenberg, S., MacLane, S.: General theory of natural equivalences. Trans. Amer. Math. Soc. **58**, 231–294 (1945).
Eilenberg, S., Nakayama, T.: On the dimension of modules and algebras, II (Frobenius algebras and Quasi-Frobenius rings). Nagoya Math. J. **9**, 1–16 (1955).
Eilenberg, S., Cartan, H. (see Cartan and Eilenberg).
[71] Eisenbud, D., Griffith, P.: Serial rings. J. Algebra **17**, 389–400 (1971).
[70] Eisenbud, D., Robson, J. C.: Hereditary noetherian prime rings. J. Algebra **16**, 86–104 (1970).
[69] Elizarov, V. P.: Quotient rings. Algebra and Logic **8**, 219–243 (1969).
[64] Faith, C.: Noetherian simple rings. Bull. Amer. Math. Soc. **70**, 730–731 (1964).
[65] Faith, C.: Orders in simple artinian rings. Trans. Amer. Math. Soc. **114**, 61–64 (1965).
[66a] Faith, C.: Rings with ascending condition on annihilators. Nagoya Math. J. **27**, 179–191 (1966).
[67a] Faith, C.: Lectures on Injective Modules and Quotient Rings. Lecture Notes in Mathematics, No. 49. Berlin/Heidelberg/New York: Springer 1967.
[67b] Faith, C.: A general Wedderburn theorem. Bull. Amer. Math. Soc. **73**, 65–67 (1967).
[71a] Faith, C.: A correspondence theorem for projective modules and the structure of simple noetherian rings. Bull. Amer. Math. Soc. **77**, 338–342 (1971).
[71b] Faith, C.: Orders in semilocal rings. Bull. Amer. Math. Soc. **77**, 960–962 (1971).
[72a] Faith, C.: A correspondence theorem for projective modules, and the structure of simple Noetherian rings. Proceedings of the Conference on Associative Algebras, Nov. 1970. Istituto Nazionale di Alta Matematica, Symposium Matematica **8**, 309–345 (1972).
[72b] Faith, C.: Modules finite over endomorphism ring. Proc. Tulane University, Symposium in Ring Theory, New Orleans. Lecture Notes in Mathematics, No. 246. Berlin/Heidelberg/New York: Springer, pp. 145–190.

[72c] Faith, C.: Galois subrings of Ore domains are Ore domains. Bull. Amer. Math. Soc. **78**, 1077—1080 (1972).

[65a] Faith, C., Utumi, Y.: On noetherian prime rings. Trans. Amer. Math. Soc. **114**, 53—60 (1965).

[65b] Faith, C., Utumi, Y.: Maximal quotient rings. Proc. Amer. Math. Soc. **16**, 1084—1089 (1965).

[67] Faith, C., Walker, E. A.: Direct sum representations of injective modules. J. Alg. **5**, 203—221 (1967).

[73/74a] Faith, C.: On a theorem of Chatters. Preprint, Rutgers University 1973/74.

[73/74b] Faith, C.: Embedding modules in projectives. Preprint, Rutgers University 1973/74.

Faith, C., Chase, S. U. (see Chase and Faith).

Faith, C., Cozzens, J. H. (see Cozzens and Faith).

[67] Feit, W.: Characters of Finite Groups. New York: Benjamin 1967.

[63] Feit, W., Thompson, J. G.: Solvability of groups of odd order. Pac. J. Math. **13**, 775—1029 (1963).

[61a] Feller, E. H., Swokowski, E. W.: Reflective N-prime rings with the ascending condition. Trans. Amer. Math. Soc. **99**, 264—271 (1961). Corrections, ibid., p. 555.

[61b] Feller, E. H., Swokowski, E. W.: Reflective rings with the ascending chain condition. Proc. Amer. Math. Soc. **12**, 651—653 (1961).

[58] Findlay, G. D., Lambek, J.: A generalized ring of quotients, I, II. Canad. Math. Bull. **1**, 77—85, 155—167 (1958).

[58] Fraenkel, A. A., Bar-Hillel, Y.: Foundations of Set Theory. Amsterdam: North Holland Publishing Comp. 1958.

[60] Freyd, P.: Functor Theory. Ph. D. Thesis, Columbia University 1970.

[64] Freyd, P.: Abelian Categories. New York: Harper Row 1964.

[72a] Formanek, E.: Central polynomials and orders. Carleton University preprint, Ottawa 1972.

[72b] Formanek, E.: A problem of Passman on semisimplicity. Carleton University preprint, Ottawa 1972.

[72] Formanek, E., Snider, R. L.: Primitive group rings. Carleton University and Miami University, preprint, Ottawa and Coral Gables 1972.

[71] Fuller, K. R.: Primary rings and double centralizers. Pac. J. Math. **34**, 379—383 (1970).

Fuller, K. R., Camillo, V. (see Camillo and Fuller).

Fuller, K. R., Dickson, S. E. (see Dickson and Fuller).

[62] Gabriel, P.: Des catégories abeliennes. Bull. Soc. Math. France **90**, 323—448 (1962).

[64] Gabriel, P., Popesco, N. (see Popesco and Gabriel).

[67] Gabriel, P., Zisman, M.: Calculus of Fractions and Homotopy Theory. New York/Heidelberg/Berlin: Springer 1967.

[60] Gentile, E.: On rings with one-sided fields of quotients. Proc. Amer. Soc. **11**, 380—384 (1960).

[64] Gödel, K.: The Consistency of the Axiom of Choice and the Generalized Continuum Hypothesis with the Axioms of Set Theory. Ann. of Math. Studies, No. 3 (Sixth Printing). Princeton: Princeton University Press 1964.

[58] Goldie, A. W.: The structure of prime rings under ascending chain conditions. Proc. Lond. Math. Soc. **VIII**, 589—608 (1958).

[60] Goldie, A. W.: Semi-prime rings with maximum condition. Proc. Lond. Math. Soc. **X**, 201—220 (1960).

[61] Goldie, A. W.: Rings with maximum condition (Lecture Notes). Yale University Math. Dept. 1961.

[62] Goldie, A. W.: Non-commutative principal ideal rings. Archiv. Math. **13**, 213—221 (1962).

[67] Goldie, A. W.: Localization in non-commutative Noetherian rings. J. Alg. **5**, 89—105 (1967).

[69] Goldie, A. W.: Some aspects of ring theory. Bull. Lond. Math. Soc. **1**, 129—154 (1969).

[51] Goldman, O.: Hilbert rings, and the Hilbert Nullstellensatz. Math. Z. **54**, 136—140 (1951).

[69] Goldman, O.: Rings and modules of quotients. J. Alg. 13, 10—47 (1969).

Goldman, O., and Auslander, M. (see Auslander and Goldman).

[68] Gorenstein, D.: Finite Groups. New York: Harper & Row 1968.

[70] Griffith, P., Robson, J. C.: A theorem of Asano and Michler. Proc. Amer. Math. Soc. **24**, 837—838 (1970).

Griffith, P., Eisenbud, D. (see Eisenbud and Griffith).

[57] Grothendieck, A.: Sur quelques points d'algèbre homologique. Tohoku Math. J. **9**, 119—221 (1957).

[65] Grothendieck, A.: Le groupe de Brauer I. Séminaire Bourbaki, exposé 290, Paris: Hermann 1965.

[60—64] Grothendieck, A., Dieudonné, J. A.: Eléments de Géometrie Algébrique, I—IV. Publications Institute des Hautes Etudes, Scientifiques. Le Bois-Marie-Bures-sur-Yvette (S.-et-O.) 1960—1964.

[66] Gupta, R. N.: Another characterisation of semi-prime rings with right Goldie conditions. J. Math. Sci. **1**, 87—89 (1966).

(68) Gupta, R. N.: Characterisation of rings whose classical quotient rings are perfect rings. Osaka J. Math. **5**, 69—87 (1968).

[67] Gupta, R. N., Saha, F.: Remark on a paper of Small. J. Math. Sci. **2**, 7—16 (1967).

[59] Hall, M. H., Jr.: The Theory of Groups. New York: Macmillan 1959.

[60] Halmos, P.: Naive Set Theory. New York: Van Nostrand 1960.

[56] Harada, M.: Note on the dimension of modules and algebras. J. Inst. Polytechnics Osaka City University, Ser. A, **7**, 17—27 (1956).

[67] Hart, R.: Simple rings with uniform right ideals. J. Lond. Math. **42**, 614—617 (1967).

[49] Hausdorff, F.: Grundzüge der Mengenlehre. Leipzig 1914, Reprinted by Chelsea, New York 1949.

[57] Hausdorff, F.: Set Theory (English translation of Hausdorff [49]). New York: Chelsea 1957.

[64] Herstein, I. N.: Topics in Algebra. New York: Harper & Row 1964.

[65] Herstein, I. N.: A counter example in Noetherian rings. Proc. Nat. Acad. Sci. (U.S.A.) **54**, 1036—1037 (1965).

[68] Herstein, I. N.: Noncommutative Rings. The Carus Mathematical Monographs, No. 15. Amer. Math. Ass. New York: Wiley 1968.

[56] Hochschild, G.: Relative homological algebra. Trans. Amer. Math. Soc. **82**, 246—269 (1956).

[58] Hochschild, G.: Note on relative homological algebra. Nagoya Math. J. **13**, 89—94 (1958).

[42] Jacobson, N.: The theory of rings. Surveys, Vol. 2. Amer. Math. Soc., Providence 1942.

[45a] Jacobson, N.: The radical and semisimplicity for arbitrary rings. Amer. J. Math. **67**, 300—342 (1945).

[45b] Jacobson, N.: The structure of simple rings without finiteness assumptions. Trans. Amer. Math. Soc. **57**, 228—245 (1945).

[45c] Jacobson, N.: Structure theory for algebraic algebras of bounded degree. Ann. of Math. **46**, 695—707 (1945).

[50] Jacobson, N.: Some remarks on one-sided inverses. Proc. Amer. Math. Soc. **1**, 352—355 (1950).

[51, 53, 64] Jacobson, N.: Lectures in Abstract Algebra, Vol. I—III. New York/Princeton: Van Nostrand 1965, 1953, 1964.

[56, 64] Jacobson, N.: Structure of Rings. Colloquium Publication, Vol. 37. Amer. Math. Soc., Providence 1956 (1964 revised).

[64] Jans, J. P.: Rings and Homology. New York: Holt 1964.

[67] Jans, J. P.: On orders in Quasi-Frobenius rings. J. Alg. **7**, 35—43 (1967).

[70] Jans, J. P.: On the double centralizer property. Math. Ann. **188**, 85—89 (1970).

[69] Janusz, G. J.: Indecomposable modules for finite groups. Ann. of Math. **89**, 209—241 (1969).

[69a] Jategaonkar, A. V.: Ore domains and free algebras. Bull. Lond. Math. Soc. **1**, 45—46 (1969).

[69b] Jategaonkar, A. V.: A counter-example in ring theory and homological algebra. J. Alg. **12**, 418—440 (1969).

[70a] Jategaonkar, A. V.: Left Principal Ideal Rings. (Lecture Notes in Mathematics, Vol. 123.) Berlin/Heidelberg/New York: Springer 1970.

[70b] Jategaonkar, A. V.: Orders in artinian rings. Bull. Amer. Math. Soc. **75**, 1258—1259 (1969).

[72] Jategaonkar, A. V.: Noetherian Prime Rings. Proceedings of Ring Theory Conference. New York: Academic Press 1972, pp. 171—229.

[41] Jennings, S. A.: The structure of the group ring of a p-group over a modular field. Trans. Amer. Math. Soc. **50**, 175—185 (1941).

[51a] Johnson, R. E.: The extended centralizer of a ring over a module. Proc. Amer. Math. Soc. **2**, 891—895 (1951).

[51b] Johnson, R. E.: Prime rings. Duke Math. J. **18**, 799—809 (1951).

[53a] Johnson, R. E.: Representations of prime rings. Trans. Amer. Math. Soc. **74**, 351—357 (1953).

[69] Johnson, R. E.: Extended Malcev domains. Proc. Amer. Math. Soc. **21**, 211—213 (1969).

[59] Johnson, R. E., Wong, E. T.: Self-injective rings. Canad. Math. Bull. **2**, 167—173 (1959).

[61] Johnson, R. E., Wong, E. T.: Quasi-injective modules and irreducible rings. J. Lond. Math. Soc. **36**, 260—268 (1961).

[58] Kan, D.: Adjoint functors. Trans. Amer. Math. Soc. **87**, 294—329 (1958).

[58a] Kaplansky, I.: Projective modules. Ann. Math. **68**, 372—377 (1958).

[58b] Kaplansky, I.: On the dimension of modules and algebras, X, A right hereditary ring which is not left hereditary. Nagoya Math. J. **13**, 85—88 (1958).

[59] Kaplansky, I.: Homological dimension of rings and modules, Mimeographed notes. University of Chicago 1969.

[66] Kaplansky, I.: The homological dimension of a quotient field. Nagoya Math. J. **27**, 139—142 (1966).

[69a] Kaplansky, I.: Fields and Rings, Chicago Lectures in Mathematics. Chicago/London: University of Chicago Press 1969.

[69b] Kaplansky, I.: Infinite Abelian Groups, 2nd Ed. Ann Arbor: University of Michigan Press 1969.

[57] Kasch, F., Kneser, M., Kuppisch, H.: Unzerlegbare modulare Darstellungen endlicher Gruppen mit zyklischer p-Sylow-Gruppe. Arch. Math. **8**, 320—321 (1957).

[55] Kelley, J.: General Topology. Princeton: Van Nostrand 1955.

Kneser, M., Kuppisch, H., Kasch, F. (see Kasch, Kneser, and Kuppisch).

[70] Knus, M. A.: Algèbres d'Azumaya et modules projectifs. Comment. Math. Helv. **45**, 372—383 (1970).

[72] Knus, M. A., Ojanguren, M.: On the torsion of the Brauer group. Preprint, Math. Inst. E.T.H., Zürich 1972.

[66] Koh, K.: On simple rings with maximal annihilator right ideal. Canad. Math. Bull. **9**, 667—668 (1966).

[53] Kolchin, E. R.: Galois theory of differential fields. Amer. J. Math. **75**, 753—824 (1953).

[48] Krull, W.: Idealtheorie. New York: Chelsea 1948.

Kuppisch, H., Kasch, F., Kneser, M. (see Kasch, Kneser, and Kuppisch).

[55, 56] Kurosch, A. G.: Theory of Groups, Vols. 1 and 2. New York: Chelsea 1955, 1956.

[63] Kurosch, A. G.: General Algebra. New York: Chelsea 1963.

[63] Lambek, J.: On Utumi's ring of quotients. Canad. J. Math. **15**, 363—370 (1963).

[64a] Lambek, J.: A module is flat if and only if its character module is injective. Canad. Math. Bull. **7**, 237—243 (1964).

[64b] Lambek, J.: Goursat's theorem and homological algebra. Canad. Math. Bull. **7**, 597—607 (1964).

[66] Lambek, J.: Rings and Modules. New York: Blaisdell 1966.

[71] Lambek, J.: Torsion Theories, Additive Semantics, and Rings of Quotients. Lecture Notes in Mathematics, No. 177. Berlin/Heidelberg/New York: Springer 1971.

Lambek, J., Findlay, G. D. (see Findlay and Lambek).

[69] Lanski, C.: Nil subrings of Goldie rings are nilpotent. Canad. J. Math. **21**, 904—907 (1969).

[65] Lang, S.: Algebra. Cambridge: Addison-Wesley 1965.

[64] Lazard, D.: Sur les modules plats, C.R. Acad. Sci. Paris **258**, 6313—6316 (1964).

[57] Leavitt, W.: Modules without invariant basis number. Proc. Amer. Math. Soc. **7**, 322—328 (1957).

[62] Leavitt, W.: The module type of a ring. Trans. Amer. Math. Soc. **103**, 113—130 (1962).

[71] Lenagan, T. H.: Bounded Asano orders are hereditary. Bull. Lond. Math. Soc. **3**, 67—69 (1971).

[69] Lenzing, H.: Endlich präsentierbare Moduln. Arch. Math. **20**, 262—266 (1969).

[58] Lesieur, L., Croisot, R.: Théorie noethérienne des anneaux, des demi-groupes et des modules dans le cas non commutatif II. Mat. Ann. **134**, 458—476 (1958).

[59] Lesieur, L., Croisot, R.: Sur les anneaux premiers noetheriens à gauche. Ann. Sci. École Norm. Sup. **76**, 161—183 (1959).

[45] Levitzki, J.: Solution of a problem of G. Koethe. Amer. J. Math. **67**, 437—442 (1945).

[64] Levitzki, J.: On nil subrings. (Posthumous paper edited by S. A. Amitsur.) Israel J. Math. **1**, 215—216 (1963).

[63a] Levy, L.: Unique direct sums of prime rings. Trans. Amer. Math. Soc. **106**, 64—76 (1963).

[63b] Levy, L.: Torsionfree and divisible modules over non-integral domains. Can. J. Math. **15**, 132—151 (1963).

[33] Littlewood, D. E.: On the classification of algebras. Proc. Lond. Math. Soc. **(2) 35**, 200—240 (1933).

[60] Lubkin, S.: Imbedding of abelian categories. Trans. Amer. Math. Soc. **97**, 410—417 (1960).

[50] MacLane, S.: Duality in groups. Bull. Amer. Math. Soc. **56**, 485—516 (1950).

[63] MacLane, S.: Homology. Berlin/Göttingen/Heidelberg: Springer 1963.
[71] MacLane, S.: Categorical Algebra and set-theoretical foundations. Amer. Math. Soc. Proceedings of Symposia in Pure Math. Vol. XIII, Part I, 231—240 (1971).
[72] MacLane, S.: Categories for the working mathemacian. Berlin/Heidelberg/New York: Springer 1972.
MacLane, S., Eilenberg, S. (see Eilenberg and MacLane).
[36] Malcev, A. I.: On the immersion of an algebraic ring into a field. Math. Ann. **113**, 686—691 (1936).
[98] Maschke, H.: Über den arithmetischen Charakter der Coefficienten der Substitutionen endlicher linearer Substitutionsgruppen. Math. Ann. **50**, 483—498 (1898).
[69] Mewborn, A. C., Winton, C. N.: Orders in self-injective semi-perfect rings. J. Algebra **13**, 5—9 (1969).
[69a] Michler, G.: On quasi-local noetherian rings. Proc. Amer. Math. Soc. **20**, 222—224 (1969).
[69b] Michler, G.: Structure of semiperfect hereditary Noetherian rings. J. Algebra **13**, 327—344 (1969).
[69c] Michler, G.: Idempotent ideals in perfect rings. Canad. J. Math. **21**, 301—309 (1969).
[69d] Michler, G.: Asano orders. Proc. Lond. Math. Soc. **19**, 421—443 (1969).
[64] Mitchell, B.: The full embedding theorem. Amer. J. Math. **86**, 619—637 (1964).
[65] Mitchell, B.: Theory of Categories. New York: Academic Press 1965.
[72] Mitchell, B.: Rings with several objects, Advances in Math. **8**, 1—161 (1972).
[58] Morita, K.: Duality for modules and its applications to the theory of rings with minimum condition. Sci. Reports, Tokyo Kyoiku Daigaku **6**, 83—142 (1958).
[70, 71] Morita, K.: Localizations in categories of modules, I; III. Math. Z. **114**, 121—144 (1970); **119**, 313—320 (1971).
[71a] Morita, K.: Flat modules, injective modules, and quotient rings. Math. Z. **120**, 25—40 (1971).
[62] Nagata, M.: Local Rings. New York: Interscience Publishers 1962.
Nakayama, T., Eilenberg, S. (see Eilenberg and Nakayama).
Nesbitt, E., Thrall, R., Artin E. (see Artin, Nesbitt, and Thrall).
[60] Northcott, D. G.: Homological Algebra. Cambridge: Cambridge University Press 1960.
[31] Ore, O.: Linear equations in non-commutative fields. Annals of Math. **32**, 463—477 (1931).
[33] Ore, O.: On a special class of polynomials. Trans. Amer. Math. Soc. **35**, 559—584 (1933).
[67] Ornstein, A. J.: Rings with restricted minimum condition. Ph. D. Thesis, Rutgers University 1967.
[68a] Osofsky, B. L.: Endomorphism rings of Quasi-injective modules. Canad. J. Math. **20**, 895—903 (1968).
[68b] Osofsky, B. L.: Noncommutative rings whose cyclic modules have injective hulls. Pac. J. Math. **25**, 331—340 (1968).
[68c] Osofsky, B. L.: Homological dimensions and the continuum hypothesis. Trans. Amer. Math. Soc. **132**, 217—230 (1968).
[68d] Osofsky, B. L.: Upper bounds on homological dimensions. Nagoya Math. J. **32**, 315—322 (1968).
[69] Osofsky, B. L.: A commutative local ring with finite divisors. Trans. Amer. Math. Soc. **141**, 377—385 (1969).

[70] Osofsky, B. L.: Homological dimension and cardinality. Trans. Amer. Math. Soc. **151**, 641—649 (1970).

[42] Perlis, S.: A characterization of the radical of an algebra. Bull. Amer. Math. Soc. **48**, 128—132 (1942).

[39] Pontryagin, L.: Topological Groups. Princeton: Princeton University Press 1939.

[64] Popesco, N., Gabriel, P.: Caractérisations des catégories abeliennes avec générateurs et limites inductives exactes. C. R. Acad. Sci. Paris **258**, 4188—4190 (1964).

Preston, G. B., Clifford, A. H. (see Clifford and Preston).

[63] Procesi, C.: On a theorem of Goldie concerning the structure of prime rings with maximal condition (in Italian). Acad. Naz. Lincei Rend. **34**, 372—377 (1963).

[65] Procesi, C.: On a theorem of Faith and Utumi (in Italian). Rend. Math. e. Appl. **24**, 346—347 (1965).

[65] Procesi, C., Small, L.: On a theorem of Goldie. J. Algebra **2**, 80—84 (1965).

[67] Puppe, D.: Über die Axiome für abelsche Kategorien. Arch. Math. **7**, 217—222 (1967).

[72] Rao, M. L. R.: Azumaya, semisimple, and ideal algebras. Bull. Amer. Math. Soc. **78**, 588—592 (1972).

Reiner, I., Curtis, C. W. (see Curtis and Reiner).

[65] Richman, F.: Generalized quotient rings. Proc. Amer. Math. Soc. **16**, 794—799 (1965).

[65] Rieffel, M. A.: A general Wedderburn theorem. Proc. Nat. Acad. Sci. USA **54**, 1513 (1965).

Ringel, C. M., Dlab, V. (see Dlab and Ringel).

[67a] Robson, J. C.: Artinian quotient rings. Proc. Lond. Math. Soc. **17**, 600—616 (1967).

[67b] Robson, J. C.: Pri-rings, and Ipri-rings. Quarterly J. Math. **18**, 125—145 (1967).

[68] Robson, J. C.: Non-commutative Dedekind rings. J. Algebra **9**, 249—265 (1968).

Robson, J. C., Eisenbud, D. (see Eisenbud and Robson).

Robson, J. C., Griffith, P. (see Griffith and Robson).

[68] Roiter, A. V.: Unboundedness of the dimensions of indecomposable representations of algebras having infinitely many indecomposable representations. Izv. Akad. Nauk S.S.S.R. Ser. Math. **32**, 1275—1282 (1968).

[65] Roos, J. E.: Caractérisation des catégories qui sont quotients des çatégories des modules par des sous-catégories bilocalisantes. C. R. Acad. Sci. Paris **261**, 4954—4957 (1965).

[67] Roos, J. E.: Sur la condition AB 6 et ses variantes dans les catégories abéliennes. C. R. Acad. Sci. Paris **264**, 991—994 (1967).

[68] Roos, J. E.: Sur l'anneau maximal de fractions des AW-algebras et des anneaux de Baer. C. R. Acad. Sci. Paris **266**, 120—123 (1968).

[65] Rotman, J.: The Theory of Groups. Boston: Allyn & Bacon 1965.

Rutter, E. A., Colby, R. R. (see Colby and Rutter).

Saha, F., Gupta, R. N. (see Gupta and Saha).

[64] Sandomierski, F.: Relative injectivity and projectivity. Ph. D. Thesis, Penna, State University, University Park 1964.

[67] Sandomierski, F. L.: Semisimple maximal quotient rings. Trans. Amer. Math. Soc. **128**, 112—120 (1967).

[68] Sandomierski, F. L.: Nonsingular rings. Proc. Amer. Math. Soc. **19**, 225—230 (1968).

Schopf, A., Eckmann, B. (see Eckmann and Schopf).

[71] Scott, D. (editor): Axiomatic Set Theory, Proceedings of Symposia in Pure Mathematics, Vol. 13, Part I. Amer. Math. Soc., Providence 1971.

[58] Seshadri, C.: Trivality of vector bundles over the affine space K^2. Proc. Nat. Acad. Sci. USA **44**, 456—458 (1958).

[51] Shepherdson, J. C.: Inverse and zero divisors in matrix rings. Proc. Lond. Math. Soc. **61**, 71—85 (1951).

[71a] Shock, R. C.: Injectivity, annihilators, and orders. J. Alg. **19**, 96—103 (1971).

[71b] Shock, R. C.: Nil subrings in finiteness conditions. Amer. Math. Monthly **78**, 741—748 (1971).

[72a] Shock, R. C.: Orders in self injective cogenerator rings, preprint.

[72b] Shock, R. C.: Polynomial rings over finite dimensional rings, preprint. Southern Illinois University, Carbondale.

[58] Sierpinski, W.: Cardinal and Ordinal Numbers. Mathematical Monographs of the Polish Academy. Warszawa 1958.

[67] Silver, L.: Noncommutative localizations and applications. J. Algebra **7**, 44—76 (1967).

[65] Small, L. W.: An example in Noetherian rings. Proc. Nat. Sci. USA **54**, 1035—1036 (1965).

[66a] Small, L. W.: Hereditary rings. Proc. Nat. Acad. Sci. USA **55**, 25—27 (1966).

[66b] Small, L. W.: Orders in Artinian rings. J. Algebra **4**, 13—41 (1966). Corrections and addendum. Ibid. **4**, 505—507 (1966).

[66c] Small, L. W.: Remarks on the homological dimension of a quotient field. Mimeographed Notes. University of California, Berkeley 1966.

[68] Small, L. W.: Orders in artinian rings, II. J. Algebra **9**, 266—273 (1968).

Small, L., Procesi, C. (see Procesi and Small).

[50, 51, 52] Snapper, E.: Completely Primary rings, I—IV. Ann. Math. **52**; **53**; **55**; 666—693; 125—142; 207—234; 46—64 (1950—1952).

Snider, R., Formanek, E. (see Formanek and Snider).

[63] Solovay, R.: Independence results in the theory of cardinals (Preliminary Report). Notices of the Amer. Math. Soc. **10**, 595 (1963).

[50] Spekcer, E.: Additive Gruppen von Folgen ganzer Zahlen. Portugaliae Math. **9**, 131—140 (1950).

[48] Steinitz, E.: Algebraische Theorie der Körper. New York: Chelsea 1948.

[71] Stenstrom, B.: Rings and modules of quotients. Lecture Notes in Mathematics, No. 237. Berlin/Heidelberg/New York: Springer 1971.

[59] Swan, R.: Projective modules over finite groups. Bull. Amer. Math. Soc. **65**, 365—367 (1959).

[68] Swan, R. G.: Algebraic K-Theory. Lecture Notes in Mathematics, No. 76. Berlin/Heidelberg/New York: Springer 1968.

[50] Szele, T.: Ein Analogon der Körpertheorie für abelsche Gruppen. J. reine u. angew. Math. **188**, 167—192 (1950).

[71] Takeuchi, M.: A simple proof of Gabriel and Popesco's theorem. J. Algebra **18**, 112—113 (1971).

[63] Talintyre, T. D.: Quotient rings of rings with maximal condition for right ideals. J. Lond. Math. Soc. **38**, 439—450 (1963).

[66] Talintyre, T. D.: Quotient rings with minimum condition on right ideals. J. Lond. Math. Soc. **41**, 141—144 (1966).

Thompson, J. G., Feit, W. (see Feit and Thompson).

Thrall, R., Artin, E., Nesbitt, E. (see Artin, Nesbitt, and Thrall).

[56] Utumi, Y.: On quotient rings. Osaka Math. J. **8**, 1—18 (1956).

[63a] Utumi, Y.: A theorem of Levitzki. Math. Assoc. of Amer. Monthly **70**, 286 (1963).

[63b] Utumi, Y.: On rings of which any one-sided quotient ring are two-sided Proc. Amer. Math. Soc. **14**, 141—147 (1963).

[63c] Utumi, Y.: A note on rings of which any two-sided quotient rings are two-sided. Proc. Japan Acad. **39**, 287—288 (1963).

Utumi, Y., Faith, C. (see Faith and Utumi).

[48] Van der Waerden, B. L.: Modern Algebra, Vols. 1 and 2. New York: Ungar 1948.

[69] Vasconcelos, W. V.: On finitely generated flat modules. Trans. Amer. Math. Soc. **138**, 505—512 (1969).

[70a] Vasconcelos, W. V.: Flat modules over commutative noetherian rings. Trans. Amer. Math. Soc. **152**, 137—143 (1970).

[70b] Vasconcelos, W. V.: Simple flat extensions. J. Alg. **16**, 106—107 (1970).

[36] Von Neumann, J.: On regular rings. Proc. Nat. Acad. Sci. **22**, 707—713 (1936).

[60] Von Neumann, J.: Continuous Geometry. Princeton Mathematical Series No. 25. Princeton: Princeton University 1960.

[66] Walker, C. L.: Relative homological algebra and abelian groups. Ill. J. Math. **10**, 186—209 (1966).

[66] Walker, E. A., Walker, C. L.: Quotient Categories of Modules. Proceedings of the Conference on Categorical Algebra (La Jolla, 1965). Berlin/Heidelberg/New York: Springer 1966.

[72] Walker, E. A., Walker, C. L.: Quotient categories and rings of quotients. Rocky Mountain J. Math. **2**, 513—55 (1972).

Walker, E. A., and Faith, C. (see Faith and Walker).

[60] Watts, C.: Intrinsic characterizations of some additive functors. Proc. Amer. Math. Soc. **11**, 5—8 (1960).

[70] Webber, D. B.: Ideals and modules of simple Noetherian hereditary rings. J. Algebra **16**, 239—242 (1970).

[08] Wedderburn, J. H. M.: On hypercomplex numbers. Proc. Lond. Math. Soc. (2) **6**, 77—117 (1908).

[09] Wedderburn, J. H. M.: On the direct product in the theory of finite groups. Ann. Math. **10**, 173—176 (1909).

[14] Wedderburn, J. H. M.: A type of primitive algebra. Trans. Amer. Math. Soc. **15**, 162—166 (1914).

[37] Wedderburn, J. H. M.: A note on algebras. Ann. Math. **38**, 854—856 (1937).

[39] Weyl, H.: The Classical Groups (revised). Princeton: Princeton University Press 1946.

[44] Weyl, H.: David Hilbert and his mathematical work. Bull. Amer. Math. Soc. **50**, 612—654 (1944).

Whaples, G. (see Artin and Whaples).

[44] Whitney, H.: Topics in the theory of abelian groups. I. Divisibility of homomorphisms. Bull. Amer. Math. Soc. **50**, 129—144 (1944).

[65] Willard, E.: Properties of generators. Math. Ann. **158**, 352—364 (1965).

Winton, C. N., Mewborn, A. C. (see Mewborn and Winton).

Wong, E. T., Johnson, R. E. (see Johnson and Wong).

[71] Woods, S. M.: On perfect group rings. Proc. Amer. Math. Soc. **27**, 49—52 (1971).

[300 B.C.] Xerxes, C.: The Athenian plateau problem. Archiv. Manuskriptus **1**, 1—100 (300 B.C.).

[54] Yoneda, N.: On the homology theory of modules. J. Fac. Sci. Univ. Tokyo Sect. I 193—227 (1954).

[58, 60] Zariski, O., Samuel, P.: Commutative Algebra, Vol. I and II. Princeton/New York: Van Nostrand 1958, 1960.

[58] Zassenhaus, H.: The Theory of Groups, 2nd Ed. New York: Chelsea 1958.

[67] Zassenhaus, H.: Orders as endomorphism rings of modules of the same rank. J. Lond. Math. Soc. **42**, 180—182 (1967).

[67] Zelmanowitz, J. M.: Endomorphism rings of torsion-less modules. J. Algebra **5**, 325—341 (1967).

[69] Zelmanowitz, J.: A shorter proof of Goldie's theorem. Canad. Math. Bull. **12**, 597—602 (1969).

[04] Zermelo, E.: Beweiss, daß jede Menge wohlgeordnet kann. Math. Ann. **59**, 514—516 (1904).

[08a] Zermelo, E.: Neuer Beweis für die Möglichkeit einer Wohlordnung. Math. Ann. **65**, 107—128 (1908).

[08b] Zermelo, E.: Untersuchungen über die Grundlagen der Mengenlehre I. Math. Ann. **65**, 261—281 (1908).

Zisman, M., Gabriel, P. (see Gabriel and Zisman).

Index

Although the main headings are alphabetical, the (indented) subheadings are listed in order of appearance in the text.

Die Grundlehren der mathematischen Wissenschaften in Einzeldarstellungen mit besonderer Berücksichtigung der Anwendungsgebiete

Eine Auswahl

---

67. Byrd/Friedman: Handbook of Elliptic Integrals for Engineers and Scientists. 2nd edition
68. Aumann: Reelle Funktionen. 2. Aufl.
74. Boerner: Darstellungen von Gruppen. 2. Aufl.
76. Tricomi: Vorlesungen über Orthogonalreihen. 2. Aufl.
77. Behnke/Sommer: Theorie der analytischen Funktionen einer komplexen Veränderlichen. Nachdruck der 3. Aufl.
78. Lorenzen: Einführung in die operative Logik und Mathematik. 2. Aufl.
86. Richter: Wahrscheinlichkeitstheorie. 2. Aufl.
87. van der Waerden: Mathematische Statistik. 3. Aufl.
94. Funk: Variationsrechnung und ihre Anwendung in Physik und Technik. 2. Aufl.
97. Greub: Linear Algebra. 3rd edition
99. Cassels: An Introduction to the Geometry of Numbers. 2nd printing
104. Chung: Markov Chains with Stationary Transition Probabilities. 2nd edition
107. Köthe: Topologische lineare Räume I. 2. Aufl.
114. MacLane: Homology. Reprint of 1st edition
116. Hörmander: Linear Partial Differential Operators. 3rd printing
117. O'Meara: Introduction to Quadratic Forms. 2nd printing
120. Collatz: Funktionalanalysis und numerische Mathematik. Nachdruck der 1. Aufl.

121./122. Dynkin: Markov Processes

123. Yosida: Functional Analysis. 3rd edition
124. Morgenstern: Einführung in die Wahrscheinlichkeitsrechnung und mathematische Statistik. 2. Aufl.
125. Itô/McKean jr.: Diffusion Processes and Their sample Paths
126. Lehto/Virtanen: Quasiconformal Mappings in the plane
127. Hermes: Enumerability, Decidability, Computability. 2nd edition
128. Braun/Koechler: Jordan-Algebren
129. Nikodým: The Mathematical Apparatus for Quantum Theories
130. Morrey jr.: Multiple Integrals in the calculus of Variations
131. Hirzebruch: Topological Methods in Algebraic Geometry. 3rd edition
132. Kato: Perturbation Theory for Linear Operators
133. Haupt/Künneth: Geometrische Ordnungen
134. Huppert: Endliche Gruppen I
135. Handbook for Automatic Computation. Vol. 1/Part a: Rutishauser: Description of ALGOL 60
136. Greub: Multilinear Algebra
137. Handbook for Automatic Computation. Vol. 1/Part b: Grau/Hill/Langmaack: Translation of ALGOL 60
138. Hahn: Stability of Motion
139. Doetsch/Schäfke/Tietz: Mathematische Hilfsmittel des Ingenieurs. 1. Teil
140. Collatz/Nicolovius/Törnig: Mathematische Hilfsmittel des Ingenieurs. 2. Teil
141. Mathematische Hilfsmittel des Ingenieurs. 3. Teil
142. Mathematische Hilfsmittel des Ingenieurs. 4. Teil
143. Schur: Vorlesungen über Invariantentheorie
144. Weil: Basic Number Theory

152. Hewitt/Ross: Abstract Harmonic Analysis. Vol. 2: Structure and Analysis for Compact Groups, Analysis on Locally Compact Abelian Groups
154. Singer: Bases in Banach Spaces I
155. Müller: Foundations of the Mathematical Theory of Electromagnetic Waves
156. van der Waerden: Mathematical Statistics
157. Prohorov/Rozanov: Probability Theory
158. Constantinescu/Cornea: Potential Theory on Harmonic Spaces
159. Köthe: Topological Vector Spaces I
160. Agrest/Maksimov: Theory of Incomplete Cylindrical Functions and Their Applications
161. Bhatia/Szegö: Stability Theory of Dynamical Systems
162. Nevanlinna: Analytic Functions
163. Stoer/Witzgall: Convexity and Optimization in Finite Dimensions I
164. Sario/Nakai: Classification Theory of Riemann Surfaces
165. Mitrinović: Analytic Inequalities
166. Grothendieck/Dieudonné: Eléments de Géométrie Algebriqué I
167. Chandrasekharan: Arithmetical Functions
168. Palamodov: Linear Differential Operators with Constant Coefficients
169. Rademacher: Topics in Analytic Number Theory
170. Lions: Optimal Control of Systems Governed by Partial Differential Equations
171. Singer: Best Approximation on Normed Linear Spaces by Elements of Linear Subspaces
172. Bühlmann: Mathematical Methods in Risk Theory
173. F. Maeda/S. Maeda: Theory of Symmetric Lattices
174. Stiefel/Scheifele: Linear and Regular Celestial Mechanics. Perturbed two-body Motion—Numerical Methods—Canonical Theory
175. Larsen: An Introduction of the Theory of Multipliers
176. Grauert/Remmert: Analytische Stellenalgebren
177. Flügge: Practical Quantum Mechanics I
178. Flügge: Practical Quantum Mechanics II
179. Giraud: Cohomologie non abélienne
180. Landkof: Foundations of Modern Potential Theory
181. Lions/Magenes: Non-Homogeneous Boundary Value Problems and Applications I
182. Lions/Magenes: Non-Homogeneous Boundary Value Problems and Applications II
183. Lions/Magenes: Non-Homogeneous Boundary Value Problems and Applications III
184. Rosenblatt: Markov Processes. Structure and Asymptotic Behavior
185. Rubinowicz: Sommerfeldsche Polynommethode
186. Wilkinson/Reinsch: Handbook for Automatic Computation II, Linear Algebra
187. Siegel/Moser: Lectures on Celestial Mechanics
188. Warner: Harmonic Analysis on Semi-Simple Lie Groups I
189. Warner: Harmonic Analysis on Semi-Simple Lie Groups II
190. Faith: Algebra: Rings, Modules, and Categories I
191. Faith: Algebra: Rings, Modules, and Categories II
192. Mal'cev: Algebraic Systems
193. Pólya/Szegö: Problems and Theorems in Analysis. Vol. 1
194. Igusa: Theta Functions
195. Berberian: Baer*-Rings
196. Athreya: Branching Processes
197. Benz: Vorlesungen über Geometrie der Algebren
200. Dold: Lectures on Algebraic Topology